Microbiology of Drinking Water Production and Distribution

Microbiology of Drinking Water Production and Distribution

Editor:

Patricia Marques, Ph.D.

Delve Publishing

www.delvepublishing.com

Microbiology of Drinking Water Production and Distribution

Editor: Patricia Marques, Ph.D.

708 3rd Avenue, 6th Floor
New York, NY 10017
United States of America

www.delvepublishing.com

ISBN: 978-1-68095-829-4

Library of Congress Control Number: 2016951571

About the Editor

Patricia Marques, Ph.D.

Patricia obtained her PhD form University College Dublin in 2010. Her interests are on Microbiology and Parasitology. She is currently working as a Postdoc at University of Maryland Baltimore, USA on Chlamydia infections

Contents

Section 6 - Biological Treatment and Biostability of Drinking Water

Section 7 - Bioterrorism and Drinking Water Safety

Section 8 - Water Treatment Technologies for Developing Countries

Section 9 - Bottled Water Microbiology

List of Contributors

V. Goncharuk
A.V. Dumansky Institute of Colloid Chemistry and Chemistry of Water, National Academy of Sciences, 42 Acad. Vernadskoho Boulevard, Kiev 03680, Ukraine

Joe Brown
Department of Disease Control, Faculty of Infectious and Tropical Diseases, London School of Hygiene and Tropical Medicine, London, United Kingdom

Thomas Clasen
Department of Disease Control, Faculty of Infectious and Tropical Diseases, London School of Hygiene and Tropical Medicine, London, United Kingdom

João P. S. Cabral
Center for Interdisciplinary Marine and Environmental Research (C. I. I. M. A. R.), Faculty of Sciences, Oporto University, Rua do Campo Alegre

Nicholas J. Ashbolt
School of Public Health, University of Alberta, Edmonton, Room 3-57D, South Academic Building, Alberta, T6G 2G7 Canada

Pavan Kumar Pindi
Department of Microbiology, Palamuru University, Mahabubnagar, Andhra Pradesh 509001, India

P. Raghuveer Yadav
Department of Microbiology, Palamuru University, Mahabubnagar, Andhra Pradesh 509001, India

A. Shiva Shanker
Department of Microbiology, Palamuru University, Mahabubnagar, Andhra Pradesh 509001, India

EMRAH CAYLAK
Department of Biochemistry, School of Health, Karatekin University, 18200 CANKIRI, TURKEY

MUSTAFA TOKAR[b]
Department of Microbiology, School of Health, Karatekin University, 18200 CANKIRI, TURKEY

Aimee M. Gall
Department of Civil and Environmental Engineering, University of Illinois at Urbana-Champaign, Urbana, Illinois, United States of America, Safe Global Water Institute, University of Illinois at Urbana-Champaign, Urbana, Illinois, United States of America,

Benito J. Mariñas
[1]Department of Civil and Environmental Engineering, University of Illinois at Urbana-Champaign, Urbana, Illinois, United States of America, Safe Global Water Institute, University of Illinois at Urbana-Champaign, Urbana, Illinois, United States of America,

Yi Lu
Safe Global Water Institute, University of Illinois at Urbana-Champaign, Urbana, Illinois, United States of America, Department of Chemistry, University of Illinois at Urbana-Champaign, Urbana, Illinois, United States of America

Joanna L. Shisler
Safe Global Water Institute, University of Illinois at Urbana-Champaign, Urbana, Illinois, United States of America, Department of Microbiology, University of Illinois at Urbana-Champaign, Urbana, Illinois, United States of America, College of Medicine, University of Illinois at Urbana-Champaign, Urbana, Illinois, United States of America

Suma George Mulamattathil
Department of Water and Sanitation, University of Limpopo, Turfloop Campus, Private Bag X 1106, Sovenga 0727, South Africa

Carlos Bezuidenhout
School of Biological Science, North-West University, Potchefstroom Campus, Private Bag X 6001, Potchefstroom 2520, South Africa

Moses Mbewe
Department of Water and Sanitation, University of Limpopo, Turfloop Campus, Private Bag X 1106, Sovenga 0727, South Africa

Collins Njie Ateba
Department of Biological Sciences, School of Environmental and Health Sciences, North-West University, Mafikeng Campus, Private Bag X2046, Mmabatho 2735, South Africa

G. Liu
Section Sanitary Engineering, Department of Water Management, Faculty of Civil Engineering and Geosciences, Delft University of Technology, P.O. Box 5048, 2600 GA Delft, The Netherlands

E. J. Van der Mark
Dunea Water Company, Plaza of the United Nations 11-15, P.O. Box 756, 2700 AT Zoetermeer, The Netherlands

J. Q. J. C. Verberk
Section Sanitary Engineering, Department of Water Management, Faculty of Civil Engineering and Geosciences, Delft University of Technology, P.O. Box 5048, 2600 GA Delft, The Netherlands

J. C. Van Dijk
Section Sanitary Engineering, Department of Water Management, Faculty of Civil Engineering and Geosciences, Delft University of Technology, P.O. Box 5048, 2600 GA Delft, The Netherlands

A. A. Al-Sulami
College of Education, University of Basrah, Ashar P.O. Box 2108, Ashar, Basrah, Iraq

T. A. A. Al-Edani
Department of Biology, College of Sciences, University of Basrah, Ashar P.O. Box 2108, Ashar, Basrah, Iraq

A. A. Al-Abdula[2]
Department of Biology, College of Sciences, University of Basrah, Ashar P.O. Box 2108, Ashar, Basrah, Iraq

Ian R. Falconer
Department of Clinical and Experimental Pharmacology, University of Adelaide Medical School, Adelaide, South Australia

Andrew R. Humpage
Department of Clinical and Experimental Pharmacology, University of Adelaide Medical School, Adelaide, South Australia
C. R. C. for Water Quality and Treatment, Australian Water Quality Centre, Salisbury, South Australia.

Gongduan Fan
College of Civil Engineering, Fuzhou University, Fuzhou, Fujian 350116, China

Zhaoyue Su
College of Civil Engineering, Fuzhou University, Fuzhou, Fujian 350116, China

Rujing Lin
College of Civil Engineering, Fuzhou University, Fuzhou, Fujian 350116, China

Xiuyong Lin
College of Civil Engineering, Fuzhou University, Fuzhou, Fujian 350116, China

Renxin Xu
College of Civil Engineering, Fuzhou University, Fuzhou, Fujian 350116, China

Wei Chen
College of Civil Engineering, Fuzhou University, Fuzhou, Fujian 350116, China

Alina Nescerecka
Department of Water Engineering and Technology, Riga Technical University, Riga, Latvia,
Department of Environmental Microbiology, Eawag, Swiss Federal Institute for Aquatic Science and Technology, Du¨bendorf, Switzerland

Janis Rubulis
Department of Water Engineering and Technology, Riga Technical University, Riga, Latvia,

Marius Vital
Department of Environmental Microbiology, Eawag, Swiss Federal Institute for Aquatic Science and Technology, Du¨bendorf, Switzerland

Talis Juhna
Department of Water Engineering and Technology, Riga Technical University, Riga, Latvia,

Frederik Hammes
Department of Environmental Microbiology, Eawag, Swiss Federal Institute for Aquatic Science and Technology, Du¨bendorf, Switzerland

J. Anthony Byrne
Nanotechnology and Integrated BioEngineering Centre, University of Ulster, Newtownabbey BT37 0QB, Northern Ireland, UK

Pilar A. Fernandez-Ibañez,
Plataforma Solar de Almería (CIEMAT), Crta Senés km 4, 04200 Tabernas, Spain

Patrick S. M. Dunlop
Nanotechnology and Integrated BioEngineering Centre, University of Ulster, Newtownabbey BT37 0QB, Northern Ireland, UK

Dheaya M. A. Alrousan
Nanotechnology and Integrated BioEngineering Centre, University of Ulster, Newtownabbey BT37 0QB, Northern Ireland, UK

Jeremy W. J. Hamilton
Nanotechnology and Integrated BioEngineering Centre, University of Ulster, Newtownabbey BT37 0QB, Northern Ireland, UK

Jeffery Deal
Health Studies at Water Missions International, 2049 Savannah Highway, Charleston, SC 29407, USA

Vessoni Thereza Christina Penna
Department of Biochemical and Pharmaceutical Technology, School of Pharmaceutical Sciences, University of São Paulo, São Paulo, Brazil

Silva Alzira Maria Martins
Department of Biochemical and Pharmaceutical Technology, School of Pharmaceutical Sciences, University of São Paulo, São Paulo, Brazil

Priscila Gava Mazzola
Department of Biochemical and Pharmaceutical Technology, School of Pharmaceutical Sciences, University of São Paulo, São Paulo, Brazil

Huirong Lin
Institute of Urban Environment, Chinese Academy of Science, Xiamen 361021, China

Shuting Zhang

Song Gong
Institute of Urban Environment, Chinese Academy of Science, Xiamen 361021, China

Shenghua Zhang
Institute of Urban Environment, Chinese Academy of Science, Xiamen 361021, China

Xin Yu
Institute of Urban Environment, Chinese Academy of Science, Xiamen 361021, China

John J. Kelly[1] *
Department of Biology, Loyola University Chicago, Chicago, Illinois, United States of America,

Nicole Minalt[1]
Department of Biology, Loyola University Chicago, Chicago, Illinois, United States of America,

Alessandro Culotti
Department of Civil and Environmental Engineering, Northwestern University, Evanston, Illinois, United States of America,

Marsha Pryor
Pinellas County Utilities Laboratory, Largo, Florida, United States of America

Aaron Packman
Department of Civil and Environmental Engineering, Northwestern University, Evanston, Illinois, United States of America,

Virgínia M. Siqueira
Institute for Biotechnology and Bioengineering, Centre of Biological Engineering, University of Minho, Campus de Gualtar, (IBB), 4710-057 Braga, Portugal

Nelson Lima
Institute for Biotechnology and Bioengineering, Centre of Biological Engineering, University of Minho, Campus de Gualtar, (IBB), 4710-057 Braga, Portugal

Virgínia M. Siqueira [1]
IBB-Institute for Biotechnology and Bioengineering, Centre of Biological Engineering, University of Minho, Campus de Gualtar, 4710-057 Braga, Portugal

Helena M. B. Oliveira
Department of Antibiotics, Federal University of Pernambuco, Av. Prof. Morais Rego, 1235, 50670-901, Recife, Pernambuco, Brazil

Cledir Santos
IBB-Institute for Biotechnology and Bioengineering, Centre of Biological Engineering, University of Minho, Campus de Gualtar, 4710-057 Braga, Portugal

R. Russell M. Paterson
IBB-Institute for Biotechnology and Bioengineering, Centre of Biological Engineering, University of Minho, Campus de Gualtar, 4710-057 Braga, Portugal

Norma B. Gusmão
Department of Antibiotics, Federal University of Pernambuco, Av. Prof. Morais Rego, 1235, 50670-901, Recife, Pernambuco, Brazil

Nelson Lima
IBB-Institute for Biotechnology and Bioengineering, Centre of Biological Engineering, University of Minho, Campus de Gualtar, 4710-057 Braga, Portugal
Maria S Gião
School of Biological Sciences, Life Sciences Building, Highfield Campus, University of Southampton
Institute for Biotechnology and Bioengineering, Centre of Biological Engineering, Universidade do Minho

Nuno F Azevedo
School of Biological Sciences, Life Sciences Building, Highfield Campus, University of Southampton
Institute for Biotechnology and Bioengineering, Centre of Biological Engineering, Universidade do Minho
LEPAE, Department of Chemical Engineering, Faculty of Engineering, University of Porto

Sandra A Wilks
School of Biological Sciences, Life Sciences Building, Highfield Campus, University of Southampton

Maria J Vieira
Institute for Biotechnology and Bioengineering, Centre of Biological Engineering, Universidade do Minho

Charles W Keevil
School of Biological Sciences, Life Sciences Building, Highfield Campus, University of Southampton

Giorgio Liguori
Department of Public, Clinical and Preventive Medicine, Second University of Naples

Ivan Cavallotti
Department of Public, Clinical and Preventive Medicine, Second University of Naples

Antonio Arnese
Department of Public, Clinical and Preventive Medicine, Second University of Naples

Ciro Amiranda
Department of Public, Clinical and Preventive Medicine, Second University of Naples

Daniela Anastasi
Department of Public, Clinical and Preventive Medicine, Second University of Naples

Italo F Angelillo
Department of Public, Clinical and Preventive Medicine, Second University of Naples

Ma José Figueras
Unit of Microbiology, School of Medicine, IISPV, Universitat Rovira i Virigili, Sant Llorenç 21, 43201-Reus, Spain

Juan J. Borrego
Department of Microbiology, Faculty of Sciences, University of Malaga, Campus Universitario Teatinos, 29071-Malaga, Spain

Nicholas E. Kman
Ohio Task Force 1-FEMA Urban Search and Rescue, Department of Emergency Medicine, The Ohio State University Medical Center, 4813 Cramblett Hall, 456 West Tenth Avenue, Columbus, OH 43210, USA

Daniel J. Bachmann
Emergency Medicine, The Ohio State University Medical Center, 4734 Cramblett Hall, 456 West Tenth Avenue, Columbus, OH 43210, USA

Steven E. Schutzer
Department of Medicine, New Jersey Medical School, University of Medicine and Dentistry of New Jersey, Newark, New Jersey, United States of America

Bruce Budowle
Laboratory Division, Federal Bureau of Investigation, Quantico, Virginia, United States of America

Ronald M. Atlas
graduate school of University of Louisville, Louisville, Kentucky, United States of America

Rachel L. Pullan
Faculty of Infectious and Tropical Diseases, London School of Hygiene & Tropical Medicine, London, United Kingdom,

Matthew C. Freeman
Department of Environmental Health, Rollins School of Public Health, Emory University, Atlanta, Georgia, United States of America,

Peter W. Gething
Spatial Ecology and Epidemiology Group, Department of Zoology, University of Oxford, Oxford, United Kingdom,

Simon J. Brooker
Faculty of Infectious and Tropical Diseases, London School of Hygiene & Tropical Medicine, London, United Kingdom, Kenya Medical Research Institute (KEMRI), Nairobi, Kenya

Robert Bain
The Water Institute, University of North Carolina at Chapel Hill, Chapel Hill, North Carolina, United States of America,

Ryan Cronk
The Water Institute, University of North Carolina at Chapel Hill, Chapel Hill, North Carolina, United States of America,

Jim Wright
University of Southampton, Southampton, United Kingdom,

Hong Yang
University of Southampton, Southampton, United Kingdom,

Tom Slaymaker
WaterAid UK, London, United Kingdom

Jamie Bartram
The Water Institute, University of North Carolina at Chapel Hill, Chapel Hill, North Carolina, United States of America,

Hiroshi Sakai
Department of Urban Engineering, The University of Tokyo, 7-3-1 Hongo, Bunkyo-ku, Tokyo 113-8656, Japan

Yatsuka Kataoka
Institute for Global Environmental Strategies, 2108-11 Kamiyamaguchi, Hayama, Kanagawa 240-0115, Japan

Kensuke Fukushi
Todai Institute for Advanced Study, Integrated Research System for Sustainability Science, The University of Tokyo, 7-3-1 Hongo, Bunkyo-ku, Tokyo 113-8654, Japan

KYLE ONDA
The Water Institute, Gillings School of Global Public Health, University of North Carolina at Chapel Hill, 135 Dauer Drive, CB #7431, Chapel Hill, NC 27599, USA;

JOE LOBUGLIO
The Water Institute, Gillings School of Global Public Health, University of North Carolina at Chapel Hill, 135 Dauer Drive, CB #7431, Chapel Hill, NC 27599, USA;

JAMIE BARTRAM
The Water Institute, Gillings School of Global Public Health, University of North Carolina at Chapel Hill, 135 Dauer Drive, CB #7431, Chapel Hill, NC 27599, USA;

Mark R Riley
The University of Arizona, Tucson, AZ, 85721, USA

Charles P Gerba
The University of Arizona, Tucson, AZ, 85721, USA

Menachem Elimelech
Yale University, New Haven, CT 06520-8286, USA

Sophie Boisson
Department of Infectious and Tropical Diseases, London School of Hygiene and Tropical Medicine, London, United Kingdom,

Mbela Kiyombo
Ecole de Sante´ publique de Kinshasa, Kinshasa, Re´publique De´mocratique du Congo,

Larry Sthreshley
Communaute´ Presbyte´rienne de Kinshasa, Kinshasa, Re´publique De´mocratique du Congo,

Saturnin Tumba
Zone de Sante´ de Bibanga, Province du Kasaı¨ Oriental, Re´publique De´mocratique du Congo

Jacques Makambo
Communaute´ Presbyte´rienne de Kinshasa, Kinshasa, Re´publique De´mocratique du Congo,

Thomas Clasen
Department of Infectious and Tropical Diseases, London School of Hygiene and Tropical Medicine, London, United Kingdom,

Daniel Ma¨usezahl
Department of Public Health and Epidemiology, Swiss Tropical Institute (STI), University of Basel, Switzerland,

Andri Christen
Department of Public Health and Epidemiology, Swiss Tropical Institute (STI), University of Basel, Switzerland,

Gonzalo Duran Pacheco
Department of Public Health and Epidemiology, Swiss Tropical Institute (STI), University of Basel, Switzerland,

Fidel Alvarez Tellez
Project Concern International (PCI), Cochabamba, Bolivia,

Mercedes Iriarte
Centro de Aguas y Saneamiento Ambiental (CASA), Facultad de Tecnologı´a, Universidad Mayor de San Simo´n (UMSS), Cochabamba, Bolivia,

Maria E. Zapata
Instituto de Investigaciones Biome´dicas (IIBISMED), Facultad de Medicina, Universidad Mayor de San Simon, Cochabamba, Bolivia,

Myriam Cevallos
Department of Public Health and Epidemiology, Swiss Tropical Institute (STI), University of Basel, Switzerland,

Jan Hattendorf
Department of Public Health and Epidemiology, Swiss Tropical Institute (STI), University of Basel, Switzerland,

Monica Daigl Cattaneo
Department of Public Health and Epidemiology, Swiss Tropical Institute (STI), University of Basel, Switzerland,

Benjamin Arnold
Centers for Occupational and Environmental Health and Family and Community Health, School of Public Health, Berkeley, University of California, Berkeley, California

Thomas A. Smith
Department of Public Health and Epidemiology, Swiss Tropical Institute (STI), University of Basel, Switzerland,

John M. Colford Jr
Centers for Occupational and Environmental Health and Family and Community Health, School of Public Health, Berkeley, University of California, Berkeley, California

Abdullah Ali Halage
Department of Disease Control and Environmental Health, School of Public Health, College of Health Sciences, Makerere University, P.O. Box 7072, Kampala, Uganda

Charles Ssemugabo
Department of Disease Control and Environmental Health, School of Public Health, College of Health Sciences, Makerere University, P.O. Box 7072, Kampala, Uganda

David K. Ssemwanga
Department of Disease Control and Environmental Health, School of Public Health, College of Health Sciences, Makerere University, P.O. Box 7072, Kampala, Uganda

David Musoke
Department of Disease Control and Environmental Health, School of Public Health, College of Health Sciences, Makerere University, P.O. Box 7072, Kampala, Uganda

Richard K. Mugambe
Department of Disease Control and Environmental Health, School of Public Health, College of Health Sciences, Makerere University, P.O. Box 7072, Kampala, Uganda

David Guwatudde
Department of Epidemiology and Biostatistics, School of Public Health, College of Health Science, Makerere University, P.O. Box 7072, Kampala, Uganda

John C. Ssempebwa
Department of Disease Control and Environmental Health, School of Public Health, College of Health Sciences, Makerere University, P.O. Box 7072, Kampala, Uganda

Zhihua Hu
Department of Sociology, Iowa State University, 103 East Hall, Ames, IA 50011, USA

Lois Wright Morton
Department of Sociology, Iowa State University, 103 East Hall, Ames, IA 50011, USA

Robert L. Mahler
PSES Department, University of Idaho, P.O. Box 442339, Moscow, ID 83844, USA

Lorna A Ward
Medical Student, University of Birmingham, Birmingham, B15 2TT, UK

Owen L Cain
Medical Student, University of Birmingham, Birmingham, B15 2TT, UK

Ryan A Mullally
Medical Student, University of Birmingham, Birmingham, B15 2TT, UK

Kathryn S Holliday
Medical Student, University of Birmingham, Birmingham, B15 2TT, UK

Aaron GH Wernham
Medical Student, University of Birmingham, Birmingham, B15 2TT, UK

Paul D Baillie
Medical Student, University of Birmingham, Birmingham, B15 2TT, UK

Sheila M Greenfield
Primary Care Clinical Sciences, University of Birmingham Medical School, Birmingham, B15 2TT, UK

Preface

Although there is no universal method for concentrating all enteric viruses from water, some methods are quite effective for some of the enteric viruses in some types of water. Flowthrough-filter adsorption-elution systems and ultrafiltration methods with anisotropic, polymeric membranes in tangential-flow systems are the best for recovery of small quantities of enterovirus in large volumes of treated or finished water. Virus concentration from water by adsorption to precipitable salts or polyelectrolytes, aqueous polymer two-phase separation, reverse osmosis, hydroextraction, continuous-flow ultracentrifugation, and forced-flow electrophoresis methods can generally be applied only to samples of a few liters. Therefore, they are not suited to recovery of a low number of infectious units of virus from large samples of water.

The principal microbiological contaminants found in drinking water are bacteria, viruses, and pathogenic protozoa. Routine microbiological monitoring of water supplies is primarily based on tests for coliforms. These organisms have been used successfully for indicating contamination and treatment failures, but may need supplementing with other organisms to further improve confidence in water supply hygiene. Novel techniques may greatly alter the water microbiology laboratory. Microbiological contaminants, such as fungi and algae, do not seem to be important causes of waterborne disease, although they are sometimes associated with undesirable tastes and odors. This book entitled Microbiology of Drinking Water Production and Distribution presents the public health facets of drinking water treatment and distribution. It clarifies the different water treatment processes and their effects on waterborne microbial pathogens and parasites. It highlights several methodologies to gauge the bacterial growth potential in water distribution systems. Microbiology of Drinking Water Production and Distribution starts by introducing the new standards for drinking water and other public health aspects of drinking water treatment and distribution (Section 1). Section 2 and 3 talk about the microbiology of drinking water. Section 4 gives some examples of the different water disinfection and treatment processes, and their impacts on waterborne microbial pathogens and parasites. Drinking water quality may be degraded in water distribution systems—microorganisms form biofilms within distribution systems that allow them to flourish (Section 5). The monitoring and stability of drinking water biology is addressed on Section 6. Very few studies are available about bioterrorism on drinking water. So Section 7 reviews the general biosurveillance in which drinking water is included. In addition, the Section 8 reviews the research on drinking water quality in developing countries. Section 9 examines the microbiological water quality of bottled water. Although there is no universal method for concentrating all enteric viruses from water, some methods are quite effective for some of the enteric viruses in some types of water. Flowthrough-filter adsorption-elution systems and ultrafiltration methods with anisotropic, polymeric membranes in tangential-flow systems are the best for recovery of small quantities of enterovirus in large volumes of treated or finished water. Virus concentration from water by adsorption to precipitable salts or polyelectrolytes, aqueous polymer two-phase separation, reverse osmosis, hydroextraction, continuous-flow ultracentrifugation, and forced-flow electrophoresis methods can generally be applied only to samples of a few liters. Therefore, they are not suited to recovery of a low number of infectious units of virus from large samples of water. Drinking water quality may be degraded in water distribution systems—microorganisms form biofilms within distribution systems that allow them to flourish. Various methodologies have been proposed to assess the bacterial growth potential in water distribution systems. Microbiology of Drinking Water Production and Distribution gives guidance on appropriate application of monitoring parameters for ensuring the safety of drinking water and to inform risk management decisions, with an emphasis on control of facial contamination.

Editor
Patricia Marques, Ph.D.

INTRODUCTION

Microbiology of Drinking Water Production and Distribution, the public health aspects of drinking water treatment and distribution. It explains the different water treatment processes, such as pretreatment, coagulation, flocculation, sedimentation, filtration, disinfection, and their impacts on waterborne microbial pathogens and parasites. Drinking water quality may be degraded in water distribution systems—microorganisms form biofilms within distribution systems that allow them to flourish. Various methodologies have been proposed to assess the bacterial growth potential in water distribution systems. Microbiology of Drinking Water Production and Distribution also places drinking water quality and public health issues in context; it addresses the effect of bioterrorism on drinking water safety, particularly safeguards that are in place to protect consumers against the microbial agents involved.

Due to the increase in the demand for drinking water, more and more types of hydraulic resources are used. This means that to obtain water fit for human consumption, it is possible to come across many different situations and characteristics as regards the water to be treated. In Drinking Water Treatment Plants (DWTP) the unprocessed water is submitted to a series of processes to eliminate substances that are harmful to human health. In conventional plants, these processes consist basically of:

- **Regulation**: A set of mechanisms which regulate the water flowing into the plant, to adapt it to the plant's treatment capacity.
- **Flocculation**: A process which, by means of chemical reagents, creates floccules which, when moving slowly through the decanters, are deposited on the bottom of those tanks, and then removed and eliminated in the appropriate manner.
- **Filtration**: Normally, this is done by forcing the decanted water through a bed of sand, where the particles not separated during the flocculation phase are retained.
- **Disinfection**: A chlorination process to guarantee a residual chlorine load in the distribution pipes.

Certain processes are specifically designed to remove unwanted elements such as iron, manganese, ammonium, fluorides, arsenic, pesticides, nitrates and heavy metals. These processes can be carried out at a facility or in metal package units using Ferazur Mangazur technology that simultaneously reduces iron, manganese and arsenic.

Because of the poor quality of raw water, dissolved organic matter and micro-pollutants must be thoroughly removed. This requires performing clarification under optimal conditions, such as oxidation by ozone or by introducing additional steps, for instance adsorption on activated carbon using a Pulsazur lamellar clarifier.

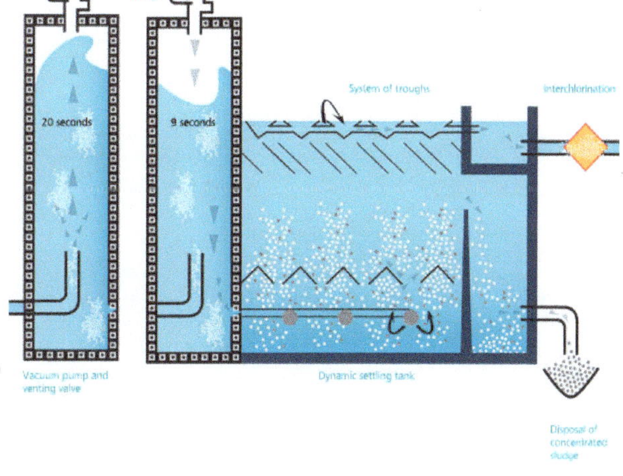

Water treatment

Water treatment is any process that makes water more acceptable for a specific end-use. The end use may be drinking, industrial water supply, irrigation, river flow maintenance, water recreation or many other uses including being safely returned to the environment. Water treatment removes contaminants or reduces their concentration so that the water becomes fit for its desired end-use.

Treatment for drinking water production

Water purification

Treatment for drinking water production involves the removal of contaminants from raw water to produce water that is pure enough for human consumption without any short term or long term risk of any adverse health effect. Substances that are removed during the process of drinking water treatment include suspended solids, bacteria, algae, viruses, fungi, and minerals such as iron and manganese. The processes involved in removing the contaminants include physical processes such as settling and filtration, chemical processes such as disinfection and coagulation and biological processes such as slow sand filtration. Measures taken to ensure water quality not only relate to the treatment of the water, but to its conveyance and distribution after treatment. It is therefore common practice to keep residual disinfectants in the treated water to kill bacteriological contamination during distribution.

Processes

A combination selected from the following processes is used for municipal drinking water treatment worldwide:

- Pre-chlorination for algae control and arresting biological growth
- Aeration along with pre-chlorination for removal of dissolved iron and manganese
- Coagulation for flocculation or slow-sand filtration
- Coagulant aids, also known as polyelectrolytes – to improve coagulation and for thicker floc formation
- Sedimentation for solids separation that is removal of suspended solids trapped in the floc
- Filtration to remove particles from water
- Disinfection for killing bacteria viruses and other pathogens.

Technologies for potable water treatment are well developed, and generalized designs are available that are used by many water utilities (public or private). In addition, a number of private companies provide patented technological solutions. Automation of water and waste-water treatment is common in the developed world. Capital costs, operating costs available quality monitoring technologies, locally available skills typically dictate the level of automation adopted.

Polluted water treatment

Wastewater treatment is the process that removes the majority of the contaminants from wastewater or sewage and produces both a liquid effluent suitable for disposal to the natural environment and a sludge. Biological processes can be employed in the treatment of wastewater and these processes may include, for example, aerated lagoons, activated sludge or slow sand filters. To be effective, sewage must be conveyed to a treatment plant by appropriate pipes and infrastructure and the process itself must be subject to regulation and controls. Some wastewaters require different and sometimes specialized treatment methods. At the simplest level, treatment of sewage and most wastewaters is carried out through separation of solids from liquids, usually by sedimentation. By progressively converting dissolved material into solids, usually a biological floc, which is then settled out, an effluent stream of increasing purity is produced.

Industrial water and wastewater treatment

Two of the main processes of industrial water treatment are boiler water treatment and cooling water treatment. A lack of proper water treatment can lead to the reaction of solids and bacteria within pipe work and boiler housing. Steam boilers can suffer from scale or corrosion when left untreated. Scale deposits can lead to weak and dangerous machinery, while additional fuel is required to heat the same level of water because of the rise in thermal resistance. Poor quality dirty water can become a breeding ground for bacteria such as Legionella causing a risk to public health.

With the proper treatment, a significant proportion of industrial on-site wastewater might be reusable. This can save money in three ways: lower charges for lower water consumption, lower charges for the smaller volume of effluent water discharged and lower energy costs due to the recovery of heat in recycled wastewater.

Domestic water treatment

Water supplied to domestic properties may be further treated before use, often using an in-line treatment process. Such treatments can include water softening or ion exchange. Many propriety systems also claim to remove residual disinfectants and heavy metal ions.

Functions of a distribution system

The purpose of distribution system is to deliver water to consumer with appropriate quality, quantity and pressure. Distribution system is used to describe collectively the facilities used to supply water from its source to the point of usage. The basic function of a water distribution system is to transport the water from the treatment facility to the customer. In addition, distribution systems may also provide storage, as well as provide flow and pressure adequate for fire protection. Typically, the size of water mains, pump station capacity, and storage reservoir volume is determined by fire suppression needs. The volume of water necessary to fight fires is much greater than domestic and commercial water demand in most communities.

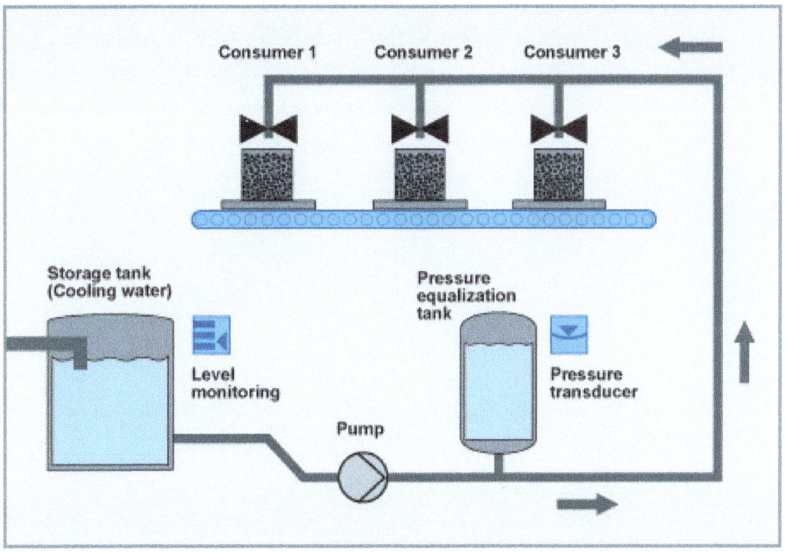

Microbial hazards

Most microorganisms present in drinking-water distribution systems are harmless. However, infectious microorganisms may enter the distribution system and can survive and in some cases grow in the distribution system, increasing the potential for waterborne disease outbreaks. Faecal contamination is a common source of infectious microorganisms. These include bacteria, viruses and parasites that occur naturally in the gut of humans and other warm blooded animals. Faecal contamination of distribution system water may occur when a pathway – such as broken mains, intrusion, cross-connections or openings in storage tanks – has been established for faecal contaminant entry.

Microorganisms that grow in the environment may enter the drinking-water distribution system and attach to and grow on drinking-water pipes and other surfaces, forming biofilms, particularly at the ends of distribution systems where flows can be low, Biofilms contain many sorption areas that can bind and accumulate organic and inorganic contaminants, as well as particulate and colloidal matter. These substances, also known as extracellular polymeric substances, protect microorganisms from biological, physical, chemical and environmental stresses, including predation, desiccation, flux changes and disinfectants.

For example, although viruses and protozoan cysts or oocysts need a suitable warmblooded host in which to grow, incorporation into biofilms could prolong their survival. Biofilms may develop within pipes and on other components, such as outlets, mixing valves and washers. Biofilms are extremely difficult to remove from all parts of the system once they are established, and they can be resistant to disinfectants. Hence, the goal is to control biofilm growth and not to eliminate biofilms totally.

Although most free-living heterotrophic bacteria, fungi, protozoa, nematodes and crustaceans in biofilms are not likely to be of risk to the drinking-water consumers, their activities can produce taste and odour problems, increase disinfectant demand and contribute to corrosion.

Chemical hazards

There are many chemical hazards that could contaminate water in distribution systems. Some of the common chemical contaminants are unwanted residues of chemicals used in water treatment, chemicals coming from materials or reacting with materials in the distribution systems, chemicals accumulated in and then released from scales and deposits, and chemicals entering the distribution system through faults and breaks

Disinfection by-products Disinfection by-products are produced by the reaction between chemical disinfectants and naturally occurring organic material in the source water.

Chemicals from pipe materials and fittings

Materials that are in contact with drinking-water in public water distribution systems may leach agents into the water that are hazards to public health. Pipes, fittings, lubricants, oranges, pipe and storage tank coatings, and mechanical devices have demonstrated the potential to leach contaminants with health concerns. The most significant contributor of lead to drinking-water in many countries remains the presence of lead service lines in the distribution system. Lead may leach into potable water from lead pipes in old water mains, lead service lines, lead in pipe jointing compounds and soldered joints, lead in brass and bronze plumbing fittings, and lead in goosenecks, valve parts or gaskets used in water treatment plants or distribution mains.

Physical hazards

Physical hazards refer to contamination affecting physical properties of water, such as color, odor and turbidity. Drinking-water that suddenly changes color may indicate a hazard that should be addressed. Consumer complaints of green or blue water are typically associated with copper contamination from plumbing supplies. New copper installations and the use of excess flux during soldering are sometimes associated with the problem. Copper solubility is increased at low pH as well as high alkalinity, and corrosion control chemicals such as orthophosphate can decrease the leaching of copper.

Section I - Introduction

CHAPTER 1 THE NEW STANDARD FOR DRINKING WATER

V. Goncharuk

INTRODUCTION

Drinking water, that is, the water whose organoleptic, physicochemical, and biological properties meet human biological needs and has neither color nor smell, should have a taste determined by the presence of physiologically necessary salts of calcium, magnesium, sodium, and potassium in corresponding concentrations without which metabolism in the human organism is impossible.

What optimal parameters should the water possess to be completely safe from the biological and physiological points of view?

For the first time in the world in 1853, in Brussels, at the International Congress of Hygienists, an attempt was made to set a standard for drinking water, which would include only nine controllable components. Among them there were vitally important substances such as magnesium, calcium, total mineralization (determined mainly by the presence of salts of sodium and potassium), and the content of oxygen in the water, sulfates, chlorides, nitrates, and ammonium. However, the age of industrialization, a fast development of the agroindustrial complex and the emergence of megapolises with compact settlement of people, whose number dramatically increased, started on the planet. A further development of civilization became impossible without the creation of civilized systems of water supply and a complex of water disposal facilities. In connection with the worsening of the ecological state of surface and underground sources the issue of water quality control, used by people for drinking purposes, became more acute. As a result of a low quality of drinking water real threats to the sanitary-epidemiological situation in various regions of the planet cropped out. And only early in the 20th century the first standard for drinking water appear in the USA. After almost a century after adoption in Brussels of the first recommendation for drinking water quality a document of the World Health Organization (WHO) was published, which today, at the beginning of the 21st century, proposes to bring under regulation 95 indices of which 26 are toxic substances, whose presence in drinking water is undesirable. The US standard regulates 102 indexes in drinking water, where complete absence is required for as many as 35 toxic indexes.

A high level of technogenic load on water bodies, the use of imperfect technologies of water conditioning, and secondary contamination of the water in distribution networks result in the ingress to the drinking water of a substantial amount of inorganic and organic pollutants whose joint effect on the human organism causes the effect of synergism known in chemistry and biology, posing real threat to human health. In the situation that has taken shape there is in principle no

Citation: Goncharuk V (2013) The New Standard for Drinking Water. Journal of Chemistry. Volume 2013, Article ID 321609, 7 pages doi: http://dx.doi.org/10.1155/2013/321609.

possibility to provide the population of any country or continent with quality and safe for human health drinking water [1–4].

An increase of the quality of the water being controlled by a national standard in various countries of the world does not solve the issue of obtaining safe drinking water at the centralized water conditioning stations either. All this calls for the ever more expensive equipment and complication of technological processes.

Against the background of such catastrophic changes in the quality of the environment the monitoring of several tens of substances in drinking water causes, to say the least, a surprise. A surprise is also caused by a lack of understanding of the fact that in pursuit of "technical progress," even increasing the amount of controllable components in drinking water several times as much, one cannot give any guarantees that it is in fact drinking water and safe for human health. Even the use of underground water at present does not always guarantee the quality of the drinking water obtained, especially if one takes into account the fact that an absolute majority of organic matter in the environment under the effect of various physicochemical and ecological factors undergoes biotransformation into unknown and, most often, into unpredictable compounds. According to the aforementioned synergism the effects of two or more compounds of toxic nature of each individually taken substance increase in that case if this substance initially is not toxic. Hence the senselessness of using classic approaches to the assessment of drinking waters follows. The notion itself of "normalizing maximum allowable concentrations in drinking water of different toxicants" purportedly safe for human health is, as a matter of fact, immoral.

The proposed new standard include fundamentally new, very effective integrated methods of monitoring water quality, which are designed for revealing acute toxicity at the level of organism and finding chronic toxicity at the cellular level using cytogenetic methods on biological objects for this purpose. These methods are universal for all types and kinds of toxic compounds irrespective of their origin and the nature of their effect.

The new Standard for Drinking Water includes three different regulatory documents [5]:(i)GOST (state all-union standard) for tap water (conditionally drinking);(ii)GOST for drinking water of increased quality (the water absolutely safe for human health);(iii)GOST for bottled water.

A new concept of providing the population with quality drinking water along with the general provisions and regulation requirements of the quality of drinking water accepted in the European Union, the WHO, and the Codex Alimentarius [6–10] has been taken into account in the development of the standard [11–15].

Such a concept provides for the following. Water safety in terms of toxicological and microbiological indices is prepared at waterworks. This water is suitable for household or domestic use and is fed to the distribution networks. In this case it is obtained mainly from the water of surface sources.

High-quality drinking water in the amount sufficient for satisfying human physiological needs is prepared at the consumption site in the plant of water pump rooms. It is then not fed to the distribution network but is supplied to the consumer by other methods (by gravitation flow or bottling). Such drinking water is obtained mainly from underground water protected against anthropogenic influence or from tap water.

The fundamental feature of the new standard is the regulatory justification of a new conceptual approach to water supply of population with quality drinking water. It provides economic water supply by utilizing the differentiating consumption of drinking water depending on physiological, sanitary-hygienic, and economic needs of human over one 24-hour period in a specific populated area. The necessity of such an approach is determined by the following objective reasons: unjustified use of quality drinking water for sanitary-hygienic needs; unsatisfactory state of transportation and distribution networks of the system of centralized drinking water supply; discrepancy of the technology of drinking water treatment in centralized systems of drinking water supply to the water quality in a water source; unjustified high economic costs for reconstruction and modernization of the systems of centralized drinking water supply for the preparation of high-quality drinking water in a volume substantially exceeding physiological and sanitary-hygienic human needs.

The second fundamental distinction of the new standard is the fact that the main accent was made on new approaches to the assessment of the quality of drinking water in terms of integral indexes of water toxicity, which is set by the methods of biotesting according to standardized techniques. This makes it possible to obtain objective information on the quality of drinking water before the conduction of a detailed analysis by all indices. Such an analysis is expedient in the case of finding toxicity and the necessity of establishing cause of its emergence.

The third fundamental distinction of the new standard is the requirement of complete absence of all toxic chemical and biological pollutants in drinking water designed for human consumption.

The employment of the methods of assessment of the drinking water quality by the integral indexes makes it possible within short terms and at the lowest economic costs to assess the suitability of drinking water for human consumption.

Unlike the methods of chemical analysis adopted today in the world (more than 100 indexes of water quality), the employment of the integrated methods of biotesting makes it possible to substantially (~ by 3-4 times) reduce the amount of controlled indexes.

It is a well-known fact that in the modern period in all countries of the world units, devices of various powers, and miniplants for add-on treatment of drinking water from various sources of water supply were used on a wide scale including from drinking water of centralized water supply and the center of realization or the pouring of tertiary treated drinking water to the containers of the consumers. An indispensable condition of achieving all regulatory indexes of drinking water safety is the setting of certain requirements of drinking water of both centralized and uncentralized systems of water supply.

By microbiological indexes drinking water should conform to the requirements given in Table 1.

Table 1: Microbiological indicators of drinking water quality.

Indicator	Measurement units	Norm, not more than	
		Water of systems of centralized drinking water supply	Water of uncentralized drinking water supply additionally purified (unbottled, bottled)
Number of bacteria in 1 cm^3 of water under study (TMN) at 37°C	CFU/cm^3	100[1*]	20[1*]
Number of bacteria in 1 cm^3 of water under study (TMN) at 22°C	CFU/cm^3	100[1*]	20[1*]
Number of bacteria of the colibacillus group (coliform microorganisms) in 1 dm^3 of water under study (BCG index)	CFU/dm^3	3[2*]	Absence[2*]
Number of thermostable colibacillus (fecal coliform—FC index) in 100 cm^3 of water under study	CFU/100 cm^3	Absence[3*]	Absence[3*]
Number of pathogenic microorganisms in 1 dm^3 of water under study	CFU/dm^3	Ditto[3*]	Ditto[3*]
Number of coliphages in 1 dm^3 of water under study	PFU/dm^3	" "	" "
Spores of sulfate-reducing clostridia	Presence (amount)/20 cm^3	Absence[4*]	Absence[4*]
Blue pus bacillus *(Pseudomonas aeruginosa)*	CFU/dm^3	Not determined	Absence

[1*] An excess of the norm shall not be allowed for 95% of water sample in the water supply network that is under investigation during the year.
[2*] An excess of the norm shall not be allowed for 98% of water samples in the water supply network that is under investigation during the year. In the event the BCG index is exceeded at the stage of colony identification investigations shall be carried out for the presence of fecal coliforms.
[3*] In the event of the presence of coliform bacteria or coliphages in a water sample their amount shall be immediately determined in repeatedly taken water samples. If in these samples generally coliform bacteria are found in the amount of >2/100^3 and (or) thermostable coliform bacteria and (or) coliphages, pathogenic bacteria of the colibacillus group and (or) enteroviruses shall be determined. The investigation of drinking water for the presence of pathogenic bacteria of the colibacillus group and enteroviruses shall also be carried out following the decision of the organs of sanitary supervision in the event of the emergence of the epidemic situation.
[4*] The control shall be exercised at the outlet from the drinking water conditioning station in the case of using surface sources of water supply or underground ones, which have the hydraulic connection with the surface water body; during the transition period annually the index of the spores of sulfate-reducing clostridia with the norm "absence/20 cm^3" shall be controlled.

By mycological indexes (micromycetes) drinking water should be in line with the norm given in Table 2.

Table 2: Mycological indicators of drinking water quality.

Indicator	Measurement units	Norm	
		Water of systems of centralized drinking water supply	Water of uncentralized drinking water supply additionally purified (unbottled, bottled)
Micromycetes	CFU/100 cm^3	Absence	Absence*

**Aspergillus fumigatus, Aspergillus niger*, and *Penicillium expansum* must not be present at all.

Toxicity of drinking water of uncentralized water supply is an integral (express) quality index of drinking water in the case of suspicion of contamination of a water source or a distribution network with toxic compounds. The list of indices, test objects, and norms for determination of toxicity obtained from the results of biotesting are given in Table 3.

Table 3: Toxicity level of drinking water.

Indicator	Measurement units	Norm: Water of systems of centralized drinking water supply	Norm: Water of uncentralized drinking water supply additionally purified (unbottled, bottled)
Chronic toxicity on *Ceriodaphnia affinis*	Number of dead individuals and (or) a decrease of the number of newly born individuals in an experiment compared with the reference during 7 ± 1 days	Not determined	Absence of chronic toxicity
Genotoxicity on *Drosophila melanogaster* Mg	Rate of occurrence of dominant lethal mutations in an experiment compared with the reference during 72 h	Ditto	Genotoxicity absent
Toxicity on *Tetrahymena pyriformis*	A reduction of the increment coefficient of the number of infusorians in an experiment compared with the reference during the time set—24 hours (a short-term biotesting) or 96% (a long-term biotesting)	" "	Toxicity absent
Toxicity on *Vibrio fischeri*	Reduction of the level of luminescence of bacteria in an experiment compared with the reference over 30 min	" "	Toxicity absent
Genotoxicity on *Salmonella typhimurium*	A deviation of the total mutagenic activity of umuC-gene of bacteria *Salmonella typhimurium* in an experiment compared with a reference over 4 hours	" "	Genotoxicity absent

Radiation safety of drinking water shall be determined by the next acceptable levels: total volumetric activity ofα-radiators ($\Sigma\alpha$-activity) not exceeding 0.1 Bq/dm^3 and total volumetric activity of β-radiators ($\Sigma\beta$-activity)—1.0 Bq/dm^3.

By organoleptic and chemical quality indexes, which affect organoleptic properties, drinking water should correspond to norms given in Tables 4 and 5.

Table 4: Organoleptic indicators of drinking water quality.

Indicator	Measurement units	Norm, not more than: Water of systems of centralized drinking water supply	Norm, not more than: Water of uncentralized drinking water supply additionally purified (unbottled, bottled)
		Organoleptic indicators of quality	
Smell at 20°C	Point	2	0
Smell during heating to 60°C	Point	2	1
Smell and taste	Point	2	0
Color	Deg	20	5
Turbidity	NUT	2.5	0.5

Table 5: Chemical indicators of quality, which affect the organoleptic properties of drinking water.

Indicator	Measurement units	Norm, not more than: Water of systems of centralized drinking water supply	Norm, not more than: Water of uncentralized drinking water supply additionally purified (unbottled, bottled)
		Inorganic components	
Hydrogen index (pH)	pH unit	6.5–8.5	6.5–8.5
Dry residue (total mineralization) optimal content	mg/dm^3	1000 (1500)*	1000 100–400
Total hardness, optimal value	mmol/dm^3	7 (10)*	7 1.5–4.0
Total alkalinity, optimal value	mmol/dm^3	6.5	6.5 0.5–5.0
Sulfates	mg/dm^3	250 (500)*	150
Chlorides	mg/dm^3	250 (350)*	150
Total iron (Fe)	mg/dm^3	0.2	Absence
Manganese (Mn)	mg/dm^3	0.05	Ditto
Copper (Cu)	mg/dm^3	1.0	" "
Zinc (Zn)	mg/dm^3	5.0	" "
Calcium (Ca), optimal value	mg/dm^3	100	100 20–60
Magnesium (Mg), optimal value	mg/dm^3	30	30 6.0–15
Sodium (Na), optimal value	mg/dm^3	200	200 20
Potassium (K), optimal value	mg/dm^3	20	20 2.0–20
		Organic components	
Methyl-*tert*-butyl ether	mg/dm^3	0.015	Absence
Oil products	mg/dm^3	0.05	Ditto
Chlorophenols	mg/dm^3	0.0003	" "

*The value in the brackets may be established following the ruling by the Chief State Sanitary Physician in the corresponding area for the specific system of the drinking water supply based on the evaluation of the sanitary-epidemiological state in a town and the technology of drinking water treatment used in a case, when other sources of drinking water supply are inaccessible.

According to toxicological indices of harmlessness of the chemical composition drinking water should be in line with norms given in Table 6 and the EC directives [6].

Table 6: Toxicological indicators of the harmlessness of drinking water chemical composition.

Indicator	Measurement units	Norm, not more than	
		Water of systems of centralized drinking water supply	Water of uncentralized drinking water additionally purified (unbottled, bottled)
		Inorganic components	
Aluminum (Al)	mg/dm^3	0.2 (0.5)[2*]	Absence
Ammonia (by NH_4^+)	mg/dm^3	0.5 (1.5)[2*]	Ditto
Barium (Ba)	mg/dm^3	0.1	0.1
Beryllium (Be)	mg/dm^3	0.0002	Absence
Boron (B)	mg/dm^3	0.5	Ditto
Cadmium (Cd)	mg/dm^3	0.001	« »
Arsenic (As)	mg/dm^3	0.01	« »
Nickel (Ni)	mg/dm^3	0.1	« »
Nitrates (by NO_3^-)	mg/dm^3	45	5
Nitrites (by NO_2^-)	mg/dm^3	0.1	0.02
Perchlorates (ClO_4^-)	mg/dm^3	0.01	Absence
Mercury (Hg)	mg/dm^3	0.0005	Ditto
Leas (Pb)	mg/dm^3	0.01	« »
Selenium (Se)	mg/dm^3	0.01	« »
Strontium (Sr)	mg/dm^3	7	2
Stibium (Sb)	mg/dm^3	0.005	Absence
Thallium (T1)	mg/dm^3	0.0001	Ditto
Fluorides (F^-) for climatic region:	mg/dm^3		
II		1.5	1.5
III		1.2	1.2
IV		0.7	0.7
Total chromium (CCr)	mg/dm^3	0.05	Absence
Cyanides (CN^-), in particular cyanogen chloride	mg/dm^3	0.05	Ditto
		Organic components	
Benz(a)pyrene	mg/dm^3	0.00001	Absence
Benzol	mg/dm^3	0.001	Ditto
Pesticides (sum)[3*]	mg/dm^3	0.0005	« »
Synthetic anionoactive surface-active substances (ASAS)	mg/dm^3	0.1	« »
Trichloroethylene and tetrachloroethylene (sum)	mg/dm^3	0.01	« »
Tetrachloride carbon	mg/dm^3	0.002	« »
		Integral indicators	
Permanganate oxidizability	$mg\ O/dm^3$	2	0.75
Total organic carbon	$mg\ C/dm^3$	4	1.5

[1*]When in drinking water several chemical substances are detected, which refer to the 1st and 2nd classes of safety, and have sanitary-toxicological indicators of hazard, the sum of relationships of concentrations of every one of them determined in water to the standard should not exceed 1.
[2*]The value in brackets may be set only by the ruling of the Chief State Sanitary Physician in the corresponding area for the specific system of drinking water supply based on the evaluation of the sanitary-epidemiological state in a town and the technology of drinking water treatment used with the account of the specific situation.
[3*]Pesticides (sum): organic insecticides, herbicides, fungicides, nematicides, acaricides, algicides, bactericides, virucides, rodenticides, slimicides, bound products (in particular growth regulators), and then metabolites and degradation products. The program of control involves only these pesticides, which, most likely, may be contained in this water.

For the water supply system, which uses reagent water treatment methods prior to its delivery to the distribution network, during pouring out, transportation, and storage for a set time of suitability and in conducting research the indexes given in Table 7 are additionally taken into account. The content of substances that are formed and enter the drinking water during its treatment should not exceed the norms given in Table7.

Table 7: Substances formed and getting into drinking water during water conditioning.

Indicator[1*]	Measurement units	Norm, not more than Water of systems of centralized drinking water supply	Water of uncentralized drinking water supply additionally purified (unbottled, bottled)
Acrylamide[2*]	mg/dm^3	0.001	Absence
Bromates	mg/dm^3	0.1	Ditto
Residual chlorine dioxide	mg/dm^3	0.2–0.8	" "
Residual ozone, within	mg/dm^3	0.1–0.3	" "
Residual polyphosphates (PO_4^{3-})	mg/dm^3	3.5	" "
Trihalogenmethanes: chloroform, bromoform, dibromochloromethane, bromodichloromethane (total)	mg/dm^3	0.1	" "
Formaldehyde	mg/dm^3	0.05	" "
Residual free chlorine, within	mg/dm^3	0.3–0.5	" "
Residual bound chlorine, within	mg/dm^3	0.8–1.2	" "
Chlorate ion	mg/dm^3	0.7	" "
Chlorite ion	mg/dm^3	0.7	" "
Chloroform	mg/dm^3	0.06	" "
Dibromochloromethane	mg/dm^3	0.03	" "

[1*] The program of control involves specific indicators with the account of the technology of water conditioning from the cited list.
[2*] The program of control involves specific indicators of using respective reagents and calculating them based on the analysis of the content of the monomer in the commodity flocculant. The reagent doses and the level of the monomer should correspond to the following requirements: at the polyacrylamide dose 1 mg/dm^3 the content of acrylamide in it should not exceed 0.05%.

Reagents, introduced as preservatives for drinking water, shall be in accordance with Table 8. Other substances and methods for preservation of add-on purified bottled water from a noncentralized drinking water supply avoid the introduction of carbon dioxide and silver ions.

Table 8: Reagents permitted as preservatives for drinking water.

Preservatives	Measurement units	Permitted mass fraction of a preservative in water, not more than
Silver (Ag)	mg/dm^3	0.025 (0.05)[1*]
Carbon dioxide (CO_2)	%	0.2–0.6

[1*] For waters containing chloride ions >50 mg/dm^3.

Thus, the approval and introduction of the new standard will allow us:(i)to substantially reduce capital and operational costs for obtaining high-quality drinking water in amounts sufficient for meeting physiological needs of the population;(ii)to reduce costs for exercising control over the quality of drinking water intended for human consumption as a result of using integral methods of assessing water quality without reequipment of chemical-analytical laboratories with expensive equipment;(iii)to improve the state legislative database for providing the population with good-quality drinking water, safe for human health.

Designations and Abbreviations

BCG:	Groupofcolibacillusbacteria
Bq:	Becquerel
PFU:	Plaque-formingunits
GMN:	Generalmicrobialnumber—numberofsaprophytemicroorganisms
CFU:	Colony-formingunits
NUT:	Nephelometricunitsofturbidity
ASAS:	Anionoactivesurface-activesubstances
SSAS:	Syntheticsurface-activesubstances
TCB:	Thermostablecolibacillusbacteria.

REFERENCES

1. M. F. Falcone-Dias, I. Vaz-Moreira, and C. M. Manaia, "Bottled mineral water as a potential source of antibiotic resistant bacteria," Water Research, vol. 46, no. 11, pp. 3612–3622, 2012.
2. M. José Figueras and J. J. Borrego, "New perspectives in monitoring drinking water microbial quality,"International Journal of Environmental Research and Public Health, vol. 7, no. 12, pp. 4179–4202, 2010.
3. M. A. Shannon, P. W. Bohn, M. Elimelech, et al., "Science and technology for water purification in the coming decades," Nature, vol. 19, pp. 295–2301, 2008.
4. A. Bosch, S. Guix, D. Sano, and R. M. Pinto, "New tools for the study and direct surveillance of viral pathogens in water," Current Opinion in Biotechnology, vol. 452, pp. 301–310, 2008.
5. V. V. Goncharuk, "SOS: drinking water," Journal of Water Chemistry and Technology, vol. 32, no. 5, pp. 255–283, 2010.
6. "The Directive of the European Union Council 98/82/EU concerning water quality designed for human consumption," The Official Journal of the European Communities, pp. 2–5, 1998.
7. Guidelines for Drinking—Water Quality, vol. 1 of Recommendations, World Health Organization, Geneva, Switzerland, 3rd edition, 2004.
8. CODEX STAN 227-2991, General Standard for Drinking Waters Bottled/Packed (different of mineral waters).
9. U.S. Environmental Protection Agency, "National primary drinking water regulations, disinfectants and disinfection byproducts, final rule," Federal Register, vol. 63, no. 241, Article ID 69389, 1998.
10. Directive of the EU 98/83/EU of November 3, 1998, Drinking Water. Regulatory Documents. Reference Book [in Ukrainian].—L'viv, Leonorm, Part 2.—323 P., 2001.
11. V. V. Goncharuk, "A new concept of supplying the population with a quality drinking water," Journal of Water Chemistry and Technology, vol. 30, no. 3, pp. 129–136, 2008.
12. O. P. Oksiyuk and V. M. Zhukiskii, "Environmental water quality standards for the Ros' River,"Hydrobiological Journal, vol. 35, no. 6, pp. 16–21, 1999.
13. V. V. Goncharuk, A. V. Rudenko, E. Z. Koval, et al., "The problem of water infection with pathogens of mycoses and prospects of resolving this problem," Journal of Water Chemistry and Technology, vol. 26, no. 2, pp. 1–18, 2004.
14. M. Fielding and M. Farrimond, Disinfection by-Products in Drinking Water: Current Issues, Royal Society of Chemistry, Cambridge, UK, 1999.
15. V. V. Goncharuk, N. A. Klimenko, L. A. Savchina, T. L. Vrubel, and I. P. Kozyatnik, "Water treatment and demineralization technology: current issues in the technology of drinking water conditioning," Journal of Water Chemistry and Technology, vol. 28, no. 1, pp. 2–49, 2006.

CHAPTER 2 WATER AS THE EARTH'S BUFFER AND IMMUNE SYSTEM

Vladyslav V. Goncharuk

WATER RESOURCES OF THE PLANET

The bulk of the planet's water is concentrated in the seas and oceans and constitutes around 1350 million km^3. This is salt water, with an average mineralization of 35.0 g/L, which makes it unsuitable not only for drinking but also for agriculture needs. Fresh water is localized mainly in the North and South poles and in the mountains in the form of glaciers. This resource constitutes 30–50 million km^3. However, this source of fresh water is almost unobtainable for human use.

The earth's rivers and lakes contain about 0.4 million km^3 of fresh water that is accessible for mankind. Subterranean regions of the planet have substantial stocks of water, mainly saline and salt ones. At a depth of 800 m and up to 1600 m, the water resource constitutes around 4 million km^3.

Thus, our planet has only 3% of fresh water of its total amount, with its bulk in the pack ice of the Arctic and Antarctic. The resources of fresh water accessible to human constitute only 0.06%, 0.8 million km^3.

At present, more than 40 countries of the world experience an absolute water deficiency (e.g., the Near East, Africa, Indochina, and Australia). A fifth of the population in Europe and the Americas drinks contaminated water, which does not meet international standard criterion. According to the World Health Organization (WHO) official data, around 80% of human diseases around the globe are related to the consumption of low-quality drinking water [1].

The first attempts to establish priorities for conservation and functioning resilient natural resources were defined in the documents of the United Nations (UN) conference in Rio de Janeiro (June, 1992) and the World Summit on Sustainable Development in Johannesburg "Johannesburg Agreement on Sustainable Development" (August 2002) [2]. "Every inhabitant of the planet has the right for clean drinking water," is the formula declared by the UN [3, 4]. This humane statement is unfortunately not supported by any specific actions. An ill-advised economy, the pursuit of profits without accounting for ecological problems, and the harsh competitive struggle of monopolies have resulted in the majority of surface and underground drinking water supply sources being depleted and heavily polluted. As a result, polluted water penetrates rivers, seas, and oceans. A fourfold increase of the world population within the 20th century alone—from 1.5 to 6 billion people—has dramatically exacerbated this deficiency of fresh water, which is used not only for drinking purposes but also for industrial activity. All of this combined has resulted in the worsening of the global ecological situation.

Citation: Vladyslav V. Goncharuk, "Water as the Earth's Buffer and Immune System," Journal of Chemistry, vol. 2013, Article ID 472323, 14 pages, 2013. doi:10.1155/2013/472323

Artesian groundwater is a source of drinking water that is reliably protected from human impacts. We currently know of more than 150 different types of freshwater and saline drinking water treatments.

The most abundant impurities in underground waters are two-valence iron and manganese, fluorine ions, nitrates, ammonia, hydrogen sulfide, hard salts, and an increased salt content. Normally, these impurities substantially exceed maximum allowable concentrations for drinking water. Their presence is caused by natural factors of a geological nature. At the same time, one has to remember that high-quality water for human health should contain a wide spectrum of micro-impurities, vitally important, biologically active elements, and natural organic compounds. It is these compounds dissolved in water that provide its taste, smell, transparency and physiological properties.

Surface-level fresh waters are the second most common source of the drinking water supply. However, the global nature of human activities in the 20th century in terms of industrial development, progress in agriculture, transportation, urban utilities, the formation of megalopolises, and the growth of cities and settlements resulted in the wide-scale pollution and contamination of surface waters. The composition of wastewaters is constantly changing due to the synthesis of new chemical substances, often possessing toxic, carcinogenic, and mutagenic properties.

One can identify the following most hazardous types of pollutants entering the environment which eventually affect the water supply:(i)chemical pollutants:(1)inorganic compounds:(a)heavy metal ions,(b)salts,(c)toxic, biologically active substances,(2)organic compounds:(a)oil products,(b) phenols,(c)pesticides,(d)surface active substances,(e)chlorine-containing compounds,(f)xenobiotics;(ii)bactericidal and viral pollutants;(iii)radioactive substances of natural and anthropogenic originisotopes of elements;(iv)mutagenic compounds of organic and inorganic origin;(v)mycotic pollutants.

Seas and oceans may also serve as the source of drinking water supply. It is known that the average salt content in water constitutes about 35 g/l. It is natural that such water cannot be used for drinking. There are various approaches intended to desalinate water. In practice, desalinization technologies such as distillation, membrane, and electrochemical methods are used.

Advantages and Drawbacks of Water Desalination by Reverse Osmosis for Drinking Water Purposes

Desalination of sea (ocean) water is one of the fastest growing sectors of the economy. The introduction of reverse osmosis technology has had a big impact on the rapid development of desalination. Reverse osmosis is used for water treatment and purification of drinking water from heavy metals, nitrates, nitrites, surfactants, phenols, hardness, organic and microbiological contaminants, and organochlorine compounds.

At present, this method is one of the most popular methods of purifying drinking water, including artesian well water for the industrial production of bottled drinking water. The reverse osmosis method of water purification produces water which contains virtually no contaminants. This method has more economical operation and maintenance than any other technology. The effectiveness of using reverse osmosis technology for the desalination of sea water is three times more effective in terms of energy consumption and the degree of purification than the other methods.

At the same time, the obvious disadvantage of reverse osmosis is the lack of selectivity of the membranes to remove contaminants. This means that not only toxic impurities and microorganisms are removed, but also vital minerals and microelements. Will this water fulfill physiological requirements? Of course not! After all, purity of drinking water is not the only criteria of water quality. Also important is the extent to which drinking water is the source of elements necessary for normal body functioning. Water obtained by the reverse osmosis method cannot be called drinking water; one gets somewhat distilled water.

The results of studies on biotesting demineralized (distilled) water show that it affects both the physiological and intracellular processes in the body.

The required presence in the optimal drinking water content of calcium and magnesium is confirmed in Figure1. Regions supplied by water with a low concentration of calcium and magnesium exhibit a significant (30–40%) increase in the incidence of hypertonic disease (see Figure 1).

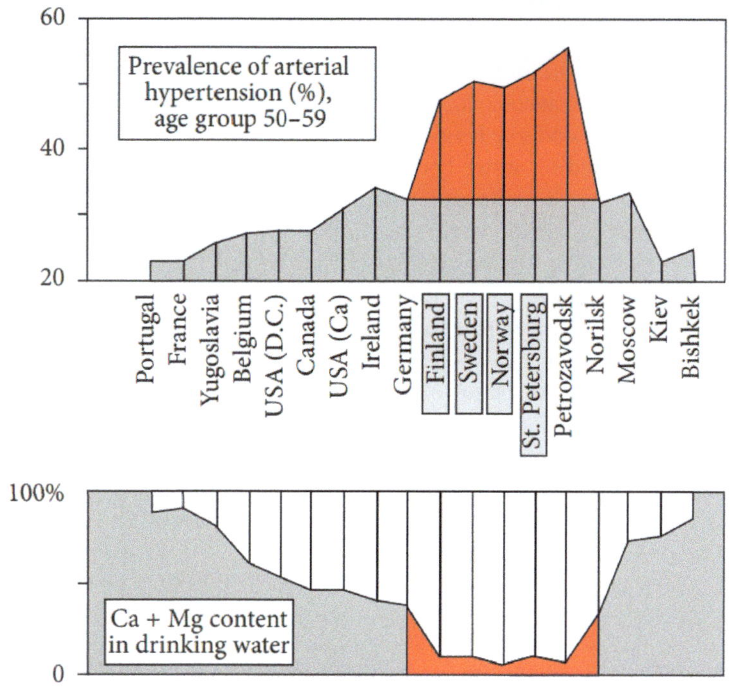

Figure 1: Effect of calcium and magnesium in the drinking water on the incidence of hypertonic disease [5].

It is necessary to do adjustments to the salt composition in water obtained by reverse osmosis [6]. This is achieved by the introduction of salts required for normal human activity: calcium, magnesium, sodium, and potassium. Taking into account this fact, factories using the reverse osmosis method need to provide special mineralizers, in order to restore the optimal mineral content of water. Thus, an artificial formation of water for drinking purposes is created. This is not the best option, but is vitally necessary where there is no other source of fresh water.

The Proof of the Failure of World's Approach to Assessing the Quality of Drinking Water and New Scientifically-Based Approaches to Assessing Its Quality

Life on earth is only possible as a result of the presence of water. The emergence and development of biological diversity on our planet is due to the presence of water, which has unique physical and chemical properties. Almost a hundred years it is known that water—not a simple compound where apart of protium there are also heavy isotope patterns however no one has ever thought about why they exist and what their role.

The unique properties of water is determined by its isotope composition and, first of all, by the ratio of protium and deuterium. The natural concentration of deuterium in the world's oceans, in salt and fresh waters varies within the interval 90–180 ppm. The stable optima, concentration of deuterium in water, constitutes 150 ppm. It is within this concentration interval that our ordinary water possesses maximum biological activity both in sea and in fresh water bodies. A decrease or an increase of the deuterium concentration in water results in a radical change of its physicochemical and biological properties [7–9].

Water is currently viewed from fundamentally new positions. The comprehensive study of water has yielded quite new results [10]. It has been found that if the deuterium concentration in water is adjusted, all ideas about its fundamental properties accepted today are changed drastically. Both the melting point and the freezing point, and other water properties, are substantially changed depending on the relation of the isotope composition of hydrogen. If deuterium is completely removed from water, then it will not exist on our planet in its liquid state. Consequently, water structure, just as all physicochemical parameters, is determined by the presence of deuterium. The results regarding the influence of deuterium on water properties became the basis for the discovery of a fundamentally new method for monitoring water quality.

Here, questions arise. What is drinking water? How can the quality of drinking water be assessed? What are the optimal parameters that water should possess in order to be completely safe according to biological and physiological points of view?

The development and the setting of regulations and standards for drinking water quality in different countries since the late 19th and early 20th centuries have changed from simply regulating the macrocomponents of natural water to more insightful knowledge about the impact of anthropogenic contaminants and toxic microcomponents of water on human organisms [11].

In connection with the worsening of the ecological state of surface and underground sources, the issue of water quality control, used by people for drinking purposes, became more acute [12, 13]. As a result of the low quality of drinking water, real threats to the sanitary-epidemiological situation in various regions of the planet crop out.

An increase of the quality of the matter being controlled in national standards of various countries of the world does not solve the issue of obtaining safe drinking water at centralized water-treatment stations either. All this calls for ever more expensive equipment and the complication of technological processes. Special attention should be paid to the fact that over the last two to three decades, an uncontrolled sharp increase of chemical compounds in the environment took place [11].

While the environment experiences these catastrophic changes, the quality control of dozens of substances in tap water does not guarantee that it really is drinkable and safe for human health. A high level of a technogenic load on water bodies, the use of imperfect technologies of water treatment, and secondary contamination of the water in distribution networks result in the ingress to the drinking water of a substantial amount of inorganic and organic pollutants, whose joint effect on the human organism causes the effect of synergism known in chemistry and biology, threatening human health. In this situation there is no possibility to provide the population of any country and continent with quality drinking water that is safe for human health. In addition, drinking water obtained from surface sources is unsafe due to the presence microtoxins in it of. As a result of our research, we found the presence of toxic micromycetes in surface waters and distribution systems. These micromycetes are not disinfected by current techniques used at centralized water treatment stations, even when chlorinated with high doses.

Hence, classic approaches to the assessment of drinking waters are useless. The notion of "normalizing maximum allowable concentrations in drinking water of different toxicants" purportedly safe for human health, is, as a matter of fact, immoral.

QUALITY OF DRINKING WATER: PROBLEMS AND SOLUTION

The harmonic development of Homosapiens concurrently with the planet's biosphere calls for the immediate prevention of environmental pollution. Our generation has witnessed qualitative changes of natural and drinking waters. For the first time in the history of mankind, the necessity of introducing quality standards for drinking water has appeared. These standards over time were improved, and new ones were introduced. Substances both of natural and anthropogenic origins were hazardous for human health and were strictly controlled. General human and worldwide approaches for drinking water quality evaluation were established. The first countries in the world that developed state standards for drinking water quality were the USA and the USSR. Owing to a high level of bacterial pollution of surface waters, drinking waters were disinfected with chlorine. This was one of the greatest mistakes of mankind. Since water always contains organic compounds, its chlorination inevitably results in the formation of very toxic, mutagenic, and carcinogenic organochlorine compounds. For the first time in the history of mankind, people began to drink chlorinated water hazardous for human health—technogenic water!

At the beginning of the 20th century, the level of surface-water contamination with chemical and bacterial components did not achieve a critical state. Therefore, small doses of chlorine, used for disinfection of water, do not lead to the formation of substantial concentrations of hazardous organochlorine compounds. However, the turbulent development of industry, agriculture, the formation of megapolises, and a dramatic increase of the population size of the world (a fourfold rise over the 20th century) resulted in a catastrophic level of bacterial and chemical contamination of water sources used for conditioning drinking water. It entailed the necessity of using large doses of chlorine for the disinfection of water. As a result, there was an increase in the concentration of organochlorine compounds in drinking water. Given such a dangerous factor, new, more sophisticated technologies of water treatment began to be developed. These technologies include preliminary filtration from suspended particles, primary chlorination, and chemical water treatment by coagulation concurrent with aluminum, iron, and flocculants of organic and inorganic origin, then filtration on sand and carbon filters. For suppressing the development of microorganisms in pipelines, the water again is treated with chlorine of such a concentration that at the outlet of the faucet of every consumer the content of residual activated chlorine is within the range of 0.3–0.5 mg/dm^3.

Schematic chlorine dioxide, possessing a higher oxidizing potential, is used instead of chlorine in a number of countries, according to regulations. Potentials of some oxidants are given in Table 1.

Table 1

Chemical substance and its formula		Oxidizing potential, eV
Chlorine	Cl_2	1.36
Chlorine dioxide	ClO_2	1.57
Hydrogen peroxide	H_2O_2	1.78
Ozone	O_3	2.07
Atomic oxygen	O	2.42
Hydroxyl radical	$HO^{\bullet}$	2.80

The current technology has its advantages and disadvantages. Advantages include the fact that the decontamination process occurs more effectively. However, it also means that chlorine dioxide may produce a broader set of organochlorine compounds.

An especially negative side of the abovementioned technologies is the use of coagulants containing aluminum. On the one hand, this final introduction of many types of contaminants into the initial water is a necessary stage of treatment. On the other hand, it introduces a new, very dangerous contamination of drinking water with residual aluminum compounds. It was known for a long time that aluminum ions contained in drinking water are exceptionally toxic and affect human health. Several publications deal with this research. This is why the World Health Organization from year to year toughens the requirements regarding the concentration of residual aluminum in drinking water.

Based on these two important prerequisites, namely, the presence in drinking water of organochlorine compounds and aluminum, I allow myself to discuss drinking water not as tap water but as technogenic water. Never before has mankind consumed water for drinking purposes which contained very toxic organochlorine compounds and also aluminum ions, which are byproducts of the modern water treatment technology.

Contaminations of the initial water reached such a level that water treatment facilities built in compliance with the effective world standards already are not capable of preventing the ingress of substances to drinking water. In this regard, their joint effect became a real threat to human health [14].

All this results in the necessity of a search for new alternative water treatment technologies to obtain safe water. Here, first and foremost, one should pay attention to natural, ecologically safe oxidants, which could be used in advanced technologies for obtaining high-quality drinking water from the severely fouled sources.

In the last century, for the first time, Russian scientists proposed a fundamentally new approach in water treatment technology—the use of ozone instead of chlorine. The world first ozonizer was developed in Lomonosov Moscow State University, and the first ozonation station for water treatment was built and commissioned in St. Petersburg before World War I. Since then, many years passed before the world community started to treat this technology with due attention.

Below are specified the potential oxidants used today in water treatment, which are the most promising for creating new technologies.

The analysis of the cited data is evidence of the fact that the maximum reactivity is possessed by a hydroxyl radical, which is capable of oxidizing any organic compound into carbon dioxide and water. It is formed in interaction of the ozone with water or hydrogen peroxide with water by the following mechanism:

$$O_3 + H_2O \longrightarrow H_2O_2 + O_2$$

$$H_2O \xrightarrow{h\nu} HO^\bullet + HO^\bullet$$

$$O_3 + HO^\bullet \longrightarrow HO_2^\bullet + O_2$$

$$O_3 + HO_2^\bullet \longrightarrow HO^\bullet + 2O_2$$

$$O_3 + H_2O_2 \xrightarrow{\text{Very slowly}} HO_2^\bullet + O_2 + HO^\bullet$$

$$O_3 + HO_2^- \xrightarrow{\text{Quickly}} HO^\bullet + O_2^{\bullet-} + O_2$$

$$O_3 + H_2O_2 \longrightarrow HO_3^\bullet + O_2^{\bullet-} + H^+ \tag{1}$$

In aqueous alkaline solutions ozone is decomposed with the formation of $O_3^{\bullet-}$ according to the following reaction:

$$O_3 + HO^- \longrightarrow O_3^{\bullet-} + OH^\bullet \tag{2}$$

Then the process follows the schematic with separation of the anion radical of atomic oxygen $O_2^{\bullet-}$ superoxide anion radical $O_2^{\bullet-}$, and then molecular oxygen:

$$HO^\bullet + OH^- \longrightarrow O^{\bullet-} + H_2O$$

$$2O^{\bullet-} + H_2O \longrightarrow O + 2OH^-$$

$$O^{\bullet-} + O \longrightarrow O_2^{\bullet-}$$

$$2O_2^{\bullet-} + H_2O \longrightarrow HO_2^{\bullet-} + OH^- + O_2 \tag{3}$$

Acid conjugate to anion radical $O_2^{\bullet-}$ that is, radical $HO_2^{\bullet-}$ is a weaker donor of electrons than $O_2^{\bullet-}$, while in the system as following:

$$O_3^{\bullet-} = O_2 + O^{\bullet-} \tag{4}$$

The most active particle is the anion radical of atomic oxygen. Therefore, many reactions of $O_3^{\bullet-}$ are actually reactions of $O^{\bullet-}$. In an aqueous solution, radicals $O_3^{\bullet-}$, $O_2^{\bullet-}$, and $O^{\bullet-}$ are also formed in photolysis or γ-radiolysis.

According to reactivity the active particles may be arranged in the following series:

$$HO^\bullet > O_2^{\bullet-} > O_3^{\bullet-} > O^- > HO_2^\bullet \tag{5}$$

It has been found that impurities of hydrogen peroxide produce a rather strong catalytic effect on ozone decomposition. Schematically it may be displayed as the following:

$$HO_3^{\bullet} > HO^{\bullet} + O_2$$

$$O_3 + HO_2^{\bullet} > HO^{\bullet} + 2O_2$$

$$O_3 + O_2^{\bullet -} > O_3^{\bullet -} + O_2$$

$$O_3^{\bullet -} + H_2O > HO^{\bullet} + HO^{-} + O_2 \tag{6}$$

The results of our research confirmed that the identified processes are substantially accelerated in the presence of catalysts (both homogenous and heterogeneous) and under ultraviolet radiation.

It is exactly these results that were used for creating fundamentally new water treatment technologies for any types of pollutants in combination with a set of physical-chemical and biological methods. The research of the kinetics and the mechanism of all processes taking place in aqueous systems provided us with the possibility to develop not only new complex technologies, but new equipment for water treatment. These methods may help obtain high quality drinking water from effectively any source of drinking water supply as far as pollution is concerned. They have an exceptionally great significance for solving the issue of water disinfection from any microbiological and chemical source of pollution.

We offer three ways to solve these problems in the selection of the appropriate parameters—concentration, temperature, pH environment, and other factors.

The first is a simple water treatment with ozone or hydrogen peroxide.

The second is photooxidative disinfection:

$$O_3 + H_2O \xrightarrow{h\nu}$$

$$H_2O_2 + H_2O \xrightarrow{h\nu}$$

$$O_3 + H_2O_2 + H_2O \xrightarrow{h\nu} \tag{7}$$

The third is the combinated effect of three factors simultaneously:

(i) photocatalytic disinfection with ozone

$$O_3 + H_2O \xrightarrow{h\nu + cat} \tag{8}$$

(ii) photocatalytic disinfection with hydrogen peroxide:

$$H_2O_2 + H_2O \xrightarrow{h\nu + cat} \tag{9}$$

(iii) photocatalytic disinfection using the combined effects of ozone and hydrogen peroxide:

$$O_3 + H_2O_2 + H_2O \xrightarrow{h\nu + cat.} \tag{10}$$

(iv) photocatalytic disinfection with chlorine:

$$Cl_2 + H_2O \xrightarrow{h\nu + cat.} \tag{11}$$

To date, we are unaware of more efficient and environmentally friendly methods of water treatment.

WATER AND HUMAN HEALTH

The wide-scale use of chlorine in the technology for centralized drinking water supply stations is determined by the following three main reasons. The use of chlorine ensures:(i)primary decontamination of raw water;(ii)substantial improvement of the coagulation process whereby phyto- and zooplankton and many organic and inorganic substances are removed from water;(iii)prevention of the biological fouling of treatment facility tanks, filtering media, water system networks and equipment.

However, as was indicated above, water treatment with chlorine is accompanied by the formation of a whole range of highly toxic halogen-containing organic compounds such as chlorinated methanes, phenols, aldehydes, ketones, acids and especially hazardous polychlorinated biphenyls up to dioxins—the most toxic matter known on earth. Many years of medical-biological research demonstrated that the long-term use of such drinking water results in an emergence of neurotoxins, cardiovascular diseases, and oncological diseases of liver, kidneys, and the hematopoietic system of human and produces mutagenic effects. It is these compounds that lead to the development of impotence. Therefore, the technology of water ozonation has been used increasingly in water treatment systems all over the world. Ozone as an alternative reagent simultaneously plays both the role of a disinfectant and an oxidant. In addition, as was noted above, it possesses a higher oxidizing potential, especially the products of its interaction with water. Therefore, the rate of its interaction with all classes of organic compounds is very high. It decolorizes water very well, eliminates fetid smells and odors, removes iron and manganese, suppresses the growth of algae, and so forth. At the same time, one should remember that the use of ozone in the technology of water treatment is possible only in conjunction with other physicochemical methods.

Ecology—The Health of Human

Having addressed all the above problems, the world-wide issue of determining the degree of risk in the ecology-health system becomes a high-priority one [4]. The hygienic standards and regulations for the levels of environmental pollutants effective today are mostly of a declarative nature. In each specific country, in every region of the world, one needs to assess the priorities of pollutants by the risk levels to human health.

As sources of the risk I would separate four factors playing the most important part in the life of human:(1)drinking water,(2)air,(3)soil,(4)foods.

It is necessary to carry out a complex monitoring of all environmental aspects. Why is monitoring needed? First of all, for finding priority pollutants in each specific region that produces maximum negative impact on human health. Certainly, one may follow the path of eliminating all types of pollutions, but this requires fantastically large financial costs.

The most expedient formula of the ecological policy in every country should be the following: minimum costs → maximum results, starting from that of the highest priority, the most dangerous pollutants to human health.

One should isolate such main priorities of risk levels:(i) chemical pollutants,(ii)bacterial pollutants,(iii)radioactive and isotopic pollutants,(iv)mutagenic pollutants.

One should not forget that complex pollution of the environment with different classes of toxic substances very often results in synergism of their effects, when each toxicant's individual effect is much weaker than a blend of different toxicants. The most glaring among the known examples of the synergetic effect is a joint presence of radioactive matter and asbestos, which strengthen the negative effect of each other sixfolds.

As risk indicators let us point out the following five factors whose presence may help unambiguously assess the state of the environment:(i)genetic violations,(ii)disease incidence,(iii) birth rate,(iv)death rate,(v)average longevity.

Bottled Water: Issue of Disinfection and Preservation

Over the last two decades, the technology of water bottling and its sale through a wide network of shops has become very popular [14, 15]. There are just a few who would think about the enormous scientific and technical issues of preserving the quality of such waters. First of all, the water being subjected to bottling should be immaculate in terms of chemical, biological, and organoleptic indicators and in addition should preserve its qualities in a closed state.

We talked about the issue of water disinfection above, but the issue of its preservation requires a more detailed consideration. Carbon dioxide is the most common preservative, which diluting in water, forms carbonic acid. The taste of such acid water is experienced by everybody who drank aerated water in their lifetime. This water is used for extreme situations of quenching one's thirst rather than for its systematic consumption.

The other effective reagent preserving water is silver or, to be more accurate, silver ions. Since ancient times, the technology of water storage in silverware has been known. Such water can be drunk without worry since the concentration of silver ions is meaningless, but the disinfection process is rather lengthy. It can be amplified a thousand times using the electrochemical technology of silvering water. However, the excessive silver concentration is dangerous. Over time, silver ions are adsorbed on the walls of the vessel and rather quickly lose their disinfecting properties. Under these conditions, secondary bacterial contamination is possible.

Unscrupulous companies may use as preservatives special chemical substances whose functional properties have effects similar to those of antibiotics, which suppress or destroy microorganisms. These substances present enormous hazard for human health. The consumption of such water by drinking leads to the sterilization of the digestive tract causing a known disease—dysbacteriosis.

The chemical nature of the containers used is a very serious issue in the process of water bottling. There are a sufficiently large number of investigations, whose results are an evidence of the toxicity of polymer containers. The safest containers are glassware, glass containers, and enameled and ceramic ware.

"GREEN" CHEMISTRY

The contemporary world early in the 20th century found itself on the verge of the global ecological crisis, which threatens the very existence of civilization. Industrial production still remains high-waste. That is why a new science has appeared and continues to develop effectively. This is the science of ecological economy, which tries to unite social, economic, and ecological systems. Ecological economy should take into account ecological costs when assessing the economic performance of production. A modern economy should develop with the account of ecological factor.

It is quite clear that the correction of the current unfavorable situation does not look to be an easy or quickly resolved issue. This results from the impossibility of allocating sufficient investments for the solution of the ecological issues.

In the world science, this area of science and technical investigations has been referred to as "green chemistry".

When we talk about drinking water quality we, first of all, should think and care about the quality of drinking water supply sources, and this is almost the entire water basin of the planet. The solution to the existing issue of protecting the whole ecosystem from the products of mankind's life activity lies in the quest for fundamentally new technical solutions. These solutions aim not only to overcome the consequences, but should prevent the causes, leading to the unfavorable ecological aftermath.

The most widely known water contamination, referred to as water "color," is caused by blue-green algae—cyanobacteria. An important adverse influence on the water basin is related to their ability to synthesize biologically active substances possessing very high toxicity. The ingress to water of such biogenic elements as phosphorus, nitrogen, potassium, sulfur, and also organic matter from industrial, domestic, rainfall, and waste waters of the agricultural production created the prerequisites for intensive development of algae of various systematic groups, which results in the formation of water "color".

The facts of the formation of toxins by algae have been known for a rather long time, as early as the 1960s, while the systematic research of the causes of this phenomenon and factors increasing the toxicity started about 20 years ago. The most serious attention at present has been drawn to the toxicant microcystine, produced by cosmopolite cyanobacteria, causing water "blooming" in water bodies of virtually all countries of the world. In terms of biological activity, toxins of cyanobacteria surpass by many times such known substances as strychnine, curare, a range of fungi toxins, and potassium cyanide. They exceed the toxicity of potassium cyanide by as much as 50–1000 times. Such a strong poison as curare exceeds them by least 10 times. Hence, it is quite obvious that such water presents a serious threat to the life of fish, birds, and humans. Even swimming in "blooming" water pools is dangerous for human health, not to mention the use of this water as a source of drinking water supply.

All these facts indicate the necessity of forming the Global World Program for the Protection of the Planet's Water Basin against Pollution.

THE BIOSPHERE AND CIVILIZATION—THE ISSUE OF COMPATIBILITY

The biosphere in its classical state no longer exists. We are witnessing the apocalypse of the biosphere, which started in the middle of the 20th century. Transforming the environment—the biosphere, whose part is human himself—society underestimated the fact that creating the technogenic environment by its existence violates the basic laws of the biosphere, the laws of the universe.

Nowadays we live in the transition period of the coexistence of two worlds—the biosphere and the noosphere. There is no return to the biosphere. Man as a creature of the biosphere origin should find new ways for his survival and development. The gene pool of the biosphere has started to intensively rearrange itself—the turbulent process of evolution has begun, which has given birth to ecological changes on a global scale. Mass extinction of the existing species and emergence of new ones are observed across the board. Especially intensive is the evolution of infections. This is evidenced by a widely known fact of the adaptation of pathogenic microflora to antibiotics. This is an illustrative manifestation of mutagenesis—the emergence of a new, earlier absent quality—resistance to antibiotics. Complex genomes are created by nature during millions of years, while in modern laboratories they are created within a year. Man has designed a new artificial habitat, the TechnoSphere, kept safe from all biological processes of the biosphere, without connection with which he simply cannot exist.

The issues of nuclear energy and its impact on the environment bear a complex nature. Since the radioactivity phenomena have been discovered, the irreversible processes of dramatic increase of background radiation all over the planet have been seen [16].

The intensive development of the nuclear industry led to the emergence of a considerable number of objects presenting potential radiation danger: nuclear power plants with reactors of various types, chemical mining plants, and plants for processing nuclear fuel. The radionuclide compositions of pollutants emitted to the environment during the operation and also in emergency situations substantially differ, which makes the use of a single flow chart of purification impossible in all mentioned cases.

Among the numerous aspects of protecting the water basin from technogenic pollutants, one of the most important aspects is achieving the effective purification of various types of pollutants, including radioactive pollutants. Achievement of this task is hampered by the general worsening of the ecological situation, accompanied by the appearance in the waters of an ever larger amount of various toxicants of organic and inorganic nature. Together with the necessity of development and the employment of more complex and sophisticated technologies to achieve purification levels to meet the existing standards, this begs the question of the applicable range of the standards themselves. Here the ever increasing restraint, which currently manifests itself despite the sufficient theoretical and practical substantiation at the moment of adoption, is determined by insufficient accounting of the possibility for aggravation of the adverse influence of toxicants on live organism in their synergetic effect. The synergetic effect of pollutants of various types and radionuclides seems to be one of the most dangerous for human health.

The adverse effect of radioactive elements of the environment is of a complex nature. Despite the fact that it is well studied, the issue of the processing and burial of radioactive waste, which today like "a nuclear gene" are either on the sea bottom or in the mines under earth and are waiting for their time, is still outstanding.

There is one more aspect of this issue which has not been sufficiently studied yet. I mean the issue of the impact of nonradioactive isotopes on the development of biological objects and, first of all, hydrogen isotopes, in particular, deuterium. The impact of the isotopic composition of substances on kinetic characteristics of numerous reactions has been known for a long time.

It turns out that deuterium strongly affects the chemical parameters of chemical and biological processes since it forms a stronger bond with oxygen in a water molecule, unlike light hydrogen—protium. It is also known that an increased concentration of deuterium in drinking water results in acceleration of the organism aging process. Deuterium manifests toxic properties with respect to living organisms [17].

Nowadays, one of the most serious issues is increased levels of background radioactivity and an increase of the concentration of heavy isotopes with respect to their light prototypes. First of all, it concerns vitally important elements such as hydrogen, oxygen, and carbon, which are a basis of protein life in our biosphere and affect the mechanism of biological evolution.

Thus, ecology is a science studied, strictly speaking, the state of the biosphere, where human is its indispensable part. However, today, when it is quite obvious that the biosphere is being transformed, in an increasing degree, to the noosphere transforming into the TechnoSphere, one cannot help taking into consideration that the world has really changed. It is necessary to have a clear notion of the place of human in a quickly changing world. Ecology transforms into a new geological phenomenon, into a new branch of science investigating the interaction of three globally coexisting constituent components of the universe: biosphere, noosphere, and TechnoSphere. The decisive role in their interaction is played by human. The future of our civilization is determined by the level of the awareness of the changes that have taken place in the environment.

A NEW APPROACH TO THE "ENVIRONMENT-HUMAN HEALTH" PROBLEM

Nature built up self renewing systems in the process of evolution supporting life activity of all live bodies. Gastrointestinal tract (GI) is the most impressive system of people and animals. Actually this is a flow-through bioreactor inhabited by many microorganisms. Total number of microorganisms exceeds number of macroorganism's own cells and makes up about 10^{14}–10^{15} cells [18]. Total weight of microorganisms amounts to 3-4 kg. Intestinal tract of healthy people contains more than 500 types of microorganisms. In the different gastrointestinal tract segments, number of bacteria is different and correlate with the state of digestive system (Table 2).

Table 2: Composition of the intestinal organisms of the human [19].

Types of bacteria	Average concentration of microorganisms (as per 1 mL or 1 g)			
	Intestine	Small intestine	Twisted intestine	Colon
Total number	$0–10^3$	$0–10^5$	$10^2–10^7$	$10^{10}–10^{12}$
Anaerobes				
Bacteroides	Seldom	$0–10^3$	$10^3–10^7$	$10^{10}–10^{12}$
Bifidobacteria	Seldom	$0–10^4$	$10–10^9$	$10^8–10^{12}$
Enterococci	Seldom	$0–10^3$	$10^2–10^6$	$10^{10}–10^{12}$
Clostridia	Seldom	Seldom	$10^2–10^6$	$10^6–10^3$
Eubacteria	Seldom	Seldom	Seldom	$10^9–10^{12}$
Aerobes				
Enterobacteria	$0–10^2$	$0–10^3$	$10^2–10^7$	$10^4–10^{10}$
Streptococci	$0–10^2$	$0–10^4$	$10^2–10^6$	$10^5–10^{10}$
Staphylococci	$0–10^2$	$0–10^3$	$10^2–10^5$	$10^4–10^9$
Lactobacteria	$0–10^2$	$0–10^4$	$10^2–10^5$	$10^4–10^{10}$
Fungi	$0–10^2$	$0–10^2$	$10^2–10^4$	$10^4–10^6$

Set of genes being a part of bacterial metagenome exceeds approximately 100 times set of genes of human organism [20]. Number of bacteria differs in the various segments of gastrointestinal tract. Number of microorganisms in the mouth cavity in acid environment conditions is little and makes up from 0 to 10^3 CGU (colonies of generatrices units) as per one milliliter of environment. In the lower sections of gastrointestinal tract number of microorganisms is much higher. The main factors determining reproduction of bacteria in the environment in the upper sections of gastrointestinal tract is a fast movement of food, secretion of bile and pancreatic juice. Environmental conditions in the colon are diametrically opposite, that is why number of bacteria in this section of gastrointestinal tract amounts to 10 CGU per one milliliter. Bifidobacteria andBacteroids dominate among several hundreds of bacteria in the intestinal tract which share constitutes 25 and 30%, correspondingly, with regard to the total number of anaerobic bacteria. Intestinal tract of a child contains no bacteria before his birth. Fast innidiation of child's intestine with bacteria which are a part of the composition of intestinal and vaginal flora of the mother takes place at the moment of birth. As a consequence it is formed complicated community of microorganisms consisting of Bifidobacteria, Lactobacteria,Enterobacteria, Clostridia, and Gram-positive cocci.

Then composition of microbial population is subject to changes due to the influence of several factors of the environment, the most important of which is child's feeding. As early as 1900 Tissier proved that the main components of intestine microflora of the breastfed children are Bifidobacteria. Such bifidodominant microflora exercises protective functions and is instrumental in maturation of the mechanisms of child's immune response. On the other hand, artificially fed children have much less Bifidobacteria in the colon, and species' composition of intestine microflora is less diversified.

Composition of intestine microflora is fairly individual and is formed from the first days of child's life, nearing indices of an adult person by the end of the 1st-2nd year of life, and undergoes some changes at the elderly age (Table 3).

Table 3: Composition of intestine microflora in various age groups.

Name of the group of microorganisms	Children of the first year of life	Adult people	People at the elderly age
Bifidobacteria	10^{10}–10^{11}	10^{9}–10^{10}	10^{9}–10^{10}
Lactobacteria	10^{6}–10^{7}	10^{7}–10^{8}	10^{6}–10^{7}
Bacteroides	10^{7}–10^{9}	10^{9}–10^{10}	10^{10}–10^{11}
Fusobacteria	$<10^{6}$	10^{8}–10^{9}	10^{8}–10^{9}
Veillonella	$<10^{5}$	10^{5}–10^{6}	10^{5}–10^{6}
Eubacteria	10^{6}–10^{7}	10^{9}–10^{10}	10^{9}–10^{10}
Peptostreptococcus	$<10^{5}$	10^{9}–10^{10}	10^{10}
Clostridium (lecitinpositive)	$<10^{5}$	$<10^{5}$	$<10^{6}$
Clostridium (lecitinnegative)	$<10^{7}$	10^{7}–10^{8}	10^{8}–10^{9}
Enterobacteria: *Escherichia coli* (with typical properties)	10^{7}–10^{8} $<10^{4}$	10^{7}–10^{8} $<10^{7}$	10^{7}–10^{8} $<10^{7}$
Enterococci	10^{6}–10^{7}	10^{7}–10^{8}	10^{6}–10^{7}
Staphylococci: Staphylococci epidermidis; Staphylococci aureus	$<10^{5}$ $<10^{1}$	$<10^{4}$ $<10^{2}$	$<10^{3}$ $<10^{2}$
Aerobic bacteria	$<10^{3}$	$<10^{5}$	$<10^{5}$
Fungi pertaining to *Candida* genus	$<10^{3}$	$<10^{4}$	$<10^{4}$

Numerous research works revealed that there is a connection between composition of intestine microflora and children's immune resistance [21–24]. Distribution of microorganisms in the gastrointestinal tract has quite exact laws. Most microorganisms (about 90%) are presented constantly in one or another section of gastrointestinal tract and represent the main (resistant) microflora, facultative (supernumerary, accompanying) microflora (makes up 10%), and 0.1–0.02% is accounted for accidental (transit or residual) microorganisms. Thus, the basic microflora in the colon is represented by anaerobic microorganisms, while aerobic bacteria make up an accompanying microflora. Staphylococci, Clostridium, Proteus, and Fungi belong in the colon to the residual microflora (Tables 2 and 3). Besides, there are about 10 intestine viruses and certain representatives of the nonpathogenic protozoa in the intestine. The quantity of strict and facultative anaerobes is always one order more than aerobes in the colon, and obligate anaerobes are immediately adhesived on epithelial cells, facultative anaerobes are located higher, followed by aerobic microorganisms. Thus, anaerobic bacteria (mainlyBifidobacteria and Bacteroides), which on the whole constitute about 60% of the total number of anaerobic bacteria, represent the most permanent and numerous group of intestine microflora carrying out the basic functions.

The whole complex of microorganisms and macroorganisms creates specific symbiosis. Recently some microbiologists declared a man and his microbes as one whole symbiotic superman. Progress in the development of methods of molecular biology thrust the scientists forward to the new level of understanding of processes of symbiosis of a man and his microflora. There seemed that processes of symbiosis of a man and his microflora were well studied and their further study will not give any specific surprise. However, quick growth of the speed with simultaneous fall of cost of methods of sequence analysis for deoxyribonucleic acid (DNA) (determining its nucleotide sequence), simultaneous growth of PC capacity, and development of Internet made it possible to analyze information about large areas of genomes. New (population) approach appeared in genetics of microorganisms after decoding of chromosomes of hundreds of types of certain bacteria: analysis of genes of all bacteria inhabiting a certain natural area at once. The population of "human bioreactor" is of course one of the most important for the investigation of microbial population. The first study made to look in a new way at the intestinal microbiota was published in 1999 by the group of scientists from the National Institute for Agronomic Research (France) and University of Reading (Great Britain). The authors decided to apply method of sequence analysis of genes of 16S RNA for the investigation of microbial population of intestines (16S RNA is certificate of identity of bacteria). Since Pasteur times the first and obligatory stage of identification of microorganisms was their cultivation in culture media. But a great number of important (useful and pathogenic) microbes do not grow in any media. Due to the development of bioinformatics and appearance of modern methods of molecular biology, it has become possible to study before inaccessible and uncultivated bacteria and to begin to put in order ultraentangled classification of already known prokaryotes [20].

It was established using new methods that only 24% of obtained sequences of 16S RNA belonged to the already known microorganisms. Three fourths of microflora in the intestine of each person avoided attention of the researchers armed by the methods of classic microbiology for more than one-hundred years! The scientists just could not select conditions for cultivation of these bacteria because the most capricious inhabitants of intestine refused to grow in traditional microbiological media. Just the same situation is observed in the chlorinated water of the water supply system. For the time being it was established by means of molecular method that in the microbiota of an adult person are presented 10 out of 70 large bacterial taxons. About 90% of our microbes belong to Firmicutes type (to them belong, e.g., known by everybody as Lactobacteria responsible for milk souring) and Bacteroidetes—strict anaerobes (organisms able to live only in absence of oxygen). The last are often used as indicators of contamination of natural waters by sewage runoff. Remaining populations (10%) are divided between taxa Proteobacteria (among others intestinal bacteria (Escherichia coli) belong to them), Actinobacteria (from one of actinomycetes types was obtained antibiotic streptomycin), Fusobacteria (usual inhabitants of mouth cavity and are frequent cause of parodontosis), Verrucomicrobia (not long ago a type of these microbes was found in geothermal spring feeding methane which is more than enough in intestine due to the vital functions of other microorganisms), Cyanobacteria (they are often called "blue-green algae"), Spirochaetes, Synergistes, and VadinBE97.

Feeding of microorganisms inhabiting intestine are provided at the cost of so-called nutrients coming from superposed sections of gastrointestinal tract, which are not digested with their own enzymatic systems and are not soaked in the thin intestine. Proteins and carbohydrates not soaked in the thin intestine are exposed to deeper bacterial decomposition in the blind intestinal—mainly by intestinal bacteria and anaerobic microorganisms. These agents are needed to provide energy and flexible requirements of microorganisms. Ability to use nutrients for its vital functions depends on the enzymatic systems of various bacteria. According to this, the following types of bacteria are conditionally marked out:(i) mostly with saccharolytic activities, carbohydrates are basic energy substrates which is characteristic for the saprophyte flora;(ii)mostly with proteolytic activities which are used with proteins for energy goals (representatives of pathogenic and conditionally pathogenic flora);(iii)bacteria with mixed activities.

Accordingly, predominance in food of one or another nutrient and violation of their digestion will stimulate growth of various microorganisms [25, 26]. Carbohydrate nutrients are especially needed for the normal vital functions of intestinal microflora. They were considered as "ballast", but later it was found out their importance for the intestinal microflora and people's health on the whole. According to the modern definition, prebiotics are partly or fully indigestible food which selectively stimulates growth and metabolism of one or several groups of microorganisms, providing normal composition of microbiocenosis [27–30].

Microflora of the gastrointestinal tract performs the following functions [31–35].(i)Morphokinetic and energy effects (energy supply of epithelium, regulation of intestinal peristalsis, organism's thermal supporting, regulation of the differentiation, and regeneration of epithelial fabrics).(ii)Forming of protective barrier of mucous membrane of intestine, suppression of growth of pathogenic microflora.(iii) Immunogenic role (stimulation of immune system, stimulation of tissue immunity including generation of immunoglobulins).(iv)Modulation of functions of cytochromes P450 in the liver and generation of P450-like cytochromes.(v)Detoxication of toxic substances and compounds.(vi)Production of various biologically active compounds, promotion of certain medicaments.(vii)Mutagenic/antimutagenic activity (growth of resistance of increase of epithelial cells to the mutagens (carcinogens), destruction of mutagens).(viii)Regulation of cavities' gas composition.(ix)Regulation of behavioral reaction.(x)Regulation of replication and gene expression of prokaryotic and eukaryotic cells.(xi)Regulation of the programmed death of eukaryotic cells (apoptosis).(xii)Depository of microbial genetic material.(xiii)Participation in etiopathogenesis of the diseases.(xiv)Participation in water-salt metabolism, sustention of organism's ionic homeostasis.(xv)Forming immunological tolerance to the food and microbial antigens.(xvi)Participation in colonization resistance.(xvii)Providing homeostasis of symbiotic relations of prokaryotic and eukaryotic cells.(xviii) Participation in the metabolic activity: metabolism of proteins, lipo (delivery of lipogenesis substrates), and carbohydrates (delivery of gluconeogenesis substrates) and regulation of bile acids, steroids, and other macromolecules.

Thus, gastrointestinal tract bioreactor is represented with sequencing circuit of microorganisms' specific communities each of which performs certain functions and transfers to the following community substances required for its life activity and functioning. Gastrointestinal tract bioreactor is now the most advanced advice providing complete destruction of the wide range of organic and inorganic matters.

Aerotanks and digestors are used in the practice of water treatment where aerobic and aerobic processes take place. However, each reactor works individually and, what is extremely important, a limited number of microorganisms participate in their work.

Creation of communities' chain of microorganisms where conditions necessary for their normal life activity (gastrointestinal tract type) will be will allow in the visible future to develop an universal bioreactor for destruction of the wide spectrum of organic and inorganic matters in the waste waters.

TAP WATER IS THE MEDIUM FOR MULTIRESISTANT MICROORGANISMS—MUTANTS

As it is known, the main task of chlorine treatment of the drinking water is water sterilization against pathogenic microorganisms. However, this task is not fulfilled to the full extent because of high adaptive capacities of single-celled animals (protozoa).

Scientists of the Swiss Federal Institute for water supply problems found out with the new method of microbiologic control based on flow cytometry that the content of microorganisms 100–1000 times higher in the sanitarily pure drinking water than the standard method. What is happening with the chlorinated water in the obsolete water distribution systems of the centralized water supply of the people? Under these at-first-sight aseptic conditions, the conglomerate of various microorganisms multiresistant to chlorine and its compounds is developed. One-celled organisms, having large adaptive flexibility, completely adapt to the habitation in the tap water. They are developing and propagating themselves in the water and educe their metabolic by-products having toxicant properties into aqueous media. Thus, in the water supply lines of the water distribution systems is forming their own "microcosmos" of the communities of multiresistant microorganisms, including pathogenic ones. They die in the dechlorinated aqueous media.

Emergence of adaptation by combinative way sometimes results in the new mutations which being included into system of genotype changes a phenotype. This path is called the adaptation of a combinative. The use of sublethal concentrations of disinfectants in water treatment processes to the emergence of strains of microorganisms, resistant to chlorine and chlorine compounds.

Adaptation emerging in the combinatorial way sometimes leads to the new mutation which, being included in the genotype system, changes the phenotype. Application of sublethal concentrations of disinfecting agents used during drinking-water treatment contributes to the appearance of microorganism strains, resistant to chlorine and its compounds. This phenomenon is proved by the studies of Mokiyenko and colleagues [36, 37], who considered very important medical problem "drinking water—water-caused infections." The authors concern themselves with the nature of bacteria resistance to the biocides effect, including chlorine and its compounds. In this connection they take notice of deterioration of the situation with infectious diseases in Ukraine on account of the drinking water of poor quality [38]; they make a hypothesis of chlorine resistance of bacteria which rests on very complicated process. It consists of two stages: information-spatial interaction of the receptive body and a substrate [39, 40].

According to this hypothesis the problem of bacteria's adaptive multiresistance to chlorine is closely connected with hormesis. This phenomenon is determined with two-phase effect of chemical matters directed to the water disinfection. Small doses produce stimulating effect, and large doses result in the inhibition of biological processes. hormesis connections "dose-effect" are observed in the biota representatives of all levels of organization starting from viruses and bacteria and ending with endotherms and human beings within the wide range of doses [41, 42].

In the papers [43, 44], investigating dynamics of microbial cenosis in the bottled slightly carbonated and strongly carbonated mineral waters, the scientists come to conclusion as to inadequacy of the carbonation methods in question. It was noted substantial increase of saprophyte microorganisms even after the first month of storage of these waters. Saturation, according to the authors, does not guarantee sanitarian safety of the waters. The data is provided which testifies about adaptation of microorganisms to the high concentrations of carbon dioxide.

The mechanism of adaptation of microorganisms to the aggressive environment is based on the process of adaptation development. Process of adaptation development, as any other process, is inconsistent internally. It consists of interaction of two counter-parties: of partial disorganization of already fully formed norm of organism's and species' reaction and its next new organization [45, 46].

In the process of adaptation of microorganisms, takes place a change in the structure and physiological processes under the influence of the environmental factor. Degree of adaptation of microorganisms to the new conditions depends on resistance of this culture to the intensity of impact of external environmental factors. Some species of microorganisms, for example, acid-alcohol- alkali-resistant, diphtheritic group and fungi, are changeable somewhat less than typhous group, round bacteria, and anaerobes. The last adapt easier to the disturbing factors of the environment. Adaptation of the microorganisms is expressed mostly in the attitude to the temperature and availability of oxygen. The better and more perfect adaptation is realized, the slowly and gradually increases impact of the new factors [47].

New conditions can make microorganisms to become less demanding to the environment, limit their physiological requirements or sporulate. Thus, the morphology of microbical cell and its structure sometimes is changes.

Mutations lead to the emergence of bacteria strains with new properties. Mutagenic effect takes place often in case of encountering harmful chemical compounds in the environment where microorganisms usually reproduce themselves; a part of them die and the most stable survive and produce multiresistant strains, sometimes even towards antibiotics and immune serum.

The primary effect of mutagenic agent does not necessary lead to the true mutation. A new phenotype shows up only when changed gene starts to function. There is a possibility to accumulate and to highlight mutants with various defects: with violation of processes of transport or usage of the substrate, with the defects of intermediate metabolism, hypersensirivity to the temperature, and so on [48].

Theoretically, mutations provoked by radiations, chemical matters or other factors could lead to the extinction of bacterial population; however, there are biochemical mechanisms in any living cell that are able to fully or partially restore parent structure of DNA. Complex of ferments catalyzing reactions of corrections of DNA damage make up systems of reparation. They differ fundamentally as to their biochemical mechanism of genome reconstruction. There are three known main mechanisms of correction of DNA defects: immediate reversion from DNA damage to the initial structure, excision (falling out) of damaged initial structure with the further reconstruction, and activation of mechanisms providing immunity to damage.

Thus, there is a high-scale possibility of formation in the tap water of multiresistant populations to the chloride and its compounds. It is largely due to the fact that prokaryotes are the most ancient living organisms on the Earth. They successfully existed during various eras owing to their unique adaptability to the drastic change of physical-chemical factors of media. They are widely presented in these latter days which are described dangerously increasing anthropogenic pollution of the environment.

Microorganisms adapt easily to the aggressive media because they have high mutagenic properties. They are agamic and are reproducing themselves by cell division that is also a positive factor for their adaptive possibilities. Thus, if in the changed media conditions survives at least one cell-mutant, in case of intensive division it forms a new population of microorganisms with changed phenotypic characteristics. Animal unit in such population of microorganisms differ from the natural forms by the appearance, structure of colonies, and physiological characteristics.

An important factor influencing the adaptation of microbiotas to biocides as well as to chloride is hormesis. It allows microorganisms to stimulate its biological activity on exposure of small doses of decontaminant. Such "training" increases immunity of the cells to the higher doses and increases number of mutations in the population holding out hope to the survival of microbiota in the aggressive media.

It is necessary to change from time to time decontaminant substances to make the process of adaptation of microorganisms to the tap water more difficult or impossible. For this purpose, it is necessary to use a set of biocides differing as to chemical composition and mechanism of action. But realization of such technological process of decontamination of drinking water at the water treatment works of the centralized water supply is not a trivial task.

TOXIC INFLUENCE OF BACTERIA ON TEST ORGANISMS IN THE DRINKING WATER

The influence of Escherichia coli bacteria and products of their metabolism on the toxic level of the drinking water was not studied before. As a result of our last research work [49] it was found that the drinking water including bottled water contains colibacillus. Standard research as to the usage of Escherichia coli bacteria as an index of sanitary-hygienic condition of the water does not include toxic evaluation of the water bacterized withEscherichia coli.

It is established that metabolites of Escherichia coli bacteria separated in the aqueous media considerably deteriorate its quality, inhibit roots' growth on the plant body, demonstrate chronic toxicity on hydra, and provoke acute toxicity on Ceriodaphnia.

The obtained results on the cellular level showed that deviations from control were observed starting from the concentration of microorganisms in the amount of 10^2. Even in case of finding few bacteria, the inflammatory processes affect hematological indices in the leucocytes of fish blood.

The leucocytes play a very important role for the protection of the organism against bacterial and mycotic infections. Growth of neutrocytosis in the blood is a response of the organism to the bacterial and many other infections. Lymphopenia (decrease of the number of lymphocytes) is typical for the initial stage of infectious-toxic process and is related with their migration from the vessels to the tissues (inflammatory tissue).

In the studied samples of the bottled drinking water with different concentrations of bacteria, significant increase of nucleus rate of the fish tissues is observed. Structural and quantitative changes of cell nucleus were observed in case of bacteria concentration equal to Escherichia coli 10^2. Changes in the components of cell nucleus which are carriers of genetic information leads to the mutation of cells. This can be a reason for error exclusion of one or another disease and lead to the proliferation of oncologic cells.

According to WHO acting hygienic requirements to the water quality from the centralized drinking water supply systems, total number of E. coli in 1 mL of water should not exceed 3 CGU. Our research proved that existence of bacteria E. coli in the drinking water is inadmissible.

REFERENCES

1. World Health Organization, Guidelines on Drinking Water Quality Control. Recommendation, vol. 1, World Health Organization, Geneva, Switzerland, 2004.
2. United Nations, Report of the World Summit on Sustainable Development, United Nations, New York, NY, USA, 2002.
3. World Water Day, "Clean Water for a Healthy World," 2010, http://www.unepcom.ru/?go=razdel&level=2&cid=66.
4. Global Environment Outlook, GEO4. Environment For Development. United Nations Environment Programme, Prog Press, Valletta, Malta, 2007.
5. "How to turn water into the water," Mineral additive, "SEVERYANKA", 2011, http://www.koshcheev.ru/2011/10/20/severyanka/.
6. V. V. Goncharuk, "SOS: drinking water," Journal of Water Chemistry and Technology, vol. 32, no. 5, pp. 255–283, 2010.
7. V. N. Lobyshev and L. P. Kalinichenko, Izotopnyye Effekty D_2O V Biologicheskikh Sistemakh (Isotope Effects of D_2O in Biological Systems), Nauka, Moscow, Russia, 1978.
8. G. Somlyai, G. Laskay, T. Berkenyi et al., "The biological effects of deuterium-depleted water, a possible new tool in cancer therapy," Zeitschrift fur Onkologie, vol. 30, no. 4, pp. 91–94, 1998.
9. V. V. Goncharuk, A. A. Kavitskaya, I. Y. Romanyukina, et al., "Revealing water's secrets: deuterium depleted water," Chemistry Central Journal, vol. 7, no. 1, pp. 103–108, 2013.
10. V. V. Goncharuk, V. B. Lapshin, T. N. Burdeinaya et al., "Physicochemical properties and biological activity of the water depleted of heavy isotopes," Journal of Water Chemistry and Technology, vol. 33, no. 1, pp. 8–13, 2011.
11. M. J. Figueras and J. J. Borrego, "New perspectives in monitoring drinking water microbial quality,"International Journal ofEnvironmental Research and Public Health, vol. 7, pp. 4179–4202, 2010.
12. M. A. Shannon, P. W. Bohn, M. Elimelech, J. G. Georgiadis, B. J. Marĩas, and A. M. Mayes, "Science and technology for water purification in the coming decades," Nature, vol. 452, no. 7185, pp. 301–310, 2008. ·
13. C. A. Stedmon, B. Seredyńska-Sobecka, R. Boe-Hansen, N. Le Tallec, C. K. Waul, and E. Arvin, "A potential approach for monitoring drinking water quality from groundwater systems using organic matter fluorescence as an early warning for contamination events," Water Research, vol. 45, no. 18, pp. 6030–6038, 2011.
14. S.-L. Loo, A. G. Fane, W. B. Krantz, and T. Lim, "Emergency water supply: a review of potential technologies and selection criteria," Water Research, vol. 46, no. 10, pp. 3125–3151, 2012.
15. C. Reimann, M. Ayräs, A. Chkushin, et al., Environmental Geochemical Atlas of the Central Barents Region. Special Publication of the Central Kola Expedition, Geological Survey of Finland and Geological Survey of Norway, Schweizerbart Science, Stuttgart, Germany, 1998.
16. E. V. Sobotovich and V. V. Dolin, Eds., Tritij Y Biosferi (Tritium in the Biosphere), Nauk Dumka, Kyiv, Ukraine, 2012.

17. In Search of the Fountain of Youth, "Preliminary analysis of deuterium's role in DNA degradation," 2003,http://www.springboard4health.com/notebook/health_deuterium_analysis.
18. V. V. Goncharuk, Nauka O Vode, Naukova Dumka, Kiev, Ukraine, 2010.
19. I. A. Meshkov, M. M. Anikhovskaya, M. Y. Yakovleva, et al., "Intestinal endotoxin in regulation of hemostasis activity and in pathogenesis of the DIC syndrome," Human Physiology, vol. 131, no. 6, pp. 91–96, 2005.
20. G. A. Zavarzin, Bacteria and Composition of Atmosphere, Nauka, Moscow, Russia, 1984.
21. S. V. Belmer, T. V. Gasilina, and A. V. Malkoch, "Znacheniye prebiotikov dlya funktsionirovaniya kishechnoi microflory (Importance of prebiotics for functioning of intestinal microflora)," Russkiy Meditsinskiy Zhurnal, vol. 3, pp. 151–153, 2008.
22. S. V. Belmer and A. V. Malkoch, "Kishechnaia microflora y znacheniye prebiotikov dlya eyo funktsionirovaniya (Intestinal microflora and importance of prebiotics for its functioning)," Lechfshchiy Vrach, vol. 4, pp. 25–29, 2006.
23. V. Poroyko, "Glubokouvazhayemy microb (Dear microbe)," Populyarnaya Mechanika, vol. 4, pp. 44–47, 2008.
24. P. L. Stark and A. Lee, "The microbial ecology of the large bowel of breast-fed and formula-fed infants during the first year of life," Journal of Medical Microbiology, vol. 15, no. 2, pp. 189–203, 1982.
25. S. Salminen, C. Bouly, R. M. C. Boutron, et al., "Functional food science and gastrointestinalphysiology and function," British Journal of Nutrition, vol. 80, supplement 1, pp. 147–171, 1998.
26. A. I. Khavkin, "Microbiotsenos kishechnika y immunitet (Intestine's microbiocenosis and immunity),"Russkiy Meditsinskiy Zhurnal, vol. 11, no. 3, pp. 1–16, 2003.
27. G. R. Gibson and M. B. Roberfroid, "Dietary modulation of the human colonic microbiota: introducing the concept of prebiotics," Journal of Nutrition, vol. 125, no. 6, pp. 1401–1412, 1995.
28. S. Salminen, R. Tanaka, H. Takayama, et al., "Effects of administration of TOS and Bifidobacterium breve 4006 on the human faecal Hoga," Bifidobacter Microflora, vol. 2, pp. 17–24, 1983.
29. B. Björkstén, E. Sepp, K. Julge, T. Voor, and M. Mikelsaar, "Allergy development and the intestinal microflora during the first year of life," Journal of Allergy and Clinical Immunology, vol. 108, no. 4, pp. 516–520, 2001.
30. S. V. Belmer and T. V. Gasilina, "Ratsionalnoye pitaniye y sostav kishechnoy microflory (Balanced feeding and composition of intestine microflora)," Voprosy Detskoy Diyetologii, vol. 1, no. 5, pp. 17–20, 2003.
31. A. F. Voronin and B. A. Tenderov, Funktsionalnoye Pitaniye (Balanced Feeding), Koloss-Press, Moscow, Russia, 2002.
32. S. V. Belmer, A. V. Malkoch, and O. N. Minushkin, "Znacheniye prebiotikov dlya funktsionirovaniya kishechnoy mikroflory (Importance of prebiotics for the functioning of intestine microflora)," inProceedings of the 13th Congress Detskikh Gastroenterologov, Actualniye voprosy abdominalnoy patologii u ditey, Moscow, Russia, March 2006 2006.
33. R. D. Rolfe, "Interactions among microorganisms of the indigenous intestinal flora and their influence on the host," Reviews of Infectious Diseases, vol. 6, supplement 1, pp. S73–S79, 1984.
34. D. A. Jenkins, C. W. C. Kendall, and V. Vuksan, "Inulin, oligofructose and intestinal function," Journal of Nutrition, vol. 129, no. 7, pp. 1431–1433, 1999.
35. C. A. Edwards and A. M. Parrett, "Intestinal flora during the first months of life: new perspectives,"British Journal of Nutrition, vol. 88, supplement 1, pp. S11–S18, 2002.
36. A. V. Mokiyenko, A. I. Gozhenko, N. F. Petrenko, et al., "Water and water caused infections Voda y vodnoobuslovlennyye," Odessa, ART-B, 2: 288, 2008.
37. A. V. Mokiyenko, A. I. Gozhenko, and N. F. Petrenko, "Water chlorination: disinfectation or adaptiveness, inactivation or stimulation?" Vesnik NAN Ukrainy, vol. 1, pp. 32–40, 2012.
38. M. A. Anfriychik, "Infectious diseases and demographic crisis in Ukraine," Journal of Academy of Medical Sciences, vol. 13, no. 3, pp. 533–542, 2007.
39. A. V. Mokiyenko, A. I. Gozhenko, and N. F. Petrenko, "Mechanism of formation of adaptive multiresistance of bacteria to biocides from the point of view of contribution of fundamental basis of supramolecular chemistry," Vesnik NAN Ukrainy, vol. 8, pp. 49–56, 2010.
40. A. V. Mokiyenko, A. I. Gozhenko, and N. F. Petrenko, "Adaptive multiresistance of bacteria to the analysis of contribution to the evolution of epidemiological process," Preventive Medicine, vol. 2, pp. 90–95, 2011.
41. L. M. Shafran, A. I. Gozhenko, N. F. Petrenko, et al., "For the substantiation of hormesis as a fundamental medical paradigm (review of literature and results of own research)," Present Day Problems of Toxicology, vol. 2, no. 3, pp. 13–23, 2011.
42. A. V. Mokiyenko and N. F. Petrenko, "hormesis as trigger mechanism of regulatory mutations and its role in the formation of multiresistance bacteria," Contemporary Problems of Industrial Toxicology, vol. 5, pp. 47–52, 2011.
43. O. M. Khmelevska, A. V. Mokiyenko, and S. I. Nikolayenko, "Hygienic evaluation of dynamics of microbal coenosis of the bottled slightly carbonated water," Medicine Today and Tomorrow, vol. 4, no. 49, pp. 161–165, 2010.
44. O. M. Khmelevska, A. V. Mokiyenko, and S. I. Nikolayenko, "Hygienic evaluation of illustratory microorganisma from the bottled on-carbonated, slightly carbonated and highly carbonated mineral waters of Ukraine," Hygiene of Population Centers, no. 58, pp. 122–126, 2011.

45. E. V. Ermilova, Molecular Aspects of Adaptation of Prokaryote, Khimizdat, Sankt-Peterburg, Russia, 2012.
46. V. L. Petukhov, V. S. Korotkevich, and S. Z. Stambekov, "Genetics. Sem.GPI, Novosibirsk," 2007.
47. L. S. Buzoleva, Adaptation of Pathogenic Bacteria to the Environmental Abiotic factors, Press, Vladivostok, Russia, 2001.
48. S. N. Schelkunov, Genetic Engineering, Siberian University Edition, Novosibirsk, Russia, 2004.
49. V. V. Goncharuk and M. R. Vergolyas, "Toxic influence of bacteria on test organisms in the drinking water," Journal of Water Chemistry and Technology, 2013.

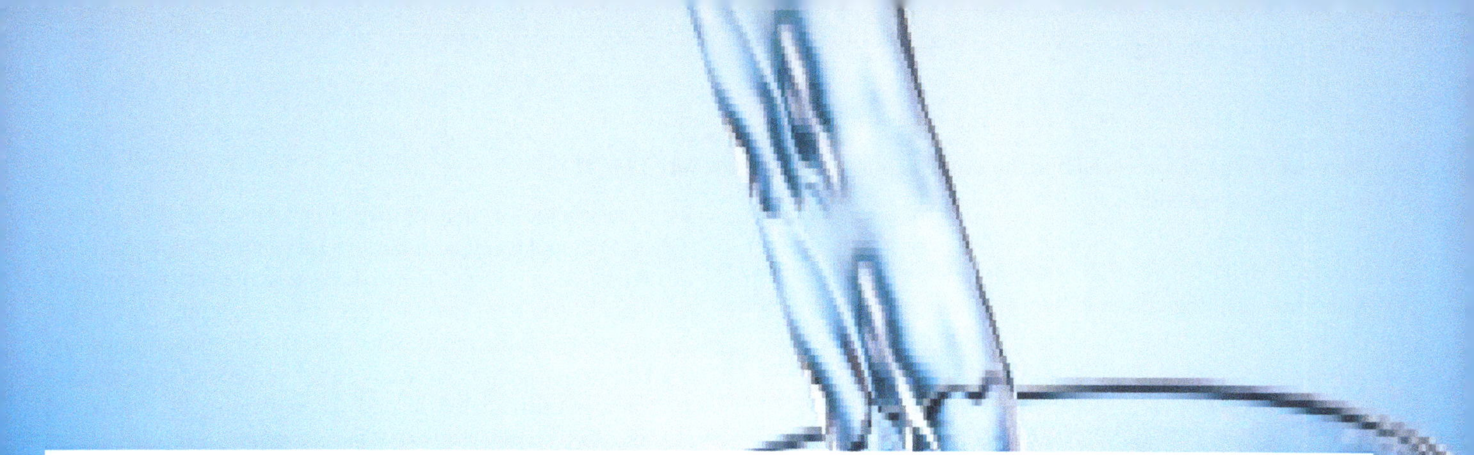

CHAPTER 3

HIGH ADHERENCE IS NECESSARY TO REALIZE HEALTH GAINS FROM WATER QUALITY INTERVENTIONS

Joe Brown , Thomas Clasen

INTRODUCTION

Over 780 million people now lack access to an "improved" water source [1], and one study has estimated the number of people who rely on microbiologically or chemically unsafe water to be 1.8 billion, or about 28% of the global population [2]. Unsafe drinking water is a leading cause of preventable disease, with the burden borne primarily by children in low and middle-income countries. Pathogens transmitted in drinking water account for an unknown but presumed significant percentage of the estimated 4 billion cases and 1.9 million deaths from diarrhoeal diseases each year [3], [4]. According to WHO and UNICEF monitoring data from 2004, diarrhoeal diseases are the second most common cause of deaths among children under 5 years of age worldwide, or 16% of the total [5].

Providing safe, reliable, piped-in water to every household is an essential goal, yielding optimal health gains while contributing to the Millennium Development Goal targets for poverty reduction, nutrition, childhood survival, school attendance, gender equity and environmental sustainability. Recent research has shown, however, that even such improved water supplies are often subject to faecal contamination [2] and that even occasional interruptions in the quality of water from piped water supplies can undermine the health benefits from safe drinking water [6]. While careful not to encourage diversion of resources away from expansion of safe piped water, public health officials have called for alternative approaches that may provide some of the health benefits of safe drinking-water as progress is made in improving infrastructure services [7].

One such alternative is household water treatment and safe storage (HWT) [8]. In many settings, both rural and urban, populations may have access to sufficient quantities of water, but that water may be unsafe for consumption as a result of microbial or chemical contamination. Effective HWT, such as boiling, filtration and chlorination, has been shown significantly improve microbial water quality [9], [10], [11], [12]. Moreover, by focusing on the point of use rather than the point of delivery, treating water at the household level minimizes the risk of recontamination [13] and the risk of contamination associated with poorly functioning or intermittent water supply systems. There is evidence that HWT can be effective in preventing self-reported diarrhoeal disease [14], [15], [16], [17], although systematic reviews have noted that the limited number of placebo-controlled trials have not reported a statistically significant reduction in risk and estimates of effect from unblinded trials may be subject to a number of well-known biases [15], [18]. Like other environmental health interventions such as sanitation or hygiene, blinding intervention studies of HWT has proved to be challenging, but additional blinded

Citation: Brown J, Clasen T (2012) High Adherence Is Necessary to Realize Health Gains from Water Quality Interventions. PLoS ONE 7(5): e36735. doi:10.1371/journal.pone.0036735.

trials may be useful in evaluating the evidence for this strategy of safe water provision.

One of the other challenges that promoters of HWT have reported is lack of adherence—*correct, consistent and sustained use*, sometimes referred to as *compliance*. Unlike centrally treated, piped-in water supplies, HWT is normally a batch process that must be undertaken by the end users on a daily or other frequent basis in order to provide consistent protection against waterborne pathogens. A systematic review of chlorine-based HWT, where adherence can be objectively assessed by residual free chlorine in treated water, found several studies with adherence under 70% [14]. Even when uptake of an HWT filter intervention was high when measured by current use, 83% of adults and 95% of children were reported to also consume untreated water [19]. A number of other studies of HWT reported reduced use of interventions over time, raising questions about whether low adherence may limit the usefulness of HWT as a strategy for securing access to safe water [19], [20], [21], [22], [23], [24].

To date, however, there is limited evidence characterizing the impact of adherence on health gains attributable to HWT. A systematic review of water quality interventions to prevent diarrhoea reported a pooled risk ratio of 0.46 (95% CI: 0.25–0.84) among 16 trials reporting adherence over 50%, and 0.75 (95% CI: 0.63–0.90) among 5 trials reporting adherence of less than 50% [14]. In a review of chlorine interventions, Arnold and Colford (2007) found that the reduction of diarrhoeal disease was greater in studies that reported a higher percentage of samples with detectable levels of chlorine at the point of use [14]. Despite the limited evidence from reviews, the relationship between adherence and health impact has not been clearly characterised in field trials [16]. This is partly due to the difficulty of measuring adherence in a field setting for many technologies, especially at the level of the individual. Adherence has not been measured or reported in the majority of field trials of HWT to date.

In this paper, we present a basic quantitative microbial risk model examining the influence of pre-treatment water quality, treatment effectiveness, and adherence on predicted health gains. We seek to estimate the magnitude of the effect of variable adherence on health impacts achievable through safe water interventions, particularly HWT.

METHODS

We constructed a quantitative microbial risk assessment (QMRA) model using assumptions for ranges of untreated and treated water quality, published dose-response relationships for reference pathogens, per-case severity weighting, population susceptibility, adherence, and per-capita consumption of drinking water. In most cases, we used the default assumptions and methods articulated in and recommended by the World Health Organization (WHO)'s*Guidelines for Drinking-water Quality (GDWQ)*, 4th Edition [25], where the rationale, caveats, and methodological issues that apply to this approach to estimating health risks are described in greater detail. We briefly describe the method here and provide a summary of calculations.

We report model output in disability-adjusted life years (DALYs) averted per 100,000 people per year attributable to HWT under the conditions we describe. DALYs can be used to compare health outcomes using severity and duration estimates of disease(s), combining quality and quantity of life factors [25], [26]. This measure has been extensively described previously [27]and is used here for ease of comparison with other health impact estimates.

Reference pathogens

We used *Campylobacter jejuni*, rotavirus, and *Cryptosporidium* as reference pathogens in this risk assessment because of their relatively low infectious doses, high per-case DALY severity weighting, and the fact that we have published dose-response models for them [25]. The DALY estimates we calculated were intended to estimate the DALY contributions from bacteria, viruses, and protozoa (as classes of pathogens) that may be found in untreated water, following the approach of the WHO recommended model [25]. It is important to note that the use of these specific microbes is not a statement about their relative importance as aetiologic agents of diarrhoeal diseases globally, or their importance as specifically waterborne pathogens. Rotavirus, for example, is thought to be more frequently transmitted via pathways other than drinking water. They are all known and relatively common causes of gastrointestinal illness. We list assumptions for pathogen-specific per-case DALY weighting, population susceptibility, and risk of illness following infection in Table 1 and present an overview of dose-response calculations below.

Table 1. Overview of assumptions used in calculating DALYs averted, adapted from Table 7.6, Guidelines for Drinking-water Quality, 4th Edition [24].

	Units	Probability density function (pdf) for independent variables	Values (independent variables) or formulae (dependent variables)
Untreated water quality (C_R)	Mean organisms per litre	Log-normal (mean = standard deviation)	Means of 0.0001, , 0.001, 0.01, 0.1, or 1
Log_{10} reduction	Unitless	Uniform between bound specified	1–2, 2.01–3, or 3.01–6
Drinking-water quality (C_D)	Organisms per litre		$C_R \times (1 - \%$ reduction)
Consumption of drinking-water (V)	Litres per person per day	Mean (point estimate)	1, 2, or 5
Adherence or consistency of drinking water treatment (A)	Percent of water consumed that is treated (%)	Uniform between bound specified	50–70%, 71–90%, or 91–100%
Exposure by drinking-water (D)	Organisms per day ingested	-	Equation (1)
Risk of infection ($P_{inf,d}$)	Per day	-	Equations (2, 3)
Risk of infection ($P_{inf,y}$)	Per year	-	Equation (4)
Risk of diarrhoeal illness given infection ($P_{ill\|inf}$)		Mean (point estimate)	*Campylobacter*: 0.3, rotavirus: 0.5, *Cryptosporidium*: 0.7
Risk of diarrhoeal illness (P_{ill})	Per year	-	Equation (5)
Disease burden (DB)	DALYs per case	Uniform between bound specified (for rotavirus only)	*Campylobacter*: 4.6×10^{-3}, rotavirus: 0.014–0.48, *Cryptosporidium*: 1.5×10^{-3}
Susceptible fraction (S)	Percentage of population	Mean (point estimate)	*Campylobacter*: 100%, rotavirus: 6%, *Cryptosporidium*: 100%
Disease burden per 100,000 persons	DALYs per year per 100,000	-	Equation (6)

Water quality assumptions

The use of QMRA models to predict reductions in waterborne disease risk requires estimation of pre- and post-treatment concentrations of microbes for which we have dose-response models. Faecal-oral pathogen occurrence and distribution are variable in drinking water sources and depend on many factors, including infection (and shedding) rates of specific pathogens in a given population, sanitation and hygiene conditions that may affect excreta containment, zoonotic reservoirs, seasonality and weather events, and other context-specific variables. No sufficiently detailed, country-level or international data exist for waterborne pathogens that would permit the meaningful estimation of means to be used in a generalizable and scientifically credible risk assessment model for drinking water risk. We have used assumptions for microbial counts in pre-treatment waters over a range from 0.0001 per litre ("low risk") to 1 per litre ("high risk"). For the purposes of this modelling exercise, we have assigned a log-normal probability density function to water quality assumptions, which is one distribution that has been observed to characterize microbial density in water [28], [29].

Treatment effectiveness

We modelled DALYs averted for improved drinking water quality under several different assumptions about microbiological effectiveness, including: (1), a hypothetical "best case" technology that would be able to reduce bacteria, viruses, and protozoa by 3.01–6 log_{10}(99.9%–99.9999%), typical of the most advanced single or multiple-barrier technologies for household water treatment [30], [31], [32] and consistent with well operated conventional or advanced water treatment; (2), a mid-range treatment option that could reduce all classes of microbes by 2.01–3 log_{10} (99%–99.9%), in the range of basic drinking water treatment and consistent with effective HWT options; and (3), a basic level of protection corresponding to poorly performing water treatment systems, systems without adequate disinfection, or some single-barrier HWT options (1–2 log_{10} reduction, or 90%–99%) [25]. In the probabilistic model, we assumed a uniform probability density function between the upper and lower log_{10} bounds we specify.

Adherence

For safer water to translate into health gains, the percentage of treated water that is consumed is an important factor. For piped-in, domestic water service, this might be called *consistency of treatment*. In household water treatment, this may be called *adherence* and sometimes*compliance*. For the purposes of this modelling exercise, we use adherence and define it as the percentage of water an individual uses that is treated (0%–100%). We have divided adherence into *high* (91%–100%), *medium* (71%–90%), and *low* (50%–70%), with an assumed uniform probability density function describing the likelihood of values within this range in the probabilistic model. We assume here that the balance of the drinking water consumed by the user is untreated water. In reality, individuals and families may rely on a number of drinking water sources and treatment options, and these options may be a function of location, season, preferences, beliefs, knowledge, convenience, infrastructure, cost, previous household investment (e.g., a well or rainwater storage), or other factors. Our simplified assumption is that users have two options: treated water and untreated water that result in microbial risks that vary according to the assumptions we have specified.

Model structure

We used Oracle Crystal Ball, Fusion Edition (release 11.1.2.1.000, www.oracle.com) to build a stochastic (Monte Carlo) model based on probability density functions (pdfs) specified for key high-leverage variables of interest, while conserving point estimates for other variables in the base model [25] and using a single-hit version for sensitivity analysis. Model output was collected following 10,000 simulations of each model configuration and descriptive statistics were computed in Crystal Ball. Model predictive and structural validity were tested using both the single-hit and probabilistic models using a structured comparison of model output across a range of default assumptions.

Table 1 describes the basic model structure. Pre-treatment water quality is assumed with means of 0.0001–1 per litre, distributed log-normally (C_R). Based on effectiveness of treatment in reducing reference pathogens, treated drinking-water quality is calculated (C_R). We then used assumed consumption of water per day (V) and adherence or consistency of water treatment (A) to calculate the predicted exposure by drinking-water (d) as the number of microbes ingested per day.

$$d = C_D VA + C_R(1 - A) \quad (1)$$

We used the exponential (*Cryptosporidium*) and the Beta-Poisson (rotavirus, *Campylobacter*) models to estimate the risk of infection from exposure to microbial pathogens [26]. We calculated the risk of infection of a single microbe (d) and then calculated the annual risk from ingestion of multiple microbes over the period of a year.

The exponential model is:

$$P_{inf}(d) = 1 - e^{-rd} \quad (2)$$

Where *P(d)* is the probability of an individual becoming infected after ingesting a dose (d) and r is an infectivity constant. The value of r is 0.2 for *Cryptosporidium* (Table 1) [25].

The equation for the Beta-Poisson model is:

$$P_{inf}(d) = 1 - \left(1 + \frac{d}{N_{50}}\left(2^{\frac{1}{\alpha}} - 1\right)\right)^{-\alpha} \quad (3)$$

N_{50} is the median infective dose, which has been experimentally derived as the dose estimated to cause infection in 50% of individuals exposed, and α is a dimensionless infectivity constant. The values for N_{50} and α are 896 and 0.145 for *Campylobacter*, respectively, and 6.17 and 0.253 for rotavirus [25], [26].

Yearly infection risk is then calculated as:

$$P_{inf.y} = 1 - (1 - P_{inf.d})^{365} \quad (4)$$

The risk of illness is calculated as:

$$P_{ill.y} = P_{inf.d} \times P_{ill|inf} \quad (5)$$

where the probability of illness given infection ($P_{ill|inf}$) is 0.3 for *Campylobacter*, 0.5 for rotavirus, and 0.7 for *Cryptosporidium* [24].

To calculate the disease burden in DALYs associated with yearly risk of illness, we used the following formula for each reference microbe:

$$DALYs = P_{ill} \times DB \times S \quad (6)$$

Where DB is the estimated per-case disease burden in DALYs and S is the susceptible fraction of the population for the outcome of interest. Per-case DALY weights are 4.6×10^{-3}(campylobacteriosis), 0.014–0.48 (rotavirus), and 1.5×10^{-3} (cryptosporidiosis). The model further assumes that 100% of the population is susceptible to cryptosporidiosis and campylobacteriosis, but that only 6% is susceptible to illness associated with rotavirus infection (Table 1). These DALYs weights and estimates for susceptible fraction are those used in the base model as described in the WHO *Guidelines for Drinking-water Quality* [25]. To calculate the estimated disease burden among 100,000 person-years across all three reference pathogens, we multiplied the microbe-specific disease burden by 100,000 and summed estimates for *Campylobacter*, *Cryptosporidium*, and rotavirus. Figures for averted DALYs were then calculated as the difference between DALYs before and after treatment, assuming that the only alternative to treated water was untreated source water.

We conducted a four-way sensitivity analysis to determine the relative sensitivity of mean DALYs averted to the key independent variables of background water quality, litres consumed per day, treatment effectiveness, and consistency/adherence. For the sensitivity analysis, we used the single-hit model using means for all inputs to examine variability in output resulting from changes to independent variables of interest.

RESULTS

Predicted health impacts associated with water quality interventions under pre-treatment water quality, technology effectiveness, and adherence assumptions are presented in Table 2 as DALYs averted per 100,000 people per year. For some combinations of parameters resulting in high impact (moderate pre-treatment water quality, medium to high treatment effectiveness, and high adherence), point estimates for mean DALYs averted per 100,000 person-years are over 500. With lower levels of adherence, DALYs averted decrease dramatically when untreated water is of high risk, since adherence is of greater importance when untreated water is a source of disease. The range of estimates of effect when pre-treatment water is of moderate to high risk indicate that the potential health gains are very high as adherence approaches 100% (up to 1840 DALYs per 100,000 per year).

Table 2. Probabilistic model output with estimated mean DALYs averted per 100,000 persons pear year as a function of pre-treatment water quality; treatment effectiveness against bacteria, viruses, and protozoa; and adherence.

Pre-treatment water quality (log-normal distribution)	Treatment effectiveness: bacteria, viruses, and protozoa	Adherence, percentage of total water consumed that is treated	Mean DALYs averted per 100,000 persons (standard deviation)	Range	Mean standard error
High risk:	High	High (91–100%)	196 (178)	3.54–1,570	1.78
reference	(3.01–6 log_{10})	Medium (71–90%)	50.4 (28.7)	0–523	0.29
pathogens		Low (50–70%)	21.5 (16.8)	0–129	0.17
present,	Medium	High (91–100%)	173 (140)	0–1,740	1.40
mean:	(2.01–3 log_{10})	Medium (71–90%)	49.2 (28.2)	0–494	0.28
1 per litre		Low (50–70%)	20.8 (16.7)	0–86.0	0.17
	Low	High (91–100%)	104 (62.0)	0–1,010	0.62
	(1–2 log_{10})	Medium (71–90%)	43.3 (26.1)	0–347	0.26
		Low (50–70%)	19.4 (15.9)	0–85.8	0.16
Moderate	High	High (91–100%)	520 (326)	29.5–1,840	3.26
high risk:	(3.01–6 log_{10})	Medium (71–90%)	174 (149)	4.91–1,130	1.49
reference		Low (50–70%)	68.1 (62.6)	3.95–659	0.63
pathogens	Medium	High (91–100%)	498 (316)	36.0–1,890	3.16
present,	(2.01–3 log_{10})	Medium (71–90%)	170 (146)	6.82 (1,110)	1.46
mean:		Low (50–70%)	69.0 (63.5)	1.14–676)	0.63
0.1 per litre	Low	High (91–100%)	363 (257)	20.0–1,460	2.57
	(1–2 log_{10})	Medium (71–90%)	147 (127)	5.41–998)	1.27
		Low (50–70%)	63.7 (57.6)	2.93–604	0.58
Moderate	High	High (91–100%)	533 (313)	26.9–1,630	3.13
risk:	(3.01–6 log_{10})	Medium (71–90%)	369 (213)	19.3–1,180	2.13
reference		Low (50–70%)	216 (131)	6.56–682	1.31
pathogens	Medium	High (91–100%)	527 (304)	31.4–1,650	3.04
present,	(2.01–3 log_{10})	Medium (71–90%)	365 (211)	18.6–1,140	2.11
mean:		Low (50–70%)	215 (129)	6.69–724	1.29
0.01 per litre	Low	High (91–100%)	483 (277)	24.5–1,520	2.77
	(1–2 log_{10})	Medium (71–90%)	338 (196)	17.9–1,140	1.96
		Low (50–70%)	206 (124)	8.23–676	1.24
Moderate	High	High (91–100%)	133 (127)	2.15–1,100	1.27
low risk:	(3.01–6 log_{10})	Medium (71–90%)	109 (102)	3.14–966	1.02
reference		Low (50–70%)	78.5 (71.3)	2.29–557	0.71
pathogens	Medium	High (91–100%)	133 (125)	3.49–1,180	1.25
present,	(2.01–3 log_{10})	Medium (71–90%)	109 (101)	2.80–865	1.01
mean:		Low (50–70%)	79.3 (70.5)	2.20–615	0.71
0.001 per	Low	High (91–100%)	129 (121)	2.06–1,070	1.21
litre	(1–2 log_{10})	Medium (71–90%)	105 (98.6)	2.55–885	0.99
		Low (50–70%)	74.9 (67.0)	1.82–511	0.67
Low risk:	High	High (91–100%)	18.3 (15.9)	0–231	0.18
reference	(3.01–6 log_{10})	Medium (71–90%)	15.5 (13.3)	0–231	0.15
pathogens		Low (50–70%)	11.5 (9.92)	0–212	0.12
present,	Medium	High (91–100%)	19.3 (15.8)	0–546	0.19
mean:	(2.01–3 log_{10})	Medium (71–90%)	16.1 (13.3)	0–311	0.16
0.0001 per		Low (50–70%)	11.9 (9.94)	0–221	0.12
litre	Low	High (91–100%)	18.5 (15.5)	0–319	0.18
	(1–2 log_{10})	Medium (71–90%)	15.2 (12.9)	0–377	0.15
		Low (50–70%)	11.3 (9.60)	0–157	0.11

Results suggest that when pre-treatment water is of moderate to high risk, adherence is more important than treatment effectiveness. To take one example from Table 2, when pre-treatment water is of moderate high risk, the mean estimated DALYs averted per 100,000 person-years for a technology with low effectiveness (1–2 log_{10} reduction of bacteria, viruses, and protozoa) but that is used very consistently (91–100% adherence) is 363, compared with 68.1 for a technology that is more effective (3.01–6 log_{10} reduction of bacteria, viruses, and protozoa) but that is used less consistently (50–70% adherence).

Assuming lower risk untreated water, both the potential health gains from water treatment and the sensitivity to adherence are reduced, suggesting that the importance of water treatment and high adherence are elevated in high-risk waters and reduced in low-risk waters. We used a uniform probability density function in the model to indicate "high" adherence between 91%–100% (individual simulation estimates being equally likely to be assigned a value between these in individual simulations), a range which probably represents highest likely achievable adherence *in situ*.

Figure 1 presents a graphical illustration of how the predicted health gains vary as a function of both adherence and pre-treatment water quality. Calculation of DALYs averted in Figure 1 assumed a 2 log_{10} reduction across all reference pathogens and 1 litre per day consumption. When pre-treatment water was of high risk [A] and [B], there is a dramatic decline in predicted health gains with decreasing adherence. Water that is of moderate risk [C] is less affected by reduced adherence, although sharp declines are observed. Waters that are already low risk [D] and [E] do not show a strong association between DALYs averted and adherence because the risk of drinking untreated water is already low.

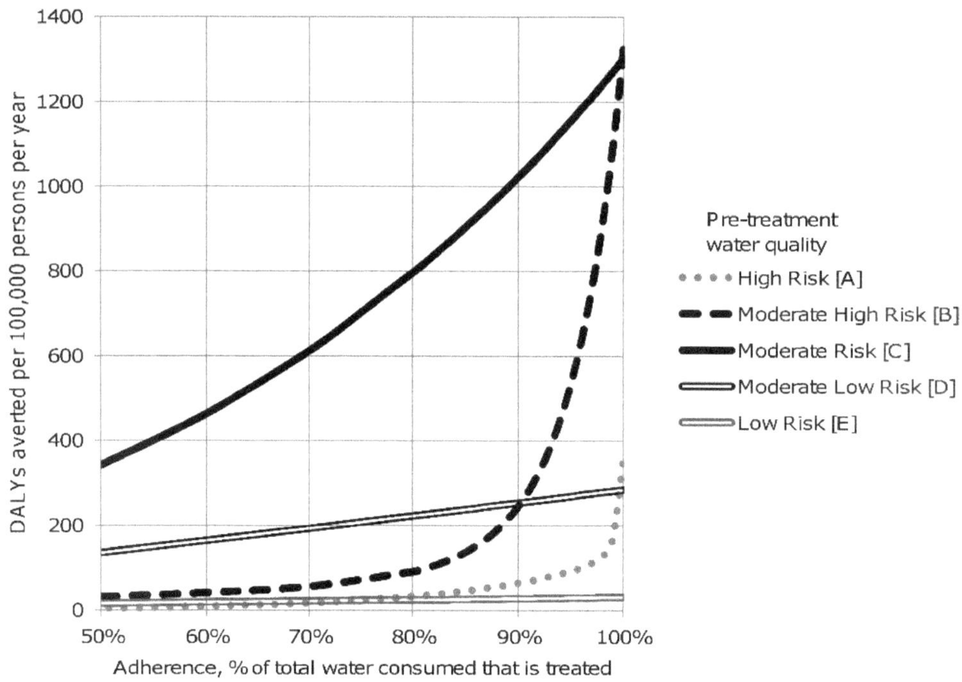

Figure 1. DALYs averted per 100,000 population, per year, based on assumptions about technology effectiveness (2 $\log_{10}$ **reduction in each pathogen class), adherence, and background water quality assumptions from Table 2.**

Results of the sensitivity analysis (Table 3) indicate that model output is highly sensitive to pre-treatment water quality and adherence, moderately sensitive over the range of assumptions to treatment effectiveness against waterborne pathogens, and relatively insensitive to volume consumed per person per day over realistic values (1–5 litres per day per person). A straightforward comparison of point estimates across changes in water quality, adherence, effectiveness and consumption suggest that, within pre-treatment water risk categories, DALYs averted estimates are most sensitive to adherence. Reducing adherence from 100% to 90% reduces the number of DALYs averted by over 90% under some model assumptions and up to 96%, with sharper declines when pre-treatment water quality is higher risk (Table 3 and Figure 1). The model was also very sensitive to the per-case DALY disease burden estimates (data not shown), which have been derived elsewhere and are key assumptions included in the base model we used [25], [26] (Table 1).

Table 3. Sensitivity analysis across a range of pre-treatment microbial counts of reference pathogens and adherence, at "high" and "low" effectiveness (6 $\log_{10}$and 2 $\log_{10}$, respectively) and at 2 levels of water consumption (1 litre per person per day and 5 litres per person per day).

Consumption, litres per person per day	Treatment effectiveness: bacteria, viruses, and protozoa	Adherence, percentage of total water consumed that is treated	DALYs averted per 100,000 people per year over pre-treatment water risk categories				
			High: 1 per litre	Moderate high: 0.1 per litre	Moderate: 0.01 per litre	Moderate low: 0.001 per litre	Low: 0.0001 per litre
1	6 $\log_{10}$ (99.9999%)	100%	1680	1618	1339	290	32
		90%	69	278	1049	258	29
		80%	35	96	815	227	25
		70%	18	57	626	196	22
		60%	9	41	473	166	19
		50%	5	31	349	137	16
1	2 $\log_{10}$ (99%)	100%	347	1324	1302	285	31
		90%	65	245	1022	254	28
		80%	33	91	796	223	25
		70%	17	56	612	193	22
		60%	9	41	463	164	19
		50%	5	31	342	135	16
5	6 $\log_{10}$ (99.9999%)	100%	1679	1612	1333	288	32
		90%	69	278	1045	257	28
		80%	35	96	812	226	25
		70%	18	57	624	195	22
		60%	9	41	471	166	19
		50%	5	31	348	136	16
5	2 $\log_{10}$ (99%)	100%	100	626	1181	272	30
		90%	50	157	934	243	27
		80%	26	75	732	213	24
		70%	14	50	567	185	21
		60%	7	38	432	157	18
		50%	4	29	321	129	15

DISCUSSION

When and where water is an important source of pathogen exposure, water quality interventions can reduce exposure to pathogens and result in improved health. Our results suggest, however, that the potential health gains are reduced sharply with even occasional consumption of untreated drinking water.

Our results are consistent with the findings of a similar QMRA analysis by Hunter et al. (2009)[6]. This study concluded that the health benefits of improved quality drinking water (as delivered by centralised treatment and distribution) were limited if even a small percentage of overall water consumed was of lower quality, for example during interruptions of service in piped water supply of when the alternative drinking-water source was surface water. Hunter *et al.* concluded as we have that the overall risk attributable to drinking water is controlled by those periods of higher exposure risk when no quality protection is in place, reducing overall impacts of water quality improvements significantly if the intervention is not present a high percentage of the time. We propose that our results are applicable to water quality interventions generally, including centralized water treatment.

There is reason to believe, however, that there is even greater risk among the estimated 1.2 billion who report treating water in the home. First, most of those that practice HWT reside in low-income settings where the risk of highly contaminated water is greatest. Second, unlike residents of higher-income settings who may receive warning when their water treatment systems fail, these individuals are unable to judge the quality of their drinking water and must follow the practice daily. Third, those who consistently treat their water at home are likely to be exposed to untreated drinking water when they are away from the home—a fundamental shortcoming of HWT compared to effective community-wide improvements in water quality.

Reconciling these results with the existing evidence base for water quality interventions raises certain questions. Our findings suggest that the dramatic protective effects against diarrhoeal disease reported in recent systematic reviews of household water treatment [14], [15], [16], [17]and other water quality interventions [33], [34] would seem to be unlikely in the absence of high adherence. Unfortunately, even studies that report on coverage and overall use rarely report on the extent to which participants consume untreated water. Nevertheless, our findings are consistent with two recent studies of the same water filter among young children in low-income settings. While investigators found no protective effect from an intervention in which most participants acknowledged regularly drinking untreated water [19], the intervention was highly protective against diarrhoea when few participants consumed untreated water (Rachel Peletz, personal communication).

Another potential explanation is that the large protective effects reported by open trials of water quality interventions are exaggerated due to reporting bias [15], [18]. Correct, consistent use has been an ongoing challenge for HWT, and few studies appear to reach the levels of uptake that our results suggest would be necessary to achieve the reported effects on diarrhoea. In this respect, our results lend support to the assertion that published health impact estimates of water quality interventions may have been overestimated. An analysis of reported DALYs averted by HWT interventions and a comparison with this and other modelling efforts may be a useful next step in contextualizing the evidence base for water quality interventions.

Another possibility that may explain higher than expected estimates of health effects from trials with low apparent adherence is that water quality interventions may have been more likely to have been used consistently when untreated water was perceived to be of higher risk due to aesthetic or other indicators, and those perceptions may have correlated with actual levels of pathogens (i.e., that adherence is higher when it is more important, when water is of higher risk). More research is needed on whether and how HWT behaviours are driven by perceptions of water safety, and whether and how perceptions of water safety are related to actual water safety.

Our results may also help explain the diminished health impact that has been observed from longer-term evaluations of HWT interventions [14], [16]. A number of studies of HWT have shown reduced use of HWT interventions over time, raising questions about the potential for sustained use [19], [20], [21], [22], [23], [24]. Technologies with a high user burden, recurrent costs, or those that involve substantial behaviour change may be especially challenging to achieve the continued high adherence that our results indicate is necessary to sustain health impact [35]. Boiling has overcome many of these same obstacles and become the leading HWT method worldwide [36]. However, studies of the effectiveness of boiling demonstrate the challenge of achieving consistently safe drinking water even from this highly normative practice.

Comparing predicted health gains resulting from two levels of microbiological treatment effectiveness (2 $\log_{10}$ and 6 $\log_{10}$ reduction of reference pathogens) suggests decreasing marginal health gains with increasing efficiency (Table 3). A focus on increasing microbiological effectiveness of water quality interventions, for example from 99% to 99.9999%, may not result in proportionately lower risk, especially if adherence will be less than 100%; the swamping effect of lower than 100% adherence essentially makes the real differences between microbial reduction efficiencies meaningless with respect to outcomes as predicted in this modelling exercise. We acknowledge that the value of requiring higher efficiencies for water treatment technologies is due to the higher efficiencies that may be required under outbreak scenarios, when untreated water may be of higher risk, or if there are no alternatives to high risk source waters. Given the fact that some technologies may be adopted more readily and used more consistently than other technologies (even those that may be more effective in reducing microbes), these results would seem to suggest that a focus on promoting adherence may be more critical in delivering health gains than focusing exclusively on increasing microbiological effectiveness [31].

The health impact estimates we report here should be interpreted in light of the uncertainty of the assumptions and necessary simplifications used to produce them. Pre-treatment water quality, for example, is likely to be highly variable in any given setting. Dose-response models for reference pathogens have been derived from few studies, mostly using data from healthy adults in wealthier countries. As an approach to estimating health risk, QMRA incorporates a number of assumptions whose values and ranges are uncertain. Where possible, we have attempted to use assumptions that would tend to result in conservative estimates of health impacts, and realistic ranges of pre-treatment water quality, adherence, treatment effectiveness in reducing microbes, and other key variables. QMRA is a quickly evolving approach and models like the one we have used will benefit from further refinement of methods and assumptions.

The greatest risks of waterborne disease globally are from microbial pathogens, although chemical contaminants are locally significant risks to public health. This risk assessment does not address the potential problem of anthropogenic or naturally occurring radiological or chemical contaminants that may be present in drinking-water, including but not limited to pesticides, arsenic, fluoride, heavy metals, nitrate, excess salts, disinfection by-products, pharmaceuticals, or others. Our study also omits helminthic and other diseases that may be transmitted by drinking water. Including these additional factors would require additional data and different modelling approaches.

Current guidelines supporting HWT emphasize the need for correct, consistent and sustained use [31] and our results underscore the need to make high adherence an essential priority in HWT design and program implementation. In order to maximise health gains from interventions, it is necessary to provide HWT solutions that people will use exclusively or nearly so. This may require not only intensified behaviour change interventions, but also process improvements in HWT products and technologies that minimize the likelihood of non-adherence. In order to monitor and evaluate these and other water quality interventions, it is also important to develop and improve tools for monitoring and to ensure that field studies report adherence. Given the challenges of achieving this high level of adherence through point-of-use water treatment, these results also confirm the need to continue efforts to provide safe, reliable, piped-in water to household taps in order to realize the full promise of health gains from water quality interventions.

ACKNOWLEDGMENTS

The authors gratefully acknowledge input by colleagues at the Hygiene Centre, LSHTM, and thoughtful critique from anonymous reviewers of this manuscript.

AUTHOR CONTRIBUTIONS

Conceived and designed the experiments: JB TC. Performed the experiments: JB. Analyzed the data: JB. Contributed reagents/materials/analysis tools: JB. Wrote the paper: JB TC.

REFERENCES

1. WHO and UNICEF (2012) Progress on Sanitation and Drinking Water: 2012 Update. Geneva: WHO Press. Available: www.wssinfo.org. Accessed 2012 March 15.
2. Onda K, LoBuglio J, Bartram J (2012) Global Access to Safe Water: Accounting for Water Quality and the Resulting
3. WHO (2005) World Health Report 2005. Geneva: World Health Organization.
4. Blakely T, Hales S, Kieft C, Wilson N, Woodward A (2005) The global distribution of risk factors by poverty level. Bulletin of the World Health Organization 83: 118–126.
5. UNICEF (2009) Diarrhea: why children are still dying and what can be done. New York: United Nations Children's Fund, Geneva: World Health Organization.
6. Hunter PR, Zmirou-Navier D, Hartemann P (2009) Estimating the impact on health of poor reliability of drinking water interventions in developing countries. Science of the Total Environment 407(8): 2621–2624.
7. Mintz ED, Reiff FM, Tauxe RV (1995) Safe water treatment and storage in the home: a practical new strategy to prevent waterborne disease. JAMA 273(12): 948–953.
8. WHO (2007) Combating waterborne disease at the household level. Geneva: World Health Organization.
9. 9.Luby S, Agboatwalla M, Raza A, Sobel J, Mintz E, et al. (2001) A low-cost intervention for cleaner drinking water in Karachi, Pakistan. International Journal of Infectious Diseases 5: 144–150.
10. Clasen TF, Thao do H, Boisson S, Shipin O (2008) Microbiological effectiveness and cost of boiling to disinfect drinking water in rural Vietnam. Environmental Science and Technology 42(12): 4255–60.
11. Sobsey MD (2002) Managing Water in the Home: Accelerated Health Gains from Improved Water Supply. Geneva: World Health Organization.
12. Brown J, Sobsey MD, Loomis D (2008) Drinking water filters reduce diarrheal disease in Cambodia: a randomized, controlled trial of locally made ceramic filters. American Journal of Tropical Medicine and Hygiene 79(3): 394–400.
13. Wright J, Gundry S, Conroy R (2004) Household drinking water in developing countries: a systematic review of microbiological contamination between source and point-of-use. Tropical Medicine and International Health 9(1): 106–117.
14. Arnold B, Colford J (2007) Treating water with chlorine at point-of-use to improve water quality and reduce child diarrhea in developing countries: A systematic review and meta-analysis. American Journal of Tropical Medicine and Hygiene 76(2): 354–364.
15. Clasen T, Roberts I, Rabie T, Schmidt W, Cairncross S (2006) *Interventions to improve water quality for preventing diarrhoea (A Cochrane Review).*
16. Waddington H, Snilstveit , White H, Fewtrell L (2009) Water, sanitation and hygiene interventions to combat childhood diarrhoea in developing countries. Delhi: International Initiative for Impact Evaluation.
17. Fewtrell L, Kaufmann RB, Kay D, Enanoria W, Haller L, et al. (2005) Water, sanitation, and hygiene interventions to reduce diarrhea in less developed countries: a systematic review and meta-analysis. Lancet Infectious Diseases 5: 42–52.
18. Schmidt WP, Cairncross S (2009) Household water treatment in poor populations: is there enough evidence for scaling up now? Environmental Science and Technology 43(4): 986–992.
19. Boisson S, Kiyombo M, Sthreshley L, Tumba S, Makambo J, et al. (2010) Field assessment of a novel household-based water filtration device: a randomised,

placebo-controlled trial in the Democratic Republic of Congo. PLoS One 10;5(9): e12613.

20. Mausezahl D, Christen A, Duran Pacheco G, Tellez FA, Iriarte M, et al. (2009) Solar Drinking Water Disinfection (SODIS) to Reduce Childhood Diarrhoea in Rural Bolivia: A Cluster-Randomized, Controlled Trial. PLoS Med 6(8): e1000125. doi:10.1371/journal.pmed.1000125.
21. Mong Y, Kaiser R, Ibrahim D, Razafimbololona L, Quick RE (2001) Impact of the safe water system on water quality in cyclone-affected communities in Madagascar. American Journal of Public Health 91(10): 1577–1579.
22. Brown J, Sobsey M, Proum S (2007) Use of Ceramic Water Filters in Cambodia. Washington, DC: WSP-World Bank Field Note.
23. Colindres RE, Jain S, Bowen A, Mintz E, Domond P (2007) After the flood: an evaluation of in-home drinking water treatment with combined flocculent-disinfectant following Tropical Storm Jeanne – Gonaives, Haiti, 2004. Journal of Water & Health 5: 367–374.
24. Luby S, Agboatwalla M, Raza A, Sobel J, Mintz E, et al. (2001) A low-cost intervention for cleaner drinking water in Karachi, Pakistan. International Journal of Infectious Diseasess 5: 144–150.
25. WHO (2011) Guidelines for Drinking-water Quality, 4th Edition. Geneva: World Health Organization.
26. Havelaar A, Melse JM (2003) *Quantifying Public Health Risk in the WHO GDWQ*. RIVM report 734301022/2003.
27. Gold MR, Stevenson D, Fryback DG (2002) HALYS and QALYS and DALYS, Oh My: similarities and differences in summary measures of population Health. Annual Review of Public Health 23: 115–134.
28. Haas CN (1996) How to average microbial densities to characterize risk. Water Research 30(4): 1036–1038.
29. Christian RR, Pipes WO (1983) Frequency distribution of coliforms in water distribution systems 45(2): 603–609.
30. EPA (United States Environmental Protection Agency) (1987) Guide Standard and Protocol for Testing Microbiological Water Purifiers. Office of Drinking Water.
31. NSF (National Sanitation Foundation, now NSF-International) (2003) NSF P231- Microbiological Water Purifiers. Ann Arbor, USA: NSF International.
32. WHO (2011) Evaluating Household Water Treatment Options: Health-Based Targets and Performance Specifications. Geneva: World Health Organization.
33. Esrey SA, Feachem RG, Hughes JM (1985) Interventions for the control of diarrheal diseases among young children: improving water supplies and excreta disposal facilities. Bulletin of the World Health Organization 63: 757–72.
34. Esrey SA, Potash JB, Roberts L, Shiff C (1991) Effects of improved water supply and sanitation on ascariasis, diarrhoea, dracunculiasis, hookworm infection, schistosomiasis, and trachoma. Bulletin of the World Health Organization 69(5): 609–621.
35. Brown J, Outlaw T, Clasen T, Wu J, Sobsey M (2009) Safe Water for All: Harnessing the Private Sector to Reach the Underserved. Washington, D.C.: World Bank/ International Finance Corporation.
36. Rosa G, Clasen T (2010) Estimating the scope of household water treatment in low- and medium-income countries. American Journal of Tropical Medicine and Hygiene 2010 Feb;82(2): 289–300.

CHAPTER 4

WATER MICROBIOLOGY. BACTERIAL PATHOGENS AND WATER

João P. S. Cabral

Keywords: Drinking water, cholera, typhoid fever, bacillary dysentery, fecal indicator bacteria, coliforms; ammonia

DRINKING WATER AS A VEHICLE OF DISEASES

Water is essential to life. An adequate, safe and accessible supply must be available to all. Improving access to safe drinking-water can result in significant benefits to health. Every effort should be made to achieve a drinking water quality as safe as possible [1].

Many people struggle to obtain access to safe water. A clean and treated water supply to each house may be the norm in Europe and North America, but in developing countries, access to both clean water and sanitation are not the rule, and waterborne infections are common. Two and a half billion people have no access to improved sanitation, and more than 1.5 million children die each year from diarrheal diseases [2]. According to the WHO, the mortality of water associated diseases exceeds 5 million people per year. From these, more that 50% are microbial intestinal infections, with cholera standing out in the first place.

In general terms, the greatest microbial risks are associated with ingestion of water that is contaminated with human or animal feces. Wastewater discharges in fresh waters and costal seawaters are the major source of fecal microorganisms, including pathogens [1–4].

Acute microbial diarrheal diseases are a major public health problem in developing countries. People affected by diarrheal diseases are those with the lowest financial resources and poorest hygienic facilities. Children under five, primarily in Asian and African countries, are the most affected by microbial diseases transmitted through water [5].

Microbial waterborne diseases also affect developed countries. In the USA, it has been estimated that each year 560,000 people suffer from severe waterborne diseases, and 7.1 million suffer from a mild to moderate infections, resulting in estimated 12,000 deaths a year [6]. The most important bacterial diseases transmitted through water are listed in Table 1.

Citation: João P. S. Cabral, Water Microbiology. Bacterial Pathogens and Water, doi:10.3390/ijerph7103657.

Table 1. The main bacterial diseases transmitted through drinking water.

Disease	Causal bacterial agent
Cholera	*Vibrio cholerae*, serovarieties O1 and O139
Gastroenteritis caused by vibrios	Mainly *Vibrio parahaemolyticus*
Typhoid fever and other serious salmonellosis	*Salmonella enterica* subsp. *enterica* serovar Paratyphi
	Salmonella enterica subsp. *enterica* serovar Typhi
	Salmonella enterica subsp. *enterica* serovar Typhimurium
Bacillary dysentery or shigellosis	*Shigella dysenteriae*
	Shigella flexneri
	Shigella boydii
	Shigella sonnei
Acute diarrheas and gastroenteritis	*Escherichia coli*, particularly serotypes such as O148, O157 and O124

CHOLERA

The Genus Vibrio

Vibrio are small, curved-shaped Gram-negative rods, with a single polar flagellum. Vibrios are facultative anaerobes capable of both fermentative and respiratory metabolism. Sodium stimulates growth of all species and is an absolute requirement for most. Most species are oxidase-positive and reduce nitrate to nitrite. Cells of certain species (V. cholerae, V. parahaemolyticus and V. vulnificus) have pili (fimbriae), structures composed of protein TcpA. TcpA formation is co-regulated with cholera toxin expression and is a key determinant of in vivo colonization (see below) [7,8].

Several Vibrio species can infect humans (Table 2). V. cholerae is, by far, the most important of these species. V. alginolyticus has been isolated from several types of soft tissue infections.

Table 2. Main species of Vibrio and their occurrence in human clinical specimensa.

Main species	Occurrence in human clinical specimens	
	Intestinal	Extra-intestinal
Vibrio alginolyticus	+	++
Vibrio cholerae O1 and O139	+++++	+
Vibrio cholerae non O1 or O139	++	++
Aliivibrio fischeri (*Vibrio fischeri*)	–	–
Vibrio fluvialis	++	–
Vibrio furnissii	++	–
Vibrio harveyi	–	+
Grimontia hollisae (*Vibrio hollisae*)	++	–
Vibrio mimicus	++	+
Vibrio natriegens	–	–
Vibrio parahaemolyticus	++++	+
Vibrio vulnificus	+	+++

[a]Adapted from [7,8]. Nomenclature according to [9]. The symbols give the relative frequency of each organism in human clinical specimens, and apply to the whole World, rather than to a particular country.

V. fluvialis, Grimontia hollisae (V. hollisae), and V. mimicus can cause diarrhea or infections of the gastrointestinal tract. V. furnissii has been isolated from a few individuals with diarrhea, but there is no evidence that it can actually cause this pathology. V. parahaemolyticus is a well-documented causal agent of acute food-borne gastroenteritis, particularly in Japan and South East Asia. Cases are associated with the consumption of raw or undercooked shellfish such as oysters, shrimp, crabs, and lobster. V. vulnificus is an important cause of (often fatal) septicemia and wound infections. Other vibrios, namely Allivibrio fischeri (Vibrio fischeri) and V. natriegens, have no relation with humans [7,8].

Vibrios are primarily aquatic bacteria. Species distribution depends on sodium concentration and water temperature. Vibrios are very common in marine and estuarine environments, living free or on the surfaces and in the intestinal contents of marine animals. Species with a low sodium requirement are also found in freshwater habitats [7,8].

The Species Vibrio Cholerae

Vibrio cholerae cells can grow at 40 °C with pH 9–10. The growth is stimulated by the presence of sodium chloride. Vibrio cholerae is a very diverse bacterial species (Table 3). It is divided in ca. 200 serovarieties, characterized by the structure of the lipopolysaccharide (LPS) (O antigens). Only serovarieties O1 and O139 are involved in "true" cholera. Some other serovarieties can cause gastroenteritis, but not cholera. The distinction between Classical and El Tor biotypes is based on biochemical and virological characteristics [1,7,8,10,11].

Table 3. Subdivision of Vibrio cholerae below the species levela.

Serovariety	Serotype	Biotype
O1	Inaba	Classical
		El Tor
	Ogawa	Classical
		El Tor
	Hikojima	
O139		
others		

[a]Adapted from [8].

Cholera

Characterization of the disease

The incubation period for cholera is ca. 1–3 days. The disease is characterized by an acute and very intense diarrhea that can exceed one liter per hour. Cholera patients feel thirsty, have muscular pains and general weakness, and show signs of oliguria, hypovolemia, hemoconcentration, followed by anuria. Potassium in blood drops to very low levels. Patients feel lethargic. Finally, circulatory collapse and dehydration with cyanosis occurs [7].

The severity of the disease depends on several factors: (1) personal immunity: this may be conferred by both previous infections and by vaccines; (2) inoculum: the disease only occurs after ingestion of a minimum amount of cells, ca. 10^8 [1,7,8,10,11]; (3) The gastric barrier: V. cholera cells likes basic media and therefore the stomach, normally very acidic, is an adverse medium for bacterial survival. Patients consuming anti-acidic medications are more susceptible to infection than healthy people; (4) blood group: for still unknown reasons, people with O-group blood are more susceptible than others [1,7,8,10,11].

In the absence of treatment, the mortality of cholera-patients is ca. 50%. It is mandatory to replace not only lost water but also lost salts, mainly potassium. In light dehydrations, water and salts can be orally-administered, but in severe conditions, rapid and intravenous-administration is obligatory. The most efficient antibiotic is currently doxicyclin. If no antibiotic is available for treatment, the administration of water with salts and sugar can, in many cases, save the patient and help in the recovery [1,7,8,10,11].

There are two main determinants of infection: (1) the adhesion of the bacterial cells to the intestinal mucous membrane. This depends on the presence of pili and adesins at the cell's surface; (2) the production of cholera toxin [1,7,8,10,11].

Cholera toxin

Cholera toxin is an exotoxin with a very precise action on target cells. The toxin attaches to a specific receptor (ganglioside Gl) on the cell membrane of intestinal cells and activates the enzyme adenylate cyclase. This results in a non-stop degradation of internal ATP, with release of cAMP and inorganic phosphate. The rise in the internal concentration of cAMP causes an efflux of water, sodium, potassium, chloride and carbonate ions from the cells of the mucous membrane, and this is the main cause of diarrhea [7].

Cholera pandemics and the emergence of El Tor biotype and O139 serovariety. New facts about cholera epidemiology

Cholera has been a well known disease since the 19th century. In the 19th and 20th centuries, seven major pandemics are recognized. The first six pandemics occurred during the following periods: 1st: 1816–1826, 2nd: 1829–1851, 3rd: 1852–1860, 4th: 1863–1875, 5th: 1881–1896, 6th: 1899–1923. These pandemics all started in Asia, passed through Europe and then reached South America. The Classical biotype was involved. The seventh pandemic, still in course, started in 1961 in the Celebes Isles, in Asia. In the 1960s, the disease spread through Asia, in the 1970s reached the Middle East and Africa, and in 1991 streaked violently across South America. Now El Tor has replaced the Classical biotype. El Tor biotype had been detected before, in 1905, but only in the development of the seventh pandemic did this biotype replace the Classical one and become dominant [1,7,8,10,11].

In 1992, a new serovariety (O139), which was coined the Bengal serovariety, was detected for the first time in Bangladesh. This new serovariety quickly spread to India and to southeastern Asia, displacing O1. Although serovariety O1 El Tor has reappeared in 1994 and 1995, the Bengal serovariety still remains the dominant one. The illness caused by serovarieties O139 and O1 are indistinguishable [8,12,13].

In 1991, the seventh pandemic entered South America through the coastal area of Peru. On 23 January, in Chancay, north Peru, Vibrio cholerae O1 El Tor was isolated from patients with cholera symptoms, confirming the disease. In this region, between 24 January and 9 February, 1,859 people were hospitalized and 66 died. From Peru, the disease spread rapidly to other countries in South America. Two routes have been proposed for the entrance of the bacterium in Peru: (1) ballast water from a boat coming from Asia; (2) the El Niño current may have transported zooplankton harboring V. cholerae cells. Shellfish and fish nourishing on this zooplankton became contaminated and the bacterium was transmitted to humans who ate these marine foods [14–17].

The misfortune of people who died in the first months of this disastrous South American cholera epidemic appeared to have unleashed scientists to study the disease harder and, indeed, important epidemiological studies were carried out during this outbreak. These studies confirmed that contaminated uncooked food and beverages can also be a vehicle for transmission of cholera [18].

Genes for toxin and pili protein production

The genes responsible for toxin production are harbored in the CTXΦ segment (7–9.7 kb) of the chromosome (only in toxigenic strains). The CTXΦ segment carries at least six genes. In addition to the gene encoding cholera toxin production, this segment (virulence cassette) include an accessory cholera toxin (ace), a zonula occludens toxin (zot), core encoded pilin (cep), and an open reading frame of unknown function. During the replication of the chromosome, the CTXΦ fragment can form an autonomous copy and this can constitute an independent plasmid. The plasmid can give rise to virus-like particles—CTXΦ bacteriophages, which can infect non-toxigenic strains. The CTXΦ segment incorporates into the chromosome of the infected cells which became toxigenic. This process was demonstrated in vitro in cell suspensions and in vivo in the gut of the rat [8,13,19,20].

Epidemic and pandemic strains of V. cholerae contain another chromosomal segment designated as VPI. VPI is 39.5 kb in size and contains two ToxR-regulated genes: a regulator of virulence genes (ToxT) and a gene cluster containing colonization factors, including the toxin co-regulated pili (TCP). The tcp gene encodes for the 20.5-kDa TcpA pili protein. This VPI segment appears to be transferable from V. cholerae O1 to non-O1 strains. V. cholerae O139 strains, like O1, carry the structural genes encoded by the CTX operon and TCP. V. cholerae strains non-O1 or O139 normally lack cholera toxin genes and have never been found to carry TCP [8].

Ecology of the bacterium and the cycle of the disease

V. cholerae non-O1 or O139 strains are common in the environment, especially in estuaries. They have been isolated from many estuarine animals such as birds, frogs, fishes and shellfish, and survive and multiply on the surface of phytoplankton and zooplankton cells [8,21].

V. cholerae O1 and O139 strains are isolable from the environment only in epidemic areas. They survive in the cultivable state in water and aquatic and marine organisms for a considerable period of time [8,12,22–24]. When V. cholerae cells face adverse environmental conditions, they reduce cell size, became coccoid and enter a dormant stage inside exopolysaccharide biofilms. Cells display a certain metabolism, but are not able to growth and multiply on the surface of agarized media and give rise to colonies. Cells in this viable but non-culturable state retain viability as well as the potential for pathogenicity for significant periods of time [25–27].

Viable but non-culturable cells can leave their dormant stage and multiply again, resulting in an explosion of their concentration in the environment. Since the presence of non-toxigenic strains is common in aquatic milieu, especially in estuaries, if a horizontal transfer of cholera exotoxin producing genes occurs between toxigenic and non-toxigenic strains, the number of toxigenic cells in the environment can rise rapidly and pronouncedly. The episodic nature and the sudden appearance of violent cholera outbreaks, followed by a rapid slowing down, are probably related with these phenomena.

SALMONELLOSIS

The Genus Salmonella. Pathogenicity of Main Serovars

The genus Salmonella was designated by Lignières in 1900 [28,29]. Antigenic analysis began when Castellani described, in 1902, a method for absorbing antisera. The first antigenic scheme forSalmonella was published by White in 1926, and subsequently developed extensively by Kauffmann, in two classical works published in 1966 and 1978 [28,29]. The Kauffmann-White antigenic scheme contained, by 1988, about 2,250 different serovars [28,29].

The genus Salmonella, a member of the family Enterobacteriaceae, include Gram-negative motile straight rods. Cells are oxidase-negative and catalase-positive, produce gas from D-glucose and utilize citrate as a sole carbon source. Salmonellae have several endotoxins: antigens O, H and Vi [28,29].

The concept "one serovar-one species", in use for many years, is no longer acceptable. The taxonomy and nomenclature of the genus Salmonella has been subject of debate since Le Minor and Popoff proposed changes in a paper published in 1987. The issue was settled by a decision of the International Committee on the Systematics of Prokaryotes and published in 2005. The current taxonomy of the genus is presented in Table 4. According to the rules of bacterial nomenclature, the names of the serovars are not italicized and the first letter must be a capital [28–30].

Table 4. Current taxonomy and nomenclature of the genusSalmonella. Habitat and pathogenicity of main serovarsa.

Species	Sub-species	Main serovars (from a total of ca. 1,443)	Habitat and pathogenicity
Salmonella enterica	*Salmonella enterica* subsp. *enterica*	Abortusovis	Pathogenic to sheeps.
		Choleraesuis	Pathogenic to humans and animals.
		Enteritidis	Ubiquitous and frequently the cause of infections in humans and animals. Very frequent agent of gastroenteritis in humans.
		Gallinarum	Isolated chiefly from chickens and other birds. Causal agent of fowl thyphoid.
		Paratyphi A	Pathogenic only to humans. Causes paratyphoid fever.
		Paratyphi B	Causes paratyphoid fever in humans and very rarely infects animals.
		Paratyphi C	Causes paratyphoid fever in humans.
		Typhi	Pathogenic only to humans, causing typhoid fever. Transmitted by water and food contaminated with feces.
		Typhimurium	Ubiquitous and frequently the cause of infections in humans and animals. Very frequently, the causal agent of gastroenteritis in humans.
		Typhisuis	Pathogenic to swines.
	Salmonella enterica subsp. *arizonae*	At least 94 serovars.	
	Salmonella enterica subsp. *diarizonae*	At least 323 serovars.	
	Salmonella enterica subsp. *houtenae*	At least 70 serovars.	Isolated mainly from cold-blooded animals and from the environment. Not pathogenic to humans.
	Salmonella enterica subsp. *indica*	At least 11 serovars.	
	Salmonella enterica subsp. *salamae*	At least 488 serovars.	
Salmonella bongori		At least 20 serovars.	

[a]Adapted from [29]. Nomenclature according to [9].

Table 5. Current taxonomy and nomenclature of the genusShigella. Habitat and pathogenicity of speciesa.

Species	Main serotypes	Habitat and pathogenicity
Shigella dysenteriae	15 serotypes.	Intestinal pathogens of humans and primates, causing bacillary dysentery.
Shigella flexneri	8 serotypes 9 subserotypes	Humans are the primary reservoir. A long-term carrier state occurs in few cases.
Shigella boydii	19 serotypes	*Shigella dysenteriae* serotype 1 causes more severe disease then other serotypes and produces a potent exotoxin (Shiga toxin). Large epidemics in developing countries are commonly caused by serotype 1. Diseases caused by other serotypes may be mild or severe.
Shigella sonnei	1 serotype	*Shigella sonnei* illness is usually milder than that caused by other *Shigella* species.

[a]Adapted from [34]. Nomenclature according to [9].

S. enterica subsp. enterica serovar Enteritidis is the most frequently isolated serovar from humans all over the world. However, locally, other serovars can be predominant. In the period 1994–2004, Tunisia was exposed to salmonellosis outbreaks in 1997, 1999, 2002 and 2004. In 1997, salmonellosis outbreak was caused by serovar Mbandaka. In 1999, three salmonellosis outbreaks were reported from hospitals located in three different regions. Each outbreak was associated with a different serotype: Mbandaka, Livingstone and Typhi Vi+. In 2002, a S. enterica subsp. enterica serovar Livingstone infection occurred in the same hospital that reported an outbreak caused by serovar Typhi Vi+ in 1999, but in a different unit. In that year, the Livingstone serovar jumped to the first position in human infection in Tunisia. In 2004, a second outbreak by serovar Typhi Vi+ was reported. The source of isolation was a fermented juice traditionally extracted from palm-tree [31].

Characterization of the Diseases

Salmonellae pathogenic to humans can cause two types of salmonellosis: (1) typhoid and paratyphoid fever (do not confuse with typhus, a disease caused by a rickettsia); (2) gastroenteritis [28]. Low infective doses (less than 1,000 cells) are sufficient to cause clinical symptoms. Salmonellosis of newborns and infants presents diverse clinical symptoms, from a grave typhoid-like illness with septicemia to a mild or asymptomatic infection. In pediatric wards, the infection is usually transmitted by the hands of staff [29].

Food-borne Salmonella gastroenteritis are frequently caused by ubiquitous Salmonella serovars such as Typhimurium. About 12 h following ingestion of contaminated food, symptoms (diarrhea, vomiting and fever) appear and last 2–5 days. Spontaneous cure usually occurs. Salmonella may be associated with all kinds of food. Prevention of Salmonella food-borne infection relies on avoiding contamination (improvement of hygiene), preventing multiplication of Salmonella in food (constant storage of food at 4 °C), and use of pasteurization (milk) or sterilization when possible (other foods). Vegetables and fruits may carry Salmonella when contaminated with fertilizers of fecal origin, or when washed with polluted water [28].

The incidence of typhoid fever decreases when the level of development of a country increases (i.e., controlled water sewage systems, pasteurization of milk and dairy products). Where these hygienic conditions are missing, the probability of fecal contamination of water and food remains high and so is the incidence of typhoid fever [29].

Ecology of Salmonellae and the Cycle of Salmonellosis

The principal habitat of Salmonella is the intestinal tract of humans and animals [28]. Salmonellae are constantly found in environmental samples, because they are excreted by humans, pets, farm animals, and wild life. Municipal sewage, agriculture pollution, and storm water runoff are the main sources of these pathogens in natural waters [1,32]. Salmonellae do not seem to multiply significantly in the natural environment, but they can survive several weeks in water and in soil if conditions of temperature, humidity, and pH are favorable [28].

Salmonellae isolated from environmental sources are predominantly non-Typhi or Paratyphi serovars. In a study carried out in Tunisia during 1994–2004, S. enterica subsp. enterica serovars Anatum, Enteritidis and Corvallis were the most common serotypes isolated from food. The great majority of the strains were isolated from poultry, red meat, milk and dairy products, vegetables and fruits. From environmental sources, 73% of the isolates were from tap water. Serovars Corvallis, Enteritidis, and Anatum were the commonest [31]. Arvanitidou et al. [32] reported a comparative study carried out in Rivers Aliakmon and Axios, in northern Greece, during a 1-year period, from May 2002 to April 2003. A total of 29 Salmonella species were recovered from the water samples. Many of the isolated Salmonella serovars were of non-human animal origin such as Mbantaka, Virchow, Hadar, Infantis and Senftenberg, commonly isolated from poultry farm.

Unlike cholera, humans infected with salmonellae can carry the bacteria in the gut without signs of disease. Infected humans can harbor the bacteria for considerable periods of time. About 5% of patients clinically cured from typhoid fever remain carriers for months or even years. These people can be chronic holders of the bacterium in the gut, and constitute the main reservoir of the bacteria in the environment [29].

The salmonellosis cycle in the environment can involve shellfish. Salmonellae survive sewage treatments if suitable germicides are not used in sewage processing. If effluent from the sewage plant passes into a coastal area, edible shellfish (mussels, oysters) can become contaminated. Shellfish concentrate bacteria as they filter several liters of water per hour. Ingestion by humans of these seafoods (uncooked or superficially cooked) may cause typhoid fever or other salmonellosis. Evidence of such a cycle has been obtained by the use of strain markers, including phage typing [29].

SHIGELLOSIS OR BACILLARY DYSENTERY

The Genus Shigella

Shigella are Gram-negative, non-sporeforming, non-motile, straight rod-like members of the family Enterobacteriaceae. Cells ferment sugars without gas production. Salicin, adonitol and myo-inositol are not fermented. Cells do not utilize citrate, malonate and acetate as sole carbon source and do not produce H_2S. Lysine is not decarboxylated. Cells are oxidase-negative and catalase-positive. Members of the genus have a complex antigenic pattern, and taxonomy is based on their somatic O antigens [1,33,34].

Characterization of the Disease

The incubation period is 1–4 days. The disease usually begins with fever, anorexia, fatigue and malaise. Patients display frequent bloody stools of small volume (sometimes grossly purulent) and abdominal cramps. Twelve to 36 hours later, diarrhea progresses to dysentery, blood, mucus and pus appearing in feces that decreases in volume (no more than 30 mL of fluid per kg per day) [34–36].

Although the molecular basis of shigellosis is complex, the initial step in pathogenesis is penetration of the colonic mucosa. The resulting focus of Shigella infection is characterized by degeneration of the epithelium and by an acute inflammatory colitis in the lamina propria. Ultimately, desquamation and ulceration of the mucosa cause leakage of blood, inflammatory elements, and mucus into the intestinal lumen. Under these conditions the absorption of water by the colon is inhibited and the volume of stool is dependent upon the ileocecal flow. As a result, the patient will pass frequent, scanty, dysenteric stools [37,38].

In order for Shigella to enter an epithelial cell, the bacterium must first adhere to its target cell. Generally, the bacterium is internalized via an endosome, which it subsequently lyses to gain access to the cytoplasm where multiplication occurs [37,38].

Virulence Factors

S. dysenteriae serotype 1 produces high levels of a cytotoxic Shiga toxin. S. sonnei and S. flexneri produce much lower amounts of this toxin. Shiga toxin binds to Galotl-4Galp (galabiose) glycolipid receptors and inhibits mammalian protein synthesis by cleaving the N-glycosidic bond at adenine 4324 in 28S rRNA. The toxic mechanism is identical to that of the plant toxin ricin, produced byRicinus communis. Shigella also release a LPS endotoxin (O antigens), that cause an inflammatory response [37,38].

Shigella 180- to 230-kb plasmids encode genes essential for virulence, namely for: production of adhesins involved in the adherence of bacteria onto the surface of target epithelial cells; production of invasion plasmid antigens (Ipa) that have a direct role in the Shigella invasion process; transport or processing functions that ensure the correct surface expression of the Ipa proteins; induction of endocytic uptake of bacteria and disruption of endocytic vacuoles; regulation of plasmid-encoded virulence genes [37,38].

Shigella emerged from E. coli during evolution. The acquisition and evolution of the pathogenicity island which encodes all of the genes required for cell invasion and phagolysosomal lysis, permitted a major alteration in pathogenesis [37,38].

Risk Factors

Many studies have identified risk factors and protective effects for shigellosis incidence and fatality. Despite gradual improvements in water supply, shigellosis continues to be endemic among the disadvantaged populations living in the tropics, often among displaced populations following natural disasters and political crises. In Guatemala, young children, the elderly, and 15–44-year-old males were found to be most susceptible to S. dysenteriae serotype 1. In Sierra Leone, the attack rate was higher among children younger than 5 years of age than in the rest of the population. In rural Bangladesh, shigellosis was most common in children aged 1–2 years and in people 60 years or older. In Dhaka, Bangladesh, it was found that shigellosis mortality was most common in severely malnourished people of all ages, in children under 2 who were not being breastfed, and in all children under 1. In a 3-year study carried out in Matlab, Bangladesh, during 1992 to 1994, it was found that the incidence of S. dysenteriae serotype 1 and S. flexneri was highest in children under 2 followed by children from 2 to 5. The location of S. dysenteriae serotype 1 risk varies in time but S. flexneri risk areas were persistent in time. Neighborhoods near bazaars with many non-septic latrines were at highest risk for S. dysenteriae serotype 1. S. flexneri was most common in flood-controlled areas. It was concluded that S. dysenteriae serotype 1 risk was more related to hygiene and sanitation whereas S. flexneri was more related to the environment [35].

Shigellosis through the World

The total number of Shigella episodes that occur each year throughout the World is estimated to be 164.7 million, including 163.2 million cases in developing countries, 1.1 million of which result in death. Children under 5 account for 61% of all deaths attributable to shigellosis [35,36].

Shigella species are not uniformly distributed in the world. S. dysenteriae is usually found in densely populated areas of South America, Africa and Asia. Infections usually result in significant epidemic outbreaks. Serotype 1 has been distinguished by both its virulence and its ability to produce ravaging epidemics. It predominates in India, Malaysia and Guatemala. Serotype 2 predominates in Yemen and Nigeria. S. flexneri is usually found in areas where endemic shigellosis occurs. S. boydii occurs sporadically, except in the Indian subcontinent where it was first identified.S. sonnei usually occurs in Western developed countries, such as France and USA [35,36].

Important epidemics were reported in the last decades: (1) in 1970 in Central America where 112,000 people were affected and 13,000 died; (2) in 1985, in Texas (USA), 5,000 people became infected after ingestion of contaminated lettuce; (3) in May–June 1994, domestic cases of S. sonneiinfection were detected in several European countries, including Norway, Sweden, and the United Kingdom. Epidemiological evidence incriminated imported iceberg lettuce as the vehicle of transmission; (4) in 1996, in Paris, with 153 reported patients [33].

Ecology of Shigellae and the Cycle of Shigellosis

Shigella is typically an inhabitant of the intestinal tract of humans and other primates [1,33,34,36,39]. It is typically spread by fecal-contaminated drinking water or food, or by direct contact with an infected person. In water, shigellae can survive for at least six months at room temperature, and this high survival favors transmission through water. Flies have been implicated on the transmission of Shigella cells from human feces to foods. The hand is an important vehicle for transmission of shigellosis, since S. dysenteriae serotype 1 cells survives for up to one hour on a human's skin and a very small inoculum is required to unchain infection and disease. Indeed, studies on American volunteers experimentally infected with Shigella have shown that as few as one hundred Shigella cells given orally cause the disease in 25–50% of the cases. Resistance ofShigella to gastric juice certainly accounts, although not exclusively, for this high infectivity [36,40]. Asymptomatic and inappropriately-treated patients with shigellosis can harbor the bacteria in the gut and these appear to be the main reservoirs of the bacteria in the environment [41].

Recent reports on the ecology of shigellae have brought new elements for the understanding of the cycle of the disease. In environmental waters of regions with high numbers of shigellosis' cases, it has been found that, although numbers of cultivable cells were low, genetic elements such as plasmids and genetic fragments with bacteriophage origin, could be detected. Many of the genes that code for exotoxin production are precisely found in these genetic elements. These results suggest that the sudden rise of the number of virulent strains in the environment can result from the incorporation, by cells with reduced virulence, of this type of genetic elements present in the waters. If this hypothesis is confirmed, there is a certain similarity between the cholera and the shigellosis cycles in the environment. It remains to be elucidated if shigellae can also exist in environmental waters in a viable but non-culturable state, as vibrios [41].

PATHOGENIC ESCHERICHIA COLI STRAINS

E. coli strains isolated from intestinal diseases have been grouped into at least six different main groups, based on epidemiological evidence, phenotypic traits, clinical features of the disease and specific virulence factors. From these, enterotoxigenic (ETEC, namely O148), enterohemorrhagic (EHEC, namely O157) and enteroinvasive serotypes (EIEC, namely O124) are of outstanding importance and can be transmitted through contaminated water [42,43].

Enterotoxigenic E. coli (ETEC) Strains

Enterotoxigenic E. coli (ETEC) serotypes can cause infantile gastroenteritis. The number of reports of their occurrence in developed countries is comparatively small, but it is an extremely important cause of diarrhea in the developing world, where there is no adequate clean water and poor sanitation. In developing countries, these strains are the most commonly isolated bacterial enteropathogen in children below 5 years of age, and account for several hundred million cases of diarrhea and several ten of thousand deaths each year [42–44].

Disease caused by ETEC follows ingestion of contaminated food or water and is characterized by profuse watery diarrhea lasting for several days that often leads to dehydration and malnutrition in young children [42–44]. ETEC also are the most common cause of "travelers' diarrhea" that affects individuals from industrialized countries travelling to developing regions of the World [42–44].

Enterohemorrhagic E. coli (EHEC) Strains

Reported outbreaks had been associated mainly with the consumption of contaminated foods, such as raw or undercooked ground meat products and raw milk. The primary reservoir of this bacterium has been found to be healthy cattle [42,45,46].

E. coli serotype O157:H7 causes abdominal pain, bloody diarrhea, and hemolytic uremic syndrome. This bacterium produces Shiga-like toxins. The incubation period is 3–4 days, and the symptoms occur for 7–10 days. It is estimated that 2–7% of E. coli O157:H7 infections result in acute renal failure [42,45,46].

Although E. coli O157:H7 is not usually a concern in treated drinking water, outbreaks involving consumption of drinking water contaminated with human sewage or cattle feces have been documented. An increasing number of outbreaks are associated with the consumption of fruits and vegetables (sprouts, lettuce, coleslaw, salad) contaminated with feces from domestic or wild animals at some stage during cultivation or handling. EHEC has also been isolated from bodies of water (ponds, streams), wells and water troughs, and has been found to survive for months in manure and water-trough sediments [45,46].

Person-to-person contact is an important mode of transmission through the oral-fecal route. An asymptomatic carrier state has been reported, where individuals show no clinical signs of disease but are capable of infecting others [45,46].

Enteroinvasive E. coli (EIEC) Strains

Enteroinvasive E. coli (EIEC) behave in many respects like shigellae. They are capable of invading and multiplying in the intestinal epithelial cells of the distal large bowel in humans. The illness is characterized by abdominal cramps, diarrhea, vomiting, fever, chills, a generalized malaise, and the appearance of blood and mucus in the stools of infected individuals. [42,43,47].

EIEC strains were isolated, for instance, from 28 subjects in the Jesreel district of Israel during a peak period for dysentery. An investigation in Croatia showed that E. coli O124 could frequently be isolated from cases of gastroenteritis, enterocolitis, and dysentery. The dysentery was more common among the older age groups, while the two other types of disease occurred equally in all age groups. A 1985 survey was carried out in Bankok, Thailand in which 410 children with diarrhea and an equal number of control children without diarrhea were examined for the presence of strains of Shigella, EIEC, and other pathogens. It was found that 17 of the children with diarrhea and six without yielded EIEC [42,43].

Any food contaminated with human feces from an ill individual, either directly or via contaminated water, could cause disease in others. Outbreaks have been associated with hamburger meat and unpasteurized milk [47].

EMERGING WATERBORNE BACTERIAL PATHOGENS

The emerging pathogenic bacteria of concern outlined here have the potential to be spread through drinking water, but they do not correlate with the presence of E. coli or with other commonly used drinking water quality indicators, such as coliform bacteria. In most cases, there are no satisfactory microbiological indicators of their presence. More studies are needed in order to understand the real significance and dimension of the diseases caused by water contaminated with these bacteria, and the ecology of these pathogens [45].

Mycobacterium Avium Complex (Mac)

The Mycobacterium avium complex (Mac) consists of 28 serovars of two distinct species:Mycobacterium avium and Mycobacterium intracellulare. The importance of Mac organisms was recognized with the discovery of disseminated infection in immunocompromised people, particularly people with HIV and AIDS. Members of MAC are considered opportunistic human pathogens [45,48].

Mac organisms have been identified in a broad range of environmental sources, including marine waters, rivers, lakes, streams, ponds, springs, soil, piped water supplies, plants, and house dust. Mac organisms have been isolated from natural water and drinking water distribution systems in the USA [45,49,50].

The ubiquitous nature of Mac organisms results from their ability to survive and grow under varied conditions. Mac organisms can proliferate in water at temperatures up to 51 °C and can grow in natural waters over a wide pH range [45]. These mycobacteria are highly resistant to chlorine and the other chemical disinfectants used for the treatment of drinking-water. Standard drinking-water treatments will not eliminate Mac organisms but, if operating satisfactorily, will significantly reduce the numbers that may be present in the source water to a level that represents a negligible risk to the general population. The entryway of these mycobacteria in distribution systems is through leaks. Growth of Mac organisms in biofilms is probably important for their continuous presence in distribution systems. Slow growing mycobacteria can be found at densities greater than 4,000 per cm^2 in the surface biofilm, creating a potentially high level of exposure [48].

The symptoms encountered with Mac infections result from colonization of either the respiratory or the gastrointestinal tract, with possible dissemination to other locations in the body. Exposure to Mac organisms may occur through the consumption of contaminated food, the inhalation of air with contaminated soil particles, or contact with or ingestion, aspiration, or aerosolization of potable water containing the organisms [45].

With respect to water supplies, infection with M. avium and M. intracellulare has been well documented. Unlike gastrointestinal pathogens, where E. coli can be used to indicate their potential presence, no suitable indicators have been identified to signal increasing concentrations of Mac organisms in water systems [45].

Helicobacter Pylori

Helicobacter pylori has been cited as a major etiologic agent for gastritis and has been implicated in the pathogenesis of peptic and duodenal ulcer disease and gastric carcinoma. However, most individuals that are infected by this pathogen remain asymptomatic [45].

Using culture-based methods, H. pylori has not been isolated from environmental sources, including water [45,51]. On the contrary, molecular methods have been successful in detecting this pathogen. Fluorescence in situ hybridization has been successfully used to detect this pathogen in drinking water distribution systems and other water bodies. Polymerase chain reaction has also been used to detect the presence of H. pylori DNA in drinking water, especially associated with biofilms [45,51,52]. In drinking-water biofilms, H. pylori cells rapidly lose culturability, entering a viable but non-culturable state. In these biofilms, cells can persist for more than one month, with densities exceeding 10^6 cells per cm^2 [51].

How the organism is transmitted is still not fully understood. However, the fact that it has been recovered from saliva, dental plaques, the stomach, and fecal samples strongly indicates oral-oral or fecal-oral transmission. Water and food appear to be of lesser direct importance, but they can still play a significant role in situations with improper sanitation and hygiene [45].

Aeromonas Hydrophyla

In recent years, A. hydrophila has gained public health recognition as an opportunistic pathogen. It has been implicated as a potential agent of gastroenteritis, septicemia, meningitis, and wound infections. It can play a significant role in intestinal disorders in children under five years old, the elderly, and immunosuppressed people. [45,53,54].

Aeromonas hydrophila are Gram-negative, non-sporeforming, rod-shaped, facultative anaerobic bacilli belonging to the family Aeromonadaceae. Although A. hydrophila is usually the dominant species, other aeromonads, such as A. caviae and A. sobria, have also been isolated from human feces and from water sources [45,54].

Aeromonas species, including A. hydrophila, are ubiquitous in the environment. It is frequently isolated from food, drinking water, and aquatic environments [45,53,54]. In clean rivers and lakes, concentrations of Aeromonas spp. are usually around 10^2 colony-forming units (CFU)/mL. Groundwaters generally contain less than 1 CFU/mL. Drinking water immediately leaving the treatment plant has been found to contain between 0 and 10^2 CFU/mL. Drinking water in distribution systems can display higher Aeromonas concentrations, due to the growth in biofilms [45,55]. Aeromonas spp. have been found to grow between 5 °C and 45 °C [44,54]. A. hydrophila is resistant to standard chlorine treatments, probably surviving inside biofilms [56].

The common routes of infection suggested for Aeromonas are the ingestion of contaminated water or food or contact of the organism with a break in the skin. Drinking or natural mineral water can be a possible source of contamination for humans. No person-to-person transmission has been reported [45,54].

MICROBIOLOGICAL WATER ANALYSIS

The Rationale of the Use of Fecal Indicator Bacteria

The most important bacterial gastrointestinal diseases transmitted through water are cholera, salmonellosis and shigellosis. These diseases are mainly transmitted through water (and food) contaminated with feces of patients. Drinking water can be contaminated with these pathogenic bacteria, and this is an issue of great concern. However, the presence of pathogenic bacteria in water is sporadic and erratic, levels are low, and the isolation and culture of these bacteria is not straightforward. For these reasons, routine water microbiological analysis does not include the detection of pathogenic bacteria. However, safe water demands that water is free from pathogenic bacteria [57].

The conciliation of the two needs was met by the discovery and testing of indicator bacteria. Water contaminated with pathogenic species also has the normal inhabitants of the human intestine. A good bacterial indicator of fecal pollution should fulfill the following criteria: (1) exist in high numbers in the human intestine and feces; (2) not be pathogenic to humans; (3) easily, reliably and cheaply detectable in environmental waters. Additionally, the following requisites should be met if possible: (4) does not multiply outside the enteric environment; (5) in environmental waters, the indicator should exist in greater numbers than eventual pathogenic bacteria; (6) the indicators should have a similar die-off behavior as the pathogens; (7) if human fecal pollution is to be separated from animal pollution, the indicator should not be very common in the intestine of farm and domestic animals [1,4,6,57,58]. The usefulness of indicator bacteria in predicting the presence of pathogens was well illustrated in many studies, namely by Wilkes et al.[59].

The Composition of Human and Animal Feces

Microbiological analysis of the human feces was important in order to structure and validate the use of fecal indicator bacteria in environmental waters. Bacteria present in feces are naturally derived from the microbiota of the human gastrointestinal tract.

Although bacteria are distributed throughout the human gastrointestinal tract, the major concentration of microbes and metabolic activity can be found in the large intestine. The upper bowel (stomach, duodenum, and jejunum) has a sparse microbiota with up to 10^5 CFU/ml of contents. From the ileum on, bacterial concentrations gradually increase reaching in the colon $10^{10}$to 10^{11} CFU/g [60].

It has been estimated that at least 500–1,000 different microbial species exist in the human gastrointestinal microbiota, although on a quantitative basis 10–20 genera usually predominate (Table 6). The total number of microbial genes in the human gastrointestinal tract has been estimated as 2–4 million. This represents an enormous metabolic potential which is far greater than that possessed by the human host [60,64].

Table 6. Total viable count in feces of healthy humans (children, adults and elderly)a.

Microbial group	Log_{10} CFU/g feces
Bacteroides	11.3*
Eubacterium	10.7*
Bifidobacterium	10.2*
Ruminococcus	10.2*
Peptostreptococcus	10.1*
Peptococcus	10.0*
Clostridium	9.8*
Lactobacillus	9.6*
Propionobacterium	9.4*
Actinomyces	9.2*
Methanobrevibacter	8.8*
Desulphovibrio	8.4*
Fusobacterium	8.4*
Enterococci	3.5–7.2**
Enterobacteriaceae	5.9–8.0**
Escherichia coli	7.5–7.7**
Citrobacter	3.3**
Klebsiella	2.4**
Yeasts	1.0–2.5**

[a]Adapted from [61–63]** and [64]*. *Values expressed as dry weight. **Values expressed as wet weight.

The composition of feces from an individual is stable at genus level, but the species composition can vary markedly from day to day. The relative proportion of intestinal bacterial groups can vary between individuals [60,64].

The microflora of the human gastrointestinal tract is dominated by obligate anaerobes, which are ca. 10^3 more abundant than facultative anaerobes. The main anaerobic genera are Bacteroides,Eubacterium and Bifidobacteria. These organisms account for ca. 90% of the cultivable human fecal bacteria. Bacteroides (mainly B. thetaiotaomicron and B. vulgatus) are the most abundant organism in the human feces and account for 20–30% of cultivable bacteria. The most abundant facultative anaerobes are Enterococci and Enterobacteriaceae. The main Enterobacteriaceae genera are Escherichia,Citrobacter, Klebsiella, Proteus and Enterobacter. Citrobacter and Klebsiella are present in most individuals although in low numbers. Proteus and Enterobacter are only present in a minority of humans [64].

A variety of molecular techniques have been used to study the microbial composition of the human gastrointestinal tract. Results yielded by these studies have shown that many microbes detected by molecular techniques are not isolable by conventional culture-based methods. The presence of high proportions of bifidobacteria detected by culture-based methods is not supported by the results of molecular-based studies. However, the results of molecular-based approaches support many of the findings derived from culture-based methods: the dominance of the obligate anaerobes over facultative anaerobes; the presence of high counts of Bacteroides, Clostridium andEubacterium [64].

Anaerobic bacteria such as Bacteroides and Eubacterium are not easily cultured by conventional techniques since require incubation chambers with nitrogen atmosphere. Bifidobacterium andLactobacillus tolerate some oxygen but are fastidious bacteria growing very slowly in culture media. Therefore, these four genera are not adequate to be used as indicators of fecal pollution (the introduction of molecular techniques may improve the situation). Citrobacter, Klebsiella andEnterobacter are present in low numbers in the human intestine and are widespread in environmental waters, and therefore are also not suitable as indicators of fecal pollution. Clostridium, Streptococcus and Escherichia do not suffer from these drawbacks. Therefore, their suitability as fecal indicators has been tested since several decades.

Fecal Bacteria in Their Hosts and in the Environment

Bacteroides

The traditional genus Bacteroides included Gram-negative, non-sporeforming, anaerobic pleiomorphic rods. Many species have been transferred to other genera—Mitsuokella, Porphyromonas, Prevotella, Ruminobacter. Bacteroides are the most abundant bacteria in human feces. In animal feces, on the contrary, Bacteroides are present at low numbers. Although anaerobic,Bacteroides are among the most tolerant to oxygen of all anaerobic human gastrointestinal species.B. thetaiotaomicron is one of the most abundant species in the lower regions of the human gastrointestinal tract. Bacteroides have a high pathogenic potential and account for approximately two-thirds of all anaerobes isolated from clinical specimens. The most frequently isolated species has been B. fragilis. The survival of Bacteroides in environmental waters is usually much lower than the survival of coliforms [64,65].

Eubacterium

The traditional genus Eubacterium included anaerobic non-sporeforming Gram-positive rods. Some species have been transferred to other genera—Actinobaculum, Atopobium, Collinsella, Dorea,Eggerthella, Mogibacterium, Pseudoramibacter and Slackia. Cells are not very aerotolerant. Species isolated from the human gastrointestinal tract include: E. barkeri, E. biforme, E. contortum, E. cylindrioides, E. hadrum, E. limosum, E. moniliforme, E. rectal and E. ventricosum [64].

Bifidobacterium

Bifidobacteria are Gram-positive, non-sporeforming, pleiomorphic rods. Bifidobacteria are anaerobic (some species tolerate oxygen in the presence of carbon dioxide) or facultative anaerobic. The optimum growth temperature is 35–39 °C. The genus Bifidobacterium contains ca. 25 species, most of which have been detected in the human gastrointestinal tract [64–66].

Bifidobacteria are present in high numbers in the feces of humans and some animals. SeveralBifidobacterium species are specific either for humans or for animals. B. cuniculi and B. magnum have only been found in rabbit fecal samples, B. gallinarum and B. pullorum only in the intestine of chickens and B. suis only in piglet feces. In human feces, the species composition changes with the age of the individual. In the intestine of infants B. breve and B. longum generally predominate. In the adult, B. adolescentis, B. catenulatum, B. pseudocatenulatum and B. longum are the dominant species. In both human and animal feces, bifibobacteria are always much more abundant than coliforms [64–66].

Bifidobacteria have been found in sewage and polluted environmental waters, but appears to be absent from unpolluted or pristine environments such as springs and unpolluted soil. This results from the fact that upon introduction into the environment, bifidobacteria decrease appreciably in numbers, probably due to their stringent growth requirements. Bifidobacteria grow poorly below 30 °C and have rigorous nutrient requirements. Reports on the survival of bifidobacteria in environmental waters indicate that their survival is lower than that of coliforms [64–66].

The presence of bifidobacteria in the environment is therefore considered an indicator of fecal contamination. Since some species are specific for humans and animals, the identification ofBifidobacterium species present in the polluted water could, in principle, provide information on the origin of fecal pollution [64–66].

A study carried out in a highly contaminated stream near Bologna, Italy, revealed that B. adolescentis, B. catenulatum, B. longum, B. pseudocatenulatum and B. thermophilum were the most representative species, whereas B. angulatum, B. animalis subsp. animalis (B. animalis), B. breve, B. choerinum, B. minimum, B. pseudolongum subsp. globosum (B. globosum) and B. subtile occurred only in low numbers [66].

Bifidobacteria are the less studied of all fecal bacteria, due to the technical difficulties in their isolation and cultivation. Other Gram-positive bacteria, such as Streptococcus and Lactobacillus, which may occur in higher numbers than bifidobacteria, can inhibit their growth. Although selective media has been designed for the isolation of bifidobacteria from environmental waters, the outcome is still unsatisfactory, with appreciable numbers of false positives and low recovery percentages [64–66].

Clostridia

The genus Clostridium is one of the largest genera of the prokaryotes containing 168 validly published species. From these, 77 (including C. perfringens) are considered to belong to a united group—Clostridium sensu stricto [64,67,68].

Clostridia are Gram-positive rods, forming endospores. Most of the clostridial species are motile with peritrichous flagellation. Cells are catalase-negative and do not carry out a dissimilatory sulphate reduction. Clostridia usually produce mixtures of organic acids and alcohols from carbohydrates and proteins. Many species are saccharolytic and proteolytic. Some species fix atmospheric dinitrogen [64,67,68].

The genus Clostridium includes psychrophilic, mesophilic, and thermophilic species. The major role of these organisms in nature is in the degradation of organic material to acids, alcohols, CO_2, H_2, and minerals. Frequently, a butyric acid smell is associated with the proliferation of clostridia. The ability to form spores that resist dryness, heat, and aerobic conditions makes the clostridia ubiquitous [64,67,68].

Most species are obligate anaerobic, although tolerance to oxygen occurs. Oxygen sensitivity restricts the habitat of the clostridia to anaerobic areas or areas with low oxygen tensions. Growing and dividing clostridia will, therefore, not be found in air saturated surface layers of lakes and rivers or on the surface of organic material and soil. Clostridial spores, however, are present with high probability in these environments, and will germinate when oxygen is exhausted and when appropriate nutrients are present [64,67,68].

C. perfringens ferment lactose, sucrose and inositol with the production of gas, produce a stormy clot fermentation with milk, reduce nitrate, hydrolyze gelatin and produce lecithinase and acid phosphatase. The species is divided into five types, A to E, on the basis of production of major lethal toxins [68,69].

C. perfringens appears to be a universal component of the human and animal intestine, since has been isolated from the intestinal contents of every animal that has been studied. Humans carry C. perfringens as part of the normal endogenous flora. The main site of carriage is the distal gastrointestinal tract. The principal habitats of type A are the soil and the intestines of humans, animals, and birds. Types B, C, D, and E appears to be obligate parasites of animals and occasionally are found in humans [68,69].

Clostridium perfringens is the most frequently isolated Clostridium in clinical microbiology laboratories, although it seldom causes serious infections. C. perfringens is isolated from infections in humans and the organism most commonly found in gas gangrene in humans. C. perfringens is most commonly isolated from infections derived from the colonic flora, namely peritonitis or abdominal abscess [68,69].

This organism is a common cause of food poisoning due to the formation of the enterotoxin in the intestine. C. perfringens food poisoning is seldom fatal, being marked by diarrhea and nausea, with no vomiting and no fever [68,69].

Sources yielding C. perfringens include soil and marine sediment samples worldwide, clothing, raw milk, cheese, semi-preserved meat products, and venison. Like E. coli, C. perfringens does not multiply in most water environments and is a highly specific indicator of fecal pollution. Berzirtzoglou et al. [70] reported a comparative study on the occurrence of vegetative cells and spores of Clostridium perfringens in a polluted station of the lake Pamvotis, in rural North-West Greece. The numbers of C. perfringens varied according to the water depth. Sporulated forms were found in all sampling sites with the exception of the surface sampling.

Lactobacillus

Lactobacilli are non-sporeforming Gram-positive long rods. There are more than thirty species in the genus. Most are microaerophillic, although some are obligate anaerobes. Cells are catalase-negative and obtain their energy by the fermentation of sugars, producing a variety of acids, alcohol and carbon dioxide. Lactobacilli have complex nutritional requirements and in agarized media may need the supplementation with aminoacids, peptides, fatty-acid esters, salts, nucleic acid derivatives and vitamins. Lactobacilli very rarely cause infections in humans [64].

Enterococci

Enterococci are Gram-positive, non-sporeforming, catalase-negative ovoid cells. Cells occur singly, in pairs or short chains. Optimal growth for most species is 35–37 °C. Some will grow at 42–45 °C and at 10 °C. Growth requires complex nutrients but is usually abundant on commonly used bacteriological media. Cells are resistant to 40% bile, 0.4% azide, 6.5% sodium chloride, have β-glucosidase and hydrolyze esculin. The enterococci are facultative anaerobic but prefer anaerobic conditions [64,71].

The genus was separated from Streptococcus in the 1980s. Enterococci form relatively distinct groups. Members of such groups exhibit similar phenotypic characteristics and species delimitation can be difficult. The E. faecalis group contains, among others, E. faecalis. The E. avium group contains, among others, E. avium. The E. faecium group contains, among others, E. faecium, E. duransand E. hirae. The E. gallinarum group contains, among others, E. gallinarum [64,71].

Most species are part of the intestinal flora of mammals, reptiles, birds, and other animals. In the human digestive tract, E. faecalis is the prevailing species, although in particular situations, E. faecium may predominate. In poultry, E. cecorum, E. durans, E. faecalis, E. faecium and E. hirae and dominate the intestinal flora [64,71].

Enterococci have been increasingly isolated from a variety of nosocomial and other infections, mainly from the urinary tract and wound infections, bacteremias, and endocarditis [64,71].

Although enterococci are considered only a temporary part of the microflora of plants, in optimal conditions, cells can proliferate on their surfaces. E. casseliflavus, E. faecalis,

E. faecium, E. hirae, E. mundtii and E. sulfureus have been isolated from plants. They are generally isolated more often from flowers than from buds or leaves [64,71].

Enterococci are naturally present in many kinds of foods, especially those of animal origin such as milk and milk products, meat and fermented sausages. Enterococci are usually considered secondary contaminants of food, although they often play a positive role in ripening and aroma development of some types of cheeses [64,71]. Although soil is not a natural habitat for enterococci, cells can be found in this habitat due to the transport by rain [64,71].

Environmental waters are not a natural habitat for enterococci and their presence in this milieu is considered the result of fecal pollution. The most common species found in environmental waters are E. durans, E. faecalis, E. faecium and E. hirae, and less commonly, E. avium, E. cecorum, E. columbaeand E. gallinarum. However, pristine waters in Finland have been reported to contain E. casseliflavus[64,71].

In environmental samples (compost, sewage effluent, harbor sediments, brackish water and swimming pool water), Pinto et al. [72] reported the isolation of E. casseliflavus, E. durans, E. faecalis,E. faecium, E. gallinarum and E. hirae. E. durans, E. faecium and E. hirae were isolated from all sources except from harbor sediments. E. raffinosus was only isolated from compost and swimming pool water. E. faecalis and E. faecium accounted for the vast majority of enterococcal strains.

Escherichia

Escherichia, a member of Enterobacteriaceae, are oxidase-negative catalase-positive straight rods that ferment lactose. Cells are positive in the Methyl-Red test, but negative in the Voges-Proskauer assay. Cells do not use citrate, do not produce H_2S or lipase, and do not hydrolyze urea [73]. E. coliis a natural and essential part of the bacterial flora in the gut of humans and animals. Most E. colistrains are nonpathogenic and reside harmlessly in the colon. However, certain serotypes do play a role in intestinal and extra-intestinal diseases, such as urinary tract infections [43]. In a study of the enteric bacteria present in the feces of Australian mammals, Gordon and FitzGibbon [74] reported that E. coli was the commonest species, being isolated from nearly half of the species studied.

Citrobacter

Citrobacter, a member of Enterobacteriaceae, are motile straight rods. Cells are oxidase-negative, catalase-positive and positive in the Methyl-Red test. Cells use citrate, are negative in the Voges-Proskauer test and do not decarboxylate lysine [73].

In a study of the enteric bacteria present in the feces of Australian mammals, Gordon and FitzGibbon [74] reported the isolation of C. amalonaticus, C. freundii and C. koseri (C. diversus).Citrobacter species can be isolated from different clinical sites. In particular, C. freundii is an intestinal inhabitant of humans that may sometimes have—or acquire—the ability to produce an enterotoxin and thus become an intestinal pathogen. Citrobacter is reported to occur in environments such as water, sewage, soil and food [75,76].

Klebsiella and Raoultella

Klebsiella and Raoultella are Enterobacteriaceae, oxidase-negative catalase-positive non-motile straight rods, surrounded by a capsule. Cells decarboxylate lysine, but are ornithine and arginine dihydrolase negative. Cells grow on KCN, do not produce H_2S and ferment most carbohydrates [73].

In humans, K. pneumoniae is present as commensal in the nasopharynx and in the intestinal tract.Klebsiella spp. can cause human diseases, ranging from asymptomatic colonization of the intestinal, urinary, or respiratory tract to fatal septicemia. Klebsiella are mostly considered nosocomial pathogens. K. pneumoniae and Enterobacter aerogenes (K. mobilis) are most frequently involved, although K. oxytoca and R. planticola, and rarely R. terrigena, can be found. In the hospital, the principal reservoir of K. pneumoniae is the gastrointestinal tract of patients. The principal vectors are the hands of personnel [77,78]. In a study of the enteric bacteria present in the feces of Australian mammals, Gordon and FitzGibbon [74] reported the isolation of K. pneumoniae and K. oxytoca.

Klebsiellae are ubitiquous in the environment. They have been found in a variety of environmental situations, such as soil, vegetation, or water, and they influence many biochemical and geochemical processes. They have been recovered from aquatic environments receiving industrial wastewaters, plant products, fresh vegetables, food with a high content of sugars and acids, frozen orange juice concentrate, sugarcane wastes, living trees, and plants and plant byproducts. They are commonly associated with wood, sawdust, and waters receiving industrial effluents from pulp and paper mills and textile finishing plants (see below). Klebsiella have been isolated from the root surfaces of various plants. K. pneumoniae, K. oxytoca, and R. planticola are all capable of fixing dinitrogen [77,78].

Enterobacter

Enterobacter a member of Enterobacteriaceae, are motile straight rods. Cells are positive in the Voges-Proskauer test VP and in Simmons citrate agar. Cells do not decarboxylate lysine, but are ornithine positive. Malonate is usually utilized and gelatin is slowly liquefied. Cells do not produce H_2S, deoxyribonuclease and lipase [73].

In a study of the enteric bacteria present in the feces of Australian mammals, Gordon and FitzGibbon [74] reported the isolation of Enterobacter cloacae subsp. cloacae (E. cloacae), E. cancerogenus (E. taylorae) and E. aerogenes (Klebsiella mobilis).

Before the widespread use of antibiotics, Enterobacter species were rarely found as pathogens, but these organisms are now increasingly encountered, causing nosocomial infections such as urinary tract infections and bacteremia. In addition, they occasionally cause community-acquired infections [79,80].

In the USA, the Surveillance and Control of Pathogens of Epidemiological Importance project analyzed 24,179 nosocomial bloodstream infections, from 1995–2002. Enterobacter species were the second most common gram-negative organism, behind Pseudomonas aeruginosa. Both bacteria were reported to each represent 4.7% of bloodstream infections in intensive care units. Enterobacterspecies represented 3.1% of bloodstream infections in non-intensive care units. Of nearly 75,000 gram-negative organisms collected from intensive care units' patients in the USA, between 1993 and 2004, Enterobacter species comprised 13.5% of the isolates. Multidrug resistance increased over time, especially in infections caused by E. cloacae [81].

In the USA, the National Healthcare Safety Network reported a study on healthcare-associated infections between 2006 and 2007. They found Enterobacter species to be the eighth most common cause of healthcare-associated infections (5% of all infections) and the fourth most common gram-negative cause of these infections [82].

Enterobacter cloacae subsp. cloacae (E. cloacae) occurs in the intestinal tracts of humans and animals, in hospital environments, the skin, in water, sewage, soil, meat. Nitrogen-fixing strains have been isolated from the roots of rice plants. E. amnigenus has been mostly isolated from water, but some strains were isolated from clinical specimens from the respiratory tract, wounds and feces. E. asburiae strains were isolated from clinical specimens, mostly urine, respiratory tract, feces, wounds, and blood [79,80].

Origin of the Use of Fecal Indicator Bacteria

Historically, the design and use of indicators of fecal pollution comes from the end of the 19th to beginning of the 20th century. In 1880, von Fritsch described Klebsiella pneumoniae and K. pneumoniae subsp. rhinoscleromatis (Klebsiella rhinoscleromatis) as micro-organisms characteristically found in human feces [83]. In 1885, Escherich described several microorganisms in the feces of newborn and suckling babies. This included a motile, rod-shaped microorganism that caused milk to clot, which was named "Bacterium coli commune". He observed that within a few weeks after birth, this bacterium became the dominant organism in the infant colon [6]. Also in 1885, Percy and Grace Frankland started the first routine bacteriological examination of water in London, using Robert Koch's solid gelatin media to count bacteria [83]. In 1891, Percy and Grace Frankland came up with the concept that organisms characteristic of sewage must be identified to provide evidence of potentially dangerous pollution [83]. In 1892, Schardinger proposed that since "Bacterium coli" was a characteristic component of the fecal flora, its presence in water could be taken as an indication of the presence of fecal pollution and therefore of the potential presence of enteric pathogens [6]. Soon after the description of "Bacterium coli", other bacteria were isolated from stools and water—Klebsiella in 1882 and Enterobacter in 1890 [6]. By 1893, the "Wurtz method" of enumerating "Bacterium coli", by direct plating water samples on litmus lactose agar, was being used by sanitary bacteriologists. This was based on the concept of acid and gas production (detected by the Durham tube) from lactose as a diagnostic feature [6]. In 1905, MacConkey described his now famous MacConkey's broth, which was diagnostic for lactose-fermenting bacteria tolerant of bile salts. Coliforms were already considered to be a heterogeneous group of organisms, many of which were not of fecal origin. The origins of the critical observation that "Bacterium coli" was largely fecal in origin while other coliforms were not, could be claimed by Winslow and Walker in 1907 [83].

Various classification schemes for coliforms have emerged. The earliest were those of MacConkey in 1909, which recognized 128 different coliform types, while Bergey and Deehan in 1908, identified 256. By the early 1920s, differentiation of coliforms had come to a series of correlations that suggested that indole production, gelatin liquefaction, sucrose fermentation and the Voges–Proskauer reaction were among the more important tests for determining fecal contamination. These developments culminated in the IMViC (Indole, Methyl Red, Voges–Proskauer and Citrate) tests for the differentiation of so-called fecal coliforms, soil coliforms and intermediates [83].

Fecal Indicator Bacteria

Coliforms

Total coliforms are Gram-negative, oxidase-negative, non-sporeforming rods, that ferment lactose with gas production at 35–37 °C, after 48h, in a medium with bile salts and detergents [1,4,6,57,84]. When the test of coliforms is carried out with environmental waters, several species of the four Enterobacteriaceae genera Escherichia, Klebsiella, Enterobacter and Citrobacter give positive results and therefore are coliforms according to this definition. However, the environmental significance of these four genera is very disparate as discussed in the present text. Therefore, total coliform counts are not necessarily a measure of fecal pollution and indeed can have no relation with this cause [1,4,6,84].

Fecal coliforms (or thermotolerant coliforms) are traditionally defined as coliforms that ferment lactose at 44.5 °C in a medium with bile salts [1,4,57,84]. The range of species detected by the experimental procedure is much lower than that of total coliforms. With environmental polluted waters, only E. coli, and K. oxytoca and K. pneumoniae gave positive results in the test [85].

Traditional tests for total and fecal coliforms are carried out either by the multiple-tube fermentation technique or by filtration through membrane. The multiple-tube fermentation technique is used for medium or highly contaminated waters, and the filtration through membrane for low or very low contaminated waters. Filtration through membrane is a very sensitive technique since can detect one (culturable) cell in 500 or even 1,000 mL of water. However, both methods take several days to complete and do not detect viable but non-culturable bacteria [3,57,86]. These limitations stimulate the discovery of alternative methods, faster and, if possible, less prone to false negative results such as those caused by the viable but non-culturable bacteria.

The detection of β-D-galactosidase activity (at 37 °C) is usually a good marker for total coliforms in environmental waters, since most of these bacteria display this enzymatic activity [1,3,57,87–91]. Most Escherichia, Citrobacter, Enterobacter, Klebsiella and Raoultella strains have galactosidase. Hafnia,Serratia and Yersinia also possess this enzymatic activity. Most Proteus, Salmonella and Edwardsiellastrains do not display β-galactosidase [92–95]. Ca. 10% of the coliform strains isolated from the environment do not have an active formic hydrogenolyase (cleaves formate with the formation of CO_2) and therefore do not produce gas being undetected by the traditional techniques but are detected by the assay of β-galactosidase activity [57,96,97].

β-galactosidase cleaves lactose in glucose and galactose, and can be detected by using colored or fluorescent markers that change color after enzyme action, such as XGAL (5-bromo-4-chloro-3-indol-β-galactopyranoside) and ONPG (O-nitrophenyl-β-D-galactopyranoside) or MUGAL (4-methylumbelliferyl-β-D-galactopyranoside), respectively [57,96,97].

In environmental waters, the presence of Aeromonas or Vibrio cholerae can be a source of false positives in the β-D-galactosidase assay, since these bacteria have galactosidase, but are not coliforms [93,95,96,98]. Additionally, in particular environments, such as estuaries, β-galactosidase activity can overestimate total coliform count due to UV-stimulated enzymatic activity in certain bacteria such as E. coli [86].

The detection of β-D-glucuronidase activity (at 44.5 °C) is, generally, a good marker for fecal coliforms in environmental polluted waters and very specific for E. coli [13,85,87–91,97,99–101]. In Gram-negative bacteria, this enzymatic activity if found in most E. coli strains and in someSalmonella and Shigella strains [92–95,97,102]. Aeromonas, Citrobacter, Enterobacter, non-coli Escherichia, Hafnia, Klebsiella, Proteus, Serratia, Vibrio, Yersinia, and most Salmonella strains do not display β-glucuronidase activity [93–95,101,102].

β-D-glucuronidase activity can be detected by using colored or fluorescent markers that change color after enzyme action, such as XGLUC (5-bromo-4-chloro-3-indoxyl-β-d-glucuronide), IBDG (indoxil-β-glucuronide), and MUGLU (4-methylumbelliferyl-β-d-glucuronide), respectively [57,97,100].

The presence of this enzyme in some strains of Bacteroides, Flavobacterium, Staphylococcus,Streptococcus, in anaerobic corynebacteria and Clostridium, has also been reported [93,95–97,102]. β-D-glucuronidase activity in fecal bacteria other then E. coli (Bacteroides, bifidobacteria, clostridia, enterococci and Lactobacillus) is very limited [61]. Although all these glucuronidase positive bacteria could lead to false positive detections in the fecal coliform test, experimental results for environmental polluted waters indicate a significant correlation between fecal coliform detection using conventional techniques and the glucuronidase assay, suggesting that false positives are not significant [85,96].

The detection of total coliforms and fecal coliforms by enzymatic methods are much less time consuming than traditional techniques. With fluorescent markers and the use of a spectrofluorimeter the detection of coliforms can be performed in minutes [57,101]. However, in very low contaminated waters, enzymatic methods might not be able to detect coliform cells. Moreover, on-line monitoring of glucuronidase activity is currently too insensitive to replace culture based detection of E. coli. Nevertheless, on-line enzymatic methods can be a valuable complementary tool for high temporal resolution monitoring. More research is needed in order to enhance sensitive and lower detection limits of available on-line glucuronidase techniques.

The seminal work of Leclerc et al. [103] clarified the diversified roles that coliforms have in the environment and the real meanings of the tests on total coliforms and fecal coliforms. It was shown that Enterobacteriaceae encompass three groups of bacteria with very different roles in the environment. Group I harbored only E. coli. Since this species usually do not survive for long periods outside this environment (but see topic 10), it was considered a good and reliable indicator of fecal pollution (both animal and human). Group II, the "ubiquitary" group, encompassed several species of Klebsiella (K. pneumoniae and K. oxytoca), Enterobacter (Enterobacter cloacae subsp. cloacae, E. aerogenes) and Citrobacter (C. amalonaticus, C. koseri and C. freundii). These bacteria live in the animal and human gut, but also in the environment, and are easily isolated from the soil, polluted water and plants. Their presence in polluted waters does not necessarily indicate fecal pollution. Finally Group III was composed of Raoultella planticola, R. terrigena, Enterobacter amnigenus and Kluyvera intermedia (Enterobacter intermedius), Serratia fonticola, and the genera Budvicia, Buttiauxella, Leclercia,Rahnella, Yersinia, and most species of Erwinia and Pantoea. These bacteria live in fresh waters, plants and small animals. They grow at 4 °C, but not at 41 °C. They are not indicators of fecal pollution, although can be detected in the total coliform test. Leclerc et al. concluded that: (1) in the enterobacteria, E. coli is the only true and reliable indicator of fecal pollution in environmental waters; (2) the traditional total coliform test should be abandoned, since it can detect bacteria that have no connection with fecal pollution; (3) the detection of fecal coliforms must be carried out at 44.5 °C, and positive results confirmed by identification to species levels in order to exclude false positives such as K. pneumoniae.

Streptococci and Enterococci

Fecal streptococci also belong to the traditional indicators of fecal pollution. Fecal streptococci are Gram-positive, catalase-negative, non-sporeforming cocci that grow at 35 °C in a medium containing bile salts and sodium azide. Cells hydrolyze esculin [1,4,57]. Azide is a strong inhibitor of the respiratory chain. Since streptococci are one of the very few bacteria that have no respiratory chain, the test is very specific for this group, and false positives are rarely found [104,105].

Fecal enterococci (E. faecalis, E. faecium, E. avium and E. gallinarum) are fecal streptococci that grow in the presence of 6.5% NaCl at 45 °C. Selective media use these particular characteristics in order to separate enterococci from the other streptococci [104,105].

Several studies [104,106] have reported on the microbiological composition of human and animal (cattle, chicken, deer, dog, fowl, goose, and swine) feces. E. faecalis and E. faecium were present in human and animal feces. However, whereas human feces almost have only these two enterococci, in the animals others species co-occur, like E. avium, E. cecorum, E. durans, E. gallinarum and E. hirae. It was concluded that in urban areas where contamination with dog and chicken feces is not likely, the best marker for human fecal pollution was E. faecalis.

The intestinal enterococci group has been used as an index of fecal pollution. In human feces, the numbers of intestinal enterococci are generally about an order of magnitude lower than those of E. coli (Table 6). Most species do not grow in environmental waters. In this milieu, fecal enterococci are able to survive longer, are more resistant to drying and chlorination, than E. coli [1,84].

However, caution should be taken with interpreting the results obtained by the enterococci procedure in water analysis. Enterococci and other group D-streptococci are present in many foods, especially those of animal origin. The isolation of E. faecalis and E. faecium was used to indicate fecal contamination of food. However, enterococci are now also considered as normal parts of the food microflora and not only as indicators for poor hygiene [104]. In addition, agricultural soils and crops with added with manure also harbor enterococci [105].

The use of ratios between indicator counts

The ratio of counts of fecal coliforms to fecal streptococci has been proposed as a means to differentiating between contamination from human and animal sources. Ratios greater than 4 have been suggested to indicate a human source whereas ratios less than 0.7 suggest an animal source. This results from the fact that streptococcal concentrations in human feces are generally less than coliforms (Table 6). In contrast, in animal feces fecal streptococci generally outnumber fecal coliforms (Table 7). In urban sewage, fecal streptococci tend to be present in concentrations 10–100 times less than fecal coliforms [65].

Table 7. Bacteria in the feces of farm and domestic warm-blooded animalsa.

Animal	Log$_{10}$ cells/g wet weight feces		
	Fecal coliforms	Fecal streptococci	*Clostridium perfringens*
Chicken	5.4	6.1	2.3
Duck	7.5	7.7	-
Horse	4.1	6.8	< 0
Pig	6.5	7.9	3.6
Sheep	7.2	7.6	5.3
Turkey	5.5	6.4	-
Cat	6.9	7.4	7.4
Dog	7.1	9.0	8.4

[a]Adapted from [83].

Geldreich [107] summarized the information available on the fecal coliforms to fecal streptococci ratios in the feces of warmblooded animals, and reported the following values: human feces, 4.3; cattle, sheep, and poultry, from 0.104 to 0.421; and wild animals (including rabbits, field mice, chipmunks, and birds), 0.0008 to 0.043. Fecal coliforms to fecal streptococci ratios for the feces of wild animals appear to be at least 10-fold lower than those of domestic livestock.

Doran and Linn [108] reported a study of the runoff from a cow-calf pasture in eastern Nebraska (USA), monitored during a three-year period. It was concluded that the fecal coliforms to fecal streptococci ratio in pasture runoff was useful in identifying the relative contributions of cattle and wildlife and in evaluating the effects of cattle management and distribution on runoff water quality. Ratios below 0.05 were indicative of wildlife sources and ratios above 0.1 were characteristic of grazing cattle. Fecal coliforms to fecal streptococci ratios of diluted cattle waste in excess of 1 were interpreted as the result of differential aftergrowth and die-off between fecal coliforms and fecal streptococci. Ratios between 0.7 and 4.0 may indicate situations where cattle are localized close to sampling or outflow points.

However, the interpretation of this ratio should be cautious. It has been observed a shift in the ratio with time and distance from the fecal pollution source. This resulted from the fact that both in surface and groundwaters, fecal streptococci are more persistent than fecal coliforms. Therefore increasing the distance from the pollution point and with passing time, the ratio tends to decrease without a change in the nature of the pollution source [65]. For these reasons, this ratio has been considered by some authors as too unreliable to be useful in characterizing pollution sources [65,84].

The ratio of fecal enterococci to fecal streptococci differs among vertebrate species. Humans have a predominance of enterococci, whereas animals contain appreciable amounts of streptococci. However, since enterococci are also present in animals and are more persistent in the environment than other fecal streptococci, the identification of the enterococci and streptococci species present in polluted waters, and the concomitant calculation of this ratio is generally considered unreliable as an indicator of the source of fecal pollution [65].

Limitations of coliform and enterococcus counts as indicator of fecal pollution

An extreme case of uselessness of the determination of total and fecal coliforms and enterococci in the assessment of fecal pollution has been demonstrated by several authors studying the microbiology of pulp and paper mill effluents.

Caplenas and Kanarek [109] reported a study of pulp and paper mills located in Wisconsin (USA). Fresh water supplies, re-cycled water within mills, treated effluent wastewater and waters receiving effluent wastes downstream, were assessed for the presence of fecal coliforms andKlebsiella. Wastewaters prior to treatment contained fecal coliforms and Klebsiella. Up to 84% of the fecal coliforms (detected by the standard test procedure) were indeed Klebsiella. In treated effluent wastewaters this value reached 90%. Treatment of the wastewater lowered the concentration of "true" fecal bacterial contamination, but since Klebsiella grew rapidly in the wastewaters, fecal coliform counts were high, although no true fecal contamination was involved. The source ofKlebsiella was traced to the early pulping stages in the mills. Klebsiella maintains a wood, bark or soil reservoir. It was concluded that: (1) Klebsiella are ubiquitous in the pulp and paper mill industry processing stages; (2) the standard procedure for fecal coliform estimation is useless to assess the microbiological quality of the effluents of these industries; (3) The assay of E. coli should replace the fecal coliform detection procedure.

Gauthier et al. [110] and Gauthier and Archibald [58] reported two studies of seven pulp and paper mills in Ontario and Quebec, Canada. Total and fecal coliforms and enterococci were detected in nearly all the biotreatment, biosolids (sludges), and in-mill water system samples. In the mill samples, the majority of the fecal coliforms (detected by the standard test procedure) wereK. pneumoniae, Raoultella terrigena and Raoultella planticola, with E. coli in minority. E. faecalis and E. faecium were detected in relatively large numbers in most samples from all of the seven mills examined. Other coliforms such as Enterobacter spp. and Citrobacter freundii were occasionally recovered from total and fecal coliform tubes. Biofilms established in the piping, tanks, and machinery where thermal and pH conditions permit were the most likely source of these bacteria. Analyses using two independent Salmonella detection/enumeration methods showed no detectableSalmonella cells in the sludges and final effluents of the five mills tested. It was concluded that for these particular systems, the determination of total and fecal coliforms and enterococci is useless and have no relationship with real fecal pollution. These studies also demonstrated the importance of checking the identities of bacteria causing the positive results in the tests. Both Escherichia andKlebsiella can give positive results in the fecal coliform test, but their ecological meaning is opposite.

Another important case of failure of the use of coliforms to detect fecal pollution was the 1993Cryptosporidium outbreak in Milwaukee (USA).

Cryptosporidium parvum, a protozoan parasite that causes gastrointestinal illness, is transmitted by ingestion of oocysts excreted in human or animal feces. Typical modes of transmission include person to person, animal to person, by exposure to contaminated food or water [111,112].

From 1990 to 2000, at least 10 cryptosporidiosis outbreaks associated with contaminated drinking water were reported in the USA. In 1993, an estimated 403,000 residents of the greater Milwaukee area (Wisconsin, population, ca. 1.61 million) became ill when an ineffective filtration process led to the inadequate removal of Cryptosporidium oocysts in one of two municipal water-treatment plants [111,112].

It was the largest waterborne disease outbreak in documented USA history. Over the span of approximately two weeks, people became ill with stomach cramps, fever, diarrhea and dehydration caused by the pathogen. More than half the people who received residential drinking water from the southern water-treatment plant became ill, which was twice the rate of illness among people whose residential drinking water came mainly from the northern water-treatment plant. Over 54 deaths were attributed to this outbreak, mostly among the elderly and immunocompromised people, such as AIDS patients [111,112].

Standard microbiological water analysis was ineffective in detecting this parasite. Indeed, throughout the period from February to April, samples of treated water from both plants were negative for coliforms. The origin of the contamination was determined as water from Lake Michigan. No specific source of the Cryptosporidium was ever identified but runoff from abnormally heavy spring rains most likely carried the parasite to the lake [111,112].

Clostridium perfringens

Sulphite-reducing clostridia, namely Clostridium perfringens, are spore-forming Gram-positive, non-motile, anaerobic, sulfite-reducing rods. C. perfringens is present in higher numbers in the feces of some animals, such as dogs, than in the feces of humans and less often in the feces of many other warm-blooded animals. The numbers excreted in feces are normally substantially lower than those of E. coli.

Clostridium spores are exceptionally resistant to unfavorable conditions in water environments, including UV irradiation, temperature and pH extremes, and disinfection processes, such as chlorination. Although clostridia probably do not growth in surface waters, the high resistance of their spores makes their presence ubiquitous in environmental waters [1,4,84,113].

The presence of chlorine in water rapidly inactivates indicator bacteria such as E. coli and coliforms, but it leaves the most resistant pathogens almost unaffected for several hours. This creates a false sense of security by providing negative coliform and negative E. coli results to authorities responsible for water testing. Giardia cysts, Crystosporidium oocysts, and human enteric viruses all have higher resistance to disinfectants and constitute a major public health risk if distribution system integrity is breached. C. perfringens spores are less affected by the residual concentrations of chlorine. Testing for the spores of this bacterium can probably provide an added margin of safety in the evaluation of treatment [114].

Correlations between parameters used to assess fecal pollution

In environmental waters, several studies have reported significant correlations between indicators of fecal pollution and between indicators and pathogenic gastrointestinal bacteria.

Charriere et al. [115] reported a study of deep aquifer waters (raw waters and piped chlorinated waters) in Normandy, France. In heavily contaminated raw waters and in slightly contaminated treated waters, fecal coliforms and enterococci were correlated.

Martins et al. [116] reported a study of 60 public outdoor swimming pools in Sao Paulo city, Brazil. Total coliforms, fecal coliforms and fecal streptococci levels increased with number of bathers and water temperature, and decreased with chlorine levels. All these indicators were significantly correlated with each other.

Ferguson et al. [117] reported a study carried out in Georges River, in the Sydney region, Australia. In the water column, concentrations of fecal coliforms, fecal streptococci, C. perfringensspores were all positively correlated with each other. Isolation of Salmonella spp. were most frequent during rainfall and sewage overflow events. In the water column, 55% of samples contained Salmonella when fecal coliform densities exceeded 2,000 CFU/100 mL.

Medema et al. [118] reported a study of seven different fresh water sites normally used for triathlon competitions. Sites were small rivers, channels, lakes and harbors, and were influenced by sewage effluents and agricultural run-off. When data from all triathlons were pooled, geometric mean densities of fecal coliforms and E. coli, of E. coli and fecal enterococci and of fecal coliforms and fecal enterococci, were significantly correlated.

Polo et al. [119] reported a study of water samples obtained from 213 beaches, eight rivers and 14 freshwaters in north-eastern Spain. In freshwaters and heavily contaminated seawaters, Salmonellaand fecal coliforms were correlated, while in less contaminated seawaters, the highest correlation was with between Salmonella and fecal streptococci.

Byamukama et al. [99] reported a study of the microbiology of Nakivubo channel, Uganda. This channel receives raw sewage from slums, industrial effluents, and discharges from a sewage treatment plant and from a complex of slaughterhouses. Water from eight sampling sites was assessed for the presence of total and fecal coliforms, E. coli and sulphite-reducing clostridia. All microbiological parameters were significantly correlated.

Noble et al. [120] reported a comparative determination of total coliforms, fecal coliforms and enterococci in 108 sites along the southern California coastline, USA. Results by traditional and enzymatic methods and from all three parameters were correlated.

Harwood et al. [121] reported a study on indicator and pathogenic microorganisms carried out in six wastewater reclamation facilities in the USA. Data from disinfected effluent (reclaimed water) samples were analyzed separately (by facility) and as a pooled data set (all facilities). Significant correlations between indicator organism concentrations were observed in the pooled data sets, namely for total and fecal coliforms.

Cabral and Marques [85] reported a microbiological study of a polluted river (Febros) in the Great Oporto area, northwest Portugal. Total and fecal coliforms, fecal streptococci and enterococci were all significantly correlated with each other.

Touron et al. [122] reported a study carried out in the Seine estuary, France. Water was sampled at nine stations (along an upstream/downstream transect of 156 km), during nine years, for fecal coliforms, E. coli, enterococci and Clostridium perfringens spores. At the upstream part of the estuary (at Poses), Salmonella and fecal coliforms, and E. coli and enterococci counts were correlated. At the mouth of the estuary (at Honfleur), significant correlation was found for Salmonella and enterococci counts. No significant correlation between concentrations of any combination of indicator organism and pathogen was observed.

Wilkes et al. [59] reported a comparative study on the presence and concentration of several pathogenic and indicator bacteria in the surface water of a Canadian river. Surface water was collected within the South Nation River basin in eastern Ontario, from the river proper, and from several lower stream order tributaries. Using data aggregated during the entire multi-year study, significant correlations were found among all indicator bacteria - total and fecal coliforms, E. coli,Enterococcus, and C. perfringens.

However, in others studies no correlation was found between the different fecal indicator bacteria. Garrido-Pérez et al. [123] reported a study of the bathing seawater quality in 18 Spanish beaches near the Strait of Gibraltar. Sample locations were selected as a single point located in the area of highest bather density of each beach. No significant correlation was found between fecal coliforms and Clostridium perfringens counts in the bathing seawater.

FECAL INDICATOR CHEMICAL COMPOUNDS

Several chemical substances have been used as markers of fecal pollution in environmental waters. Caffeine is present in several beverages and in many pharmaceutical products. It is excreted in the urine of individuals who have ingested the substance. The main source of caffeine in domestic wastewaters is excretion following consumption of coffee, tea, soft drinks, or medication. Levels of caffeine in domestic wastewater have been measured to be between 20 and 300 μg/L. Levels in receiving waters are much lower due to significant dilution. Due to its high solubility, low octanol-water partition coefficient, insignificant volatility and clear anthropogenic origin, the presence of caffeine in environmental waters can be a good marker for human fecal pollution [124–129].

However, relationships between fecal indicator bacteria and caffeine are variable. Wu et al. [129] reported a study carried out in the Rochor Canal and Marina Bay, Singapore. In Rochor Canal, the highest concentration of caffeine (1.35 ng/mL) was found at downstream, and the lowest (0.68 and 0.37 ng/mL) were determined at middle and upstream points. At Marina Bay, the concentration of caffeine was in the range of 0.41–0.96 ng/mL. Fecal coliform concentrations were very high, exceeding 5,000 CFU/100 mL. Caffeine and fecal coliform concentrations were significantly correlated in Rochor Canal samples, but no significant correlation was observed in Martina Bay water samples.

The use of caffeine as marker of fecal pollution has additional important limitations. Caffeine is often present in the urban environment from numerous plant species debris as well as from human "dumping" of coffee wastes. In addition, the current analytical methods used are relatively complex and expensive [124].

Coprostanol (5β-cholestan-3β-ol) is a fecal stanol that is formed by indigenous bacteria present in the gut of humans and higher animals, during catabolism of cholesterol. It is the main stanol present in human feces (24 to 89% of total steroids) and in domestic wastewater. Based on these facts, it has been proposed as a chemical indicator of human fecal pollution. Feces from pigs and cats also contain coprostanol, but at much lower levels. Additional fecal stanols, such as 24-ethylcoprostanol, were found to be predominant in herbivores, such as cows, horses, and sheep, suggesting potential use of this chemical as an indicator of fecal pollution from these sources. Reported half-lives of coprostanol in aerobic conditions are generally lower than 10 days at 20 °C. Thus, the presence of coprostanol in an aerobic environment can be considered an indication of recent fecal input to the waters. [124,128,130].

Isobe et al. [130] reported a study carried out in the Mekong Delta (Vietnam) and in the Tokyo metropolitan area. During the wet season in the Mekong Delta, higher bacterial densities were observed in rivers, probably due to the higher bacterial inputs from soil particles with runoff. In Tokyo, higher bacterial densities were usually observed during summer, followed by those in the typhoon aftermath and winter. Significant correlations between the concentrations of E. coli and coprostanol (log scale) were found in all surveys. It was concluded that the determination of coprostanol can improve standard microbiological assays of fecal pollution.

Reports from several regions throughout the world indicate variable quantitative relationships between fecal coliform densities and coprostanol concentrations. In the Derwent Estuary and Sydney region (Australia), coprostanol concentration of 400 ng/L corresponded to 1,000 CFU of fecal coliforms/100 mL. In the Mekong Delta, during the wet season, this coliform density corresponded to 30 ng coprostanol/L, and in the dry seasons, 100 ng/L. In Tokyo metropolitan area, these values were 30 ng coprostanol/L, in summer, and 100 ng coprostanol/L, in a typhoon aftermath. These differences were interpreted as a result of differences in water temperature and soil particle concentration [130].

Fecal sterol analysis, although expensive and complex, has resolved problems of source attribution in urban and rural environments not possible with use of traditional fecal indicator bacteria [124]. These chemical indicators are especially useful in environments which allow survival and growth of fecal bacteria. For instance, in tropical regions, characterized by high temperatures and frequent rainstorms that facilitate erosion of soils, fecal bacteria can proliferate in environmental waters reaching densities that are not representative of real sewage inputs in the environment [130].

SOURCES OF FECAL BACTERIAL POLLUTION OF ENVIRONMENTAL WATERS

Sources of Surface and Groundwater Contamination

Determinations carried out in the sewage systems of urbanized areas have confirmed the presence of high numbers of intestinal bacteria. Treatment of sewage reduces the concentration of these bacteria by 1–2 logs, but effluent still contains high levels of intestinal bacteria (Table 8). Effluents from sewage treatment plants can be a source of contamination of surface waters with fecal bacteria.

Table 8. Typical concentrations of selected bacteria in raw and treated domestic wastewatera.

Bacterial group	Raw sewage (cells/ml)	Treated effluent (cells/ml)
Salmonella	$10^{-1} - 10^{1}$	$10^{-1} - 10^{1}$
Total coliforms	$10^{4} - 10^{6}$	$10^{3} - 10^{5}$
Fecal coliforms	$10^{3} - 10^{5}$	$10^{2} - 10^{4}$
Enterococci	$10^{3} - 10^{4}$	$10^{1} - 10^{3}$
Clostridium perfringens	$10^{2} - 10^{3}$	$10^{1} - 10^{2}$

[a]Adapted from Medema *et al.* [131].

Septic tanks, cesspools, latrines and other on-site systems are widely used for wastewater storage and treatment. The water percolating from these facilities contains bacteria that may contaminate groundwater supplies.

Many farmers use cellars, tanks or landfills to store manure. Water leaching from these storage sites may also contaminate groundwater, especially during periods of rainfall. The application of animal manure to agricultural lands as fertilizer is common practice throughout the world. Bacteria present in the manure may leach into the groundwater.

The potential for bacteria present in human and animal wastes to contaminate water in nearby wells needs special attention [131]. An important source of contamination of surface and ground waters is runoff water from agricultural and pasture lands, and urban areas.

Fecal bacteria enter surface water by direct deposit of feces and by overland runoff. The movement of animal wastes into surface waters can be a major factor contributing to the pollution of available water in many regions. Over one-third of the land area of USA is used for grazing livestock and receives 50% of all livestock wastes.

In a study reported by Doran and Linn [108], runoff from a cow-calf pasture in eastern Nebraska was monitored during a three-year period. Rainfall runoff from the grazed area contained 5 to 10 times more fecal coliforms than runoff from the fenced, ungrazed area. However, fecal streptococci counts were higher in runoff from the ungrazed area and reflected the contributions from wildlife.

Urban and suburban areas are dominated by impervious cover. During storms, rainwater flows across these impervious surfaces, mobilizing contaminants. The pollutants carried in runoff originate from a variety of urban and suburban nonpoint sources. Contaminants commonly found in stormwater runoff include fecal and pathogenic bacteria. Stormwater transports pollutants to water bodies such as lakes and streams [132].

Enterococci and E. coli can be found in high numbers in most storm drains and creeks. In Southern California (USA), Ferguson et al. [133] found high levels of enterococci (Enterococcus faecalis, Enterococcus faecium, Enterococcus hirae, Enterococcus casseliflavus and Enterococcus mundtii) in intertidal sediments in a seasonal river, and near a storm drain outlet.

Survival in Surface Water

Most intestinal bacteria that contaminate environmental waters are not able to survive and multiply in this environment. Survival rates vary widely among fecal bacteria introduced in environmental waters. Pathogenic enteric bacteria and E. coli display low survival rates (Table 9).

Table 9. Reduction times for fecal bacteria in surface watersa.

Bacterial group	Time for 50% reduction in concentration (days)
Total coliforms	0.9
E. coli	1.5–3
Enterococci	0.9–4
Clostridium perfringens	60 – > 300
Salmonella	0.1–0.67
Shigella	1

[a]Adapted from Medema *et al.* [131].

Table 10. Disappearance rates of fecal bacteria in groundwatersa.

Bacterial group	Disappearance rate (per day)
E. coli	0.063–0.36
Fecal streptococci	0.03–0.24
Clostridium bifermentans spores	0.00
Salmonella enterica subsp. *enterica* serovar Typhimurium	0.13–0.22

[a]Adapted from Medema *et al.* [131].

The ability of fecal bacteria to survive in environmental waters generally increases as the temperature decreases. Others factors that influence survival include dissolved organic carbon concentration, sunlight intensity and the ability to enter the viable but non-culturable state [131].

In a comparative study on the survival of 10 different coliform species (E. coli, Citrobacter freundii,Citrobacter youngae, Klebsiella pneumoniae, K. oxytoca, Enterobacter amnigenus, Enterobacter cloacaesubsp. cloacae, and Pantoea agglomerans (Enterobacter agglomerans)) inoculated in sterilized river water with different concentrations of dissolved organic carbon, Boualam et al. [134] found that only C. freundii, K. pneumoniae and E. cloacae subsp. cloacae remained cultivable after 96 hours of incubation. In a posterior study, using the same bacteria and medium, Boualam et al. [135] found that after 28 days, only C. freundii and E. cloacae subsp. cloacae survived.

Baudišová [136] reported a comparative study on the survival of total coliforms, fecal coliforms and E. coli, in sterile and non-sterile river water. In sterile water, all bacteria survived for many months. However, in non-sterile conditions (closer to true environmental conditions), the elimination rate of all bacteria was considerably faster. Total coliforms survived the longest and E. coli the shortest.

Survival in Groundwater

Survival of bacteria in groundwater is influenced by several factors, namely the survival in soil, since in order to reach the groundwater bacteria have to percolate through the soil. Generally, survival in soil (and concomitantly in groundwater) is enhanced by low temperatures, high soil humidity, neutral or alkaline soil pH and the presence of organic carbon [131].

WHICH INDICATORS OF FECAL POLLUTION SHOULD BE USED?

Several fecal indicator bacteria in environmental waters are in current use. From these stand out fecal coliforms, E. coli and enterococci [1,6,137]. In environmental waters, most fecal coliform strains are E. coli.

In particular situations, the presence of E. coli is definitively not associated with fecal pollution. These situations were firstly detected in some African countries, namely Nigeria, Ivory Coast and New Guinea (although not in others, such as Uganda) [42,99].

Recent studies carried out in temperate zones indicated that E. coli can persist in secondary, nonhost habitats, outside the hot tropical areas, and become naturalized in these habitats. Byappanahalli et al. [138] reported that E. coli could be isolated from coastal temperate forest soils in Indiana (USA). The aquatic alga Cladophora glomerata (L.) from several Lake Michigan beaches was shown to harbor high densities of E. coli [139].

Ishii et al. [140] reported a study on the survival of E. coli in temperate riverine soils of northern Minnesota (USA). Viable E. coli populations were repeatedly isolated from northern temperate soils in three Lake Superior watersheds. Seasonal variation in the population density of soilborne E. coliwas observed; the greatest cell densities were found in the summer to fall, and the lowest numbers, occurred during the winter to spring months. Horizontal, fluorophore-enhanced repetitive extragenic palindromic PCR (HFERP) DNA fingerprint analyses indicated that identical soilborneE. coli genotypes, overwintered in frozen soil and were present over time, and that these strains were different from E. coli strains obtained from wildlife commonly found in the studied habitats or river water. Soilborne E. coli strains had HFERP DNA fingerprints that were unique to specific soils and locations. In laboratory studies, naturalized E. coli strains had the ability to grow and replicate to high cell densities, in nonsterile soils when incubated at 30 or 37 °C and survived longer than 1 month when soil temperatures were lower than 25 °C. It was concluded that these E. coli strains became naturalized, autochthonous members of the soil microbial community.

In a latter paper, Ksoll et al. [141] studied epilithic periphyton communities at three sites on the Minnesota shoreline of Lake Superior (USA). Fecal coliform densities increased up to 4 orders of magnitude in early summer, and decreased during autumn. HFERP DNA fingerprint analyses indicated that waterfowl (geese, terns, and gulls) were the major primary source of periphyton E. coli strains that could be identified. Periphyton and sewage effluent were also major potential sources. Several periphyton E. coli isolates were genotypically identical, repeatedly isolated over time. Inoculated E. coli rapidly colonized natural periphyton in laboratory microcosms and persisted for several weeks, and some cells were released to the overlying water. It was concluded that E. coli had became a naturalized member of the bacterial periphyton communities.

The presence, persistence, and possible naturalization of E. coli in these habitats can confound the use of fecal coliforms as a reliable indicator of recent fecal contamination of environmental waters. Future studies should consider other nonhost habitats as potential sources of fecal coliform bacteria in aquatic environments.

Considering these limitations, it appears to be advisable, in order to check the microbiological quality of drinking water, to complement the determination of Escherichia coli with the assay of enterococci. This rationale that has been followed, for many years, in the making of the drinking-water legislation in the European Union.

However, for many developing countries, where limited financial resources are the norm and reality, the routine determination of these two parameters can be difficult to implement. In these circumstances, it appears common sense that is better to determine a (good) parameter, such asEscherichia coli, than have no analysis done.

In this context, USA legislation emerges as a pragmatic approach to the problem. According to the American legislation, total coliforms are the routine parameter to be determined. Only when these determinations are repeatedly positive, it is mandatory to assess fecal coliforms [142,143]. Although total coliforms are not necessarily fecal bacteria, the rationale behind this system is correct, since: (1) a positive test in fecal coliforms (which is our target) is necessarily positive in the total coliform procedure; (2) the inverse is not necessarily true; (3) total coliforms are easily and cheaply assayed in waters.

As an alternative to the determination of both E. coli and enterococci, the assay of ammonia in environmental waters can be useful and complement the determination of fecal coliforms.

Ammonia is one of the key molecules in the nitrogen cycle. The presence of ammonia in surface waters can be due to direct contamination by agricultural fertilizers, and/or to microbial degradation of proteins, nucleic acids and urea, implying therefore the presence of a considerable concentration of organic matter in the water. Ammonia is rapidly oxidized in the environment and is typically found in natural waters at concentrations less than 0.1 mg/L. Concentrations significantly above this indicate gross contamination by fresh sanitary waste, where ammonia levels are typically very high (tens or hundreds of mg/L) [84].

Espigares et al. [144] reported a comparative study of chemical and microbiological indicators (total and fecal coliforms, fecal streptococci and sulphite-reducing clostridia) in a stretch of the Guadalquivir River (Spain) and its affluents. Total coliforms were correlated with fecal coliforms, but were not correlated with fecal streptococci and clostridia. Fecal coliforms were correlated with the other indicators. Fecal streptococci and sulphite-reducing clostridia were correlated with the other indicators except for total coliforms. All these microbiological indicators were correlated with dissolved oxygen (negatively), dissolved organic carbon and ammonia (positively). Cabral and Marques [85] in a study of a polluted river (Febros) in the Great Oporto area (northwest Portugal) found that ammonia was significantly correlated with all the microbiological assayed parameters—total and fecal coliforms, fecal streptococci and enterococci. These correlations are most probably due to the carry-over of organic matter in wastewaters, and to a high microbial ammonification activity [85].

Simple and rapid in-field tests and automated and continuous systems are available for the assay of ammonia in environmental waters. More studies are needed in order to confirm the use of ammonia as a reliable parameter in a preliminary screening for emergency fecal pollution outbreaks.

CONCLUSIONS

- Safe drinking water for all is one of the major challenges of the 21st century.
- Microbiological control of drinking water should be the norm everywhere.
- Routine basic microbiological analysis of drinking water should be carried out by assaying the presence of Escherichia coli by the culture methods. On-line monitoring of glucuronidase activity is currently too insensitive to replace culture based detection of E. coli but is a valuable complementary tool for high temporal resolution monitoring. Whenever financial resources are available, coliform determinations should be complemented with the quantification of enterococci.
- More studies are needed in order to check if ammonia is reliable for a preliminary screening for emergency fecal pollution outbreaks.
- Financial resources should be devoted to a better understanding of the ecology and behavior of human and animal fecal bacteria in environmental waters.

REFERENCES

1. WHO (World Health Organization). Guidelines for Drinking-water Quality, Incorporating 1st and 2nd Addenda, Volume 1, Recommendations, 3rd ed; WHO: Geneva, Switzerland, 2008.
2. Fenwick, A. Waterborne Diseases—Could they be Consigned to History? Science **2006**, 313, 1077–1081.
3. George, I; Crop, P; Servais, P. Use of β-D-Galactosidase and β-D-Glucuronidase Activities for Quantitative Detection of Total and Faecal Coliforms in Wastewater. Can. J. Microbiol **2001**,47, 670–675.
4. Grabow, WOK. Waterborne Diseases: Update on Water Quality Assessment and Control.Water SA **1996**, 22, 193–202.
5. Seas, C; Alarcon, M; Aragon, JC; Beneit, S; Quiñonez, M; Guerra, H; Gotuzzo, E. Surveillance of Bacterial Pathogens Associated with Acute Diarrhea in Lima, Peru. Int. J. Infect. Dis **2000**,4, 96–99.
6. Medema, GJ; Payment, P; Dufour, A; Robertson, W; Waite, M; Hunter, P; Kirby, R; Anderson, Y. Safe drinking water: an ongoing challenge. In Assessing Microbial Safety of Drinking Water Improving Approaches and Method; WHO & OECD, IWA Publishing: London, UK, 2003; pp. 11–45.
7. Farmer, JJ; Hickam-Brenner, FW. The Genus Vibrio and Phtotobacterium. In The Prokaryotes: An Evolving Electronic Resource for the Microbiological Communityelectronic release 314,, 3th ed; Dworkin, M, Falkow, S, Rosenberg, E, Eds.; Springer-Verlag: New-York, NY, USA, 2003.
8. Farmer, JJ; Janda, JM; Brenner, FW; Cameron, DN; Birkhead, KM. Genus Vibrio. In Bergey's Manual of Systematic Bacteriology, 2nd ed; Brenner, DJ, Krieg, NR, Staley, JT, Eds.; Springer: New York, NY, USA, 2005; Volume 2, Part B, pp. 494–546.
9. DSMZ (Deutsche Sammlung von Mikroorganismen und Zellkulturen GmbH). Bacterial Nomenclature Up-to-Date (Approved Lists; Validation Lists); DSMZ: Braunschweig, Germany, 2010; List 05/2010.
10. Sack, DA; Sack, RB; Nair, GB; Siddique, AK. Cholera. Lancet **2004**, 363, 223–233.
11. Todar, K. Vibrio cholerae and asiatic cholera. Todar's Online Textbook of Bacteriology, 2009. Available online: http://www.textbookofbacteriology.net/cholera.html (assessed on 4 May 2010).
12. Ali, M; Emch, M; Yunus, M; Sack, RB. Are the Environmental Niches of Vibrio cholerae 0139 Different from those of Vibrio cholerae 01 El Tor? Int. J. Infect. Dis **2001**, 5, 214–219.
13. Ramamurthy, T; Yamasaki, S; Takeda, Y; Nair, GB. Vibrio cholerae O139 Bengal: Odyssey of a Fortuitous Variant. Microbes Infect **2003**, 5, 329–344.
14. Arbora, S; Crum, S. Medical Geography and Cholera in Perú; The Geographer's Craft Project, Department of Geography, The University of Colorado at Boulder: Boulder, CO, USA, 1996. Available online: http://www.colorado.edu/geography/gcraft/warmup/cholera/cholera_f.html (assessed on 4 May 2010).
15. Glass, RI; Claeson, M; Blake, PA; Waldmann, RJ; Pierce, NF. Cholera in Africa: Lessons on Transmission and Control for Latin America. Lancet **1991**, 338, 791–795.
16. Glass, RI; Libel, M; Brandling-Bennet, AD. Epidemic Cholera in the Americas. Science **1992**,256, 1524–1525.
17. Tamplin, ML; Parodi, CC. Environmental Spread of Vibrio cholerae in Peru. Lancet **1991**, 338, 1216–1217.
18. Steinberg, EB; Greene, KD; Bopp, CA; Cameron, DN; Wells, JG; Mintz, ED. Cholera in the United States, 1995–2000: Trends at the End of the Twentieth Century. J. Infect. Dis **2001**, 184, 799–802.
19. Meibom, KL; Blokesch, M; Dolganov, NA; Wu, C-Y; Schoolnik, GK. Chitin Induces Natural Competence in Vibrio cholerae. Science **2005**, 310, 1824–1827.
20. Waldor, MK; Mekalanos, JJ. Lysogenic Conversion by a Filamentous Phage Encoding Cholera Toxin. Science **1996**, 272, 1910–1914.
21. Hervio-Heath, D; Colwell, RR; Derrien, A; Robert-Pillot, A; Fournier, JM; Pommepuy, M. Occurrence of Pathogenic Vibrios in Coastal Areas of France. J. Appl. Microbiol **2002**, 92, 1123–1135.
22. Alam, M; Hasan, NA; Sadique, A; Bhuiyan, NA; Ahmed, KU; Nusrin, S; Nair, GB; Siddique, AK; Sack, RB; Sack, DA; Huq, A; Colwell, RR. Seasonal Cholera Caused by Vibrio choleraeSerogroups O1 and O139 in the Coastal Aquatic Environment of Bangladesh. Appl. Environ. Microbiol **2006**, 72, 4096–4104.

23. Bhanumathi, R; Sabeena, F; Isac, SR; Shukla, BN; Singh, DV. Molecular Characterization ofVibrio cholerae O139 Bengal Isolated from Water and the Aquatic Plant Eichhornia crassipes in the River Ganga, Varanasi, India. Appl. Environ. Microbiol **2003**, 69, 2389–2394.

24. Munro, PM; Colwell, RR. Fate of Vibrio cholerae O1 in Seawater Microcosms. Water Res **1996**,30, 47–50.

25. Alam, M; Sultana, M; Nair, GB; Sack, RB; Sack, DA; Siddique, AK; Ali, A; Huq, A; Colwell, RR. Toxigenic Vibrio cholerae in the Aquatic Environment of Mathbaria, Bangladesh. Appl. Environ. Microbiol **2006**, 72, 2849–2855.

26. Chaiyanan, S; Chaiyanan, S; Huq, A; Maugel, T; Colwell, RR. Viability of the NonculturableVibrio cholerae O1 and O139 System. Appl. Microbiol **2001**, 24, 331–341.

27. Molloy, S. Biofilms in the Time of Cholera. Nature Rev. Microbiol **2007**, 5, 906–907.

28. Le Minor. The genus Salmonella. In The Prokaryotes: An Evolving Electronic Resource for the Microbiological Communityelectronic release 314,, 3th ed; Dworkin, M, Falkow, S, Rosenberg, E, Eds.; Springer-Verlag: New York, NY, USA, 2003.

29. Popoff, MY; Le Minor, LE. Genus Salmonella. In Bergey's Manual of Systematic Bacteriology, 2nd ed; Brenner, DJ, Krieg, NR, Staley, JT, Eds.; Springer: New York, NY, USA, 2005; Volume 2, Part B, pp. 764–799.

30. Tindall, BJ; Grimont, PAD; Garrity, GM; Euzéby, JP. Nomenclature and Taxonomy of the Genus Salmonella. Int. J. Syst. Evolut. Microbiol **2005**, 55, 521–524.

31. Ben Aissa, R; Al-Gallas, N; Troudi, H; Belhadj, N; Belhadj, A. Trends in Salmonella entericaSerotypes Isolated from Human, Food, Animal, and Environment in Tunisia, 1994–2004. J. Infect **2007**, 55, 324–339.

32. Arvanitidou, M; Kanellou, K; Vagiona, DG. Diversity of Salmonella spp. and Fungi in Northern Greek Rivers and their Correlation to Faecal Pollution Indicators. Environ. Res**2005**, 99, 278–284.

33. Kapperud, G; Rørvik, LM; Hasseltvedt, V; Høiby, EA; Iversen, BG; Staveland, K; Johnsen, G; Leitão, J; Herikstad, H; Andersson, Y; Langeland, G; Gondrosen, B; Lassen, J. Outbreak ofShigella sonnei Infection Traced to Imported Iceberg Lettuce. J. Clin. Microbiol **1995**, 33, 609–614.

34. Strockbine, NA; Maurelli, AT. Genus Shigella. In Bergey's Manual of Systematic Bacteriology, 2nd ed; Brenner, DJ, Krieg, NR, Staley, JT, Eds.; Springer: New York, NY, USA, 2005; Volume 2, Part B, pp. 811–823.

35. Emch, M; Ali, M; Yunus, M. Risk Areas and Neighborhood-Level Risk Factors for Shigella dysenteriae 1 and Shigella flexneri. Health Place **2008**, 14, 96–105.

36. Germani, Y; Sansonetti, PJ. The Genus Shigella. In The Prokaryotes: An Evolving Electronic Resource for the Microbiological Communityelectronic release 314,, 3th ed; Dworkin, M, Falkow, S, Rosenberg, E, Eds.; Springer-Verlag: New York, NY, US, 2003.

37. Hale, TL. Genetic Basis of Virulence in Shigella Species. Microbiol. Rev **1991**, 55, 206–224.

38. Todar, K. Shigella and shigellosis. Todar's Online Textbook of Bacteriology, 2009. Available online: http://www.textbookofbacteriology.net/Shigella.html (assessed on 1 September 2010).

39. Tetteh, GL; Beuchat, LR. Survival, Growth, and Inactivation of Acid-Stressed Shigella flexnerias Affected by pH and Temperature. Int. J. Food Microbiol **2003**, 87, 131–138.

40. Chompook, P; Todd, J; Wheeler, JG; von Seidlein, L; Clemens, J; Chaicumpa, W. Risk Factors for Shigellosis in Thailand. Int. J. Infect. Dis **2006**, 10, 425–433.

41. Faruque, SM; Khan, R; Kamruzzman, M; Yamasaki, S; Ahmad, QS; Azim, T; Nair, GB; Takeda, Y; Sack, DA. Isolation of Shigella dysenteriae type 1 and S. flexneri Strains from Surface Waters in Bangladesh: Comparative Molecular Analysis of Environmental ShigellaIsolates versus Clinical Strains. Appl. Environ. Microbiol **2002**, 68, 3908–3913.

42. Bettelheim, KA. The genus Escherichia. In The Prokaryotes: An Evolving Electronic Resource for the Microbiological Communityelectronic release 314,, 3th ed; Dworkin, M, Falkow, S, Rosenberg, E, Eds.; Springer-Verlag: New York, NY, USA, 2003.

43. Scheutz, F; Strockbine, NA. Genus Escherichia. In Bergey's Manual of Systematic Bacteriology, 2nd ed; Brenner, DJ, Krieg, NR, Staley, JT, Eds.; Springer: New York, NY, USA, 2005; Volume 2, Part B, pp. 607–623.

44. World Health Organization. Enterotoxigenic Escherichia coli (ETEC). Diarrhoeal Diseases, Available online: http://www.who.int/vaccine_research/diseases/diarrhoeal/en/index4.html(assessed on 4 September 2010).

45. Health Canada. Guidelines for Canadian Drinking Water Quality: Guideline Technical Document. Bacterial Waterborne Pathogens. Current and Emerging Organisms of Concern; Health Canada: Ottawa, ON, Canada, 2006.

46. World Health Organization. Enterohaemorrhagic Escherichia coli (EHEC). Fact sheet N°125; Available online: http://www.who.int/mediacentre/factsheets/fs125/en/ (assessed on 4 September 2010).

47. Enteroinvasive Escherichia coli (EIEC); US Department of Health & Human Services, U.S. Food and Drug Administration. Available online: http://www.fda.gov/Food/FoodSafety/FoodborneIllness/FoodborneIllnessFoodbornePathogensNaturalToxins/BadBugBook/ucm071298.htm (assessed on 4 September 2010).

48. WHO (World Health Organization). Pathogenic Mycobacteria in Water: A Guide to Public Health Consequences, Monitoring and Management; Pedley, S, Bartram, J, Rees, G, Dufour, A, Cotruvo, J, Eds.; IWA Publishing: London, UK, 2004.

49. Holtzman, A; Aronson, T; Froman, S; Berlin, OGW; Dominguez, P; Kunkel, KA; Overturf, G; Stelma, G; Smith, C; Yakrus, M. The Isolation and Identification of Mycobacterium aviumComplex (MAC) Recovered from Los Angeles Potable Water, a Possible Source of Infection in AIDS Patients. Int. J. Environ. Health Res **1994**, 4, 63–72.

50. Von Reyn, CF; Maslow, JN; Barber, TW; Falkinham, JO; Arbeit, RD. Persistent Colonisation of Potable Water, a Source of Mycobacterium avium Infection in AIDS. Lancet **1994**, 343, 1137–1141.

51. Gião, MS; Azevedo, NF; Wilks, SA; Vieira, MJ; Keevil, CW. Persistence of Helicobacter pylori in Heterotrophic Drinking Water Biofilms. Appl. Environ. Microbiol **2008**, 74, 5898–5904.

52. Hulten, K; Han, SW; Enroth, H; Klein, PD; Opekun, AR; Gilman, RH; Evans, DG; Engstrand, L; Graham, DY; El-Zaatari, FA. Helicobacter pylori in the Drinking Water in Peru.Gastroenterology **1996**, 110, 1031–1035.

53. Handfield, M; Simard, P; Couillard, M; Letarte, R. Aeromonas hydrophila Isolated from Food and Drinking Water: Hemagglutination, Hemolysis, and Cytotoxicity for a Human Intestinal Cell Line (HT-29). Appl. Environ. Microbiol **1996**, 62, 3459–3461.

54. Daskalov, H. The Importance of Aeromonas hydrophila in Food Safety. Food Control **2006**, 17, 474–483.

55. Chauret, C; Volk, C; Creason, R; Jarosh, J; Robinson, J; Warnes, C. Detection of Aeromonas hydrophila in a Drinking-Water Distribution System: A Field and Pilot Study. Can. J. Microbiol**2001**, 47, 782–786.

56. Fernández, MC; Beatriz, N; Giampaolo, SB; Ibañez, M; Guagliardo, V; Esnaola, MM; Conca, L; Valdivia, P; Stagnaro, SM; Chiale, C; Frade, H. Aeromonas Hydrophila and its Relation with Drinking Water Indicators of Microbiological Quality in Argentine 2000. Genetica **2000**, 108, 35–40.

57. George, I; Servais, P. Sources et Dynamique des Coliformes dans le Basin de la Sein; Rapport de Synthèse; Programme PIREN-Seine 1998–2001, Sources et dynamique des coliformes dand le bassin de la Seine; C. N. R. S.: Paris, France, 2002.

58. Gauthier, F; Archibald, F. The Ecology of "Faecal Indicator" Bacteria Commonly Found in Pulp and Paper Mill Water Systems. Water Res **2001**, 35, 2207–2218.

59. Wilkes, G; Edge, T; Gannon, V; Jokinen, C; Lyautey, E; Medeiros, D; Neumann, N; Ruecker, N; Topp, E; Lapena, DR. Seasonal Relationships Among Indicator Bacteria, Pathogenic Bacteria, Cryptosporidium Oocysts, Giardia Cysts, and Hydrological Indices for Surface Waters Within an Agricultural Landscape. Water Res **2009**, 43, 2209–2223.

60. Saarela, M; Lähteenmäki, L; Crittenden, R; Salminen, S; Mattila-Sandholm, T. Gut Bacteria and Health Foods—The European Perspective. Int. J. Food Microbiol **2002**, 78, 99–117.

61. Hawksworth, G; Drasar, BS; Hill, MJ. Intestinal Bacteria and the Hydrolysis of Glycoside Bonds. J. Med. Microbiol **1971**, 4, 451–459.

62. Hopkins, MJ; Sharp, R; Macfarlane, GT. Variation in Human Intestinal Microbiota with Age.Digest Liver Dis **2002**, 34(Suppl.2), S12–S18.

63. Leclerc, H; Moriamez, JC. Étude Quantitative de la Flore Fécale de l'Adulte et du Nourrison Alimenté Artificiellement. Path. Bio **1980**, 28, 217–226.

64. Wilson, M. Microbial Inhabitants of Humans Their Ecology and Role in Health and Disease; Cambridge University Press: Cambridge, UK, 2005.

65. Sinton, LW; Finlay, RK; Hannah, DJ. Distinguishing Human from Faecal Contamination in Water: A Review. New Zealand J. Marine Freshwater Res **1998**, 32, 323–348.

66. Biavati, B; Mattarelli, P. The family Bifidobacteriaceae. In The Prokaryotes: An Evolving Electronic Resource for the Microbiological Communityelectronic release 314,, 3th ed; Dworkin, M, Falkow, S, Rosenberg, E, Eds.; Springer-Verlag: New York, NY, USA, 2003.

67. Hippe, H; Andreesen, JR; Gottschalk, G. The genus Clostridium—Nonmedical. In The Prokaryotes: An Evolving Electronic Resource for the Microbiological Communityelectronic release 314,, 3th ed; Dworkin, M, Falkow, S, Rosenberg, E, Eds.; Springer-Verlag: New York, NY, USA, 2003.

68. Rainey, FA; Hollen, BJ; Small, A. Genus Clostridium. In Bergey's Manual of Systematic Bacteriology, 2nd ed; De Vos, P, Garrity, GM, Jones, D, Krieg, NR, Ludwig, W, Rainey, FA, Schleifer, K-H, Whitman, WB, Eds.; Springer: New York, NY, USA, 2009; Volume 3, pp. 738–828.

69. Smith, LDS. The genus Clostridium—Medical. In The Prokaryotes: An Evolving Electronic Resource for the Microbiological Communityelectronic release 314,, 3th ed; Dworkin, M, Falkow, S, Rosenberg, E, Eds.; Springer-Verlag: New York, NY, USA, 2003.

70. Bezirtzoglou, E; Panagiou, A; Savvaidis, I; Maipa, V. Distribution of Clostridium perfringens in Polluted Lake Environments. Anaerobe **1997**, 3, 169–172.

71. Švec, P; Devriese, LA. Genus Enterococcus. In Bergey's Manual of Systematic Bacteriology, 2nd ed; DE Vos, P, Garrity, GM, Jones, D, Krieg, NR, Ludwig, W, Rainey, FA, Schleifer, K-H, Whitman, WB, Eds.; Springer: New York, NY, USA, 2009; Volume 3, pp. 594–607.

72. Pinto, B; Pierotti, R; Canale, G; Reali, D. Characterization of 'Faecal Streptococci' as Indicators of Faecal Pollution and Distribution in the Environment. Lett. Appl. Microbiol **1999**,29, 258–263.

73. Bergey's Manual of Determinative Bacteriology, 9th ed; Holt, JG, et al, Eds.; Williams & Wilkins: Baltimore, MD, USA, 1994; pp. 175–190.

74. Gordon, DM; FitzGibbon, F. The Distribution of Enteric Bacteria from Australian Mammals: Host and Geographical Effects. Microbiology **1999**, 145, 2663–2671.

75. Frederiksen, W; Søgaard, P. The genus Citrobacter. In The Prokaryotes: An Evolving Electronic Resource for the Microbiological Communityelectronic release 314,, 3th ed; Dworkin, M, Falkow, S, Rosenberg, E, Eds.; Springer-Verlag: New York, NY, USA, 2003.

76. Frederiksen, W. Genus Citrobacter. In Bergey's Manual of Systematic Bacteriology, 2nd ed; Brenner, DJ, Krieg, NR, Staley, JT, Eds.; Springer: New York, NY, USA, 2005; Volume 2, Part B, pp. 651–656.

77. Grimont, F; Grimont, PAD; Richard, C. The Genus Klebsiella. In The Prokaryotes: An Evolving Electronic Resource for the Microbiological Communityelectronic release 314,, 3th ed; Dworkin, M, Falkow, S, Rosenberg, E, Eds.; Springer-Verlag: New York, NY, USA, 2003.
78. Grimont, PAD; Grimont, F. Genus Klebsiella. In Bergey's Manual of Systematic Bacteriology, 2nd ed; Brenner, DJ, Krieg, NR, Staley, JT, Eds.; Springer: New York, NY, USA, 2005; Volume 2, Part B, pp. 685–693.
79. Grimont, F; Grimont, PAD. The Genus Enterobacter. In The Prokaryotes: An Evolving Electronic Resource for the Microbiological Communityelectronic release 314,, 3th ed; Dworkin, M, Falkow, S, Rosenberg, E, Eds.; Springer-Verlag: New York, NY, USA, 2003.
80. Grimont, PAD; Grimont, F. Genus Enterobacter. In Bergey's Manual of Systematic Bacteriology, 2nd ed; Brenner, DJ, Krieg, NR, Staley, JT, Eds.; Springer: New York, NY, USA, 2005; Volume 2, Part B, pp. 661–669.
81. Lockhart, SR; Abramson, MA; Beekmann, SE; Gallagher, G; Riedel, S; Diekema, DJ; Quinn, JP; Doern, GV. Antimicrobial Resistance Among Gram-negative Bacilli Causing Infections in Intensive Care Unit Patients in the United States between 1993 and 2004. J. Clin. Microbiol**2007**, 45, 3352–3359.
82. Hidron, AI; Edwards, JR; Patel, J; Horan, TC; Sievert, DM; Pollock, DA. NHSN Annual Update: Antimicrobial Resistant Pathogens Associated with Healthcare-Associated Infections: Annual Summary of Data Reported to the National Healthcare Safety Network at the Centers for Disease Control and Prevention, 2006–2007. Infect. Control Hosp. Epidemiol**2008**, 29, 996–1011.
83. Ashbolt, NJ; Grabow, OK; Snozzi, M. Indicators of microbial water quality. In Water Quality: Guidelines, Standards and Health; Fewtrell, L, Bartram, J, Eds.; World Health Organization (WHO), IWA Publishing: London, UK, 2001; pp. 289–316.
84. Payment, P; Waite, M; Dufour, A. Introducing parameters for the assessment of drinking water quality. In Assessing Microbial Safety of Drinking Water Improving Approaches and Method; WHO & OECD, IWA Publishing: London, UK, 2003; pp. 47–77.
85. Cabral, JP; Marques, C. Faecal Coliform Bacteria in Febros river (Northwest Portugal): Temporal Variation, Correlation with Water Parameters, and Species Identification. Environ. Monit. Assess **2006**, 118, 21–36.
86. Villarino, A; Toribio, AL; Brena, BM; Grimont, PAD; Bouvet, OMM. On the Relationship Between the Physiological State of Bacteria and Rapid Enzymatic Assays of Faecal Coliforms in the Environment. Biotechnol. Lett **2003**, 25, 1329 1334.
87. George, I; Petit, M; Servais, P. Use of Enzymatic Methods for Rapid Enumeration of Coliforms in Freshwaters. Lett. Appl. Microbiol **2000**, 88, 404–413.
88. George, I; Petit, M; Theate, C; Servais, P. Distribution of Coliforms in the Seine River and Estuary (France) Studied by Rapid Enzymatic Methods and Plate Counts. Estuaries **2001**, 24, 994–1002.
89. Nelis, H; van Poucke, S. Enzymatic Detection of Coliforms and Escherichia coli Within 4 Hours. Water, Air, and Soil Pollution **2000**, 123, 43–52.
90. Rompré, A; Servais, P; Baudart, J; de-Roubin, M-R; Laurent, P. Detection and Enumeration of Coliforms in Drinking Water: Current Methods and Emerging Approaches. J. Microbiol. Method **2002**, 49, 31–54.
91. Van Poucke, SO; Nelis, HJ. Rapid Detection of Fluorescent and Chemiluminescent Total Coliforms and Escherichia coli on Membrane Filters. J. Microbiol. Method **2000**, 42, 233–244.
92. Kämpfer, P; Rauhoff, O; Dott, W. Glycosidase Profiles of Members of the FamilyEnterobacteriaceae. J. Clin. Microbiol **1991**, 29, 2877–2879.
93. Kilian, M; Bülow, P. Rapid Diagnoses of Enterobacteriaceae. I. Detection of Bacterial Glycosidases. Acta Pathol. Microbiol. Scandinavica **1976**, 84, 245–251.
94. Muytjens, HL; van der Ros-van de Repe, J; nan Druten, HAM. Enzymatic Profiles ofEnterobacter sakazakii and Related Species with Special Reference to the α-Glucosidase Reaction and Reproducibility of the Test System. J. Clin. Microbiol **1984**, 20, 684–686.
95. Tryland, I; Fiksdal, L. Enzyme Characteristics of β-D-Galactosidase- and β-D-Glucuronidase-Positive Bacteria and their Interference in Rapid Methods for Detection of Waterborne Coliforms and Escherichia coli. Appl. Environ. Microbiol **1998**, 64, 1018–1023.
96. Köster, W; Egli, T; Ashbolt, N; Botzenhart, K; Burlion, N; Endo, T; Grimont, P; Guillot, E; Mabilat, C; Newport, L; Niemi, M; Payment, P; Prescott, A; Renaud, P; Rust, A. Analytical methods for microbiological water quality testing. In Assessing Microbial Safety of Drinking Water Improving Approaches and Method; WHO & OECD, IWA Publishing: London, UK, 2003; pp. 237–292.
97. Manafi, M; Kneifel, W; Bascomb, S. Fluorogenic and Chromogenic Substrates Used in Bacterial Diagnostics. Microbiol. Rev **1991**, 55, 335–348.
98. Geissler, K; Manafi, M; Amorós, I; Alonso, JL. Quantitative Determination of Total Coliforms and Escherichia coli in Marine Waters with Chromogenic and Fluorogenic Media. J. Appl. Microbiol **2000**, 88, 280–285.
99. Byamukama, D; Kansiime, F; Mach, RL; Farnleitner, AH. Determination of Escherichia coliContamination with Chromocult Coliform Agar Showed a High Level of Discrimination Efficiency for Differing Faecal Pollution Levels in Tropical Waters of Kampala, Uganda. Appl. Environ. Microbiol **2000**, 66, 864–868.
100. Eccles, JP; Searle, R; Holt, D; Dennis, PJ. A Comparison of Methods Used to EnumerateEscherichia coli in Conventionally Treated Sewage Sludge. J. App. Microbiol **2004**, 96, 375–383.
101. Farnleitner, AH; Hocke, L; Beiwl, C; Kavka, GG; Zechmeister, T; Kirschner, AKT; Mach, RL. Rapid Enzymatic Detection of Escherichia coli Contamination in Polluted River Water. Lett. Appl. Microbiol **2001**, 33, 246–250.

102. Edberg, SC; Kontnick, CM. Comparison of β-Glucuronidase-Based Substrate Systems for Identification of Escherichia coli. J. Clin. Microbiol **1986**, 24, 368–371.

103. Leclerc, H; Mossel, DAA; Edberg, SC; Struijk, CB. Advances in the Bacteriology of the Coliform Group: their Suitability as Markers of Microbial Water Safety. Ann. Rev. Microbiol**2001**, 55, 201–234.

104. Klein, G. Taxonomy, Ecology and Antibiotic Resistance of Enterococci from Food and the Gastro-Intestinal Tract. Int. J. Food Microbiol **2003**, 88, 123–131.

105. Kühn, I; Iversen, A; Burman, LG; Olsson-Liljequist, B; Franklin, A; Finn, M; Aarestrup, F; Seyfarth, AM; Blanch, AR; Vilanova, X; Taylor, H; Caplin, J; Moreno, MA; Dominguez, L; Herrero, IA; Möllby, R. Comparison of Enterococcal Populations in Animals, Humans, and the Environment—An European Study. Int. J. Food Microbiol **2003**, 88, 133–145.

106. Wheeler, AL; Hartel, PG; Godfrey, DG; Hill, JL; Segars, WI. Potentital of Enterococcus faecalisas a Human Fecal Indicator for Microbial Source Tracking. J. Environ. Qual **2002**, 31, 1286–1293.

107. Geldreich, EE. Fecal Coliform and Fecal Streptococcus Density Relationships in Waste Discharges and Receiving Waters. Crit. Rev. Environ. Control **1976**, 6, 349–369.

108. Doran, JW; Linn, DM. Bacteriological Quality of Runoff Water from Pasteureland. Appl. Environ. Microbiol **1979**, 37, 985–991.

109. Caplenas, NJ; Kanarek, MS. Thermotolerant Non-Fecal Source of Klebsiella pneumoniae: Validity of the Fecal Coliform Test in Recreational Waters. Am. J. Public Health **1984**, 74, 1273–1275.

110. Gauthier, F; Neufeld, JD; Driscoll, BT; Archibald, FS. Coliform Bacteria and Nitrogen Fixation in Pulp and Paper Mill Effluent Treatment Systems. Appl. Environ. Microbiol **2000**, 66, 5155–5160.

111. Corso, PS; Kramer, MH; Blair, KA; Addiss, DG; Davis, JP; Haddix, AC. Cost of Illness in the 1993 Waterborne Cryptosporidium Outbreak, Milwaukee, Wisconsin. Emerg. Inf. Dis **2003**, 9, 426–431.

112. MacKenzie, WR; Hoxie, NJ; Proctor, ME; Gradus, MS; Blair, KA; Peterson, DE; Kazmierczak, JJ; Addiss, DG; Fox, KR; Rose, JB; Davis, JP. A Massive Outbreak in Milwaukee ofCryptosporidium Infection Transmitted Through the Public Water Supply. N. Engl. J. Med **1994**,331, 161–167.

113. Vilanova, X; Manero, A; Cerdà-Cuéllar, M; Blanch, AR. The Composition and Persistence of Faecal Coliforms and Enterococcal Populations in Sewage Treatment Plants. J. Appl. Microbiol**2004**, 96, 279–288.

114. Araujo, M; Sueiro, RA; Gómez, MJ; Garrido, MJ. Enumeration of Clostridium perfringensSpores in Groundwater Samples: Comparison of Six Culture Media. J. Microbiol. Methods**2004**, 57, 175–180.

115. Charriere, G; Mossel, DAA; Beaudeau, P; Leclerc, H. Assessment of the Marker Value of Various Components of the Coli-Aerogenes Group of Enterobacteriaceae and of a Selection ofEnterococcus spp. for the Official Monitoring of Drinking Water Supplies. J. Appl. Bacteriol**1994**, 76, 336–344.

116. Martins, MT; Sato, MIZ; Alves, MN; Stoppe, NC; Prado, VM; Sanchez, PS. Assessment of Microbiological Quality for Swimming Pools in South America. Water Res **1995**, 29, 2417–2420.

117. Ferguson, CM; Coote, BG; Ashbolt, NJ; Stevenson, IM. Relationships Between Indicators, Pathogens and Water Quality in an Estuary System. Water Res **1996**, 30, 2045–2054.

118. Medema, GJ; vanAsperen, IA; Havelaar, AH. Assessment of the Exposure of Swimmers to Microbiological Contaminants in Fresh Waters. Water Sci. Technol **1997**, 35, 157–163.

119. Polo, F; Figueras, MJ; Inza, I; Sala, J; Fleisher, JM; Guarro, J. Relationship Between Presence of Salmonella and Indicators of Faecal Pollution in Aquatic Habitats. FEMS Microbiol. Lett**1998**, 160, 253–256.

120. Noble, RT; Leecaster, MK; McGee, CD; Weisberg, SB; Ritter, K. Comparison of Bacterial Indicator Analysis Methods in Stormwater-Affected Coastal Waters. Water Res **2004**, 38, 1183–1188.

121. Harwood, VJ; Levine, AD; Scott, TM; Chivukula, V; Lukasik, J; Farrah, SR; Rose, JB. Validity of the Indicator Organism Paradigm for Pathogen Reduction in Reclaimed Water and Public Health Protection. Appl. Environ. Microbiol **2005**, 71, 3163–3170.

122. Touron, A; Berthe, T; Gargala, G; Fournier, M; Ratajczak, M; Servais, P; Petit, F. Assessment of Faecal Contamination and the Relationship Between Pathogens and Faecal Bacterial Indicators in an Estuarine Environment (Seine, France). Mar. Pollut. Bull **2007**, 54, 1441–1450.

123. Garrido-Pérez, MC; Anfuso, E; Acevedo, A; Perales-Vargas-Machuca, JA. Microbial Indicators of Faecal Contamination in Waters and Sediments of Beach Bathing Zones. Int. J. Hyg. Environ. Health **2008**, 211, 510–517.

124. Ashbolt, N; Fujioka, R; Glymph, T; McGee, C; Schaub, S; Sobsey, M; Toranzos, G. Pathogen indicators, and indicators of fecal contamination. In Report of the Experts Scientific Workshop on Critical Research Needs for the Development of New or Revised Recreational Water Quality; EPA 823-R-07–006; U.S. Environmental Protection Agency, Office of Water, Office of Research and Development: Warrenton, VA, 2007; Chapter 2,; pp. 35–56.

125. Ferreira, AP. Caffeine as an Environmental Indicator for Assessing Urban Aquatic Ecosystems. Cad. Saúde Pública **2005**, 21, 1884–1892.

126. Gardinali, PR; Zhao, X. Trace Determination of Caffeine in Surface Water Samples by Liquid Chromatography-Atmospheric Pressure Chemical Ionization–Sass spectrometry (LC-APCI-MS). Environ. Int **2002**, 28, 521–528.

127. Peeler, KA; Opsahl, SP; Chanton, JP. Tracking Anthropogenic Inputs Using Caffeine, Indicator Bacteria, and Nutrients in Rural Freshwater and Urban

Marine Systems. Environ. Sci. Technol **2006**, 40, 7616–7622.

128. Scott, TM; JB; Rose, TM; Jenkins, S; Farrah, R; Lukasik, J. Microbial Source Tracking: Current Methodology and Future Directions. Appl. Environ. Microbiol **2002**, 68, 5796–5803.

129. Wu, J; Yue, J; Hu, R; Yang, Z; Zhang, L. Use of Caffeine and Human Pharmaceutical Compounds to Identify Sewage Contamination. World Acad. Sci. Eng. Technol **2008**, 44, 438–442.

130. Isobe, KO; Tarao, M; Chiem, NH; Minh, LY; Takada, H. Effect of Environmental Factors on the Relationship between Concentrations of Coprostanol and Fecal Indicator Bacteria in Tropical (Mekong Delta) and Temperate (Tokyo) Freshwaters. Appl. Environ. Microbiol **2004**,70, 814–821.

131. Medema, GJ; Shaw, S; Waite, M; Snozzi, M; Morreau, A; Grabow, W. Catchment characteristics and source water quality. In Assessing Microbial Safety of Drinking Water Improving Approaches and Method; WHO & OECD, IWA Publishing: London, UK, 2003; pp. 111–158.

132. EPA. Source Water Protection Practices Bulletin. Managing Stormwater Runoff to Prevent Contamination of Drinking Water; Office of Water (4606); United States Environmental Protection Agency (EPA): Washington, DC, USA, 2009.

133. Ferguson, DM; Moore, DF; Getrich, MA; Zhowandai, MH. Enumeration and Speciation of Enterococci Found in Marine and Intertidal Sediments and Coastal Water in Southern California. J. Appl. Microbiol **2005**, 99, 598–608.

134. Boualam, M; Mathieu, L; Fass, S; Cavard, J; Gatel, D. Relationship between Coliform Culturability and Organic Matter in Low Nutritive Waters. Water Res **2002**, 36, 2618–2626.

135. Boualam, M; Fass, S; Saby, S; Lahoussine, V; Cavard, J; Gatel, D; Mathieu, L. Organic Matter Quality and Survival of Coliforms in Low-Nutrient Waters. J. AWWA **2003**, 95, 119–126.

136. Baudišová, D. Evaluation of Escherichia coli as the Main Indicator of Faecal Pollution. Water Sci. Technol **1997**, 35, 333–336.

137. Edberg, SC; Rice, EW; Karlin, RJ; Allen, MJ. Escherichia coli: The Best Biological Drinking Water Indicator for Public Health Protection. J. Appl. Microbiol **2000**, 88, 106S–116S.

138. Byappanahalli, MN; Whitman, RL; Shively, DA; Sadowsky, MJ; Ishii, S. Population Structure, Persistence, and Seasonality of Autochthonous Escherichia coli in Temperate, Coastal Forest Soil from a Great Lakes Watershed. Environ. Microbiol **2006**, 8, 504–513.

139. Ishii, S; Yan, T; Shively, DA; Byappanahalli, MN; Whitman, RL; Sadowsky, MJ. Cladophora(Chlorophyta) spp. Harbor Human Bacterial Pathogens in Nearshore Water of Lake Michigan. Appl. Environ. Microbiol **2006**, 72, 4545–4553.

140. Ishii, S; Ksoll, WB; Hicks, RE; Sadowsky, MJ. Presence and Growth of Naturalized Escherichia coli in Temperate Soils from Lake Superior Watersheds. Appl. Environ. Microbiol **2006**, 72, 612–621.

141. Ksoll, WB; Ishii, S; Sadowsky, MJ; Hicks, RE. Presence and Sources of Fecal Coliform Bacteria in Epilithic Periphyton Communities of Lake Superior. Appl. Environ. Microbiol **2007**, 73, 3771–3778.

142. Hecq, P; Hulsmann, A; Hauchman, FS; McLain, JL; Schmitz, F. Drinking water regulations. InAnalytical Methods for Drinking Water; Quevauviller, P, Thompson, KC, Eds.; John Wiley: London, UK, 2006; Chapter 1.

143. Total Coliform Rule: A Handbook for Small Noncommunity Water Systems Serving less than 3,300 Persons; EPA 816-B-06–001; Office of Water (4606M), EPA: Washington, DC, USA, 2006.

144. Espigares, M; Coca, C; Fernández-Crehuet, M; Moreno, O; Gálvez, R. Chemical and Microbiologic Indicators of Faecal Contamination in the Guadalquivir (Spain). Eur. Water Pollut. Control **1996**, 6, 7–13.

Section II - Microbial Contaminants in Drinking Water

CHAPTER 5

MICROBIAL CONTAMINATION OF DRINKING WATER AND HUMAN HEALTH FROM COMMUNITY WATER SYSTEMS

Nicholas J. Ashbolt

Keywords: Water safety plans, HACCP, QMRA, Surrogate, Indicator, Antimicrobial resistance, Enteric pathogen, Opportunistic pathogen, Viruses, Bacteria, Parasitic protozoa, Fungi, Waterborne, Water-based, Risk management

INTRODUCTION

The provision of safe drinking water has been one of humanity's most successful public health interventions and is a defining aspect of a developed country. Nonetheless, ignorance of the potential risks and inappropriate training of staff and managers working on drinking water systems still results in unnecessary waterborne disease outbreaks in affluent communities [1•]. Furthermore, re-introduction of once-controlled diseases, such as cholera, may rapidly spread during periods of disasters when sanitation systems are non-functional and drinking water treatment is inadequate. A recent example was the Haitian epidemic [2] where, although a developing region, international aid workers introduced the outbreak strain and then tourists spread infections to more developed regions [3, 4•]. Hence, to some degree, differentiating pathogen risks between developed regions and those less developed, particularly rapidly developing regions, is artificial and not very useful. Therefore, this review is relevant to most regions with functional drinking water treatment provided through a community system. A key realization is the need for ongoing system-wide vigilance, coupled with a preventative rather than just responsive management approach. This approach is best practiced globally using principles from the food industry's Hazard Analysis Critical Control Point (HACCP) approach, described by the World Health Organization (WHO) as Water Safety Plans (WSPs) [5, 6]. In addition to existing regulatory framework constraints, this WSP approach may be particularly hard to implement in (developed) regions that have not identified major waterborne outbreaks for decades.

Even with well-operated drinking-water treatment systems, there is growing concern that aging drinking water distribution systems (DWDSs) are vulnerable to higher rates of mains breaks/repairs and related pressure losses that may lead to pathogen intrusion scenarios [7•, 8•]. Also, traditional end-of-pipe compliance monitoring practices may not identify short periods of DWDS intrusions or short periods of poorer treatment performance, such as is associated with rain-induced dirty-water events that appear to be associated with increased rates of waterborne gastrointestinal disease [9]. Generally, drinking-water gastrointestinal cases are not well quantified, even in developed regions, due to the insensitivities of surveillance and specific epidemiology studies [10, 11]. For the US (with approximately 300 million people) estimates of annual drinking-water gastrointestinal cases range from 12–19 million [12]. In addition, beyond the DWDS is a vast network of building or in-premise plumbing that, under certain conditions, allows the growth and release of water-based opportunistic pathogens, many resulting in respiratory or

Citation: Nicholas J. Ashbolt, Microbial Contamination of Drinking Water and Human Health from Community Water Systems, doi: 10.1007/s40572-014-0037-5.

skin diseases, such as from Legionella pneumophila and non-tuberculous mycobacteria (NTM) [13]. At least in the US, these water-based pathogens appear to cause a higher health burden via hospitalization than waterborne enteric pathogens [14•]. With the exception of recently enacted regulatory monitoring for Legionella in The Netherlands, France, and Germany, these water-based pathogens are neither targeted nor identified by current regulatory monitoring that focuses on fecal indicator bacteria (FIB) [e.g. Escherichia coli and enterococci]. Due to environmental growth of water-based opportunistic pathogens, quite different but familiar control strategies (elimination of stagnation zones and related temperature and disinfectant control) are required for in-premise plumbing, particularly in healthcare settings [15].

This review builds on previously conducted reviews [5, 16•, 17–19], and is organized around recent findings associated with drinking-water microbial hazards and scenarios that need to be managed as part of a WSP-like system-wide management framework to provide safe drinking water. This information should also be informative for the growing use of quantitative microbial risk assessment (QMRA) to inform WSPs [20]. As such, pathogens are grouped into waterborne (enteric viruses, bacteria, parasitic protozoa, and fungi) and water-based (environmental viruses, bacteria, free-living protozoa, and fungi) groups. Furthermore, crosscutting pathogen issues, such as antimicrobial resistance (AMR) gene transfer and the role of the host microbiome, are introduced.

WATERBORNE (ENTERIC) PATHOGENS

There are over 500 waterborne pathogens of potential concern in drinking waters, identified by the US Environmental Protection Agency (EPA) through its Candidate Contaminant List ('CCL 3 Universe' list, available at http://www.epa.gov/safewater/ccl/pdfs/report_ccl3_microbes_universe.pdf). To aid in identifying representatives within each of the microbial groups (viruses, bacteria, parasitic protozoa, and fungi), Table 1 lists current and likely members of importance to manage waterborne risks from community drinking waters. This subset contains representative members, known as reference pathogens, which are increasingly being used to support WSPs via QMRA, given their general coverage of the vast majority of human health effects associated with pathogens in drinking water [26].

Table 1 Recognized and potential enteric and water-based microbial pathogens to manage community drinking water risks

Recognized and potential enteric and water-based microbial pathogens to manage community drinking water risks

Microbial group	Enteric (waterborne)		Water-based (opportunistic)	
	Recognized	Potential	Recognized	Potential
Viral	*Adenovirus* 40 and 41 *Avipolyomavirus* [a] *Enterovirus* A–D *Hepatitis* A and E *Norovirus* G1 and G2 *Rotavirus* A *Sapovirus* G1	*Mamastrovirus* 1 *Orthoreovirus* C	None	*Mimivirus* [b] *Mamavirus* [b]
Bacterial	*Aeromonas hydrophila* [c] *Campylobacter coli* *C. jejuni* *Salmonella enterica* (non-typhoid) *Shigella sonnei* *Vibrio cholerae* [e]	*Acinetobacter baumannii* [c] *Arcobacter butzleri* *Helicobacter pylori* *Clostridium difficile* [c] *Listeria monocytogenes* *Pseudomonas aeruginosa* [c] *Staphylococcus aureus* [c] *Yersinia enterocolitica*	*Legionella longbeachеae* [d] *Legionella micdadei* [d] *Legionella pneumophila* [d] *Escherichia coli* O157:H7 [f] NTM [c,d,g] *Pseudomonas aeruginosa* [c]	*Acinetobacter baumannii* [c] *Aeromonas hydrophila* [c] ARB (*Afipia, Bosea, Parachlamydia* spp., *Coxiella burnetii*) [d] *E. coli* (toxigenic strains) *Listeria monocytogenes* [d] *Staphylococcus aureus* [c] *Stenotrophomonas maltophilia* [c,d]
Protozoan	*Cryptosporidium hominis* and *parvum* *Cyclospora cayetanensis* *Giardia intestinalis* assemblages A and B *Toxoplasma gondii*	*Blastocystis hominis*	*Acanthamoeba T4* *Balamuthia mandrillaris* *Naegleria fowleri*	*Acanthamoeba* spp. [d] *Vahlkampfia* spp. [d] *Vannella* spp. [d] *Vermamoeba vermiformis* [d]
Fungal	None Microsporidia (e.g. *Encephalitozoon bieneusi, E. intestinalis*)	*Candida albicans*	None	*Aspergillus fumigatus* *Aspergillus terreus* *Candida albicans* *Candida parapsilosis* *Exophiala dermatitidis*

AMR antimicrobial-resisting, *ARB* amoeba-resisting bacteria, *QMRA* quantitative microbial risk assessment, *VBNC* viable but non-culturable, *DWDS* drinking water distribution systems, *NTM* non-tuberculous mycobacteria

[a] Main species being *JC polyomavirus*, which is largely excreted in urine, as it infects the kidneys along with the respiratory system or brain

[b] *Acanthamoeba polyphaga mimivirus* (APMV) may cause respiratory disease and unknown health effects from *Mamavirus* [21]

[c] Most strains of species from the environment may be non-pathogenic, however there is future potential for AMR strains. For *P. aeruginosa*, most clinical disease is identified with *otitis media*, with less severe disease via drinking water aerosols leading to diffuse bronchopneumonia and more severe disease in high-risk children with cystic fibrosis; folliculitis is important directly via drinking waters used in pools/spas

[d] Largely non-pathogenic amoeba hosts containing ARB, of which many ARB are of key concern, from drinking water [22], except possibly *Coxiella burnetii*, which is of low theoretical risk estimated by QMRA [23•]

[e] Cholera may re-emerge if a major event interferes with drinking-water disinfection, and an epidemic strain is introduced from an endemic region of the world

[f] Shiga toxin and verotoxin-producing *E. coli* (and various intracellular *Salmonella* and *Listeria*) strains may grow within free-living protozoa [24], and non-pathogenic VBNC *E. coli* in DWDS biofilms [25]

[g] Various NTM, including *Mycobacterium avium* (*M. intracellulare*) complex, *M. chelonae, M. fortuitum, M. gordonae*, and *M. kansasii*

Various counties have developed treatment goals or drinking-water parameters based on microbial risk assessment [5]; however, only in the Netherlands is there a regulatory requirement for drinking-water companies to provide water that, in theory, meets an annual gastrointestinal risk of $<10^{-4}$ 95 % of the time. This means, for example, that drinking water is required to have less than one enteric virus per million liters of drinking water [27], a concentration well below the capabilities of current measurement techniques. Epidemiology studies have shown an increased gastrointestinal risk (30 %) when enteric virus concentrations were at approximately one genomic copy per liter [28], highlighting the difference between what can be measured directly compared with the low concentrations sought by regulations to control risk of gastrointestinal illness as estimated by QMRA. Hence, estimations of enteric pathogen risks are more reliant upon measuring pathogen concentrations in contaminated source water(s), and using surrogates to estimate treatment removals [29]. As such, QMRA estimates contain uncertainties associated with reference pathogen detection, their relationships to surrogates used, and relevance of limited dose–response models that may not address life-stage of most interest. Therefore, QMRA is probably better used to present relative risks to inform management about different risk scenarios in their considerations for developing WSPs, rather than trying to estimate absolute risk levels.

Key areas not currently addressed in most QMRAs of systems are the DWDS and premise plumbing risks. For both, biofilms on pipe walls and sediments within storage reservoirs and pipes [30] present a sequestering environment for various enteric pathogens, including viruses that are relatively resistant to normal disinfection treatments [31], and allow growth of water-based pathogens [32]. Limited QMRA studies of biofilm enteric pathogen risks are available but results suggest that accumulation of virions, known to occur within DWDSs [28], could present a higher level of risk than background levels when they slough off and re-enter the mass flow to customers [33]. In addition, various free-living protozoa and metazoan feeding within DWDS biofilms may also act as disinfectant-resistance transport hosts for enteric pathogens, including viruses [34•, 35]. The risks associated with water-based pathogens, of particular relevance to premise plumbing, are discussed latter.

Reference Enteric Viruses

Human noroviruses cause the most gastrointestinal illness in all regions of the world, with the vast majority thought to be acquired via person-to-person and then by food [36•], given the predominance of genogroup II strains implicated. In waterborne cases, genogroup I is normally implicated [37], presumably due to increased environmental robustness. An interesting finding with human noroviruses and the second most common cause of gastrointestinal illness, rotavirus (although greatly diminishing due to childhood vaccination programs), is the need for certain histo-blood group antigen (HBGA) receptors for these pathogens to bind to target cells [38]. Not only do certain gut bacteria have these HBGA binding sites but these bacteria may also facilitate infection, as recently demonstrated with human B cells [39•]. Therefore, one's gut microbiome and blood group impact the likelihood of infection. Furthermore, there is now optimism that a routine cell culture system for human noroviruses may be developed, which would be of particular value to the water-treatment industry. Non-human, culturable noroviruses, such as murine noroviruses among others, are used as surrogates for treatment performance (inactivation studies) but there is limited understanding of the validity of these surrogates for any human norovirus genogroup or mode of inactivation [40, 41•].

Overall, Norovirus is considered one of the most conservative virus targets for drinking-water QMRA studies (measured via reverse transcriptase-quantitative polymerase chain reaction [RT-qPCR])—conservative in that if Norovirus risks are managed, most other enteric viruses will also be controlled. However, there are limited dose–response models to interpret molecular (RT-qPCR) exposure data [42•], noting it is unknown what fraction of virions may be infectious (given that fresh suspensions of virions were used in dose–response studies versus more aged virions typically present in the environment, and that both infectious and non-infectious virions are measured by RT-qPCR). Counteracting the infectious fraction to some unknown degree is virion clumping, which would increase infection likelihood [43]. In the absence of infectivity data, the WHO, as well as various other jurisdictions, recommend culturable rotaviruses as a preferred enteric virus reference, or the environmentally hardier human adenoviruses or reoviruses, which may be in similarly high concentrations as noroviruses in sewage-contaminated waters [44–46]. The problem with using adenoviruses is that gastrointestinal illness is largely caused by serogroups 40 and 41, whereas the dose–response model available for QMRA is only for a respiratory strain, and there are limited illness or infection rates known for reoviruses (Orthoreovirus C) in humans.

Regarding virus surrogates (typically bacteriophages), there is no identified single surrogate known to mimic the various behaviors of different human enteric viruses (with respect to surface charge, hydrophobic interactions, and inactivation by sunlight, disinfectants, etc.). Human adenoviruses provide a good example of a key virus risk and considerations for what surrogate is most suited, being more resistant than Enterovirus or Rotavirus when ultraviolet (UV) disinfection is practiced. Given the expectation for a 4-$\log_{10}$ inactivation of surface waters in the US (i.e. 99.99 %), it is of concern that human adenoviruses are relatively poorly inactivated by so-called low pressure (LP) UV disinfection (monochromatic at 254 nm), but readily inactivated with medium pressure (MP), polychromatic UV treatment (including 254, 265, 280 and 295 nm wavelengths) [47•]. The differences are, in part, thought to arise due to LP-UV targeting nucleic acids, which may be repaired by the host cell, whereas MP-UV also impairs the protein coats of virions, potentially impeding cell infection.

Therefore, consideration of these different modes of inactivation is important when selecting an appropriate surrogate. For example, a commonly used disinfection bacteriophage surrogate, the F-RNA coliphage MS2, demonstrates a 4-log inactivation with 64 mJ/cm^2 LP-UV and 46 mJ/cm^2 MP-UV compared with adenovirus 2 requiring 120 and 45 mJ/cm^2, respectively [47•]. Not only do these LP- and MP-UV systems inactivate bacteriophages by different mechanisms but their host bacterial cells may also use different repair mechanisms (referred to as photoreactivation and dark repair mechanisms). In the case of MS2, there appears to be no repair by the E. coli host, whereas the dsDNA Salmonella bacteriophage PRD1, appears to be the most resistant of four bacteriophages examined, with a 4-$\log_{10}$ reduction (similar for both LP- and MP-UV) requiring 103 mJ/cm^2 with photoreactivation versus only 35 mJ/cm^2 without reactivation [47•]. In general, MS2 is a valid surrogate for chemical disinfection processes but due to its E. coli source from various warm-blooded animals, it is not a good index of human enteric virus presence in environmental waters [48].

In summary, human enteric reference viruses include species of the genera Adenovirus, Enterovirus, Norovirus, and Rotavirus, and potentially Orthoreovirus C (a reovirus). Generally, only one or two are chosen but only when there is likely to be a human source of fecal contamination yielding human enteric virus risk. In the absence of a sanitary survey to indicate possible sewage/septic seepage pollution to source waters, or to give a second opinion, human-targeted Bacteroides provide a valuable and potentially more sensitive indicator than assaying for human enteric viruses [49]. In urbanized settings, DWDS intrusions are likely to include human enteric viruses, which may dominate gastrointestinal risks via drinking water [43].

Reference Enteric Bacteria

The classic waterborne enteric pathogens include Vibrio cholerae (serogroups O1 and O139, causing cholera), Salmonella enterica (subsp. enterica ser. Typhi, causing typhoid), and Shigella spp. (four species causing dysentery), which have largely been controlled by water treatment/disinfection and are therefore rarely an issue via drinking water in developed regions. However, person-to-person and foodborne spread maintains Shigella sonnei within the sewage of developed regions, along with closely-related shiga toxin and verotoxin-producing E. coli, and pathogenic species of Campylobacter, Salmonella, Arcobacter, Helicobacter and Yersinia (Table 1). An emerging issue is that of AMR, which may occur within any of the bacterial members listed in Table 1 but is noted here by example for E. coli in well waters associated with animal production [50]. These AMR genes may horizontally transfer between commensal and enteric pathogenic bacteria, and present a higher risk due to antimicrobial treatment failures [51•]. Within healthcare facilities, there is also a considerable health burden due to the prevalence of AMR Pseudomonas aeruginosa and Clostridium difficile; with the latter being a spore-former it may persist in sewage and river waters and eventually make its way to drinking waters, and AMR-P. aeruginosa may grow post-water treatment (discussed further in the section on "Control of Water-Based Pathogens") [Table 1, potential future concern]. AMR Staphylococcus aureus is also of potential concern via companion animals to water [52] and could be considered a useful reference pathogen for AMR in the future.

The most recognized and useful reference enteric bacteria in developed regions are Salmonella enterica, Campylobacter jejuni, and E. coli O157:H7, each containing human pathogenic strains that vary by fecal source [53, 54]. However, there is limited data on the quantification of these pathogens in source drinking waters (most studies provide presence/absence data) due to the difficulties in culturing these bacteria from the environment [31, 55]. Therefore, given the general presence of FIB and their ease of culture, either E. coli or enterococci have been used as surrogates to assess enteric bacterial pathogen removals by treatment barriers. However, there are improvements in risk management if culture-free methods are used, as discussed in the next section.

Non-Culturable States of Enteric Bacteria

Historically, clinical and environmental microbiology methods have been based on culturing cells on selective media. Today this is still the general situation in clinical laboratories, although next-generation sequencing costs are so rapidly decreasing that single gene to whole metagenomic approaches are enabling rapid and broader detection of pathogens from clinical and environmental samples (Fig. 1). From a water perspective, qPCR assay for Enterococcus spp. (targeting 16S rDNA) was the first molecular method approved by the EPA (for recreational water assessment in treated/sewage-impacted water bodies [57•]). This assay of total (dead or alive) enterococci provides the best index to health risk following fresh (and marine) water exposures in epidemiology studies [58•] (gastrointestinal risk was assumed to be dominated by enteric viruses due to the presence of municipal wastewater contamination). Hence, it would seem appropriate to consider the use of qPCR for enterococci as possibly the most useful microbial index identified to date for sewage-contaminated drinking water. Furthermore, improved detection sensitivity and apparent viability appears possible by qPCR targeting the thousands of copies of 16S ribosomal RNA within viable bacterial cells rather than qPCR directed to low-copy-number DNA-based genes [59•].

As for all possible pathogen infections, we now realize that our health is also reliant upon the 'health' of our microbiomes and, in the case of the gastrointestinal tract, the gutome is being explored by next-generation sequencing approaches. Informed by these metagenomic studies and clinical samples from outbreak cases, the expectation is that many more important, currently uncultured drinking water pathogens (and pathogen-vulnerable gutomes) will be identified over the coming decades than those listed in Table 1, noting that etiologic agents are identified in less than 45 % of drinking-water outbreaks in the US [16•], in part due to the inability to culture them.

However, even well-known bacterial pathogens lose the ability to be cultured when moved from recent cell growth in the gut to the aquatic environment. Campylobacters are particularly well known for this, forming viable but non-culturable (VBNC) cells [60•]. Whereas the term active but non-culturable (ABNC) makes more sense given the term 'viable' normally refers to the ability to grow an organism, VBNC is used in this review due to the majority of papers on this topic describing these cells as such. The VBNC cell state is important to drinking water and public health due to the potential of VBNC cells to cause infection in humans [61], formed during water disinfection [62], and this has not yet been addressed by current culture-based compliance monitoring of drinking water, nor has the likely environmental biofilm niche for VBNC environmental pathogens been sampled. A further complication (in describing these pathogens as enteric and/or environmental) is that resuscitation of VBNC cells may also occur within environmental (free-living) protozoa—true for a variety of the intracellular enteric pathogens listed in Table 1 [63]. Hence, it is probably more useful to take a microbial ecology perspective and think of the VBNC state as part of the normal lifecycle of most bacteria that do not form spores. As such, various mechanisms have probably evolved over millennia, with their various interactions within aquatic (and subsequently water system) predatory eukaryotic organisms, well before they adapted to the human gut (also true for the FIB, e.g. Enterococcus faecalis [64]).

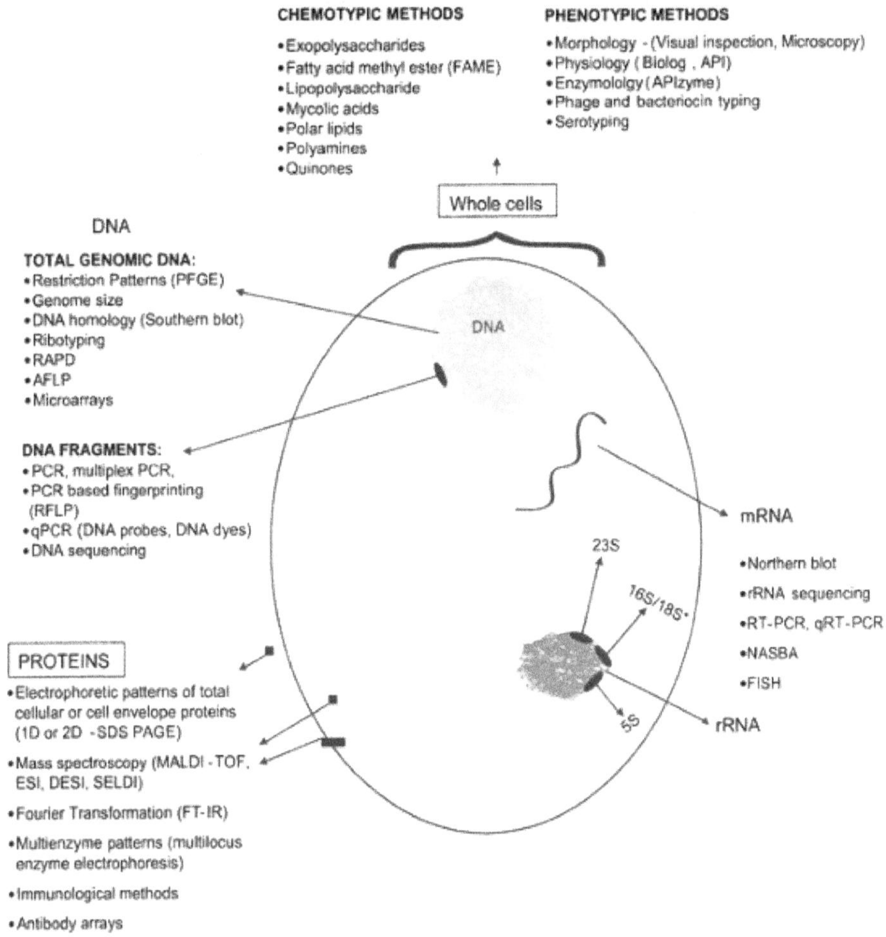

Figure 1 Various cell targets used for non-culture-based methods and culture-based phenotypic methods to detect microorganisms from water (from Sen and Ashbolt [56]). PFGE pulse field gel electrophoresis, RAPD random amplified polymorphic DNA, AFLP amplified fragment ...

Of particular relevance to drinking water disinfection, and easily confused with the VBNC state, are highly-resistant cell forms that enable bacterial survival in the presence of stressors, known as persister cells [65]. Unlike VBNC cells, there is always some fraction of persisters in a population present as a reversible non-replicating state, a particular feature of AMR strains. Persisters may represent a few percent of a bacterial population and are more common post-exponential growth [65]. Although not specifically identified, persisters may contribute to the inability to completely disinfect drinking water, seen by various tailing effects in disinfection kinetic studies [66] and the downstream presence of various enteric bacteria identified by molecular methods within drinking water/biofilms, e.g. along with VBNC forms of E. coli and Helicobacter pylori [25, 67].

H. pylori is a particularly controversial 'waterborne' pathogen due to the presence of non-culturable cells in drinking water. To date, H. pylori has only been detected by molecular methods in drinking waters [67], and it demonstrates a subpopulation able to 'survive' drinking-water chlorination treatment [68], yet there is only weak epidemiologic evidence for waterborne transmission [69]. As with many enteric pathogens, person-to-person spread is probably more important than the waterborne route, yet it remains unclear if H. pylori should even be considered waterborne in developed regions.

Reference Parasitic Protozoa and Fungi

When a parasitic protozoan agent is identified during waterborne outbreaks in the US it was most often due to the presence of cysts to human-infective Giardia intestinalis(synonyms G. lamblia and G. duodenalis) [16•], further described as assemblages A or B. However, from a European perspective, and globally in developed regions over the last decade, more disease burden has resulted from chlorine-resistant oocysts of Cryptosporidium hominis or C. parvum [70]. Considerable research and management changes have been successful in reducing waterborne cryptosporidiosis from large municipal systems, and molecular methods are now available to identify the small subset of genotypes that are likely to be human infectious [71]. Nonetheless, QMRA estimates of waterborne cryptosporidiosis from small systems in developed regions is considerably higher than giardiasis, and possibly well above levels considered acceptable [72]. Therefore, C. parvum and/or C. hominis are generally the reference parasitic protozoan used in QMRA to assess and manage drinking waters. Unfortunately, most jurisdictions do not discriminate between the genotypes that may infect humans versus those that may not (as a precautionary principle), yet that decision may have major cost ramifications, and money could be better spent elsewhere to reduce drinking-water disease burden when the two key Cryptosporidium oocyst sources (human sewage or calf feces [73]) are not likely to impact source drinking waters.

Waterborne outbreaks from other parasitic protozoa appear to be rarer (e.g. Blastocystis hominis, Cyclospora cayetanensis, and Toxoplasma gondii) [74], and attention toCryptosporidium control in watersheds, water treatment, and distribution should largely address these other members. Less well-understood are the microsporidia, once classified as parasitic protozoan and identified in some waterborne outbreaks [70]. Enterocytozoon bieneusi has been identified in source waters [73] and is considered the most common member among 17 human pathogenic microsporidian species that largely impact HIV/AIDS and immunosuppressed patients [75]. Taxonomically, the microsporidia are within the phylum Microspora, and are classified among spore-forming unicellular fungal parasites. As such, they produce smaller spores than the oo/cysts of parasitic protozoa, but due to likely low occurrence their movement through sand filtration processes [76] may be better modeled using bacterial spore surrogates, which are typically removed to a lesser degree than parasitic protozoan oo/cysts [77]. Zoonotic spread from bovines is considered important, although microsporidial infection rates in cattle are probably significantly lower than for Cryptosporidium spp. [78].

WATER-BASED (ENVIRONMENTAL) PATHOGENS

Amoeba-Resisting Bacteria

Respiratory disease caused by NTM and L. pneumophila is the dominant hospitalization cost claim in the US, and largely results from drinking-water-related aerosol exposures [14•]. However, there has been slow recognition of this fact, due, in part, to the overlap in clinical findings of the severe form of pneumonia known as Legionnaires' disease with other more common causes of community-acquired pneumonia (CAP), noting that NTM also contribute to the health burden through wound and soft tissue infections, andP. aeruginosa via AMR infections in healthcare settings and, to a lesser degree, from folliculitis via pools and spas [79]. Looking at the legionellae first, despite Legionella micdadei being identified from blood in a CAP patient in 1943 it took the 1976 Legionnaires' disease epidemic in Philadelphia to fully recognize and describe L. pneumophilaand Legionnaires' disease [80]. Indeed, retrospectively, the first documented epidemic of Legionnaires' disease was traced back to Austin, MN, USA, in the summer of 1957 [81]. However, legionellosis was only made a reportable disease in the US in 2001, with L. pneumophila causing 80–90 % of identified cases in the US [16•] and worldwide [82], and with L. longbeacheae adding 2–7 % of Legionnaires' disease cases, except in Australia where it accounted for approximately 30 % [82]. A further complication results from the difficulty in culturing L. pneumophila, which requires the reduced forms of the amino acid cysteine and ferrous iron [83], and the likelihood of VBNC and other difficult-to-culture cell forms [84] being dominant in drinking waters. One solution identified in the 1980s is to co-culture these problematic cell forms with free-living amoebae [85], as occurs in nature [86]. Follow-up identification can then use molecular methods targeting the 16S rRNA gene and PCR directed to the macrophage infectivity potentiator (mip) gene to confirm L. pneumophila, with other genes (gyrA, rpoB, rnpB) also used to describe additional species [87].

The NTM are ubiquitous in freshwaters and are often the dominant group within drinking-water pipe biofilms, probably selected for by the presence of a residual disinfectant [88•]. Opportunistic members linked to drinking water infections include the Mycobacterium avium complex (which includes M. intracellulare), M. chelonae, M. fortuitum, M. gordonae, and M. kansasii [89]. However, care is needed to resolve between clinically important strains and non-pathogenic environmental members. Environmental isolates of clinically relevant species are often not identified as the etiologic agents. However, is this just a limit of our culture-based approaches or truly a misdiagnosis of the environmental source? Next-generation sequencing not reliant on culturing is likely to resolve such controversies with these water-based pathogens.

Most important to note is that L. pneumophila, NTM, and an increasing list of other amoeba-resisting bacteria (ARB) are opportunistic pathogens growing within drinking-water biofilms (Table 1). Chronic persistence of ARB is particularly problematic to control in healthcare settings due to the ubiquitous nature of amoebae and the protection they confer on their intracellular pathogens [90]. Comparative genomics is also proving to be beneficial to identify human pathogenic members, such as key Legionella species [82].

Free-Living Amoebae and Their Viruses

Both culture- and molecular-based methods have identified the free-living amoebae Acanthamoeba, Naegleria, Protacanthamoeba spp. and Vermamoeba (formallyHartmannella) vermiformis. Less frequent detections include Echinamoeba, Vahlkampfia, and Vannella spp., amongst others, all likely environmental hosts of amoeba-resisting bacterial pathogens such as NTM and legionellae in drinking waters [22, 91•] (Table 1). Some of these free-living amoebae may be pathogens in their own right, such as those causing Acanthamoeba keratitis via drinking water, a serious eye infection primarily affecting contact lens users, although several cases have involved other genera (Vahlkampfia, Vannella, and Hartmannella spp.) [22]. More severe and often life-threatening infections affecting immunocompetent children and immunocompromised adults include encephalitis involving Acanthamoeba spp. and Balamuthia mandrillaris, and Naegleria fowleri causing meningoencephalitis [22]. Nasal irrigation with drinking water from warm climatic zones has been identified as an important source of primary amoebic meningoencephalitis (PAM) caused by growth of N. fowleri through nasal passages to the brain [92]. Luckily, N. fowleri is readily controlled in drinking water by maintaining a disinfectant residual [93], remembering the main growth is thought to occur within biofilms, therefore biofilm control through minimization of stagnation zones and monitoring is also suggested for problematic climatic zones.

Acanthamoeba polyphaga mimivirus (APMV), first misidentified as a bacterium due to its large size (>700 nm capsid), appears to be one of a group of giant viruses that infect aquatic and soil protozoa and metazoan [94]. They are environmentally very robust and some have been implicated in human cases of pneumonia [21]. Recent viral metagenomic studies have vastly increased the number of new members of the Megavirales order of giant viruses and their virophages [95], with new species of clinical significance expected.

Control of Water-Based Pathogens

A common feature of the water-based pathogens is the ability to grow to problematic concentrations within biofilms on pipe walls and sediments, particularly during periods of water stagnation and warmer conditions; therefore, control below some critical concentration is necessary to manage these environmental pathogens. For L. pneumophila, some millions of cells per liter of drinking water may be necessary so that aerosols of a respiratory size reach the alveoli and cause human infection [96]. Lesser concentrations, but some contact, are required for P. aeruginosa folliculitis [79]; the infective doses for NTM and N. fowleri are probably low but are undocumented.

Therefore, there are competing issues in managing all of these water-based pathogens, should they co-occur. Legionella and NTM of human health concern probably both grow within amoebae; therefore, limiting susceptible amoebae hosts by way of biofilm control seems logical, as would monitoring for total amoebae to evaluate control [97]. However, maintaining a high disinfectant residual may select for NTM biofilms that could contain problematic strains, although control N. fowleri. Some evidence suggests that monochloramine is effective against free and amoeba-cultured L. pneumophila, while chlorine and chlorine dioxide are less effective drinking-water disinfectants against amoeba-cultured pathogens, implying different modes of disinfection [98]. Of pipe materials, copper appears to develop less biofilm biomass, but select for VBNC L. pneumophila compared with PVC [99•, 100•], whereas cross-linked polyethylene appears to support both legionellae and M. avium complex as culturable cells at 40–55 °C [101]. Hence, keeping piped cold water cool (<20 °C) and hot water >60 °C via constant circulation seems to offer the most pragmatic control options for all of these pathogens within premise plumbing [15]. A speculative control approach is to maintain an actively antagonistic biofilm community that suppresses growth of members that may be opportunistic pathogens [102].

Opportunistic Fungal Infections

Of the filamentous fungi, Aspergillus fumigatus and A. terreus have been isolated from hospital drinking waters that caused nosocomial infections [103]. However, while fungal filaments are often observed in drinking-water biofilms, they are rarely identified. Recent next-generation sequencing of drinking-water biofilms has identified three opportunistic fungal pathogens—Candida albicans, C. parapsilosis, and Exophiala (formally Wangiella) dermatitidis [104]. It is noted that disseminated candidiasis caused byC. albicans is a leading nosocomial bloodstream infection in the US, with a high case fatality rate. E. dermatitidis is a ubiquitous black yeast that grows optimally around 40 °C and is most commonly seen in saunas, steam baths, and humidifiers [105].

CONCLUSIONS

With of advent of next-generation sequencing for routine clinical and environmental microbiology, there is renewed hope to improve upon identifying novel and currently known, but non-detected, physiological states of drinking water pathogens.

There is also intriguing new evidence that what were once thought of as strictly enteric pathogens may contain members with environmental amplification potential, ranging from human enteric viruses [34•] to E. coli O157:H7 [23•], but for now that is largely speculative. What is more concrete is the rising health burden resulting from opportunistic pathogens via drinking water [13], which are largely unregulated worldwide.

CONFLICT OF INTEREST

Nicholas J. Ashbolt has a patent method for detection of viable Cryptosporidium parvum cells—Australian patent number 96/00274; EP0840799 A1 filed 6 May 1996, issued 13 May 1998—and receives royalties.

REFERENCES

1. Hrudey SE, Hrudey EJ. Ensuring safe drinking water: learning from frontline experience with contamination. Denver (CO): American Water Works Association; 2014.
2. Grandesso F, Allan M, Jean-Simon PS, Boncy J, Blake A, Pierre R, et al. Risk factors for cholera transmission in Haiti during inter-peak periods: insights to improve current control strategies from two case–control studies. Epidemiol Infect. 2014;142(8):1625–35. doi: 10.1017/S0950268813002562.
3. Moore SM, Shannon KL, Zelaya CE, Azman AS, Lessler J. Epidemic risk from cholera introductions into Mexico. PLoS Curr. 2014;6. doi:10.1371/currents.outbreaks.c04478c7fbd9854ef6ba923cc81eb799. Edition 1.
4. 4.•. Loharikar A, Newton AE, Stroika S, Freeman M, Greene KD, Parsons MB, et al. Cholera in the United States, 2001–2011: a reflection of patterns of global epidemiology and travel. Epidemiol Infect. 2014. doi:10.1017/S0950268814001186. Cholera surveillance that provides insight into global and United States trends.
5. World Health Organization . Water safety plan manual: step-by-step risk management for drinking-water suppliers. Geneva: World Health Organization; 2009.
6. World Health Organization . Guidelines for drinking-water quality. 4. Geneva: World Health Organization; 2011.
7. Ebacher G, Besner MC, Clément B, Prévost M. Sensitivity analysis of some critical factors affecting simulated intrusion volumes during a low pressure transient event in a full-scale water distribution system. Water Res. 2012;46(13):4017–30. doi: 10.1016/j.watres.2012.05.006.
8. Lambertini E, Borchardt MA, Kieke BA, Jr, Spencer SK, Loge FJ. Risk of viral acute gastrointestinal illness from nondisinfected drinking water distribution systems. Environ Sci Technol. 2012;46(17):9299–307. doi: 10.1021/es3015925.

9. Beaudeau P, Schwartz J, Levin R. Drinking water quality and hospital admissions of elderly people for gastrointestinal illness in Eastern Massachusetts, 1998–2008. Water Res. 2014;52:188–98. doi: 10.1016/j.watres.2014.01.005.
10. Ford TE. Microbiological safety of drinking water: united states and global perspectives. Environ Health Perspect. 1999;107(Suppl. 1):191–206.
11. Hellard ME, Sinclair MI, Forbes AB, Fairley CK. A randomized, blinded, controlled trial investigating the gastrointestinal health effects of drinking water quality. Environ Health Perspect. 2001;109(8):773–8. doi: 10.1289/ehp.01109773.
12. Reynolds KA, Mena KD, Gerba CP. Risk of waterborne illness via drinking water in the United States. Rev Environ Contam Toxicol. 2008;192:117–58.
13. National Research Council. Drinking water distribution systems: assessing and reducing risks. Committee on Public Water Supply Distribution Systems: assessing and reducing risks. Water Science and Technology Board, Division on Earth and Life Studies, National Research Council of the National Academies. Washington, DC: The National Academies Press; 2006.
14. Collier SA, Stockman LJ, Hicks LA, Garrison LE, Zhou FJ, Beach MJ. Direct healthcare costs of selected diseases primarily or partially transmitted by water. Epidemiol Infect. 2012;140(11):2003–13. doi: 10.1017/S0950268811002858.
15. Sidari FP, III, Stout JE, Duda S, Grubb D, Neuner A. Maintaining Legionella control in building water systems. J Am Water Works Assoc. 2014;106(10):24–32. doi: 10.5942/jawwa.2014.106.0147.
16. Craun GF, Brunkard JM, Yoder JS, Roberts VA, Carpenter J, Wade T, et al. Causes of outbreaks associated with drinking water in the United States from 1971 to 2006. Clin Microbiol Rev. 2010;23(3):507–28. doi: 10.1128/CMR.00077-09.
17. Exner M, Kramer A, Lajoie L, Gebel J, Engelhart S, Hartemann P. Prevention and control of health care-associated waterborne infections in health care facilities. Am J Infect Control. 2005;33(5):S26–40. doi: 10.1016/j.ajic.2005.04.002.
18. Leclerc H, Schwartzbrod L, Dei-Cas E. Microbial agents associated with waterborne diseases. Crit Rev Microbiol. 2002;28(4):371–409. doi: 10.1080/1040-840291046768.
19. Szewzyk U, Szewzyk R, Manz W, Schleifer KH. Microbiological safety of drinking water. Ann Rev Microbiol. 2000;54:81–127. doi: 10.1146/annurev.micro.54.1.81.
20. Medema G, Loret J-F, Stenström TA, Ashbolt N. Quantitative microbial risk assessment in the water safety plan. Final report on the EU MicroRisk Project. Brussels: European Commission. 2006. Available at: http://www.microrisk.com/publish/cat_index_11.shtml. Accessed 24 Dec 2014.
21. Dornas FP, Silva LC, de Almeida GM, Campos RK, Boratto PV, Franco-Luiz AP, et al. Acanthamoeba polyphaga mimivirus stability in environmental and clinical substrates: implications for virus detection and isolation. PLoS One. 2014;9(2):e87811.
22. Thomas V, McDonnel G, Denyer SP, Maillard J-Y. Freeliving amoebae and their intracellular pathogenic microorganisms: risk for water quality. FEMS Microbiol Rev.2010;34:231–59. doi: 10.1111/j.1574-6976.2009.00190.x.
23. Sales-Ortells H, Medema G. Screening-level microbial risk assessment of urban water locations: a tool for prioritization. Environ Sci Technol. 2014;48(16):9780–9. doi: 10.1021/es5020407.
24. Gourabathini P, Brandl MT, Redding KS, Gunderson JH, Berk SG. Interactions between food-borne pathogens and protozoa isolated from lettuce and spinach. Appl Environ Microbiol. 2008;74(8):2518–25. doi: 10.1128/AEM.02709-07.
25. Juhna T, Birzniece D, Larsson S, Zulenkovs D, Sharipo A, Azevedo NF, et al. Detection of Escherichia coli in biofilms from pipe samples and coupons in drinking water distribution networks. Appl Environ Microbiol. 2007;73(22):7456–64.
26. Soller JA, Bartrand T, Ashbolt NJ, Ravenscroft J, Wade TJ. Estimating the primary etiologic agents in recreational freshwaters impacted by human sources of faecal contamination. Water Res. 2010;44(16):4736–47. doi: 10.1016/j.watres.2010.07.064.
27. Smeets PW, Rietveld LC, van Dijk JC, Medema GJ. Practical applications of quantitative microbial risk assessment (QMRA) for water safety plans. Water Sci Technol.2010;61(6):1561–8. doi: 10.2166/wst.2010.839.
28. Borchardt MA, Spencer SK, Kieke BA, Lambertini E, Loge FJ. Viruses in nondisinfected drinking water from municipal wells and community incidence of acute gastrointestinal illness. Environ Health Perspect. 2012;120(9):1272–9. doi: 10.1289/ehp.1104499.
29. Schijven JF, Teunis PF, Rutjes SA, Bouwknegt M, Husman AM. QMRAspot: a tool for quantitative microbial risk assessment from surface water to potable water. Water Res. 2011;45(17):5564–76. doi: 10.1016/j.watres.2011.08.024.
30. Liu G, Bakker GL, Li S, Vreeburg JH, Verberk JQ, Medema GJ, et al. Pyrosequencing reveals bacterial communities in unchlorinated drinking water distribution system: an integral study of bulk water, suspended solids, loose deposits, and pipe wall biofilm. Environ Sci Technol. 2014;48(10):5467–76. doi: 10.1021/es5009467.
31. Szabo J, Minamyer S. Decontamination of biological agents from drinking water infrastructure: a literature review and summary. Environ Int. 2014;72:124–8. doi: 10.1016/j.envint.2014.01.031.
32. Loret JF, Greub G. Free-living amoebae: biological by-passes in water treatment. Int J Hyg Environ Health. 2010;213(3):167–75. doi: 10.1016/j.ijheh.2010.03.004.
33. Storey MV, Ashbolt NJ. Enteric virions and microbial biofilms: a secondary source of public health concern. Water Sci Technol. 2003;48(3):97–104.

34. Bichai F, Dullemont Y, Hijnen W, Barbeau B. Predation and transport of persistent pathogens in GAC and slow sand filters: a threat to drinking water safety? Water Res.2014;64:296–308. doi: 10.1016/j.watres.2014.07.005.
35. Scheid P, Schwarzenberger R. Acanthamoeba spp. as vehicle and reservoir of adenoviruses. Parasitol Res. 2012;111(1):479–85.
36. Ahmed SM, Hall AJ, Robinson AE, Verhoef L, Premkumar P, Parashar UD, et al. Global prevalence of norovirus in cases of gastroenteritis: a systematic review and meta-analysis. Lancet Infect Dis. 2014;14(8):725–30. doi: 10.1016/S1473-3099(14)70767-4.
37. Matthews JE, Dickey BW, Miller RD, Felzer JR, Dawson BP, Lee AS, et al. The epidemiology of published norovirus outbreaks: a review of risk factors associated with attack rate and genogroup. Epidemiol Infect. 2012;140(7):1161–72. doi: 10.1017/S0950268812000234.
38. Tan M, Jiang X. Histo-blood group antigens: a common niche for norovirus and rotavirus. Expert Rev Mol Med. 2014;16:e5. doi: 10.1017/erm.2014.2.
39. Jones MK, Watanabe M, Zhu S, Graves CL, Keyes LR, Grau KR, et al. Enteric bacteria promote human and mouse norovirus infection of B cells. Science.2014;346(6210):755–9. doi: 10.1126/science.1257147.
40. Li D, De Keuckelaere A, Uyttendaele M. Application of long-range and binding reverse transcription-quantitative PCR to indicate the viral integrities of noroviruses. Appl Environ Microbiol. 2014;80(20):6473–9. doi: 10.1128/AEM.02092-14.
41. Cromeans T, Park GW, Costantini V, Lee D, Wang Q, Farkas T, et al. Comprehensive comparison of cultivable norovirus surrogates in response to different inactivation and disinfection treatments. Appl Environ Microbiol. 2014;80(18):5743–51. doi: 10.1128/AEM.01532-14.
42. Messner MJ, Berger P, Nappier SP. Fractional poisson: a simple dose–response model for human norovirus. Risk Anal. 2014;34(10):1820–9. doi: 10.1111/risa.12207.
43. Teunis PF, Xu M, Fleming KK, Yang J, Moe CL, LeChevallier MW. Enteric virus infection risk from intrusion of sewage into a drinking water distribution network.Environ Sci Technol. 2010;44(22):8561–6. doi: 10.1021/es101266k.
44. Grøndahl-Rosado RC, Yarovitsyna E, Trettenes E, Myrmel M, Robertson LJ. A one year study on the concentrations of Norovirus and enteric adenoviruses in wastewater and a surface drinking water Source in Norway. Food Environ Virol. 2014. doi:10.1007/s12560-014-9161-5.
45. Hata A, Katayama H, Kojima K, Sano S, Kasuga I, Kitajima M, et al. Effects of rainfall events on the occurrence and detection efficiency of viruses in river water impacted by combined sewer overflows. Sci Total Environ. 2014;468–469:757–63. doi: 10.1016/j.scitotenv.2013.08.093.
46. Sedmak G, Bina D, Macdonald J, Couillard L. Nine-year study of the occurrence of culturable viruses in source water for two drinking water treatment plants and the influent and effluent of a wastewater treatment plant in Milwaukee, Wisconsin (August 1994 through July 2003) Appl Environ Microbiol. 2005;71(2):1042–50. doi: 10.1128/AEM.71.2.1042-1050.2005.
47. Rodriguez RA, Bounty S, Beck S, Chan C, McGuire C, Linden KG. Photoreactivation of bacteriophages after UV disinfection: Role of genome structure and impacts of UV source. Water Res. 2014;55C:143–9. doi: 10.1016/j.watres.2014.01.065.
48. Ashbolt NJ, Grabow WOK, Snozzi M. Indicators of microbial water quality. In: Fewtrell L, Bartram J, editors. Water quality: guidelines, standards and health. Risk assessment and risk management for water-related infectious disease. London: IWA Publishing; 2001. pp. 289–315.
49. Ashbolt NJ, Schoen ME, Soller JA, Roser DJ. Predicting pathogen risks to aid beach management: the real value of quantitative microbial risk assessment (QMRA) Water Res. 2010;44(16):4692–703. doi: 10.1016/j.watres.2010.06.048.
50. Coleman BL, Louie M, Salvadori MI, McEwen SA, Neumann N, Sibley K, et al. Contamination of Canadian private drinking water sources with antimicrobial resistantEscherichia coli. Water Res. 2013;47(9):3026–36.
51. Ashbolt NJ, Amézquita A, Backhaus T, Borriello SP, Brandt K, Collignon P, et al. Human health risk assessment (HHRA) for environmental development and transfer of antibiotic resistance. Environ Health Perspect. 2013;121(9):993–1001.
52. Harrison EM, Weinert LA, Holden MT, Welch JJ, Wilson K, Morgan FJ, et al. A shared population of epidemic methicillin-resistant Staphylococcus aureus 15 circulates in humans and companion animals. AmBio. 2014;5(3):e00985–13.
53. Angulo FJ, Tippen S, Sharp DJ, Payne BJ, Collier C, Hill JE, et al. A community waterborne outbreak of Salmonellosis and the effectiveness of a boil water order. Am J Public Health. 1997;87(4):580–4. doi: 10.2105/AJPH.87.4.580.
54. Hrudey SE, Huck PM, Payment P, Gillham RW, Hrudey EJ. Walkerton: Lessons learned in comparison with waterborne outbreaks in the developed world. J Environ Eng Sci. 2002;1(6):397–407. doi: 10.1139/s02-031.
55. Pitkänen T. Review of Campylobacter spp. in drinking and environmental waters. J Microbiol Meth. 2013;95(1):39–47.
56. Sen K, Ashbolt NJ. Environmental microbiology: current technology and water applications. Norfolk: Caister Academic Press; 2011.
57. Haugland RA, Siefring SD, Varma M, Dufour AP, Brenner KP, Wade TJ, et al. Standardization of enterococci density estimates by EPA qPCR methods and comparison of beach action value exceedances in river waters with culture methods. J Microbiol Meth. 2014;105C:59–66. doi: 10.1016/j.mimet.2014.07.007.

58. US Environmental Protection Agency. Recreational water quality criteria. Office of Water Report 820-F-12-058. Washington, DC: US Environmental Protection Agency; 2012. Key summary of recent epidemiologic studies of bathing water gastrointestinal illness from fecally-contaminated recreational waters, and identification of qPCR as the preferred index of health outcome.

59. Pitkänen T, Ryu H, Elk M, Hokajärvi AM, Siponen S, Vepsäläinen A, et al. Detection of fecal bacteria and source tracking identifiers in environmental waters using rRNA-based RT-qPCR and rDNA-based qPCR assays. Environ Sci Technol. 2013;47(23):13611–20. doi: 10.1021/es403489b.

60. Li L, Mendis N, Trigui H, Oliver JD, Faucher SP. The importance of the viable but non-culturable state in human bacterial pathogens. Front Microbiol. 2014;5:258.

61. Bullman S, O'Leary J, Corcoran D, Sleator RD, Lucey B. Molecular-based detection of non-culturable and emerging campylobacteria in patients presenting with gastroenteritis. Epidemiol Infect. 2012;140(4):684–8. doi: 10.1017/S0950268811000859.

62. Liu Y, Wang C, Tyrrell G, Li XF. Production of Shiga-like toxins in viable but nonculturable Escherichia coli O157:H7. Water Res. 2010;44(3):711–8.

63. Vaerewijck MJM, Baré J, Lambrecht E, Sabbe K, Houf K. Interactions of foodborne pathogens with free-living protozoa: potential consequences for food safety. Compr Rev Food Sci Food Saf. 2014;13(5):924–44. doi: 10.1111/1541-4337.12100.

64. Signoretto C, Burlacchini G, Pruzzo C, Canepari P. Persistence of Enterococcus faecalis in aquatic environments via surface interactions with copepods. Appl Environ Microbiol. 2005;71(5):2756–61.

65. Lewis K. Persister cells, dormancy and infectious disease. Nat Rev Microbiol. 2007;5(1):48–56. doi: 10.1038/nrmicro1557.

66. Jensen JN. Disinfection model based on excess inactivation sites: implications for linear disinfection curves and the Chick-Watson dilution coefficient. Environ Sci Technol.2010;44(21):8162–8. doi: 10.1021/es101818z.

67. Watson CL, Owen RJ, Said B, Lai S, Lee JV, Surman-Lee S, et al. Detection of Helicobacter pylori by PCR but not culture in water and biofilm samples from drinking water distribution systems in England. J Appl Microbiol. 2004;97(4):690–8.

68. Moreno Y, Piqueres P, Alonso JL, Jiménez A, Gonzállez A, Ferrús MA. Survival and viability of Helicobacter pylori after inoculation into chlorinated drinking water. Water Res. 2007;41(15):3490–6.

69. Travis PB, Goodman KJ, O'Rourke KM, Groves FD, Sinha D, Nicholas JS, et al. The association of drinking water quality and sewage disposal with Helicobacter pyloriincidence in infants: the potential role of water-borne transmission. J Water Health. 2010;8(1):192–203.

70. Karanis P, Kourenti C, Smith H. Waterborne transmission of protozoan parasites: a worldwide review of outbreaks and lessons learnt. J Water Health. 2007;5(1):1–38. doi: 10.2166/wh.2006.002.

71. Hunter PR, de Sylor MA, Risebro HL, Nichols GL, Kay D, Hartemann P. Quantitative microbial risk assessment of cryptosporidiosis and giardiasis from very small private water supplies. Risk Anal. 2011;31(2):228–36. doi: 10.1111/j.1539-6924.2010.01499.x.

72. Xiao L. Molecular epidemiology of cryptosporidiosis: an update. Exp Parasitol. 2010;124(1):80–9. doi: 10.1016/j.exppara.2009.03.018.

73. Dowd SE, Gerba CP, Pepper IL. Confirmation of the human-pathogenic microsporidia Enterocytozoon bieneusi, Encephalitozoon intestinalis, and Vittaforma corneae in water. Appl Environ Microbiol. 1998;64(9):3332–5.

74. Yang R, Murphy C, Song Y, Ng-Hublin J, Estcourt A, Hijjawi N. Specific and quantitative detection and identification of Cryptosporidium hominis and C. parvum in clinical and environmental samples. Exp Parasitol. 2013;135(1):142–7.

75. Didier ES, Weiss LM. Microsporidiosis: not just in AIDS patients. Curr Opin Infect Dis. 2011;24(5):490–5. doi: 10.1097/QCO.0b013e32834aa152.

76. Brusseau ML, Oleen JK, Santamaria J, Cheng L, Orosz-Coghlan P, Chetochine AS, et al. Transport of microsporidium Encephalitozoon intestinales spores in sandy porous media. Water Res. 2005;39(15):3636–42.

77. Hijnen WAM, Medema GJ. Elimination of micro-organisms by drinking water treatment processes: a review. Third. London: IWA Publishing; 2010. p. 160.

78. Ma J, Li P, Zhao X, Xu H, Wu W, Wang Y, et al. Occurrence and molecular characterization of Cryptosporidium spp. and Enterocytozoon bieneusi in dairy cattle, beef cattle and water buffaloes in China. Vet Parasitol. 2014. doi:10.1016/j.vetpar.2014.10.011.

79. Roser DJ, van den Akker B, Boase S, Haas CN, Ashbolt NJ, Rice SA. Dose–response algorithms for water-borne Pseudomonas aeruginosa folliculitis. Epidemiol Infect. 2014. doi:10.1017/S0950268814002532:1%E2%80%9314.

80. Edelstein PH. Legionnaires' disease: history and clinical findings. In: Heuner K, Swanson MS, editors. Legionella: molecular microbiology. Norfolk: Caister Academic Press; 2008.

81. Osterholm MT, Chin TD, Osborne DO, Dull HB, Dean AG, Fraser DW, et al. A 1957 outbreak of Legionnaires› disease associated with a meat packing plant. Am J Epidemiol. 1983;117(1):60–7.

82. Gomez-Valero L, Rusniok C, Rolando M, Neou M, Dervins-Ravault D, Demirtas J, et al. Comparative analyses of Legionella species identifies genetic features of strains causing Legionnaires' disease. Genome Biol. 2014;15(11):505.

83. Devos L, Boon N, Verstraete W. Legionella pneumophila in the environment: occurrence of a fastidious bacterium in oligotrophic conditions. Rev Environ Sci Biotechnol. 2005;4(1):61–74.

84. Garduño RA, Garduño E, Hiltz M, Hoffman PS. Intracellular growth of Legionella pneumophila gives

rise to a differentiated form dissimilar to stationary-phase forms. Infect Immun. 2002;70(11):6273–83.

85. Rowbotham TJ. Isolation of Legionella pneumophila from clinical specimens via amoebae, and the interaction of those and other isolates with amoebae. J Clin Pathol. 1983;36(9):978–86.

86. Price CT, Richards AM, Von Dwingelo JE, Samara HA, Abu Kwaik Y. Amoeba host-Legionella synchronization of amino acid auxotrophy and its role in bacterial adaptation and pathogenic evolution. Environ Microbiol. 2014;16(2):350–8.

87. Luck PC, Jacobs E, Roske I, Schroter-Bobsin U, Dumke R, Gronow S. Legionella dresdenensis sp. nov., isolated from river water. Int J Syst Evol Microbiol. 2010;60(Pt 11):2557–62.

88. Wang H, Masters S, Edwards MA, Falkinham JO, 3rd, Pruden A. Effect of disinfectant, water age, and pipe materials on bacterial and eukaryotic community structure in drinking water biofilm. Environ Sci Technol. 2014;48(3):1426–35. doi: 10.1021/es402636u.

89. Falkinham 3rd JO. Hospital water filters as a source of Mycobacterium avium complex. J Med Microbiol. 2010;59(Pt 10):1198–202.

90. Cateau E, Delafont V, Hechard Y, Rodier MH. Free-living amoebae: what part do they play in healthcare-associated infections? J Hosp Infect. 2014;87(3):131–40. doi: 10.1016/j.jhin.2014.05.001.

91. Delafont V, Brouke A, Bouchon D, Moulin L, Hechard Y. Microbiome of free-living amoebae isolated from drinking water. Water Res. 2013;47(19):6958–65. doi: 10.1016/j.watres.2013.07.047.

92. Yoder JS, Straif-Bourgeois S, Roy SL, Moore TA, Visvesvara GS, Ratard RC, et al. Primary amebic meningoencephalitis deaths associated with sinus irrigation using contaminated tap water. Clin Infect Dis. 2012;55(9):e79–85. doi: 10.1093/cid/cis626.

93. Trolio R, Bath A, Gordon C, Walker R, Wyber A. Operational management of Naegleria spp in drinking water supplies in Western Australia. Water Sci Technol Water Supply. 2008;8(2):207–15.

94. Abrahao JS, Dornas FP, Silva LC, Almeida GM, Boratto PV, Colson P, et al. Acanthamoeba polyphaga mimivirus and other giant viruses: an open field to outstanding discoveries. Virol J. 2014;11(1):120.

95. Boughalmi M, Saadi H, Pagnier S, Colson P, Fournous G, Raoult D, et al. High-throughput isolation of giant viruses of the Mimiviridae and Marseilleviridae families in the Tunisian environment. Environ Microbiol. 2013;15(7):2000–7.

96. Schoen ME, Ashbolt NJ. An in-premise model for Legionella exposure during showering events. Water Res. 2011;45(18):5826–36.

97. Codony F, Pérez LM, Adrados B, Agustí G, Fittipaldi M, Morató J. Amoeba-related health risk in drinking water systems: could monitoring of amoebae be a complementary approach to current quality control strategies? Future Microbiol. 2012;7(1):25–31. doi: 10.2217/fmb.11.139.

98. Dupuy M, Mazoua S, Berne F, Bodet C, Garrec N, Herbelin P, et al. Efficiency of water disinfectants against Legionella pneumophila and Acanthamoeba. Water Res. 2011;45(3):1087–94.

99. Lu J, Buse H, Gomez-Alvarez V, Struewing I, Santo Domingo J, Ashbolt NJ. Impact of drinking water conditions and copper materials on downstream biofilm microbial communities and Legionella pneumophila colonization. J Appl Microbiol. 2014;117(3):905–18. doi: 10.1111/jam.12578.

100. 100.•. Buse HY, Lu J, Lu X, Mou X, Ashbolt NJ. Microbial diversities (16S and 18S rRNA gene pyrosequencing) and environmental pathogens within drinking water biofilms grown on the common premise plumbing materials unplasticized polyvinylchloride and copper. FEMS Microbiol Ecol. 2014;88(2):280–95. doi: 10.1111/1574-6941.12294.

101. Bukh AS, Roslev P. Mycobacterium avium complex in day care hot water systems, and persistence of live cells and DNA in hot water pipes. Curr Microbiol. 2014;68(4):428–39.

102. Wang H, Edwards MA, Falkinham JO, 3rd, Pruden A. Probiotic approach to pathogen control in premise plumbing systems? A review. Environ Sci Technol.2013;47(18):10117–28.

103. Anaissie EJ, Penzak SR, Dignani MC. The hospital water supply as a source of nosocomial infections: a plea for action. Arch Intern Med. 2002;162(13):1483–92. doi: 10.1001/archinte.162.13.1483.

104. Heinrichs G, Hubner I, Schmidt CK, de Hoog GS, Haase G. Analysis of black fungal biofilms occurring at domestic water taps. I: compositional analysis using Tag-Encoded FLX Amplicon Pyrosequencing. Mycopathologia. 2013;175(5–6):387–97. doi: 10.1007/s11046-013-9618-3.

105. Matos T, de Hoog GS, de Boer AG, de Crom I, Haase G. High prevalence of the neurotrope Exophiala dermatitidis and related oligotrophic black yeasts in sauna facilities. Mycoses. 2002;45(9–10):373–7.

CHAPTER 6 IDENTIFICATION OF OPPORTUNISTIC PATHOGENIC BACTERIA IN DRINKING WATER SAMPLES OF DIFFERENT RURAL HEALTH CENTERS AND THEIR CLINICAL IMPACTS ON HUMANS

Pavan Kumar Pindi, P. Raghuveer Yadav, and A. Shiva Shanker

INTRODUCTION

Human population growth wields several and various pressures on the quality and the quantity of drinking and fresh water resources and on the access to them. Safe drinking water remains inaccessible to several millions people in the globe. Safe drinking water, thus, reduces the load of waterborne diseases. Contamination of drinking water due to natural and manmade contaminants is frequently reported in developing countries where mainstream of the inhabitants survive in countryside and uptown areas with meager hygiene and waste clearance practices. In developing countries, poor water quality is the most important risk of child mortalities which are mainly through infectious diarrhea. In India, about 10% of the countryside and city populations do not have access to usual safe drinking water and several others are threatened [1].

Potable or drinking water is defined as having a satisfactory quality in terms of its physical, chemical, and bacteriological parameters so that it can be securely used for drinking and cooking. The most common and widespread health risks associated with drinking water in developing countries are of biological origin. The WHO estimates that about 1100 million people globally drink unsafe water, and the greater part of diarrheal disease in the globe (88%) is attributable to insecure water, sanitation, and hygiene [2]. Ten major waterborne diseases are responsible for over 28 billion disease episodes annually in developing countries. The quality of drinking water might be ascertained by its microbiological examination. The maximum threat from microbes (coliform) in water is related to consumption of drinking water, that is, contaminated with human or animal excreta. Coliform bacteria include members of the family Enterobacteriaceae, for example, Escherichia coli, Enterobacter aerogenes, Salmonella, and Klebsiella in water that are accountable for a variety of diseases like cholera, typhoid, dysenteries, bacillary dysentery, and so forth in human and livestock [3].

Recently, Suthar et al. [4, 5] reported a wide range of pathogenic bacteria in potable water samples from some rural habitations of Northern Rajasthan, India. A total of ten bacterial species, E. coli, Pseudomonas aeruginosa, E. aerogenes, Klebsiella sp., Proteus vulgaris, Alcaligenes faecalis, Bacillus cereus, Staphylococcus aureus, Streptococcus lactis, and Micrococcus luteus, were isolated and identified from the potable water samples from this region [4] M. luteus, S. lactis, Klebsiella sp., and E. coli were the dominant microflora (recorded from 73.1% of the villages/towns) in the water samples. The presence of coliforms shows the danger of fecal

Citation: Pindi PK, Yadav R, Shanker S (2013) Identification of Opportunistic Pathogenic Bacteria in Drinking Water Samples of Different Rural Health Centers and Their Clinical Impacts on Humans. BioMed Research International Volume 2013, Article ID 348250, 10 pages doi: http://dx.doi.org/10.1155/2013/348250

pollution and a consequent hazard of contracting a disease through pathogenic organisms. Nonetheless, the disease-causing organisms (pathogens) mostly transmitted via drinking water are predominantly of fecal origin.

The results of the recent study on the detection of potentially pathogenic bacteria in the drinking water distribution system of a hospital in Hungary by Felföldi et al. [5] had shown the presence of Legionella, Pseudomonas aeruginosa, and also several other opportunistic pathogenic bacteria, such as Escherichia albertii, Acinetobacter lwoffii and Corynebacterium tuberculostearicum, emphasizing that drinking water systems, particularly those with stagnant water sections, could be the source of nosocomial infections.

Hospitals drinking (potable) water systems are the most important and controllable as well as the most overlooked source of nosocomial pathogens. Conventional culture-based microbiological water quality monitoring techniques take a long time (several days), and usually a small volume of water is sampled (typically 100 mL), which gives rise to inadequate detection limits with regard to drinking water safety. Furthermore, the presence of some important waterborne pathogens (such as Pseudomonas aeruginosa or legionellae) shows no correlation with conventional indicator organism counts. Water-related pathogens can also find niches in water systems (i.e., an association with biofilms or free-living amoebae), rendering their observation with conventional techniques more difficult. Molecular techniques provide new and rapid facilities for the detection of pathogens involved in nosocomial infections. Five representative end-points (taps) in the drinking water system of a hospital (Budapest, Hungary) were sampled in October 2005.

MATERIALS AND METHODS

Source of Samples

Thirteen drinking water samples were collected from different government hospitals, Mahabubnagar District, Andhra Pradesh, India, on 22nd and 23rd of November, 2009. Fifty five strains were isolated from these 13 drinking water sources on nutrient agar medium at 37°C. For isolation of bacteria, 100 µL of the water sample was plated on nutrient agar medium and incubated at 37°C for 2 days. Based on the different colony morphology from each sample, a total of 55 strains were selected and characterized in the present study.

S rRNA Gene Sequencing and Phylogenetic Analysis

For 16S rRNA gene sequencing, DNA was prepared using the Mo Bio microbial DNA isolation kit (Mo Bio Laboratories Inc., Solano Beach, CA, USA) and sequenced as described previously [6]. The resultant almost complete sequence of the 16S rRNA gene contained 1502 nucleotides. The 16S rRNA gene sequence of the isolate was subjected to BLAST sequence similarity search [7] and EzTaxon [8] to identify the nearest taxa. The entire related 16S rRNA gene sequences were downloaded from the database (http://www.ncbi.nlm.nih.gov/) aligned using the CLUSTAL_X program [9] and the alignment corrected manually. Phylogenetic tree was constructed using tree-making algorithm and the Maximum likelihood (ML) using the PhyML program [10].

Biochemical Tests

The biochemical tests were performed for all the strains obtained by using the HiMViCTM Biochemical Test Kit (HIMEDIA-KB001) and HiAssorted Biochemical Test Kit (HIMEDIA-KB002).

RESULTS AND DISCUSSION

Characterization of the Bacterial Strains from Drinking Water

All these isolates were from different drinking water sources, and all are mesophilic and could grow in the temperature range of 20 to 40°C with optimum growth temperature of 37°C. All the strains could grow without NaCl in the medium, and the tolerance to NaCl varied from 1 to 2%. All strains could grow either in the pH range of 6–8.

Identification of Isolated Strains from Drinking Water

A total of 55 bacterial strains were recovered from the 13 drinking water sources (Table 1). Taxonomic analysis of all the 55 strains isolated from water samples indicated that 43 strains were Gram-negative and 12 Gram-positive. The nearest phylogenetic neighbor of all 55 isolated strains was identified following BLAST analysis of the 16S rRNA gene sequence. BLAST analysis indicated that twenty one strains belonged to the genus Acinetobacter, nine strains belonged to the genus Bacillus, eight strains belonged to the genus Pseudomonas, seven strains to the genus Aeromonas, two strains each belonged to the genera Methylobacterium and Pantoea, and one strain belonged to the genera Cupriavidus, Delftia, Exiguobacterium, Kocuria, and Lysinibacillus, respectively (Table 2). A total 27 different taxa were identified based on the BLAST analysis out of 55 strains. The most common group was similar to Acinetobacter junii (21.8%) and Acinetobacter calcoaceticus (10.9%) which were shared by 7 and 5 water samples, respectively. The phylogenetic trees constructed to determine their affiliations are shown in Figures 1 and 2. Out of 12 Gram-positive groups, 11 belonged to the phylum Firmicutes and 1 belonged to the phylum Actinobacteria (Table 2 and Figure 2). All 43 Gram-negative groups were belonged to the phylum Proteobacteria. Among the Proteobacteria, the Alphaproteobacteria, Betaproteobacteria, and Gammaproteobacteria were represented by 2, 2, and 39 groups, respectively (Table 2 and Figure 2). Phylogenetic analysis based on 16S rRNA gene sequences indicated that the 21 Acinetobacter strains belonged to 4 groups, 9 Bacillus strains belonged to 7 groups, 8 Pseudomonas strains belonged to 3 groups, 7 Aeromonas strains belonged to 5 groups, and 2 Methylobacterium strains formed two distinct groups (Figures 1 and 2). The affiliation of the strains to the nearest phylogenetic neighbor and the percentage of 16S rRNA gene sequence similarities are indicated in Table 2. Many of the strains are closely related to one another and exhibit 98.4 to 100% sequence similarity at the 16S rRNA gene sequence level like KL-1 and JD-2 which are closely related to one another (100%) and are affiliated to Acinetobacter calcoaceticus DSM 30006^{T} (98.5%).

Table 1: Bacterial abundance from the drinking water samples collected from government hospitals, Mahabubnagar.

Serial number	Sample number	Place of sample collection	Number of isolates	Isolated strains
1	AL	Alampur	3	AL1, AL3, AL4
2	AP	Achampet	3	AP1, AP2, AP4
3	D	Shadnagar	4	D12, D22, D24, D25
4	GD	Gadwal	4	GD2, GD4–D6
5	JD	Jadcherla	4	JD1–JD4
6	JU	Jurala	4	JU1–JU4
7	KL	Kalwakurthy	5	KL1–KL5
8	MB	Mahabubnagar-1	4	MB1–MB4
9	MK	Makthal	3	MK2–MK4
10	NG	Nagar Kurnool	2	NG1, NG2
11	NR	Narayanpet	3	NR2–NR4
12	R	Mahabubnagar-2	12	R1, R2, R5–R8, R21–R23, R26, R30, R31
13	WN	Wanaparthy	4	WN1, WN3–WN5

Table 2: Identification of the 55 bacterial strains isolated from the drinking water samples collected from different government hospitals, Mahabubnagar, based on BLAST analysis of the 16S rRNA gene sequences.

Sl no.	Strain no.	Nearest phylogenetic neighbor	16S rRNA gene sequence similarity %
		Gram-negative bacterial strains	
		Proteobacteria	
		Alphaproteobacteria	
1	D24	*Methylobacterium chloromethanicum* CM4^T CP001298	99.6
2	AL1	*Methylobacterium zatmanii* DSM 5688^T AB175647	99.7
		Betaproteobacteria	
3	AP1	*Cupriavidus pauculus* LMG 3413^T AF085226	100.0
4	AL3	*Delftia lacustris* DSM 21246^T EU888308	100.0
		Gammaproteobacteria	
5	JD2	*Acinetobacter calcoaceticus* DSM 30006^T X81661	98.5
6	KL1	*Acinetobacter calcoaceticus* DSM 30006^T X81661	98.5
7	KL5	*Acinetobacter calcoaceticus* DSM 30006^T X81661	98.5
8	MB1	*Acinetobacter calcoaceticus* DSM 30006^T X81661	98.4
9	MK4	*Acinetobacter calcoaceticus* DSM 30006^T X81661	98.5
10	NR4	*Acinetobacter calcoaceticus* DSM 30006^T X81661	98.4
11	GD5	*Acinetobacter johnsonii* DSM 6963^T X81663	99.9
12	NG1	*Acinetobacter johnsonii* DSM 6963^T X81663	99.5
13	AP2	*Acinetobacter junii* LMG 998^T AM410704	100.0

14	AP4	*Acinetobacter junii* LMG 998^T AM410704	100.0
15	GD2	*Acinetobacter junii* LMG 998^T AM410704	100.0
16	JD4	*Acinetobacter junii* LMG 998^T AM410704	100.0
17	KL3	*Acinetobacter junii* LMG 998^T AM410704	100.0
18	KL4	*Acinetobacter junii* LMG 998^T AM410704	100.0
19	MB3	*Acinetobacter junii* LMG 998^T AM410704	100.0
20	R1	*Acinetobacter junii* LMG 998^T AM410704	99.9
21	WN1	*Acinetobacter junii* LMG 998^T AM410704	99.9
22	WN3	*Acinetobacter junii* LMG 998^T AM410704	100.0
23	WN4	*Acinetobacter junii* LMG 998^T AM410704	100.0
24	WN5	*Acinetobacter junii* LMG 998^T AM410704	99.9
25	MB4	*Acinetobacter ursingii* LUH 3792^T AJ275038	99.6
26	JD1	*Aeromonas aquariorum* MDC47^T EU085557	100.0
27	R31	*Aeromonas aquariorum* MDC47^T EU085557	100.0
28	R30	*Aeromonas aquariorum* MDC47^T EU085557	99.9
29	GD4	*Aeromonas hydrophila* subsp. *hydrophila* ATCC 7966^T CP000462	100.0
30	JD3	*Aeromonas ichthiosmia* DSM 6393^T X71120	99.8
31	MK2	*Aeromonas ichthiosmia* DSM 6393^T X71120	99.8
32	R26	*Aeromonas punctata* subsp. *punctata* NCIMB 13016^T X60408	99.2
33	NG2	*Citrobacter freundii* DSM 30039^T AJ233408	98.3
34	R21	*Pantoea anthophila* LMG 2558^T EF688010	99.8
35	R22	*Pantoea anthophila* LMG 2558^T EF688010	99.9
36	NR2	*Pseudomonas aeruginosa* LMG 1242^T Z76651	99.9
37	R6	*Pseudomonas alcaligenes* LMG 1224^T Z76653	100.0
38	R7	*Pseudomonas alcaligenes* LMG 1224^T Z76653	99.8
39	R23	*Pseudomonas alcaligenes* LMG 1224^T Z76653	99.8
40	D22	*Pseudomonas otitidis* MCC10330^T AY953147	99.8
41	GD6	*Pseudomonas otitidis* MCC10330^T AY953147	99.8
42	MB2	*Pseudomonas otitidis* MCC10330^T AY953147	99.6
43	MK3	*Pseudomonas otitidis* MCC10330^T AY953147	99.7
		Gram-positive bacterial strains	
		Actinobacteria	
44	R8	*Kocuria rhizophila* DSM 11926^T Y16264	100.0
		Firmicutes	
		Bacilli	
45	D12	*Bacillus anthracis* ATCC 14578^T AB190217	100.0
46	KL2	*Bacillus aryabhattai* B8W22^T EF114313	98.7
47	NR3	*Bacillus aryabhattai* B8W22^T EF114313	99.7
48	R5	*Bacillus cereus* ATCC 14579^T AE016877	100.0
49	JU1	*Bacillus infantis* SMC 4352-1^T AY904032	99.5
50	JU3	*Bacillus licheniformis* ATCC 14580^T CP000002	99.9
51	JU4	*Bacillus niabensis* 4T19^T AY998119	98.9
52	AL4	*Bacillus stratosphericus* 41KF2a^T AJ831841	99.9
53	R2	*Bacillus stratosphericus* 41KF2a^T AJ831841	100.0
54	D25	*Exiguobacterium indicum HHS31* ^{T} AJ846291	99.9
55	JU2	*Lysinibacillus fusiformis* NBRC 15717^T AB271743	99.8

The accession numbers of the 55 strains are GU566304 and GU566358.

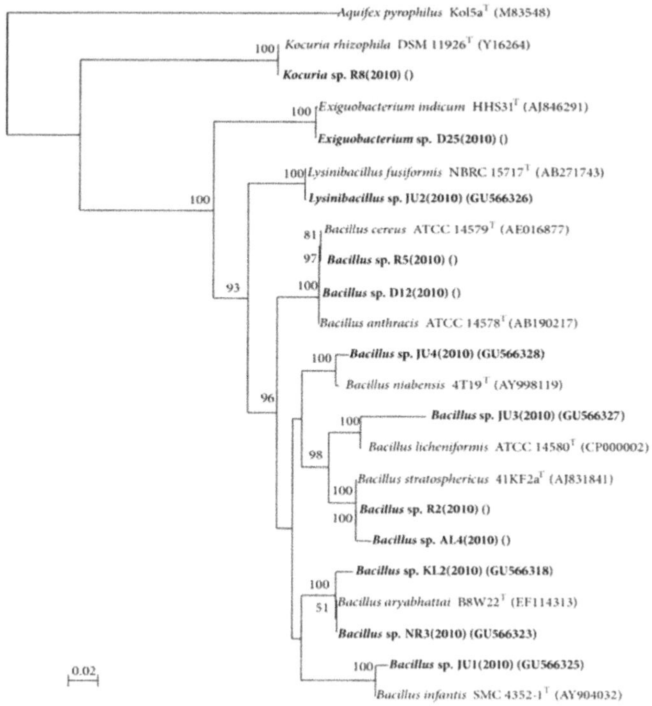

Figure 1: The affiliation of the strains to the nearest phylogenetic neighbor and the percentage of 16S rRNA gene sequence similarities.

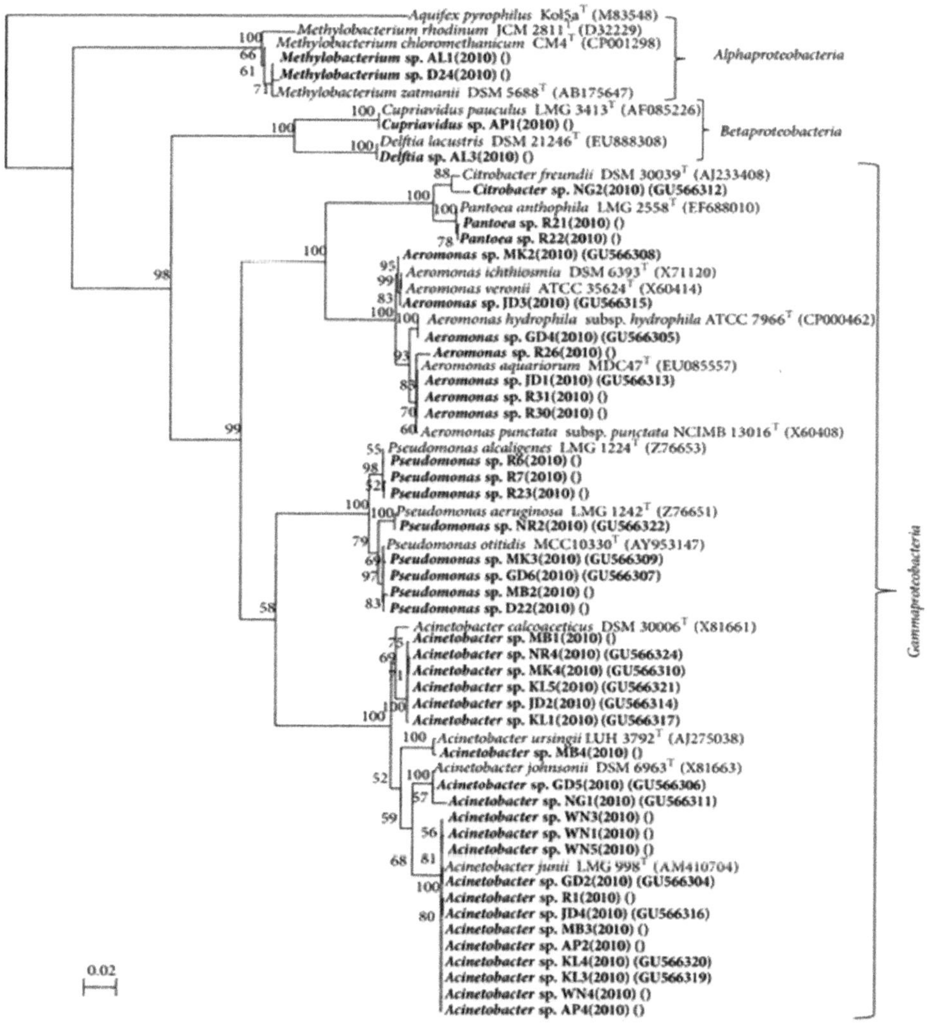

Figure 2: The affiliation of the strains to the nearest phylogenetic neighbor and the percentage of 16S rRNA gene sequence similarities.

Biochemical Assay

Biochemical analysis of the genus Bacillus gave positive results for Indole, Citrate, Oxidase, and Urease and positive results for Methyl red Voges-Proskaeur, Catalase, and Gelatinase tests. Genus Methylobacterium gave positive results for Citrate, Catalase, and Urease and negative results for Indole, Methyl Red, Voges-Proskaeur, Oxidase, and Gelatinase. Genus Cupriavidus gave positive results for Citrate, Oxidase, Catalase, and Urease and negative results for Indole, Methyl red, Voges-Proskaeur, and Gelatinase. Delftia gave positive results for Citrate and Urease and negative ones for Indole, Methyl red, Voges-Proskaeur, Oxidase, Catalase, and Gelatinase. Genus Acinetobacter gave positive results for Citrate and Gelatinas and negative results for Indole, Methyl red, Voges-Proskaeur, Oxidase, Catalase, and Urease (A. ursingii gave positive results even for Citrate and Oxidase and Catalase). Genus Aeromonas (Ar. aquariorum gave positive results for all except Catalase and Gelatinase, whereas Ar. hydrophila, Ar. Ichthiosmia, and Ar. punctata gave positive results for all tests except Methyl red). Genus Citrobacter gave positive results for Methyl red, Oxidase, Catalase, and Urease and negative results for Indole, VP, Citrate, and Gelatinase. Genus Pantoea gave positive results for VP, Citrate, and Catalase and negative results for Indole, Methyl red, Oxidase, Urease, and Gelatinase. Genus Pseudomonas gave positive results for Citrate, Oxidase, Catalase, and Gelatinase and negative ones for Indole, Methyl red, VP, and Urease. Kocuria rhizophila gave positive results for Oxidase and Catalase and negative ones for the remaining tests. Exiguobacterium indicum gave positive results for Methyl red, Citrate, and Catalase and negative ones for Indole,VP, Oxidase, Urease, and Gelatinase. Lysinibacillus fusiformis gave positive results for Oxidase and Gelatinase and negative results for the remaining tests (Table 3).

Table 3: Biochemical tests for the strains obtained.

Sl no.	Strain no.	Nearest phylogenetic neighbor	Biochemical tests							
			Ind	MR	VP	Cit	Oxi	Cat	Ure	Gel
1	D24	*Methylobacterium chloromethanicum* CM4^T CP001298	−	−	−	+	−	+	+	−
2	AL1	*Methylobacterium zatmanii* DSM 5688^T AB175647	−	−	−	+	−	−	+	−
3	AP1	*Cupriavidus pauculus* LMG 3413^T AF085226	−	−	−	+	+	+	+	−
4	AL3	*Delftia lacustris* DSM 21246^T EU888308	−	−	−	+	−	−	+	−
5	JD2	*Acinetobacter calcoaceticus* DSM 30006^T X81661	−	−	−	+	−	−	+	+
6	KL1	*Acinetobacter calcoaceticus* DSM 30006^T X81661	−	−	−	+	−	−	−	+
7	KL5	*Acinetobacter calcoaceticus* DSM 30006^T X81661	−	−	−	+	−	−	−	+
8	MB1	*Acinetobacter calcoaceticus* DSM 30006^T X81661	−	−	−	+	−	−	−	+
9	MK4	*Acinetobacter calcoaceticus* DSM 30006^T X81661	−	−	−	+	−	−	−	+
10	NR4	*Acinetobacter calcoaceticus* DSM 30006^T X81661	−	−	−	+	−	−	−	+
11	GD5	*Acinetobacter johnsonii* DSM 6963^T X81663	−	−	−	−	−	−	−	−
12	NG1	*Acinetobacter johnsonii* DSM 6963^T X81663	−	−	−	−	−	−	−	−
13	AP2	*Acinetobacter junii* LMG998^T AM410704	−	−	−	−	−	−	−	−
14	AP4	*Acinetobacter junii* LMG998^T AM410704	−	−	−	−	−	−	−	−
15	GD2	*Acinetobacter junii* LMG998^T AM410704	−	−	−	−	−	−	−	−
16	JD4	*Acinetobacter junii* LMG998^T AM410704	−	−	−	−	−	−	−	−
17	KL3	*Acinetobacter junii* LMG998^T AM410704	−	−	−	−	−	−	−	−
18	KL4	*Acinetobacter junii* LMG998^T AM410704	−	−	−	−	−	−	−	−
19	MB3	*Acinetobacter junii* LMG998^T AM410704	−	−	−	−	−	−	−	−
20	R1	*Acinetobacter junii* LMG998^T AM410704	−	−	−	−	−	−	−	−
21	WN1	*Acinetobacter junii* LMG998^T AM410704	−	−	−	−	−	−	−	−
22	WN3	*Acinetobacter junii* LMG998^T AM410704	−	−	−	−	−	−	−	−
23	WN4	*Acinetobacter junii* LMG998^T AM410704	−	−	−	−	−	−	−	−
24	WN5	*Acinetobacter junii* LMG998^T AM410704	−	−	−	−	−	−	−	−
25	MB4	*Acinetobacter ursingii* LUH 3792^T AJ275038	−	−	−	+	+	+	−	−
26	JD1	*Aeromonas aquariorum* MDC47^T EU085557	+	+	+	+	+	−	+	−
27	R31	*Aeromonas aquariorum* MDC47^T EU085557	+	+	+	+	+	−	+	−

28	R30	*Aeromonas aquariorum* MDC47^T EU085557	+	+	+	+	+	−	+	−
29	GD4	*Aeromonas hydrophila* subsp. *hydrophila* ATCC 7966^T CP000462	+	−	+	+	+	+	+	+
30	JD3	*Aeromonas ichthiosmia* DSM 6393^T X71120	+	−	+	+	+	+	+	+
31	MK2	*Aeromonas ichthiosmia* DSM 6393^T X71120	+	−	+	+	+	+	+	+
32	R26	*Aeromonas punctata* subsp. *punctate* NCIMB 13016^T X60408	+	−	+	+	+	+	+	+
33	NG2	*Citrobacter freundii* DSM 30039^T AJ233408	−	+	−	−	+	+	+	−
34	R21	*Pantoea anthophila* LMG2558^T EF688010	−	−	+	+	−	+	−	−
35	R22	*Pantoea anthophila* LMG2558^T EF688010	−	−	+	+	−	+	−	−
36	NR2	*Pseudomonas aeruginosa* LMG 1242^T Z76651	−	−	−	+	+	+	−	+
37	R6	*Pseudomonas alcaligenes* LMG 1224^T Z76653	−	−	−	+	+	+	−	+
38	R7	*Pseudomonas alcaligenes* LMG 1224^T Z76653	−	−	−	+	+	+	−	+
39	R23	*Pseudomonas alcaligenes* LMG 1224^T Z76653	−	−	−	+	+	+	−	+
40	D22	*Pseudomonas otitidis* MCC10330^T AY953147	−	−	−	+	+	+	−	+
41	GD6	*Pseudomonas otitidis* MCC10330^T AY953147	−	−	−	+	+	+	−	+
42	MB2	*Pseudomonas otitidis* MCC10330^T AY953147	−	−	−	+	+	+	−	+
43	MK3	*Pseudomonas otitidis* MCC10330^T AY953147	−	−	−	+	+	+	−	+
44	R8	*Kocuria rhizophila* DSM 11926^T Y16264	−	−	−	−	+	+	−	−
45	D12	*Bacillus anthracis* ATCC 14578^T AB190217	−	+	+	−	−	+	−	+
46	KL2	*Bacillus aryabhattai* B8W22^T EF114313	−	+	+	−	−	+	−	+
47	NR3	*Bacillus aryabhattai* B8W22^T EF114313	−	+	+	−	−	+	−	+
48	R5	*Bacillus cereus* ATCC 14579^T AE016877	−	+	+	−	−	+	−	+
49	JU1	*Bacillus infantis* SMC4352-1^T AY904032	−	+	+	−	−	+	−	+
50	JU3	*Bacillus licheniformis* ATCC 14580 CP000002	−	+	+	−	−	+	−	+
51	JU4	*Bacillus niabensis* 4T19^T AY998119	−	+	+	−	−	+	−	+
52	AL4	Bacillus stratosphericus 41KF2a^T AJ831841	−	+	+	−	−	+	−	+
53	R2	*Bacillus stratosphericus* 41KF2a^T AJ831841	−	+	+	−	−	+	−	+
54	D25	*Exiguobacterium indicum HHS31* ^{T} AJ846291	−	+	−	+	−	+	−	−
55	JU2	*Lysinibacillus fusiformis* NBRC 15717^T AB271743	−	−	−	−	+	−	−	+

Ind: Indole; MR: Methyl red; VP: Voges Proskeur; Cit: Citrate; Oxi: Oxidase; Cat: Catalase; Ure: Urease; and Gel: Gelatinase.

Coliform is the name of a test adopted in 1914 by the Public Health Service for the Enterobacteriaceae family. It is the commonly used bacterial indicator of sanitary quality of foods and water. Coliform bacteria live in soil or vegetation and in the gastrointestinal tract of animals. Coliform bacteria enter water supplies from the direct disposal of waste into streams or lakes, or from runoff, from wooded areas, pastures, feedlots, septic tanks, and sewage plants into streams or groundwater. Coliform bacteria are not a single type of bacteria, but a group of bacteria that includes many strains, such as E. coli. They are ubiquitous in nature, and many types are harmless. Coliform bacteria belong to the genera Citrobacter, Enterobacter, Escherichia, Hafnia, Klebsiella, Pantoea, and Serratia. Out of 55 isolates only 3 isolates belonged to coliform group which are NG2 which is phylogenetically close to Citrobacter freundii and R21 and R22 which are phylogenetically close to Pantoea anthophila. This shows that the water samples NG and R collected from Nagarkurnool and Mahabubnagar-2 might be contaminated with fecal pollution and a consequent hazard of contracting a disease through pathogenic organisms. Nonetheless, the disease-causing organisms (pathogens) mostly transmitted via drinking water are predominantly of fecal origin. A total of 2 and 12 strains were isolated from the samples NG and R which include 3 coliform bacteria, and out of 14 strains, 7 were likely to be pathogenic which include NG1 which is phylogenetically close to Acinetobacter johnsonii, R1 which is phylogenetically close to Acinetobacter junii, NG2 which is phylogenetically close to Citrobacter freundii, R6, 7, and 23 which are phylogenetically close to Pseudomonas alcaligenes, and R5 which is phylogenetically close to Bacillus cereus.

In the present study, species that belonged to the genera Aeromonas, Methylobacterium, Bacillus, and Pseudomonas were isolated which is supported by the different studies where several novel species which belong to the genera Aeromonas [11], Methylobacterium [12], Bacillus [13], and Pseudomonas [14] were reported from drinking water systems. Aeromonas spp. are ubiquitous bacteria found in diverse aquatic environments worldwide such as bottled water, chlorinated water, well water, and heavily polluted waters. Aeromonas hydrophila survives easily in waters polluted by feces and seems resistant to various disinfectants, insecticides, and chemicals.

Recent studies on drinking water in rural areas in Northern Rajasthan, India, [4] and in drinking water distribution system of a hospital in Hungary, [5] showed the presence of several bacteria in the drinking water which included Acinetobacter lwoffii, Alcaligenes faecalis, Bacillus cereus, Corynebacterium tuberculostearicum, Enterobacter aerogenes, Escherichia coli, Escherichia albertii, Klebsiella sp., Micrococcus luteus, Proteus vulgaris, Pseudomonas aeruginosa, Staphylococcus aureus, and Streptococcus lactis. In the present study, species belonging to the genus Acinetobacter, Bacillus cereus, and Pseudomonas aeruginosa which are opportunistic pathogens were isolated from 9 drinking water samples such as AP, GD, JD, KL, MB, NG, NR, R, and WN out of 13.

Of the 27 groups, only few (GD4, R5, AP1, and NR2) of the phylogenetic neighbors had been isolated earlier from the drinking water sources including Aeromonas hydrophila [4], Bacillus cereus [4], Cupriavidus pauculus [15], and Pseudomonas aeruginosa [4, 5]. More than half (52.7%, 29 strains) of the phylogenetic neighbors which belonged to 12 groups were reported to be pathogenic and were isolated from clinical specimens, and other 15 groups were reported to be isolated from several distinct habitats. Most of the drinking water sources studied for bacterial diversity include bacteria which are phylogenetically close to pathogenic bacteria except one sample AL from which 3 strains AL1, AL3, and AL4 were isolated which belong to the phylogenetic neighbors Methylobacterium zatmanii, Delftia lacustris, and Bacillus stratosphericus, respectively. The isolates which are likely to be pathogenic include strain AP1 which is phylogenetically close to Cupriavidus pauculus which causes human infections sporadically and were isolated earlier from a variety of human clinical sources and also from drinking water sources [15] (Table 4); strains GD5 and NG1 which are phylogenetically close to Acinetobacter johnsonii which causes vascular catheter-related bloodstream infection were isolated from clinical specimens and milk [16] (Table 4); strains AP2, AP4, GD2, JD4, KL3, KL4, MB3, R1, WN1, WN3, WN4, and WN5 which are phylogenetically close to Acinetobacter junii which causes septicemia, meningitis, peritonitis, and so forth were isolated from human clinical specimens like urine [17] (Table 4); strain MB4 which is phylogenetically close to Acinetobacter ursingii which causes bacteremia was isolated from blood [18] (Table 4); strain GD4 which is phylogenetically close to Aeromonas hydrophila subsp. Hydrophila which causes acute diarrheal disease was isolated from humans, animals, fish, and fresh water [11, 19] (Table 4); strain NG2 which is phylogenetically close to Citrobacter freundii which causes opportunistic infections was isolated from canal water [20] (Table 4); strain NR2 which is phylogenetically close to Pseudomonas aeruginosa which causes several infections in immunocompromised individuals was isolated from wide variety habitats including clinical samples and also from drinking water samples [4, 5] (Table 4); strains R6, R7, and R23 which are phylogenetically close to Pseudomonas alcaligenes which causes endocarditis occasionally were isolated from soil, water, and also from clinical samples [21] (Table 4); strains D22, GD6, MB2, and MK3 which are phylogenetically close to Pseudomonas otitidis which causes inflammation of the ear were isolated from the ear [22] (Table 4); strain D12 which is phylogenetically close to Bacillus anthracis which causes anthrax disease was isolated from sheep blood (Table 4); strain R5 which is phylogenetically close to Bacillus cereus which causes food poisoning was isolated from the soil and also food materials [23] (Table 4); and strain JU1 which is phylogenetically close to Bacillus infantis which causes sepsis was isolated from clinical specimens [24] (Table 4).

Table 4: Isolation of source of the type strains associate with disease.

Sl no.	Nearest phylogenetic neighbor	Isolation source	Associated human disease	Reference
1	*Cupriavidus pauculus* LMG 3413^T AF085226	Isolated from a variety of human clinical sources including blood, wounds, sputum, urine, eye, throat, and peritoneal fluid. In addition strains have been isolated from pool water, ground water, and bottled mineral water	Sporadically cause human infections	Vandamme et al., 1999 [15]
2	*Acinetobacter johnsonii* DSM 6963^T X81663	Isolated from clinical specimens and from eviscerated chickens and may cause ropiness in milk. Isolated from duodenum	Vascular catheter-related bloodstream infection	Seifert et al., 1993 [16]
3	*Acinetobacter junii* LMG 998^T AM410704	Isolated from human clinical specimens like urine	Septicemia, community-acquired bacterial meningitis, peritoneal dialysis-related peritonitis, and infections associated with corneal perforation	Hung et al., 2009 [17]
4	*Acinetobacter ursingii* LUH 3792^T AJ275038	Isolated from blood of a hospitalized patient with endocarditis	Bacteremia	Loubinoux et al., 2003 [18]
5	*Aeromonas hydrophila* subsp. *hydrophila* ATCC 7966^T CP000462	Isolated from humans, animals, fish, and fresh water	Acute diarrheal disease	Ljungh et al., 1977 [19]
6	*Citrobacter freundii* DSM 30039^T AJ233408	Isolated from canal water	Opportunistic infections like neonatal meningitis	Badger et al., 1999 [20]

7	*Pseudomonas aeruginosa* LMG 1242^T Z76651	Found in soil, water, skin flora, and most manmade environments throughout the world	Localized infection of eye, ear, skin, urinary, and respiratory. Bone, joint, gastrointestinal tract, central nervous system, and systemic infection with bacteremia. Secondary pneumonia. Endocarditis	
8	*Pseudomonas alcaligenes* LMG 1224^T Z76653	Common soil and water inhabitant that has rarely been proven a human pathogen	Endocarditis	Valenstein et al., 1983 [21]
9	*Pseudomonas otitidis* MCC 10330^T AY953147	Isolated from the ears of patients with acute otitis externa (inflammation of the ear)	Inflammation of the ear	Clark et al., 2006 [22]
10	*Bacillus anthracis* ATCC 14578^T AB190217	Isolated from the blood of sheep suffering from anthrax	Cutaneous anthrax, pulmonary anthrax, and gastrointestinal anthrax	
11	*Bacillus cereus* ATCC 14579^T AE016877	Isolated from soil and food materials	Food poisoning	Todar, 2008 [23]
12	*Bacillus infantis* SMC 4352-1^T AY904032	Isolated from a newborn child with sepsis	Sepsis	Ko et al., 2006 [24]

Out of 27 representative taxa are affiliated have eight representative genera in drinking water except for those affiliated with the genera *Exiguobacterium*, *Delftia*, *Kocuria*, and *Lysinibacillus*.

CONCLUSION

This study is an overview of drinking water quality in the rural and urban parts of Mahabubnagar District, India. Recent investigations on bacterial contaminations of drinking water in the hospitals suggested risks of secondary infection to the patients, staff, and also patient relatives. These observations were just a drinking water quality assessment, and no data on human health scenario in this region were recorded. Further detailed studies on the health issues of the patients and other people using this contaminated drinking water are still required to trace the impact of this water consumption on people in those hospitals. The lack of awareness about good sanitation and personal hygienic practices, among governmental rural and also urban hospitals is an important factor for poor drinking water quality in these sources.

ACKNOWLEDGMENTS

The authors are grateful to Profesor Bagyanarayana, Vice-Chancellor, and Profesor K Venkata Chalam, Registrar, Palamuru University, for their encouragement and support. The authors wish to acknowledge the Department of Science and Technology (DST) for their financial assistance for the Project "Advanced Molecular Methods for rapid detection of coliforms in potable water" DST/TM/WTI/2 K10/265.

REFERENCES

1. S. Suthar, "Contaminated drinking water and rural health perspectives in Rajasthan, India: an overview of recent case studies," Environmental Monitoring and Assessment, vol. 173, pp. 837–849, 2011.
2. WHO, Emerging Issues in Water and Infectious Disease, World Health Organization (WHO), Geneva, Switzerland, 2003.
3. N. J. Ashbolt, "Microbial contamination of drinking water and disease outcomes in developing regions," Toxicology, vol. 198, no. 1–3, pp. 229–238, 2004.
4. S. Suthar, S. Singh, V. Chhimpa et al., "The problem of safe drinking water in Northern Rajasthan, India," in Proceedings of the 16th National Symposium on Environment, Puranik, et al., Ed., pp. 491–496, Department of Environmental Science and Technology, Guru Jambheshwar University of Science and Technology, Hisar, India, 2008.
5. T. Felföldi, Z. Heéger, M. Vargha, and K. Márialigeti, "Detection of potentially pathogenic bacteria in the drinking water distribution system of a hospital in Hungary," Clinical Microbiology and Infection, vol. 16, pp. 89–92, 2010.
6. D. J. Lane, "16S/23S rRNA sequencing," in Nucleic Acid Techniques in Bacterial Systematics, E. Stackebrandt and M. Good Fellow, Eds., pp. 115–175, John Wiley & Sons, Chichester, UK, 1991.
7. S. F. Altschul, W. Gish, W. Miller, E. W. Myers, and D. J. Lipman, "Basic local alignment search tool," Journal of Molecular Biology, vol. 215, no. 3, pp. 403–410, 1990.
8. J. Chun, J. H. Lee, Y. Jung et al., "EzTaxon: a web-based tool for the identification of prokaryotes based on 16S ribosomal RNA gene sequences," International Journal of Systematic and Evolutionary Microbiology, vol. 57, no. 10, pp. 2259–2261, 2007.
9. J. D. Thompson, T. J. Gibson, F. Plewniak, F. Jeanmougin, and D. G. Higgins, "The CLUSTAL X windows interface: flexible strategies for multiple sequence alignment aided by quality analysis tools," Nucleic Acids Research, vol. 25, no. 24, pp. 4876–4882, 1997.
10. S. Guindon and O. Gascuel, "A simple, fast, and accurate algorithm to estimate large phylogenies by maximum

likelihood," Systematic Biology, vol. 52, no. 5, pp. 696–704, 2003.

11. G. Huys, P. Kämpfer, M. J. Albert, I. Kühn, R. Denys, and J. Swings, "Aeromonas hydrophila subsp. dhakensis subsp. nov., isolated from children with diarrhoea in Bangladesh, and extended description of Aeromonas hydrophila subsp. hydrophila (Chester 1901) stanier 1943 (approved lists 1980)," International Journal of Systematic and Evolutionary Microbiology, vol. 52, no. 3, pp. 705–712, 2002.
12. S. W. Lee, H. W. Oh, K. H. Lee, and T. Y. Ahn, "Methylobacterium dankookense sp. nov., isolated from drinking water," Journal of Microbiology, vol. 47, no. 6, pp. 716–720, 2010.
13. T. Zhang, X. Fan, S. Hanada, Y. Kamagata, and H. H. P. Fang, "Bacillus macauensis sp. nov., a long-chain bacterium isolated from a drinking water supply," International Journal of Systematic and Evolutionary Microbiology, vol. 56, no. 2, pp. 349–353, 2006.
14. E. P. Ivanova, N. M. Gorshkova, T. Sawabe et al., "Pseudomonas extremorientalis sp. nov., isolated from a driking water reservoir," International Journal of Systematic and Evolutionary Microbiology, vol. 52, no. 6, pp. 2113–2120, 2002.
15. P. Vandamme, J. Goris, T. Coenye et al., "Assignment of Centers for Disease Control group IVc-2 to the genus Ralstonia as Ralstonia paucula sp. nov.," International Journal of Systematic Bacteriology, vol. 49, no. 2, pp. 663–669, 1999.
16. H. Seifert, A. Strate, A. Schulze, and G. Pulverer, "Vascular catheter-related bloodstream infection due to Acinetobacter johnsonii (formerly Acinetobacter calcoaceticus var. lwoffii): report of 13 cases," Clinical Infectious Diseases, vol. 17, no. 4, pp. 632–636, 1993.
17. Y. T. Hung, Y. T. Lee, L. J. Huang et al., "Clinical characteristics of patients with Acinetobacterjunii infection," Journal of Microbiology, Immunology and Infection, vol. 42, no. 1, pp. 47–53, 2009.
18. J. Loubinoux, L. Mihaila-Amrouche, A. Le Fleche et al., "Bacteremia caused by Acinetobacter ursingii," Journal of Clinical Microbiology, vol. 41, no. 3, pp. 1337–1338, 2003.
19. Ljungh, M. Popoff, and T. Wadstrom, "Aeromonas hydrophila in acute diarrheal disease: detection of enterotoxin and biotyping of strains," Journal of Clinical Microbiology, vol. 6, no. 2, pp. 96–100, 1977.
20. J. L. Badger, M. F. Stins, and K. S. Kim, "Citrobacter freundii invades and replicates in human brain microvascular endothelial cells," Infection and Immunity, vol. 67, no. 8, pp. 4208–4215, 1999.
21. P. Valenstein, G. H. Bardy, C. C. Cox, and P. Zwadyk, "Pseudomonas alcaligenes endocarditis," American Journal of Clinical Pathology, vol. 79, no. 2, pp. 245–247, 1983.
22. L. L. Clark, J. J. Dajcs, C. H. McLean, J. G. Bartell, and D. W. Stroman, "Pseudomonas otitidis sp. nov., isolated from patients with otic infections," International Journal of Systematic and Evolutionary Microbiology, vol. 56, no. 4, pp. 709–714, 2006.
23. K. Todar, "Bacillus cereus food poisoning," in Todar's Online Textbook of Bacteriology, 2008, http://www.textbookofbacteriology.net/B.cereus.html.
24. K. S. Ko, W. S. Oh, M. Y. Lee et al., "Bacillus infantis sp. nov. and Bacillus idriensis sp. nov., isolated from a patient with neonatal sepsis," International Journal of Systematic and Evolutionary Microbiology, vol. 56, no. 11, pp. 2541–2544, 2006.

CHAPTER 7 METALLIC AND MICROBIAL CONTAMINANTS IN DRINKING WATER OF CANKIRI, TURKEY

Emrah Caylak, and Mustafa Tokar

EXPERIMENTAL

The chemical and microbiological dataset of thirty different waterworks were obtained from the Health Directorate of Cankiri for this study. Drinking water samples were collected from centre of Cankiri City and villages from near Cankiri City in August 2009 (Table 1). Samples were collected from residential tap water according to correct sampling techniques to analyze chemical and microbiological them

Table 1. Water sampling locations.

Location no	Location name
I	Karatekin park
II	100th year Orphanage
III	Superior reservoir
IV	Inferior reservoir
V	Ahlat village
VI	Akcavakif village
VII	Akoren village
VIII	Alanpinar village
IX	Asagicavus village
X	Asagipelitozu village
XI	Asagiyanlar village
XII	Ayan village
XIII	Ayan village Deliklikaya quarter
XIV	Basegmez village
XV	Civi village
XVI	Civi village Dede quarter
XVII	Degim village
XVIII	Derecati village
XIX	Germece village
XX	Hasakca village
XXI	Hidirlik village
XXII	Icyenice village
XXIII	Inanc village
XXIV	Inandik village
XXV	Konak village
XXVI	Pasa village
XXVII	Suleymanli village
XXVIII	Tuzlu village
XXIX	Tuney village
XXX	Yukaricavus village

Citation: Emrah Caylak and Mustafa Tokar, "Metallic and Microbial Contaminants in Drinking Water of Cankiri, Turkey," E-Journal of Chemistry, vol. 9, no. 2, pp. 608-614, 2012. doi:10.1155/2012/720317

The concentrations of elements As, Cd, Cr, Cu, Hg, Ni, Pb, Se, and Sn were measured using Inductively Coupled Plasma-Optical Emission Spectrophotometer (ICP-OES) (Varian Vista, Australia). As concentration in samples was measured using atomic absorption spectrophotometer (Perkin Elmer HGA 700, USA). B, F, NO_3^-, and NO_2^- were analyzed by using Ion Chromatography technique (Dionex DX-600, USA). Cn was analyzed colorimetrically (Hach) Pyridine-Pyrazalone Metot spectrofotometric kit (Merck) with Hach DR/2500 Spectrophotometer, USA. Determination in water of free Cl was also performed colorimetrically using diethyl-p-phenylenediamine. Detection and enumeration of coliform bacteria, E.coli and Enterococci was performed using membran filtration method[10] . The all data were analyzed descriptive statistically by software program (SPSS 15.0 for Windows).

RESULTS AND DISCUSSION

The chemical values of the drinking water samples including As, B, Cd, Cn, Cr, Cu, F, free Cl, Hg, Ni, NO_3^-, NO_2^-, Pb, Se, and Sn from Cankiri-Turkey were given in Table 2. Several quality assessment studies on drinking tap water have been carried out[12-14] .

Table 2. Chemical values of the drinking water samples from Cankiri, Turkey (μg/L).

Location	As TR[a] = 10 WHO[b] = 10	B TR[a] = 1000 WHO[b] = 500	Cd TR[a] = 5 WHO[b] = 3	Cn TR[a] = 50 WHO[b] = 50	Cr TR[a] = 50 WHO[b] = 50	Cu TR[a] = 2000 WHO[b] = 2000	F TR[a] = 1500 WHO[b] = 1500	Free Cl TR[a] = 500 WHO[b] = 5000	Hg TR[a] = 1 WHO[b] = 0.6	Ni TR[a] = 20 WHO[b] = 70	NO_3^- TR[a] = 50000 WHO[b] =50000	NO_2^- TR[a] = 500 WHO[b] = 200	Pb TR[a] = 25 WHO[b] = 10	Se TR[a] = 10 WHO[b] = 10	Sb TR[a] = 5 WHO[b] = 5
I	10.7	160	< 1.4	< 1.0	< 7.0	< 3.2	240	1000	< 0.2	4.90	4630	< 50	< 7.0	< 0.5	< 2.0
II	8.92	280	< 1.4	1.0	< 7.0	< 3.2	240	< 50	< 0.2	6.20	6330	< 50	< 7.0	< 0.5	< 2.0
III	8.19	360	< 1.4	< 1.0	< 7.0	11	240	< 50	< 0.2	5.10	6860	< 50	< 7.0	< 0.5	< 2.0
IV	13.1	200	1.40	1.0	< 7.0	< 3.2	300	< 50	< 0.2	5.10	5020	< 50	< 7.0	< 0.5	< 2.0
V	9.96	130	< 1.4	1.0	15.7	3.3	260	< 50	< 0.2	< 3.7	1050	< 50	< 7.0	< 0.5	3.10
VI	8.16	130	< 1.4	< 1.0	< 7.0	11	370	< 50	< 0.2	< 3.7	8850	< 50	< 7.0	< 0.5	< 2.0
VII	4.32	400	< 1.4	1.0	7.90	< 3.2	1140	< 50	< 0.2	< 3.7	26300	< 50	< 7.0	< 0.5	< 2.0
VIII	1.25	60	< 1.4	< 1.0	< 7.0	8	190	< 50	< 0.2	< 3.7	760	< 50	< 7.0	< 0.5	< 2.0
IX	7.45	130	< 1.4	1.0	10.7	5	190	< 50	< 0.2	6.10	4360	< 50	< 7.0	< 0.5	< 2.0
X	3.96	210	1.40	1.0	< 7.0	< 3.2	690	100	< 0.2	< 3.7	38400	< 50	< 7.0	< 0.5	< 2.0
XI	1.57	110	< 1.4	1.0	18.5	10	150	< 50	< 0.2	5.60	5570	< 50	< 7.0	< 0.5	< 2.0
XII	7.73	130	< 1.4	1.0	10.5	5	230	< 50	0.32	4.50	4290	< 50	< 7.0	< 0.5	< 2.0
XIII	55.7	480	< 1.4	1.0	< 7.0	7	310	< 50	< 0.2	5.10	6200	< 50	< 7.0	< 0.5	< 2.0
XIV	3.13	65	< 1.4	1.0	< 7.0	11	320	< 50	< 0.2	< 3.7	410	< 50	< 7.0	< 0.5	< 2.0
XV	1.65	100	< 1.4	1.0	< 7.0	4	150	< 50	< 0.2	5.60	11400	< 50	< 7.0	< 0.5	< 2.0
XVI	1.69	100	< 1.4	1.0	8.40	15	140	< 50	0.36	5.40	11400	< 50	< 7.0	< 0.5	< 2.0
XVII	27.6	160	< 1.4	1.0	< 7.0	5	200	< 50	< 0.2	5.10	1430	< 50	< 7.0	< 0.5	< 2.0
XVIII	4.0	< 60	< 1.4	1.0	< 7.0	3.2	90	< 50	< 0.2	< 3.7	1810	< 50	< 7.0	< 0.5	< 2.0
XIX	3.39	460	1.70	1.0	< 7.0	12	220	< 50	< 0.2	< 3.7	2130	< 50	< 7.0	< 0.5	< 2.0
XX	1.55	110	< 1.4	1.0	10.1	6	160	< 50	0.24	6.30	11500	< 50	< 7.0	< 0.5	< 2.0
XXI	5.72	680	< 1.4	< 1.0	< 7.0	24	240	< 50	< 0.2	< 3.7	3340	< 50	< 7.0	< 0.5	< 2.0
XXII	7.65	360	< 1.4	< 1.0	< 7.0	7	300	< 50	< 0.2	< 3.7	14200	< 50	< 7.0	< 0.5	< 2.0
XXIII	1.79	100	1.40	1.0	< 7.0	10	170	< 50	< 0.2	4.80	11400	< 50	< 7.0	< 0.5	< 2.0
XXIV	3,01	430	< 1.4	1.0	< 7.0	< 3.2	260	< 50	< 0.2	< 3.7	2060	< 50	< 7.0	< 0.5	< 2.0
XXV	3.87	450	< 1.4	1.0	< 7.0	29	220	< 50	< 0.2	4.10	2030	< 50	< 7.0	< 0.5	< 2.0
XXVI	5.98	230	< 1.4	1.0	< 7.0	7	280	< 50	0.20	< 3.7	3120	< 50	< 7.0	< 0.5	< 2.0
XXVII	1.52	110	1.60	1.0	17.2	34	160	< 50	< 0.2	5.70	5510	< 50	< 7.0	< 0.5	< 2.0
XXVIII	12,4	210	< 1.4	1.0	< 7.0	13	260	< 50	0.20	6.90	5060	< 50	< 7.0	< 0.5	< 2.0
XXIX	3.95	910	< 1.4	1.0	< 7.0	< 3.2	220	< 50	< 0.2	5.30	8270	< 50	< 7.0	< 0.5	< 2.0
XXX	4.06	120	< 1.4	< 1.0	< 7.0	5	150	< 50	< 0.2	< 3.7	140	< 50	< 7.0	< 0.5	< 2.0

[a] *Guide values established by Turkish legislation for drinking water*[11]. [b] *Guideline values established by the World Health Organization for drinking water*[1].

Arsenic in the investigated water samples were found in the range of 1.25-55.77 μg/L. The arsenic values in Karatekin park and inferior reservoir at the centre of Cankiri, Ayan village Deliklikaya quarter, Degim village near Cankiri were found above the Turkish and WHO recommended value for drinking water (10 μg/L)[1,11]. Concentrations of arsenic in drinking water samples were found to be ranging between 10 and 50 μg/L in 22 points and were found to be >50μg/L in 5 stations of Aksaray in Turkey[15]. The presence of arsenic in water sources resulted from leaching of arsenic from rocks and sediments. It is known that long time absorption of arsenic can cause neurological, dermal, hematopoietic, reproductive, and carcinogenic (dermis and lung) diseases[16,17]; whereas positive relationships between low-dose As exposures and cumulative incidence ratios of bladder, lung and urinary-related cancers were recently found[18] .

The boron concentrations were found in the range of The boron concentrations were found in the range of <60-480 pg/L. B in Hidirlik and Tuney village›s samples was above according to the limit of 500 pg/L permitted by WHO in drinking waters; whereas the values of stations of 7, 13, 19, 24, 25 (400-480 pg/L) were slightly below that limit. Col and Col (2003) found higher boron levels in waters of Kutahya-Turkey resulted from rich boron mine areas°. Boron is found naturally in groundwater, but its presence in surface water is frequently a consequence of the discharge of treated sewage effluent, in which it arises from use in some detergents, to surface waters. In laboratory animals, chronic low-level boron exposure has demonstrated to cause reduced growth, cutaneous disorders and suppression of male reproductive system function[20]. Fluoride in the drinking water samples were found in the range of 90-1140 pg/L. F concentrations in Akoren village sample (1140 pg/L) slightly below the Turkish regulation and WHO guidelines established for drinking water (1500 pg/L) «I. Ruiz-Payana et al.[21] found fluoride content in wells and tap water samples of Villa Ahumada-Mexico ranged from 5000 to 5700 pg/L. Fluoride is present in many waters, with higher concentrations often associated with underground sources in areas rich in fluoride-containing minerals. Epidemiological evidence about exposure to high concentrations of fluoride can carry an increasing risk of dental fluorosis, and progressively higher concentrations lead to increasing risks of skeletal fluorosis[22]

Disinfection of drinking water in developing countries is done primarily with free chlorine, commonly available as chlorine bleach. The concentration of free chlorine should normally be such that it should not give rise to unacceptable tastes and should not normally exceed Turkish regulation (500 pg/L) and WHO (5000 pg/L) guidelines values for free chlorine)» I. Removing excess chlorine is important to prevent taste problems. The levels of free CI in the samples were found in the range of <50-1000 pg/L. The value of Karatekin Park at the centre of Cankiri was above the Turkish regulation criteria. In Nakhon Pathom Municipality-China, Panyakapo et al.23 found residual concentration of chlorine in tap water and swimming pool water ranged from nondetectable to 900 pg/L and from nondetectable to 8510 pg/L, respectively. Chlorine reacts with natural organic matter and/or inorganic substances occurring in water to form various chlorination by-products (CBPs) such as trihalomethanes (THMs), haloacetic acids (HAAS) and other undesirable compounds24. Toxicological and epidemiological studies have been carried out to establish relationships between levels of CBPs in drinking water and various carcinogenic, genotoxic and adverse reproductive health effects25.

Table 3. Microbiological values of the drinking water samples from Cankiri, Turkey (cfu/100 ml).

Location	Coliform bacteria $TR^a = 0$; $WHO^b = 0$	*E.coli* $TR^a = 0$; $WHO^b = 0$	Enterococci $TR^a = 0$; $WHO^b = 0$
I	0	0	0
II	0	0	0
III	0	0	1
IV	1	1	2
V	0	0	0
VI	0	0	56
VII	0	0	0
VIII	0	0	1
IX	0	0	1
X	0	0	0
XI	0	0	1
XII	0	0	2
XIII	0	0	14
XIV	1800	1800	90
XV	0	0	0
XVI	3	3	0
XVII	0	0	2
XVIII	0	0	110
XIX	0	0	32
XX	2	0	0
XXI	0	0	1
XXII	0	0	3
XXIII	1	1	0
XXIV	100	100	126
XXV	0	0	0
XXVI	500	0	3
XXVII	600	0	116
XXVIII	0	0	0
XXIX	0	0	2
XXX	0	0	2

The nitrate levels in all stations were ranged from 410 to 38400 µg/L, The values of Akoren and Asagipelitozu village were slightly below Turkish and WHO recommended limits[1,11]. In 2003, the nitrate levels were found in the water supply Tenerife island-Spain (maximum concentration 57300 µg/L 26. The important source of human exposure in drinking water to nitrate by rain and irrigation water washing through soils that have been treated excessively with nitrated fertilizers. Nitrates are involved in the origin of highly toxic compounds like nitrites and nitrosamines. Nitrite can produce metahaemoglobinaemia, and nitrosamines are known to cause certain cancers, adverse pregnancy outcomes, diabetes, and thyroid disorders27. The antimony content of the samples was in the range of Table 3 shows microbiological contamination values found in the drinking water collected from thirty location of Cankiri. The highest E. Coli and total coliform value (1800 cfu/100 mL) were detected in drinking water collected from Basegmez village of Cankiri, while the highest Enterecocci value (126 cfu/100 mL) was found in samples collected from Inandik village of Cankiri. The presence of E. coli is indicative of a fecal contamination and the highly probable presence of pathogens29. In addition, the presence of these microbial contaminants in drinking water must at least be considered as a possible threat or indicative of microbiological water quality deterioration.

CONCLUSION

The concentrations of the investigated heavy metals (As, B, and free Cl) in thedrinking water samples from Cankiri/Turkey were found above the guidelines for drinking waters given by the Turkish legislation and WHO. However, it was concluded that drinking waters in Cankiri contain low other investigated heavy metal levels. This study, therefore, recommends the government and other responsible authorities to prevent any kind of metallic and microbial contaminants into water sources that supply domestic drinking water.

REFERENCES

1. World Health Organization (WHO), Guidelines for drinking water quality, 3rd Edition, Geneva, 2008.
2. Boffetta P, Mc Laughlin J K, la Vecchia C, Autier P and Boyle P, Carsinogenesis., 2007, 28(5), 913-915.
3. Gohar A V and Muhammadi A, J Biol Sci., 2010, 10(3), 200-208.
4. Caylak E, Aytekin M and Halifeoğlu I, Exp Toxicol Pathol., 2008, 60(4-5), 289-294.
5. Caylak E and Halifeoglu I, Cocuk Sagligi ve Hastaliklari Dergisi, 2010, 53(2), 159- 173. (in Turkish).

6. Pimparkar B D and Bhave A, J Assoc Physicians India., 2010, 58, 617-624, 629.
7. Calderon R L, Food Chemical Toxicol., 2000, 38(1 supply), 13-30.
8. Murray F J, Regul Toxicol Pharmacol., 1995, 22(3), 221-230.
9. Gasana J J, Morin J, Ndikuyeze A and Kamoso P, Environ Res Section A., 2002, 90(2), 76-88.
10. Lee R J and Cole S R, J Appl Bacteriol., 1994, 76, 270-274.
11. ITAS, Insani Tuketim Amacli Sular Hakkinda Yonetmelik, 25730 sayili Resmi Gazete, 17th February 2005. (in Turkish).
12. Bhaskar C V, Kumar K and Nagendrappa G, E-J Chem., 2010, 7(2), 349-352.
13. Nair I V, Singh K, Arumugam M and Clarson D, E-J Chem., 2011, 8(1), 257-263.
14. Kaplan O, Yildirim N C, Yildirim N and Tayhan N, E-J Chem., 2011, 8(1), 276-280.
15. Alta L, Isik M and Kavurmaci M, J Environ Manage., 2011, (E pub ahead of print).
16. Choong T S Y, Chuah T G, Robiah Y, Gregory Koay F L and Azni I, Desalination, 2007, 217,139-166.
17. Celik I, Gallicchio L, Boyd K, Lam T K, Matanoski G, Tao X, Shiels M, Hammond E, Chen L, Robinson KA, Caulfield L E, Herman J G, Guallar E and Alberg J, Environ Res., 2008, 108(1), 48-55.
18. Liao C M, Shen H H, Chen C L, Hsu L I, Lin T L, Chen S C and Chen C J, J Hazard Mater., 2009, 165(1-3), 652-663.
19. Col M and Col C, Food Chem Toxicol., 2003, 41(10), 1417-1420.
20. Benderdour M, Bui-Van T, Dicko A and Belleville F, J Trace Elem Med Biol., 1998, 12(1), 2-7.
21. Ruiz-Payan A, Ortiz M and Duarte-Gardea M, Microchem J., 2005, 656, 1-4.
22. Den Besten P K, Adv Dent Res., 1994, 8(1),105-110.
23. Panyakapo M, Soontornchai S and Paopuree P, J Environ Sci (China), 2008, 20(3), 372-378.
24. Singer P C, J Environ Eng (New York), 1994, 120(4), 727-744.
25. Legay C, Rodriguez M J, Sadiq R, Serodes J B, Levallois P and Proulx F, J Environ Manage., 2011, 92(3), 892-901.
26. Caballero Mesa J M, Rubio Armendariz C and Hardisson de la Torre A, Sci Total Environ., 2003, 302(1-3), 85-92.
27. Ward M H and Brender J D, Encyclopedia of Environmental Health., 2011, 167-178.
28. De Boeck M, Kirsch-Volders M and Lison D, Mutat Res., 2003, 533(1-2), 135-152.
29. Fewtrell L and Bartram J, Eds., Water quality:guidelines, standards and health. London, IWA Publishing, 2001, 289–316.

Section III - Microbiological Aspects of Drinking Water

CHAPTER 8

WATERBORNE VIRUSES: A BARRIER TO SAFE DRINKING WATER

Aimee M. Gall, Benito J. Mariñas, Yi Lu, Joanna L. Shisler

WHAT IS THE GLOBAL STATUS OF ACCESS TO SAFE, PATHOGEN-FREE DRINKING WATER?

Nearly 25% of the global population (1.8 billion people in 2012) is consuming fecally-contaminated water [1]. This water can contain bacteria, protozoa, and viruses that can cause a variety of diseases in humans, most notably gastroenteritis. The impact on public health is staggering. Unsafe water, inadequate sanitation, and poor hygiene are responsible for about 90% of diarrheal deaths worldwide [2]. Not surprisingly, diarrhea is the second leading cause of death for children under the age of five globally (1.2 million deaths in 2012) [2]. In addition to the human cost, the World Bank estimates that lack of access to safe water and sanitation results in a global economic loss of US$260 billion annually [3].

The lack of access to improved water disproportionally affects those living in poverty in rural, developing regions; however, even populations living in countries with state-of-the-art water and waste treatment facilities are prone to waterborne disease outbreaks. For example, there were at least 33 outbreaks associated with drinking water reported in the United States of America during 2009–2010 [4]. Regardless of the socioeconomic status of a country, illnesses due to contaminated drinking water are considered significantly underreported because people do not seek medical attention for self-limiting infections and because of the current limitations on clinical detection of virus infections [4].

ARE WATERBORNE VIRUSES A PARTICULAR CONCERN?

It is well known that bacteria are major causes of diarrhea transmitted through unsafe drinking water. What is less appreciated are viruses in these same drinking water sources and their impact on human health. Water-transmitted viral pathogens that are classified as having a moderate to high health significance by the World Health Organization (WHO) include adenovirus, astrovirus, hepatitis A and E viruses, rotavirus, norovirus and other caliciviruses, and enteroviruses, including coxsackieviruses and polioviruses [5]. Also, viruses that are excreted through urine like polyomaviruses [5] and cytomegalovirus [6] can potentially be spread through water. Other viruses, such as influenza and coronaviruses, have been suggested as organisms that can be transmitted through drinking water, but evidence is inconclusive [5].

Most of the above viruses are most commonly associated with gastroenteritis, which can cause diarrhea as well as other

Citation: Gall AM, Mariñas BJ, Lu Y, Shisler JL (2015) Waterborne Viruses: A Barrier to Safe Drinking Water. PLoS Pathog 11(6): e1004867. doi:10.1371/journal.ppat.1004867

symptoms including abdominal cramping, vomiting, and fever. It should be noted that some of these same viruses could also cause more severe illnesses including encephalitis, meningitis, myocarditis (enteroviruses), cancer (polyomavirus), and hepatitis (hepatitis A and E viruses) [5]. Hepatitis E virus can also cause a mortality rate of up to 25% in pregnant women [5]. Viral infections are usually self-limiting in healthy individuals. They can cause greater morbidity in children under the age of five, the elderly, immunocompromised people, and pregnant women. Waterborne virus-based diseases may be higher in developing regions, where there is widespread malnutrition and large populations of HIV-positive people. Regardless, there are few broad spectrum anti-viral drugs to treat these diseases.

Certainly, there is good cause for controlling these waterborne viruses. Rotavirus, for example, is the leading cause of severe acute diarrhea in children under the age of five globally, resulting in over half a million deaths annually [7]. The worldwide use of new rotavirus vaccines may make the removal of this virus from water supplies less of an issue in the future. Similarly, better coverage with vaccinations against hepatitis A and poliovirus would also greatly decrease the health risks of these viruses in drinking water.

Waterborne viruses differ in terms of their genome content and capsid proteins, but these viruses share several properties that make them of particular concern regarding the risk of disease outbreak associated with drinking water contamination. Several of these viruses have extremely low infectious doses; the probability of infection from exposure to one rotavirus particle is 31% [8]. Viruses are shed in feces in very high numbers even asymptomatically. For example, up to 10^{11} norovirus particles can be present per gram of stool [9]. In addition, non-enveloped viruses can persist in water for long periods of time [10]. When considering these characteristics, inadequate disinfection of fecally contaminated drinking water could easily lead to outbreaks of viral gastroenteritis from ingestion. Notably, drinking water can also transmit viruses via inhalation (e.g., showering) or contact with skin and eyes (e.g., swimming) causing respiratory and ocular infections.

WHAT IS THE STATE OF THE ART FOR CONTROL OF VIRUSES IN WATER?

Water treatment utilities routinely assay for the presence of fecal coliforms in water supplies, but they do not assay for the presence of infectious viruses because it is either impossible or not feasible to detect or propagate infectious virus particles in a cost-efficient and timely manner. Despite these barriers, the United States Environmental Protection Agency (USEPA) is evaluating adenovirus, caliciviruses, enteroviruses, and hepatitis A virus for potential regulatory action [11]. The current US regulations require the removal or inactivation of 99.99% of enteric viruses by approved treatment techniques, but specific virus families are not individually regulated. These approved treatment techniques are based on bench scale studies where a specific virus is exposed to a disinfectant at various environmental conditions until reaching 99.99% inactivation. By studying a range of enteric viruses, regulations are decided for each disinfectant based on an appropriate dose to adequately inactivate the most resistant enteric virus studied. Utilities must apply an appropriate disinfectant dose to meet enteric virus regulations.

Common water treatment techniques worldwide include physically removing pathogens through conventional treatment and inactivating pathogens by applying ultraviolet light or chemical oxidants such as chlorine, chloramines, ozone, and chlorine dioxide. Since viruses are so small, conventional treatment, including filtration, is ineffective at physically removing viruses. The application of disinfectants highly depends on water chemistry and local regulations. Free chlorine (i.e., sum of hypochlorous acid and hypochlorite ion formed by dissolution and hydrolysis of chlorine gas in water) is the most commonly used disinfectant worldwide and has been used to disinfect water since the early 1900s [12]. Most viruses are inactivated by this strong oxidant. However, free chlorine treatment may produce regulated toxic disinfection by-products (DBPs), and it is ineffective to control *Cryptosporidium*, a protozoan that is transmitted in water and causes diarrhea [13,14]. Thus, some drinking water utilities are moving towards using monochloramine (i.e., formed by mixing chlorine and ammonia with the latter in slight excess) to control the formation of regulated toxic DBPs, and either monochromatic (~254 nm) or polychromatic (200–300 nm) ultraviolet (UV) light for controlling both DBP formation and*Cryptosporidium* contamination. Unfortunately, these changes in disinfection practice come at a cost to virus control. For example, while adenovirus is susceptible to inactivation by free chlorine, it is highly resistant to inactivation by both monochloramine and UV light [15]. Chlorine dioxide and ozone, also strong oxidants, are both effective at controlling viruses, but they have operational challenges, such as the need for on-site generation, and the production of DBPs, including chlorite from chlorine dioxide and bromate from ozone [12].

Regardless of the disinfectant applied at a drinking water utility, as the treated water travels from the treatment plant to the tap, cross-contamination can occur throughout the miles of water distribution infrastructure due to cavitation and accidental depressurization; therefore, the use of secondary disinfectants in distribution systems is required. Unfortunately, the only two disinfectants capable of maintaining a residual in the distribution system are free chlorine and monochloramine. Although free chlorine is a stronger disinfectant with respect to pathogen inactivation, monochloramine provides a more stable residual in distribution systems, and so both are utilized.

WHAT ARE THE BARRIERS TOWARD DISINFECTING VIRUSES IN DRINKING WATER?

There are several barriers that prevent viruses in drinking water from being detected. From a technological standpoint, as compared to detecting fecal coliforms, virus propagation requires the use of tissue culture, a system that requires increased time, labor, expertise, and expensive equipment [8]. Furthermore, several of these viruses cannot be grown easily (adenovirus serotypes 40 and 41) or at all (human norovirus, hepatitis A virus) in cell culture. Consequently, traditional viral growth assays (plaque assays) are either unavailable or too lengthy in time to be practical for water treatment facilities. For example, a 10-day incubation period is required to detect replicating adenoviruses via plaque assays. While ELISA or qPCR-based technologies can be used to rapidly detect viral proteins or genomes, respectively, they do not distinguish infectious versus non-infectious viral particles. Integrated cell culture-PCR (ICC-PCR) reduces time requirements of traditional plaque assays and allows for infectious viruses to replicate in host cells, but it still employs the use of cell culture that is impractical at water treatment utilities [8]. Although there have been advances in concentrating viruses from large volumes of water [16], there has yet to be a rapid way to detect viable viruses.

In addition to these detection technology limitations, there is not one "silver bullet" water treatment that will inactivate all virus types independently of water quality. For example, human adenovirus is nearly five times more resistant to monochromatic (254 nm) UV inactivation compared to other enteric viruses [17]. While bacteriophages are often used as models to study enteric eukaryotic viruses, no bacteriophage studied to date accurately represents enteric virus behavior for all disinfectants. The scientific community does not yet understand why viruses have different profiles of resistance to different disinfectants. Regardless, from a regulatory standpoint, a major barrier is that not one disinfection method is effective against all viruses that can be applied to all water quality conditions.

Waterborne viruses have a range of genome types (e.g., DNA, RNA, linear, segmented) and capsid protein structures that contribute to their resistance or susceptibility to specific disinfectants. Many studies have determined reaction rates of disinfectants with amino acids and nucleotides, and this information can be useful in analyzing the mechanism of virus inactivation. However, these data are not always predictive due to the complex nature of viral capsid structures and secondary reactions [18]. It was initially thought that UV treatment of viruses would damage the viral genome as the mechanism for disinfection, whereas chemical oxidants like free chlorine would damage viral capsid proteins as a means for disinfection. UV light is known to be more reactive with nucleotides than amino acids; however, UV light can also damage viral proteins [19], suggesting that UV light has multiple mechanisms to disinfect viruses. For example, UV irradiation inactivates bacteriophage MS2 by both site-specific backbone cleavage of the major capsid protein resulting in genome injection inhibition and by damaging the RNA genome leading to genomic replication inhibition [19]. Whether this is true for other viruses is unknown. Chemical oxidants like free chlorine typically have higher reaction rates with amino acids, but have been shown to damage both viral proteins and genomes [19–21]. Regardless, determining precisely how a disinfectant damages and neutralizes specific viruses is necessary to fully understand the mechanisms of virus inactivation.

WHAT IS THE FUTURE OF WATERBORNE VIRUS RESEARCH?

How does the scientific community overcome the technological and knowledge barriers? First, we must increase our fundamental understanding of how individual viruses become inactivated by disinfectants on a molecular level. This will require detailed studies of the individual components of a viral particle as well as the virion as a whole (Fig 1). The state-of-the-art technique is to correlate virus disinfection with a block in the virus replication cycle as a means to pinpoint what alteration of the protein capsid and/or the viral genome by the disinfectant results in loss of infectivity [19,22]. Technologies including mass spectrometry will afford the opportunity to understand, on an amino acid level, how a disinfectant modifies viral capsid proteins, and how this may result in a non-infectious virus particle [18]. Labeling techniques can also be utilized to determine if viral capsid proteins have undergone specific modifications, such as using a biotin hydrazide that forms covalent bonds with carbonyl groups [23], one of many oxidative products formed on amino acids. The scientific field is now beginning to capitalize on the sensitivity of qPCR or RT-qPCR to detect specific regions of a viral genome that are damaged by a disinfectant. Disinfectant-induced genomic damage can block viral DNA/RNA replication in host cells or viral genome amplification in PCR reactions [24]. However, PCR cannot elucidate what type of genome modification is caused by a disinfectant (e.g., pyrimidine dimers, crosslinking to proteins, chlorine-carbon bonds) or if the host cell is able to repair the lesions [17,24]. A variety of techniques will need to be used and developed to determine on a molecular level how a disinfectant neutralizes a virus and what stage of the replication cycle is blocked. This will likely vary for different viruses and for each disinfectant.

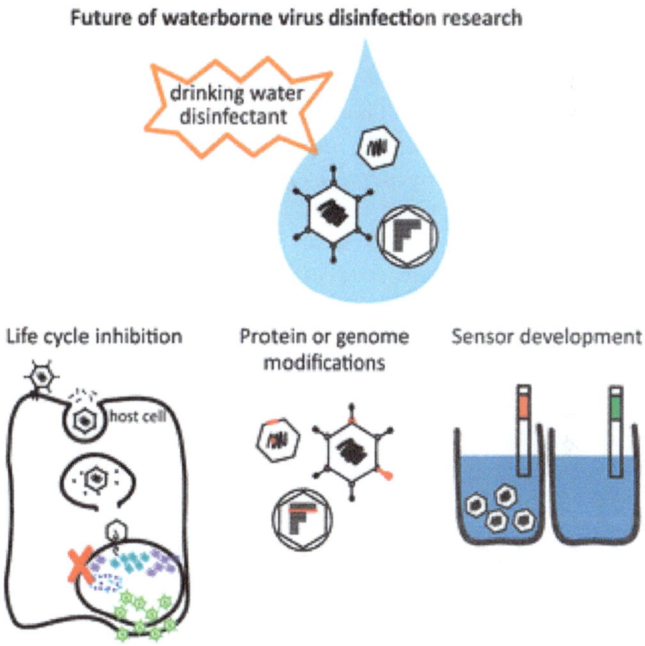

Figure 1. Future of waterborne virus research to provide safe drinking water globally. Gaining a better understanding of how viruses become inactivated by disinfectants requires detailed studies of many virus types and disinfectants to determine what stage of the virus replication cycle becomes blocked, and what modifications to the viral protein and/or genome lead to inactivation. The development of sensors to detect infectious viruses in drinking water will benefit from these studies and is also necessary to ensure safe drinking water.

A grand challenge remains to design technology to rapidly distinguish infectious from damaged or portions of viral particles. Recently developed methods combine the antibody-based capture of viral capsids and subsequent qPCR amplification of genomes as a strategy to quantify structurally intact viruses in environmental samples [25]. This immunocapture-qPCR (IC-qPCR) technique is promising to detect infectious viral particles rather than fragments of viral nucleic acid and proteins using ELISA- or qPCR-based techniques alone. Another recent technique employs dyes (ethidium monoazide and propidium monoazide) that can penetrate damaged viral capsids and intercalate into viral nucleic acids inhibiting PCR amplification [26]. This method, like IC-qPCR, requires an intact viral capsid to detect an infectious virus. Unlike IC-qPCR, these dye-based assays rely on the mechanism of inactivation to be a leaky capsid structure. Results from these assays must be interpreted carefully because it is possible that a disinfectant modifies a capsid protein responsible for a key replication cycle event, like attachment, but does not affect the capsid structure or genome. Another new, promising technology to meet this grand challenge relies on the use of aptamers to selectively detect infectious virus particles. For this to occur, there is an in vitro selection from a large library of nucleic acids of aptamers that will selectively bind surface receptors of infectious viruses, but not those from non-infectious viral particles [27]. This technology has already been used in sensors to detect toxins and metals, and many aptamers have been selected for viral proteins and whole viruses for use in antiviral agents [28]. Indeed, an aptamer-based sensor can already distinguish between viable and heat-inactivated vaccinia virus, showing the promise of this technology for detecting other viruses [29]. Because aptamers are much more stable and cheaper than antibodies used in ELISA, such a technology would allow rapid detection of infectious viruses more cost-effectively and selectively.

Once we understand how disinfectants inactivate viruses, we can develop effective treatment protocols for water utilities and the next generations of sensors that rapidly detect and quantify infectious viruses in finished drinking water. Populations most vulnerable to unsafe drinking water live in rural and periurban areas of developing countries. Thus, scientists and engineers must design protocols and sensors to be cost-effective, rugged, and easy to use for these populations. Ultimately, the fundamental knowledge of how viruses become inactivated will ensure better control of viruses in drinking water, increasing access to safer drinking water globally.

REFERENCES

1. WHO, UNICEF (2014) Progress on drinking water and sanitation: 2014 Update. Geneva, Switzerland: WHO Press.
2. UNICEF (2012) Pneumonia and diarrhoea: Tackling the deadliest diseases for the world's poorest children. New York, NY: Statistics and Monitoring Section- Division of Policy and Strategy.
3. WHO (2012) Global costs and benefits of drinking-water supply and sanitation interventions to reach the MDG target and universal coverage. Geneva, Switzerland: WHO Press, World Health Organization.
4. (2013) Center for Disase Control and Prevention. Morbidity and mortality weekly report. 62.
5. WHO (2011) Guidelines for drinking-water quality -4th ed. Geneva, Switzerland: WHO Press.
6. Cannon MJ, Hyde TB, Schmid DS (2011) Review of cytomegalovirus shedding in bodily fluids and relevance to congenital cytomegalovirus infection. Rev Med Virol 21: 240–255. doi: 10.1002/rmv.695. pmid:21674676
7. WHO (2008) Global networks for surveillance of rotavirus gastroenteritis, 2001–2008. Weekly epidemiological record 47: 421–428. pmid:19024780
8. Fong TT, Lipp EK (2005) Enteric viruses of humans and animals in aquatic environments: health risks, detection, and potential water quality assessment tools. Microbiol Mol Biol Rev 69: 357–371. pmid:15944460 doi: 10.1128/mmbr.69.2.357-371.2005
9. Hall AJ (2012) Noroviruses: the perfect human pathogens? J Infect Dis 205: 1622–1624. doi: 10.1093/infdis/jis251. pmid:22573872
10. Reynolds KA, Mena KD, Gerba CP (2008) Risk of waterborne illness via drinking water in the United States. Rev Environ Contam Toxicol 192: 117–158. pmid:18020305 doi: 10.1007/978-0-387-71724-1_4 (2009) U.S. EPA. Drinking water contaminant candidate list 3: Final. Federal Register 74: 51850–51862.
11. Crittenden JC (2005) MWH's water treatment: principles and design. Hoboken, N.J.: John Wiley and Sons.
12. (2006) U.S. EPA. 40 CFR Parts 9, 141, and 142. National primary drinking water regulations: stage 2 disinfectants and disinfection byproducts rule, final rule. Fed Regist 71: 388–493.
13. (2006) U.S. EPA. 40 CFR Parts 9, 141, and 142. National primary drinking water regulations: Long term 2 enhanced surface water treatment rule. Federal Register 71: 654–786.
14. Sirikanchana K, Shisler JL, Marinas BJ (2008) Effect of exposure to UV-C irradiation and monochloramine on adenovirus serotype 2 early protein expression and DNA replication. Appl Environ Microbiol 74: 3774–3782. doi: 10.1128/AEM.02049-07. pmid:18424543
15. Ikner LA, Gerba CP, Bright KR (2012) Concentration and recovery of viruses from water: a comprehensive review. Food Environ Virol 4: 41–67. doi: 10.1007/s12560-012-9080-2. pmid:23412811
16. Eischeid AC, Thurston JA, Linden KG (2011) UV Disinfection of Adenovirus: Present State of the Research and Future Directions. Critical Reviews in Environmental Science and Technology 41: 1375–1396. doi: 10.1080/10643381003608268
17. Wigginton KR, Kohn T (2012) Virus disinfection mechanisms: the role of virus composition, structure, and function. Curr Opin Virol 2: 84–89. doi: 10.1016/j.coviro.2011.11.003. pmid:22440970
18. Wigginton KR, Pecson BM, Sigstam T, Bosshard F, Kohn T (2012) Virus inactivation mechanisms: impact of disinfectants on virus function and structural integrity. Environ Sci Technol 46: 12069–12078. doi: 10.1021/es3029473. pmid:23098102
19. Sharma VK (2013) Halogenated species. Oxidation of amino acids, peptides, and proteins: Kinetics and mechanism. Hoboken, NJ: John Wiley & Sons, Inc. pp. 78–121.
20. Prutz WA (1996) Hypochlorous acid interactions with thiols, nucleotides, DNA, and other biological substrates. Arch Biochem Biophys 332: 110–120. pmid:8806715 doi: 10.1006/abbi.1996.0322

21. Gall AM, Shisler JL, Marinas BJ (2015) Analysis of the Viral Replication Cycle of Adenovirus Serotype 2 after Inactivation by Free Chlorine. Environ Sci Technol.

22. Sano D, Pinto RM, Omura T, Bosch A (2010) Detection of oxidative damages on viral capsid protein for evaluating structural integrity and infectivity of human norovirus. Environ Sci Technol 44: 808–812. doi: 10.1021/es9018964. pmid:20000802

23. Pecson BM, Ackermann M, Kohn T (2011) Framework for using quantitative PCR as a nonculture based method to estimate virus infectivity. Environ Sci Technol 45: 2257–2263. doi: 10.1021/es103488e. pmid:21322644

24. Ogorzaly L, Bonot S, Moualij BE, Zorzi W, Cauchie HM (2013) Development of a quantitative immunocapture real-time PCR assay for detecting structurally intact adenoviral particles in water. J Virol Methods 194: 235–241. doi: 10.1016/j.jviromet.2013.07.009. pmid:23850702

25. Leifels M, Jurzik L, Wilhelm M, Hamza IA (2015) Use of ethidium monoazide and propidium monoazide to determine viral infectivity upon inactivation by heat, UV- exposure and chlorine. Int J Hyg Environ Health.

26. Liu J, Cao Z, Lu Y (2009) Functional nucleic acid sensors. Chem Rev 109: 1948–1998. doi: 10.1021/cr030183i. pmid:19301873

27. Gopinath SC (2007) Antiviral aptamers. Arch Virol 152: 2137–2157. pmid:17851732 doi: 10.1007/s00705-007-1014-1

28. Labib M, Zamay AS, Muharemagic D, Chechik AV, Bell JC, Berezovski M. (2012) Aptamer-based viability impedimetric sensor for viruses. Anal Chem 84: 1813–1816. doi: 10.1021/ac203412m. pmid:22303883

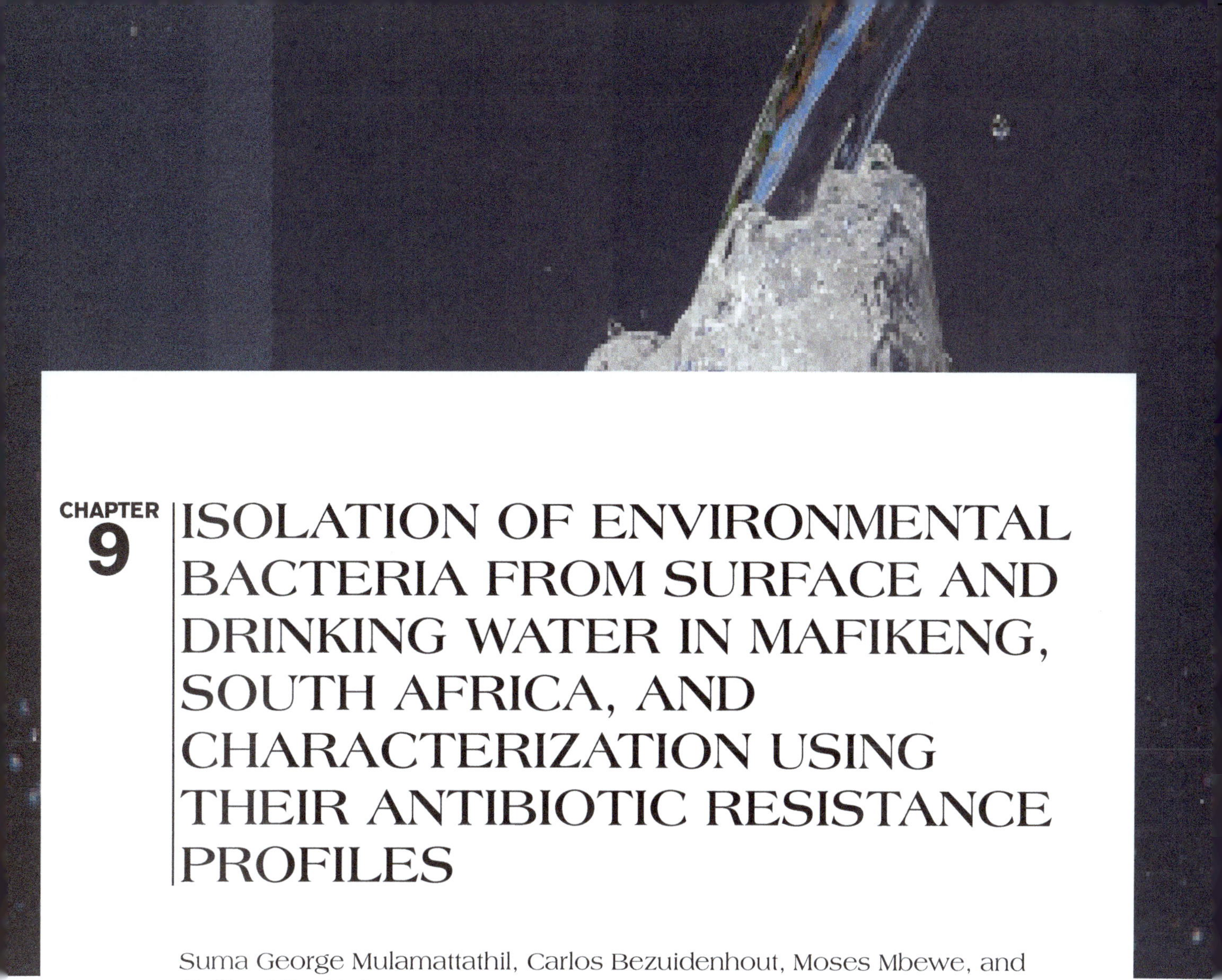

CHAPTER 9

ISOLATION OF ENVIRONMENTAL BACTERIA FROM SURFACE AND DRINKING WATER IN MAFIKENG, SOUTH AFRICA, AND CHARACTERIZATION USING THEIR ANTIBIOTIC RESISTANCE PROFILES

Suma George Mulamattathil, Carlos Bezuidenhout, Moses Mbewe, and Collins Njie Ateba

INTRODUCTION

Water is considered a vehicle for the propagation and dissemination of human associated bacteria [1]. Safe drinking water is a fundamental human right and if contaminated with opportunistic pathogenic environmental bacteria, it may have health implications for consumers [2, 3]. Human health should therefore be protected by preventing microbial contamination of water that is intended for consumption [4]. In rural communities, untreated surface water from rivers, dams, and streams is directly used for drinking and other domestic purposes [5]. These unprotected water sources can be contaminated with microbes through rainfall run-off and agricultural inputs, mixing with sewage effluents and faeces from wild life [6, 7], which render them unacceptable for human consumption. Faecal coliforms, Aeromonas and Pseudomonas, are used as indicators of faecal contamination in water [8] and the presence of these pathogens may have severe health implications on consumers especially those that are immunocompromised [5, 9, 10].

South Africa is a semiarid country with very low rainfall and high evaporation [11] and it has a scarcity of fresh water systems due to the highly variable and spatial distribution of rainfall [12]. Moreover, safe drinking water is frequently used for unsustainable nondrinking applications such as irrigation, toilet and urinary flushing [12], and general cleaning. To manage the existing water resources and to address some of the challenges associated with water shortages in South Africa and the world at large, waste water reuse can form an important component of water demand management [13], but this waste water reuse may affect the quality of drinking water if proper treatment procedures are not implemented. Whereas waste water reuse has been extensively implemented in some European and African countries [13], yet in South Africa, only a few waste water reuse schemes have been documented [14, 15] and there is limited implementation of this alternative in communities.

Alarming increases in the consumption of antibiotics through human therapy and agricultural processes have been reported [16, 17] and this extensive usage in both human and animal medicine has resulted in the development of antibiotic-

Citation: Suma George Mulamattathil, Carlos Bezuidenhout, Moses Mbewe, and Collins Njie Ateba, "Isolation of Environmental Bacteria from Surface and Drinking Water in Mafikeng, South Africa, and Characterization Using Their Antibiotic Resistance Profiles," Journal of Pathogens, vol. 2014, Article ID 371208, 11 pages, 2014. doi:10.1155/2014/371208.

resistant bacteria which affect the treatment of infections [18, 19]. Antibiotic resistance has therefore become a major public health issue [20] and its presence in waste water, surface water, and drinking water is well documented [19–22]. The hazard associated with the pathogenicity of microbes is aggravated by its ability to resist destruction by antibiotics. Biological treatment processes in the waste water treatment plants may result in a selective increase of antibiotic-resistant bacteria and therefore increase the occurrence of multidrug-resistant organisms [23]. Although microorganisms in drinking water are reduced by chlorination, they may survive the treatment process and enter the distribution system [1]. Moreover, the presence of antibiotic resistance in microorganisms has been previously reported [24–26]. Considering the fact that the public health of a community may be related to the quality of treated waste water supplied and that public health can be protected by reducing the pathogenic microorganisms in drinking water, the present study was designed to isolate environmental bacteria from surface and drinking water in Mafikeng and identify thePseudomonas and Aeromonas species using polymerase chain reaction (PCR). A further objective was to characterize the isolates using their antibiotic resistance profiles.

MATERIALS AND METHODS

Study Area

Water samples were collected from five sampling points around Mafikeng, namely, both raw and treated (drinking) water from a karstic groundwater source, the Molopo Eye, both raw water and treated (drinking) water from the Modimola dam, and finally mixed water, treated water from both sources mixed in the Signal hill reservoir and distributed to some areas in the city. These sampling points were chosen for the study because water from Molopo eye and Modimola dam, after purification, is used for human consumption, and for recreational, agricultural, and industrial purposes. As few small scale of farmers live near these two water resources and the Modimola Dam receives treated sewage effluent from the Mmabatho sewage treatment plant, which is the major source of pathogens, it is therefore important to investigate the microbiological quality of water at these points.

Sampling

Raw and treated water samples were collected during a one-year period, in February, April, July, and October, to cover the four different seasons. Water samples from the Molopo eye and Modimola dam were collected aseptically in sterile 500 mL Duran Schott glass bottles from different sampling points by directly dipping the bottles into the surface of the water. Purified water samples were collected directly in to the sterile bottles from the tap, after letting the tap run for a minute. The samples were labelled properly and transported on ice to the laboratory for analysis. Aliquots of the samples were used for selective isolation of faecal coliforms, total coliforms, Aeromonas, Pseudomonas, and heterotrophic bacteria based on standard microbiological procedures [27].

Isolation, Purification, and Characterization of Planktonic Bacteria

Isolation by Membrane Filtration

For all the samples, three volumes of 100 mL were filtered through 0.45 μm pore-sized filter (cellulose nitrate membranes, Whatman Laboratory Division, Maidstone, England) using a water pump (model Sartorius 16824). These membranes were aseptically placed up on plates with appropriate selective media ensuring that no air bubbles were trapped. The selective media used are as follows. mFC agar used as a selective medium for faecal coliforms, mEndo for total coliforms, nutrient agar for heterotrophic bacteria, and Aeromonas selective medium for Aeromonas and Pseudomonas (Biolab, Merck, South Africa). All the media were prepared according to the manufacturers' instructions (Biolab, Merck, South Africa). Each sample was analysed in triplicate. In order to isolate heterotrophic bacteria, 1 mL of the treated water samples was spread onto the nutrient agar plates. Water samples from Modimola dam and Molopo eye were serially diluted and 1 mL of the 5 fold serial dilutions was spread on to the nutrient agar plates.

The plates were incubated at 37°C except for mFC agar which were incubated at 45°C for 24 hours. The colonies were enumerated, characterized, and recorded. The results were expressed as the number of faecal coliforms, total coliforms, Pseudomonas and Aeromonas in 100 mL of water, and heterotrophic bacteria in 1 mL of water. Blue colonies from mFC agar (presumptive coliforms), metallic-sheen colonies from mEndo agar (presumptive total coliforms), and yellow (presumptive Aeromonas) and green colonies (presumptive Pseudomonas) from Aeromonas selective media were picked for further work (Biolab Catalogue).

Purification of Colonies

Colonies were purified by twice subculturing using the streaking plate method. Young cultures were used for Gram staining and all isolates were identified as Gram-negative Bacilli. All the Gram-negative isolates were subjected to primary and secondary biochemical identification tests. The bacteria that were picked to create antibiograms were streaked on to nutrient agar slants to make sample cultures and for PCR purpose.

Antimicrobial Susceptibility Testing

An antibiotic susceptibility test was performed using the Kirby-Bauer disk diffusion method [28]. The following antibiotic discs (Mast Diagnostics, UK) at the final concentrations that are indicated were used: ampicillin (AP) −10 μg, cephalothin (KF) 5 μg, streptomycin (S) 10 μg, erythromycin (E) 15 μg, chloramphenicol (C) 30 μg, neomycin (NE) 30 μg, amoxycillin (A) 10 μg, ciprofloxacin (CIP) 5 μg, trimethoprim (TM) 25 μg, kanamycin (K) 30 μg, and oxytetracycline (OT) 30 μg. These antibiotics were chosen because they are either used in both human medicine and animal veterinary practice or because previous studies have reported microbial resistance to them [29].

Three colonies were picked from each sample and each colony was transferred in to 3 mL of sterile distilled water to prepare bacterial suspension. Aliquots of 100 μL from each suspension were spread-plated on Mueller-Hinton agar plates. Antibiotic discs were applied on to the plates using sterile needles and the plates were incubated at 37°C for 24 hours [30]. After incubation, the antibiotic inhibition zone diameters (IZD) were measured. Results obtained were used to classify isolates as being resistant, intermediate resistant, or susceptible to a particular antibiotic using standard reference values according to National Committee for Clinical Laboratory Standards [30], now Clinical and Laboratory Standards Institute (CLSI). Multiple antibiotic resistance (MAR) phenotypes were generated for isolates that showed resistance to 3 or more antibiotics.

Primary Identification Tests

Triple Sugar Iron (TSI) Test

Triple sugar iron (TSI) agar is a differential medium obtained from Biolab, Merck (South Africa), with the substrates glucose, sucrose, and lactose at sample concentrations of 0.1, 1.0, and 1.0%, respectively. It can distinguish between a number of Gram-negative enteric bacteria based on their physiological ability (or lack thereof) to metabolize lactose and/or sucrose, conduct fermentation to produce acid, produce gas during fermentation, and generate H_2S. The media were prepared according to the manufacturer's instructions. Aliquots were placed in test tubes and autoclaved. Once out of the autoclave, the tubes were placed on a rack and clamped so that the tubes (with liquid medium in them) have a 3 cm slant with a 2 to 3 cm butt. They were allowed to cool down until they become solid. The slants are inoculated with a pure culture by streaking over the entire surface of the slant (zigzagging to cover the surface) and then stabbing deep into the butt. Incubation was at 37°C for 24 hours. If only glucose is fermented, acid is produced in the butt and it will turn yellow. However, if either sucrose or lactose is fermented, sufficient fermentation products will be formed to turn both the butt and the slant yellow. If gas is formed during the fermentation, it will be shown in the butt either as bubbles or as cracking of the agar. If no fermentation occurs, the slant and butt will remain red. The medium also contains ferrous sulphate. If the bacterium forms H_2S, this chemical will react with the iron to form ferrous sulphide, which is seen as a black precipitate in the butt (a black butt).

Oxidase Test

This test was performed using the test oxidase reagent (PL.390) from Mast Diagnostics (Nesto, Wirral, UK) in accordance with the manufacturer's published protocol. A well-isolated pure colony was placed on a filter paper using a sterile wire loop. A drop of test oxidase reagent was added on to it and mixed. After 30 seconds, the filter was observed for a colour change with oxidase positive isolates producing a purple colour being taken as presumptive Aeromonas and Pseudomonas isolates. Oxidase negative colonies were colourless and were presumptively considered to be E. coli.

Secondary Identification Tests

Analytical Profile Index (API) 20E Test

The API 20E test was performed in accordance with the manufacturer's protocol (BioMérieux, 69280, Marcy I'Etoile, France) and the organisms were identified to species level using API software.

Haemolysis on Blood Agar

Haemolysis, on blood agar (Biolab, Merck, SA) supplemented with 5% (v/v) sheep blood, was determined after the incubation of the plates at 37°C for 24 hours. Haemolysis is determined by streaking for isolation on a blood agar plate. After incubation overnight, the medium is inspected for signs of alpha- or beta-haemolysis. If the medium is discoloured or darkened after growth, the organism has demonstrated alpha-haemolysis. If the medium develops clear halo under growth, the organism is beta-haemolytic. No discernible change in the colour of the medium constitutes gamma haemolysis.

Extraction of Genomic DNA and PCR for the Identification of Culture Species

DNA from the isolates was extracted using the peqGOLD (PEQLAB Biotechnologie GmbH 12-3450) bacterial DNA extraction kit according to the manufacturer's protocol. The concentration of the extracted DNA in solution was determined spectrophotometrically (NanoDrop ND 1000, Thermo Scientific, USA). The integrity of the purified template DNA was assessed by conventional 1% (w/v) agarose gel.

Identification of the Isolates by PCR Assays

The identities of the presumptive Pseudomonas were confirmed through amplicons of gyrB 222 [31], toxA 367[32], and ecfX 528 [33] gene fragments, respectively. The identity of Aeromonas species was determined by screening them for the presence of specific virulent genes aerolysin (aerA) [34] and haemolysin (hylH) [35].

PCRs were performed using oligonucleotide primer combinations and cycling conditions that appear in Tables1 and 2. Amplifications were performed using a Peltier Thermal Cycler (model-PTC-220DYAD DNA ENGINE, MJ Research Inc., USA). The reactions were prepared in 25 μL volumes that constituted 1 μg/μL of the template DNA, 50 pmol of each oligonucleotide primer set, 1x PCR master mix, and RNase free water. All PCR reagents used were Fermentas, USA, products supplied by Inqaba Biotechnological Industries Pty Ltd., Pretoria, South Africa. All PCR products were stored at 4°C until used for further analysis.

Table 1: Oligonucleotide primers that were used for specific detection of Pseudomonasspecies.

Primer	Oligonucleotide sequence (5′-3′)	Target and size (pb)	PCR cycling conditions
ECF1 ECF2	ATGGATGAGCGCTTCCGTG TCATCCTTGCCTCCCTG	*ecfX*, (528)	35X 94°C for 45 s 58°C for45 s 72°C for 1 m
GyrPA-398 GyrPA-620	CCTGACCATCCGTCGCCACAAC CGCAGCAGGATGCCGACGCC	gyrB, (222)	35X 94°C for 45 s 66°C for45 s 72°C for 1 m
ETA1 ETA2	GACAACGCCCTCAGCATCACCAGC CGCTGGCCCATTCGCTCCAGCGCT	*toxA*, (367)	35X 94°C for 45 s 66°C for45 s 72°C for 1 m

Initial denaturing step of 95°C for 5 min and final strand extension of 72°C for 5 min.

Table 2: Oligonucleotide primers that were used to detect virulence genes in Aeromonasspecies.

Genes	Oligonucleotide sequence (5′-3′)	Target gene and size (bp)	PCR cycling conditions
aer A	Aer 2F: AGCGGCAGAGCCCGTCTATCCA Aer 2R: AGTTGGTGGCGGTGTCGTAGCG	*aerA* (416)	30X 95°C for 2 m 55°C for 1 m 72°C for 1 m
hyl H	Hyl 2F: GGCCCGTGGCCCGAAGATGCAGG Hyl 2R: CAGTCCCACCCACTTC	*hylH* (597)	30X 95°C for 2 m 55°C for 1 m 72°C for 1 m

Initial denaturing step of 95°C for 5 min and final strand extension of 72°C for 7 min.

Electrophoresis of PCR Products

PCR products were separated by electrophoresis on 2% (w/v) agarose gel. Electrophoresis was conducted in a horizontal Pharmacia Biotech equipment system (model Hoefer HE 99X; Amersham Pharmacia Biotech, Sweden) for 2 h at 60 V using 1x TAE buffer (40 mM Tris, 1 mM EDTA, and 20 mM glacial acetic acid, pH 8.0). Each gel contained a 100 bp DNA molecular weight marker (Fermentas, USA). The gel was stained in ethidium bromide (0.1 µg/mL) for 15 min and amplicons were visualized under UV light. A Gene Genius Bioimaging system (Syngene, Synoptics, UK) was used to capture the image using GeneSnap (version 3.07.01) software (Syngene, Synoptics, UK).

Statistical Analysis

Cluster analysis based on the antibiotic inhibition zone diameter data of different organisms isolated from different sites was determined using Wards algorithm and Euclidean distance of Statistica version 7.

RESULTS

Occurrence of Coliform Bacteria and Aeromonas and Pseudomonas Species in Water

The primary aim of this study was to determine the levels of environmental bacteria from source and drinking water from Mafikeng. Raw water from Modimola dam and Molopo eye, treated water from these two sites, and mixed water were analysed for the presence of total coliforms, faecal coliforms, heterotrophic bacteria, andAeromonas and Pseudomonas species. Table 3 shows the average number of different organisms isolated from the various sites during summer and winter. Heterotrophic bacteria were isolated from all the sampling sites in both seasons and their occurrence was high. Moreover, Pseudomonas species, faecal and total coliforms were the most prevalent during all the seasons in both raw and treated water from Modimola dam. However,Aeromonas species were isolated only from the raw water samples, not from treated water from all the sites. The numbers of the different organisms isolated were higher in summer than in winter.

Table 3: Average number of microorganisms isolated.

Site	Seasons	FC	TC	H	Ae	Ps
Molopo eye (ME)	Summer	30	40	>100	42	>100
	Winter	50	40	5	28	>100
Modimola dam (MD)	Summer	>100	>100	>100	>100	>100
	Winter	35	60	20	6	>100
Treated water ME	Summer	15	0	17	0	0
	Winter	0	0	7	0	0
Treated water MD	Summer	15	6	20	0	>100
	Winter	0	0	7	0	>100
Mixed water	Summer	5	5	10	0	0
	Winter	0	0	3	0	1

FC: faecal coliforms, TC: total coliforms, H: heterotrophic bacteria, and Ae: *Aeromonas*.

No total coliforms were observed in treated water from the Molopo eye. Pseudomonas species were isolated in large numbers from Modimola dam both in raw and treated water from this site.

Biochemical Tests Used to Identify the Isolates

API tests results are listed in Table 4. Various species were isolated from the water sources particularly from the untreated water. The diversity of bacteria from this source was greater than from the treated water.Pseudomonas spp. were isolated from the Modimola dam (untreated as well as treated) water, the Molopo eye (untreated only), and the mixed water. Aeromonas spp. were isolated from the untreated as well as the mixed water. Aeromonas and Pseudomonas species isolated in this study were -haemolytic and indicated potential pathogenicity.

Table 4: Identification of the isolates from the treated (drinking) and source water using biochemical tests.

Site	API 20E (presumptive isolates)
Modimola dam (MD) untreated water	*Pseudomonas aeruginosa* *Pseudomonas luteola* *Aeromonas hydrophila* *E. coli* *Serratia odorifera* *Serratia liquefaciens* *Proteus vulgaris* *Providencia rettgeri1* *Chryseobacterium meningosepticum*
MD treated water	*Pseudomonas aeruginosa* *Pseudomonas luteola* *Serratia odorifera* *E. coli*
Molopo eye (ME) untreated water	*Serratia odorifera* *Serratia liquefaciens* *Aeromonas salmonicida* spp. *Pseudomonas oleovorans* *Chryseobacterium meningosepticum* *Salmonella choleraesuis* *Enterobacter asburiae*
ME treated water	*Serratia liquefaciens* *Myroides* spp.
Mixed water	*Serratia liquefaciens* *Serratia odorifera* *Aeromonas salmonicida* spp. *Pseudomonas aeruginosa* *Myroides* spp.

Antibiotic-Resistant Data of Different Isolates from Drinking Water

Isolates were subjected to an antibiotic susceptibility test using 11 different antibiotics from which their antibiotic resistance profiles and multiple antibiotic resistance phenotypes were compiled. The results obtained are depicted in Tables 5, 6, 7, and 8. The results revealed that a large proportion of the environmental isolates were resistant to erythromycin, followed by trimethoprim and amoxicillin. None of the isolates were resistant to ciprofloxacin and only very few isolates from Modimola dam were resistant to streptomycin and neomycin. Despite the fact that all isolates from Molopo eye, treated water from Molopo eye and Modimola dam, and mixed water were susceptible to neomycin, only a small proportion (20%) from Modimola dam was resistant to this antibiotic. Multiple antibiotic-resistant (MAR - 3 to 8 antibiotics) bacteria were common among the isolates.

Table 5: Percentage of total coliforms resistant to various antibiotics.

Site	API 20E (presumptive isolates)
Modimola dam (MD) untreated water	*Pseudomonas aeruginosa* *Pseudomonas luteola* *Aeromonas hydrophila* *E. coli* *Serratia odorifera* *Serratia liquefaciens* *Proteus vulgaris* *Providencia rettgeri1* *Chryseobacterium meningosepticum*
MD treated water	*Pseudomonas aeruginosa* *Pseudomonas luteola* *Serratia odorifera* *E. coli*
Molopo eye (ME) untreated water	*Serratia odorifera* *Serratia liquefaciens* *Aeromonas salmonicida* spp. *Pseudomonas oleovorans* *Chryseobacterium meningosepticum* *Salmonella choleraesuis* *Enterobacter asburiae*
ME treated water	*Serratia liquefaciens* *Myroides* spp.
Mixed water	*Serratia liquefaciens* *Serratia odorifera* *Aeromonas salmonicida* spp. *Pseudomonas aeruginosa* *Myroides* spp.

Table 6: Percentage of faecal coliforms resistant to various antibiotics.

Antibiotics	KF	AP	C	E	OT	K	TM	S	A	NE	CIP
Molopo eye $N = 5$	40	60	0	100	40	20	80	0	60	0	0
Modimola dam $N = 4$	50	50	25	50	50	25	50	0	100	0	0

Table 7: Percentage of heterotrophic bacteria resistant to various antibiotics.

Antibiotics	KF	AP	C	E	OT	K	TM	S	A	NE	CIP
Molopo eye $N = 5$	80	60	40	80	20	40	80	0	80	0	0
Modimola dam $N = 5$	60	80	20	20	0	0	80	0	60	0	0
Treated Molopo eye $N = 4$	50	50	0	0	50	25	50	0	50	0	0
Treated dam $N = 3$	66.7	66.7	66.7	33.3	66.7	0	100	0	66.7	0	0
Mixed water $N = 2$	50	50	0	0	50	0	100	0	50	0	0

Table 8: Percentage of Aeromonas spp. resistant to various antibiotics.

Antibiotics	KF	AP	C	E	OT	K	TM	S	A	NE	CIP
Molopo eye $N = 4$	25	25	25	100	50	0	50	0	75	0	0
Modimola dam $N = 2$	50	100	50	100	100	50	100	50	100	50	0

KF: cephalothin, AP: ampicillin, C: chloramphenicol, E: erythromycin, OT: oxytetracycline, K: kanamycin, TM: trimethoprim, S: streptomycin, A: amoxicillin, NE: neomycin, and CIP: ciprofloxacin.

Faecal coliforms from all sites were susceptible to streptomycin, neomycin, and ciprofloxacin, respectively. The highest number of bacteria was resistant to erythromycin. However, two isolates from Modimola dam were susceptible to erythromycin. Heterotrophic bacteria were isolated from all sites and were also resistant to several antibiotics. Isolates from all sites were susceptible to streptomycin, neomycin, and ciprofloxacin.Aeromonas species isolated were all resistant to erythromycin. All of these species were susceptible to ciprofloxacin. However, isolates from Molopo eye were susceptible to kanamycin, neomycin, and streptomycin while isolates from the Modimola dam were resistant to most antibiotics. All these isolates were resistant to ampicillin, oxytetracycline, trimethoprim, and amoxicillin. There were single isolates from Pseudomonas spp. for Modimola dam, Molopo eye (untreated water sources), and the Modimola dam treated water and mixed treated water. All the isolates were resistant to up to eight antibiotics but all were susceptible to streptomycin, neomycin, and ciprofloxacin.

Predominant Multiple Antibiotic-Resistant (MAR) Phenotypes Observed among the Bacterial Groups

Different types of multiple antibiotic resistance patterns were observed amongst all the bacterial groups isolated from the various sites. In some cases, some of the isolates were resistance to up to ten different antibiotics. The predominant antibiotic-resistant phenotypes that were obtained for the groups from different sites are depicted in Table 9. Similar types of MAR phenotypes were observed in isolates from groups.

Table 9: Prevalent antibiotic resistance phenotypes observed amongst the bacterial groups.

Bacterial group	Antibiotic resistance phenotype
Faecal coliform	KF-AP-C-E-OT-K-TM-A
Total coliform	KF-AP-C-E-OT-K-TM-A
Heterotrophic bacteria	KF-AP-C-E-OT-K-TM-A
Aeromonas sp.	KF-AP-C-E-OT-K-TM-S-A-NE
Pseudomonas sp.	KF-AP-C-E-OT-K-TM-A

KF: cephalothin, AP: ampicillin, C: chloramphenicol, E: erythromycin, OT: oxytetracycline, K: kanamycin, TM: trimethoprim, S: streptomycin, A: amoxicillin, and NE: neomycin.

Cluster Analysis of the Isolates Based on Inhibition Zone Diameter Data

All the Pseudomonas, Aeromonas, and heterotrophic bacterial isolates were subjected to cluster analysis based on their inhibition zone diameter (IZD) data and a dendrogram was generated using Ward's method. This approach was used as a tool in determining the commonness and in resolving differences between the MAR phenotypes of different isolates.

Pseudomonas, Aeromonas, and heterotrophic bacteria isolated from all sites were subjected to cluster analysis and two main clusters were observed (Figure 1) with the number of isolates from different sites within the clusters being depicted in Table 10. Both clusters have subclusters. Cluster 1 is a large cluster and has a total of 21 Aeromonas and heterotrophic bacterial isolates, while cluster 2 is a mixed cluster with Pseudomonas,Aeromonas, and heterotrophic bacteria. Pseudomonas is only found in cluster 2.

Table 10: Number of heterotrophic bacteria (H), Pseudomonas (PS), and Aeromonas (AE) isolated from different sites within the various clusters.

Site	Cluster 1 (Number of isolates = 21)			Cluster 2 (Number of isolates = 8)		
	H	AE	PS	H	AE	PS
Dam	5	2	0	0	1	1
Eye	4	2	0	1	1	1
Treated water dam	2	0	0	1	0	1
Treated water eye	3	0	0	0	0	0
Mixed water	2	0	0	0	0	1

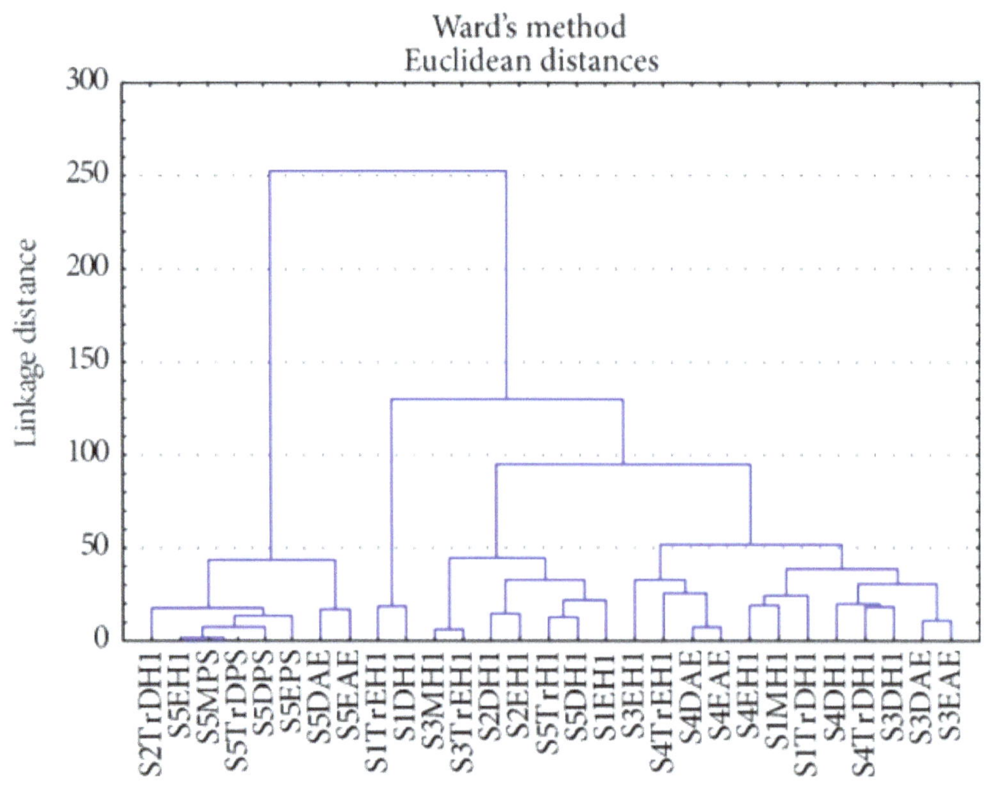

Figure 1: Cluster analysis for Pseudomonas, Aeromonas, and heterotrophic bacteria isolated from different sites within the various clusters (PS: Pseudomonas; AE: Aeromonas, H: heterotrophic bacteria).

The cluster analysis did not demonstrate groupings based on bacterial groups or sites. This suggests that all the isolates had similar chemical exposure histories.

Molecular Identification of Pseudomonas and Aeromonas Species

Specific PCR was used to determine the identities of presumptive Pseudomonas species through amplification of the gyrB, toxA, and the ecfX gene fragments. Figures 2 and 3 show agarose (1% w/v) gels indicating gyrB andecfX gene fragments generated by PCR using genomic DNA extracted from Pseudomonas species isolated from different sites. Gel electrophoresis of PCR products revealed the desired 222 bp and 528 bp fragments for thegyrB and ecfX gene fragments, respectively. A total of 61 isolates were screened and 17 were positively identified.

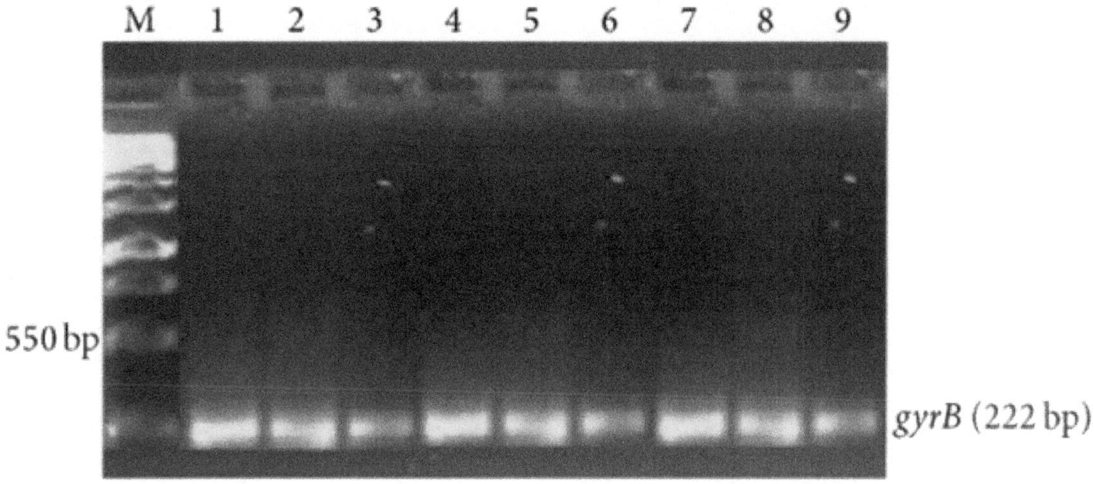

Figure 2: Image of a composite agarose (1% w/v) gel depicting DNA extracted fromPseudomonas species. Lane M (1 kb DNA Ladder); Lane 1–9 (gyrB gene fragments (222 bp) from Pseudomonas species isolated from different sites.

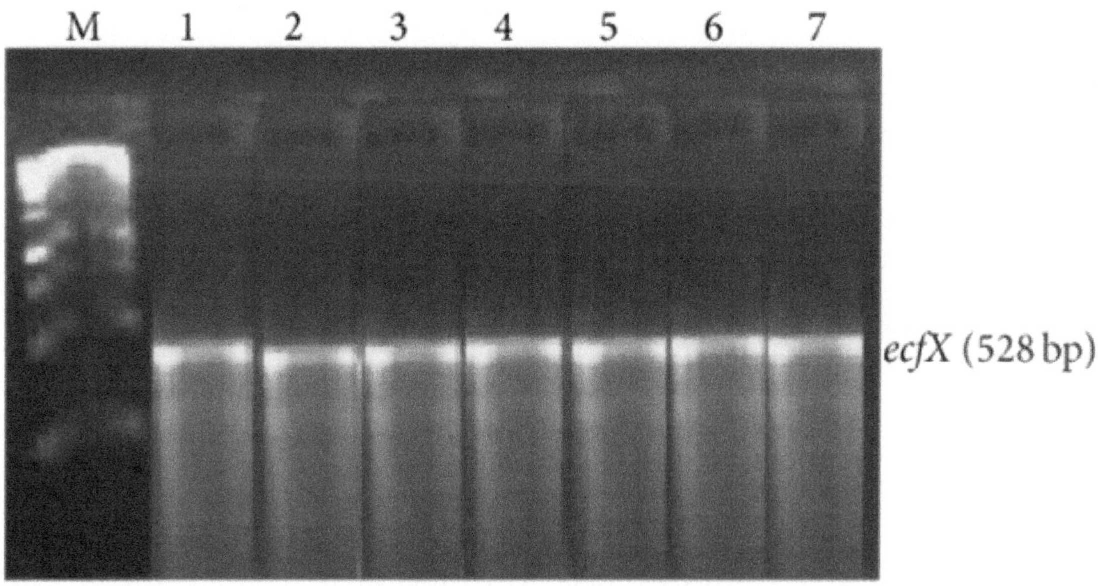

Figure 3: Image of a composite agarose (1% w/v) gel depicting genomic DNA extracted from Pseudomonas species. Lane M (1 kb DNA Ladder); Lane 1–7 ecfX gene fragments from Pseudomonas species isolated from different sites.

The identity of Aeromonas species was determined by screening them for the presence of specific virulent genes aerolysin (aerA) [34] and haemolysin (hylH) [35]. Figure 4 shows agarose (1% w/v) gels indicating hylH gene fragments generated by PCR using genomic DNA extracted from Aeromonas species isolated from different sites.

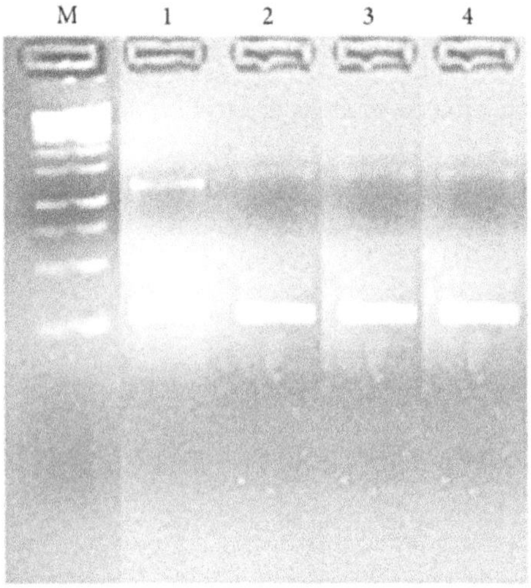

Figure 4: Image of a composite agarose (1% w/v) gel depicting the hylH gene fromAeromonas species. Lane M (1 kb DNA Ladder); Lanes 1–4 (hylH gene fragments fromAeromonas species isolated from different sites).

PCR results depicted in Figures 2–4 support the biochemical identification of the Pseudomonas spp. andAeromonas spp. Furthermore, it demonstrated that certain virulence genes were present in these isolates. This is also supported by positive -haemolytic activity indicating the pathogenic potential of these environmental isolated species.

DISCUSSION

One of the objectives of this study was to isolate environmental bacteria from surface and drinking water in Mafikeng, North West Province, South Africa. A motivation for this study was the numerous reports about the occurrence of pathogenic microorganisms in drinking water and the associated diseases [6, 36–40]. In addition to this, resistance of microorganisms to antibiotics of clinical interest has previously been reported in the area [24, 26, 29]. The study demonstrated the occurrence of total coliforms, faecal coliforms, heterotrophic bacteria, and Aeromonas and Pseudomonas in water samples analysed which indicated the incidence of water contamination as some of these species are indicators of faecal contamination [8]. These organisms may harbour potential pathogens and the presence of pathogenic organisms that can pose severe health risks to consumers in general and immunocompromised individuals in particular [5]. Reduction in the number of bacteria in the treated water could be due to the treatment process, when comparing drinking water to raw water. However, occurrence of bacteria in the water after treatment could also harbour potential pathogens and the health risk caused by these should be taken into consideration when water is distributed. This is of particular importance when the drinking water abstraction and purification facility are at a relatively short distance from the sewage treatment and effluent disposal facility. In Mafikeng, the latter is the case. One of the waste water treatment plants in Mafikeng is situated on the Modimola dam and is upstream from the drinking water purification plant. This may explain the larger number and diversity of isolates from this water. Results from this and other studies show that drinking water produced from the Modimola dam should be further analysed and that general microbiological tests that only include faecal coliforms and E. coli may be insufficient.

Molopo eye is a protected spring into which no treated sewage effluent is being discharged. However, there is housing development that is linked to septic tanks. The occurrence of bacteria at this site may be due to contamination from human activity (septic tank, diving), pets, and wild birds. The levels of bacteria from this site were lower and after treatment very low numbers of heterotrophic bacteria and no Aeromonas andPseudomonas spp. were detected. However, the overall results obtained in the present study indicated that the organisms in treated water may have survived the treatment process. When the levels of Pseudomonas spp. in the raw water were high, this increased the chances of finding this species in treated water.

A further objective of this study was to characterise the isolates using their antibiotic resistance profiles. The results revealed that a large proportion of the environmental isolates were resistant to erythromycin followed by trimethoprim and amoxicillin. The trend was in accordance with earlier studies that showed resistance towardsβ-lactam, macrolides, and phenicols [41, 42]. All the isolates from all the sites were resistant to erythromycin which was in accordance with the study conducted by [20] where most isolates were also resistant to erythromycin and tetracycline (92.9%). A large percentage (60%–100%) of isolates were also resistant to trimethoprim and ampicillin. All these results could be attributed to the overuse of these antibiotics in the clinical and veterinary setting. All isolates were found to be susceptible to streptomycin and ciprofloxacin in line with an earlier observation [43]. This is not in agreement with yet another observation [44] where increased resistance against ciprofloxacin in E. coli and Pseudomonas was observed.

Coliforms detected in all samples in the present study were resistant to 3 or more classes of antibiotics. This is supported by previous studies which obtained similar results [19, 20]. Multiple antibiotic resistance (MAR) inE. coli and Pseudomonas spp. isolated from environmental samples and hospitalized patients had previously been reported [45–49]. These studies support the findings of the present study. Antibiotic resistance patterns were generally similar in all different sites. However, the largest numbers of isolates that were resistant to the largest numbers of antibiotics were from the Modimola dam.

Although heterotrophic bacteria were isolated from all sites and exhibited antibiotic resistance, isolates from all sites were susceptible to streptomycin, neomycin, and ciprofloxacin which agree with the susceptibility to ciprofloxacin observed by Biyela et al. [5]. They found the highest resistant towards trimethoprim followed byβ-lactams and tetracycline. This was also similar to the results by Biyela et al. [5]. All isolates from treated water from Molopo eye and mixed water were sensitive to erythromycin.

High loads of heterotrophic bacteria and opportunistic pathogens that harbour multiple drug resistance determinants pose significant health hazards to consumers, especially those whose immune systems are compromised [50]. Associated virulence factors give heterotrophic bacteria the ability to act as opportunistic pathogens [10]. These bacteria (heterotrophic bacteria) are generally harmless; but some may harbour pathogenic features which may cause potential health risks to humans and animals. Thus, the concern that we report here is the high levels of heterotrophic bacteria from water sources, particularly drinking water from Mafikeng. What is also of concern are the multiple antibiotic resistance patterns observed amongst these isolates. This may be a therapeutic challenge if individuals infected by these opportunistic pathogens are to be treated by available antibiotics.

In a study conducted by Sharma et al. [7] all Aeromonas spp. isolated from fresh water were resistant to ampicillin. However, in this study resistance to tetracycline, trimethoprim, and cephalothin was observed. Furthermore, resistance of Aeromonas and Pseudomonas spp. have also been reported among isolates from municipally treated tap and raw water samples [51, 52]. These findings support the observations in the present study that shows resistance of Aeromonas and Pseudomonas spp. to a variety of antibiotics.

Wide distribution of antibiotic-resistant bacteria in surface and ground waters has been reported in previous studies [53, 54] and results of the present study is thus not uncommon. There is, however, concern about the increase in the incidence of MAR in organisms from various sources [55]. Resistance could be attributed to heavy contamination from sewage effluent, surface runoff, agricultural activities, wild life, industrial pollution, and so forth. Considering the high incidence of HIV/AIDS in South Africa, the importance of such findings cannot be overemphasized. The results of this study on bacterial resistance profiles are consistent with previous studies in other surface and drinking water systems [10, 20, 55].

Besides receiving treated sewage effluent, Modimola dam also supports small scale farming communities close to the dam. Livestock graze around the dam and their faeces is washed off into the dam during the rainy season. Therefore, the dam could serve as a reservoir of antibiotic-resistant bacteria with sewage contamination contributing to the dissemination of antibiotic-resistant bacteria in this environment. There may thus be a link between the resistant isolates from sewage treatment plant, Modimola dam, and the treated water. This aspect, however, needs to be demonstrated which could be done in follow-up studies. Cluster analysis indicated that isolates from all sites had similar antibiotic resistance profiles. This data indicates common chemical exposure histories of the various isolates. Such an approach could be used to investigate the relationship between sewage treatment plant, Modimola dam, and treated water. The results from the present study show that bacterial isolates from the Modimola dam and treated water had similar antibiotic profiles.

Another objective of this study was to identify the Pseudomonas and Aeromonas species using PCR. Isolates were identified using gyrB, ecfX, and hylH. The ecfX gene encodes an extra cytoplasmic function sigma factor, which may be involved in haem uptake or virulence [56], whereas the gyrB gene encodes the DNA gyrase subunit B, a protein which plays a crucial role in the DNA replication process [34] and toxA encoding the exotoxin A precursor [33]. PCR assay targeting the ecfX and gyrB genes is suitable for the identification of P. aeruginosa [56]. In this study, biochemical identification methods were supported by PCR assays to identify thePseudomonas spp. isolated.

Application of PCR technique to target gyrB, aerA, and hylH genes are molecular markers for identifying and screening potentially virulent Aeromonas species in food and the environment [34, 35]. In this study, this approach supported biochemical data to identify the various Aeromonas spp. The detection of Aeromonasspecies in the present study that harbour putative virulence factors and resistance to various antibiotics indicates that drinking water that is supplied to the community could serve as a source for the transmission of pathogens to humans [51, 57].

Overall, the results from the present highlight the need to conduct studies to determine the prevalence of antibiotic-resistant genes among environmental isolates and the dissemination of resistant genes among the pathogens. This would provide information about the health risks associated with the consumption of contaminated water.

CONCLUSIONS

An evaluation of the bacteriological quality of drinking water in the present study confirmed the presence of various bacterial species including opportunistic pathogens such as Aeromonas and Pseudomonas spp. These organisms were resistant to several classes of antibiotics. Undesirable properties of water quality caused by the presence of drug-resistant bacteria can pose a negative impact on human health.

The data on multiple antibiotic resistance (MAR) profiles of bacterial isolates from water and the resistance patterns of organisms in drinking water in Mafikeng suggested that there has been an indiscriminate use of the antibiotics tested. The high prevalence of multiple antibiotic-resistant organisms in the drinking water distribution system could potentially pose a threat to humans consuming this water. The presence of MAR organisms in the drinking water of Mafikeng, South Africa, is an important health concern due to the risk of developing waterborne diseases and the health risks associated with immunocompromised patients living in the area. It is therefore imperative to monitor the quality of water and strict quality control measures should be put in place to ensure the effective treatment of drinking water. Since the Modimola dam is the recipient of treated waste water as well as the water source for drinking water production, extremely strict measures should guide the waste water treatment plant. The quality of effluent leaving in this plant should be extremely high. This would decrease the load of microorganisms (and other contaminants) allowed to enter the dam. Such measures will improve the overall quality of water available for drinking water production, preventing outbreaks and spreading water borne diseases. Antibiotic resistance surveillance can be used as tool to control the problem of antibiotic resistance and to educate the public on the consequences of the misuse of antibiotics and also to regulate the usage of drugs in both human and veterinary medicine. It is also helpful to formulate guidelines for the optimal use of antibiotics. Further studies should be conducted to assess the level of antibiotics in water and the potential risks associated with human consumption of polluted water. It is also very important that findings from studies such as this one should be disseminated to the relevant stakeholders and the affected communities.

CONFLICT OF INTERESTS

The authors declare that there is no conflict of interests regarding the publication of this paper.

ACKNOWLEDGMENTS

The authors would like to thank the National Research Foundation (FA20040101000030 and FA2006040700029) for the financial support of this study. The assistance received from Mrs. Rika Huyser and the employees of Mmabatho Water Works is also hereby acknowledged.

REFERENCES

1. C. Faria, I. Vaz-Moreira, E. Serapicos, O. C. Nunes, and C. M. Manaia, "Antibiotic resistance in coagulase negative staphylococci isolated from wastewater and drinking water," Science of the Total Environment, vol. 407, no. 12, pp. 3876–3882, 2009.
2. WHO, Guidelines for Drinking Water Quality, vol. 1, World Health Organization, Geneva, Switzerland, 3rd edition, 2004.
3. J. Fawell and M. J. Nieuwenhuijsen, "Contaminants in drinking water Environmental pollution and health," British Medical Bulletin, vol. 68, no. 1, pp. 199–208, 2003.
4. S. Völker, C. Schreiber, and T. Kistemann, "Drinking water quality in household supply infrastructure—a survey of the current situation in Germany," International Journal of Hygiene and Environmental Health, vol. 213, no. 3, pp. 204–209, 2010.
5. P. T. Biyela, J. Lin, and C. C. Bezuidenhout, "The role of aquatic ecosystems as reservoirs of antibiotic resistant bacteria and antibiotic resistance genes," Water Science and Technology, vol. 50, no. 1, pp. 45–50, 2004.
6. C. L. Obi, N. Potgieter, P. O. Bessong, and G. Matsaung, "Assessment of the microbial quality of river water sources in rural Venda communities in South Africa," Water SA, vol. 28, no. 3, pp. 287–292, 2002. ·
7. A. Sharma, N. Dubey, and B. Sharan, "Characterization of aeromonads isolated from the river Narmada, India," International Journal of Hygiene and Environmental Health, vol. 208, no. 5, pp. 425–433, 2005. ·
8. L. F. Webster, B. C. Thompson, M. H. Fulton et al., "Identification of sources of Escherichia coli in South Carolina estuaries using antibiotic resistance analysis," Journal of Experimental Marine Biology and Ecology, vol. 298, no. 2, pp. 179–195, 2004.
9. S. Dumontet, K. Krovacek, S. B. Svenson, V. Pasquale, S. B. Baloda, and G. Figliuolo, "Prevalence and diversity of Aeromonas and Vibrio spp. in coastal waters of Southern Italy," Comparative Immunology, Microbiology and Infectious Diseases, vol. 23, no. 1, pp. 53–72, 2000.
10. D. Pavlov, C. M. E. de Wet, W. O. K. Grabow, and M. M. Ehlers, "Potentially pathogenic features of heterotrophic plate count bacteria isolated from treated and untreated drinking water," International Journal of Food Microbiology, vol. 92, no. 3, pp. 275–287, 2004.
11. R. Eberhard and P. Robinson, Guidelines for the Development of National Water Policies and Strategies to Support IWRM 2003 Draft, SADC Water Sector Co-Ordination Unit, Gaborone, Botswana, 2003.
12. J. R. Adewumi, A. A. Ilemobade, and J. E. Van Zyl, "Treated wastewater reuse in South Africa: overview, potential and challenges," Resources, Conservation and Recycling, vol. 55, no. 2, pp. 221–231, 2010.
13. International Water Association (IWA), 2008.
14. R. Mckenzie, J. Buckle, W. Wegelin, and N. Meyer, "Water demand management cookbook, rand water in collaboration with WRP, managing water for African Cities UN Habitat Programme, New York," Medical Bulletin, vol. 68, pp. 199–208, 2003.
15. S. Dimitriadis, Issues Encountered in Advancing Australia›s Water Recycling Schemes, Research Brief, Parliamentary Library, Parliament of Australia, Department of Parliamentary Services, 2005.
16. B. Vaseeharan, P. Ramasamy, T. Murugan, and J. C. Chen, "In vitro susceptibility of antibiotics againstVibrio spp. and Aeromonas spp. isolated from Penaeus monodon hatcheries and ponds," International Journal of Antimicrobial Agents, vol. 26, no. 4, pp. 285–291, 2005.
17. J. L. Martinez, "Environmental pollution by antibiotics and by antibiotic resistance determinants,"Environmental Pollution, vol. 157, no. 11, pp. 2893–2902, 2009.
18. M. Akram, M. Shahid, and A. U. Khan, "Etiology and antibiotic resistance patterns of community-acquired urinary tract infections in J N M C Hospital Aligarh, India," Annals of Clinical Microbiology and Antimicrobials, vol. 6, article 4, pp. 1–10, 2007.
19. A. Łuczkiewicz, K. Jankowska, S. Fudala-Ksiazek, and K. Olańczuk-Neyman, "Antimicrobial resistance of fecal indicators in municipal wastewater treatment plant," Water Research, vol. 44, no. 17, pp. 5089–5097, 2010.
20. J. E. Moore, P. J. A. Moore, B. C. Millar et al., "The presence of antibiotic resistant bacteria along the River Lagan," Agricultural Water Management, vol. 98, no. 1, pp. 217–221, 2010.
21. T. Schwartz, W. Kohnen, B. Jansen, and U. Obst, "Detection of antibiotic-resistant bacteria and their resistance genes in wastewater, surface water, and drinking water biofilms," FEMS Microbiology Ecology, vol. 43, no. 3, pp. 325–335, 2003.
22. T. I. Lobova, Y. V. Barkhatov, O. V. Salamatina, and L. Y. Popova, "Multiple antibiotic resistance of heterotrophic bacteria in the littoral zone of Lake Shira as an indicator of human impact on the ecosystem," Microbiological Research, vol. 163, no. 2, pp. 152–160, 2008.
23. Y. Zhang, C. F. Marrs, C. Simon, and C. Xi, "Wastewater treatment contributes to selective increase of antibiotic resistance among Acinetobacter spp.," Science of the Total Environment, vol. 407, no. 12, pp. 3702–3706, 2009.
24. S. G. Mulamattathil, H. A. Esterhuysen, and P. J. Pretorius, "Antibiotic-resistant Gram-negative bacteria in a virtually closed water reticulation system," Journal

of Applied Microbiology, vol. 88, no. 6, pp. 930–937, 2000.

25. J. Lin and P. T. Biyela, "Convergent acquisition of antibiotic resistance determinants amongst the Enterobacteriaceae isolates of the Mhlathuze River, KwaZulu-Natal (RSA)," Water SA, vol. 31, no. 2, pp. 257–260, 2005.

26. C. W. Kinge and M. Mbewe, "Characterisation of Shigella species isolated from river catchments in the North West province of South Africa," South African Journal of Science, vol. 106, no. 11-12, article 211, 2010.

27. SANS: South African National Standards, "Drinking water—part 1: microbiological , physical, chemical, aesthetic and chemical determinands," SABS Standards Division, Pretoria, South Africa, 2011.

28. W. Kirby, W. Bauer, J. Sherris, and M. Turck, "Antibiotic susceptibility testing by a standardized single disk method," The American Journal of Clinical Pathology, vol. 45, no. 4, pp. 493–496, 1966.

29. C. N. Ateba, M. Mbewe, and C. C. Bezuidenhout, "Prevalence of Escherichia coli O157 strains in cattle, pigs and humans in North West province, South Africa," South African Journal of Science, vol. 104, no. 1-2, pp. 7–8, 2008.

30. National Committee for Clinical Laboratory Standards, "Performance standards for antimicrobial disk and dilution susceptibility tests for bacterial isolates from animals," National Committee for Clinical Laboratory Standards, Wayne, Pa, USA, Approved standard, M31-A19, (11), 1999.

31. X. Qin, J. Emerson, J. Stapp, L. Stapp, P. Abe, and J. L. Burns, "Use of real-time PCR with multiple targets to identify Pseudomonas aeruginosa and other nonfermenting gram-negative bacilli from patients with cystic fibrosis," Journal of Clinical Microbiology, vol. 41, no. 9, pp. 4312–4317, 2003.

32. A. A. Khan and C. E. Cerniglia, "Detection of Pseudomonas aeruginosa from clinical and environmental samples by amplification of the exotoxin A gene using PCR," Applied and Environmental Microbiology, vol. 60, no. 10, pp. 3739–3745, 1994.

33. R. Lavenir, D. Jocktane, F. Laurent, S. Nazaret, and B. Cournoyer, "Improved reliability of Pseudomonas aeruginosa PCR detection by the use of the species-specific ecfX gene target," Journal of Microbiological Methods, vol. 70, no. 1, pp. 20–29, 2007.

34. E. Kaszab, S. Szoboszlay, C. Dobolyi, J. Háhn, N. Pék, and B. Kriszt, "Antibiotic resistance profiles and virulence markers of Pseudomonas aeruginosa strains isolated from composts," Bioresource Technology, vol. 102, no. 2, pp. 1543–1548, 2011.

35. A. H. Yousr, S. Napis, G. R. A. Rusul, and R. Son, "Detection of aerolysin and hemolysin genes inAeromonas spp. isolated from environmental and shellfish sources by polymerase chain reaction,"ASEAN Food Journal, vol. 14, no. 2, pp. 115–122, 2007.

36. A. I. Schäfer, H. M. A. Rossiter, P. A. Owusu, B. S. Richards, and E. Awuah, "Physico-chemical water quality in Ghana: prospects for water supply technology implementation," Desalination, vol. 248, no. 1–3, pp. 193–203, 2009.

37. P. T. Yillia, N. Kreuzinger, and J. M. Mathooko, "The effect of in-stream activities on the Njoro River, Kenya. Part I: stream flow and chemical water quality," Physics and Chemistry of the Earth, vol. 33, no. 8–13, pp. 722–728, 2008.

38. H. Schraft and L. A. Watterworth, "Enumeration of heterotrophs, fecal coliforms and Escherichia coli in water: comparison of 3M Petrifilm plates with standard plating procedures," Journal of Microbiological Methods, vol. 60, no. 3, pp. 335–342, 2005.

39. W. Germs, M. S. Coetzee, L. van Rensburg, and M. S. Maboeta, "A preliminary assessment of the chemical and microbial water quality of the Chunies River—Limpopo," Water SA, vol. 30, no. 2, pp. 267–272, 2004.

40. J. T. Carter, E. W. Rice, S. G. Buchberger, and Y. Lee, "Relationships between levels of heterotrophic bacteria and water quality parameters in a drinking water distribution system," Water Research, vol. 34, no. 5, pp. 1495–1502, 2000.

41. C. L. Obi, P. O. Bessong, M. Momba, N. Potgieter, A. Samie, and E. O. Igumbor, "Profiles of antibiotic susceptibilities of bacterial isolates and physico-chemical quality of water supply in rural Venda communities, South Africa," Water SA, vol. 30, no. 4, pp. 515–519, 2004.

42. P. Messi, E. Guerrieri, and M. Bondi, "Antibiotic resistance and antibacterial activity in heterotrophic bacteria of mineral water origin," Science of the Total Environment, vol. 346, no. 1–3, pp. 213–219, 2005. ·

43. J. Lin, P. T. Biyela, and T. Puckree, "Antibiotic resistance profiles of environmental isolates from Mhlathuze River, KwaZulu-Natal (RSA)," Water SA, vol. 30, no. 1, pp. 23–28, 2004.

44. G. J. Oudhuis, A. Verbon, J. A. A. Hoogkamp-Korstanje, and E. E. Stobberingh, "Antimicrobial resistance in Escherichia coli and Pseudomonas aeruginosa from Intensive Care Units in The Netherlands, 1998–2005," International Journal of Antimicrobial Agents, vol. 31, no. 1, pp. 58–63, 2008.

45. T. C. Siewicki, T. Pullaro, W. Pan, S. McDaniel, R. Glenn, and J. Stewart, "Models of total and presumed wildlife sources of fecal coliform bacteria in coastal ponds," Journal of Environmental Management, vol. 82, no. 1, pp. 120–132, 2007.

46. A. Sorlózano, J. Gutiérrez, J. Luna et al., "High presence of extended-spectrum β-lactamases and resistance to quinolones in clinical isolates of Escherichia coli," Microbiological Research, vol. 162, no. 4, pp. 347–354, 2007.

47. M. E. Scott, A. R. Melton-Celsa, and A. D. O›Brien, "Mutations in hns reduce the adherence of Shiga toxin-producing E. coli 091:H21 strain B2F1 to human colonic epithelial cells and increase the production of hemolysin," Microbial Pathogenesis, vol. 34, no. 3, pp. 155–159, 2003.

48. E. Tuméo, H. Gbaguidi-Haore, I. Patry, X. Bertrand, M. Thouverez, and D. Talon, "Are antibiotic-

resistant Pseudomonas aeruginosa isolated from hospitalised patients recovered in the hospital effluents?"International Journal of Hygiene and Environmental Health, vol. 211, no. 1-2, pp. 200–204, 2008.

49. E. Breidenstein, C. de la Fuente-Núñez, and R. Hancock, "Pseudomonas aeruginosa: all roads lead to resistance," Trends in Microbiology, vol. 19, no. 8, pp. 419–426, 2011.

50. M. I. Jeena, P. Deepa, K. M. Mujeeb Rahiman, R. T. Shanthi, and A. A. M. Hatha, "Risk assessment of heterotrophic bacteria from bottled drinking water sold in Indian markets," International Journal of Hygiene and Environmental Health, vol. 209, no. 2, pp. 191–196, 2006.

51. G. Emekdas, G. Aslan, S. Tezcan et al., "Detection of the frequency, antimicrobial susceptibility, and genotypic discrimination of Aeromonas strains isolated from municipally treated tap water samples by cultivation and AP-PCR," International Journal of Food Microbiology, vol. 107, no. 3, pp. 310–314, 2006. ·

52. E. E. Odjadjare, E. O. Igbinosa, R. Mordi, B. Igere, C. L. Igeleke, and A. I. Okoh, "Prevalence of multiple antibiotics resistant (MAR) Pseudomonas species in the final effluents of three municipal wastewater treatment facilities in South Africa," International Journal of Environmental Research and Public Health, vol. 9, no. 6, pp. 2092–2107, 2012.

53. S. Harakeh, H. Yassine, and E. Mutasem, "Antimicrobial-resistance patterns of Escherichia coli andSalmonella strains in the aquatic Lebanese environments," Environment Pollution, vol. 295, no. 2, pp. 503–511, 2005.

54. K. Kümmerer, "Antibiotic in the aquatic environment. A review: part 2," Chemosphere, vol. 75, pp. 435–441, 2009.

55. R. Shrivastava, R. K. Upreti, S. R. Jain, K. N. Prasad, P. K. Seth, and U. C. Chaturvedi, "Suboptimal chlorine treatment of drinking water leads to selection of multidrug-resistant Pseudomonas aeruginosa,"Ecotoxicology and Environmental Safety, vol. 58, no. 2, pp. 277–283, 2004.

56. S. N. Anuj, D. M. Whiley, T. J. Kidd et al., "Identification of Pseudomonas aeruginosa by a duplex real-time polymerase chain reaction assay targeting the ecfX and the gyrB genes," Diagnostic Microbiology and Infectious Disease, vol. 63, no. 2, pp. 127–131, 2009. ·

57. X. Li, M. Mehrotra, S. Ghimire, and L. Adewoye, "β-Lactam resistance and β-lactamases in bacteria of animal origin," Veterinary Microbiology, vol. 121, no. 3-4, pp. 197–214, 2007.

CHAPTER 10 FLOW CYTOMETRY TOTAL CELL COUNTS: A FIELD STUDY ASSESSING MICROBIOLOGICAL WATER QUALITY AND GROWTH IN UNCHLORINATED DRINKING WATER DISTRIBUTION SYSTEMS

G. Liu, E. J. Van der Mark, J. Q. J. C. Verberk, and J. C. Van Dijk

INTRODUCTION

Microbial regrowth during water distribution is undesirable since the regrowth can lead to hygienic, aesthetic, and technical problems, such as the proliferation of opportunistic pathogenic bacteria [1], an increase in taste and odor [2], and the biocorrosion of pipe material [3]. In practice, there are two approaches to limit the regrowth: maintaining disinfectant residuals and limiting the concentration of growth-promoting compounds in the treated water [4–7].

Biomass quantification is the most direct way to evaluate the microbial regrowth in distributed water. Worldwide, a heterotrophic plate count (HPC) is the primary parameter to quantify biomass [8]. The worldwide accepted guideline for HPC is 20 colony forming units per mL (CFU mL^{-1}) for treated water and 300 CFU mL^{-1} for distributed water [9]. In The Netherlands, the regulation for the quality of distributed water has been set to the geometric year mean of 100 CFU mL^{-1}. Despite its common use, the cultivation-dependent HPC method is increasingly challenged by the fact that only a small fraction (less than 1%) of the microorganisms in drinking water can be detected [5, 10, 11]. Other limitations, such as the fact that the method is time consuming and the results are influenced by various factors (e.g., temperature and needed culture time), have been well discussed previously [5, 11–14].

In the last decade, cultivation independent detection methods, such as adenosine triphosphate (ATP) [5, 13] and flow cytometry total cell count (FCM-TCC, referred to as TCC afterwards [11, 15, 16]) have gained foremost interest because these methods are accurate, rapid, quantitative, detect both cultivable and uncultivable microorganisms, and are easy to perform.

ATP is a parameter that assesses active biomass by quantifying the activated, energy-rich metabolic compounds present in all viable cells [17]. ATP has been proven to be a suitable parameter for biomass quantification in drinking water and has found a wide application including the analysis of groundwater [18, 19], assessment of biofilters in water treatment plants [20, 21], drinking water [11, 12, 22], and biofilms in drinking water distribution system (DWDS) [23, 24]. For microbial regrowth in DWDS, research evaluating the influence of water composition, distance, and season on distributed water ATP concentrations [5] has been conducted.

Citation: Liu G, Van der Mark EJ, Verberk QJC, Van Dijk JC (2013) Flow Cytometry Total Cell Counts: A Field Study Assessing Microbiological Water Quality and Growth in Unchlorinated Drinking Water Distribution Systems. BioMed Research International Volume 2013, Article ID 595872, 10 pages doi: http://dx.doi.org/10.1155/2013/595872.

As ATP is an indirect biomass quantification method, weak relationships between HPC and ATP were found in all the reported investigations.

Flow cytometry (FCM) is a routine method for characterizing cellular biology that has been used since the 1970s. One promising application of FCM is that it is able to characterize and distinguish different physiological states of microorganisms at the single cell level with high-throughput capacity [25]. As reported by Berney et al. [15], double staining with the nucleic acid-binding SYTO dyes and propidium iodide (PI) was used to distinguish between intact cells and membrane-damaged cells. Although most FCM is not used for the purpose of quantifying microorganisms, it became an attractive method for studying microbial growth in oligotrophic environments, for example, drinking water environments. It has been used for rapid detection bacteria associated with loss of disinfectant residual in drinking water distribution system (Hoefel et al., 2005) [26], Hammes et al. [11] developed an FCM quantification method to accurately count microbial cells at concentrations as low as 1,000 cells mL^{-1} and have demonstrated biomass changes during treatment. No relationship between TCC and HPC was observed. However, a good relationship between TCC and ATP was found [11, 12] and an even better relationship between intact biovolume and microbial ATP [13] in drinking water samples was seen. The changes in biomass during drinking water distribution have also been monitored by TCC [4]. However, in this research, only two locations in the distribution system were selected, which were within close proximity of the treatment plant.

The objectives of the present study were (1) to evaluate the application of TCC in DWDS, taking into account the influence of water composition, distance to the treatment plant, and water temperature; (2) to determine the relationship between TCC, ATP, Aeromonas, and HPC in distributed drinking water.

MATERIAL AND METHODS

Selected Treatment Plants and Sampling Locations

In this study, distribution systems of two treatment plants of the Dunea Water Company were selected. Both treatment plants (called, hereafter, TP1 and TP2) take source water from the Meuse River and the water is treated by similar processes. The source water, after pretreatment, is transported over 30 km to a dune area of natural lakes where it is infiltrated by gravity into the subsoil. After an average residence time of 2 months, the water is abstracted from the dunes. Afterwards, water is posttreated by softening, powdered activated carbon, aeration, rapid sand filtration (RSF), and slow sand filtration (SSF) before being pumped into the distribution system. The water is supplied without disinfectant.

A one-year-long monitoring program was carried out in the distribution systems of TP1 and TP2. Samples were taken from the supply area in the surrounding areas of the treatment plants (proximal area), the central part of the distribution systems, and a distal part of distribution systems. The average distance of the proximal, central and distal parts of the distribution system from each production plant is shown in Table 1 and Figure S1 (see supplementary materials available online at http://dx.doi.org/10.1155/2013/595872). To get representative and reliable results, the number of sampling locations was determined based on the supply connections in the area, and samples were taken randomly every month from locations on the same street. In total, 260 samples were collected in 2011. All samples from the distribution systems were taken directly from consumers' taps. Before taking the water samples, each tap was flushed until the water temperature remained stable for 30 seconds (with temperature meter with response time of 10 s, sensitivity of 0.1°C). Water samples were transported within two hours after sampling and stored at 4°C. Analyses were performed within 24 h after water sample collection.

Table 1: The source water and biological quality of treated water of selected treatment plants.

Parameters	TP1	TP2
Source water	AAR water	AAR water
AOC ($\mu g\ C\ L^{-1}$)	5 (±2)	11 (±8)
BFR ($pg\ ATP\ cm^{-2}\ day^{-1}$)	0.73	0.61
TCC ($cells\ mL^{-1}$)	120000 (±25000)	220000 (±30000)
ATP ($ng\ L^{-1}$)	0.5	0.5
Studying period	Jan. 2011–Dec. 2011	
Distance (km)		
Proximal	5 ± 0.8	4 ± 0.8
Central	11 ± 1.5	9 ± 1.1
Distal	15 ± 3	16 ± 2

*In the Table: AAR stands for artificial aquifer recharge; AOC (assimilable organic carbon), was averaged values measured from 2006 to 2012 and measured as described by van der Kooij [22]. For both treatment plants, $N = 25$. Detailed results are shown in supplementary data, Figure S3. BFR (biofilm formation rate), was measured in 2009 according to methods described by Van der Kooij [27]. ATP and TCC were measured monthly at treatment plants during this study.

Flow Cytometry Total Cell Count (TCC)

Total Cell Count and Damaged/Intact Cell Count

All collected samples were analyzed by TCC, and damaged cells and intact cells were distinguished by flow cytometer measurements. Staining and flow cytometry were done as described previously [11, 15]. In short, for working solutions, SYBR Green I (SG) (Invitrogen AG, Basel, Switzerland) was diluted 100× in anhydrous dimethylsulfoxide (DMSO) and propidium iodide (PI; 30 mM) and was mixed with the SG working solution at a ratio of 1 : 50 (SGPI). Both working solutions were stored at −20°C until use.

From every water sample analysis, two 1 mL subsamples were prepared and stained with both SG (for total cell counting) and SGPI (for intact cell counting) at 10 μL mL^{-1}. Before analysis, samples were incubated in darkness for 10 min (SG) and 15 min (SGPI). Prior to flow cytometric analysis, the water samples were diluted with 0.22 μm filtered, commercially available bottled water to 10% v/v of the initial concentration. Recalculation of the actual cell number was made based on the dilution ratio. TCC was performed with a BD ACCURI C6 flow cytometer (United States). Green fluorescence was collected at 520±10 nm, red fluorescence above 630 nm, and high-angle sideward scatter (SSC) at 488 nm.

Low Nucleic Acid Content (LNA) Bacteria and High Nucleic Acid Content Bacteria

The fluorescence intensity measured by FCM (indicated as FL1-A in Figure 1) has been used as an indicator of apparent cellular nucleic acid content [28, 29] and sideward scatter (SSC) has been applied as an indication of cellular size [28–30]. Based on the distinctly different fluorescence intensity and SSC signals detected by FCM in combination with nucleic acid stains, bacteria have been broadly classified into two groups: low nucleic acid content (LNA) bacteria and high nucleic acid content (HNA) bacteria. LNA bacteria and HNA bacteria were distinguished and quantified according to two clusters in FCM dot-plots of FL1-A and SSC-A (Figure 1 and Figure S2).

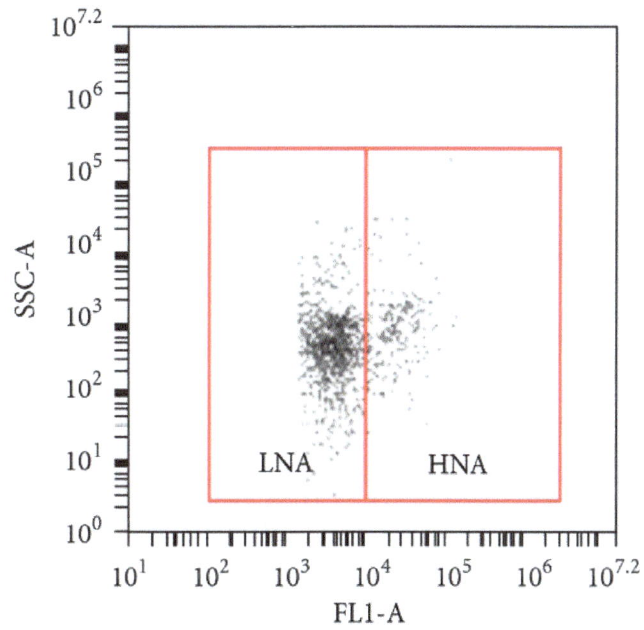

Figure 1: FCM results are represented in dot-plots of total cell counts, and low nucleic acid content (LNA) bacteria and high nucleic content acid (HNA) bacteria are indicated by two solid lines. LNA and HNA are two groups of bacteria generally classified by FCM measurements. Complete and original plot with noise background is shown in Figure S2.

Adenosine Triphosphate (ATP)

All collected samples were analyzed for ATP. Total ATP concentration was determined as described previously using the BacTiter-Glo reagent and a luminometer [20]. In short, ATP was released from suspended cells with nucleotide-releasing buffer (NRB, Celsis), and the luminometer was calibrated with solutions of free ATP (Celsis) in autoclaved tap water following the procedure as given by the manufacturer. The detection limit was 1 ng ATP L^{-1}.

Heterotrophic Plate Count (HPC)

Water samples were analyzed on HPC. The water samples were diluted tenfold in sterile drinking water and 50 μL of the appropriate dilution was spread in triplicate over the surface of R2A agar plates (Oxoid Led). Subsequently, agar plates were incubated at 25°C for 10 days. The number of colony forming units (CFUs) was determined.

Aeromonas

Samples were analyzed by Aeromonas plate counts. After filtering 100 mL water samples through a 0.45 μm filter, the filter was incubated with Aeromonas dextrin agar, and agar plates were incubated for 24 h at 30°C before the number of CFUs was determined.

Statistical Analyses

The obtained data were statistically analyzed to determine whether TCC and ATP values were significantly different. The normally distributed data were tested by the -test. The nonnormal distributed data were tested by nonparametric tests (Kruskal-Wallis tests). Differences were considered significant when the P value was lower than 0.05. Correlation analyses were done by determining Pearson's (to test for linear correlations) and Spearman's rho (to test for nonlinear correlations) correlation coefficients between paired values of TCC and ATP, TCC, and HPC. All statistical calculations were performed with Minitab 16.0.

RESULTS AND DISCUSSION

Microbiological Quality of the Water at the Two Treatment Plants

Table 1 includes the results of the microbiological parameters at TP1 and TP2. From these data it can be concluded that both distribution systems were supplied by biologically stable water.

Figure 2 gives the changes in the ATP and TCC numbers in the treatment train. It can be concluded that biomass was efficiently removed by the multiple barriers treatments: TP1 removed 62% TCC and 98% ATP; TP2 removed 47% TCC and 99% ATP. Comparing results from TP1 and TP2, the values in abstracted water from dune area of TP2 were higher than that of TP1. Since the same pretreated water is transported to the dune areas, the different values in abstracted water are probably related to the different soil compositions in the two dune areas. Another notable observation is that although ATP and TCC concentrations were generally decreasing in every treatment unit, an increase in TCC after softening in TP2 was observed. This may be related to some biological growth in TP2, possibly caused by the softening chemical. Moreover, the long-term monitoring of treated water AOC content in the two treatment plants (Figure S3 in supplementary data) revealed that TP1 has a stable performance, while some fluctuation was observed in TP2. However, these aspects were not within the scope of the current study. Detailed studies are necessary to understand the differences in performance of the two dune areas and the two treatment plants.

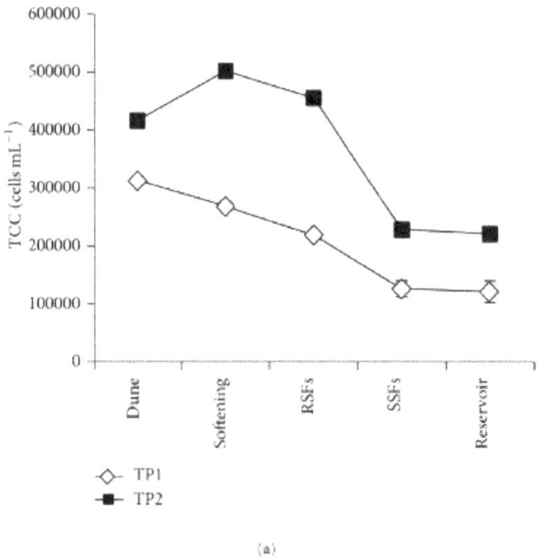

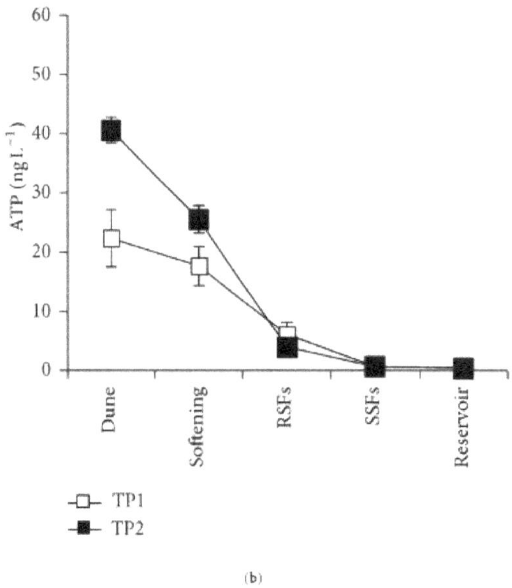

Figure 2: Changes of biomass during drinking water treatment: total cell concentration (a) (n=3) and ATP concentration (b) (n=3). RSF = rapid sand filtration; SSF = slow sand filtration.

TCC in Distributed Drinking Water

Effect of Water Composition on TCC Levels in Distributed Drinking Water

The TCC in the unchlorinated, distributed drinking water sampled from both distribution systems ranged from 0.5×10^5 cells mL^{-1} to $3.2 \times 10_5$ cells mL^{-1} (Figure 3(a), shown as boxplot). According to the literature, these values are typical for the cell concentration of unchlorinated drinking water [4, 11]. The damaged cells and intact cells were distinguished by flow cytometer measurements. It was found that most of the cells (more than 80%) were alive/intact (Figure S4 in supplementary data). That is related to the fact that no disinfectant is being used. TCC values in the TP2 system were significantly higher than those of the TP1 system ($P < 0.05$). This difference in TCC values originated at the treatment plants; it is consistent with the treated water AOC values (Table 1). The treated water of TP1 had relatively lower AOC than TP2. So it may be concluded that the water in the supply area of TP1 is slightly more biologically stable than the water from TP2. However, it should be noted that both TCC and AOC values of both systems are very low compared to studies in the literature.

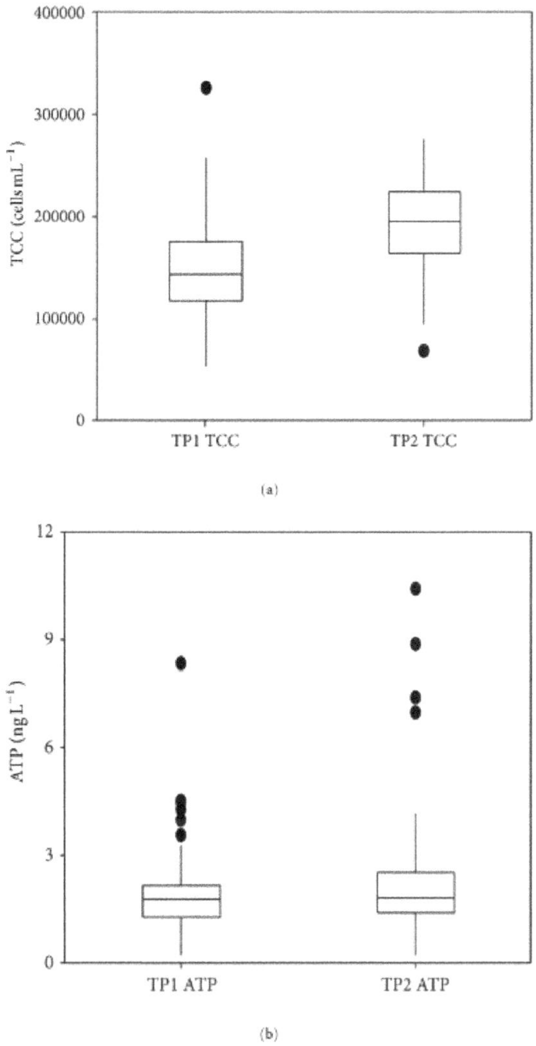

Figure 3: Boxplot of overall TCC (a) and ATP (b) results from distribution systems of TP1 (n=150) and TP2 (n=110).

The ATP concentrations from the two systems did not show significant differences (Figure 3(b)). ATP concentrations were almost all below 3 ng L^{-1} with few outliers. So, it may be concluded that the TCC analysis was better suited than the ATP method to pick up the slight changes in biomass that occurred in the distribution area of TP2.

Effect of Distance to Treatment Plant on TCC Levels in Distributed Drinking Water

The differences in TCC values in the distributed water sampled from proximal, central and distal parts of the distribution systems are shown in Figure 4(a). The TCC values in distributed water did not change significantly during transportation to the distal part of the TP1 system. For the TP2 system, a significant decrease in TCC was observed ($P<0.05$; Figure 4(a)). For both of the systems, the changes measured by ATP between proximal and central distribution areas were not significant (Figure 4(b)). These results confirm that the water quality of TP1 is slightly better than at TP2, even though both systems can be characterized as biologically stable. Stable trends of biological water quality in unchlorinated distribution systems have been reported by a TCC study of two locations within 6 km from a treatment plant (i.e., Zurich,

Switzerland: treated water AOC 32 μg l^{-1} [4]) and by an ATP study in several distribution systems (i.e., the Netherlands: treated water AOC 1.9–14.2 μg l^{-1}[5]). It is important to note that both aforementioned studies concluded that the substrate (AOC, dissolved organic carbon, and/or biodegradable organic matter) decreases corresponding to bioregrowth in distribution systems. Hammes et al. [4] found that a TCC increase corresponded to a decrease in AOC and dissolved organic carbon (DOC) during water treatment, while stable TCC values were found in distributed water where DOC and AOC changed minutely during distribution. In the last study, ATP concentrations decreased in a larger distribution system (the distal area averaged 40 km from the treatment plant, that is, more than two times farther than the two systems in the present study) when the distributed water reached the distal distribution area. The decrease has been attributed to lower biodegradable organic matter concentrations in the distal part of the distribution system due to biological processes and sedimentation during water distribution. In the present study, monitoring of straightforward parameters for substrate (DOC; Figure S5 and Figure S6, supplementary information) and biomass (TCC and ATP) showed only slight changes, suggesting that the treated water of the two selected treatment plants had a high level of biological stability.

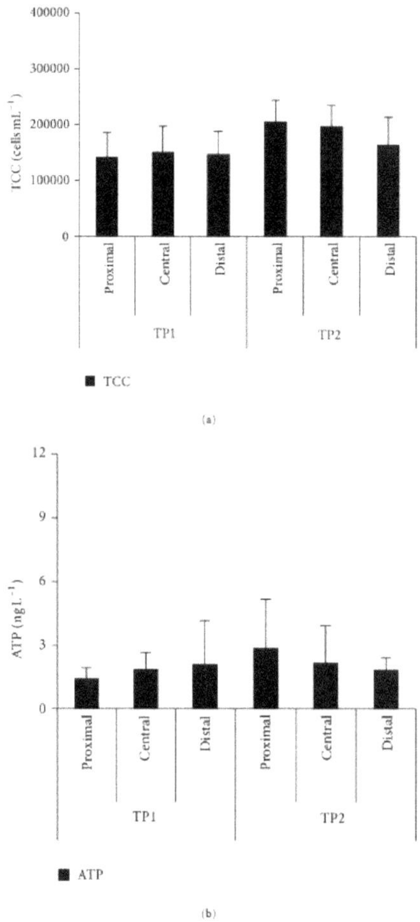

Figure 4: Average TCC (a) and ATP (b) concentrations (with standard deviations) in drinking water sampled from different areas in distribution systems of TP1 and TP2.

The processes occurring during drinking water distribution, such as biofilm formation and detachment, loose deposits accumulation and resuspension, all have an impact on biomass concentration in the bulk water and are strongly related to hydraulic conditions. It is important to mention that the lower biomass concentration measured in distributed water by TCC (in the present study) and ATP (in [5]) in the distal supply area can be caused by biomass enrichment in DWDS either by cells attached as biofilm on the pipe wall surface [31] or by cells associated with particulate matter such as suspended solids and loose deposits [31, 32]. Detailed studies on biomass in different phases in DWDS, such as pipe wall biofilm and loose deposits, are necessary before such a conclusion can be made.

Effect of Temperature on TCC Levels in Distributed Drinking Water

Higher water temperatures promote bacterial regrowth during distribution [33–35]. Over the study period, the water temperature ranged from 4.5°C to 19.5°C. It was observed that water samples could be collected with lower temperatures in the summer time and with higher temperatures in the spring. Therefore, the results in this study are being presented based on different temperatures instead of using general seasonal differences. An earlier statistical study of coliform occurrences in 31 water systems in North America [35] found significant differences in the occurrence of coliform bacteria between water samples with temperatures lower than 15°C and those with temperatures higher than 15°C.

In the present study, the same boundary temperature (15°C) was used to compare the influence of temperature on biomass in the distributed drinking water (Figure 5). The average TCC values at temperatures higher than 15°C were significantly higher than the values of samples collected with temperatures lower than 15°C (Figure5(a):$P<0.05$). Again, the same influence was not found in the ATP results, corroborating the higher sensitivity of the TCC measurements (Figure 5(b)). In a recent investigation, no correlation was found between increased temperatures and increased cell concentrations [4], but this may be caused by the limited number of sampling points (only two locations) and the limited distance from the feed reservoir (1 km and 6.1 km).

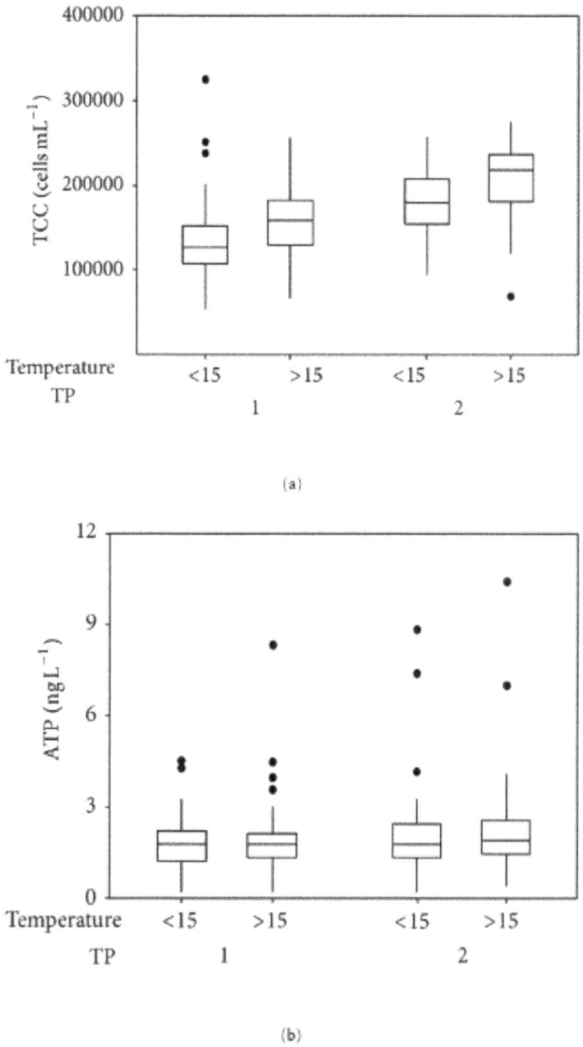

Figure 5: Boxplot of TCC (a) and ATP (b) concentrations in distributed drinking water sampled from two systems at temperatures below 15°C (<15) and above 15°C (>15).

In another investigation carried out in The Netherlands, low temperature was found to limit microbial activity as measured by ATP [5]. The different results found in this current study can be explained by the greater biological stability of the treated water. In The Netherlands, AOC and BFR tests are used together to evaluate biological stability. BFR is used as a measure for the regrowth potential of drinking water [27]. In DWDS, high BFR was found to be consistent with high concentration of ATP in distributed water [5]. With similar AOC concentrations, BFR values in the systems selected in the present study were 25–47 times lower than those of the systems in the previous study done by van der Wielen and van der Kooij [5].

Relationship between TCC and ATP, HPC, and Aeromonas in Distributed Drinking Water

All of the samples (n=260) were analyzed using ATP, HPC, Aeromonas, and TCC. The observed HPC values were within the water quality regulations. The correlation analysis between TCC, ATP, HPC, and Aeromonasdemonstrated that none of the parameters had a good correlation with each other (Figure 6). Although the TCC is a powerful technique, the weak correlation with Aeromonas indicated that it cannot replace Aeromonasmeasurements in the regular biological water quality monitoring. The weak relationship between TCC and HPC has been reported before [4, 5, 11, 12]. These weak relationships are logical, as HPC counts only cultivable cells and Aeromonas is only a limited group of the cultivable cells, whereas TCC counts all cells.

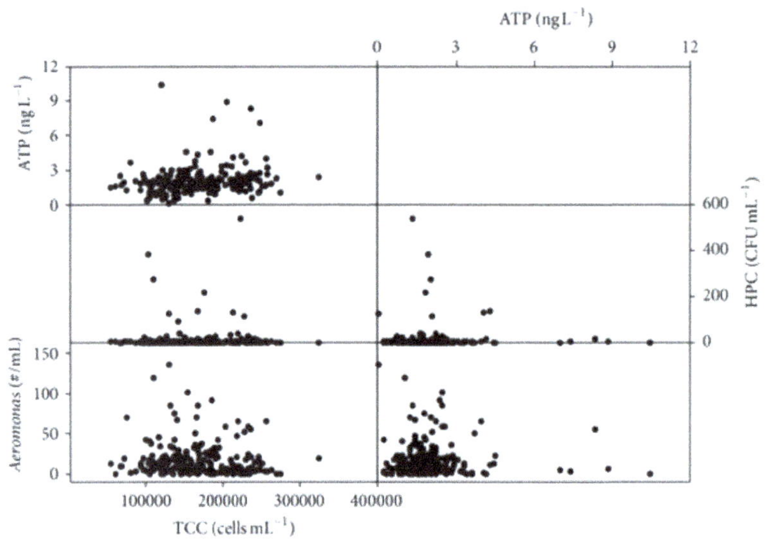

Figure 6: Matrix diagram of TCC, ATP, and Aeromonas, HPC, and ATP (n=260).

In the literature a good relation between TCC and ATP has been reported, in which TCC was measured either by FCM [12, 13] or by microscopy-based cell counts [5]. However, a good correlation between TCC and ATP was not found in the current study (Figure 6). With TCC ranging from 0.5–3.2 × 10^5 cells mL^{-1}, ATP concentrations were found to be relatively stable, mainly between 1–3 ng L^{-1} with a few outliers (Figure 3(b)). By comparing data reported in the literature [12, 13] in which strong correlations were found against the results obtained from the current study, it can be seen that only a limited number of the analyzed samples in the reported research had an ATP concentration lower than 3 ng L^{-1} (0.06 nM). For example, less than 8% of the samples in the study on different aquatic samples (n=102, R^2;=0.8 [13] and less than 10% of the samples in the study on distributed water (n=102, R^2= 0.69 [12] had ATP concentrations lower than 3 ng L^{-1}. As Figure8 shows, the current research revealed a correlation at ATP concentrations above 3 ng L^{-1}. It is possible that the correlation between TCC and ATP occurs only at ATP concentrations above 3 ng L^{-1}.

Relationship between HNA Bacteria and ATP

Theoretically, a correlation between the number of active bacteria and the ATP concentrations can be expected. It should be noticed that the intact LNA bacteria have been considered inactive [36], therefore, may have no contribution of ATP concentration. Therefore, the correlation between cell counts of HNA bacteria and ATP concentrations was studied based on dot-plots of LNA and HNA bacteria from FCM measurements (Figure 7). As shown in Figures 7(a) and 7(b), the two samples contained similar ATP concentrations, but the sample shown in Figure 7(b) had a much higher TCC than the sample shown in Figure 7(a). It is interesting to note that most of the cells in Figure 7(b) were classified as LNA bacteria. Similarly, Figures 7(b) and 7(c) show two samples with similar TCC values, while the increase of HNA bacteria was related to an increase of ATP from 2.0 ng L^{-1} to 8.4 ng^{-1}. It is, therefore, hypothesized that ATP concentrations correspond better to the number of HNA bacteria. To confirm this hypothesis, ATP was correlated to the number of HNA bacteria in all of the 260 samples. Indeed, a better correlation was found between HNA bacteria and ATP (Figure 8, R^2=0.63) than the correlation between TCC and ATP (Figure 6, R^2 =0.0002). The correlation was not as good as reported in previous research [12, 13]. Again, this may be because the ATP concentration was mainly lower than 3 ng L^{-1}.

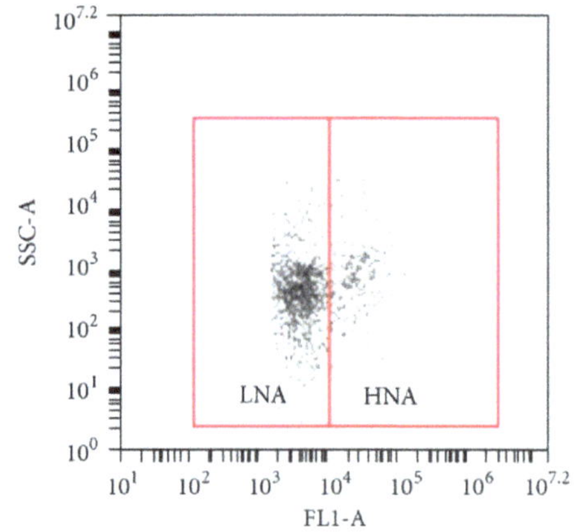

(a)

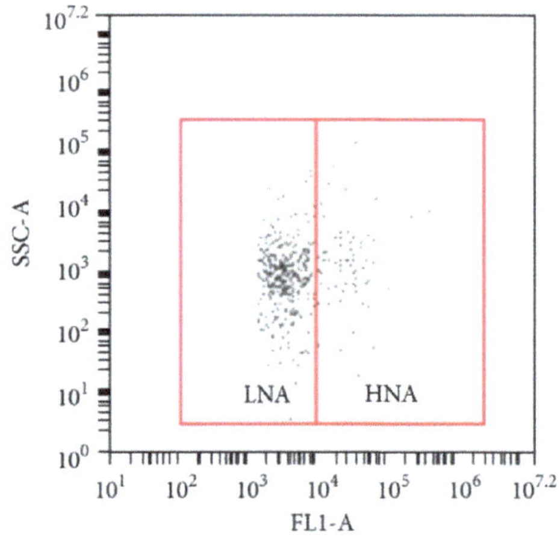

(b)

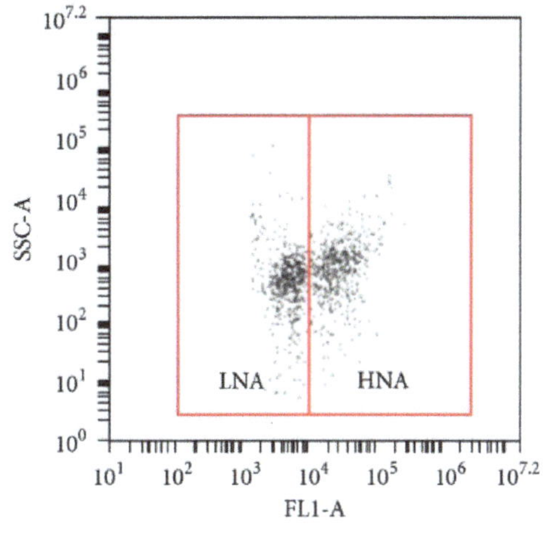

(c)

Figure 7: FCM dot-plots of total cell counts of selected FCM measurements based on ATP values. LNA and HNA are indicated by two solid lines. (a) 90,000 cells mL^{-1}, 2.0 ng ATP L^{-1}; (b) 210,000 cells mL^{-1}, 2.0 ng ATP L^{-1}; (c) 220,000 cells mL^{-1}, 8.4 ng ATP L^{-1}.

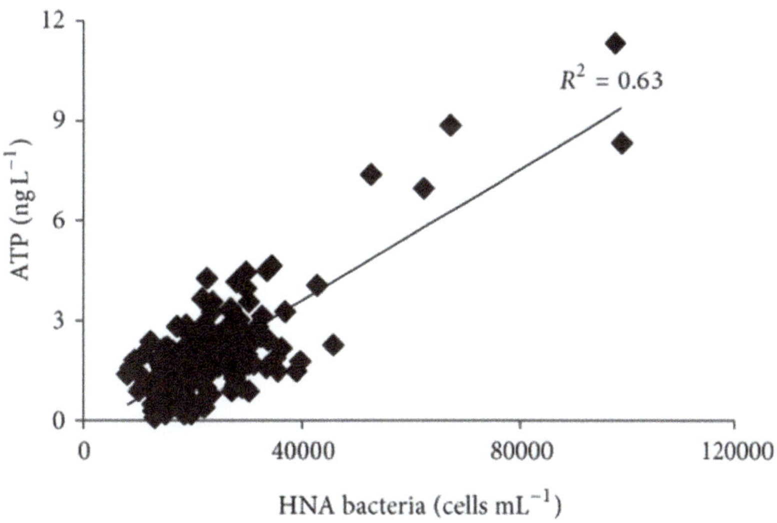

Figure 8: Correlation between HNA bacteria and ATP concentrations driven from the overall data set (n=260).

CONCLUSIONS

(i) The TCC levels were 1.2 × 10^5 cells mL^{-1} and 2.2 × 10^5 cells mL^{-1} for treated water at TP1 and TP2, respectively. Based on the levels of TCC, ATP, AOC, and BFR, it is concluded that both TP1 and TP2 produce biologically stable water. Nevertheless, TCC and AOC results demonstrated that the biological quality of treated water at TP1 is slightly better than that at TP2. The differences may result from different soils in their respective dune areas and differences in the treatments. Further investigation is needed to understand these differences.

(ii) TCC values in unchlorinated distributed water averaged between 0.5 and 3.2 × 10^5 cells mL^{-1}. In both distribution systems, the water remains biologically stable and there is no growth, as indicated by the stable levels of TCC and ATP. As revealed by TCC measurements, the biological quality of the distributed water at TP1 is slightly more stable than at TP2. TCC values were found to be higher at temperatures higher than 15°C than at temperatures lower than 15°C.

(iii) Significant changes in TCC values were observed in water with different compositions, different supply areas, and different temperatures. The changes measured by ATP were not significant. ATP and TCC analysis demonstrated that when the ATP concentrations in distributed drinking water are too low to observe significant changes, TCC is a more sensitive method to monitor the biomass changes.

(iv) There is no direct relationship between TCC and ATP, HPC, or Aeromonas numbers in distributed drinking water characterized as biologically stable with low ATP concentrations. It was found that in unchlorinated distributed drinking water, metabolic activity (as measured by ATP concentrations) showed a better relationship with the number of HNA bacteria. The extra information on subgroups of LNA and HNA is a clear advantage of TCC over HPC and ATP for bioregrowth study.

(v) Based on this investigation, it is concluded that it is useful to include TCC by FCM on a regular basis to assess the biological quality and stability of distributed drinking water. The TCC quantifies the total biomass and can detect small changes directly and sensitively. Moreover, it can determine the subgroup of HNA, which are related to the active cells and the ATP concentration.

CONFLICT OF INTERESTS

The authors declare no conflict financial of interests.

ACKNOWLEDGMENT

The authors thank the Chinese Scholarship Council for their support (2008612022).

REFERENCES

1. L. M. Feazel, L. K. Baumgartner, K. L. Peterson, D. N. Frank, J. K. Harris, and N. R. Pace, "Opportunistic pathogens enriched in showerhead biofilms," Proceedings of the National Academy of Sciences of the United States of America, vol. 106, no. 38, pp. 16393–16398, 2009.
2. R. C. Hoehn, "Biological causes of tastes and odors in drinking water supplies," Water Quality Bulletin, vol. 13, pp. 46–52, 1988.
3. S. H. Lee, J. T. O'Connor, and S. K. Banerji, "Biological mediated corrosion and its effects on water quality in distribution systems," Journal of the American Water Works Association, vol. 72, no. 11, pp. 636–645, 1980.
4. F. Hammes, C. Berger, O. Köster, and T. Egli, "Assessing biological stability of drinking water without disinfectant residuals in a full-scale water supply system," Journal of Water Supply: Research and Technology, vol. 59, no. 1, pp. 31–40, 2010.
5. P. W. J. J. van der Wielen and D. van der Kooij, "Effect of water composition, distance and season on the adenosine triphosphate concentration in unchlorinated drinking water in the Netherlands," Water Research, vol. 44, no. 17, pp. 4860–4867, 2010.
6. D. Van Der Kooij, A. Visser, and W. A. M. Hijnen, "Determing the concentration of easily assimilable organic carbon in drinking water," Journal of the American Water Works Association, vol. 74, no. 10, pp. 540–545, 1982.
7. G. Liu, M. C. Lut, Verberk, J. Q. J. C, and J. C. Van Dijk, "A comparison of additional treatment processes to limit particle accumulation and microbial growth during drinking water distribution," Water Research, vol. 47, no. 8, pp. 2719–2728, 2013.
8. J. Bartram, Heterotrophic Plate Counts and Drinking-Water Safety: The Significance of HPCs for Water Quality and Human Health, IWA Publishing, 2003.
9. WHO, Guidelines for Drinking-Water Quality (Electronic Resource): Incorporating First Addendum, World Health Organization. Water, Sanitation and Health Team, 3rd edition, 2006.
10. M. J. Allen, S. C. Edberg, and D. J. Reasoner, "Heterotrophic plate count bacteria—what is their significance in drinking water?" International Journal of Food Microbiology, vol. 92, no. 3, pp. 265–274, 2004.
11. F. Hammes, M. Berney, Y. Wang, M. Vital, O. Köster, and T. Egli, "Flow-cytometric total bacterial cell counts as a descriptive microbiological parameter for drinking water treatment processes," Water Research, vol. 42, no. 1-2, pp. 269–277, 2008.
12. E. Siebel, Y. Wang, T. Egli, and F. Hammes, "Correlations between total cell concentration, total adenosine tri-phosphate concentration and heterotrophic plate counts during microbial monitoring of drinking water," Drinking Water Engineering and Science, vol. 1, no. 1, pp. 1–6, 2008.
13. F. Hammes, F. Goldschmidt, M. Vital, Y. Wang, and T. Egli, "Measurement and interpretation of microbial adenosine tri-phosphate (ATP) in aquatic environments," Water Research, vol. 44, no. 13, pp. 3915–3923, 2010.
14. R. A. Deininger and J. Lee, "Rapid determination of bacteria in drinking water using an ATP assay," Field Analytical Chemistry and Technology, vol. 5, no. 4, pp. 185–189, 2001.
15. M. Berney, M. Vital, I. Hülshoff, H. U. Weilenmann, T. Egli, and F. Hammes, "Rapid, cultivation-independent assessment of microbial viability in drinking water," Water Research, vol. 42, no. 14, pp. 4010–4018, 2008.
16. F. Hammes and T. Egli, "Cytometric methods for measuring bacteria in water: advantages, pitfalls and applications," Analytical and Bioanalytical Chemistry, vol. 397, no. 3, pp. 1083–1095, 2010.
17. O. Holm-Hansen and C. R. Booth, "The measurement of adenosine triphosphate in the ocean and its ecological significance," Limnology and Oceanography, vol. 11, pp. 510–519, 1966.
18. B. K. Jensen, "ATP-related specific heterotrophic activity in petroleum contaminated and uncontaminated groundwaters," Canadian Journal of Microbiology, vol. 35, no. 8, pp. 814–818, 1989.
19. H. S. C. Eydal and K. Pedersen, "Use of an ATP assay to determine viable microbial biomass in Fennoscandian Shield groundwater from depths of 3-1000 m," Journal of Microbiological Methods, vol. 70, no. 2, pp. 363–373, 2007.
20. A. Magic-Knezev and D. van der Kooij, "Optimisation and significance of ATP analysis for measuring active biomass in granular activated carbon filters used in water treatment," Water Research, vol. 38, no. 18, pp. 3971–3979, 2004.
21. S. Velten, F. Hammes, M. Boller, and T. Egli, "Rapid and direct estimation of active biomass on granular activated carbon through adenosine tri-phosphate (ATP) determination," Water Research, vol. 41, no. 9, pp. 1973–1983, 2007.

22. D. van der Kooij, "Assimilable organic carbon as an indicator of bacterial regrowth," Journal of the American Water Works Association, vol. 84, no. 2, pp. 57–65, 1992.

23. R. Boe-Hansen, H. J. Albrechtsen, E. Arvin, and C. Jørgensen, "Bulk water phase and biofilm growth in drinking water at low nutrient conditions," Water Research, vol. 36, no. 18, pp. 4477–4486, 2002.

24. M. J. Lehtola, T. Juhna, I. T. Miettinen, T. Vartiainen, and P. J. Martikainen, "Formation of biofilms in drinking water distribution networks, a case study in two cities in Finland and Latvia," Journal of Industrial Microbiology and Biotechnology, vol. 31, no. 11, pp. 489–494, 2004.

25. K. Czechowska, D. R. Johnson, and J. R. van der Meer, "Use of flow cytometric methods for single-cell analysis in environmental microbiology," Current Opinion in Microbiology, vol. 11, no. 3, pp. 205–212, 2008.

26. D. Hoefel, P. T. Monis, W. L. Grooby, S. Andrews, and C. P. Saint, "Culture-independent techniques for rapid detection of bacteria associated with loss of chloramine residual in a drinking water system,"Applied and Environmental Microbiology, vol. 71, no. 11, pp. 6479–6488, 2005.

27. D. Van der Kooij, Managing Regrowth in Drinking Water Distribution Systems, IWA Publishing, London, UK, 2003.

28. Y. Wang, F. Hammes, N. Boon, M. Chami, and T. Egli, "Isolation and characterization of low nucleic acid (LNA)-content bacteria," ISME Journal, vol. 3, no. 8, pp. 889–902, 2009.

29. P. Lebaron, P. Servais, A. C. Baudoux, M. Bourrain, C. Courties, and N. Parthuisot, "Variations of bacterial-specific activity with cell size and nucleic acid content assessed by flow cytometry," Aquatic Microbial Ecology, vol. 28, no. 2, pp. 131–140, 2002.

30. M. Felip, S. Andreatta, R. Sommaruga, V. Straskrábová, and J. Catalan, "Suitability of flow cytometry for estimating bacterial biovolume in natural plankton samples: comparison with microscopy data," Applied and Environmental Microbiology, vol. 73, no. 14, pp. 4508–4514, 2007.

31. D. S. Herson, D. R. Marshall, K. H. Baker, and H. T. Victoreen, "Association of microorganisms with surfaces in distribution systems," Journal of the American Water Works Association, vol. 83, no. 7, pp. 103–106, 1991.

32. G. Liu, F. Ling, A. Magic-Knezev et al., "Quantification and identification of particle associated bacteria in unchlorined drinking water from three treatment plants by cultivation-independent methods," Water Research, vol. 47, no. 10, pp. 3523–3533, 2013.

33. A. Kerneïs, F. Nakache, A. Deguin, and M. Feinberg, "The effects of water residence time on the biological quality in a distribution network," Water Research, vol. 29, no. 7, pp. 1719–1727, 1995.

34. W. Uhl and G. Schaule, "Establishment of HPC(R2A) for regrowth control in non-chlorinated distribution systems," International Journal of Food Microbiology, vol. 92, no. 3, pp. 317–325, 2004.

35. M. W. LeChevallier, N. J. Shaw, and D. B. Smith, Factors Limiting Microbial Growth in Distribution Systems: Full-Scale Experiments, American Water Works Association, 1996.

36. P. Lebaron, P. Servais, H. Agogué, C. Courties, and F. Joux, "Does the high nucleic acid content of individual bacterial cells allow us to discriminate between active cells and inactive cells in aquatic systems?" Applied and Environmental Microbiology, vol. 67, no. 4, pp. 1775–1782, 2001.

CHAPTER 11

CULTURE METHOD AND PCR FOR THE DETECTION OF HELICOBACTER PYLORI IN DRINKING WATER IN BASRAH GOVERNORATE IRAQ

A. A. Al-Sulami, T. A. A. Al-Edani, and A. A. Al-Abdula

INTRODUCTION

Helicobacter pylori is recognized as the major cause of gastritis and peptic ulcer and gastric mucosa-associated lymphoid tissue (MALT) gastric lymphoma [1]. The mechanism of H. pylori pathogenic effect is unclear but is believed to be related to host bacterial interactions initiated by virulence genes, and it is possible that these effects are enhanced by invasiveness of the bacterium [2–5]. H. pylori changes from the normal spiral-shaped bacillary form into the coccoid form when it is exposed to water or to other adverse conditions [6]. Hence attempts have been made to develop artificial media to achieve better culture recovery results than those obtained from traditional Columbia blood agar [7, 8]. Polymerase chain reaction (PCR) methods have also been used to detect H. pylori as its 16SrRNA gene sequence analysis unambiguously differentiated theHelicobacter genus from the closely related Campylobacter genus and other Helicobacter species [9]. The presence of H. pylori in drinking water which was detected by PCR has been reported from several countries [10–12]. Hegarty et al. [13] also demonstrated the presence of respiring H. pylori from US surface water. The prevalence of disease attributed to H. pylori in Iraq is not available despite of its commonality.

Basrah Governorate, where Basrah city is located, has a population of about three millions; its water supply is mainly derived from three sources, Shatt-Al-Arab River, Tigris River, and Bada lake. Water from these sources is treated at 22 treatment works and distributed through approximately 13,000 Km pipe network.

Since the 1980s there has been a general marked deterioration in water quality in Iraq, reflecting the environmental degradation of the country caused by successive armed conflicts.

The aim of this study was isolating H. pylori from drinking water in Basrah, Iraq, on modified Columbia urea agar (MCUA) and HP media using MDCS method [7] and then confirming that by conventional biochemical tests and 16SrRNA PCR.

Citation: A. A. Al-Sulami, T. A. A. Al-Edani, and A. A. Al-Abdula, "Culture Method and PCR for the Detection of Helicobacter pylori in Drinking Water in Basrah Governorate Iraq," Gastroenterology Research and Practice, vol. 2012, Article ID 245167, 5 pages, 2012. doi:10.1155/2012/245167.

METHODS

Sample Collection and Culturing

266 samples of tap water and 205 samples from tankers supplying Reverse osmosis (RO) were collected from 24 districts covering more than 90% of Basrah Governorate during the period from February 2008 to December 2009. Samples of 500 mL water each were collected in sterile glass flasks and examined for chlorine concentration using o-toluidine. Samples were transferred within 1-2 hr. to the laboratory and filtered through 0.22 μm Millipore filter membrane. Each membrane was then immersed into 2 mL of tryptic soy broth (TSB) for 1 h. After that each 2 mL TSB was taken and placed at the lower portion of the slanted MCUA tube. Each tube was tilted a few times to allow the added broth to spread bacteria on the upper part of the slant. Slanted MCUA tube, was incubated microaerophically at 37°C for 1-2 days, after which color changes from orange to pink in the solid phase, indicating urease activity. The resulting system is a simple monophasic-diphasic culture setup (MDCS), a diphasic solid liquid environment at the lower part of the test tube and a monophasic solid one above it [7]. From the bottom and the upper portions of the slanted MCUA tube subcultures were done on plates of MCUA and HP media for purification.

No controls were used in the isolation of the strains and also in PCR as they are out of reach for us in Iraq.

Primary Diagnosis of H. pylori

The suspected purified colonies were chosen according to the Gram staining and cultural characteristics.

Biochemical Tests

Biochemical tests include production of catalase, oxidase, urease, and H_2S, nitrate reduction, growing in 3.5% NaCl, growing with 1% glycine, and growing at different temperatures.

Antibiotic Sensitivity Test

The method of Piddock [14] was used to test the sensitivity of 14 isolates of H. pylori from drinking water to seven types of antibiotics, kanamycin 30 μg, erythromycin 15 μg, tetracycline 30 mg, ampicillin 10 μg, rifampicin 5 μg, amoxicillin 30 μg and gentamycin 30 μg (Bioanalyse, Turkey).

16SrRNA Identification of Isolates

All isolates from tap and RO water samples which gave positive results by biochemical tests as H. pylori and (100) samples which were H. pylori negative by culture method were further confirmed by using primers specifically designed for the identification of H. pylori based on 16SrRNA sequence [15]. The primers for 500 bp product of the 16SrRNA sequence are represented by the forward primer sequence: 5 GCT AAG AGA TCA GCC TAT GTC C3 and the reverse one: 5 TGG CAA TCA GCG TCA GGT AAT G3.

Preparation of Bacterial Genomic DNA

Genomic DNA from each isolate was prepared by vortex after suspending a loopful of colonies in 1 mL of phosphated-buffer saline (PBS) 7.6, centrifuging at 14000×g for 2 min, and boiling the pellet in 1 mL of distilled water for 1 min [16]. The samples were then centrifuged at 12000×g for 4 min at 4°C and the supernatants were stored in sterile vials at −70°C until they were used as PCR templates. Genomic DNA from water samples, which have been cultured but did not give isolates for H. pylori, were prepared by centrifuging 1 mL of the liquid portion of slant MCUA tube at 14000×g for 2 min and washed with 1 mL of PBS to be completed by the same steps for H. pylori isolates. Concentration and purity were measured spectrophotometrically at OD_{260} and OD_{280} respectively, to exclude any possible contamination, and a gel of 0.8% agarose was used for electrophoresis.

PCR Amplification of 16SrRNA for H. pylori

Amplification was carried out in a 25 μL of reaction mixture containing 12.5 μL master mix, 0.5 μL forward primer, 0.5 μL reverse primer, 5 μL DNA samples, 6.5 μL distilled water and 25 μL mineral oil. PCR conditions for 16SrRNA include: denaturation step at 95°C for 5 min, followed by 39 cycles at 94°C for 1 min, annealing at 55°C for 1 min and extension at 72°C for 2 min, and an additional extension step at 72°C for 7 min. PCR products were electrophoresed in 2% agarose.

ureA Gene for H. pylori and PCR Amplification

All isolates which were confirmed by 16SrRNA have been tested for the presence of the ureA gene of H. pylori. The primer for 411 bp product of the ureA sequence represented by the forward primer sequence: 5 GCC AAT GGT AAA GCC TTA GTT3 and the reverse one: 5 CTC CTT AAT TGT TTT TAC 3 [17]. Amplification was carried out in a 25 μL of reaction mixture containing 12.5 μL master mix, 0.5 μL forward primer, 0.5 μL reverse primer, 5 μL DNA samples, 6.5 μL distilled water and 25 μL mineral oil. PCR conditions for ureA gene include: denaturation step at 95°C for 5 min, followed by 35 cycles at 94°C for 1 min, annealing at 45°C for. 1 min and, extension at 72°C for 1 min and an additional extension step at 72°C for 7 min. PCR products were electrophoresed in 2% agarose.

RESULTS

Culture Results

Out of 471 water samples, 14 (2.76%) isolates of H. pylori were isolated from samples taken from 14 districts by culture method and identified by biochemical tests. They consist of 11 (4.13%) H. pylori that have been isolated and diagnosed from 266 samples of tap water and 3 ones (1.46%) from 205 RO samples.

The modified Columbia urea agar using MDCS method preliminarily revealed the presence of H. pylori in water samples, correlated with the change in the color of the slant MCUA tube from orange to pink that occurred at the same time thus giving an additional evidence for the presence of H. pylori in the samples (Figure 1).

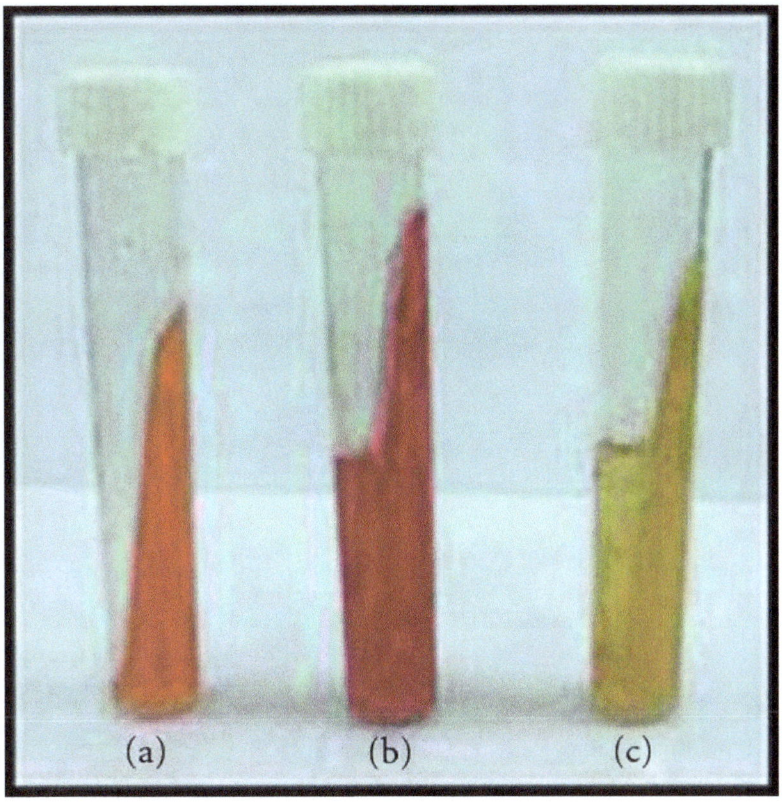

Figure 1: Change in color of slant MCUA tube, (a) slant MCUA tube only, (b) positive slant MCUA tube, culture, (c) negative slant MCUA tube, culture.

The isolation rate upon subculturing on HP medium was 14/471 (2.76%) isolates of H. pylori, while on MCUA medium was 6/471 (1.2%) isolates included in the 14 isolates of H. pylori.

On MCUA medium, the colonies of the isolated H. pylori were small to middle in size, rounded, and creamy in color, while, on HP medium, the isolated H. pylori were small in size, rounded, and transparent. Both the MCUA and HP media showed change in color from yellow/orange to red.

All H. pylori isolates were Gram-negative spiral to coccobacilli and shared the characteristic catalase, urease, and oxidase production, but differ slightly with respect to other tests (Table 1). Collectively, 3 isolates are being positive in nitrate reduction, 2 in being able to grow at 42°C, and 9 negatives in both traits.

Table 1: Results of biochemical tests characterizing H. pylori isolates from 14 districts.

District no	Catalase	Oxidase	Urease	Nitrate reduction	H_2S	Growth with 3.5% NaCl	Growth on 1% glycin	Growth at 42°C	Growth at 25°C
1	+	+	+	–	–	–	–	–	–
2	+	+	+	–	–	–	–	+	–
3	+	+	+	+	–	–	–	–	–
4	+	+	+	–	–	–	–	–	–
5	+	+	+	–	–	–	–	+	–
6	+	+	+	+	–	–	–	–	–
7	+	+	+	–	–	–	–	–	–
8	+	+	+	–	–	–	–	–	–
9	+	+	+	–	–	–	–	–	–
10	+	+	+	–	–	–	–	–	–
11	+	+	+	+	–	–	–	–	–
12	+	+	+	–	–	–	–	–	–
13	+	+	+	–	–	–	–	–	–
14	+	+	+	–	–	–	–	–	–

Antibiotics Susceptibility

For H. pylori isolates from drinking water, tetracycline was found to be the most effective antibiotic, 71% of the tested isolates were sensitive to tetracycline followed by kanamycin 57% and gentamycin 36%, ampicillin 14%. Rifampicin and amoxicillin were shown to be the least effective ones (7%) against H. pylori isolated from drinking water,while erythromycin was a non effective antibiotic, as shown in (Figure 2) and in reference to interpretive chart of zone sizes.

Figure 2: Antibiotic effects on H. pylori isolated from drinking water.

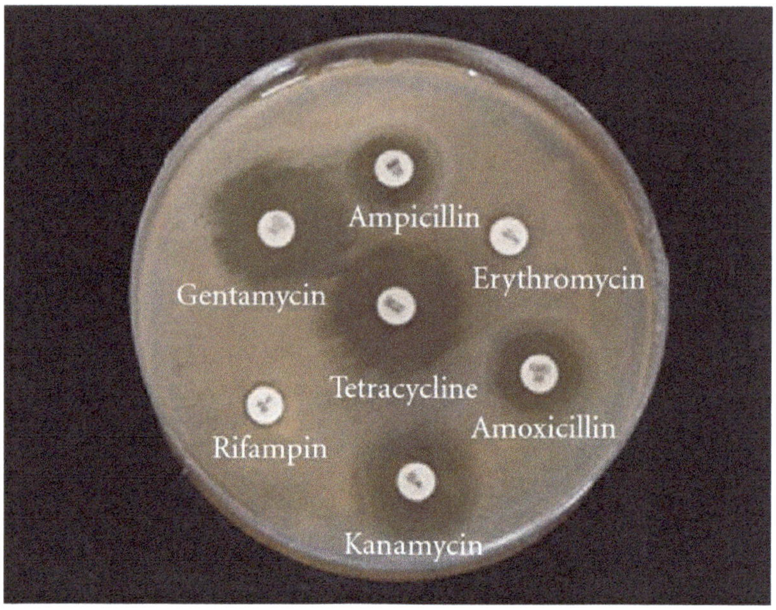

PCR Results

Only 6 out of 11 (54.5%) H. pylori morphologically and biochemically identified isolates from tap water were found to harbor 16SrRNA gene and of the 3 R.O isolates only one (33.3%) isolate gave positive results for 16SrRNA gene by PCR. Thus leaving out 50% of the conventionally identified isolates as false positive. From the100 samples negative for H. pylori by culturing, only 4 (4%), gave positive results for 16SrRNA.

PCR products for 16SrRNA based primers gave bands on agarose gel corresponding to a 500 base pair product when compared to the molecular ladder, thus identifying the isolates as H. pylori as shown in (Figure 3).

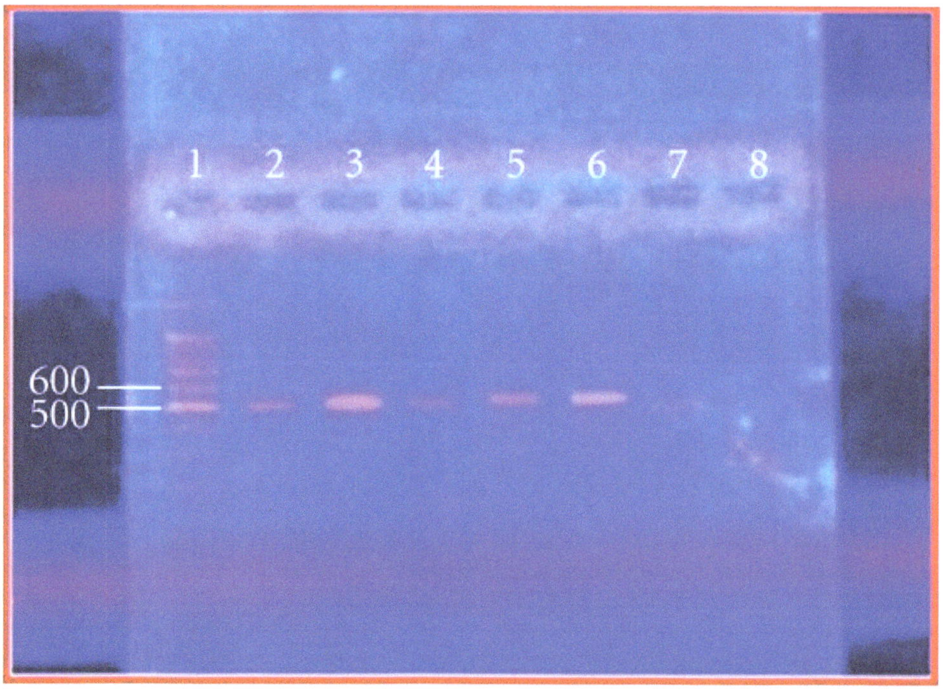

Figure 3: PCR products for 16SrRNA-based primers gave band on agarose gel corresponding to a 500 base pair product when compared to the molecular ladder. Lane 1, molecular ladder (1500–100) bp, lane (2–6) bands of PCR products for H. pylori with 16SrRNA.

ureA Gene for H. pylori and PCR Amplification

All isolates of H. pylori which have been confirmed by 16SrRNA, did not give specific results to ureA (Figure 4), only products of 100 bp have been obtained and also a much larger bands.

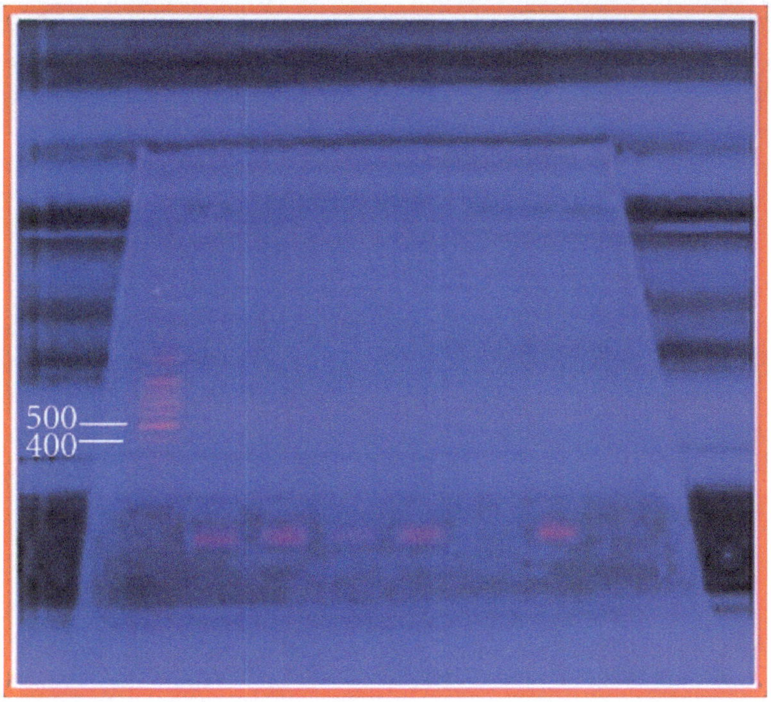

Figure 4: PCR product for H. pylori with ureA gene based primers. Lane 1, molecular ladder (1500–100) bp, no band of PCR product for H. pylori with ureA gene have been obtained.

DISCUSSION

Natural habitat of H. pylori is in the human stomach, other sources of H. pylori and its mode of transmission are unknown [18]. In this study, H. pylori has been isolated and diagnosed from drinking water by culture method and a combination of biochemical and PCR test. The first indication for the presence of H. pylori in water came from AL-Sulami et al. [8] in which 10 isolates were identified as H. pylori by biochemical tests. That finding has been confirmed by current study using the same method and a combination of conventional and PCR tests in identifying recovered H. pylori.

A low recovery of a pathogen is not surprising considering various factors affecting its survival in water. Upon primary, isolation there were 78 urease-positive isolates obtained from 266 tap water samples and 43 urease-positive ones from RO water samples. The numbers were reduced to 11 and 3 isolates, respectively, after subjecting them to conventional tests leaving 67 and 40 false-positive ones. Urease-negative isolates were not considered. Other bacteria were mainly pseudomonads.

So far there is no published paper proving the viability of coccoid form or the possibility that coccoid form transforms to spiral bacillary form. Our results indicate that some H. pylori are still viable and appear as spiral bacillary after Gram staining smears from colonies on MCUA; others are not and only can be detected by PCR.

Based on the assumption that all H. pylori in drinking water are coccoid [6], the results implicitly indicate the possibility of the transformation of some coccoid form to spiral bacillary form.

It is difficult to compare our data with those published, because each author has used a distinct method to detect the bacterium, and all attempts to culture the organism directly from water samples [18, 19] have been unsuccessful. This may be due to the fact that overgrowth by other microorganisms on the rich media led to the difficulty of isolation of H. pylori from water, and another reason for the lack of recovery of H. pylori from the environment is the fastidious nature of H. pylori which has a polymorphisms phenomenon. Under these circumstances, the organism would not be recovered by traditional culture techniques; hence in our study we developed a different protocol for culturing H. pylori from water. The importance of this method is to provide a possibility of successful culture method for H. pylori.

In general, high-resistance profile to the tested antibiotics is apparent on these isolates as indicated by H. pyloriresistance for tetracycline in 29% of the isolates, also in case of kanamycin H. pylori resistance of 43% which is less than Al-Sulami et al. [8] result of 60%. Amoxicillin which represents active antibiotics in treatment of this bacterium was ineffective with a resistance in 93% of the isolates.

16SrRNA for H. pylori Detection by PCR

In this study, this is the first report on using 16SrRNA amplification and confirmation of H. pylori isolates from environmental samples in Iraq. The 16SrRNA was chosen for detection of H. pylori because it exhibits a high degree of functional and evolutionary homology within all bacteria [9]. Only 7 isolates, out of 14 morphologically and biochemically identified H. pylori, were confirmed by 16SrRNA as they gave positive results for 16SrRNA. The prevalence of false positive isolates by conventional tests indicates a nonspecific approach. Meanwhile, in 100 drinking water samples in which no H. pylori was detected by culture method, 4 samples produced positive results by 16SrRNA. This means that cells of H. pylori that are not detected by culture method can be done by PCR, and hence, the MDCS provides the opportunity for simultaneous detection of both culturable and nonculturable forms.

Results of PCR products of 16SrRNA gene amplification revealed the presence of 500 base pair sequence of the gene coded the 16SrRNA molecule, and this result agrees with that of [15]. The size of PCR product was determined by comparing it with a DNA ladder, which contains DNA fragments of known size (1500–100) base pairs. Our results may shed additional light on the evidence supporting water-borne transmission which emanates from the fact that there is a direct recovery of H. pylori from tap water and R.O water concomitantly confirmed by PCR.

ureA Gene for H. pylori Detection by PCR

The ureA genotype was expected to be present in all Helicobacter positive strains. However, our study was unable to detect the ureA gene in the isolates of H. pylori already confirmed by 16SrRNA. This result agrees with Tiveljung et al. [20] who used ureA gene and were unable to detect it in H. pylori strain regarded as normal control.

CONCLUSION

The isolation of H. pylori from drinking water, tap and R.O, by culture method and consequent identification by biochemical tests and PCR represents a clear signal for the presence of this dangerous but illusive pathogen in our consumable water. It, certainly, will impact our search for a better epidemiological understanding and measures of control.

REFERENCES

1. J. C. Atherton, "The pathogenesis of Helicobacter pylori-induced gastro-duodenal diseases," Annual Review of Pathology, vol. 1, pp. 63–96, 2006.
2. Y. Elitsur and J. Yahav, "Helicobacter pylori infection in pediatrics," Helicobacter, Supplement, vol. 10, no. 1, pp. 47–53, 2005.
3. A. Dubois and T. Boren, "Helicobacter pylori is invasive and it may be a facultative intracellular organism," Cellular Microbiology, vol. 9, no. 5, pp. 1108–1116, 2007.
4. M. R. Amieva and E. M. El-Omar, "Host-Bacterial Interactions in Helicobacter pylori Infection,"Gastroenterology, vol. 134, no. 1, pp. 306–323, 2008.
5. V. Necchi, M. E. Candusso, F. Tava et al., "Intracellular, intercellular, and stromal invasion of gastric mucosa, preneoplastic lesions, and cancer by Helicobacter pylori," Gastroenterology, vol. 132, no. 3, pp. 1009–1023, 2007.
6. N. F. Azevedo, C. Almeida, L. Cerqueira, S. Dias, C. W. Keevil, and M. J. Vieira, "Coccoid form ofHelicobacter pylori as a morphological manifestation of cell adaptation to the environment," Applied and Environmental Microbiology, vol. 73, no. 10, pp. 3423–3427, 2007.
7. A. A. Al-Sulami, H. S. Al-Kiat, L. K. Bakker, and H. Hunoon, "Primary isolation and detection ofHelicobacter pylori from dyspeptic patients: a simple, rapid method," Eastern Mediterranean Health Journal, vol. 14, no. 2, pp. 268–275, 2008.
8. A. A. Al-Sulami, A. M. R. Al-Taee, and M. G. Juma'a, "Isolation and identification of Helicobacter pylorifrom drinking water in Basra governorate, Iraq," Eastern Mediterranean Health Journal, vol. 16, no. 9, pp. 810–816, 2010.
9. H. Liu, A. Rahman, C. Semino-Mora, S. Q. Doi, and A. Dubois, "Specific and sensitive detection of H. pylori in biological specimens by real-time RT-PCR and in situ hybridization," PLoS ONE, vol. 3, no. 7, Article ID e2689, 2008.
10. K. Hulten, S. W. Han, H. Enroth et al., "Helicobacter pylori in the drinking water in Peru,"Gastroenterology, vol. 110, no. 4, pp. 1031–1035, 1996.
11. J. A. Benson, K. A. Fode-Vaughan, and M. L. P. Collins, "Detection of Helicobacter pylori in water by direct PCR," Letters in Applied Microbiology, vol. 39, no. 3, pp. 221–225, 2004.
12. M. Mazari-Hiriart, Y. López-Vidal, and J. J. Calva, "Helicobacter pylori in water systems for human use in Mexico City," Water Science and Technology, vol. 43, no. 12, pp. 93–98, 2001.
13. J. P. Hegarty, M. T. Dowd, and K. H. Baker, "Occurrence of Helicobacter pylori in surface water in the United States," Journal of Applied Microbiology, vol. 87, no. 5, pp. 697–701, 1999.
14. L. J. V. Piddock, "A review techniques used for the determination of antimicrobial resistance and sensitivity in bacteria," Journal of Applied Bacteriology, vol. 68, no. 4, pp. 307–318, 1990.
15. T. Falsafi, R. Favaedi, F. Mahjoub, and M. Najafi, "Application of stool-PCR test for diagnosis ofHelicobacter pylori infection in children," World Journal of Gastroenterology, vol. 15, no. 4, pp. 484–488, 2009.
16. R. M. Peek Jr., G. G. Miller, K. T. Tham, et al., "Detection of Helicobacter pylori gene expression in human gastric mucosa," Journal of Clinical Microbiology, vol. 33, no. 1, pp. 28–32, 1995.
17. M. Notarnicola, F. Russo, A. Cavallini et al., "PCR identification of Helicobacter pylori DNA in faeces from patients with gastroduodenal pathology," Medical Science Research, vol. 24, no. 11, pp. 785–787, 1996.
18. N. F. Azevedo, C. Almeida, I. Fernandes et al., "Survival of gastric and enterohepatic Helicobacter spp. in water: implications for transmission," Applied and Environmental Microbiology, vol. 74, no. 6, pp. 1805–1811, 2008.
19. M. S. Giao, N. F. Azevedo, S. A. Wilks, M. J. Vieira, and C. W. Keevil, "Persistence of Helicobacter pyloriin heterotrophic drinking-water biofilms," Applied and Environmental Microbiology, vol. 74, no. 19, pp. 5898–5904, 2008.
20. A. Tiveljung, K. Borch, J. Jonasson, S. Mårdh, F. Petersson, and H. J. Monstein, "Identification of Helicobacter in gastric biopsies by PCR based on 16S rDNA sequences: a matter of little significance for the prediction of H. pylori-associated gastritis?" Journal of Medical Microbiology, vol. 47, no. 8, pp. 695–704, 1998.

CHAPTER 12

HEALTH RISK ASSESSMENT OF CYANOBACTERIAL (BLUE-GREEN ALGAL) TOXINS IN DRINKING WATER

Ian R. Falconer and Andrew R. Humpage

Keywords: Health risk, cyanobacteria, toxins, drinking water

INTRODUCTION

Cyanobacterial toxins are well recognized as a cause of livestock poisoning, which has been extensively reported in the Americas, Europe, Asia and Australasia [1]. Livestock are inevitably vulnerable to poisoning as they are restricted in access to water by topography and by fences, and hence may have no choice but to drink water infested by toxic cyanobacteria. Most stock deaths have resulted from the formation of cyanobacterial water-blooms in ponds and lakes on farms, but several major poisoning events were through water blooms on rivers and drinking water reservoirs [2, 3].

Human poisoning has also occurred, but the reports are less well documented. The symptoms of poisoning by the main toxic cyanobacteria in drinking water reservoirs overlap with a range of other gastrointestinal illnesses, largely caused by infectious disease organisms. As a consequence during an outbreak of enteric disease the pathogens are investigated first, as the most probable cause, and only after exhaustive exploration are toxins of any type evaluated. Agricultural chemicals and industrial pollutants such as heavy metals are likely to be next suspected, with cyanobacterial toxins ignored until well after the event [4].

Epidemiological data for human poisoning by cyanobacterial toxins only exists for a small number of events. The most well characterized case was the poisoning of renal dialysis patients in a clinic in Caruaru, Brazil, in 1996. In this instance the patients treated in a dialysis clinic during one week suffered severe illness following perfusion, with hepatic failure and, in more than 50 cases, death. Investigation of the water treatment unit at the clinic found contamination of the filters by two types of cyanobacterial toxin, microcystins and cylindrospermopsins [5, 6]. Microcystins were detected in the blood and liver of poisoned individuals [7]. Because of the severity of the poisoning a thorough investigation was carried out, which showed up major defects in the operation of the water treatment unit at the clinic. Exposure to toxins through renal dialysis is a particularly potent route of poisoning, equivalent to an intravenous injection in the case of water soluble toxins. The volume of water used in perfusion is large, about 120L, so that the total amount of toxin to which a dialysis patient is exposed is much greater than possible through drinking water.

Exposure to cyanobacterial toxins through consumption of contaminated drinking water has however also resulted in poisoning. The earliest demonstration of this was in 1983, when the population of a rural town in Australia was supplied

Citation: Falconer IR, Humpage AR (2005) Health Risk Assessment of Cyanobacterial (Blue-green Algal) Toxins in Drinking Water. Int. J. Environ. Res. Public Health 2005, 2(1), 43-50 doi:10.3390/ijerph2005010043.

with drinking water from a reservoir carrying a dense water bloom of a toxic species of cyanobacterium, Microcystis aeruginosa. The toxicity of this water bloom was being monitored in the reservoir. The controlling authority dosed the reservoir with copper sulphate to destroy the cyanobacteria, which caused the cells to lyse and release toxin into the water. Epidemiological data for liver injury in the affected population, a control population and comparison of the time periods before the bloom, during the bloom and lysis, and afterwards, showed clearly that liver damage had occurred only in the exposed population and only at the time of the water bloom [8]. In another less well characterized event, about 140 children and 10 adults were hospitalized, after the water supply authorities treated a cyanobacterial bloom in a small drinking water supply reservoir with copper sulphate, to resolve taste and odour problems. Within a week severe hepatoenteritis was apparent in the population, with about 20 cases requiring intravenous therapy. No-one died though several children were placed in intensive care [9]. Subsequent investigation demonstrated a "new" toxic cyanobacterial species in the reservoir, with a potent general toxin [10]. Later work on this strain of cyanobacterium lead to the identification of an alkaloid cytotoxin, with considerable liver toxicity [11].

How Abundant are Cyanobacterial Toxins?

Cyanobacteria are a normal component of the worldwide biota, with a wide tolerance of climatic conditions and environment. As a very ancient life-form they occupy every conceivable ecological niche, and their abundance is limited by nutrient and light availability [12]. In aquatic systems cyanobacteria are always present, through the population density varies from very small numbers to more than 10^6 organisms/mL. There is a strong relationship between phosphorus concentration in the water and cyanobacterial numbers and also a similar though less linked relationship between dissolved nitrate/ammonia and cyanobacteria [13]. Thus in general toxic cyanobacterial species will be presentin all water bodies, with numbers dependent on the available nutrients and light.

As human population density rises the inflow of nutrients into water bodies increases through agricultural fertilizer use, urban run-off and sewage discharge. This increase in aquatic nutrients is termed eutrophication, and it is observed worldwide. Phytoplankton in general becomes more abundant, and among these organisms are the cyanobacteria. Cyanobacteria can utilize nutrients competitively with eukaryotic phytoplankton, and will proliferate more successfully at lower nutrient concentrations than the green algae. As a result many rivers, lakes and reservoirs worldwide develop high cyanobacterial cell concentrations, especially in the summer months, which appear as greenish suspensions in the water. Some species float to the surface under warm, still conditions, forming scums with extreme cell concentrations above 1×10^6 cells/ml. Dried scums often appear blue-green or red through liberation of phycocyanin pigment, leading to the common name of these organisms - blue-green algae.

The scum-forming cyanobacterial species are largely toxic, and the majority of domestic animal poisonings have occurred from the animals drinking scum [12]. Cell populations carry over from year to year, and once a reservoir or lake has an established water bloom of cyanobacteria in summer, it is very difficult to reverse this phenomenon. Cyanobacteria proliferate in warmer weather, and often form extensive blooms in late summer. With an increase in global temperatures, cyanobacterial populations are likely to increase also. With the growth in human populations, demand for drinking water has resulted in water is being drawn from water bodies carrying substantial cyanobacterial populations, thus presenting a risk to human populations.

Annual or even permanent blooms of toxic cyanobacteria are becoming increasingly common in drinking water reservoirs. To give an illustration the three main reservoirs supplying Brisbane in Australia all carry substantial populations of the toxic Cylindrospermopsis raciborskii. This cyanobacterium forms dense layers 5–10m below the surface, so that the first indication of the proliferation of the organism may be the blocking of filters in the drinking water treatment plant. Other examples are the main drinking water supply reservoirs for the cities of Sao Paulo in Brasil and Lodz in Poland, which contain heavy blooms of the toxic Microcystis aeruginosa in summer.

Cyanobacterial Toxins

Neurotoxins form one of the major groups of cyanobacterial toxins. They are produced by several genera of cyanobacteria growing in freshwater which have the capacity to form dense waterblooms and floating scums at the edge of lakes and rivers. The neurotoxins are alkaloid compounds, fast acting and have caused many deaths of dogs and livestock [1]. Three types of alkaloid have so far been described (Figure 1).

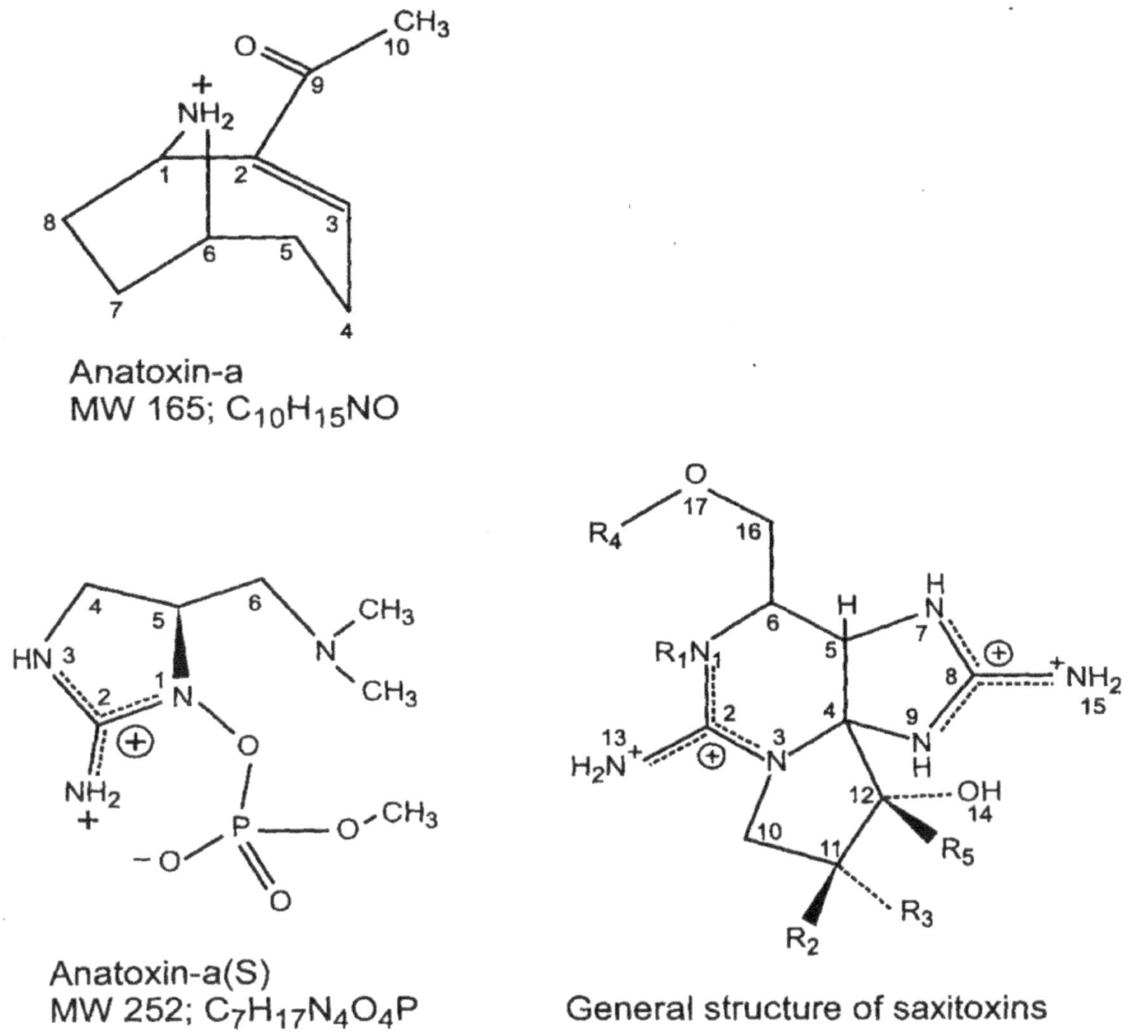

Figure 1. Structures of cyanobacterial neurotoxic alkaloids.

The first to be characterized was anatoxin-a (Fig. 1), a neuromuscular blocking agent which causes death by respiratory paralysis [14]. This toxin has been found in three common genera of cyanobacteria, Anabaena, Aphanizomenon and Planktothrix, all filamentous planktonic organisms capable of high cell concentrations and potential scum formation. There has been no clear evidence of human poisoning from these organisms, though a coroner in Wisconsin in 2003 resolved that the death of a male teenager who was diving and playing in a pond containing neurotoxic Anabaenahad died as a consequence of ingestion of these cyanobacteria [15] To verify this cause of death, evidence of toxin in the gastrointestinal tract or tissues was required, however there has been no published report of the presence of toxin. By contrast, dogs poisoned by anatoxin-a have shown the toxin in stomach contents [16]. The compound is stable in the environment, as exhibited by the dogs having died after eating decaying lumps of cyanobacteria on the lakeside. Anatoxin-a has been identified in the water of lakes in North America and in Europe, which are largely used for recreation. There is the possibility of consumption of moderate quantities of water during swimming and especially water skiing, and hence a risk exists for anatoxin-a poisoning of recreational water users [17]. Many authorities in developed countries have warning procedures for cyanobacterial blooms at popular recreational areas, to reduce risk to water users [18]. There has been little attention paid to the assessment of risk to drinking water consumers from anatoxin-a, largely because of the rapid excretion of the toxin from the body, no evidence of residual effects and low free-water concentrations in lakes.

Anatoxin-a(s) is much less common in cyanobacterial waterblooms, though it was first identified following cattle deaths in the USA[19]. The alkaloid closely resembles an organophosphorus insecticide (Figure 1), and acts as an anticholinesterase. The characteristic feature of this poisoning is excessive salivation, which is the reason for the designation (s). The compound is highly unstable and unlikely to persist in water supplies, and as a result is also unlikely to present any risk.

The saxitoxin-type neurotoxins are well known as the cause of paralytic shellfish poisonings, which have resulted in many hundreds of human deaths worldwide [20]. As a result, legislation controls the allowable concentration of saxitoxins in shellfish harvested for human consumption (80µg/100g fresh shellfish tissue) and there is a substantial monitoring program in many countries. Saxitoxins are however not limited to marine waters, and also occur in freshwater cyanobacteria.Anabaena, Aphanizomenon and Lyngbya genera of cyanobacteria have species that produce saxitoxins [21–23]. The massive waterbloom of Anabaena circinalis on 1,000km of the Darling River in Australia in 1991 killed a large number of sheep and cattle, and also resulted in detectable neurotoxicity in town water supplies [24]. Saxitoxins are heat-stable molecules, which are not easily removed in conventional water treatments unless pH and chlorine residuals are carefully controlled, but can be effectively removed by ozone or activated carbon [25].

The toxicity of saxitoxin is considerable, as the alkaloid blocks sodium conduction in axons preventing nerve impulse transmission, leading to paralysis. The oral LD_{50} in mice is about 260µg/kg bodyweight [26]. Acute poisoning in humans is unlikely to occur from contaminated water supplies, as the human body can tolerate about 100µg of saxitoxin without ill effect [20], which translated to drinking water is 50µg/L assuming 2L water drunk per day. No cumulative effects have been demonstrated, though there is limited evidence of resistance to toxicity in exposed human populations [26]. New Zealand is considering a Maximum Acceptable Value in drinking water of 3µg/L of saxitoxin equivalents in their new drinking water guidelines, which should be ratified shortly [27]. This value has also been proposed in Australia [28]. The data on which this value was based were the intraperitoneal toxicity of saxitoxin to mice, and incorporated a safety factor of 1,000. This issue will be discussed further in the section on hepatotoxins (microcystins). There are no data for the concentrations of saxitoxin-type neurotoxins in drinking water, and there were no reports of neurotoxic symptoms in the town population when neurotoxicity was detected in the drinking water supply [29].

Hepatotoxins

These toxins have received the greatest attention, as they are the source of the most likely risk to consumers of drinking water. The predominant genera of cyanobacteria forming the peptide toxins called microcystins are Microcystis, Planktothrix and Anabaena. Species from these genera are common in Europe, the Americas, Africa and Asia, and poisoning of domestic animals has been widely reported [12]. Only two epidemiological investigations have so far shown human injury from microcystin in drinking water, one in Australia [8] and one in China (unpublished). As a consequence of the frequency of cyanobacterial blooms containing hepatotoxins in drinking water reservoirs, the WHO carefully examined the need for the major toxins, the microcystins, to be included in the drinking water guidelines. An 'expert group' was established to examine the whole issue of cyanobacterial toxins in drinking water, which resulted in a comprehensive assessment of the risks involved [30]. The outcome was a recommendation that the microcystins should be included among the chemicals for which Guideline Values be determined. These peptide toxins are cyclic, and contain a majority of D-amino acids (Figure 2). The positions shown as [X] and [Y] are L-amino acids, and are variable between species and strains of cyanobacteria. The most abundant variant has L-leucine (L) and L-arginine (R) respectively at [X] and [Y] (microcystin-LR). The amino acid at the left of the molecule is unique, is connected into the ring through an amino group at the β-carbon atom, and has the trivial name of ADDA.

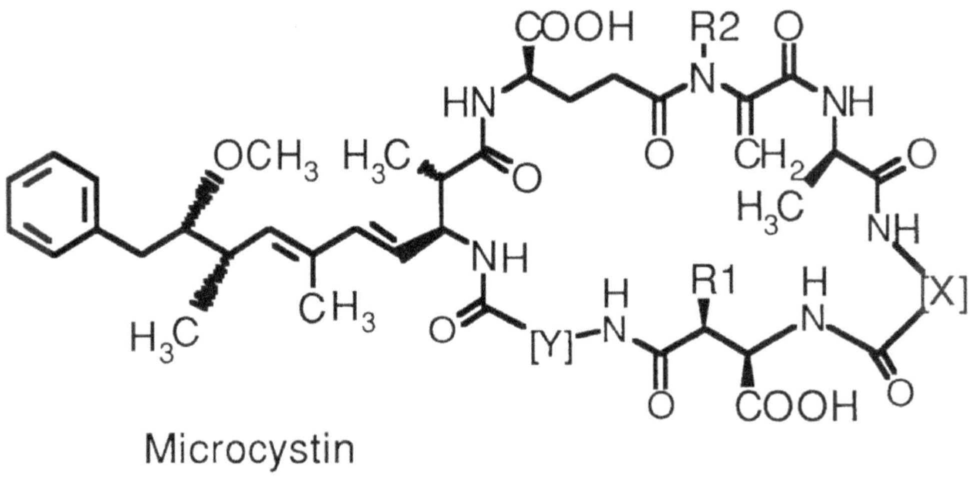

Figure 2. Structure of the peptide hepatotoxin microcystin, first isolated from the cyanobacterium Microcystis aeruginosa.

Microcystins are resistant to digestion in the gastrointestinal tract of eukaryotes, as peptide bonds linking to the D-amino acids are not susceptible to normal hydrolytic enzymes. The toxins are concentrated into the liver by an active transport system, similar to the bile acid transporter [31]. Microcystins specifically inhibit protein phosphatases 1 and 2A, which have a vital role in cell control and in intracellular structure [12]. Acute poisoning is through destruction of the liver architecture, leading to blood loss into the liver and hemorrhagic shock [32]. Later death is through liver failure with massive destruction of hepatocytes, seen in large animal deaths and human fatalities [5, 33]. Chronic exposure to these toxins in drinking water led to ongoing active liver injury in mice [34].

There is experimental evidence for tumour promotion by microcystins, and limited data for carcinogenesis in rodents [35]. In rural areas in Southern China some villages showed hyper-endemic rates of hepatocellular carcinoma, which have been shown to be linked to hepatitis, aflatoxin in the food, and drinking surface water. Microcystins in the ponds and ditches used as water sources were suspected of contributing to the cancer rates [36, 37].

These toxins are highly stable in water and are resistant to boiling. Hence they present a risk to consumers in less developed regions and countries who are collecting water from surface sources to drink. Many lakes, ponds, ditches and streams in rural and outer urban areas suffer from eutrophication through excessive nutrient leaching from housing, sewage, and intensive agricultural use, leading to cyanobacterial proliferation. In tropical and temperate regions of the world the genus Microcystis is the most abundant cyanobacterium forming toxic blooms, with toxin concentrations sufficient to poison domestic animals. If these contaminated water sources are used for human consumption, there is a risk of human poisoning. Conventional Western drinking water treatment may not be effective under bloom conditions in removing microcystins from drinking water and hence there is a risk to consumers. Advanced water treatment using ozone and activated carbon will reliably remove microcystins [12].

WHO have carried out an assessment of the safe level of microcystins in drinking water, based on data from a subchronic toxicity trial in mice, with supporting data from growing pigs [38]. The calculation used the No Observed Adverse Effect Level for male mice during a 13 week oral toxicity trial, of 40μg of microcystin-LR/Kg/day [39]. This was used to calculate a Tolerable Daily Intake (TDI) for safe human consumption, by the incorporation of uncertainty or safety factors. While these are subjective, a factor of 10 for interspecies uncertainty between rodents and humans, a further 10 for variability in sensitivity between people, and an uncertainty of 10 for inadequate data, possible tumour promotion and lack of lifetime exposure are generally accepted.

Thus the:

$$\mathrm{TDI} = \frac{40}{10 \times 10 \times 10} = 0.04\mu\mathrm{g/Kg/day}$$

From this value the Guideline Value (also called the reference dose and the maximum acceptable concentration) was calculated from the standard bodyweight of 60Kg, an assumption of the proportion of the dose from drinking water of 0.8 (some may come from food and particularly blue-green algal diet supplements) and a standard water consumption of 2L/day.

$$\mathrm{GV} = \frac{0.04 \times 60 \times 0.8}{2} = 0.96$$
$$= 1\mu\mathrm{g/L\ of\ microcystin-LR\ in\ drinking\ water.}$$

This value was determined from the toxicity of microcystin-LR, so the WHO Chemical Safety Committee set the Guideline Value for microcystin–LR. Since there are some 60 variants of the molecule, and some highly toxic blooms do not contain any of the –LR variant, it is necessary to interpret this as toxicity equivalent to microcystin-LR. Where this Guideline Value has been adopted as the basis for national legislation, the need for monitoring of all the toxin variants has been recognized and the equivalent total toxicity calculated [40]. Individual countries have also adopted a higher standard bodyweight, and a different proportion of the consumption from drinking water. All of the Guideline Values adopted so far lie between 1 and 2μg/L of microcystin equivalents, which for practical purposes are the same.

A larger potential adjustment to this value may result from re-classification of microcystin as a carcinogen, rather than a non-carcinogenic poison. While there is experimental evidence for tumour promotion by microcystin in liver, skin and colon, the only data indicating carcinogenesis have been obtained by continued very high intraperitoneal doses of toxin in mice which cause extensive liver damage [1]. In China there is ongoing investigation into the relationship between surface water consumption and cancer of liver and colon [41]. This issue is discussed in detail elsewhere [12], concluding that there is insufficient evidence at present to determine that microcystin is a probable carcinogen but the possibility requires continual evaluation.

Cylindrospermopsins

These alkaloid cytotoxins were relatively recently discovered, following the widespread human poisoning at Palm Island, Australia due to contamination of the water supply [9]. Cyanobacteria from the supply reservoir were collected, cultured and evaluated for toxicity, showing potent toxicity to liver, kidney, adrenals, lymphoid cells and other tissues in mice [10]. Subsequent investigation of the oral toxicity of the cyanobacterium responsible, Cylindrospermopsis raciborskii, further demonstrated the tissue damage caused by the toxin [42, 43]. The toxic alkaloid was isolated and identified as a potent inhibitor of protein synthesis [11, 44]. There are on-going investigations into the mechanism of cylindrospermopsin toxicity, which may involve activated metabolites of the alkaloid [44, 45]. The alkaloid has several reactive groups, including a hydroxymethyl uracil, which may be vulnerable to biological oxidation (Figure 3).

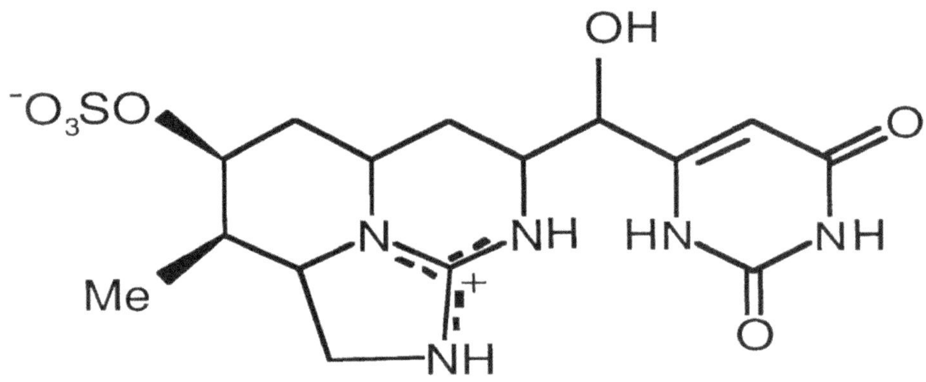

Figure 3. Molecular structure of the cyanobacterial alkaloid toxin cylindrospermopsin. The bridging hydroxyl group may be in either stereochemical position, and may also be replaced by hydrogen.

Cylindrospermopsin has been found in water bodies that have blooms of the cyanobacterial species Aphanizomenon ovalisporum[46], and Umezakia natans[47] as well as those withCylindrospermopsis[48, 49], and recently in Germany, in the absence of any of those species [50]. It is apparent from this data that cylindrospermopsin is likely to occur widely in freshwater sources, and that only a beginning has been made in identifying species producing this toxin. Monitoring of water supplies for cylindrospermopsin has found concentrations in natural water bodies, reservoirs and in drinking water that are above 10μg/L, which is a cause for concern [51, 52].

On the basis of the experimental toxicity of cylindrospermopsin, the reported human poisoning associated with Cylindrospermopsis and the toxin concentrations measured in water bodies, it was apparent that risk assessment for this toxin in drinking water was required. A subchronic oral exposure trial of cylindrospermopsin in male mice provided a No Observed Adverse Effect Level of 30μg of cylindrospermopsin/Kg/Day. On the assumption of standard uncertainty factors and the total intake arising from drinking water, a Guideline Value of 1μg/L resulted [53].

Examination of the molecular structure of cylindrospermpsin indicated that it may be able to interact with DNA or RNA in cells, through the uracil group, assisted by the planar shape of the molecule. If this proves to be the case, then evaluation of the possible carcinogenicity of the molecule is required. Preliminary data indicated that cylindrospermopsin may form DNA adducts [54], and there is evidence for clastogenicity and micronucleus formation in a cultured human white cell line incubated with cylindrospermopsin [55]. A preliminary trial of carcinogenicity in mice indicated the presence of excess tumors, providing support for a more definitive carcinogenicity trial [56]. It is premature at present to attempt to classify cylindrospermopsin as a possible human carcinogen, because of the very limited current data. Further experimental and epidemiological research on this toxin is required to clarify these issues, and cylindrospermopsin is now on the 'Candidate Contaminant List' of the US Environmental Protection Agency.

Evaluation of the Risks to the Population from Cyanobacterial Toxins

There are two areas in which more data is necessary to make a clear case for national action on minimizing health risks from cyanobacterial toxins. The first is the need for widespread monitoring for the presence of toxic cyanobacterial species and toxins in drinking water sources, to identify the abundance of locations of potential risk. This is in progress in Europe, and an initial survey has been carried out through the American Water Works Association in the USA. From data arising from these surveys, the extent of the problem has become apparent, and the location of a proportion of the populations at most risk.

The second and more difficult aspect is the need for epidemiological studies on at-risk populations to quantify the adverse health effects. Exposure biomarkers will have to be developed, in addition to quantitating the toxin concentrations in tap water. The commonly used clinical measures of liver and kidney function, and clinical records for hepatoenteritis, provide relevant health information. The earlier data on population injury from microcystins indicates the clinical parameters of particular interest [8].

CONCLUSION

The cyanobacterial toxins provide a risk to human health when the population of toxic cyanobacteria in drinking water sources rises to bloom proportions. The present assessments of Guideline Values for these toxins as chemical, non-carcinogenic contaminants indicate that a safe concentration in drinking water is in the region of 1μg/L, a concentration that has been exceeded in numerous water storages. Carcinogenicity of these toxins is not yet established, though both microcystins and cylindrospermopsin have caused excess tumors in rodent experiments. With the increased eutrophication of water supplies and global warming, cyanobacterial populations and hence toxic risks are likely to rise in the immediate future. The extent and potential severity of the risks need further evaluation.

REFERENCES

1. Falconer, I. R. Cyanobacterial Toxins of Drinking Water Supplies: Cylindrospermopsins and Microcystins. Chapter 5 Cyanobacterial poisoning of livestock and people; CRC Press: Boca Raton, Fl., 2005.
2. Bowling, L. The cyanobacterial (blue-green algal) bloom in the Darling/Barwon River system, November-December, 1991; Technical Services Division, New South Wales Department of Water Resources: Australia, Sydney, 1992.
3. Steyn, D. G. Poisoning of animals and human beings by algae. South African Journal of Science**1945**, 41, 243–244.
4. Teixera, M. G. L. C.; Costa, M. C. N.; Carvalho, V. L. P.; Pereira, M. S.; Hage, E. Gastroenteritis epidemic in the area of the Itaparica Dam, Bahia, Brazil. Bulletin of the Pan-American Health Organisation **1993**, 27, 244–253.
5. Jochimsen, E. M.; Carmichael, W. W.; An, J. S.; Cardo, D. M.; Cookson, S. T.; Holmes, C. E. M.; Antunes, M. B. D.; Demelo, D. A.; Lyra, T. M.; Barreto, V. S. T.; Azevedo, S. M. F. O.; Jarvis, W. R. Liver failure and death after exposure to microcystins at a hemodialysis center in Brazil. The New England Journal of Medicine **1998**, 338, 873–878.
6. Carmichael, W. W.; Azevedo, S. M. F. O.; An, J. S.; Molica, R. J. R.; Jochimsen, E. M.; Lau, S.; Rinehardt, K. L.; Shaw, G. R.; Eaglesham, G. K. Human Fatalities from Cyanobacteria: Chemical and Biological Evidence for Cyanotoxins. Environmental Health Perspectives **2001**,109, 663–668.
7. Azevedo, S. M. F. O.; Carmichael, W. W.; Jochimsen, E. M.; Rinehart, K. L.; Lau, S.; Shaw, G. R.; Eaglesham, G. K. Human intoxication by microcystins during renal dialysis treatment in Caruaru - Brazil. Toxicology **2002**, 181, 441–446.
8. Falconer, I. R.; Beresford, A. M.; Runnegar, M. T. Evidence of liver damage by toxin from a bloom of the blue-green alga Microcystis aeruginosam. Med. J. Aust **1983**, 1, 511–4.
9. Byth, S. Palm Island Mystery Disease. Medical Journal of Australia **1980**, 2, 40–42.
10. Hawkins, P. R.; Runnegar, M. T. C.; Jackson, A. R. B.; Falconer, I. R. Severe hepatotoxicity caused by the tropical cyanobacterium (blue-green alga) Cylindrospermopsis raciborskii(Woloszynska) Seenaya and Subba Raju isolated form a domestic supply reservoir. Applied and Environmental Microbiology **1985**, 50, 1292–1295.
11. Ohtani, L.; Moore, R. E.; Runnegar, M. T. C. Cylindrospermopsin: a potent hepatotoxin from the blue-green alga Cylindrospermopsis raciborskii. J. of the American Chem. Soc **1992**, 114, 7941–7942.
12. Falconer, I. R. Cyanobacterial Toxins of Drinking Water Supplies: Cylindrospermopsins and Microcystins; CRC Press: Boca Raton, 2005.
13. Mur, L. R.; Skulberg, O. M.; Utkilen, H. Cyanobacteria in the Environment. In Toxic Cyanobacteria in Water. A Guide to their Public Health Consequences, Monitoring and Management; Chorus, I., Bartram, J., Eds.; E & FN Spon on behalf of WHO: London, 1999; pp. 15–40.
14. Carmichael, W. W.; Mahmood, N. A. Toxins from freshwater cyanobacteria. In Seafood toxins; American Chemical Society: Washington, DC, 1984; pp. 377–389.
15. Behm, D. Coroner cites algae in teen's death; Milwaukee Journal Sentinel: Milwaukee, 2003.
16. Edwards, C.; Beattie, K. A.; Scrimgeour, C. M.; Codd, G. A. Identification of anatoxin-a in benthic cyanobacteria (blue-green algae) and in associated dog poisonings at Loch Insh, Scotland. Toxicon **1992**, 30, 1165–1175.
17. Chorus, I.; Falconer, I. R.; Salas, H. J.; Bartram, J. Health Risks Caused by Freshwater Cyanobacteria in Recreational Waters. J. of Toxicol. and Environ. Health, Part B **2000**, 3, 323–347.
18. Fromme, H.; Kohler, A.; Krause, R.; Fuhrling, D. Occurrence of cyanobacterial toxins – microcystins and anatoxin-a - in Berlin water bodies with implications to human health and regulations. Environ. Toxicol **2000**, 75, 120–130.
19. Mahmood, N. A.; Carmichael, W. W. Anatoxin-a(s), an anticholinesterase from the cyanobacterium Anabaena flos-aquae NRC-525-17. Toxicon **1987**, 25, 1221–1227.
20. Falconer, I. Algal Toxins in Seafood and Drinking Water; Academic Press: London, 1993.
21. Velzeboer, R. M. A.; Baker, P. D.; Rositano, J.; Heresztyn, T.; Codd, G. A.; Raggett, S. L. Geographical patterns of occurrence and compositions of saxitoxins in the cyanobacterial genus Anabaena (Nostocales Cyanophyta) in Australia. Phycologia **2000**, 39, 395–407.
22. Ferreira, F. M. B.; Soler, J. M. F.; Fidalgo, M. L.; Fernandez-vila, P. PSP toxins fromAphanizomenon flos-aquae (cyanobacteria) collected in the Crestuma-Lever reservoir (Douro River, Northern Portugal). Toxicon **2001**, 39, 757–761.
23. Carmichael, W. W.; Evans, W. R.; Yin, Q. Q.; Bell, P.; Moczydlowski, E. Evidence for paralytic shellfish poisons in the freshwater cyanobacterium Lyngbya wollei (Farlow ex Gomont) comb. nov. Applied and Environmental Microbiology **1997**, 65, 3104–3110.
24. Bartram, J.; Burch, M.; Falconer, I. R.; Jones, G.; Kuiper - Goodman, T. Situation Assessment Planning and Management. In Toxic Cyanobacteria in Water. A Guide to their Public Health Consequences, Monitoring and Management; Chorus, I., Bartram, J., Eds.; E & FN Spon on behalf of WHO: London, 1999; pp. 179–09.
25. Falconer, I. R.; Runnegar, M. T. C.; Buckley, T.; Huyn, V. L.; Bradshaw, P. Use of Powdered and Granular Activated Carbon to Remove Toxicity from Drinking Water Containing Cyanobacterial Toxins. J. American Water Works Association **1989**, 18, 102–105.
26. Kuiper-Goodman, T.; Falconer, I.; Fitzgerald, J. Human Health Aspects. In Toxic Cyanobacteria in Water. A Guide to their Public Health Consequences, Monitoring and Management; Chorus, I., Bartram, J., Eds.; E & FN Spon on behalf of WHO: London, 1999; pp. 113–153.

27. New Zealand Ministry of Health, Guidelines for Drinking Water Quality Management for New Zealand; Ministry of Health: Wellington, New Zealand.

28. Fitzgerald, D. J.; Cunliffe, D. A.; Burch, M. D. Development of health alerts for cyanobacteria and related toxins in drinking-water in South Australia. Environmental Toxicology **1999**, 14, 203–209.

29. Bartram, J.; Vapnek, J. C.; Jones, G.; Bowling, L.; Falconer, I.; Codd, G. A. Implementation of Management Plans. In Toxic Cyanobacteria in Water. A Guide to their Public Health Consequences, Monitoring and Management; Chorus, J Bartram, Ed.; E & FN Spon: London, 1999; pp. 211–234.

30. Chorus, I.; Bartram, J. Toxic cyanobacteria in water. In A Guide to their Public Health Consequences, Monitoring and Management; E & FN Spon (on behalf of World Health Organisation): London, 1999.

31. Runnegar, M.; Berndt, N.; Kaplowitz, N. Microcystin uptake and inhibition of protein phosphatases: effects of chemoprotectants and self-inhibition in relation to known hepatic transporters. Toxicol. App. Pharmacol **1995**, 134, 264–72.

32. Falconer, I. R.; Jackson, A. R. B.; Langley, J.; Runnegar, M. T. C. Liver pathology in mice in poisoning by the blue-green alga in Microcystis aeruginosa. Australian J. of Biol. Sci **1981**, 34, 179–187.

33. Jackson, A. R. B.; McInnes, A.; Falconer, I. R.; Runnegar, M. T. C. Clinical and pathological changes in sheep experimentally poisoned by the blue-green alga Microcystis aeruginosa.Veterinary Pathology **1984**, 21, 102–113.

34. Falconer, I. R.; Smith, J. V.; Jackson, A. R.; Jones, A.; Runnegar, M. T. Oral toxicity of a bloom of the cyanobacterium Microcystis aeruginosa administered to mice over periods up to 1 year.J. Toxicol Environ Health **1988**, 24, 291–305.

35. Ito, E.; Kondo, F.; Terao, K.; Harada, K. L. Neoplastic nodular formation in mouse liver induced by repeated intraperitoneal injections of microcystin-LR. Toxicon **1997**, 55, 1453–1457.

36. Yu, S. Z. Primary prevention ofhepatocellular carcinoma. J. of Gastroenterology and Hepatology**1995**, 10, 674–682.

37. Ueno, Y.; Nagata, S.; Tsutsumi, T.; Hasegawa, A.; Watanabe, M.; Park, H.; Chen, G. C.; Chen, G.; Yu, S. Detection of microcystins, a blue-green algal hepatotoxin, in drinking water sampled in Haimen and Fusui, endemic areas of primary liver cancer in China, by highly sensitive immunoassay. Carcino genesis **1996**, 17, 1317–1321.

38. Falconer, I.; Bartram, J.; Chorus, I.; Kuiper Goodman, T.; Utkilen, H.; Burch, M.; Codd, G. A. Safe Levels and Safe Practices. In Toxic Cyanobacteria in Water. A Guide to their Public Health Consequences, Monitoring and Management; Chorus, I., Bartram, J., Eds.; E & FN Spon on behalf of WHO: London, 1999; pp. 55–178.

39. Fawell, J. K.; James, C. P.; James, H. A. Toxins from blue-green algae: Toxicological assessment of microcystin-LR and a method for its determination in water; WRc pic: Medmenham, 1994.

40. Australian National Health and Medical Research Council, Drinking Water Guidelines; NH & MRC: Canberra, Australia, 2004.

41. Zhou, L.; Yu, H.; Chen, K. Relationship between microcystin in drinking water and colorectal cancer. Biomedical and Environmental Science **2002**, 75, 166–171.

42. Falconer, I. R.; Hardy, S. J.; Humpage, A. R.; Froscio, S. M.; Tozer, G. J.; Hawkins, P. R. Hepatic and renal toxicity of the blue-green alga (cyanobacterium) Cylindrospermopsis raciborskii in male Swiss Albino mice. Environmental Toxicology **1999**, 14, 143–150.

43. Seawright, A. A.; Nolan, C. C.; Shaw, G. R.; Chiswell, R. K.; Norris, R. L.; Moore, M. R.; Smith, M. J. The oral toxicity for mice of the tropical cyanobacterium Cylindrospermopsis raciborskii(Woloszynska). Environmental Toxicology **1999**, 14, 135–142.

44. Froscio, S. M. Investigation of the mechanisms involved in cylindrospermopsin toxicity: Hepatocyte culture and reticulocyte lysate studies. In PhD Thesis, Department of Clinical and Experimental Pharmacology; University of Adelaide: Adelaide, 2002; p. 139.

45. Runnegar, M. T.; Kong, S. M.; Zhong, Y. Z.; Lu, S. C. Inhibition of reduced glutathione synthesis by cyanobacterial alkaloid cylindrospermopsin in cultured rat hepatocytes.Biochemical Pharmacology **1995**, 49, 219–225.

46. Banker, P.; Carmeli, S.; Hadas, O.; Telsch, B.; Porat, R.; Sukenik, A. Identification of cylindrospermopsin in Aphanizomenon ovalisporum (Cyanophyceae) isolated from Lake Kinneret, Israel. J. of Applied Phycology **1997**, 33, 613–616.

47. Harada, K.; Ohtani, I.; Iwamoto, K.; Suzuki, M.; Watanabe, M. F.; Watanabe, M.; Terao, K. Isolation of Cylindrospermopsin from a Cyanobacterium Umezakia Natans and Its Screening Method. Toxicon **1994**, 32, 73–84.

48. Li, R.; Carmichael, W. W.; Brittain, S.; Eaglesham, G. K.; Shaw, G. R.; Mahakhant, A.; Noparatnaraporn, N.; Youngmanichai, W.; Kaya, K.; Watanabe, M. M. Isolation and identification of the cyanotoxin cylindrospermopsin and deoxy-cylindrospermopsin from a Thailand strain of Cylindrospermopsis raciborskii (Cyanobacteria). Toxicon **2001**, 39, 973–980.

49. Stirling, D. J.; Quilliam, M. A. First report of the cyanobacterial toxin cylindrospermopsin in New Zealand. Toxicon **2001**, 39, 1219–1222.

50. Fastner, J.; Heinze, R.; Humpage, A. R.; Mischke, U.; Eaglesham, G. K.; Chorus, I. Cylindrospermopsin occurrence in two German lakes and preliminary assessment of toxicity and toxin production of Cylindrospermopsis raciborskii (Cyanobacteria) isolates. Toxicon **2003**,42, 313–321.

51. Griffiths, D. J.; Saker, M. L. The Palm Island Mystery Disease 20 Years on: A Review of Research on the

Cyanotoxin Cylindrospermopsin. Environmental Toxicology **2003**, 18, 78–93.

52. Flewelling, L.; Pawlowicz, M.; Carmichael, W.; Williams, C. D.; Burns, J.; Chapman, A. Assessment of Cyanotoxins in Florida's lakes, reservoirs, and rivers. In Cyanobacteria Survey Project; Harmful Algal Bloom Task Force; St. John's River Water Management District: Palatka, Florida, 2001.
53. Humpage, A. R.; Falconer, I. R. Oral toxicity of the cyanobacterial toxin cylindrospermopsin in male Swiss albino mice: Determination of No Observed Adverse Effect Level for Deriving a Drinking Water Guideline Value. Environmental Toxicology **2003**, 18, 94–103.
54. Shaw, G. R.; Seawright, A. A.; Moore, M. R.; Lam, P. K. S. Cylindrospermopsin, a cyanobacterial alkaloid: evaluation of its toxicologic activity. Therapeutic Drug Monitoring**2000**, 22, 89–92.
55. Humpage, A. R.; Fenech, M.; Thomas, P.; Falconer, I. R. Micronucleus induction and chromosome loss in WIL2-NS cells exposed to the cyanobacterial toxin, cylindrospermopsin.Mutation Research **2000**, 472, 155–161.
56. Falconer, I. R.; Humpage, A. R. Preliminary Evidence for In-Vivo Tumour Initiation by Oral Administration of Extracts of the Blue-Green Alga Cylindrospermopsis raciborskii containing the toxin Cylindrospermopsin. Environmental Toxicology **2001**, 16, 192–195.

Section IV - Drinking Water Disinfection and Treatment

CHAPTER 13

INFLUENCE OF MEMBRANE MATERIALS AND OPERATIONAL MODES ON THE PERFORMANCE OF ULTRAFILTRATION MODULES FOR DRINKING WATER TREATMENT

Gongduan Fan, Zhaoyue Su, Rujing Lin, Xiuyong Lin, Renxin Xu, and Wei Chen

INTRODUCTION

Drinking water safety problem is still not optimistic in China at present [1, 2]. With the development of industry and agriculture, and the improvement of the people's standard of living, pollution of industrial wastewater and domestic sewage has become more and more serious. Polluted water which carries huge amount of toxic and harmful substances is dispersed into the water body, causing ever-decreasing self-purification capacity of natural water. The restriction of conventional water treatment process in waterworks is becoming apparent [3]. Ultrafiltration (UF), as the third technology of purifying drinking water, is widely used in the field of drinking water treatment because of its peculiar advantage with the improvement in membrane quality and decrease in membrane costs [4, 5]. UF has been reported to have the advantages of effectively removal of suspended matters, colloid, pathogenic microorganism and macromolecule organic matters [6], ease of operation, and small footprint [7, 8]. UF membranes materials can be divided into organic and inorganic materials; common commercial ones are organic polymer materials such as polyethersulfone (PES), polyvinylidene fluoride (PVDF), polyacrylonitrile (PAN), polysulfone (PS), and polyvinyl chloride (PVC) [9,10]. The performance of UF largely depends on the properties of membrane materials and operational modes. C. Jönsson and A.-S. Jönsson [11] indicated that the pore size and the hydrophilic/hydrophobic properties of the UF membrane material were correlated to the extent of the flux reduction. Li et al. [12] showed that improving hydrophilicity of the membrane could reduce membrane fouling in a certain extent and the improvement of membrane hydrophilicity and the increase of the efficient filtration area through modification resulted in better permeation performance. The permeations of thermally stable membranes were also found to have received significantly enhanced interest because of their advantage to increase permeate flux with an increasing operating temperature compared to the common membranes which need to be operated using cooling systems [13]. What is more, the operational modes of UF can be sorted into inside-out and outside-in UF modes [14]. Inside-out UF mode means that the separation layer is the inner surface of fibers and the feed water flows inside the fibers, while the outside-in UF mode means that separation layer is on the outside surface of the fiber and the feed water flows outside the fibers. The differences between operational modes of UF modules result

Citation: Gongduan Fan, Zhaoyue Su, Rujing Lin, Xiuyong Lin, Renxin Xu, and Wei Chen, "Influence of Membrane Materials and Operational Modes on the Performance of Ultrafiltration Modules for Drinking Water Treatment," International Journal of Polymer Science, vol. 2016, Article ID 6895235, 8 pages, 2016. doi:10.1155/2016/6895235

in different suitable membrane materials, different washing methods, and different application conditions. The inside-out UF module needs to use materials with higher stiffness and good acid and caustic resistance, such as PES, for adapting its washing method of using the strong flow of water to flush and the acid-base staggered method to carry out chemical enhanced backwash (CEB); its poorer capacity of receiving pollutants makes the inside-out UF module match the field of drinking water treatment. The outside-in UF module needs to use materials with higher strength, such as PVDF and PAN, for adapting to the way of gas washing; its lower fouling load caused by larger flow channel extends its applications in water and wastewater treatment. Therefore, different membrane materials and operational modes will inevitably result in different adaptability in special condition of raw water quality. In this pilot study, three kinds of hollow fiber UF modules made by different materials and operated under different operational modes were chosen to purify micropolluted source water of Minjiang River, and in order to reduce the membrane fouling [15] and enhance removal of organics [16, 17], integrative devices of in-line coagulation/UF were proposed. Their treatment performance, operation stability, clean methods, and fresh water recovery rate were assessed. The objective of this paper is to evaluate the influence of different membrane materials and operational modes on the performance of UF modules for drinking water treatment.

MATERIALS AND METHODS

Integrative Devices

Three types of hollow-fiber membranes which are widely used for drinking water production were used in this study; virgin PES, PVDF, and PAN were prepared to construct combined processes of in-line coagulation/UF. The main membranes characteristics and operation parameters of these three membrane systems are shown in Table 1 and the schematic diagram of the UF pilot testing systems is shown in Figure 1, respectively. The coagulant was added (at a concentration of 4 mg/L) to raw water entering into the water plant from Minjiang River and rapidly mixed by the tubular static mixer and mechanical mixer in conventional coagulation and flocculation unit. The water was then pumped into the feed tank of three UF systems, respectively. In this pilot study, the inside-out hollow fiber UF membrane was made by PES (Module A), and two other outside-in hollow fiber UF membranes were made by PVDF (Module B) and PAN (Module C), respectively. Each UF system had its own filtration and restorative cleaning mode. The transmembrane pressure (TMP) and membrane permeate flux were continuously monitored by pressure gauges and electronic flowmeters, respectively. For each filtration and forward flushing procedure, water from the feed tank was pumped by feed pump to the module, and the permeate from the module overflowed its permeate tank and then was rejected back to the draining tank. For each backwash and CEB procedure, filtrate from the permeate tank was pumped by the backwashing pump to the module. The CEB was carried out every 24 hours for each module to eliminate microorganism and the pollutants that are hard to remove completely by backwashing.

Table 1: Membranes characteristics and operation parameters of the UF modules.

Items	Module A	Module B	Module C
Membrane material	PES	PVDF	PAN
Operation mode	Inside-out	Outside-in	Outside-in
Filtration mode	Dead-end	Dead-end	Dead-end
Membrane effective area, m^2	55	77	45
Average pore size, nm	20	30	50
Contact angle	44	66	52
Restorative flush mode	BW + FF	Air + BW + FF	Air + BW
Filtration duration, min	40	30	30
Designed permeate flux, LMH	60~120	50~80	44~66
Feed pressure, MPa	0.1~0.3	<0.6	<0.15
TMP, MPa	0.01~0.10	<0.15	<0.08
Chemical of CEB	HCl/(NaClO + NaOH)	NaClO + NaOH	NaClO + NaOH
Frequency of CEB, h	24	24	24

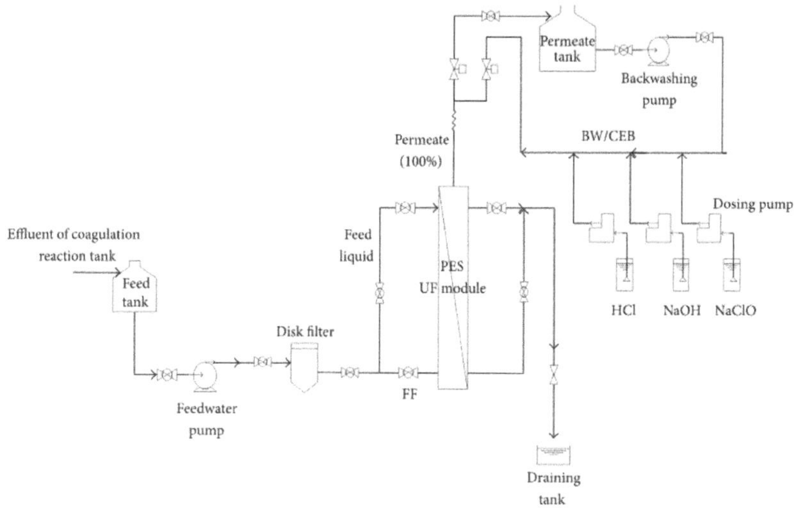

(a) The integrative device of in-line coagulation and inside-out hollow fiber UF modules

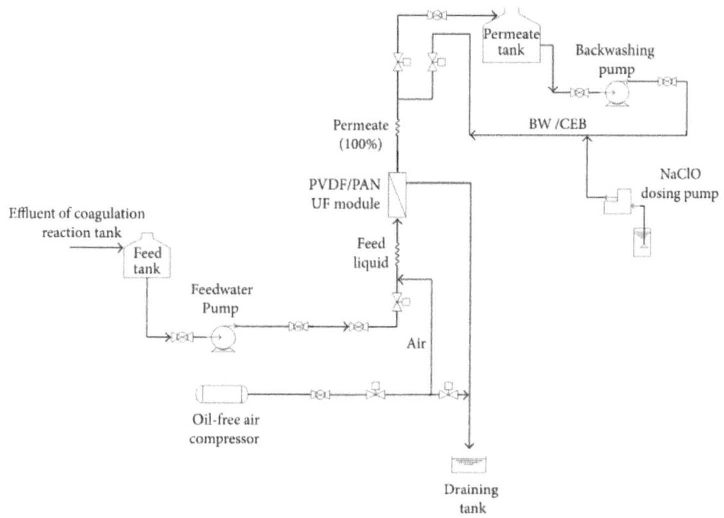

(b) The integrative device of in-line coagulation and outside-in hollow fiber UF modules

Figure 1: The schematic diagram of the three UF pilot testing systems.

Raw Water Quality

The pilot test was carried out in a water plant in Fuzhou Coastal Section of Minjiang River. For the water quality status of Minjiang River, it was seen that its water quality deteriorates rapidly in the past few years as discharge from factory effluent and domestic wastewater increase. On the other hand, for some self-reasons of some enterprises, such as the unadvanced equipment, mismanagement, and low technology level, emissions of pollutants cannot reach the national draining standard, which results in seasonal and unusual phenomena of the water quality of Minjiang River. According to the statistics, turbidity, organic content, and ammonium nitrogen (NH_4-N) concentrations are abnormally high in late spring and early summer each year regularly; therefore, assessment of purifying effects of different models is based on these three water quality indexes in this study. Source water from the drinking water treatment plant in this pilot test had pH 6.80~7.66, turbidity 14.0~80.0 NTU, COD_{Mn}, value 1.80~5.13 mg/L, and NH_4-N concentration 0.08~0.68 mg/L, respectively.

RESULTS AND DISCUSSION

Removal Abilities of Turbidity by Various UF Modules

The values of turbidity of permeate was measured during the long run in the pilot test, which is shown in Figure 2. The average turbidity of the permeation of PES, PVDF, and PAN module was 0.097, 0.109, and 0.101 NTU, respectively, and the average removal ratios of turbidity by these three modules were all higher than 99.5%. Various UF modules did not result in significant difference of the effluent turbidity. Notably, each UF module could reduce the influent turbidity (7.2~54.6 NTU) which fluctuated drastically to about 0.1 NTU steadily because of the powerful fractionation performance of the special tiny holes of UF membranes. It should be noted that an abnormal peak at 0.19 NTU was found during the testing phase, which was due to the colored NaClO dissolved in the filtrate. Actually, a short while before filtrate sampling, the CEB process of PVDF module has just been completed, which was likely to result in NaClO residual problems on the pipeline system.

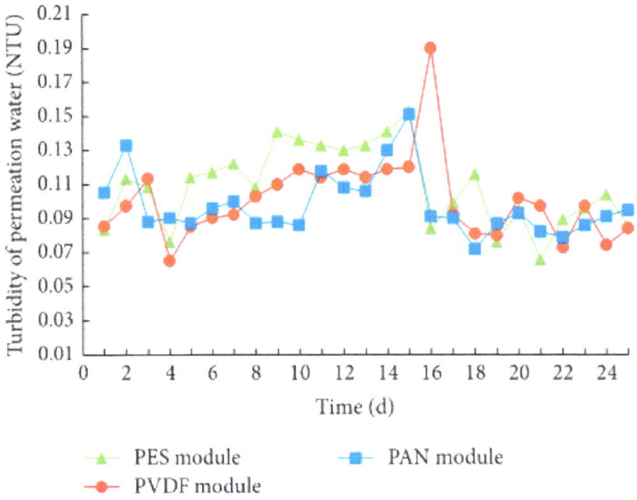

Figure 2: Turbidity trend of UF filtrate during the test period.

Removal Abilities of by Various UF Modules

COD_{Mn} value can be regarded as an index reflecting the content of organic substances in aqueous solution. The result from COD_{Mn} measurements of the UF filtrate is presented in Figure 3. The average COD_{Mn} concentration in the permeation of PES, PVDF, and PAN module was 1.37, 1.31, and 1.33 mg/L, respectively. The average removal ratio of COD_{Mn} by these three modules was 55.2%, 57.1%, and 56.5%, respectively, from which we can find that there was little difference between the three UF modules in removal capacity of COD_{Mn}. The coating layer which was formed on the membrane surface by flocs that hydrophobic macromolecular organic compounds had adsorbed on had an excellent retain capacity to micromolecular organics which can hardly be entrapped by UF membranes because of the relatively large pores of UF membranes [18], causing high removal efficiency of COD_{Mn}. Therefore, during the removal process, characteristics of membrane materials and the operational modes of the UF modules had little influence on the removal efficiency of organics. It can thus be seen that the removal ratio of COD_{Mn} by the combined process of in-line coagulation and UF is only determined by raw water quality.

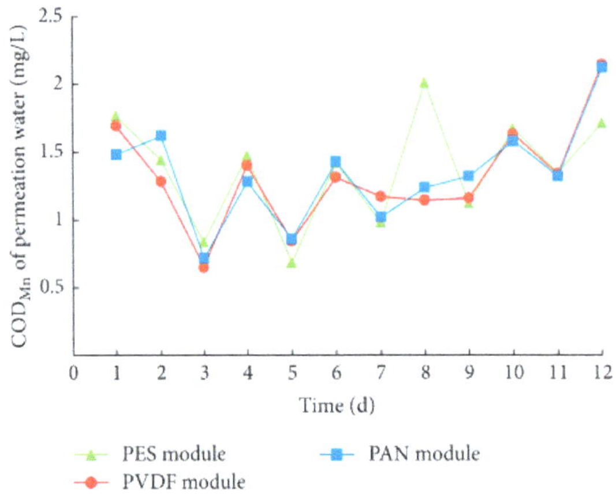

Figure 3: trend of UF filtrate during the test period.

Removal Abilities of NH_4-N Concentrations by Various UF Modules

The NH_4-N concentration of UF influent and effluent during operation is shown in Figure 4. The average water NH_4-N concentration from the permeation of PES, PVDF, and PAN module was 0.102, 0.100, and 0.093 mg/L, respectively, under the same influent quality. The NH_4-N concentrations of the effluent of different UF modules were more or less the same, indicating that UF membrane materials and operational modes have no apparent influence on the removal rate of NH_4-N concentration by the process. In general, we can also find that all effluent NH_4-N concentrations measured in this pilot test regardless of UF modules or influent quality were almost the same. In fact, UF membranes no matter which operational modes they were operated on or which material they were made from, limited by the pore size, are not able to retain NH_4-N which is in the form of ammonium salt (NH_4^+) and free ammonia (NH_3) in water [19]; only when UF combined with other process, such as oxidation, adsorption, or biological process, can NH_4-N removal be effectively accomplished [20]. In this pilot test, a fairly small amount of NH_4-N attached to suspensions and colloid substances and then was entrapped by UF membranes, thus proving the ability of NH_4-N removal to some extent, during this process. It must be pointed out that the removal of high concentration NH_4-N by the combined process of in-line coagulation/UF is poor, and process improvement should be conducted if necessary.

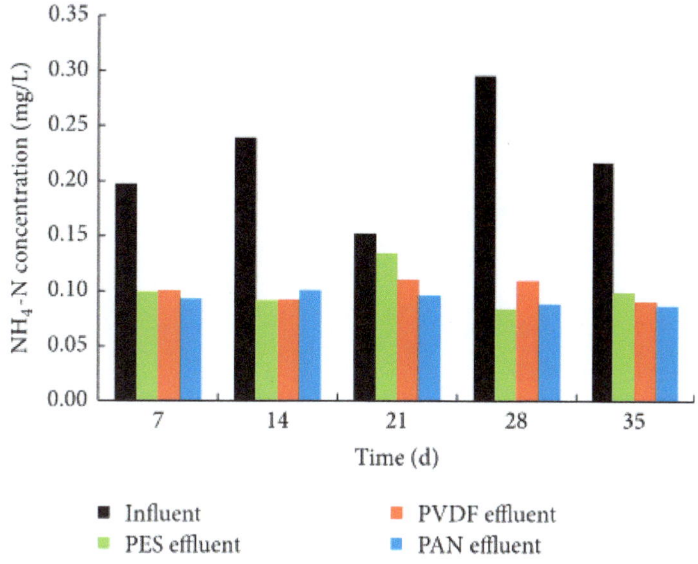

Figure 4: NH_4-N concentration trend of UF filtrate during the test period.

TMP, Membrane Permeate Flux, and Specific Flux (SF) Trends of Various UF Modules

The basic parameters which are shown in Figure 5, such as TMP, membrane permeate flux, and SF (corrected for temperature of the liquid), were monitored during the testing period. SF (also called permeability) which reflects the membrane resistance can be calculated according to the following equations:

$$SF = \frac{Q}{A \cdot TMP \cdot TCF}, \quad L{\cdot}h^{-1}{\cdot}m^{-2}{\cdot}MPa^{-1}\ (LMH),$$

$$TCF = \frac{1855 - 5596 \cdot 10^{-2} \cdot T_{reference} + 6533 \cdot 10^{-4} \cdot T^2_{reference}}{1855 - 5596 \cdot 10^{-2} \cdot T_{measured} + 6533 \cdot 10^{-4} \cdot T^2_{measured}}, \quad (1)$$

where Q is permeate flow rate of UF membrane, A is the membrane area, TCF is the temperature calibration factor, is the reference temperature (20°C), and $T_{measured}$ is the measured temperature in the pilot test, respectively.

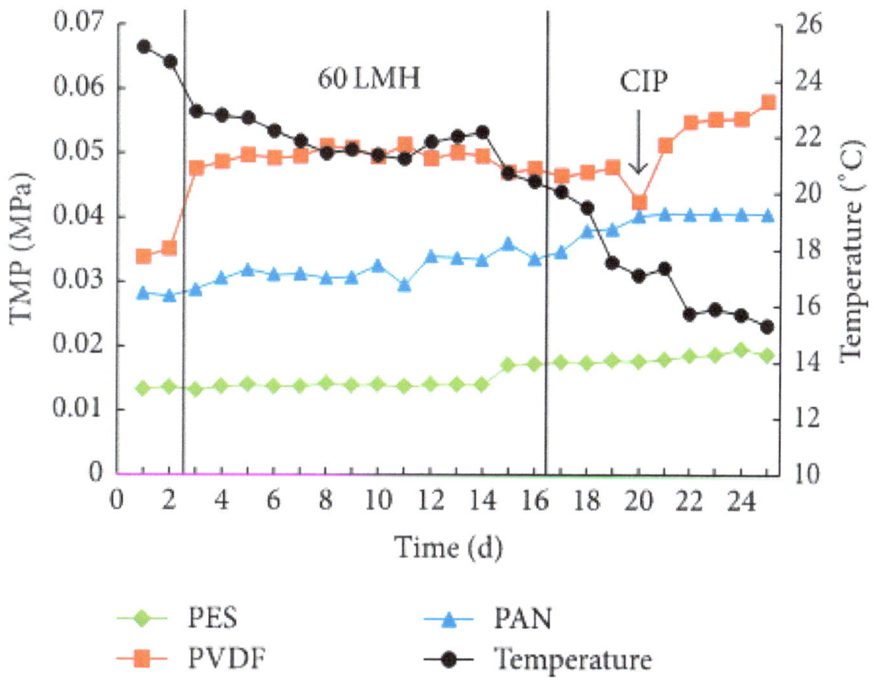

(a) TMP trend of UF modules and pilot test temperature.

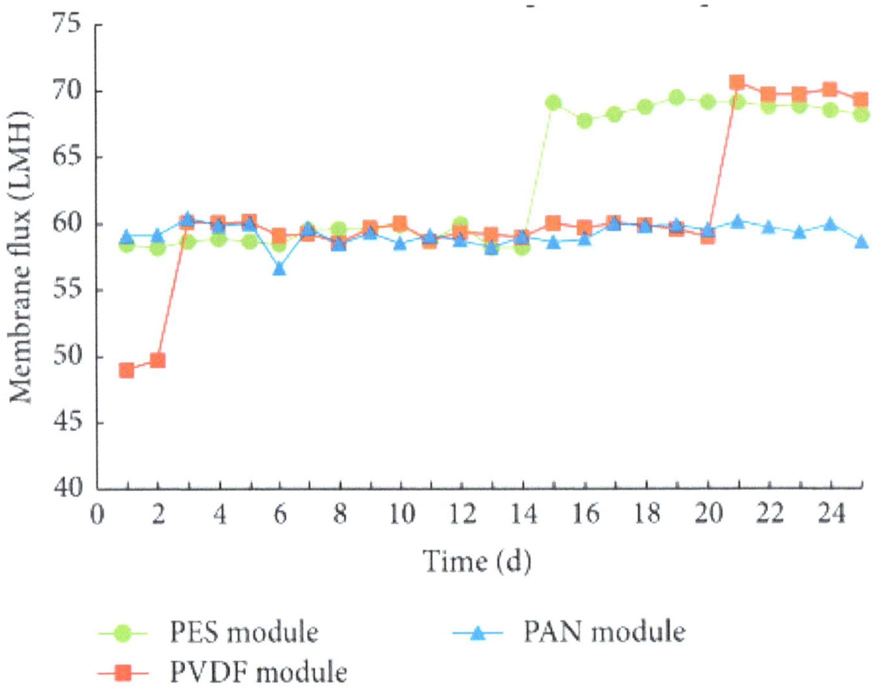

(b) Membranes flux adjustment of UF modules

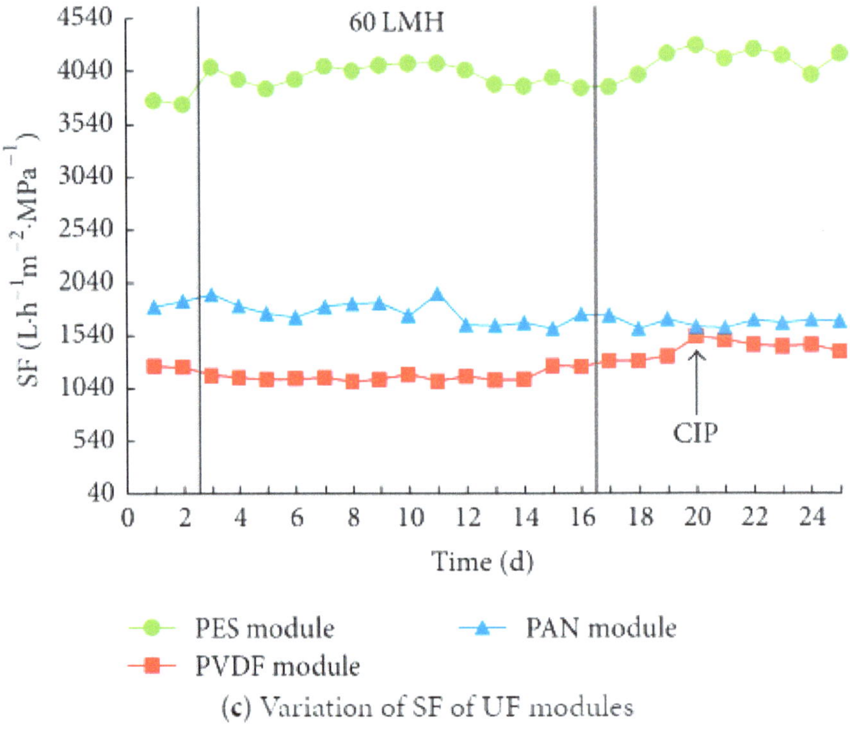

(c) Variation of SF of UF modules

Figure 5: Tendency of basic parameters of UF modules during runtime.

As shown in Figure 5(c), SF trends of PES and PVDF modules remained stable during continuous, synchronous, and fixed flux operation, while SF of PAN membrane presented obvious downtrend, which indicated that the pollution degree of the PAN membrane was greater than that of PES and PVDF membranes. As can be seen from Figure 5(a), when these modules were operated at flux of 60 LMH, the TMP of PAN module showed the most obviously ascending trend with the decrease of temperature, while PVDF and PES modules were relatively stable, and the increase of TMP of PVDF membrane was somewhat greater than that of PES. Actually, the TMP of UF membranes was enhanced in inverse proportion to the operating temperature for the reason that the activity of the solute in the mixture of feed solution had a tendency to increase but the viscosity of the feed solution decreased as the temperature increased [21, 22]. Furthermore, the increase of TMP during the running process of the system not only was caused by the descent of temperature, but also was associated with the increased intensity of membrane fouling. In conjunction with Figure 5(c), we can see that, after being corrected for temperature, SF of PAN module did not keep stable like PES and PVDF modules but presented a downward tendency to some extent. Therefore, we have every reason to believe that the increase of TMP of PAN module was mainly due to membrane fouling, while the slight increase of TMP of PVDF membrane was caused by the decrease of temperature. From the view of membrane material, it is worth noticing that the well-distributed pore size, better hydrophilicity, and higher thermal stability of the PES membrane made its SF more than double those of the other two membranes and the TMP of PES module appeared the least influenced by temperature. In fact, according to measurement results of contact angle from Table 1, the hydrophilicity of PES was the strongest, the second was PAN, and PVDF was the weakest in this pilot test [19, 20]. In addition, the fine and uniformity pores of PES membrane lead to higher adaptability in the drinking water treatment of which the demand of raw water quality and effluent water quality are relatively high.

Before the process of CIP (clean in place) in PVDF module which was operated at flux of 60 LMH, we noticed that the SF of PVDF and PES modules which was operated in synchronization and under the same flux remained stable. However, when membrane flux of PVDF module was changed to 70 LMH after CIP, with the running time growth, SF of PVDF presented obvious downtrend compared to PES module operated at flux of 70 LMH, which indicated that the flux of PVDF membrane had exceeded the critical point, called critical flux [23]. In fact, the better hydrophilicity and well-distributed pore size of PES membrane help lower TMP and operating pressure to some extent; therefore, the TMP of PES module was still far from design value, the SF has almost no attenuation in system running phase.

As shown in Figure 5(c), when these modules were operated at flux of 60 LMH, the SF of PES was the maximum one, the second was PAN, and PVDF was the minimum in this pilot test. It can be seen that the SF was consistent with the hydrophilicity of UF membranes. For PAN and PVDF modules, as the outside-in UF modules, the decreasing rate of permeability of the former was larger than that of the latter, which indicated that the hydrophilicity was not conclusive factor with the critical flux.

SF Trends of Various UF Modules in Single Filtration Duration

Figure 6 shows the SF of these three UF modules operated at flux of 60 LMH in four filtration durations. The data of SF were recorded every five-minute interval. The decay rate of SF of PES, PVDF, and PAN modules was 81.65, 44.56, and 49.74 $L{\cdot}h^{-1}{\cdot}m^{-5}{\cdot}MPa^{-1}$, respectively. As shown in Figure 7, the flow channel of the capillary inside-out UF module (PES module) is much smaller than outside-in UF module (PVDF and PAN modules), which resulted in poorer capacity of receiving pollutants. With the extension of operating time and increase of the amount of the contaminant attaching onto the membrane surface, inside-out module with higher fouling load presented higher decay rate of SF under fixed flux operation compared with outside-in modules in single filtration duration. The permeability of these three modules can be resumed to the initial filter competence through their own restorative flush mode, but some particulate fouling was still present after hydraulic clean [24, 25], which resulted in the decay of permeability of PAN module in the overall system operation process. Moreover, the pretreatment of inside-out module should be noted according to raw water quality due to its poorer dint capacity.

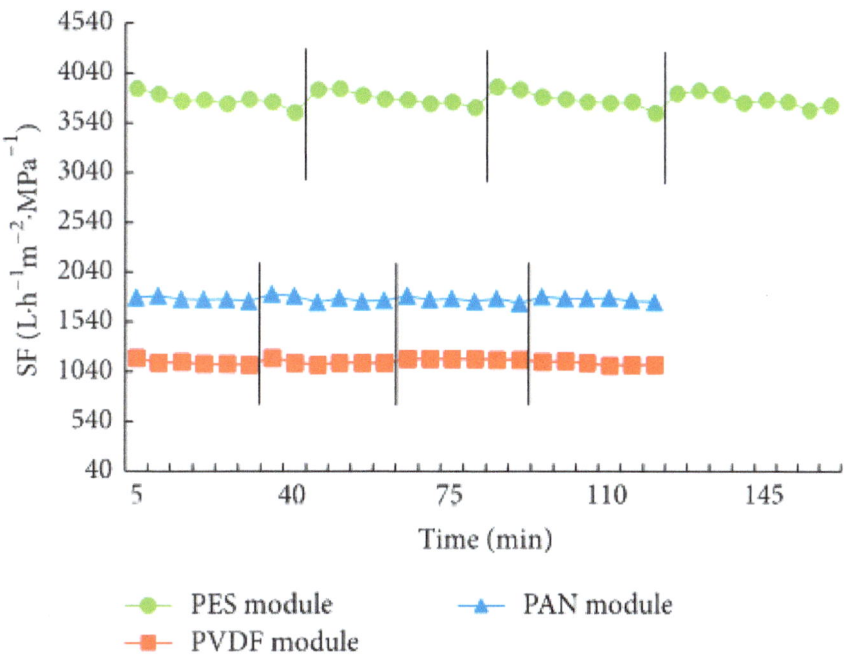

Figure 6: SF trends of various UF modules in four filtration durations.

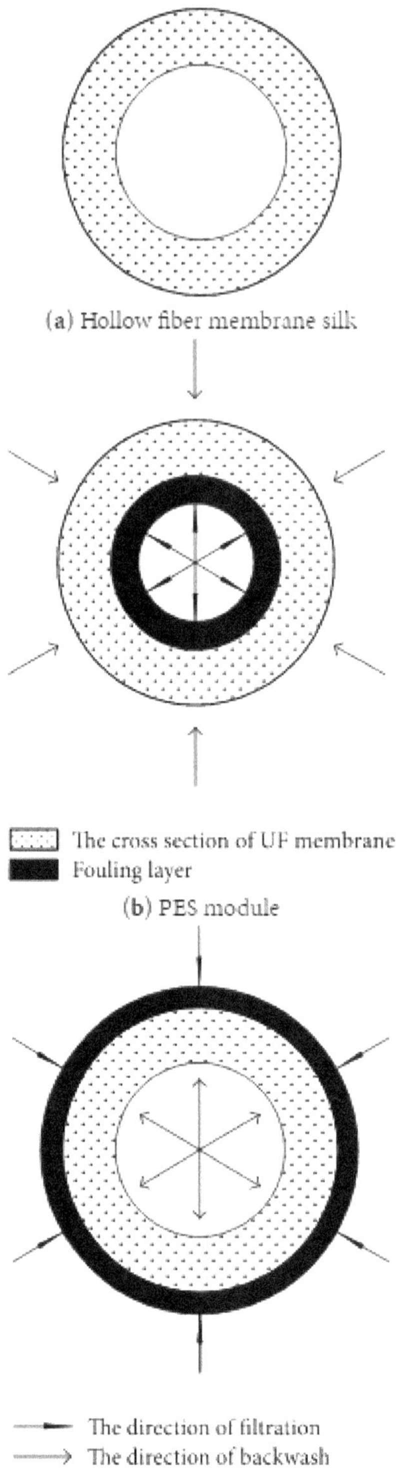

Figure 7: Flow channel of various UF modules.

Comparison of the Washing Methods of Membranes

Each UF module has its own washing methods whose appropriateness mainly depends upon their operational modes. In this pilot study, the washing methods of these three UF modules proved to be effective. The outside-in UF module made by PVDF or PAN had backwashing "dead angles" because of the interspaces of membrane silk; in view of this problem, the use of a strong flow of water to backwash is not a good choice for the reason that the backwashing channel is much smaller than that of the inside-out UF module, as shown in Figure 7(c), resulting in strength problem for backwashing with large pressure and current. Therefore, the way of gas washing is chosen to be a supplement to backwashing by relatively small water flow, causing the improvement of the water production rate and increase of the power consumption to some extent. Operational experience has shown that air-water washing is suitable for flux recovery of outside-in UF modules. The recovery of PES, PVDF, and PAN modules operated at flux of 60 LMH was 93.5%, 95.3%, and 95.9%, respectively, and the product water recovery of the former was lower than that of the latter two by 1%. It is worth noticing that there is no question of the so-called backwashing "dead angles" of inside-out module for the reason that the voids between its membranes silks were filled with filtered water instead of the accumulation of pollutants solids, compared to outside-in modules. As shown in Figure 7(b), the backwashing channel of inside-out module was large enough to withstand water backwashing with large pressure and current; therefore, it would require no gas washing and also would not fit to gas washing because of certain restrictions on the physical strength of the membrane, which reduced energy consuming and recovery rate of water production. It should be noted that acid and alkaline resistance of inside-out membrane made by PES is appropriate for the acid-base staggered method to CEB, which can intensify the effects of backwashing by water through inorganic and organic foulants cleaning [26] and therefore replace the way of gas washing of the outside-in UF module made by PVDF and PAN to some extent. In that sense, the selection of membrane material is dependent on the operational mode of UF module in a certain degree.

CONCLUSIONS

The water purification effect of UF modules made by PES, PVDF, and PAN was equivalent. PES module presented outstanding advantages for higher permeability and lower energy consuming because of the better hydrophilicity, uniform distribution porosity, and higher thermal stability of PES membrane. However, due to the higher fouling load caused by the smaller flow channel compared to outside-in module (PAN and PVDF modules), the pretreatment of inside-out module (PES module) should be noted according to raw water quality. The decreasing rate of permeability of PAN module was larger than that of PVDF module, indicating that the hydrophilicity was not conclusive factor with the critical flux. A strong flow of water to flush and the acid-base staggered method to CEB of the inside-out module increased the chemicals consumption and makes the product water recovery lower than that of outside-in modules by 1%, while air-water washing is suitable for flux recovery of outside-in UF modules, resulting in the improvement of the water production rate and the increase of the power consumption. The selection of membrane material is dependent on the operational mode of UF module to some extent.

Competing Interests

The authors declare that there is no conflict of interests regarding the publication of this paper.

ACKNOWLEDGMENTS

This work was supported by the National Natural Science Foundation of China (no. 51308123), China Postdoctoral Science Foundation (2014M561856), Housing and Urban-Rural Development of Fujian (2014-K-05), and Opening Measuring Fund of Large Precious Apparatus of Fuzhou University (2016T041).

REFERENCES

1. J. Qu and M. Fan, "The current state of water quality and technology development for water pollution control in China," Critical Reviews in Environmental Science and Technology, vol. 40, no. 6, pp. 519–560, 2010.
2. X. Zhang, Q. Guo, X. Shen, S. Yu, and G. Qiu, "Water quality, agriculture and food safety in China: current situation, trends, interdependencies, and management," Journal of Integrative Agriculture, vol. 14, no. 11, pp. 2365–2379, 2015.
3. R. Li, L. Zhang, and P. Wang, "Rational design of nanomaterials for water treatment," Nanoscale, vol. 7, no. 41, pp. 17167–17194, 2015.
4. V. Discart, M. R. Bilad, S. Van Nevel, N. Boon, J. Cromphout, and I. F. J. Vankelecom, "Role of transparent exopolymer particles on membrane fouling in a full-scale ultrafiltration plant: feed parameter analysis and membrane autopsy," Bioresource Technology, vol. 173, pp. 67–74, 2014.
5. M. Yao, J. Nan, Q. Li et al., "Effect of under-dosing coagulant on coagulation-ultrafiltration process for treatment of humic-rich water with divalent calcium ion," Journal of Membrane Science, vol. 495, Article ID 13888, pp. 37–47, 2015.
6. X. Du, F. Qu, H. Liang et al., "Removal of antimony (III) from polluted surface water using a hybrid coagulation-flocculation-ultrafiltration (CF-UF) process," Chemical Engineering Journal, vol. 254, pp. 293–301, 2014.
7. K. Kimura, Y. Hane, Y. Watanabe, G. Amy, and N. Ohkuma, "Irreversible membrane fouling during ultrafiltration of surface water," Water Research, vol. 38, no. 14-15, pp. 3431–3441, 2004.
8. E. Zuriaga-Agustí, E. Alventosa-deLara, S. Barredo-Damas, M. I. Alcaina-Miranda, M. I. Iborra-Clar, and J. A. Mendoza-Roca, "Performance of ceramic ultrafiltration membranes and fouling behavior of a dye-polysaccharide binary system," Water Research, vol. 54, pp. 199–210, 2014.
9. A. L. Ahmad, A. A. Abdulkarim, B. S. Ooi, and S. Ismail, "Recent development in additives modifications of polyethersulfone membrane for flux enhancement," Chemical Engineering Journal, vol. 223, pp. 246–267, 2013.
10. J. Gu, Y. Bai, L. Zhang et al., "VTOS cross-linked PDMS membranes for recovery of ethanol from aqueous solution by pervaporation," International Journal of Polymer Science, vol. 2013, Article ID 529474, 7 pages, 2013.
11. C. Jönsson and A.-S. Jönsson, "Influence of the membrane material on the adsorptive fouling of ultrafiltration membranes," Journal of Membrane Science, vol. 108, no. 1-2, pp. 79–87, 1995. View at Publisher · View at Google Scholar · View at Scopus
12. Y. S. Li, L. Yan, C. B. Xiang, and L. J. Hong, "Treatment of oily wastewater by organic-inorganic composite tubular ultrafiltration (UF) membranes," Desalination, vol. 196, no. 1–3, pp. 76–83, 2006.
13. J.-Y. Jeon, "Permeation properties of thermo-chemically stable ultrafiltration membranes of copoly(bis[4-(3-aminophenoxy)phenyl]sulfone/3,3',4,4'-benzophenonetetracarboxyl/pyromellite) imides,"Desalination, vol. 296, pp. 46–52, 2012.
14. J. Xu, G. Ruan, X. Gao, X. Pan, B. Su, and C. Gao, "Pilot study of inside-out and outside-in hollow fiber

UF modules as direct pretreatment of seawater at low temperature for reverse osmosis," Desalination, vol. 219, no. 1–3, pp. 179–189, 2008.

15. E. Barbot, S. Moustier, J. Y. Bottero, and P. Moulin, "Coagulation and ultrafiltration: understanding of the key parameters of the hybrid process," Journal of Membrane Science, vol. 325, no. 2, pp. 520–527, 2008.

16. M. Kabsch-Korbutowicz, "Impact of pre-coagulation on ultrafiltration process performance,"Desalination, vol. 194, no. 1–3, pp. 232–238, 2006.

17. C. Guigui, J. C. Rouch, L. Durand-Bourlier, V. Bonnelye, and P. Aptel, "Impact of coagulation conditions on the in-line coagulation/UF process for drinking water production," Desalination, vol. 147, no. 1–3, pp. 95–100, 2002.

18. P.-K. Park, C.-H. Lee, S.-J. Choi, K.-H. Choo, S.-H. Kim, and C.-H. Yoon, "Effect of the removal of DOMs on the performance of a coagulation-UF membrane system for drinking water production,"Desalination, vol. 145, no. 1–3, pp. 237–245, 2002.

19. S. Montalvo, J. S. Martin, C. Huiliñir, L. Guerrero, and R. Borja, "Assessment of a UASB reactor with high ammonia concentrations: effect of zeolite addition on process performance," Process Biochemistry, vol. 49, no. 12, pp. 2220–2227, 2014.

20. Y. Chen, Y. Tu, Y. Bai, M. Tang, W. Liu, and H. Wu, "Combined adsorption/ultrafiltration of secondary effluents using powdered zeolites," Chemical Engineering & Technology, vol. 39, no. 2, pp. 285–292, 2016.

21. Y. Su, X. G. Jian, S. H. Zhang, and C. Yan, "Preparation of novel PPES-B UF membrane with good thermal stability: the effect of additives on membrane performance and cross-section morphology,"Journal of Membrane Science, vol. 271, no. 1-2, pp. 205–214, 2006.

22. J. Cai, E. Perfect, C.-L. Cheng, and X. Hu, "Generalized modeling of spontaneous imbibition based on hagen-poiseuille flow in tortuous capillaries with variably shaped apertures," Langmuir, vol. 30, no. 18, pp. 5142–5151, 2014.

23. Y.-T. Chiou, M.-L. Hsieh, and H.-H. Yeh, "Effect of algal extracellular polymer substances on UF membrane fouling," Desalination, vol. 250, no. 2, pp. 648–652, 2010.

24. Y. Bessiere, C. Guigui, P. J. Remize, and C. Cabassud, "Coupling air-assisted backwash and rinsing steps: a new way to improve ultrafiltration process operation for inside-out hollow fibre modules," Desalination, vol. 240, no. 1–3, pp. 71–77, 2009.

25. P. J. Remize, C. Guigui, and C. Cabassud, "From a new method to consider backwash efficiency to the definition of remaining fouling," Desalination, vol. 199, no. 1-3, pp. 86–88, 2006.

26. Y. C. Woo, J. J. Lee, L. D. Tijing, H. K. Shon, M. Yao, and H.-S. Kim, "Characteristics of membrane fouling by consecutive chemical cleaning in pressurized ultrafiltration as pre-treatment of seawater desalination," Desalination, vol. 369, pp. 51–61, 2015.

CHAPTER 14 BIOLOGICAL INSTABILITY IN A CHLORINATED DRINKING WATER DISTRIBUTION NETWORK

Alina Nescerecka, Janis Rubulis, Marius Vital, Talis Juhna and Frederik Hammes

INTRODUCTION

The goal of public drinking water supply systems is to produce water of acceptable aesthetic and hygienic quality and to maintain that quality throughout distribution until the point of consumption. From a microbiological perspective, the quality of treated water can deteriorate as a result of excessive bacterial growth, which can lead to problems such as a sensory deterioration of water quality (e.g. taste, odor, turbidity, discoloration) as well as pathogen proliferation [1]–[10]. To avoid this, biological stability during distribution can be achieved by maintaining sufficient residual disinfectants in the water, and/or through nutrient limitations [3],[7], [11], [12]. However, drinking water systems should not be viewed as sterile; complex indigenous bacterial communities have been shown to inhabit both chlorinated and non-chlorinated drinking water distribution systems [5], [13]–[17].

The concept of biological stability and its impact on a system's microbiology has been discussed extensively in the framework of non-chlorinated drinking water distribution systems[3], [7], [17]–[20]. However, many treatment plants worldwide employ a final disinfection step to ensure that no viable bacteria enter the distribution system. The latter is often achieved by oxidative disinfection, usually by chlorination [21]. Disinfection has a number of implications for a biological system. During chlorination, one can expect that a considerable fraction of bacteria in the water are killed or damaged, while some residual chlorine may remain in the water (Figure 1). This could be visible through numerous microbial monitoring methods. For example, the number of cultivable bacteria, measured with heterotrophic plate counts, would reduce dramatically [22], [23]. Secondly, bacteria cells are likely to display measurable membrane damage irrespective of their cultivability [24], though the rate and extent of damage may differ between different communities. This would be detectable with several staining techniques coupled with epifluorescence microscopy or flow cytometry (FCM). Also, adenosine tri-phosphate (ATP), often used as a cultivation-independent viability method [19], [22], [25] will be severely affected. Based on data from Hammes and co-workers [4] one may reasonably expect increased levels of extracellular ATP (so-called free ATP) and decreased concentrations of intracellular ATP (bacterial ATP) following oxidative disinfection. Irrespective of the detection method, the overall consequence of disinfection is a considerable decrease in the viable biomass, potentially opening a niche for microorganisms to occupy downstream of the treatment process. Following initial disinfection, residual chlorine

Citation: Nescerecka A, Rubulis J, Vital M, Juhna T, Hammes F (2014) Biological Instability in a Chlorinated Drinking Water Distribution Network. PLoS ONE 9(5): e96354. doi:10.1371/journal.pone.0096354

might provoke undesirable changes during drinking water distribution. Disinfectants target not only bacteria, but it also react with natural organic matter, pipe surfaces and particles in the network, thus potentially forming/releasing assimilable organic carbon (AOC) [26]–[30]. AOC can easily be consumed by bacteria, and is therefore seen as a main contributor to biological instability. Moreover, chlorine decay within the network negatively affects its ability to inhibit microbial growth at the far ends of the network [12]. If all factors were considered, the presence of nutrients, a reduction in the number of competing bacteria, and the lack of residual disinfectant would potentially lead to biological instability in the distribution network, manifesting in a subsequent bacterial growth (Figure 1). Besides the importance of nutrients, the extent of bacterial growth will be influenced by a number of factors. For example, increased water temperature can accelerate chlorine decay and favor bacteria growth [19], [31], while changes in hydraulic conditions can alter nutrient supply for microorganisms in biofilms and/or bacteria detachment from the pipe surfaces [32], [33]. Finally, the quality of materials in contact with drinking water, as well as the presence of sediments and loose deposits, can both affect the general microbial quality of the water [6], [34], [35].

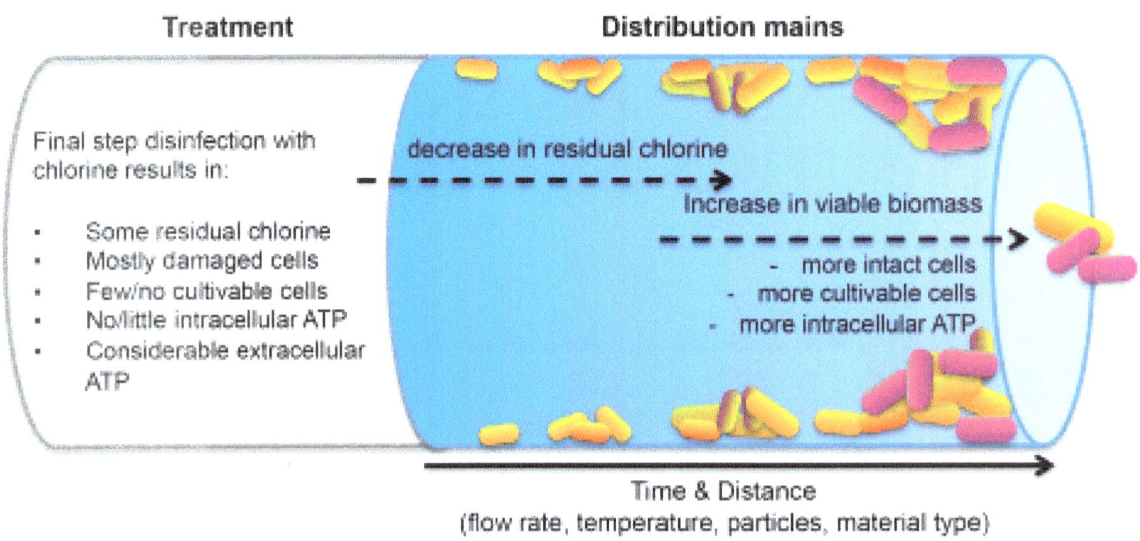

Figure 1. A worst-case-scenario in an unstable, chlorinated distribution network.

Prediction of changes in the microbiological state of the water due to the depletion of residual chlorine and the concomitant growth of bacteria, potentially resulting in hygienic and sensory deterioration of the water quality.

In the present study we examined some of the above-discussed concepts in a full-scale, chlorinated distribution system in the city of Riga (Latvia) with a number of microbiological methods. The purpose was a detailed investigation of the entire city's distribution network, asking the basic question whether evidence of spatial and/or temporal biological instability exists, and if so, to which degree. Additional goals were to evaluate the use of fluorescent staining coupled with FCM, as well as ATP analysis, for the assessment of chlorinated drinking water in a distribution network with disinfectant residuals.

MATERIALS AND METHODS

Ethics statement

Permission for sampling at all locations in the present study was obtained from the local water utility (Rīgas Ūdens).

Description of study site

Sampling was performed in the full-scale distribution network of Riga (Latvia) with a total length of about 1400 km. The city is supplied with drinking water from six water treatment plants (WTP) produced from both surface and groundwater (150 000 m^3 d^{-1}). Only the three major WTP, which are continuously operated, were included in the sampling campaign. Average WTP effluent water quality parameters for each treatment plant are shown in Table 1. The distribution network mainly consists of cast iron (80%) and unlined iron (15%) pipes as old as 50 years. The diameters of pipes ranged from 100 to 1200 mm. Three reservoirs are operated in the network to compensate for fluctuations in the daily water demand, while four high-pressure zones are maintained in some distal areas of the network. The high-pressure zones were excluded from the present study. A total of 49 sampling sites were selected across the city to cover the network broadly and to include both proximal and distal zones relative to the treatment plants. The sampling sites were selected according to the approximate water retention times obtained from a validated hydraulic model made in EPANET 2.0 [36], [37]based on a total length of 538 km (39% of the total length of the network). Apart from the effluents of the three treatment plants, the sampling sites were in all cases fire hydrants in order to attain some degree of reproducibility between sampling and to avoid localized effects (e.g. household growth). The exact locations of sampled fire hydrants can be obtained from the authors after agreement from the local water utility.

	WTP 1	WTP 2	WTP 3
Source water	surface water	artificially recharged groundwater	groundwater
Final treatment step[a]	Cl_2 (0.5–3 mg L^{-1})	Cl_2 (ca. 1.5 mg L^{-1})	N.A.
Residual chlorine (mg L^{-1})[b]	0.44±0.11	0.51±0.01	0.42±0.26
Total organic carbon (TOC) (mg L^{-1})[a]	6±1	9±3	3
Assimilable organic carbon (AOC) (μg L^{-1})[a]	213±37	209±59	N.A.
Total cell concentration (cells mL^{-1})[b]	$5.31\pm0.97\times10^5$	$5.45\pm0.47\times10^5$	$1.69\pm0.18\times10^5$
Intact cell concentration (cells mL^{-1})[b]	$1.83\pm1.18\times10^4$	$1.4\pm0.86\times10^4$	$1.03\pm0.68\times10^4$
Total ATP (nM)[b]	0.015±0.005	0.029±0.004	0.011±0.002
Intracellular ATP (nM)[b]	0.007±0.003	0.000±0.004	0.001±0.002
HPC 22°C (CFU mL^{-1})[b]	23±24	4±2	4
HPC 36°C (CFU mL^{-1})[b]	16±16	4±2	1
Conductivity (μS cm^{-1}), 25°C[a]	468±101	625±4	272±25
pH[a]	6.63±0.18	7.41±0.04	7.5±0.05

Table 1. Average water quality parameters for the final effluents of the the three main treatment plants of Riga (Latvia).

Sampling protocol

A specific sampling protocol was designed and followed in order to avoid artifacts due to water stagnation in unused fire hydrants. Each hydrant was pre-flushed at a high velocity (never exceeding 1.6 m s^{-1}) for no more than 60 s, then immediately adjusted to a low velocity of 0.015–0.25 m s^{-1} and connected to an online system for monitoring pH, temperature, redox potential, electro-conductivity and turbidity. The low sampling velocity was specifically used to ensure a minimal possible impact of cell wall erosion and detachment from biofilms on the samples and measurements. Readings of all parameters were taken at 5–10 minute intervals, and water was only sampled for microbiological analysis once all of the parameters stabilized. The impact of this hydrant flushing is demonstrated in an example in Figure 2 and discussed in detail in the results section. Samples were kept in cold storage (≈5°C) and analyzed within four hours of sampling.

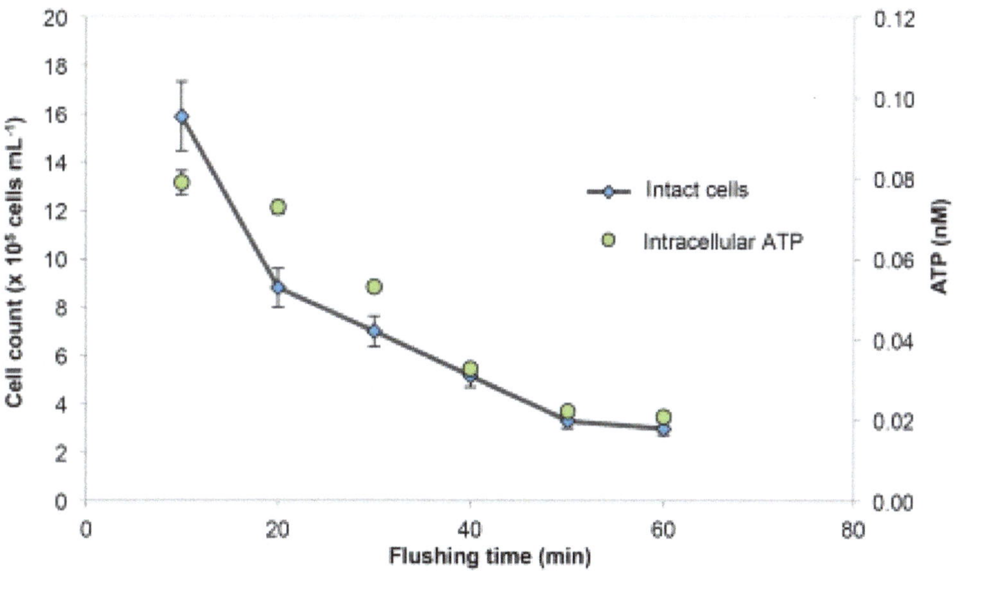

Figure 2. The impact of low velocity flushing on the water quality in a newly-opened fire hydrant.

Intracellular adenosine tri-phosphate (ATP) data points were derived from duplicate measurements of extracellular and total ATP concentrations. FCM intact cells (after SYBR Green I and propidium iodide staining) were single measurements, with a relative standard deviation of 9% calculated from all data in the present study.

Chemical analysis

Determination of free chlorine was performed according to standard method EN ISO 7393-1, based on the direct reaction with N,N-diethyl-1,4-phenylenediamine (DPD) and subsequent formation of a red compound at pH 6.2–6.5. Afterwards titration by means of a standard solution of ammonium iron (III) sulfate until disappearance of the red color was performed. Determination of total chlorine was performed according to EN ISO 7393-1, based on the reaction with DPD in the presence of an excess of potassium iodide, and then titration as described above.

Fluorescent staining and flow cytometry (FCM) of water samples

Staining and FCM analysis was done as described previously [4], [38]. In short, for a working solution, SYBR® Green I (SG) stock (Invitrogen AG, Basel, Switzerland) was diluted 100x in anhydrous dimethylsulfoxide (DMSO) and propidium iodide (PI; 30 mM) was mixed with the SYBR® Green I working solution to a final PI concentration of 0.6 mM. This working solution was stored at −20°C until use. From every water sample, 1 mL was stained with SGPI at 10 µL mL^{-1}. Before analysis, samples were incubated in the dark for 15 minutes. Prior to FCM analysis, the water samples were diluted with 0.22 µm filtered bottled water (Evian, France) to 10% v/v of the initial concentration. FCM was performed using a Partec CyFlow SL instrument (Partec GmbH, Münster, Germany), equipped with a blue 25 mW solid state laser emitting light at a fixed wavelength of 488 nm. Green fluorescence was collected at 520±10 nm, red fluorescence above 630 nm, and high angle sideward scatter (SSC) at 488 nm. The trigger was set on the green fluorescence channel and data were acquired on two-parameter density plots while no compensation was used for any of the measurements. The CyFlow SL instrument is equipped with volumetric counting hardware and has an experimentally determined quantification limit of 1000 cells mL^{-1} [4].

Adenosine tri-phosphate (ATP) analysis

Total ATP was determined using the BacTiter-Glo reagent (Promega Corporation, Madison, WI, USA) and a luminometer (Glomax, Turner Biosystems, Sunnyvale, CA, USA) as described elsewhere [25]. A water sample (500 µl) and the ATP reagent (50 µl) were warmed to 38°C simultaneously in separate sterile Eppendorf tubes. The sample and the reagent were then combined and then the luminescence was measured after 20 s reaction time at 38°C. The data were collected as relative light units (RLU) and converted to ATP (nM) by means of a calibration curve made with a known ATP standard (Promega). For extracellular ATP analysis, each sample was filtered through a 0.1 µm sterile syringe filter (Millex-GP, Millipore, Billerica, MA, USA), followed by analysis as described above. The intracellular ATP was calculated by subtracting the extracellular ATP from the total ATP for each individual sample. ATP was measured in duplicate for all samples.

Heterotrophic plate counts

To obtain heterotrophic plate counts (HPC), samples were serially diluted in sterile distilled water and then inoculated onto nutrient yeast agar plates using the spread plate technique. All plates were incubated in dark at 22°C or 36°C for 3 and 7 days, respectively. Results were expressed as colony forming units (CFU) per ml of water sample.

Statistical analysis

Statistical data evaluation was performed with the MS Excel Data Analysis tool (Descriptive statistics, Regression). The reproducibility for indirect/calculated data (e.g., intracellular ATP) was calculated by a propagation-of-uncertainty method. FCM data was not always measured in duplicate, due to practical constraints. In these cases, a 9% error (average coefficient of variation (CV) (n=39)) was applied for representing FCM data. The residual chlorine concentration distribution box plot was created using on-line calculator onhttp://www.physics.csbsju.edu/stats/.

RESULTS AND DISCUSSION

The importance of correct sampling

Sample collection during this study elucidated some of the problems specific for this network and highlighted the broader importance of correct sampling procedures. Fire hydrants were selected as sampling points to enable direct access to the distribution network and avoid potential household effects [39]. We opted for a low velocity water flow in combination with online monitoring to achieve comparable samples. In some cases, the water initially emerging from the fire hydrants were visibly turbid and/or discolored (data not shown). Turbid water is clearly unwanted and serves as a first visual confirmation of some form of system failure. In this regard, a recent study in the Netherlands has established an important link between suspended solids and microbial growth and biological instability [6]. Hence in some instances continuous low velocity flushing of up to 60 minutes was required before stable values for chemical and physical parameters as well as microbiological parameters were obtained (Figure 2; Table S1). The data in Figure 2 demonstrate clearly the need for a carefully planned sampling protocol when assessing full-scale systems. It should be noted that Figure 2represents an example of some of the worst sampling points in the system. Data from other hydrants often showed less fluctuation during flushing (Figure S1). One potential problem during the sampling procedure is the re-suspension of sediments/particles and sloughing of biofilms from the pipes, causing artifacts in the measurements. In this respect, we specifically employed a low velocity (0.015–0.25 m s^{-1}) pre-sampling flushing procedure. The latter differs from extreme flushing applied for network cleaning, which is operated with high velocities of 1.5–1.8 m s^{-1} [40], [41]. According to Antoun and co-workers [40] low-veocity flushing (below 0.3 m s^{-1}) does not cause any scouring actions. However, it should be considered that part of the samples, especially during the first minutes of the flushing, can cointain biofilm bacteria detatched in a result of pre-flushing [35].

The concept of detecting instability: a single point in the distribution network

In the introduction we proposed the straightforward hypothesis that biological parameters would show an increase between the point of treatment and a point during distribution in case of biological instability (Figure 1). Before the relation between different parameters and the impact on the entire network are discussed in detail below, a single sampling point is compared to its source water as an example to illustrate the concept (Figure 3A). The point was selected on the basis of (1) hydraulic data linking it with a specific WTP, (2) its medial distance from WTP (neither too close and nor too far from the WTP) and, (3) the fact that all microbiological parameters (FCM, ATP and HPC) as well as residual chlorine measurements were performed on this sample. For the purpose of clarity, the data was normalized to the values of the treated water and expressed as the relative change (the raw data and standard deviations for the data in Figure 3A are shown in Figure S2). Evidently the data from Figure 3A supports the basic hypothesis. The microbial parameters such as intact cell concentration, ATP and colony forming units all show a considerable increase in their values. Simultaneously, only 12% (0.06 mg L^{-1}) of the initial residual chlorine concentration (0.5 mg L^{-1}) was left in the water sample. The data suggests that the residual chlorine in the network was not sufficient to inhibit microbial growth, concurring with earlier report from Prévost and colleagues [42] showing increased HPC, total direct and direct viable bacteria counts in a distribution network coinciding with chlorine depletion. Other studies also showed the presence of viable bacteria in water with chlorine concentration lower than 0.1 mg L^{-1} [23] and that residual chlorine levels below 0.07 mg L^{-1} allows bacterial growth [12]. Data of residual chlorine concentrations in the drinking water network is summarized in Figure S3. Evidently a considerable fraction of samples (18%) had residual chlorine concentrations below 0.1 mg L^{-1}.

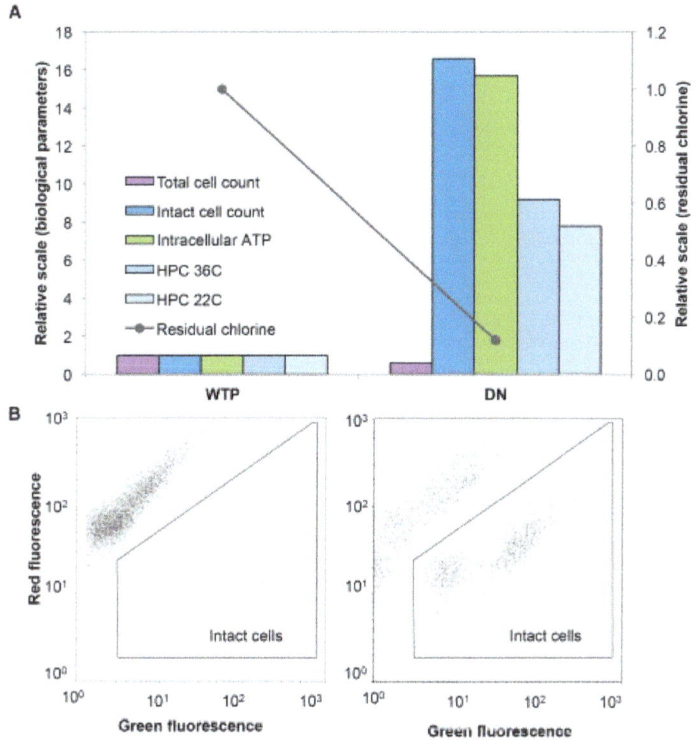

Figure 3. Changes in bacterial parameters between water treatment plant (WTP) and distribution network (DN) sampling points.

(A) For comparison, all values at the WTP were set to 1, and values in the DN were expressed relative to their values at the WTP. The original raw data for these samples are shown in Figure S2. Data points are average values for duplicate FCM and ATP measurements and triplicate HPC measurements; (B) Flow cytometric density plots of samples stained with SYBR Green I and propidium iodide, showing the intact cell concentration at the plant and in the specific network point.

Staining of bacteria with fluorescent dyes was previously suggested as a way to distinguish between viable and damaged bacteria in real water samples [43], [44], and the application of this approach has been successfully demonstrated in laboratory scale chlorination studies[24], [45]. One focus point of the present study was to determine whether FCM combined with viability staining can be used for a fast and meaningful assessment of viable bacteria in chlorinated drinking water systems. The same samples from Figure 3A, stained with SYBR Green and propidium iodide (SGPI), are shown as density plots obtained with FCM (Figure 3B). The theory behind the staining method and the interpretation of such data are discussed in detail elsewhere [20], [24], [38], [43], [46]. In the treatment plant sample, where the water was recently exposed to chlorine, 98% of all cells were measured as membrane compromised, seen by absence of events inside the gated area of the plot (Figure 3B). In the distribution network (DN) sample, a high concentration of intact cells appeared (Figure 3B). Since these intact cells were clearly not present in the influent, the plausible conclusion is that the bacterial growth occurred during distribution.

Detailed assessment of dynamic changes in a single point

High frequency monitoring of a single sampling point revealed temporal instability in the distribution network. We monitored the effluent of one treatment plant and one point in the network with 1-hour intervals during a day (ca. 21 h). The sampling was arranged in such a way that the network sampling started 15 hours after the treatment plant sampling, which corresponded with the estimated water residence time (WRT) for this location. Figure 4displays the changes of intracellular ATP and intact cell concentrations in the network and the water treatment plant. Values for both parameters were low in the water samples from the treatment plant (n=19): intracellular ATP varied from 0.0025 nM to 0.0096 nM (mean=0.0061±0.002 nM) and the intact cell concentration amongst 19 samples varied from $7.5\times10^3$to 6.3×10^4 cells

mL^{-1} (mean=$1.6\times10^4\pm1.2\times10^4$ cells mL^{-1} in average). In turn, the values from the distribution network point (n=23) were significantly higher: intact cell concentrations ranged from 1.37×10^5 to 4.66×10^5 cells mL^{-1} (mean=$2.5\times10^5\pm9.9\times10^4$ cells mL^{-1}), and the ATP concentrations from 0.021 to 0.063 nM (mean=0.038±0.012 nM). Moreover, a distinct pattern was apparent in the distribution network data, with values peaking at about 05:00–07:00 and again at 12:00–13:00. During both these events, the intracellular ATP data followed a similar pattern as the intact cell concentration data, with a good overall correlation (R^2=0.81; $p<0.005$). Although it is not evident exactly why the bacterial concentrations peaked at these specific time periods, a plausible explanation is a change in the flow velocity due to diurnal changes in water consumption by both industrial and domestic consumers. It was previously shown in laboratory scale experiments that increased flow velocity could lead to increased bacterial detachment from biofilms and a re-suspension of loose deposits, thus leading to an increase in suspended cell concentrations [32], [33], [47]. In addition, it is possible that lower water consumption overnight resulted in considerably reduced flow rates, and consequently a faster decay of chlorine and increased bacterial growth [42], [48].

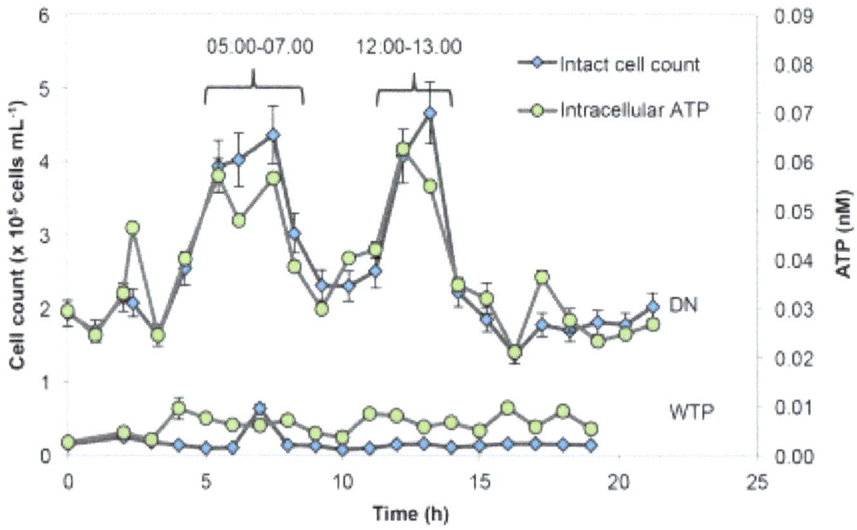

Figure 4. Diurnal changes in bacterial parameters of WTP and DN points.

Intensive sampling of one WTP (n=19) and one point in the DN (n=23) during 21 hours reveals steady cell concentrations at the treatment plant but clear variations in the distribution network. Intracellular adenosine tri-phosphate (ATP) data points were derived from duplicate measurements of extracellular and total ATP concentrations. FCM intact cells (after SYBR Green I and propidium iodide staining) were single measurements, with a relative standard deviation of 9% calculated from all data in the present study.

Detailed data sets of diurnal changes in the microbial quality of water mains, such as Figure 4, are particularly scarce in literature. Importantly, this clearly demonstrated temporal instability in the network for which the exact cause remains uncertain. Moreover, it shows that the absolute cell concentrations at any sampling point may be influenced by the time of sampling.

Instability data for the entire network

Full-scale distribution networks are complicated systems, not restricted to a single source or a straight distribution line [17]. The Riga distribution network is supplied with drinking water from several separate treatment plants (Table 1). One plant treats surface water from the Daugava River (WTP 1) and the others supply natural groundwater (WTP 3) and artificially recharged groundwater (WTP 2). Chlorination is applied as the final disinfection step at all plants, resulting in low concentrations of intact cells, intracellular ATP and cultivable bacteria in the effluents (Table 1). A large fraction of the active chorine is rapidly consumed due to relatively high levels of organic matter. Despite the fact that the purpose of chlorination and residual chlorine is to limit microbial growth during distribution, a considerable increase in the concentration of intact cells was detected throughout the distribution network. Figure 5A shows the range of intact cell concentrations arranged in ascending order. Treated water contained between 1.84×10^5–5.63×10^5 total cells mL^{-1} and between 9.7×10^3–2.13×10^4 intact cells mL^{-1}(hence 2–5% intact cells) depending on WTP. The data confirms effective final disinfection in all treatment plants. The total cell concentration values of the drinking water samples from the distribution network (n=49) varied from 1.62×10^5 cells mL^{-1} to 1.07×10^6 cells mL^{-1} and the range of the intact cell concentration was from 5.28×10^3 cells mL^{-1} to 4.66×10^5 cells mL^{-1} (3–59% intact cells). Notably, 50% of all samples contained more than 1.06×10^5 intact cells mL^{-1}corresponding to an increase of at least one order of magnitude in those samples compared to effluent water, which clearly shows that bacterial growth in the distribution network was not an isolated occurrence. The observed increase in intact cell concentration is likely related to the presence of assimilable organic carbon (AOC) in the distributed water. While AOC was not measured in the present study, previous data for two of the treatment plants were high (in the range of 200 μg L^{-1}; Table 1), and nutrient availability in the water is generally regarded as a key factor that promotes microbial growth [29], [49]. It cannot be excluded that some variability in the data resulted from bacteria detached from biofilms or re-suspended from sediments during the fire hydrant sampling procedure. However, the potential adverse impact of this was minimized by the low velocity sampling protocol (see above), while the systematic increase in cell concentrations in the network clearly suggests the occurrence of biological instability rather than sampling artifacts. In contrast to these findings, several studies analyzing drinking water distribution systems without any additional residual disinfectants showed no (or only minute) changes in bacterial parameters during distribution [17], [19], [20]. These distributions systems rely on nutrient limitation to achieve biological stability, and while intact cell concentrations are often relatively high (ca. 1×10^5 cells mL^{-1}) [17], [20], changes during distribution tend to be negligible.

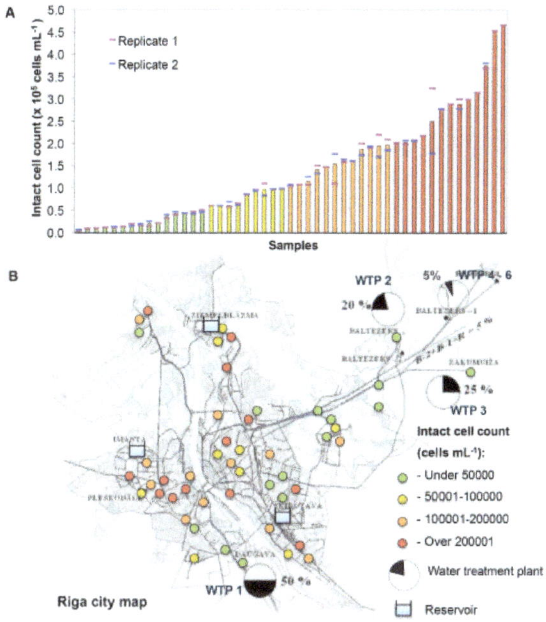

Figure 5. Intact cell concentrations of all samples measured from the distribution network (n=49).

(A) Intact cell concentrations arranged in ascending order and categorized into four main classes (colored bars) according to increasing concentrations. Data points are average values of duplicate measurements. Blue and purple stripes above and below data bars show the measured values. (B) Actual distribution of the classes of intact cells (colored circles) throughout the drinking water distribution network. WTP 1, WTP 2, WTP 3 represent location and productivity of the main water treatment plans supplying the city: WTP 1 operates using surface water, WTP 2 – artificially recharged ground water, WTP 3 – natural groundwater. WTP 4 – 6 indicates on other three pump stations with less significance for the city water supply.

To examine the spatial distribution of the growth/instability in the network, the data was divided into four broad categories based on the extent of growth (Figure 5A). These were visualized on the sampling map (Figure 5B). The sampling points with the lowest intact cell concentration (less than 5×10^4 cells mL^{-1}) are marked with green bullets. Yellow and orange colored bullets indicate higher concentrations, while the points with the highest values (over 2×10^5 cells mL^{-1}) are shown as red bullets. As could be expected, the map shows that the points with the lowest cell concentrations are mostly concentrated in areas close to the water treatment plants. Low intact cell concentrations in those areas could be ascribed to (1) disinfection during treatment and (2) growth inhibition from sufficient residual chlorine. Also the flow rate in the outgoing pipes closest to the treatment plants is high, which likely prevents water stagnation, sedimentation and cell adhesion on the pipe surface, and, consequently, biofilm formation and further bacterial growth. A different situation is observed in the distant areas from the water treatment plants and particularly in the so-called mixing zones, where the water from three different water treatment plants potentially mix. The map displays different color points spread in these zones without any visible order. The prevalence of the samples with higher cell concentrations there compared to the areas close to WTPs also corroborates the argument that increasing distance and water residence time could lead to chlorine decay with concomitant oxidation of dissolved organic matter; both these events would favor bacterial growth. Moreover, mixing zones are potential hot-spots for bacterial growth, as one water might well contain the nutrients that are growth limiting in the other.

The uneven spatial distribution of the samples with different intact cell concentrations is noteworthy, highlighted for example by the three points in upper-left corner of the map. Based on the long distance from the WTPs, high intact cell concentrations were expected, but the samples taken from the hydrants located in this small area rather show variability (respectively 1.82×10^4, 1.87×10^5, 2.51×10^5 intact cells mL^{-1}). Such different intact cell concentrations could be due to several reasons: the time the samples were taken, which is linked to water consumption and the potential impact of which is shown in Figure 4, the condition of the pipes in this specific area (unknown), the way water flows from the treatment plant, and/or the relative proximity of these sample points to one of the reservoirs (Figure 5B), etc. Other authors showed a decrease in AOC [13] and ATP [19] in the some distal points of the distribution networks. Decrease of AOC concentration was explained by its consumption by bacteria within the network. These authors argued that an insufficient amount of nutrients led to starvation and a decrease in bacterial parameters at the end of the pipelines. However, it is an unlikely reason in the present study, because this phenomenon seems more occasional than systematic.

The combined data demonstrates clearly biological instability throughout the distribution network. However, despite the relative simplicity of the concept (Figure 1; Figure 3A), a complex interplay of chemical, physical and biological parameters and hydraulic conditions should be taken into account for characterization of each particular case of instability.

Comparison of different microbiological parameters

FCM and ATP data showed clear correlations, but these data did not correlate well with conventional HPC data. A total amount of 49 different samples was measured in duplicate with ATP (total and extracellular) and FCM (total and intact cell concentration) analyses, while 38 of those samples were further analyzed with HPC. The significant linear correlation (R^2=0.77; n=49) between intracellular ATP and FCM intact cell concentration is shown in Figure 6A. This corroborates previous studies that showed good results comparing total ATP with total cell concentration [19], [50] and intracellular ATP with intact cell count as well [20], [25]. The strong correlation is encouraging, since FCM and ATP analysis are independent viability parameters – integrity of the cell membrane (FCM) and cellular energy (ATP). A correlation between these parameters during disinfection is not necessarily a given fact. The membrane integrity based PI staining method implies that PI positive cells are damaged and thus considered as inactive, yet extreme examples where living cells became permeable for propidium iodide have been described [46]. In turn, Nocker and co-workers [51] showed that after UV-C exposure cells became inactivated, while their membranes remained essentially intact. Discrepancies between intracellular ATP and intact cell concentration can also result from cell morphology, bacterial species and physiological state, that was discussed in detail previously [25]. The results provided by FCM provide information on single cell level, whereas during ATP analyses the values are evaluated per volume. Hence, intracellular ATP-per-cell was calculated for characterization of biomass activity. In the present study intracellular ATP-per-cell ranges from zero (no cell-bound ATP observed) to 5.92×10^{-10} nM $cell^{-1}$ (=3×10^{-17} g $cell^{-1}$) with the average value of 1.68×10^{-10} nM $cell^{-1}$ (=8.52×10^{-18} g $cell^{-1}$) (stdev=9.58×10^{-11} nM $cell^{-1}$, n=49). The result is in the same range as ATP-per-cell values obtained from various water sources, which were analyzed with the same methods [20], [25]. This suggests that bacterial activity (ATP values) in the intact cells was not affected by any remaining chlorine residuals, and that membrane damage (SGPI values) was in this case reflective of viability in the sample. The good correlation between these two independent parameters is an optimistic prospect for applying these methods for chlorinated water analyses in future studies.

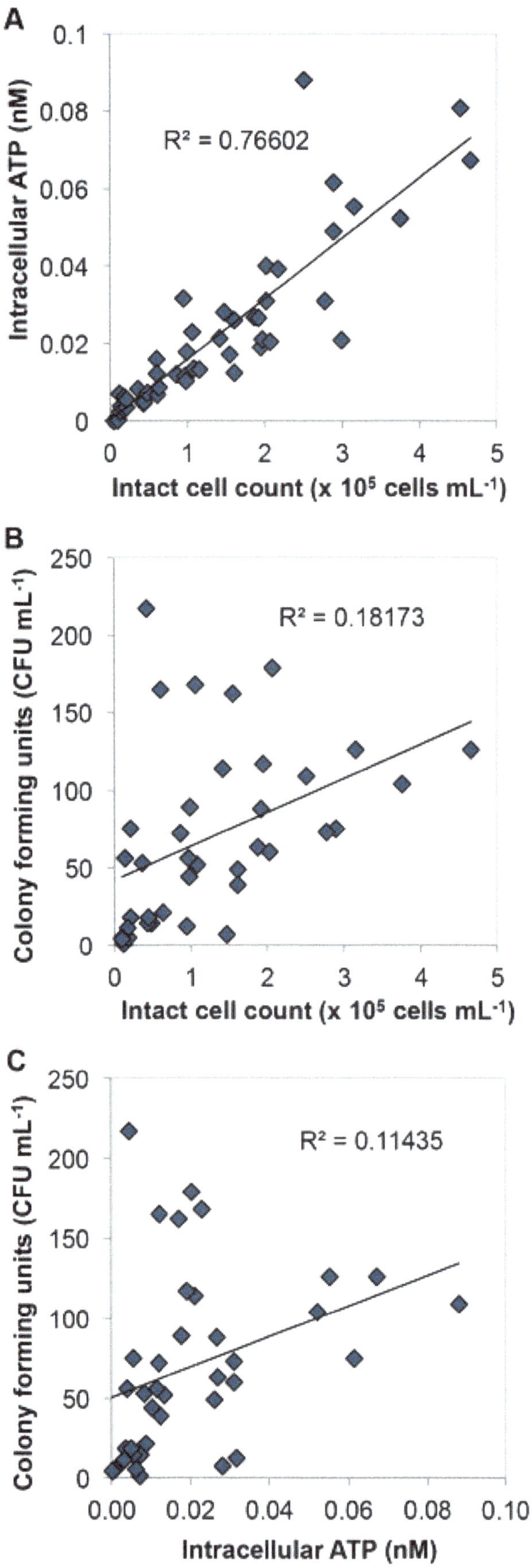

Figure 6. Comparison of the various microbiological parameters.

Clear correlations were observed between intact cells and intracellular ATP (n=49) (A), but no obvious correlations between these two parameters and heterotrophic plate counts at 22°C (n=38) (B) (C).

The conventional HPC results were compared with the FCM intact cell concentration values. A weak correlation (R^2=0.18, n=38) was observed between HPC (at 22°C) and intact cell concentrations (Figure 6B), similar to reports in previous studies [38], [50]. It could be explained by the often described phenomenon, that less than 1% of drinking water bacteria are cultivable on conventional agar plates [25], [50], [52]. In addition, Mezule and co-workers [53] demonstrated evidence of the presence of so called viable-but-not-cultivable (VNBC) bacterial state, in both drinking water and biofilms for the network investigated here, thus indicating further limitations in the HPC method. Since intracellular ATP showed a good correlation with intact cell concentration, but intact cell count correlated weakly with HPC, it was expected that intracellular ATP and HPC would not correlate well (Figure 6C; R^2=0.11, n=38). Various studies were performed to compare ATP and HPC parameters from water samples, but good correlations were never observed e.g., R^2=0.20 [19], R^2=0.36 [22] and R^2=0.31 [50]. Our results combined with those from previous studies cast further doubts on the value of using the HPC method for general microbiological drinking water quality control. In our opinion, the clear correlation between two methodologically independent viability parameters (intracellular ATP and FCM intact cell counts), and the absence of any correlations with two different HPC methods, renders the former methods more meaningful for assessing and understanding biological instability, particularly in chlorinated environments.

Importance of measuring extracellular ATP

Arguments for and against the concept and importance of measuring extracellular ATP have been made [4], [19], [20], [25], [54], [55]. To understand this better, we arranged our data according to increasing intracellular ATP concentrations, after which the measured extracellular ATP values were added to each corresponding sample (Figure 7). It is evident that extracellular ATP constitutes a considerable fraction of the total ATP amount in some samples – varying from 3% up to 100% – with an average contribution of 36% (n=49). Moreover, 33% of the samples contain more that 50% of extracellular ATP. This data supports other studies, where analyses showed high extracellular ATP in drinking water samples from the distribution networks [20], [25]. Interestingly, the highest extracellular ATP ratio is mostly observed in the samples with relatively low intracellular ATP, in this case samples with close proximity to the treatment plant. In the case of chlorinated water, this could potentially be explained by the oxidative effect of chlorine on bacterial cells. Previous studies have shown extensive damage to bacterial membranes during chlorination [24], [45], after which a release of extracellular ATP from the damaged bacteria can occur. This membrane damage was also clearly detected in the present study (e.g., Figure 3B). Although, there is lack of detailed data considering the release of extracellular ATP in water samples affected by chlorination, strong evidence of ATP release during oxidation was presented in previous studies [4], [20]. Both these works showed a significant decrease in cell concentrations and intracellular ATP after ozonation, whereas extracellular ATP comprised 83–100% of the total ATP. Moreover, Figure 7 shows that samples with increased intracellular ATP concentrations, which we linked to bacterial growth during distribution, often had considerably less extracellular ATP in relation to total ATP. This could be due to the fact that extracellular ATP can be biodegraded by bacteria or extracellular enzymes in the network [54], [56]–[58]. However, it cannot be excluded that a decrease in extracellular ATP during distribution occurs due to oxidation by residual chlorine present in the network.

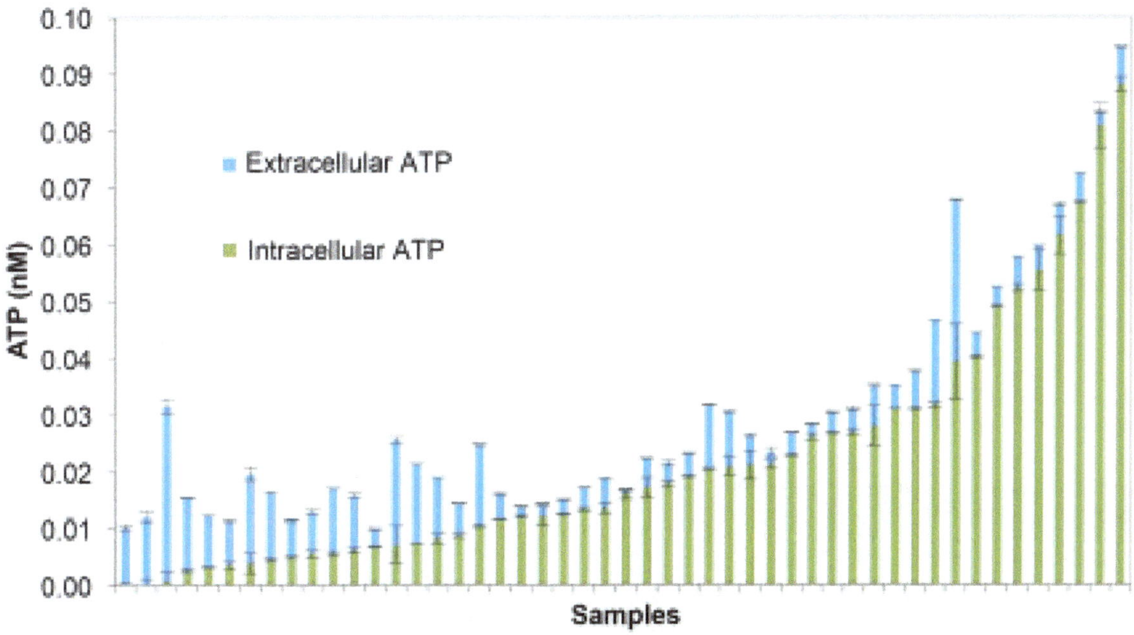

Figure 7. Distribution of intracellular and extracellular ATP in the water samples.

In general, higher concentrations and relative percentages of extracellular ATP were measured in samples that exhibited lower intracellular ATP concentrations (n=49).

CONCLUSIONS

- An investigation of a full-scale chlorinated drinking water distribution network with various microbiological methods clearly demonstrated both spatial and temporal biological instability in the network.
- Fluorescent staining with SGPI in combination with ATP measurements provided reliable and descriptive information about bacterial density and viability in chlorinated drinking water samples.
- A good correlation was observed between intracellular ATP and intact cell counts (R^2=0.77), whereas HPC showed poor correlations with both parameters (R^2=0.18 with intact cell concentration and R^2=0.11 with intracellular ATP).
- Extracellular ATP constituted on average 36% of total ATP in the present study, which confirms the necessity of extracellular ATP subtraction from total ATP measurements during chlorinated drinking water analyses.
- Overall the results raise questions with respect to the offset between increased biological safety gained from disinfection opposed to increased risk from instability (uncontrolled bacterial growth). While an improvement of the chlorination procedure could be a solution, the data suggests looking beyond only disinfection for achieving biological stability of drinking water.

REFERENCES

1. Bartram J, Cotruvo J, Exner M, Fricker C, Glasmacher A (2003) Heterotrophic plate count and drinking-water safety: The significance of HPCs for water quality and human health. World Health Organization. 244 p.
2. Boe-Hansen R, Albrechtsen H-J, Arvin E, Jørgensen C (2002) Bulk water phase and biofilm growth in drinking water at low nutrient conditions. Water Res 36: 4477–4486. doi:10.1016/S0043-1354(02)00191-4.
3. Hammes F, Berger C, Ko¨ster O, Egli T (2010) Assessing biological stability of drinking water without disinfectant residuals in a full-scale water supply system. J Water Supply Res Technol 59: 31. doi:10.2166/aqua.2010.052.
4. Hammes F, Berney M, Wang Y, Vital M, Ko¨ster O, et al. (2008) Flowcytometric total bacterial cell counts as a descriptive microbiological parameter for drinking water treatment processes. Water Res 42: 269–277. doi:10.1016/ j.watres.2007.07.009.
5. Juhna T, Birzniece D, Larsson S, Zulenkovs D, Sharipo A, et al. (2007) Detection of Escherichia coli in biofilms from pipe samples and coupons in drinking water distribution networks. Appl Environ Microbiol 73: 7456–7464. doi:10.1128/AEM.00845-07.
6. Liu G, Lut MC, Verberk JQJC, Van Dijk JC (2013) A comparison of additional treatment processes to limit particle accumulation and microbial growth during drinking water distribution. Water Res 47: 2719–2728. doi:10.1016/j.watres.2013.02.035.
7. Van der Kooij D (2000) Biological stability: A multidimensional quality aspect of treated water. In: Belkin S, editor. Environmental Challenges. Springer Netherlands. pp. 25–34.
8. Vital M, Stucki D, Egli T, Hammes F (2010) Evaluating the growth potential of pathogenic bacteria in water. Appl Environ Microbiol 76: 6477–6484. doi:10.1128/AEM.00794-10.
9. Vital M, Fu¨chslin HP, Hammes F, Egli T (2007) Growth of Vibrio cholerae O1 Ogawa Eltor in freshwater. Microbiology 153: 1993–2001. doi:10.1099/mic.0.2006/005173-0.
10. Vital M, Hammes F, Egli T (2008) Escherichia coli O157 can grow in natural freshwater at low carbon concentrations. Environ Microbiol 10: 2387–2396. doi:10.1111/j.1462-2920.2008.01664.x.
11. LeChevallier MW, Schulz W, Lee RG (1991) Bacterial nutrients in drinking water. Appl Environ Microbiol 57: 857–862.
12. Niquette P, Servais P, Savoir R (2001) Bacterial dynamics in the drinking water distribution system of Brussels. Water Res 35: 675–682.
13. Liu W, Wu H, Wang Z, Ong SL, Hu JY, et al. (2002) Investigation of assimilable organic carbon (AOC) and bacterial regrowth in drinking water distribution system. Water Res 36: 891–898.
14. Eichler S, Christen R, Ho¨ltje C, Westphal P, Bo¨tel J, et al. (2006) Composition and dynamics of bacterial communities of a drinking water supply system as assessed by RNA- and DNA-based 16S rRNA gene fingerprinting. Appl Environ Microbiol 72: 1858–1872. doi:10.1128/AEM.72.3.1858-1872.2006.
15. Hong P-Y, Hwang C, Ling F, Andersen GL, LeChevallier MW, et al. (2010) Pyrosequencing analysis of bacterial biofilm communities in water meters of a drinking water distribution system. Appl Environ Microbiol 76: 5631–5635. doi:10.1128/AEM.00281-10.
16. Pinto AJ, Xi C, Raskin L (2012) Bacterial community structure in the drinking water microbiome Is governed by filtration processes. Environ Sci Technol 46: 8851–8859. doi:10.1021/es302042t.
17. Lautenschlager K, Hwang C, Liu W-T, Boon N, Ko¨ster O, et al. (2013) A microbiology-based multi-parametric approach towards assessing biological stability in drinking water distribution networks. Water Res 47: 3015–3025. doi:10.1016/j.watres.2013.03.002.
18. Rittmann BE, Snoeyink VL (1984) Achieving biologically stable drinking water. J - Am Water Works Assoc 76: 106–114.
19. Van der Wielen PWJJ, van der Kooij D (2010) Effect of water composition, distance and season on the adenosine triphosphate concentration in unchlorinated drinking water in the Netherlands. Water Res 44: 4860–4867. doi:10.1016/j.watres.2010.07.016.
20. Vital M, Dignum M, Magic-Knezev A, Ross P, Rietveld L, et al. (2012) Flow cytometry and adenosine tri-phosphate analysis: Alternative possibilities to evaluate major bacteriological changes in drinking water treatment and distribution systems. Water Res 46: 4665–4676. doi:10.1016/j.watres.2012.06.010.

21. LeChevallier MW, Au K-K (2004) Water treatment and pathogen control: Process efficiency in achieving safe drinking-water. IWA Publishing. 136 p.

22. Delahaye E, Welte´ B, Levi Y, Leblon G, Montiel A (2003) An ATP-based method for monitoring the microbiological drinking water quality in a distribution network. Water Res 37: 3689–3696. doi:10.1016/S0043-1354(03)00288-4.

23. Francisque A, Rodriguez MJ, Miranda-Moreno LF, Sadiq R, Proulx F (2009) Modeling of heterotrophic bacteria counts in a water distribution system. Water Res 43: 1075–1087. doi:10.1016/j.watres.2008.11.030.

24. Ramseier MK, von Gunten U, Freihofer P, Hammes F (2011) Kinetics of membrane damage to high (HNA) and low (LNA) nucleic acid bacterial clusters in drinking water by ozone, chlorine, chlorine dioxide, monochloramine, ferrate(VI), and permanganate. Water Res 45: 1490–1500. doi:10.1016/ j.watres.2010.11.016.

25. Hammes F, Goldschmidt F, Vital M, Wang Y, Egli T (2010) Measurement and interpretation of microbial adenosine tri-phosphate (ATP) in aquatic environments. Water Res 44: 3915–3923. doi:10.1016/j.watres.2010.04.015.

26. LeChevallier MW, Welch NJ, Smith DB (1996) Full-scale studies of factors related to coliform regrowth in drinking water. Appl Environ Microbiol 62: 2201–2211.

27. Polanska M, Huysman K, van Keer C (2005) Investigation of assimilable organic carbon (AOC) in flemish drinking water. Water Res 39: 2259–2266. doi:10.1016/j.watres.2005.04.015.

28. Ramseier MK, Peter A, Traber J, von Gunten U (2011) Formation of assimilable organic carbon during oxidation of natural waters with ozone, chlorine dioxide, chlorine, permanganate, and ferrate. Water Res 45: 2002–2010. doi:10.1016/ j.watres.2010.12.002.

29. Van der Kooij D (1990) Assimilable organic carbon (AOC) in drinking water. In: McFeters GA, editor. Drinking Water Microbiology. New York, NY: Springer New York. pp. 57–87.

30. Weinrich LA, Jjemba PK, Giraldo E, LeChevallier MW (2010) Implications of organic carbon in the deterioration of water quality in reclaimed water distribution systems. Water Res 44: 5367–5375. doi:10.1016/j.watres.2010.06.035.

31. Jjemba P (2010) Guidance document on the microbiological quality and Biostability of reclaimed water following storage and distribution. WateReuse Research Foundation.

32. Lehtola MJ, Laxander M, Miettinen IT, Hirvonen A, Vartiainen T, et al. (2006) The effects of changing water flow velocity on the formation of biofilms and water quality in pilot distribution system consisting of copper or polyethylene pipes. Water Res 40: 2151–2160. doi:10.1016/j.watres.2006.04.010.

33. Manuel CM, Nunes OC, Melo LF (2007) Dynamics of drinking water biofilm in flow/non-flow conditions. Water Res 41: 551–562. doi:10.1016/j.watres.2006.11.007.

34. Bucheli-Witschel M, Ko¨tzsch S, Darr S, Widler R, Egli T (2012) A new method to assess the influence of migration from polymeric materials on the biostability of drinking water. Water Res 46: 4246–4260. doi:10.1016/j.watres.2012.05.008.

35. Douterelo I, Husband S, Boxall JB (2014) The bacteriological composition of biomass recovered by flushing an operational drinking water distribution system. Water Res 54: 100–114. doi:10.1016/j.watres.2014.01.049.

36. Rossman LA (2000) EPANET 2 Users manual. National Risk Management Research Laboratory. U.S. Environmental Protection Agency, Cincinatti, Ohio.

37. Rubulis J, Dejus S, Meksa R (2011) Online measurement usage for predicting water age from tracer tests to validate a hydraulic model American Society of Civil Engineers. pp. 1488–1497. doi:10.1061/41203(425)133.

38. Berney M, Vital M, Hu¨lshoff I, Weilenmann H-U, Egli T, et al. (2008) Rapid, cultivation-independent assessment of microbial viability in drinking water. Water Res 42: 4010–4018. doi:10.1016/j.watres.2008.07.017.

39. Lautenschlager K, Boon N, Wang Y, Egli T, Hammes F (2010) Overnight stagnation of drinking water in household taps induces microbial growth and changes in community composition. Water Res 44: 4868–4877. doi:10.1016/ j.watres.2010.07.032.

40. Antoun EN, Dyksen JE, Hiltebrand DJ (1999) Unidirectional flushing: A powerful tool. J - Am Water Works Assoc 91: 62–71.

41. Friedman M, Kirmeyer GJ, Antoun E (2002) Developing and implementing a distribution system flushing program. J - Am Water Works Assoc 94: 48–56.

42. Pre´vost M, Rompre´ A, Coallier J, Servais P, Laurent P, et al. (1998) Suspended bacterial biomass and activity in full-scale drinking water distribution systems: Impact of water treatment. Water Res 32: 1393–1406. doi:10.1016/S0043- 1354(97)00388-6.

43. Berney M, Hammes F, Bosshard F, Weilenmann H-U, Egli T (2007) Assessment and interpretation of bacterial viability by using the LIVE/DEAD BacLight Kit in combination with flow cytometry. Appl Environ Microbiol 73: 3283–3290. doi:10.1128/AEM.02750-06.

44. Gre´gori G, Citterio S, Ghiani A, Labra M, Sgorbati S, et al. (2001) Resolution of viable and membrane-compromised bacteria in freshwater and marine waters based on analytical flow cytometry and nucleic acid double staining. Appl Environ Microbiol 67: 4662–4670. doi:10.1128/AEM.67.10.4662-4670.2001.

45. Lisle JT, Pyle BH, McFeters GA (1999) The use of multiple indices of physiological activity to access viability in chlorine disinfected Escherichia coli O157:H7. Lett Appl Microbiol 29: 42–47. doi:10.1046/j.1365-2672.1999.00572.x.

46. Shi L, Gu¨nther S, Hu¨bschmann T, Wick LY, Harms H, et al. (2007) Limits of propidium iodide as a cell viability indicator for environmental bacteria. Cytometry A 71A: 592–598. doi:10.1002/cyto.a.20402.

47. Tsai Y-P (2005) Impact of flow velocity on the dynamic behaviour of biofilm bacteria. Biofouling 21: 267–277. doi:10.1080/08927010500398633.

48. Srinivasan S, Harrington GW, Xagoraraki I, Goel R (2008) Factors affecting bulk to total bacteria ratio in drinking water distribution systems. Water Res 42: 3393–3404. doi:10.1016/j.watres.2008.04.025.

49. Van der Kooij D (1992) Assimilable organic carbon as an indicator of bacterial regrowth. J - Am Water Works Assoc 84: 57–65.

50. Siebel E, Wang Y, Egli T, Hammes F (2008) Correlations between total cell concentration, total adenosine triphosphate concentration and heterotrophic plate counts during microbial monitoring of drinking water. Drink Water Eng Sci Discuss 1: 71–86. doi:10.5194/dwesd-1-71-2008.

51. Nocker A, Sossa KE, Camper AK (2007) Molecular monitoring of disinfection efficacy using propidium monoazide in combination with quantitative PCR. J Microbiol Methods 70: 252–260. doi:10.1016/j.mimet.2007.04.014.

52. Van der Kooij D, Vrouwenvelder JS, Veenendaal HR (2003) Elucidation and control of biofilm formation processes in water treatment and distribution using the Unified Biofilm Approach. Water Sci Technol J Int Assoc Water Pollut Res 47: 83–90.

53. Mezule L, Larsson S, Juhna T (2013) Application of DVC-FISH method in tracking Escherichia coli in drinking water distribution networks. Drink Water Eng Sci 6: 25–31. doi:10.5194/dwes-6-25-2013.

54. Cowan DA, Casanueva A (2007) Stability of ATP in Antarctic mineral soils. Polar Biol 30: 1599–1603. doi:10.1007/s00300-007-0324-9.

55. Venkateswaran K, Hattori N, La Duc MT, Kern R (2003) ATP as a biomarker of viable microorganisms in clean-room facilities. J Microbiol Methods 52: 367– 377. doi:10.1016/S0167-7012(02)00192-6.

56. Azam F, Hodson RE (1977) Dissolved ATP in the sea and its utilisation by marine bacteria. Nature 267: 696–698. doi:10.1038/267696a0.

57. Mempin R, Tran H, Chen C, Gong H, Ho KK, et al. (2013) Release of extracellular ATP by bacteria during growth. BMC Microbiol 13: 301. doi:10.1186/1471-2180-13-301.

58. Riemann B (1979) The occurrence and ecological importance of dissolved ATP in fresh water. Freshw Biol 9: 481–490. doi:10.1111/j.1365-2427.1979.tb01532.x.

CHAPTER 15

PHOTOCATALYTIC ENHANCEMENT FOR SOLAR DISINFECTION OF WATER: A REVIEW

J. Anthony Byrne, Pilar A. Fernandez-Ibañez, Patrick S. M. Dunlop, Dheaya M. A. Alrousan and Jeremy W. J. Hamilton

INTRODUCTION

Water is the most important natural resource in the world and availability of safe drinking water is a high priority issue for human existence and quality of life. Unfortunately, water resources are coming under increasing pressure due to population growth, over-use and wastage. The World Health Organization (WHO) estimates that 884 million people lack access to improved water supplies. Many more are forced to rely on sources that are microbiologically unsafe, resulting in a higher risk of waterborne disease transmission, including typhoid, hepatitis and cholera [1–3].

As ever, the poor are the worst affected and, in developing countries, 50% of the population are exposed to polluted water sources which, along with inadequate supplies of water for personal hygiene and poor sanitation, are the main contributors to an estimated 4 billion cases of diarrhea each year. These factors result in an estimated 2.2 million deaths each year, the majority of which are children under the age of five [1].

The provision of piped-in water supplies is an important long-term goal; however, the WHO and the United Nations Children's Fund (UNICEF) acknowledge that we are unlikely to meet the Millennium Development Goal (MDG) target of halving the proportion of the people without sustainable access to safe drinking water and basic sanitation by 2015. While conventional interventions to improve water supplies at source (point of distribution) have long been recognized as effective in preventing diarrhoea, more recent reviews have shown household-based (point-of-use) interventions to be significantly more effective than those at the source. As a result, there is increasing interest in such household-based interventions that deliver the health benefits associated with consumption of safe drinking water via low cost technologies [4].

In 2008, Clasen and Haller reported on the cost and cost effectiveness of house-hold based interventions to prevent diarrhoea [4]. They compared the following: chlorination using sodium hypochlorite following the "Safe Water System" (SWS) developed and promoted by the US Centers for Disease Control and Prevention (CDC), gravity filtration using either commercial "candle" style gravity filters or locally fabricated pot-style filters developed by Potters for Peace, solar disinfection following the "SODIS" method in which clear 2 L PET bottles are filled with raw water and then exposed to the sun for 6–48 h, and flocculation disinfection using Procter & Gambles PUR sachets, which combine an iron-based flocculant with a chlorine-based disinfectant and treat water in 10 L

Citation: Byrne JA, Fernandez-Ibañez PA, Dunlop PSM, Alrousan DMA, Hamilton JWJ, Photocatalytic Enhancement for Solar Disinfection of Water: A Review, doi: http://dx.doi.org/10.1155/2011/798051

batches. They concluded that household-based chlorination was the most cost-effective. Solar disinfection (SODIS) was only slightly less cost-effective, owing to its almost identical cost but marginally lower overall effectiveness. Given that household-based chlorination requires the distribution of sodium hypochlorite, solar disinfection has a major advantage in terms of nonreliance on chemical distribution. Sunlight is widely and freely available on Earth and the combined effects of IR, visible and UV energy from the sun can inactivate pathogenic organisms present in water. There are a number of parameters which affect the efficacy of the SODIS process, including the solar irradiance, the quality of the water to be treated, and the nature of the contamination—as some pathogens are more resistant to SODIS than others. Furthermore, SODIS enhancement technologies may improve the process effectiveness without substantially increasing the cost. One such approach is the use of semiconductor photocatalysis.

Semiconductor photocatalysis uses light along with a semiconductor material to produce highly oxidative species that destroy organic pollutants in water and inactivate pathogenic microorganisms [5–10]. The process occurs at atmospheric pressure and ambient temperature, without the requirement for consumable chemicals (except for oxygen from the air). Photocatalysis may be able to provide a low cost and simple solution to the purification of water in developing regions where solar irradiation can be employed.

This paper presents a critical review concerning semiconductor photocatalysis as a potential enhancement technology for the solar disinfection of water. The purpose of this review is to inform the nonexpert with respect to solar disinfection, and photocatalytic disinfection. For more detailed information the reader is referred to the scientific papers cited.

SOLAR DISINFECTION OF WATER (SODIS)

SODIS is a simple and low cost technique used to disinfect contaminated drinking water. Transparent bottles (preferably PET) are filled with contaminated water and placed in direct sunlight for a minimum of 6 hours. Following exposure, the water is safe to drink as the viable pathogen load can be significantly decreased. Simple guidance for the use of SODIS is given in Figure 1. SODIS is used by an estimated 4.5 million regular users worldwide, predominately in Africa, Latin America, and Asia, and is recognised and promoted by the WHO [11, 12].

Figure 1: SODIS process—www.sodis.ch.

SODIS harnesses light and thermal energy to inactivate pathogens via a synergistic mechanism [13]. Around 4–6% of the solar spectrum reaching the surface of the Earth is in the UV domain, with maximum reported value of around 50 W/m^2 [14]. UV radiation (200–400 nm) can be classified as UVA (320–400 nm), UVB (280–320 nm), and UVC (200–280 nm). UVC is absorbed by the ozone layer along with a proportion of the UVB; therefore UVA represents the main fragment of solar ultraviolet radiation reaching the earth's surface.

Disinfection of water using solar energy has been carried out since Egyptian times. The process was first studied and reported in scientific literature by London-based scientists Downes and Blunt in the late 1870s [15] and was effectively rediscovered as a low-cost water disinfection method by Acra et al. in the late 1970s [16, 17]. The main findings of this work were that Escherichia coli was more resistant to SODIS than other organisms tested, and as such E. coli should be used as an indicator organism for SODIS efficiency, akin to the presence of viable faecal coliforms as an indicator of efficacy for conventional disinfection processes. Furthermore, it is the UV component of sunlight, and to a lesser extent the blue end of the visible spectrum, that is mainly responsible for the biocidal action observed during SODIS. Wegelin et al. from the Swiss Federal Institute of Aquatic Science and Technology (Eawag) reported on SODIS in terms of the scope of the process and the dose of radiation required [18]. They concurred with the conclusions of Acra et al. and reported that to obtain a 3-log reduction in the viable numbers of E. coli a cumulative exposure dose of 2000 kJ/m^2 (350–450 nm) was required. The same dose was found to inactivate bacteriophage f2 whereas picornavirus required twice this dose. Water temperatures above 50°C significantly increased the rate of bacterial inactivation. The research team at Eawag have significantly contributed to the scientific development, standardisation, and promotion of the SODIS process through the development of the online information network and the publication and distribution of the SODIS Manual and associated education resources [12].

Laboratory studies have demonstrated the effects of key operational parameters such as light intensity and wavelength, solar exposure time, availability of oxygen, turbidity, and temperature [19, 20]. The SODIS mechanism is understood to involve a number of biocidal pathways based upon absorption of UVA radiation and thermal inactivation. Direct UVA exposure can induce cellular membrane damage and delay microbial growth [21]. The biocidal action of UVA has also been attributed to the production of reactive oxygen species (ROS) which are generated from dissolved oxygen in water [22] and the photosensitisation of molecules in the cell, and/or any naturally occurring dissolved organic matter that can absorb photons of wavelengths between 320–400 nm, to induce photochemical reactions [23]. The thermal effect has been attributed to the high absorption of red and infrared photons by water. At temperatures below 40°C, the thermal effect is negligible with UVA inactivation mechanisms dominating the inactivation process. Significant bactericidal action is evident at temperatures above 40–45°C with a synergistic SODIS process observed at temperatures above 45°C [13, 19, 23–25]. Detailed genetic assessment has also been used to probe the biocidal mechanism of SODIS [26,27].

Researchers have shown SODIS to be effective against a wide range of microorganisms responsible for diarrheal illness [28–30]. The inactivation of resistant protozoa has also been reported [31–33]. Field trials have demonstrated a significant health benefits from consumption of SODIS treated water [34, 35]. The effectiveness of SODIS against cholera was also demonstrated in a Kenyan health impact assessment, where an 86% reduction cholera cases was observed in households regularly using SODIS [36].

Studies to improve the efficiency of the SODIS process using low-cost, commonly available materials have been conducted [37–40]; however, the simple approach of exposing a 2 L PET bottle to full sun for a minimum of 6 hours is the most commonly promoted and practiced method.

ENHANCEMENT TECHNOLOGIES FOR SODIS

There are several drawbacks of "conventional" SODIS technology. The use of PET bottles allows for only small volumes to be treated (2-3 L), and the process efficiency is dependent on a range of environmental parameters including the solar irradiance (which depends on the latitude, time of day, and atmospheric conditions), the initial water quality for example organic loading, turbidity, level, and nature of the bacterial contamination. The resistance that microorganisms display to solar disinfection leads to variation in treatment times. Malato et al. reviewed the reported inactivation time required for a range of microorganisms using SODIS under ca. 1 kW/m^2 global irradiance [10]. These vary enormously from 20 min for Campylobacter jejuni to 8 h forCryptosporidium parvum oocysts. For Bacillus subtilis endospores, no inactivation was observed after 8 hours of SODIS treatment. SODIS is user dependent in that it requires the user to 'time' the exposure and as such there is no quality assurance in the process and as such, lack of compliance with the recommended protocol is a major issue.

There are a number of ways to improve or enhance the conventional SODIS process, including the design of SODIS bags where the solar dose per volume is increased, the use of UVA dosimetric sensors which indicate to the user when the desired dose has been received by the water, the design of customised SODIS treatment systems which maximise the solar dose using compound parabolic collectors (CPC) and may include UVA feedback sensors for automated control, and the use of semiconductor photocatalysis to enhance the treatment efficacy.

SODIS Bags

The SODIS method can be enhanced by use of a personal SODIS-reactor which would maximise the area for photon collection and minimise the path length for light penetration through the water to be treated. Furthermore, these SODIS bags could be deployed in emergency situations where access to drinking water is an immediate issue for example flooding or earthquake. The latter application can be limited due to the lack of PET bottles in disaster areas whereas bags have the advantage that they can easily be transported and stored in large quantities. Eawag has launched a project to develop specific SODIS bags, and several bag models are being developed in collaboration with partners from the private sector. A prototype is currently being field tested together with local organisations in Bolivia, DR Congo, Kenya, Nepal, and Nicaragua [41].

UV Dosimetric Indicators

Professor Mills' group at the University of Strathclyde has been investigating the use of photocatalytic systems as intelligent inks [42, 43]. As titanium dioxide (TiO_2) is excited by UVA light, these systems may be used as UVA dosimetric indicators. At the University of Ulster, we have tested a simple prototype system to measure UVA dose utilising the photocatalytic reduction of methylene blue (MB) to leuco-Methylene Blue (LMB) (method adopted from Mills, Lee, and Sheridan [43]).

In the presence of the hole scavenger (glycerol), methylene blue is photocatalytically reduced to colorless leuco-methylene blue, and the rate of decolouration is dependent on the UVA intensity (Figure 2). In the dark and in the presence of oxygen, the reduced form is reoxidised back to methylene blue. This results in a simple reversible UVA dosimetric indicator which could be utilised to provide some quality assurance for the SODIS user. There are a wide range of dyes or inks that could be utilised in these systems to provide simple dosimetric indicators for corresponding to UVA doses required for SODIS.

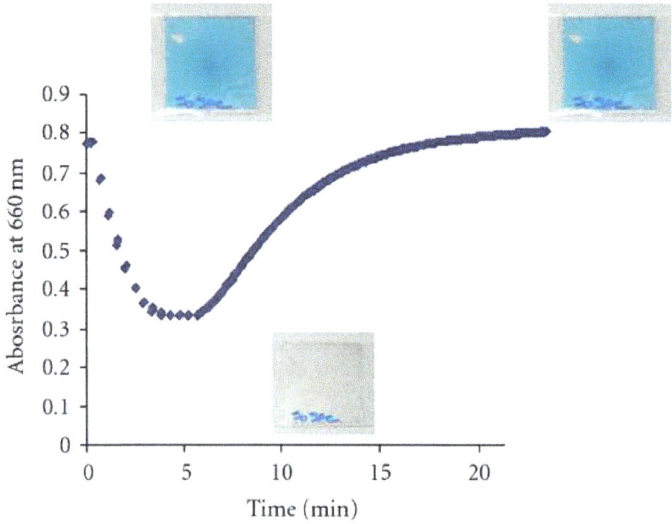

Figure 2: Colourimetric UVA dosimetric indicators—UVA exposure photo-reduced MB (blue) the LMB (white) (UVA dose = 9.0 kJ/m^2), MB subsequently reformed via oxidation in the dark.

Compound Parabolic Collectors

Fernandez-Ibanez and Malato at the Plataforma Solar de Almeria (PSA) in Spain have focused on the use of nonconcentrating solar collectors for the enhancement of solar disinfection. Compound Parabolic Collectors (CPC's) are nonimaging systems which collect diffuse radiation. The collected energy is homogeneously distributed across the absorber surface. CPCs do not rely solely on direct solar radiation and are therefore effective even on cloudy days. In addition, they collect radiation independently of the direction of sunlight and do not require sun tracking.

The SODIS process relies heavily on the solar UVA which, as received at sea level, is composed of roughly similar portions of both direct and diffuse electromagnetic radiation. Given the diffuse nature of the UVA and the cylindrical shape of SODIS bottles, the use of sunlight collecting systems based on nonimaging optics has obvious potential. Navntoft et al. demonstrated the use of CPC technology to enhance the efficacy of SODIS on sunny and cloudy days [25]. The major advantage of CPC technology is that the concentration factor remains constant for all values of sun zenith angle within the acceptance angle limit; therefore, it is theoretically possible to design larger volume SODIS systems. For example, recent work showed the effective inactivation of E. coliK12 in well water using a 25 L SODIS reactor with CPC [44]. Studies using CPC solar reactors including titanium dioxide as a photocatalyst have shown increased disinfection in comparison to SODIS alone [45–47].

Ubomba-Jaswa et al. investigated the effect of UVA dose on the inactivation efficiency of 3 types of SODIS reactors: borosilicate glass tubes (static batch), borosilicate glass tubes with compound parabolic collector (recirculating batch), and borosilicate glass tubes with CPC (recirculating batch) (Figures 3 and 4) [48]. They used E. coli K12 as the model microorganism suspended in natural well water and demonstrated inactivation of approximately 6-log colony forming units (CFU) mL^{-1} following receipt of a total uninterrupted minimum dose of >108 kJ/m^2 (295–385 nm).

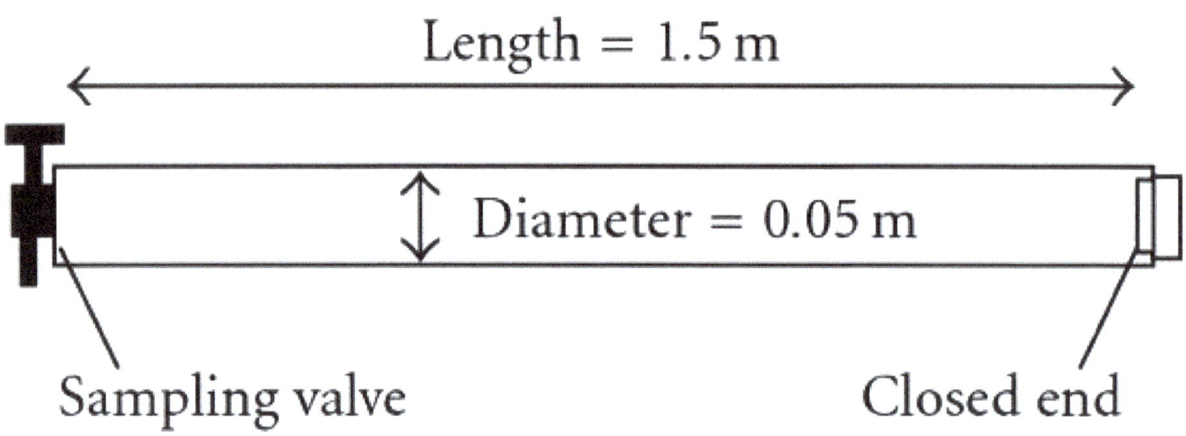

(a)

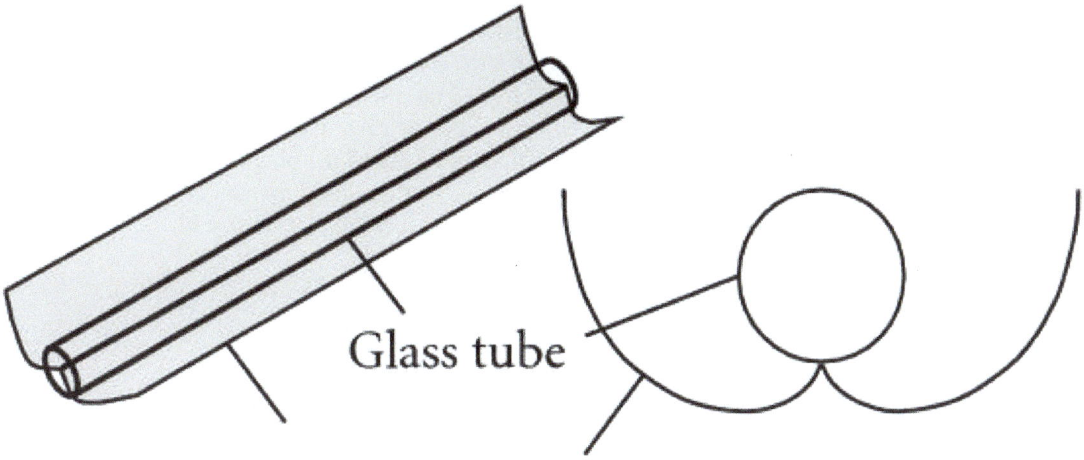

(b)

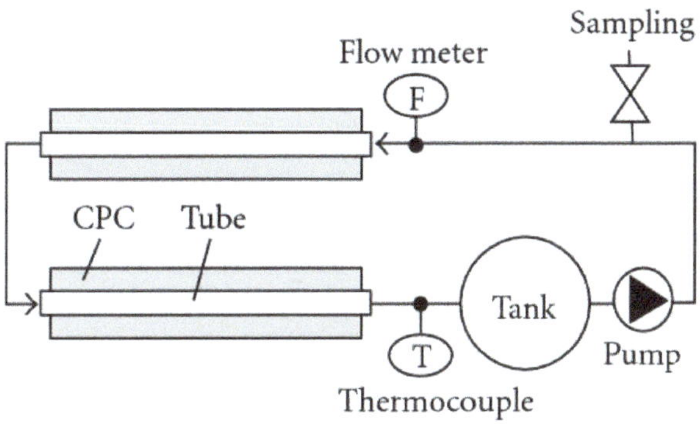

(c)

Figure 3: (a) Glass tube configuration. (b) Tube with CPC collector configuration. (c) Flow diagram of the solar CPC reactor (with permission from [48]).

Figure 4: Borosilicate glass tubes with CPC.

SEMICONDUCTOR PHOTOCATALYSIS

Mechanism of Photocatalysis

When a semiconductor is irradiated with light of wavelength equal or greater than its band gap, energy is absorbed resulting in the promotion of electrons from the valence band to the conduction band, and the formation of electron-hole pairs (e^- and h^+) [5]. These charge carriers can recombine, with the energy being reemitted as light or heat, or they may migrate to the catalyst surface where they can participate in redox reactions at the particle-solution interface [8] (Figure 5). The reactive oxygen species (ROS) produced, including the hydroxyl radical, are very active, indiscriminate oxidants [49] destroying a large variety of chemical contaminants in water and also causing fatal damage to microorganisms [50]. The final products of the photocatalytic degradation of pollutants (given long enough treatment time) are CO_2, H_2O, and respective inorganic acids or salts.

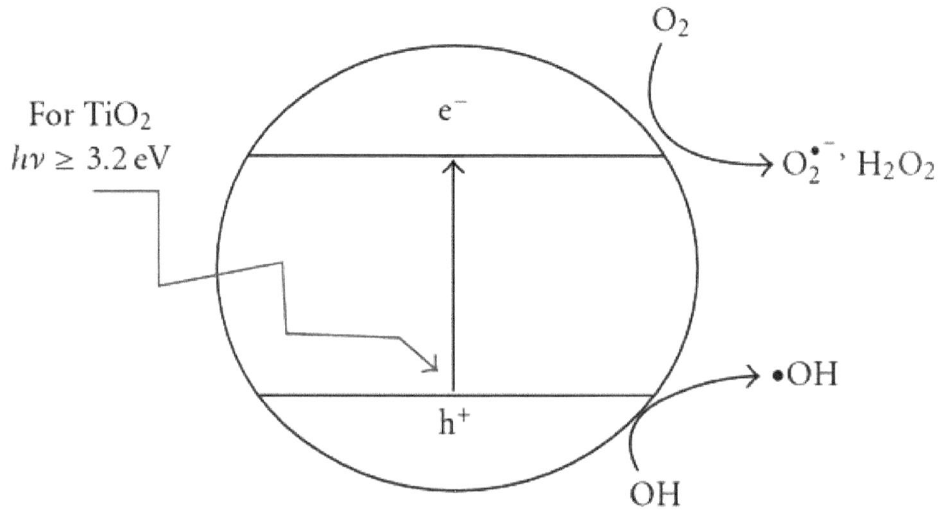

Figure 5: Schematic representation of the mechanism of photocatalysis on titanium dioxide particles.

For more detailed information on the mechanisms of semiconductor photocatalysis, the reader is referred to one of the many excellent reviews [5–10].

Photocatalytic Materials

Several compounds have been investigated as potential semiconductor photocatalysts, including metal oxides (TiO_2, ZnO, ZrO_2, V_2O_5, Fe_2O_3, SnO_2) and metal sulphides (CdS, ZnS) [51, 52]. Amongst these, the most popular photocatalyst for use in water treatment applications is titanium dioxide. TiO_2 is a wide band semiconductor (band gap=3.2 eV for anatase); therefore, it requires UV excitation. To improve efficiency in solar applications, visible light active materials are desirable. However, while absorbing a greater number of solar photons, the smaller band gap gives a narrower voltage window to drive the redox reactions at the particle-solution interface. Metal sulphide semiconductors, which absorb in the visible region of the spectrum, tend to undergo photo-anodic corrosion [53]. Considering cost, chemical and photochemical stability, availability, and lack of toxicity, the most suitable catalyst reported to date for the treatment of water is TiO_2[54].

Immobilised versus Suspended Photocatalyst

The photocatalyst can be utilised in aqueous suspension or it may be immobilised on a supporting solid substrate. Most studies have reported that suspension reactors are more efficient due to large surface area available for redox reactions [55], however; the main drawback of using nano or micro-particles in suspension is the requirement for posttreatment separation and recycling of the catalyst, potentially making the treatment more complex and expensive. Therefore, treatment reactors utilising immobilised TiO_2 have gained attention. There are a wide range of methodologies available for the preparation of immobilised photocatalyst films on a range of supporting substrates, and a careful reactor design is required to prevent efficiency loss due to a reduction in catalyst surface area and poor mass transfer of reactants to the photocatalyst surface [56].

Photocatalytic Disinfection of Water

Matsunaga et al. reported the first application of TiO_2 photocatalysis for the inactivation of bacteria in 1985 [57]. Since then, there have been a large number of research publications dealing with the inactivation of microorganisms including bacteria, viruses, protozoa, fungi, and algae. Blake et al. carried out an extensive review of the microorganisms reported to be inactivated by photocatalysis [51]. In 2007, McCullagh et al. reviewed the application of photocatalysis for the disinfection of water contaminated with pathogenic microorganisms [58]. In 2009, Malato et al. published an extensive review on the decontamination and disinfection of water by solar photocatalysis [10] and, in 2010, Dalrymple et al. reviewed the proposed mechanisms and kinetic models widely used in photocatalytic disinfection studies [59].

In most photocatalytic disinfection studies, the hydroxyl radical is suggested to be the primary species responsible for microorganism inactivation; however, some papers do report involvement of other ROS, such as H_2O_2, $O_2^{\bullet-}$ [60–63]. These reactive species can cause fatal damage to microorganisms by disruption of the cell membrane or by attacking DNA and RNA [51]. Other modes of action TiO_2 photocatalysis have been proposed, including damage to the respiratory system within the cells [64] and loss of fluidity and increased ion permeability in the cell membrane [62]. Detailed spectroscopy-based studies attributed cell death to lipid peroxidation of bacterial cell membrane [61–63]. The peroxidation of the unsaturated phospholipids contained in the bacterial cell membrane results in loss of respiratory activity [65] and/or leads to a loss of fluidity and increased ion permeability [62]. It has also been suggested that cell membrane damage can open the way for further oxidative attack of internal cellular components, ultimately resulting in cell death [65].

Research at the University of Ulster has been mainly concerned with the use of immobilised TiO_2 films prepared by the deposition of Evnoik (Degussa) P25 onto a range of supporting substrates, including borosilicate glass, ITO glass, and titanium metal. For example, Alrousan et al. reported on the photocatalytic inactivation of E. coli in surface water using immobilised nanoparticle TiO_2 films [66]. In this work, P25 was immobilized onto borosilicate glass using a dip coating method. The photoreactor was a custom-built stirred tank reactor which had excellent mixing and good mass transfer properties. The catalyst was irradiated in a back-face configuration that is, the light passed through the glass to excite the photocatalyst on the surface. It was found that the rate of photocatalytic inactivation of E. coli was more efficient with UVA-TiO_2 than direct photolytic inactivation with UVA alone, for E. coli suspended in both distilled water and surface water (Figure6). The optimum catalyst loading for the inactivation was determined to be 0.5 mg/cm², approximately half that reported for photocatalytic degradation of formic acid and atrazine in the same reactor, under the same incident light intensity.

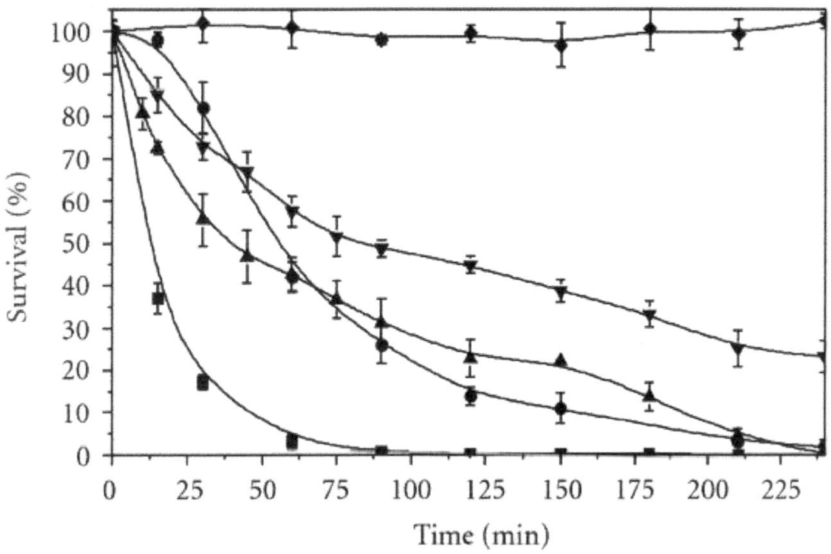

Figure 6: Comparison between photocatalytic and photolytic inactivation of *E. coli* in surface and distilled water (TiO_2, UV, Distilled water) ■, (TiO_2, UV, Surface water) ●, (UV, Distilled water) ▲, (UV, Surface water) ▼, Dark control ◆ (with permission, from [66]).

The organic and inorganic content of the surface water led to a reduction in the rate of photocatalytic disinfection in comparison to that observed in distilled water. The effect of selected individual constituents present in the surface water was examined to identify the main constituent responsible for the reduction in the rate of photocatalytic disinfection. The presence of inorganic ions, that is, sulphate and nitrate reduced the rate of photocatalytic inactivation, with sulphate having a more pronounced effect than nitrate. The presence of organic matter was found to be the dominating parameter responsible for the decrease in the rate of photocatalytic disinfection. The presence of inorganic ions will lead to a reduction in efficiency either by absorption of light, competing for ROS, or by adsorption to the catalyst surface. Organic matter will compete for ROS and may compete for photon absorption. The efficacy of photocatalytic disinfection will therefore be strongly dependent on the initial water quality. Where the water to be treated is of poor quality, a pre-treatment stage, for example, simple filtration or settling, may be desirable prior to photocatalytic treatment.

Bacterial cells have been described as a rather easy target for disinfectants, with bacterial spores and protozoa suggested as more robust target organisms. Clostridium perfringens spores have been reported to be chlorine resistant at levels used in potable water supplies. Dunlop et al. reported on the photocatalytic inactivation ofClostridium perfringens spores on TiO_2 electrodes [67]. The TiO_2 electrodes were made using electrophoretic immobilisation of commercially available TiO_2 powders onto conducting supports, that is, indium-doped tin oxide-coated glass, titanium metal, and titanium alloy. The photocatalytic inactivation of E. coli and C. perfringens spores in water was observed on all immobilised TiO_2 films under UVA irradiation. The rate of photocatalytic inactivation of E. coli was found to be one order of magnitude greater than that of C. perfringensspores, demonstrating the greater resistance of the spores to environmental stress. In this work, it was shown that the application of an external electrical bias (electrochemically assisted photocatalysis) significantly increased the rate of photocatalytic disinfection of C. perfringens spores. The effect of incident light intensity and initial spore loading were investigated, and disinfection kinetics were determined to be pseudo-first order. This work demonstrated that UVA photocatalysis is effective against bacterial spores which are more resistant to environmental stress, including UVA irradiation, than bacterial cells.

Cryptosporidium species are waterborne, protozoan parasites that infect a wide range of vertebrates. The life cycle involves the production of an encysted stage (oocyst) which is discharged in the faeces of their host. The disease, Cryptosporidiosis, in humans usually results in self-limited watery diarrhoea but has far more devastating effects on immunocompromised patients (e.g., AIDS patients) and can be life-threatening as a result of dehydration caused by chronic diarrhoea. Owing to their tough outer walls, the oocysts are highly resistant to disinfection and can survive for several months in standing water. Cryptosporidium oocysts therefore present as an excellent challenge for disinfection technologies. Sunnotel et al. reported on the photocatalytic inactivation of Cryptosporidium parvum oocysts on nanostructured titanium dioxide films (Figure 7) [68].

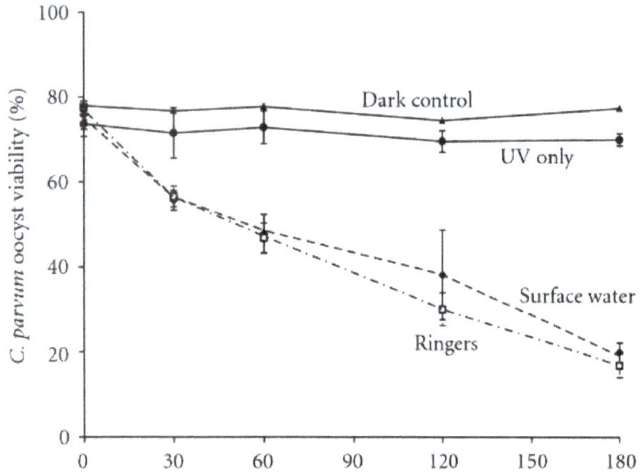

Figure 7: Photocatalytic inactivation of C. parvum oocysts suspended in Ringer's solution and surface water (with permission from [68]).

The photocatalytic inactivation of the oocysts was shown to occur in Ringer's buffer solution (78.4% after 180 min) and surface water (73.7% after 180 min). Scanning electron microscopy (SEM) confirmed cleavage at the suture line of oocyst cell walls, revealing large numbers of empty (ghost) cells after exposure to photocatalytic treatment. No significant inactivation was observed in the oocysts exposed to UVA radiation alone demonstrating the substantial benefit in the addition of TiO_2 as a photocatalyst.

It is clear from the literature, that in lab scale reactors, photocatalysis under UVA irradiation is more efficient than UVA irradiation alone for the disinfection of water contaminated with pathogenic microorganisms. Therefore, it is essential that photocatalytic reactors are tested for disinfection under real sun conditions either on small scale batch (personal use) or at pilot scale (aimed at household or small community use), and using real water sources.

Gelover et al. studied small-scale batch disinfection in plastic bottles containing spring water naturally polluted with coliform bacteria, with and without the addition of TiO_2 [69]. The bottles were mounted in simple home-made solar collectors (Figure 8). Two litre PET bottles were filled with spring water and exposed to direct sunlight. The microbial loading was typically 2.5×10^3 MPN/100 mL (most probable number) of total coliforms and 9.0×10^2 MPN/100 mL of faecal coliforms.

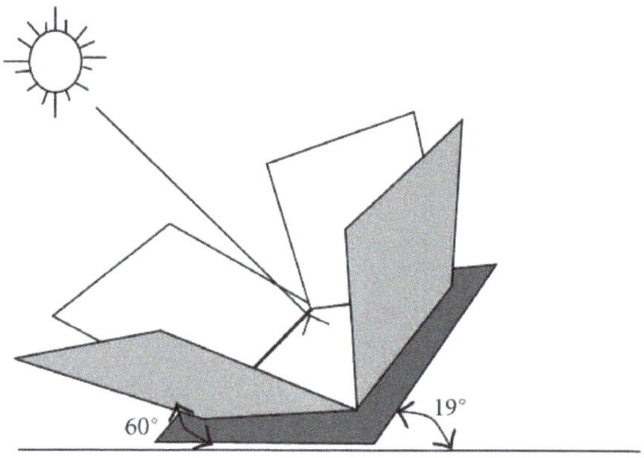

Figure 8: Simple solar collector made from 5 wooden sheets covered with aluminium foil (with permission from [69]).

TiO_2 was coated onto small Pyrex-glass cylinders, using a sol-gel method, and these were placed inside each bottle. It was found that photocatalytically enhanced SODIS was by far more effective than SODIS alone for the inactivation of both the total coliforms and the faecal coliforms (Figure 9). They measured bacterial regrowth following treatment and found that regrowth was observed with SODIS alone, but not with photocatalytically enhanced SODIS. This is an important finding, as bacteria have repair mechanisms which allow recovery following stress/injury and demonstrates the differences in the kill mechanisms involved in SODIS and photocatalytic disinfection.

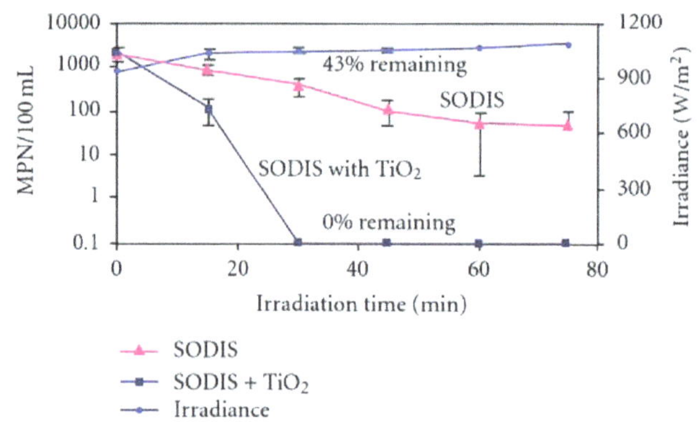

Figure 9: Decrease in faecal coliforms in natural spring water exposed to direct sunlight, comparing SODIS to TiO_2-enhanced SODIS (with permission from [69]).

With respect to larger-scale systems, Fernandez-Ibanez et al. reported on the pilot-scale photocatalytic disinfection of water under real sun conditions using a photoreactor with CPC [70]. The experiments were carried out under sunlight at the Plataforma Solar de Almeria in southern Spain using compound parabolic collectors. The pilot plant consisted of three CPC collectors, each containing two Pyrex tubes (50 mm inner diameter, 1.5 m long), with a collector surface area of

$0.25\,m^2$ and a total reactor volume treated of 11 L. Experiments were performed with suspended TiO_2 and, following modification to the reactor, with supported/immobilised TiO_2. In both cases, Degussa, P25 was employed as the photocatalyst. In the immobilised study, the catalyst was fixed onto glass fibre using SiO_2 as an inorganic binder. E. coli was used as the model microorganism suspended in distilled water. The photocatalytic suspension reactor was the most efficient, followed by the immobilised photocatalytic reactor, with SODIS being the least efficient treatment (Figure 10).

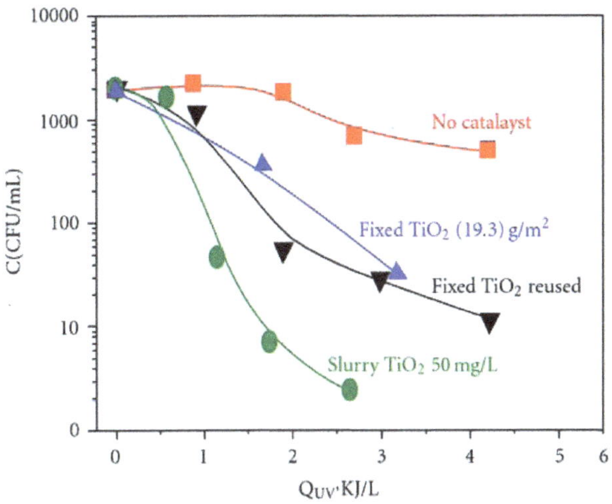

Figure 10: E. coli inactivation ($C_0 = 1 \times 10^4$ CFU/mL) versus Q_{UV} (energy per unit of volume, kJ/L) in the CPC solar photoreactor. Comparison of TiO_2 slurry (50 mg/L), with supported photocatalyst KN47 (19.3 g $TiO_2\,m^{-2}$) and SODIS. With permission from [70].

Further work undertaken by the group at Plataforma Solar de Almeria investigated the effect of UV intensity and dose on SODIS and photocatalytically enhanced SODIS [71]. The aim of the work was to study the dependence on solar irradiation conditions under natural sunlight using three microorganisms, E. coli K-12 culture and two wild strains Fusarium solani and Fusarium anthophilum. Photocatalytic disinfection experiments were carried out with TiO_2 supported on a paper matrix held concentrically within the CPC glass tube reactors described above, and with TiO_2 in the form of a slurry in Pyrex glass bottle reactors. The experiments were performed with different illuminated reactor surfaces, in different seasons of the year, and under changing weather conditions (i.e., cloudy and sunny days). Photocatalysis did not increase the rate of disinfection following receipt of the minimum solar dose; however, the solar-only disinfection was more susceptible to changes in solar irradiation than photocatalytic disinfection, that is, where light intensities were generally low or there was greater availability of diffuse UVA, the inclusion of the photocatalyst provided a significant benefit.

ISSUES TO BE ADDRESSED

Photoreactor Design

The SODIS process depends mainly on the UVA wavelengths present in sunlight. Solar UV at sea level is composed of roughly similar portions of both direct and diffuse electromagnetic radiation. Without cloud cover, the solar UVA spectrum is ca. 60% direct and ca. 40% diffuse. Therefore, the use of collecting systems based on nonimaging optics has a clear potential, compared to expensive imaging/concentrating optic-based systems. The major advantage with compound parabolic collectors is that collection factor remains constant for all values of sun zenith angle within the acceptance angle limit. Therefore, CPC enhancement can be utilised in the design of larger-scale solar disinfection systems which can be used for household or small community use. A photoreactor for use in developing countries should have the following attributes; high illuminated volume to total volume ratio; ability to operate under a low flow rate when utilising immobilised photocatalyst to maximise the residence time in flow systems; UVA dosimetric indicator (as both SODIS and photocatalytic disinfection are dose dependant); CPC of good quality that is, high UVA reflectivity (for aluminium this is typically 87–90%); reaction vessel with high (90%) UVA transmission (e.g., borosilicate glass); robust under potentially harsh environmental conditions; minimal life-cycle cost; low environmental impact; low maintenance requirements and ease of access to replacement parts; and minimal external power requirement.

If the system is to include a photocatalyst, the reactor design must also ensure the sufficient supply of an electron acceptor, typically dissolved oxygen from the air. In static batch systems, the concentration of dissolved oxygen will be rapidly consumed in the initial stages of the reaction and as water temperature increases the solubility of oxygen in water will be further reduced. For applications in developing countries, this is of paramount importance as the temperature within SODIS reactors can reach 55°C. An alternative may be to introduce other oxidants for example, H_2O_2; however, this would give rise to a dependence on consumable chemicals which is undesirable.

Photocatalyst Longevity

For application in remote locations and developing regions, the treatment system must be robust, noncomplex, and require only low-level maintenance. Therefore, photocatalyst regeneration is undesirable and more research is required to understand and predict photocatalyst longevity under typical working conditions. To reduce the complexity of the treatment system, immobilised photocatalyst systems are preferred; however, catalyst stripping may be a problem if the immobilisation protocol does not produce a robust hard wearing coating. Also, catalyst fouling by inorganic species present in the water can lead to a reduction in the photocatalytic efficiency over time. Miranda-García et al. investigated the degradation of 15 emerging contaminants (ECs) in a photocatalytic pilot plant utilising TiO_2 immobilised onto glass beads [72]. The CPC plant consisted of two modules of 12 Pyrex glass tubes mounted on a fixed platform tilted 37° (local latitude). Two of the glass tubes were packed with TiO_2-coated glass spheres (coated with a titania sol–gel). The total illuminated area was $0.30\,m^2$ and total volume was 10 L, of which 0.96 L was constantly irradiated. The system operated in recirculating batch mode with a flow rate of $3.65\,L\,min^{-1}$. They found that the degradation of the organic pollutants in distilled water was achieved under solar irradiation, and importantly, after 5 cycles of photocatalysis, the photocatalyst activity was not decreased significantly. Experiments using real water may present catalyst fouling problems.

Visible Light Active Photocatalyst Materials

The overall efficiency of TiO_2 under natural sunlight is limited to the UV-driven activity (for anatase), accounting only to ca. 4% of the incoming solar energy on the Earth's surface. Therefore, there has been substantial research effort towards shifting the absorption spectrum of TiO_2 towards the visible region of the electromagnetic spectrum. Different approaches have been attempted including doping the TiO_2 with metal ions [73]. According to the literature, one of the more promising approaches to achieve visible light activity is doping with nonmetal elements including N and S. Since Asahi et al. [74] reported the visible-light photo-activity of TiO_2 with nitrogen doping, many groups have demonstrated that anion doping of TiO_2 extends the optical absorbance of TiO_2 into the visible-light region. However, the number of publications concerning the photocatalytic activity of these materials for the inactivation of microorganisms is limited. Li et al. reported the inactivation of MS2 phage under visible light irradiation using a palladium-modified nitrogen-doped titanium oxide (TiON/PdO) photocatalytic fiber, synthesized on a mesoporous-activated carbon fiber template by a sol-gel process [75]. Dark adsorption led to virus removal, and subsequent visible light illumination (wavelengths greater than 400 nm and average intensity of 40 mW/cm^2) resulted in additional virus removal of 94.5–98.2% within 1 h of additional contact time. By combining adsorption and visible-light photocatalysis, TiON/PdO fibers reached final virus removal rates of 99.75–99.94%. EPR measurements confirmed the production of •OH radicals by TiON/PdO under visible light illumination. Wu et al. also reported visible-light-induced photocatalytic inactivation of bacteria by composite photocatalysts of palladium oxide and nitrogen-doped TiO_2 [76]. The PdO/TiON catalysts were tested for visible-light-activated photocatalysis using gram-negative organisms that is, E. coli and Pseudomonas aeruginosa and the gram-positive organism Staphylococcus aureus.Their disinfection data showed that the PdO/TiON photocatalysts had a much better visible photocatalytic activity than either palladium-doped (PdO/TiO_2) or nitrogen-doped titanium oxide (TiON). The light source used in this work was a metal halogen desk lamp with a low UV output (<0.01 mW/cm^2 for $\lambda \leq 400$ nm). While these photocatalysts show promise, there was no comparison with undoped TiO_2 and solar-simulated light (which has around 5% UV) was not used.

In many cases, the UV activity of undoped TiO_2 is much greater than the visible light activity of the doped material. Therefore, for solar applications, the photocatalysts should be tested under simulated solar irradiation or under real sun conditions. Indeed, Rengifo-Herrera and Pulgarin recently reported on the photocatalytic activity of N, S codoped and N-doped commercial anatase (Tayca TKP 102) TiO_2 powders towards phenol oxidation and E. coli inactivation under simulated solar light irradiation [77]. However, these novel materials did not present an enhancement for the photocatalytic degradation of phenol or the photocatalytic inactivation of E. coli under simulated solar light, as compared to Degussa P25. They suggest that while the N, or N-S co-doped TiO_2, may show a visible light response, the localized states responsible for the visible light absorption do not play an important role in the photocatalytic activity.

More research is required to determine if visible light active materials can deliver an increase in the efficiency of photocatalysis under solar irradiation.

CONCLUSIONS

SODIS is a simple and low cost technique used to disinfect contaminated drinking water. Transparent bottles (preferably PET) are filled with contaminated water and placed in direct sunlight for a minimum of 6 hours. Following exposure, the water is safe to drink as the viable pathogen load can be significantly decreased. The process has approximately 4.5 million regular users, predominately in Africa, Latin America, and Asia and is recognised and promoted by the WHO. However, there are several drawbacks with "conventional" SODIS technology. The use of PET bottles allows for only small volumes to be treated (2-3 L) and the process efficiency is dependent on a range of environmental parameters including the solar irradiance (which depends on the latitude, time of day, and atmospheric conditions), the initial water quality for example, organic loading, turbidity, and level and nature of the bacterial contamination. There are a number of ways to improve or enhance the conventional SODIS process and these include the design of SODIS bags where the solar dose per volume is increased, the use of UV dosimetric indicators which measure the UV dose and indicate to the user when the desired dose has been received by the water, design of customised SODIS treatment systems which maximise the solar dose and the inclusion of UV feedback sensors for automated control, and the use of semiconductor photocatalysis to enhance the treatment efficacy. Semiconductor photocatalysis has been shown to be effective for the inactivation of a wide range of microorganisms at lab scale and under real sun conditions for both small-scale and large-scale applications. The use of CPC reactors enhances the efficiency of solar disinfection and photocatalytically enhanced solar disinfection. Nevertheless, there are a number of issues to be addressed before photocatalytically enhanced solar disinfection can be effectively deployed in developing regions. These include improvements in photoreactor design and the assessment of photocatalyst longevity under real operating conditions. Future developments in relation to visible light active photocatalytic materials may lead to more efficient solar photocatalysis for the disinfection of water.

ACKNOWLEDGMENTS

The authors wish to thank the European Commission for financial support in relation to the SODISWATER project (INCO-CT-2006-031650), the Spanish Ministry of Science and Innovation under the Consolider-Ingenio 2010 programme (Project CSD2006-00044 TRAGUA), and the Department of Employment and Learning (Northern Ireland) for funding DMA Alrousan, and the Department of Employment and Learning (Northern Ireland) for funding DMA Alrousan, and DEL and Invest Northern Ireland for funding under the US-Ireland Initiative for "Degradation Mechanism of Cyanotoxins Using Novel Visible Light-Activated Titania (TiO_2) Photocatlysts".

REFERENCES

1. T. Clasen and P. Edmondson, "Sodium dichloroisocyanurate (NaDCC) tablets as an alternative to sodium hypochlorite for the routine treatment of drinking water at the household level," International Journal of Hygiene and Environmental Health, vol. 209, no. 2, pp. 173–181, 2006. View at Publisher · View at Google Scholar · View at Scopus
2. WHO, "Economic and health effects of increasing coverage of low cost household drinking-water supply and sanitation interventions to countries off-track to meet MDG target 10," 2007. View at Google Scholar
3. J. D. Burch and K. E. Thomas, "Water disinfection for developing countries and potential for solar thermal pasteurization," Solar Energy, vol. 64, no. 1–3, pp. 87–97, 1998. View at Publisher · View at Google Scholar · View at Scopus
4. T. F. Clasen and L. Haller, Water Quality Interventions to Prevent Diarrhoea: Cost and Cost-Effectiveness, WHO, 2008.
5. A. Mills and S. Le Hunte, "An overview of semiconductor photocatalysis," Journal of Photochemistry and Photobiology A, vol. 108, no. 1, pp. 1–35, 1997. View at Google Scholar · View at Scopus
6. D. A. Tryk, A. Fujishima, and K. Honda, "Recent topics in photoelectrochemistry: achievements and future prospects," Electrochimica Acta, vol. 45, no. 15-16, pp. 2363–2376, 2000. View at Google Scholar ·View at Scopus
7. D. Bahnemann, "Photocatalytic water treatment: solar energy applications," Solar Energy, vol. 77, no. 5, pp. 445–459, 2004. View at Publisher · View at Google Scholar · View at Scopus
8. V. Augugliaro, M. Litter, L. Palmisano, and J. Soria, "The combination of heterogeneous photocatalysis with chemical and physical operations: a tool for improving the photoprocess performance," Journal of Photochemistry and Photobiology C, vol. 7, no. 4, pp. 127–144, 2006. View at Publisher · View at Google Scholar · View at Scopus
9. A. Fujishima, X. Zhang, and D. A. Tryk, "TiO_2 photocatalysis and related surface phenomena," Surface Science Reports, vol. 63, no. 12, pp. 515–582, 2008. View at Publisher · View at Google Scholar · View at Scopus
10. S. Malato, P. Fernández-Ibáñez, M. I. Maldonado, J. Blanco, and W. Gernjak, "Decontamination and disinfection of water by solar photocatalysis: recent overview and trends," Catalysis Today, vol. 147, no. 1, pp. 1–59, 2009.
11. "SODIS Newsletter 2010 No. 1," Swiss Federal Institute of Aquatic Science and Technology, 2010.
12. SODIS, August 2010, http://www.sodis.ch/index.
13. K. G. McGuigan, T. M. Joyce, R. M. Conroy, J. B. Gillespie, and M. Elmore-Meegan, "Solar disinfection of drinking water contained in transparent plastic bottles: characterizing the bacterial inactivation process,"Journal of Applied Microbiology, vol. 84, no. 6, pp. 1138–1148, 1998.
14. Y. D. Goswami, F. Kreith, and J. F. Kreider, Principles of Solar Engineering, Taylor and Francis, Philadelphia, Pa, USA, 2nd edition, 2000.
15. A. Downes and T. P. Blunt, "The influence of light upon the development of bacteria," Nature, vol. 16, pp. 402–218, 1877.
16. A. Acra, Y. Karahagopian, Z. Raffoul, and R. Dajani, "Disinfection of oral rehydration solutions by sunlight," The Lancet, vol. 2, no. 8206, pp. 1257–1258, 1980.
17. A. Acra, Z. Raffoul, and Y. Karahagopian, Solar Disinfection of Drinking Water and Oral Rehydration Solutions, UNICEF, Paris, France, 1984.
18. M. Wegelin, S. Canonica, K. Mechsner, T. Fleischmann, F. Pesaro, and A. Metzler, "Solar water disinfection: scope of the process and analysis of radiation experiments," Journal of Water Supply Research and Technology—Aqua, vol. 43, no. 4, pp. 154–169, 1994.
19. M. Berney, H. U. Weilenmann, A. Simonetti, and T. Egli, "Efficacy of solar disinfection of Escherichia coli, Shigella flexneri, Salmonella Typhimurium and Vibrio cholerae," Journal of Applied Microbiology, vol. 101, no. 4, pp. 828–836, 2006.
20. R. H. Reed, "The inactivation of microbes by sunlight: solar disinfection as a water treatment process,"Advances in Applied Microbiology, vol. 54, pp. 333–365, 2004.
21. A. Hamamoto, M. Mori, A. Takahashi et al., "New water disinfection system using UVA light-emitting diodes," Journal of Applied Microbiology, vol. 103, no. 6, pp. 2291–2298, 2007.
22. R. Khaengraeng and R. H. Reed, "Oxygen and photoinactivation of Escherichia coli in UVA and sunlight," Journal of Applied Microbiology, vol. 99, no. 1, pp. 39–50, 2005.
23. P. M. Oates, P. Shanahan, and M. F. Polz, "Solar disinfection (SODIS): simulation of solar radiation for global assessment and application for point-of-use water treatment in Haiti," Water Research, vol. 37, no. 1, pp. 47–54, 2003.
24. J. Blanco, S. Malato, P. Fernández-Ibañez, D. Alarcón, W. Gernjak, and M. I. Maldonado, "Review of feasible solar energy applications to water processes," Renewable and Sustainable Energy Reviews, vol. 13, no. 6-7, pp. 1437–1445, 2009.
25. C. Navntoft, E. Ubomba-Jaswa, K. G. McGuigan, and P. Fernández-Ibáñez, "Effectiveness of solar disinfection using batch reactors with non-imaging aluminium reflectors under real conditions: natural well-water and solar light," Journal of Photochemistry and Photobiology B, vol. 93, no. 3, pp. 155–161, 2008.
26. M. Berney, H. U. Weilenmann, and T. Egli, "Flow-cytometric study of vital cellular functions in Escherichia coli during solar disinfection (SODIS)," Microbiology, vol. 152, no. 6, pp. 1719–1729, 2006.View at Publisher · View at Google Scholar · View at Scopus

27. M. Berney, H. U. Weilenmann, and T. Egli, "Gene expression of Escherichia coli in continuous culture during adaptation to artificial sunlight," Environmental Microbiology, vol. 8, no. 9, pp. 1635–1647, 2006.
28. S. C. Kehoe, M. R. Barer, L. O. Devlin, and K. G. McGuigan, "Batch process solar disinfection is an efficient means of disinfecting drinking water contaminated with Shigella dysenteriae type I," Letters in Applied Microbiology, vol. 38, no. 5, pp. 410–414, 2004.
29. J. Lonnen, S. Kilvington, S. C. Kehoe, F. Al-Touati, and K. G. McGuigan, "Solar and photocatalytic disinfection of protozoan, fungal and bacterial microbes in drinking water," Water Research, vol. 39, no. 5, pp. 877–883, 2005.
30. M. Boyle, C. Sichel, P. Fernández-Ibáñez et al., "Bactericidal effect of solar water disinfection under real sunlight conditions," Applied and Environmental Microbiology, vol. 74, no. 10, pp. 2997–3001, 2008.
31. K. G. McGuigan, F. Méndez-Hermida, J. A. Castro-Hermida et al., "Batch solar disinfection inactivates oocysts of Cryptosporidium parvum and cysts of Giardia muris in drinking water," Journal of Applied Microbiology, vol. 101, no. 2, pp. 453–463, 2006.
32. W. Heaselgrave, N. Patel, S. Kilvington, S. C. Kehoe, and K. G. McGuigan, "Solar disinfection of poliovirus and Acanthamoeba polyphaga cysts in water—a laboratory study using simulated sunlight,"Letters in Applied Microbiology, vol. 43, no. 2, pp. 125–130, 2006.
33. S. Mtapuri-Zinyowera, N. Midzi, C. E. Muchaneta-Kubara, T. Simbini, and T. Mduluza, "Impact of solar radiation in disinfecting drinking water contaminated with Giardia duodenalis and Entamoeba histolytica/dispar at a point-of-use water treatment," Journal of Applied Microbiology, vol. 106, no. 3, pp. 847–852, 2009.
34. R. M. Conroy, M. Elmore-Meegan, T. Joyce, K. G. McGuigan, and J. Barnes, "Solar disinfection of drinking water and diarrhoea in Maasai children: a controlled field trial," The Lancet, vol. 348, no. 9043, pp. 1695–1697, 1996.
35. R. M. Conroy, M. Elmore Meegan, T. Joyce, K. McGuigan, and J. Barnes, "Solar disinfection of water reduces diarrhoeal disease: an update," Archives of Disease in Childhood, vol. 81, no. 4, pp. 337–338, 1999.
36. R. M. Conroy, M. E. Meegan, T. Joyce, K. McGuigan, and J. Barnes, "Solar disinfection of drinking water protects against cholera in children under 6 years of age," Archives of Disease in Childhood, vol. 85, no. 4, pp. 293–295, 2001.
37. B. Sommer, A. Mariño, Y. Solarte et al., "SODIS—an emerging water treatment process," Journal of Water Supply Research and Technology—Aqua, vol. 46, no. 3, pp. 127–137, 1997.
38. S. C. Kehoe, T. M. Joyce, P. Ibrahim, J. B. Gillespie, R. A. Shahar, and K. G. McGuigan, "Effect of agitation, turbidity, aluminium foil reflectors and container volume on the inactivation efficiency of batch-process solar disinfectors," Water Research, vol. 35, no. 4, pp. 1061–1065, 2001.
39. M. B. Fisher, C. R. Keenan, K. L. Nelson, and B. M. Voelker, "Speeding up solar disinfection (SODIS): effects of hydrogen peroxide, temperature, pH, and copper plus ascorbate on the photoinactivation of E. coli," Journal of Water and Health, vol. 6, no. 1, pp. 35–51, 2008.
40. S. K. Mani, R. Kanjur, I. S. B. Singh, and R. H. Reed, "Comparative effectiveness of solar disinfection using small-scale batch reactors with reflective, absorptive and transmissive rear surfaces," Water Research, vol. 40, no. 4, pp. 721–727, 2006.
41. M. Saladin, "SODIS bags for humanitarian aid," March 2010, http://www.sodis.ch/index.
42. S. K. Lee, M. Sheridan, and A. Mills, "Novel UV-activated colorímetric oxygen indicator," Chemistry of Materials, vol. 17, no. 10, pp. 2744–2751, 2005.
43. A. Mills, S. K. Lee, and M. Sheridan, "Development of a novel UV indicator and dosimeter film," Analyst, vol. 130, no. 7, pp. 1046–1051, 2005.
44. E. Ubomba-Jaswa, P. Fernández-Ibáñez, C. Navntoft, M. Inmaculada Polo-Lópezb, and K. G. McGuigana, "Investigating the microbial inactivation efficiency of a 25 L batch solar disinfection (SODIS) reactor enhanced with a compound parabolic collector (CPC) for household use," Journal of Chemical Technology and Biotechnology, vol. 85, no. 8, pp. 1028–1037, 2010.
45. A. Vidal, A. I. Díaz, A. El Hraiki et al., "Solar photocatalysis for detoxification and disinfection of contaminated water: pilot plant studies," Catalysis Today, vol. 54, no. 2-3, pp. 283–290, 1999.
46. O. A. McLoughlin, P. Fernández-Ibáñez, W. Gernjak, S. Malato Rodriguez, and L. W. Gill, "Photocatalytic disinfection of water using low cost compound parabolic collectors," Solar Energy, vol. 77, no. 5, pp. 625–633, 2004.
47. C. Navntoft, P. Araujo, M. I. Litter et al., "Field tests of the solar water detoxification SOLWATER reactor in Los Pereyra, Tucumán, Argentina," Journal of Solar Energy Engineering, vol. 129, no. 1, pp. 127–134, 2007.
48. E. Ubomba-Jaswa, C. Navntoft, M. I. Polo-López, P. Fernandez-Ibáñez, and K. G. McGuigan, "Solar disinfection of drinking water (SODIS): an investigation of the effect of UV-A dose on inactivation efficiency," Photochemical and Photobiological Sciences, vol. 8, no. 5, pp. 587–595, 2009.
49. C. Lee, Y. Lee, and J. Yoon, "Oxidative degradation of dimethylsulfoxide by locally concentrated hydroxyl radicals in streamer corona discharge process," Chemosphere, vol. 65, no. 7, pp. 1163–1170, 2006.
50. D. M. Blake, P. C. Maness, Z. Huang, E. J. Wolfrum, J. Huang, and W. A. Jacoby, "Application of the photocatalytic chemistry of titanium dioxide to disinfection and the killing of cancer cells," Separation and Purification Methods, vol. 28, no. 1, pp. 1–50, 1999.
51. J.-M. Herrmann, "Heterogeneous photocatalysis: fundamentals and applications to the removal of various

types of aqueous pollutants," Catalysis Today, vol. 53, no. 1, pp. 115–129, 1999.

52. C. Karunakaran, S. Senthilvelan, S. Karuthapandian, and K. Balaraman, "Photooxidation of iodide ion on some semiconductor and non-semiconductor surfaces," Catalysis Communications, vol. 5, no. 6, pp. 283–290, 2004.

53. K. G. Kanade, J. O. Baeg, U. P. Mulik, D. P. Amalnerkar, and B. B. Kale, "Nano-CdS by polymer-inorganic solid-state reaction: visible light pristine photocatalyst for hydrogen generation," Materials Research Bulletin, vol. 41, no. 12, pp. 2219–2225, 2006.

54. M. I. Litter, "Heterogeneous photocatalysis: transition metal ions in photocatalytic systems," Applied Catalysis B, vol. 23, no. 2-3, pp. 89–114, 1999.

55. J. A. Byrne, B. R. Eggins, N. M. D. Brown, B. McKinney, and M. Rouse, "Immobilisation of TiO_2 powder for the treatment of polluted water," Applied Catalysis B, vol. 17, no. 1-2, pp. 25–36, 1998.

56. T. A. McMurray, J. A. Byrne, P. S. M. Dunlop, J. G. M. Winkelman, B. R. Eggins, and E. T. McAdams, "Intrinsic kinetics of photocatalytic oxidation of formic and oxalic acid on immobilised TiO_2 films,"Applied Catalysis A, vol. 262, no. 1, pp. 105–110, 2004.

57. T. Matsunaga, R. Tomoda, T. Nakajima, and H. Wake, "Photoelectrochemical sterilization of microbial cells by semiconductor powders," FEMS Microbiology Letters, vol. 29, no. 1-2, pp. 211–214, 1985.

58. C. Mccullagh, J. M. C. Robertson, D. W. Bahnemann, and P. K. J. Robertson, "The application of TiO_2photocatalysis for disinfection of water contaminated with pathogenic micro-organisms: a review,"Research on Chemical Intermediates, vol. 33, no. 3-5, pp. 359–375, 2007.

59. O. K. Dalrymple, E. Stefanakos, M. A. Trotz, and D. Y. Goswami, "A review of the mechanisms and modeling of photocatalytic disinfection," Applied Catalysis B, vol. 98, no. 1-2, pp. 27–38, 2010.

60. Z. Huang, P. C. Maness, D. M. Blake, E. J. Wolfrum, S. L. Smolinski, and W. A. Jacoby, "Bactericidal mode of titanium dioxide photocatalysis," Journal of Photochemistry and Photobiology A, vol. 130, no. 2-3, pp. 163–170, 2000.

61. M. Wainwright, "Methylene blue derivatives—suitable photoantimicrobials for blood product disinfection?" International Journal of Antimicrobial Agents, vol. 16, no. 4, pp. 381–394, 2000.

62. K. Sunada, T. Watanabe, and K. Hashimoto, "Studies on photokilling of bacteria on TiO_2 thin film,"Journal of Photochemistry and Photobiology A, vol. 156, no. 1–3, pp. 227–233, 2003. View at Publisher ·View at Google Scholar · View at Scopus

63. M. Cho, H. Chung, W. Choi, and J. Yoon, "Linear correlation between inactivation of E. coli and OH radical concentration in TiO_2 photocatalytic disinfection," Water Research, vol. 38, no. 4, pp. 1069–1077, 2004. View at Publisher · View at Google Scholar · View at Scopus

64. A. G. Rincón, C. Pulgarin, N. Adler, and P. Peringer, "Interaction between E. coli inactivation and DBP-precursors—dihydroxybenzene isomers—in the photocatalytic process of drinking-water disinfection with TiO_2," Journal of Photochemistry and Photobiology A, vol. 139, no. 2-3, pp. 233–241, 2001. View at Google Scholar · View at Scopus

65. A. G. Rincón and C. Pulgarin, "Field solar E. coli inactivation in the absence and presence of TiO_2 : is UV solar dose an appropriate parameter for standardization of water solar disinfection?" Solar Energy, vol. 77, no. 5, pp. 635–648, 2004. View at Publisher · View at Google Scholar · View at Scopus

66. D. M. A. Alrousan, P. S. M. Dunlop, T. A. McMurray, and J. A. Byrne, "Photocatalytic inactivation of E. coli in surface water using immobilised nanoparticle TiO_2 films," Water Research, vol. 43, no. 1, pp. 47–54, 2009. View at Publisher · View at Google Scholar · View at Scopus

67. P. S. M. Dunlop, T. A. McMurray, J. W. J. Hamilton, and J. A. Byrne, "Photocatalytic inactivation ofClostridium perfringens spores on TiO_2 electrodes," Journal of Photochemistry and Photobiology A, vol. 196, no. 1, pp. 113–119, 2008. View at Publisher · View at Google Scholar · View at Scopus

68. O. Sunnotel, R. Verdoold, P. S. M. Dunlop et al., "Photocatalytic inactivation of Cryptosporidium parvum on nanostructured titanium dioxide films," Journal of Water and Health, vol. 8, no. 1, pp. 83–91, 2010. View at Publisher · View at Google Scholar · View at Scopus

69. S. Gelover, L. A. Gómez, K. Reyes, and M. Teresa Leal, "A practical demonstration of water disinfection using TiO_2 films and sunlight," Water Research, vol. 40, no. 17, pp. 3274–3280, 2006. View at Publisher ·View at Google Scholar · View at Scopus

70. P. Fernández, J. Blanco, C. Sichel, and S. Malato, "Water disinfection by solar photocatalysis using compound parabolic collectors," Catalysis Today, vol. 101, no. 3-4, pp. 345–352, 2005. View at Publisher ·View at Google Scholar · View at Scopus

71. C. Sichel, J. Tello, M. de Cara, and P. Fernández-Ibáñez, "Effect of UV solar intensity and dose on the photocatalytic disinfection of bacteria and fungi," Catalysis Today, vol. 129, no. 1-2, pp. 152–160, 2007.View at Publisher · View at Google Scholar · View at Scopus

72. N. Miranda-García, M. I. Maldonado, J. M. Coronado, and S. Malato, "Degradation study of 15 emerging contaminants at low concentration by immobilized TiO_2 in a pilot plant," Catalysis Today, vol. 151, no. 1-2, pp. 107–113, 2010. View at Publisher · View at Google Scholar · View at Scopus

73. J. W. J. Hamilton, J. A. Byrne, C. McCullagh, and P. S. M. Dunlop, "Electrochemical investigation of doped titanium dioxide," International Journal of Photoenergy, vol. 2008, Article ID 631597, 2008. View at Publisher · View at Google Scholar · View at Scopus

74. R. Asahi, T. Morikawa, T. Ohwaki, K. Aoki, and Y. Taga, "Visible-light photocatalysis in nitrogen-doped titanium oxides," Science, vol. 293, no. 5528, pp. 269–271,

2001. View at Publisher · View at Google Scholar · View at Scopus

75. Q. Li, M. A. Page, B. J. Mariñas, and K. S. Jian, "Treatment of coliphage MS2 with palladium-modified nitrogen-doped titanium oxide photocatalyst illuminated by visible light," Environmental Science and Technology, vol. 42, no. 16, pp. 6148–6153, 2008. View at Publisher · View at Google Scholar · View at Scopus

76. P. Wu, R. Xie, J. A. Imlay, and J. K. Shang, "Visible-light-induced photocatalytic inactivation of bacteria by composite photocatalysts of palladium oxide and nitrogen-doped titanium oxide," Applied Catalysis B, vol. 88, no. 3-4, pp. 576–581, 2009. View at Publisher · View at Google Scholar · View at Scopus

77. J. A. Rengifo-Herrera and C. Pulgarin, "Photocatalytic activity of N, S co-doped and N-doped commercial anatase TiO_2 powders towards phenol oxidation and E. coli inactivation under simulated solar light irradiation," Solar Energy, vol. 84, no. 1, pp. 37–43, 2010. View at Publisher · View at Google Scholar · View at Scopus

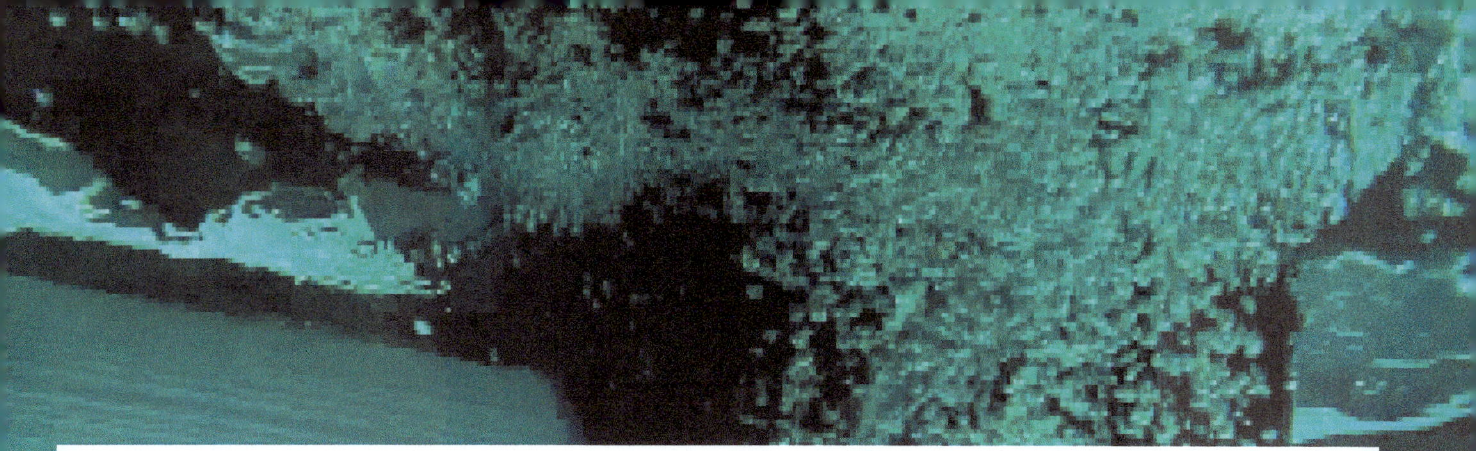

CHAPTER 16

HEALTH IMPACT OF COMMUNITY-BASED WATER TREATMENT SYSTEMS IN HONDURAS

Jeffery Deal

INTRODUCTION

A key component of the United Nations Millennium Development Goal Number 7 states "halve, by 2015 the proportion of the population (global) without sustainable access to safe drinking water and basic sanitation." Most waterborne diseases result in diarrhea which continues to be a leading cause of morbidity and mortality worldwide. According to World Health Organization data, using existing technologies approximately ten percent of the worldwide burden of disease would be removed by the water supply, sanitation, hygiene, and management of water resources, making water-related diseases arguably the most manageable set of health problems affecting humans [1].

A great deal of work has been done attempting to measure the impact of interventions to provide improved water sources at the household level, and less frequently at the community level. The overwhelming majority of these studies have also used either key informant or self-reporting of diarrhea (defined as three or more loose stools per day) as the measure of disease burden [2–4]. Reliance upon such nonobjective measures introduces a host of potentially confounding variables [2, 5, 6] and yet appears to have been used in all of the 2,120 published studies reviewed in a far reaching meta-analysis produced for the World Bank on diarrhea and water interventions [7]. While some of these deficiencies may be reduced by shortening recall time to seventy-two hours or less, potentially profound observer effects remain. Estimates of disease load changes are further impeded by researchers' concentration on known users of water systems rather than measurements on community disease levels of disease changes regardless of compliance, thus making extrapolations of disease rate changes inappropriate.

In 2006, Water Missions International (WMI) received a grant from the Pentair Foundation to provide improved water source access and toilet systems to all of the people in the district of Colon, Honduras, an area that contains approximately 340,000 people. The goal of 100% coverage utilizes a combination of solutions including home-based filtration systems (provided by a different NGO) for communities with less than 300 people, and a variety of high-capacity treatment systems for larger communities. In the initial phase of the study, all 604 water sources for the control and test communities were tested by pressure filtration methods, and all failed quality testing by being positive for coliform bacterial growth. For all households that lacked adequate sanitation facilities and agreed to assist in installation, sanitary pit latrines with

Citation: Jeffery Deal, "Health Impact of Community-Based Water Treatment Systems in Honduras," Journal of Anthropology, vol. 2011, Article ID 929860, 5 pages, 2011. doi:10.1155/2011/929860

pour-flush toilet (toilets with water traps and no reservoir that are flushed by pouring a bucket of water into the basin) were also provided. Water treatment systems and pour-flush toilets deployed during the present study were developed and manufactured by Water Missions International, a nonprofit organization based in Charleston, SC, that works to provide sustainable water treatment capacities and sanitation facilities for people in developing countries. The technology uses a combination of multimedia, multistage filters, and chlorination to provide treated water for drinking and cooking that meets the most stringent class of WHO drinking water standards as well as passed present US standards from the EPA—specifically, tests on treated water grew no coliform bacteria on repeated, monthly testing during the period of this study. In addition, WMI provides community development programs that include education and microenterprise strategies to assure sustainability of these interventions.

In the baseline study phase of the project, water sources for 613 communities were identified. Various water quality tests were conducted on the community water sources by standard membrane filter test. High counts of coliform bacteria indicative of fecal contamination were found in 100% of the water sources. As previously published [8], initial stool tests also showed that 29.3% (53 of 181) of the volunteer subjects from the twelve communities that had been randomly selected from the state of Colon carried at least one of the three tested protozoan parasites. Also as reported previously, the prevalence of positive protozoan parasite levels was significantly lower in the intervention groups as compared to the test group. Even greater reductions of disease rates (at least 52%) were noted in the medical chart reviews of visits to the local health facility for diarrhea and dysentery as well as self-reporting of the same diseases via ethnographic interviews. Ethnographic data further suggested widespread acceptance and community-wide reproduction of the awareness of health benefits derived from consuming treated water. The present paper is a followup to that report primarily looking at the development of positive parasite stool tests in the same communities twelve months after the initial round of tests with an expanded study population.

METHODS

Twelve communities from three different categories were randomly selected from the Colon district. Four communities had not yet obtained a water treatment system and were used collectively as the Control Group. Four other communities where water systems (Water Only Group) had been deployed were entered into the study as were four more communities where both water systems and sanitary flush latrines had been installed (Water and Sanitation Group). For both the initial tests in 2009 and the final round of tests in 2010, volunteers were recruited by poster advertising and via community leaders. Methods of recruitment were identical in all communities. All subjects who tested positive for protozoan antigens in either study (2009 or 2010) were treated with an appropriate dose of tinidazole (500 mgs per day for three days for adults with weight adjusted dosing for children). Subjects were directly observed to take the initial dose, which in management of Giardiahas been shown to be highly effective without the additional doses. No side effects of treatment were reported. One hundred and sixty-three of the 200 subjects in the final round of testing in 2010 had also participated in the prior studies conducted 12 months earlier.

The effect of previous medical treatment of subjects upon the present study is to provide a population that either tested negative or were given highly effective treatment in 2009—thus providing a subpopulation which was believed to have begun the 12-month test period free of any of the three tested protozoans. The results among the subjects tested both times, therefore, may be regarded as the rates of reinfection over the 12-month period (or a one-year incidence rate) for all three groups.

Recent advances in highly sensitive and specific rapid immunoassays for waterborne parasite diseases have made field testing of individual fecal specimens now possible [9, 10]. These devices test for species-specific antigens of common parasites known to be primarily waterborne. The device chosen for this study tested for three protozoan parasites: Giardia lamblia (now widely known as Giardia intestinalis), Entamoeba histolytica/Entamoeba dispar, and Cryptosporidium parvum antigens. Previous work has shown these tests to have both specificity and sensitivity in excess of 96% for the aforementioned pathogens.

Immunoassay of stool for these waterborne parasites was used as an indicator that the subject had been exposed to waterborne pathogens and was therefore at risk of these and other infectious waterborne illnesses. A separate subset of 163 subjects from the three groups who also gave specimens twelve months earlier was analyzed as a separate subgroup. Given the highly effective cure rates of tinidazole for Giardia and Entamebaand the fact that the vast majority of non-immunocompromised subjects will clear Cryptosporidium infections spontaneously, this subset is thought to represent recolonization rates with waterborne protozoan during the year after the initial round of testing and treatment. All specimens were tested within 12 hours of collection using the Triage Micro Parasite Panel manufactured by Biosite Incorporated.

Further information regarding diarrhea and dysentery rates was obtained by reviewing medical records from a public health clinic in a community where a water treatment system had been previously installed. Ethnographic data was collected using a combination of KAP surveys and guided interviews. The medical records review and the majority of the ethnographic data have been previously reported in this journal [8].

ROLE OF THE FUNDING SOURCE AND ETHICS REVIEW

Water Missions International maintains a country program in Honduras whose staff provided support and significant amount of labor for this project. The study design, collection, and analysis of data and interpretation of the data were the sole responsibility of the author.

Prior to initiation of the study, the Colon Minister of Health and the Institutional Review Board of Water Missions International reviewed and gave approval and consent for the project and study. Consistent with this review, no information from medical chart reviews which could identify subjects of the study was retained outside of the local healthcare facility. Under the supervision of a licensed physician, all individuals in whom potential pathologic parasites were found were given free treatment with regimens previously approved by the Colon Minister of Health. The control communities where no water treatment or sanitation facilities existed were selected from a preexisting construction queue and intervention was not withheld as a result of this study. Verbal consent was obtained and recorded from all subjects.

RESULTS

Age distributions within the three groups are seen in Tables 1 and 2. Gender distribution is seen in Table 3. Parasite test results for the control group compared to the combined water only group and water and sanitation group are seen in Table 4. Giardia and Entameba accounted for all but one positive test in all categories. Giardiaaccounted for 46% and Entameba 48% of the positive tests while Cryptosporidium remained rare at 6% of the totals.

Table 1: Age distribution of all subjects in all groups.

	Min.	Max.	Mean	Std. deviation
Age	2	82	23.06	19.143

Table 2: Mean age distribution by group.

	Mean (years)	N	Std. deviation
Control group	22.67	67	18.806
Water only group	24.79	68	20.662
Water and sanitation group	21.66	65	17.959

Table 3: Gender by intervention group.

	Control group	Water only Group	Water and sanitation group	Totals
Female	44	42	39	125
Male	23	26	26	75
Total	67	68	65	200

Table 4: 2010 data comparing control group to the combination of water only group and water and sanitation group.

	Controls	Combined water only and water and sanitation	Totals
Negative for any parasite	50	126	176
Positive for any parasite	17	7	24
Total	67	133	200

Subjects living in communities that did not have access to water systems had significantly higher rates of positive tests than subjects who had either access to water or who had access to water and flush toilets both in the initial survey of 2009 and in the 2010 followup (). These finding are summarized in Figure 1.

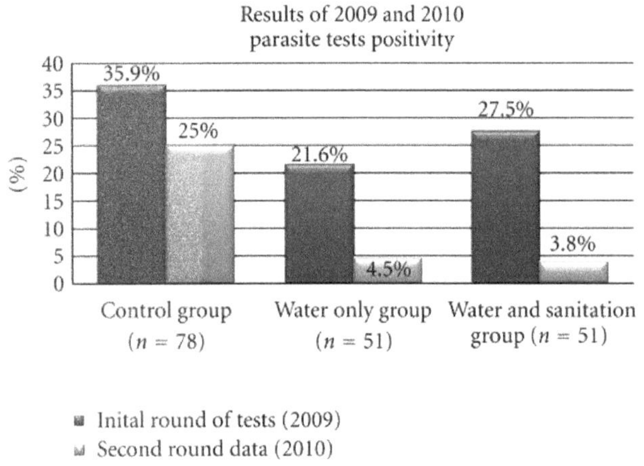

Figure 1

In 2009, a comparison of the rate of positive parasite tests appeared to demonstrate that, while access to a water treatment system reduced parasite levels, communities that had both treatment systems and installed flush toilets demonstrated a higher rate of positive parasite tests. These findings show a distinct gender bias toward women and are summarized in Figure 2. Additional ethnographic witnessing suggested that this finding may possibly be explained by the fact that women exclusively cleaned the toilets, often with inadequate supplies and protection. Water Missions International responded to this suggestion with additional training and supplies. In the followup parasite survey of 2010, what had appeared to have been a negative effect of the toilet systems was no longer present, as seen in Figure 2.

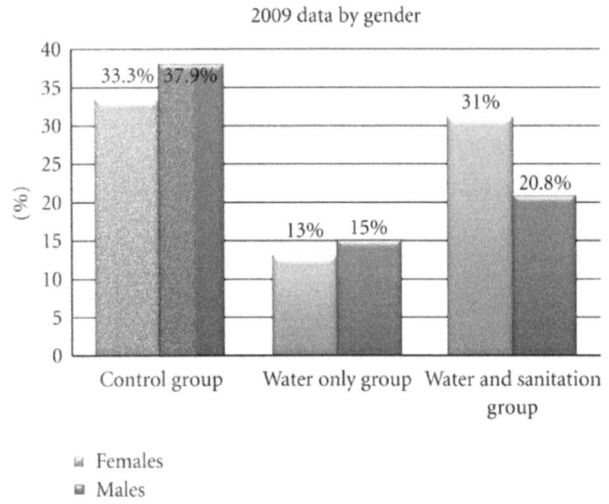

Figure 2

A separate analysis of parasite test results including only subjects who were available in both 2009 and 2010 is summarized in Figure 3. Of interest is that the apparent negative effect of the toilets in the 2009 data (Figure 1) completely vanishes when subjects lost to follow up in 2010 are removed from the 2009 analysis.

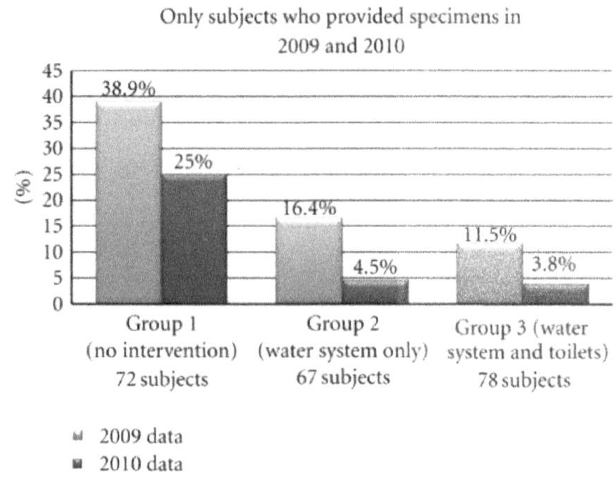

Figure 3

DISCUSSION

The present paper is a followup of the research previously reported [8] and adds support to the conclusion that access to community-based water treatment systems and flush toilets reduced the disease load in this region of Honduras. To our knowledge, these are the first studies to combine self-reported data, medical chart review, and stool immunoassays as an indicator of exposure to potential waterborne pathogens. The triangulation of these methodologies provides powerful support to what are otherwise strongly subjective and questionable measures of disease loads from waterborne pathogens.

The previously reported ethnographic data from these communities suggests a high level of understanding of the causes and prevention of diarrhea among the communities studied. The overwhelming majority (130 of 142) of the people interviewed attributed the majority of their diarrheal diseases to water and sanitation issues and improvement of the condition to improved water sources and access to flush toilets. There were also significant signs of a shift of ideations regarding drinking untreated water toward an appreciation of the importance of purified water and prohibitions against drinking untreated water with the addition of water treatment facilities and community education.

Ethnographic data found during this study suggests that, consistent with other similar work, the availability of improved water is felt by its recipients to improve a general sense of health and well-being. High levels of knowledge related to water issues exist in this area of Honduras which could be attributed to many factors including sophisticated public health efforts, high literacy rates comparable to the region, widespread health education in public schools, and the training offered by Water Missions International and other NGOs. Local indigenous belief systems appear uncommonly (mentioned in only 6 of 46 individual interviews and in none of the four focus groups) and for no one were they the basis for a preferred method of treatment.

Immunoassay evidence of decreased prevalence of waterborne parasites strongly supports the contention that community-based water treatment facilities reduce the overall stool parasite load, at least of the three protozoan species tested. Since all subjects who tested positive for either Giardia or Entameba were treated with three doses of tinidazole adjusted for age and weight, the follow-up study of all subjects who submitted stool samples initially is felt to represent the reinfection rates, in essence eliminating concerns regarding residual colonization from exposures that occurred prior to initiation of this study. Previous treatment of subjects who tested positive for any parasite creates the potential for a Hawthorne effect; however, the elevated number of positive tests initially found within the control communities would potentially bias this group more than the test groups and would tend to lessen rather than strengthen the differences found.

Also, as previously reported [8], when parasite antigens were detected in stool samples of individuals who had access to improved water sources, ethnographic investigation revealed lapses of behavior in spite of the high level of understanding of the risks associated with drinking from untreated sources. Further analysis of the interviews of subjects whose stool was positive for potential waterborne parasites suggests that risk and time management decisions rather than cultural or knowledge-based differences accounted for lapses in behavior and willingness to drink untreated water. Subjects reported that the time required to obtain treated water, sometimes a difference of only a minute or less, was too great to overcome their concerns with potential health risks associated with untreated tap water.

Multiple pathogens and inflammatory conditions cause diarrhea, making monitoring this symptom alone an inexact measure of the disease load related to water quality. Worldwide, the most common causes of diarrhea are viral infections such as the rotavirus, an ubiquitous infection that may be transmitted by personal contact. Food contamination and noninfectious inflammatory diseases add to the diarrhea prevalence. Though not precisely known, the number of diarrhea cases unrelated to waterborne pathogens is likely substantial. This means that the 52% drop in diarrhea rates noted in the previously reported community chart reviews may represent an even greater majority of the cases that could possibly be related to potential water and sanitation issues. This follow-up study strongly supports this finding and suggests that the effect where water treatment systems are maintained may even increase over time.

A comparison of the 2009 parasite test data excluding those lost to follow up a year later was dramatically different from when these individual tests were included (Figure 3). This suggests that the population that was lost to follow up significantly added to the initial rate of positive tests. This phenomenon deserves future scrutiny and incorporation into discussions of the idea of community as a social construct. If community is defined as those present at a given point in time, we see a negative impact from presence of the toilets. When we define community as those who reside in the geographic area for at least one year we find the exact opposite as this apparent negative impact is not detected. As this finding demonstrates, the unit of analysis remains paramount in such studies.

This combination of qualitative data, health records reviews, and immunoassays provides compelling evidence that community-based water treatment facilities with or without providing flush toilets significantly reduced the burden of diseases in the communities of Colon, Honduras. We further validate with objective measures prior work based upon self-reporting of diarrhea rates. Finally, this data suggests that interventions to provide potable water access on a community level when combined with community development efforts and sanitation can play a significant role in the reduction of mortality and morbidity from waterborne diseases and associated comorbidities.

CONCLUSION

Providing access to water treatment or water treatment and flush toilets significantly reduced the one-year incidence of positive test results for the three protozoan species tested. This finding combined with the previously reported ethnographic and medical chart review data provides compelling evidence that such interventions significantly reduce the disease load from waterborne pathogens within this population. Furthermore, the finding that initial results are significantly different, even in the initial round of testing, if individuals who are not followed up are eliminated from the analysis has profound methodological implications which warrant further investigation.

Adding a temporal definition of community resulted in a completely different finding regarding the impact of supplying flush toilets, demonstrating the need for precise definitions of community in future studies.

The method used here where objective measurements of health effects are coupled with more traditional anthropological tools may serve as a template for future studies in medical anthropology.

REFERENCES

1. A. Prüss-Üstün, R. Bos, F. Gore, and J. Bartram, Safer Water, Better Health: Costs, Benefits and Sustainability of Interventions to Protect and Promote Health, World Health Organization, Geneva, Switzerland, 2008.
2. S. Boisson, W. P. Schmidt, T. Berhanu, H. Gezahegn, and T. Clasen, "Randomized controlled trial in rural Ethiopia to assess a portable water," Environmental Science and Technology, vol. 43, no. 15, pp. 5934–5939, 2009.
3. R. J. Gelting and L. Ortolano, "A model describing performance of rural drinking water systems in Honduras," International Journal of Water Resources Development, vol. 14, no. 2, pp. 199–215, 1998.
4. U. D. Parashar, E. G. Hummelman, J. S. Bresee, M. A. Miller, and R. I. Glass, "Global illness and deaths caused by rotavirus disease in children," Emerging Infectious Diseases, vol. 9, no. 5, pp. 565–572, 2003.
5. D. Blum, R. N. Emeh, S. R. Huttly et al., "The Imo State (Nigeria) Drinking Water Supply and Sanitation Project, 1. Description of the project, evaluation methods, and impact on intervening variables,"Transactions of the Royal Society of Tropical Medicine and Hygiene, vol. 84, no. 2, pp. 309–315, 1990.
6. T. Clasen, I. Roberts, T. Rabie, W. Schmidt, and S. Cairncross, "Interventions to improve water quality for preventing diarrhoea," Cochrane Database of Systematic Reviews, vol. 3, p. CD004794, 2006.
7. L. Fewtrell and J. M. Colford Jr., "Water, sanitation, and hygiene interventions to reduce diarrhoea in less developed countries: a systematic review and meta-analysis," The Lancet Infectious Diseases, vol. 5, no. 1, pp. 42–52, 2005.

8. Jeffery L. Deal, et al., "A multidimensional measure of diarrheal disease load changes resulting from access to improved water sources in honduras," Practicing Anthropologist, vol. 32, no. 1, pp. 15–20, 2010.
9. L. S. Garcia, R. Y. Shimizu, and C. N. Bernard, "Detection of Giardia lamblia, Entamoeba histolytica/Entamoeba dispar, and Cryptosporidium parvum antigens in human fecal specimens using the triage parasite panel enzyme immunoassay," Journal of Clinical Microbiology, vol. 38, no. 9, pp. 3337–3340, 2000.
10. S. E. Sharp, C. A. Suarez, Y. Duran, and R. J. Poppiti, "Evaluation of the Triage Micro Parasite Panel for detection of Giardia lamblia, Entamoeba histolytica/ Entamoeba dispar, and Cryptosporidium parvum in patient stool specimens," Journal of Clinical Microbiology, vol. 39, no. 1, pp. 332–334, 2001.

CHAPTER 17

IDENTIFICATION OF BACTERIA IN DRINKING AND PURIFIED WATER DURING THE MONITORING OF A TYPICAL WATER PURIFICATION SYSTEM

Vessoni Thereza Christina Penna, Silva Alzira Maria Martins and Priscila Gava Mazzola

Keywords: Water Purification System, Pseudomonas, Minimal Inhibitory Concentration (MIC), Gram-negative non-fermenting bacteria

BACKGROUND

Water is one of the most important elements for all forms of life and is indispensable in the maintenance of life on Earth and essential for the composition and renewal of cells. Water represents 70% of our body, participates in the composition of our tissues, and transports the most diverse substances throughout our organism. Notwithstanding, human beings increasingly continue to pollute the reserves which still remain, provoking illnesses that can jeopardize the population [1].

Bioaccumulation, sewerage, agricultural, industrial, radioactive, and thermal residues are the principal polluters of water on our planet.

Purified Water

Water is essential for industrial, pharmaceutical and hospital purposes, in the preparation and processing of medicines and other health products. In the majority of cases, water is an input, which should be incorporated into the product during processing. At other times, even if it is not present in the preparation, it is especially used for cleaning and hygiene purposes. It is recognized that the greatest demand on water is destined for human consumption, its quality being relatively guarantied up to the point where the pipe transportation network terminates.

Although potable water is suitable for human consumption, it does not guarantee that the so-called potable quality could also be used in industrial installations, equipment, preparation of medicines, foodstuffs, cosmetics, chemopharmaceutical materials, or health center units for the cleaning and washing of semi-critical areas or devices prior to the application of disinfection or sterilization procedures.

For this reason, every industrial or pharmaceutical plant related to health products must rely on appropriate water purification systems, allowing it to meet its particular requirements, especially as to the problems related to storage and internal distribution. This procedure must guarantee supply according to the volume required and pursuant to the demanded quality consumption points.

Purified water is used as an expedient in the production of official preparations, pharmaceutical (industrial, and in health center units) applications such as the cleaning of semi-critical devices, areas and equipment, as well as in the preparation of pharmaceutical chemicals and bacteriological media. It is also used as the principal component in peritoneal dialyses solutions in hospitals, in nutrient solutions (including nursing bottles) and liquid enteral nutrient solutions, prepared in the hospital

Citation: Vessoni Thereza Christina Penna,,Silva Alzira Maria Martins and Priscila Gava Mazzola, Identification of bacteria in drinking and purified water during the monitoring of a typical water purification system, DOI: 10.1186/1471-2458-2-13

lactarium and administered to children and weak patients who can eat regular solid food.

Purified water is obtained from drinking water [1] through a typical water purification system of unit operations, meeting the standards set forth by the 1978/1990 directives issued by the Brazilian Ministry of Health [2]. Purified water systems must be validated in order for water to meet the requirements [3]: for the purity of ionic (conductivity ≤ 1.3 μS/cm, at 25°C) and for total organic compounds (TOC < 0.5 mg/L), and must even be protected from microbial proliferation (total enumeration < 100 CFU/mL), preventing pyrogen formation. The bacteriological standard ≤ 1.0 Endotoxin Unit (EU/mL) is required for sterile purified water used in the washing of critical devices before autoclaving.

The purified water system that produces, stores, and circulates water under background conditions is «susceptible to the establishment of adhesive biofilms or microorganisms», which can be the source of undesirable levels of viable microorganisms or endotoxins in the effluent water. Recent studies have shown that nearly all large water purification systems can cause the formation of biofilm in the piping. This biofilm can spread microorganisms within the system and contribute to an increase in particles, bacteria, and the level of total organic carbon (TOC). Contamination can affect the whole process in the pharmaceutical industry or hospital environment. These systems require frequent sanitation and microbiological monitoring to ensure water of the appropriate microbiological quality (microbial limit at the points of use) [3].

Monitoring data should be analyzed on an ongoing basis so as to ensure that the process continues to be performed within acceptable limits. It should be recognized that the microbial alert and action levels established for any pharmaceutical water system are obligatorily linked to the monitoring method chosen. Using the recommended methodologies generally considered, appropriate action levels are 500 colony-forming units(CFU) per mL of drinking water, 100 CFU/mL of purified water, and 10 CFU/mL of water for injection (WFI); the limits for pyrogen are < 1.0 Endotoxin Unit/mL of purified water, < 0.5 EU/mL of WFI and < 0.25 EU/mL of sterile WFI [2, 3]. For gram-negative fermenting bacteria in drinking water, the standards show that total coliforms must be fewer than 10^{-2} CFU/mL to that of drinking water. Neither the Brazilian Federal standards nor the USP 24 [3] includes levels for gram-negative non-fermenting bacteria, such as the*Pseudomonas* species, which are the principal cause of biofilms and enterotoxins [4] in purified water.

During the performance of a typical water purification system, the purposes of the present work were: (i) to examine the efficiency of each treatment stage; (ii) to characterize the isolated bacteria at each stage, from the storage tank of publicly supplied water up to the final point of purified water consumption. The minimum inhibitory concentration (MIC) of the main chemical agents used in the disinfection of the water purification system over the isolated and identified bacteria was also studied.

METHODS

Water Samples

Samples from thirteen points of a typical water purification system (10,000 L/day) were analyzed in triplicate. Three analysts collected samples of water from every point indicated in the Flow Sheet of a typical USFilter water purification system [5], placed in the Edwards Lifesciences Macchi, SP, Br.

Before sampling the water, every sampling tap (point) was made sanitary with ethyl 70% alcohol, the valve was opened and the water was allowed to flow (approximately 3 L/min) freely for about 60 seconds, followed by the taking of a sample of 100 mL of the water in a sterile polyethylene bag (Millipore, Bedford, MA, USA).

Flow sheet of a typical water purification system

Point 1: Storage Tank (feed water)

Point 2: Two multimedia Filters (primary filtration)

Point 3: Two water softeners (hardness reduction)

Point 4: One filter of activated carbon (chlorine removal)

Point 5: One 5.0 μm filter (removal of particle materials)

Point 6: One reverse osmosis membrane system (removal of organic and inorganic substances)

Point 7: One continuous deionization column (removal of dissolved minerals and salts)

Point 8: One storage tank (treated water)

Point 9: Light UV: 254 nm (reduce TOC)

Point 10: Three 0.05 μm filters in parallel (removal of particles and bacteria)

Points 11, 12, 13: Loop of distribution of purified water for consumption

The water purification system presented in the Flow Sheet [5] included the following stages and subsequent apparatus, from where the water points were sampled:

Point 1: The storage tank was fed with drinking water distributed by SABESP [1] (Basic Sanitation Company for the State of São Paulo). The polyethylene storage tank has a closed cover and a capacity of 10,000 L H_2O designed with a siphon valve fitted at the bottom of the tank outlet. It is also fitted with a centrifugal pressure pump (stainless steel, 40 L/min/ 414 kPa), which pressurizes the water to point 02 across the multimedia filters.

Point 2: Two multimedia filters parallel to each other (US Filter, rlzms12 fxxfa, Warrendale, PA, USA) offers a highly efficient removal of suspended fragmented matter from the water. The three layers of media (sand, anthracite and quartz) are selected in accordance with their particular size, specific gravity, and ability to trap particles of specific size ranges (≥ 10 micra). As the water flows downwards through the bed, it finds a layer of media with decreasing porousness/permeability so that successively smaller particles are trapped in each layer, providing depth filtration.

Point 3: Two water softeners of an alternated sodium resin (US Filter, rlzsd12fxxya) which remove hard minerals from the water. Ion exchange water softening exchanges the calcium and magnesium cations from water with an equivalent number of sodium cations.

Point 4: One filter of activated carbon (US Filter relzcs12fxxfa) is used to remove chlorine, chloramines, and dissolved organic substances from the water. Carbon filters are frequently used as the pretreatment to osmosis membranes and ion exchange resins, avoiding damage by oxidant substances, such as chlorine.

Point 5: One 5.0 micra polyethylene microporous depth screen filter (US Filter, fcrof2005) is often used ahead of other water purification operations, such as deionization, and reverse osmosis, as a polishing filter for removing resin, carbon fine colloids, and microorganisms.

Point 6: One reverse osmosis membrane system (polyamide polymers, US Filter, rsolv204021133)- The natural process of osmosis occurs when a solution with different concentrations of salts is separated by a semi-permeable membrane. As osmotic pressure drives the water through the membrane, the water dilutes more concentrated solutions, until an equilibrium is achieved. The permeate pure water is collected on the downstream side of the membrane. Reverse osmosis removes 90% – 99% of particles, colloids, bacteria, pyrogens, dissolved organic and inorganic substances greater than 200–300 molecular weight (MW) range or larger than the membrane's pore size of 150 to 200 angstroms. The conductivity of the water at the inlet is 150 μS and at the outlet is 5 μS. Four membranes (10.16 cm × 101.6 cm) with an outlet capacity of 16 L/min are used. The membrane›s shelf life is between 2 and 3 years and sanitation is carried out through the association of peracetic acid with a hydrogen peroxide solution.

Point 7: One continuous deionization (US Filter, cdi = ccdis20 × 1) column that removes dissolved minerals and salts, as well as some dissolved organic matter, from the water stream crossing ion exchange resins. The ion exchange operation removes positively charged cations, such as calcium, magnesium, and sodium from the water by cation exchange resins, which are replaced by hydrogen ions. Negatively charged anions such as chloride, nitrate, and silica are removed from the water by strong based anion exchange resins, and hydroxide ions then form water molecules. The water stream passes through a mixed bed of cation and anion exchange resins, which produces a very high quality of water with a resistance of up to 1.3 μs/cm (18.3 megohm-cm) at 25°C.

Point 8: A storage polyethylene tank (1500 L) with the treated water, of which the reservoir has a closed cover design with a spray ball at the top to spread the water, keeping it in continuous movement so as to prevent any hazard of contamination. The tank is also provided with a siphon valve fitted at the bottom of the tank outlet, a pressure relief, and a 0.2 μm ventilation filter. The loop of distribution for the points of use is accomplished through two pumps.

Point 9: Ultraviolet Light (λ = 254 nm, US Filter, SL-1) is used as a final step in the treatment for the purpose of preventing the growth of microorganisms, and reducing total organic carbon (TOC).

Point 10: Three 0.05 μm filters are set parallel to each other (US Filter, zha 153107). Microporous filters are used to remove particles, and bacteria, ranging from 0.05 to 0.5 μm contaminants, which would not ordinarily be removed by depth filtration.

Points of use 11,12, and 13: Every point of use is provided with 3 filters of 0.05 μm set parallel to each other. From those points, the purified water for consumption is provided by a loop of distribution and is used for the cleaning of critical devices (membrane oxygenators and PVC tubes), the washing of semi critical areas, the preparation of chemical solutions, and the culture bacteriological media.

Culture media, enzymatic and biochemical reactions efficiency test

The culture media efficiency test was carried out with standard strains: *Aspergillus niger* ATCC 16404, *Bacillus subtilis* ATCC 6633, *Candida albicans* ATCC 10231, and *Pseudomonas aeruginosa* ATCC 27853, *Escherichia coli*ATCC 25922, pursuant to USP 24 [3]. The evaluation of the enzymatic and biochemical reactions, performed in the identification of the isolated colonies, were carried out with standard strains of *Pseudomonas aeruginosa*ATCC 27853, *Aeromonas hydrophila* ATCC 35654, *Alcaligenes faecalis* ATCC 35655, and *Flavobacterium multivorum* ATCC 35656.

Filtration of sample

The samples were submitted to the filtration method by 0.45 μm membranes in a class 100 laminated flow chamber. This was followed by an incubation period of 72 hours at 30–35°C (Tryptic Soy Agar, TSA, Difco, Detroit, Michigan, USA) for total bacteria enumeration (heterotrophic), and for 22 hours at 34.5–35.5°C (m Endo Agar Less, Difco) for total coliform enumeration, in accordance with the United States Pharmacopoeia National Formulary, in the chapter: «Water For Pharmaceutical Purposes USP 24/NF 2000» [3]. Every sampling tap (point) was analyzed in triplicate for heterotrophic bacterial counts and total coliforms. In order to stimulate the growth of stressed and chlorine-tolerant bacteria, we just started using R2A Agar to improve the enumeration of heterotrophic organisms in treated potable water.

Isolation of colonies

From each TSA culture, the colonies were transferred to the surface of the Cetrimide (Difco) in plates and incubated at 30–35°C for 18–24 h. The developed colonies were then submitted to oxidase and indol tests, followed by biochemical genera and species identification tests. In every plate, all the developed colonies were isolated for identification. The criteria used were colony morphology and colony color. The total number of colonies found at each sampling point (CFU/100 mL) and picked for identification, are shown in Tables 01&2.

Table 1 Total number (and percentage) of identified genera and species of gram-negative bacteria through the percentage of the positive reactions from the isolated colonies.

Microorganisms	**strains**	**% of identification**
Pseudomonas aeruginosa	25	32.05
Pseudomonas picketti	18	23.08
Pseudomonas vesiculares	10	12.82
Pseudomonas diminuta	09	11.54
Flavobacterium aureum	05	6.42
Pseudomonas fluorescens	04	5.13
Acinetobacter lwoffi	02	2.56
Pseudomonas putida	02	2.56
Pseudomonas alcaligenes	01	1.28
Pseudomonas paucimobilis	01	1.28
Flavobacterium multivorum	01	1.28
Enumeration	78	100.0

Table 2 Total enumeration (CFU/100 mL) and identification of Gram-negative non-fermenting bacteria from the genera:*Pseudomonas, Flavobacterium, Acinetobacter* identified by the BBL Crystal kit (Becton Dickinson) and API 20 NE kit (bioMérieux), at every sampling point.

Sampling points/Flow sheet	***CFU/100 mL Average ± SD***	***Colonies***		***identification***	
		Morphology & color	***Number of isolates***	***BBL Crystal***	***API 20 NE***
Point 1 Feed Water (storage)	507 ± 50	Shining beige	02	*F. multivorum*	*P. vesiculares*
		Circular Brown	02	*P. aeruginosa*	*P. alcaligenes*
		Circular pumpkin	02	*P. diminuta*	*P. diminuta*
Point 2 Two multimedia Filters	33 ± 12	Circular pumpkin	02	*P. diminuta*	*P. diminuta*
		Circular pumpkin	02	*P. diminuta*	*P. diminuta*
		Circular pumpkin	02	*P. diminuta*	*P. diminuta*
Point 3 Two water softeners	173 ± 50	Circular pink	02	*F. aureum*	*P. vesiculares*
		Circular pink	02	*F. aureum*	*P. vesiculares*

		Circular pink	02	*F. aureum*	*P. vesiculares*
Point 4 Activated carbon filter	897 ± 823	*Circular brown*	02	*P. aeruginosa*	*P. pcketti*
		Circular white	02	*F. aureum*	*P. vesiculares*
		Circular pink	02	*P. vesiculares*	*P. vesiculares*
Point 5 One 5 micron filter	100 ± 20	Circular white	02	*F. aureum*	*P. vesiculares*
		Circular brown	02	*P. aeruginosa*	*P. picketti*
		Circular brown	02	*P. aeruginosa*	*P. aeruginosa*
Point 6 Reverse Osmosis	8 ± 2	Circular beige	02	*P. vesiculares*	*P. vesiculares*
		Circular brown	02	*P. aeruginosa*	*P. picketti*
		Circular brown	02	*P. aeruginosa*	*P. picketti*
Point 7 Continuous deionization	653 ± 344	Circular brown	02	*P. aeruginosa*	*P. fluorescens*
		Circular brown	02	*P. aeruginosa*	*P. picketti*
		Circular brown	02	*P. aeruginosa*	*P. aeruginosa*
Point 8 Storage tank for the purified water.	40 ± 35	Circular brown	02	*P. aeruginosa*	*P. picketti*
		Pointed yellow	02	*A. lwoffi*	*P. diminuta*
		Pointed cream	02	*P. fluorescens*	*P. putida*
Point 9 Light UV	27 ± 12	Circular brown	02	*P. aeruginosa*	*P. picketti*
		Pointed cream	02	*P. paucimobilis*	*P. putida*
		Circular brown	02	*P. aeruginosa*	*P. picketti*
Point 10 0.05 micron Filter	87 ± 31	Circular brown	02	*P. aeruginosa*	*P. picketti*
		Circular brown	02	*P. aeruginosa*	*P. picketti*
		Circular brown	02	*P. aeruginosa*	*P. picketti*
Point 11: use	27 ± 12	Circular brown	02	*P. aeruginosa*	*P. fluorescens*
		Circular brown	02	*P. aeruginosa*	*P. picketti*
		Circular brown	02	*P. aeruginosa*	*P. picketti*
Point 12: use	27 ± 12	Circular brown	02	*P. aeruginosa*	*P. picketti*
		Circular brown	02	*P. aeruginosa*	*P. picketti*
		Circular brown	02	*P. aeruginosa*	*P. picketti*
Point 13: use	27 ± 12	Circular brown	02	*P. aeruginosa*	*P. picketti*
		Circular brown	02	*P. aeruginosa*	*P. picketti*
		Circular brown	02	*P. aeruginosa*	*P. picketti*

SD = standard deviation ($n = 3$; $p < 0.05$)

Identification tests

The identification tests for genera and species were performed by using two standardized micro-method systems: (i) API 20 NE kit from bioMérrieux that was used for the identification of non-fastidious gram-negative rods [6] not belonging to the *Enterobacteriacea* family; (i) the BBL crystal enteric/non-fermenting ID kit from Becton & Dickinson that was used for the identification of aerobic gram-negative bacteria that belong to the family of *Enterobacteriacea* as well as some of the more frequently isolated glucose fermenting and non-fermenting gram-negative bacilli.

The micro-methods employed are modifications of the classical methods and the basic principles of the reactions follow the biochemical and enzymatic standards described for the enteric/non fermenting bacteria by Murray *et al.*[4].

The more specific API 20 NE kit is comprised of 20 microtubes containing dehydrated media and/or substrates, combining 8 conventional tests and 12 assimilation tests. The interpretation of the reactions were made according to the reading tables using the identification software with BBL crystal or API 20 NE data base.

The biochemical tests were employed in the bioMerieux identification system and the respective reactions were: (a) potassium nitrate (NO_3) for the reduction of nitrate to nitrite, (b) tryptophan (trp) for indol production, and (c) glucose (glu) for medium acidification. The substrates used for enzymatic hydrolysis reactions were: (d) arginine (adh) for arginine dihydrolase, (e) urea (ure) for urease, (f) esculin (esc) for β-glucosidase, (g) gelatin (gel) for protease, and (h) p-nitrophenyl-β-D-galactopyranoside (PNPG). For assimilation tests, the substrates employed were the following: glucose (GLU), arabinose (ARA), mannose (MNE), mannitol (MAN), N-acetyl-glucosamine (NAG), maltose (MAL), gluconate (GNT), caprate (CAP), adipate (ADI), malate (MLT), citrate (CIT), and phenyl-acetate (PAC). When the substrate was assimilated by the microorganism, its growth was detected by visible turbidity in the corresponding medium. The reactions were visualized by the addition of reactive agents and a change in the indicator.

The BBL crystal kit is a miniature method of identification that comprises modified conventional and chromogenic substrates, which are distributed in 30 dehydrated enzymatic and biochemical substrata in panels. The biochemical tests used the following carbohydrates: glucose, arabinose, mannose, sucrose, melibiose, rhamnose, sorbital, mannitol, and inositol. The utilization of carbohydrate results in lower pH and a change in the indicator. Chromagenic tests, in turn, used nitrophenyl phosphate, proline nitroanilide, nitrophenyl xyloside, arabinoside, glucoronide and glucosamine, esculin, phenylalanine, arginine, lysin, urea, glycine, citrate, and malonate.

Minimal Inhibitory Concentration (MIC test)

The reference bacteria used were *Pseudomonas aeruginosa, Pseudomonas fluorescens, Pseudomonas alcaligenes, Pseudomonas picketti, Flavobacterium aureum,* and *Acinetobacter lowffi* (Tables 1 &2), which were isolated from the selected points of the typical water purification system, and maintained on a tryptic soy agar slant (TSA, Difco) at 4°C, with monthly transfers. The 24-hour cultures developed on TSA at 30–35°C were harvested in a tryptic soy broth (TSB, Difco), centrifuged (1000 g/ 15 min/ 4°C), and suspended in saline (0.95% NaCl plus 0.1% peptone) to a final population (by pour plate) of 10^6 CFU/mL, and the suspensions were applied to the MIC tests.

The minimal inhibitory concentration (MIC) was determined by using the two-fold broth dilution method [7]. Starting from a chemical agent solution, serial dilutions were prepared in TSB inoculated with the test bacterial populations between 1.2 and 3.0×10^6 CFU/mL. The MIC was identified as having the lowest concentration of the chemical agent, which resulted in the confirmed inhibition of the growth of the tested microorganism, after 24 h of optimum incubation conditions. The chemical agent solutions, started concentrations and the pH values employed are shown in Table 03.

Table 3 The minimum inhibitory concentrations (MIC) for chemical agents for the reduction of bacteria populations over 6-log_{10}.

Bacteria	***Pseudomonas aeruginosa***		***Pseudomonas diminuta***		***Pseudomonas fluorescens***		***Pseudomonas alcaligenes***		***Pseudomonas picketti***		***Flavobacterium aureum***		***Acinetobacter lowffi***	
MIC test	**mg/L**	**%**	**mg/L**	**%**	**mg/L**	**%**	**mg/L**	**%**	**mg/L**	**%**	**mg/L**	**%**	**mg/L**	**%**
Agent (initial concentration & pH)														
Ethanol (70%; pH = 7.2)	175000	17.5	87500	8.75	87500	8.75	87500	8.75	87500	8.75	87500	8.75	87500	8.75
Sodium Hypochlorite (0.5%; pH = 11.9)	2500	0.25	2500	0.25	2500	0.25	2500	0.25	2500	0.25	2500	0.25	2500	0.25
PAA (0.45%) + H_2O_2 (2.2%) (pH = 2.3)	1125	0.11	562	0.056	140	0.014	280	0.028	1125	0.11	1125	0.11	1125	0.11
	5500	0.55	2750	0.275	680	0.068	1368	0.14	5500	0.55	5500	0.55	5500	0.55
Sodium Hydroxide (0.4%; pH = 12.8)	3000	0.3	1500	0.15	1500	0.15	750	0.075	3000	0.3	3000	0.3	4000	0.4
Citric acid (0.5%; pH= 2.5)	2500	0.25	2500	0.25	2500	0.25	2500	0.25	5000	0.5	2500	0.25	600	0.06
Hydrochloric Acid (0.3%; pH= 0.3)	1560	0.16	1560	0.16	780	0.078	1560	0.16	3125	0.31	1560	0.16	390	0.039
Sodium Bisulfite (1.0%; pH= 4.0)	780	0.078	780	0.078	780	0.078	390	0.039	780	0.078	780	0.078	780	0.078

PAA – peracetic acid H_2O_2 – hydrogen peroxide

Chemical agents, respective solutions, and concentrations used in the MIC test

1. Ethyl alcohol 70 % v/v (Ferreira, SP, Br). 70% Ethanol solution (pH = 7.2) was used as the desinfectant on the external surfaces of sampling taps and valves.
2. Sodium Hypochlorite NaOCl (Nuclear, SP, Br). Since the concentration of commercial sodium hypochlorite is variable, in order to calculate the amount of sodium hypochlorite (H) to be diluted in purified water and to perform approximately 0.5% v/v, the following formula was applied: H = (5.4 * V) / (% NaOCl concentration); where H is the volume of NaOCl at 10% (w/v) to be diluted and V is the final desired volume of the NaOCl solution at 0.5% (approximately 1% of the volume of the tank); and NaOCl is at an initial concentration of 10%. The concentration of the total available chlorine was determined by the iodometric method [7]. The solution of NaOCl at 0.5% (pH= 11.9) for 60 min exposure is used in the clearance of the tanks of feeding water (point 01), storage (point 08) and in the loop of distribution (before points 11, 12 &13).
3. A solution of 4.5 % (v/v) peracetic acid and 22% (v/v) hydrogen peroxide plus 10 mg/L acetic acid (Minncare, pH = 1.3, Minntech Corporation, Minneapolis, MN, USA) was used by dissolving Minncare into purified water, performing 0.45% of PAA and 2.2% of H_2O_2, reaching a final pH of between 2.0 and 2.3 in the tank. This 1% Minncare solution (PAA + H_2O_2) is applied to the cleaning of reverse osmosis membranes (point 6) and the continuous deionization (point 7) apparatus for three hours, so as to obtain purified water, and 18 hours for WFI which is used to prepare parenteral solutions, including peritoneal dialysis solutions.
4. A sodium hydroxide (Nuclear, SP. Br) solution was prepared by dissolving 200 g of NaOH in 50 liters of purified water, performing 0.4 % (w/v) with a final pH of 12.8 in the tank. A solution of 0.4% sodium hydroxide is used to adjust the pH value of acidic water in the reverse osmosis (point 06) and the continuous deionization (point 07).
5. A citric acid (Ciro, SP, Br) solution was prepared by dissolving 225 g of citric acid in 45 liters of purified water, performing 0.5 % (w/v) with a final pH of between of 2.0 and 2.5 in the tank. A solution of 0.5% citric acid is used to acidify the water in the reverse osmosis membranes (point 06), before the application of 1% of the (PAA+H_2O) solution.
6. A solution of hydrochloric acid (Synth, SP, Br) was prepared by adding 5.5 liters of hydrochloric acid at 37% v/v to 40 liters of purified water, performing 5.0% (pH = 0.3). The final solution of 0.3% HCl is applied in the pH adjustment of water in deionization columns (point 7).
7. A solution of sodium bisulfite (Ciro, SP, Br) was prepared by dissolving 1 kg of sodium bisulfite in 100 liters of purified water, attaining 1.0% w/v (pH of 4.0). The solution of sodium bisulfite at 0.5% concentration is used to preserve the multimedia filters (point 02), water softeners (point 03) and in the dechlorinating of carbon bed filters (point 04).

RESULTS AND DISCUSSION

Isolation of colonies

Of the thirteen points, which were analyzed in triplicate, seventy eight (78) colonies were isolated, with different aspects: circular viscosity, sharp pointed and brilliant, opaque, from light brown to dark brown, pumpkin, pink, white and cream colored. All the colonies were confirmed as bacteria of the non-fermentative gram-negative bacilli (NFGNB) group, strictly aerobic, positive for indole and negative for oxidase, with the exception of the *Acinetobacter lwoffi*, which had a negative reaction for oxidase.

The gram-negative non-fermenting rods isolated and identified from the genera *Pseudomonas, Flavobacterium, and Acinectobacter* are considered opportunistic pathogens and very common in nature: soil, water, plants (including fruits and vegetables), animals and organic material in decomposition (sewage). They are also frequently found in water treatment systems, demonstrating an adaptation to environments with a low concentration of nutrients, and to a large range of temperature, from 4°C to 42°C [4, 8].

According to the miniature kits used in the identification of gram-negative non-fermenting bacteria (Tables 1&2), there was a prevalence of isolation by *P. aeruginosa*, 25 strains (32.05%); *P. picketti (Ralstonia picketti)*, 18 strains (23.08%); *P. vesiculares*, 10 strains (12.82%); *P. diminuta*, 09 strains (11.54%); *F. aureum*, 5 strains (6.42%); *P. fluorescens*, 4 strains (5.13%); *A. lowffi*, 02 strains (2.56%); *P. putida*, 02 strains (2.56%); *P. alcaligenes*, 01 strain (1.28%); *P. paucimobilis*, 01 strain (1.28%); and *F. multivorum*, 01 strain (1.28%). The two identification systems complemented each other, since *P. aeruginosa* was identified in all stages by the Becton and Dickinson kit, whereas *P. picketti* was identified by the Bio-Mérieux kit.

The chemical and microbiological characteristics of the water from the sampling points were in compliance with the relevant laws and standards [1, 2, 3]. For the points 1, 4, 11, 12 and 13, respectively, the pH values were 7.15, 7.39, 5.60, 5.60, 5.70; conductivity was 155.2 µs, 148.2 µs, 1.2 µs, 0.8 µs, 0.9 µs. The total organic compounds (TOC) were < 0.5 mg/L for the points 11, 12 and 13.

The enumeration of heterotrophic bacteria was observed to increase from one to two logarithmic cycles for the water crossing through points 3 & 4, respectively, the two resins of softeners and the bed of activated carbon filter (Table 2). This increase in total bacteria enumeration cancelled out the multimedia (point 02) effect by reducing one cycle of the initial population (average of 507 CFU/ 100 mL) from public drinking water (point 01). The reverse osmosis (point 06) showed the best filtration reduction to a maximum residual population of 10 CFU/100 mL, although gram-negative non-fermenting bacteria were still there crossing and reaching the next stage of the continuous deionization. Although in the storage tank (point 08) the total enumeration was maintained at around 40 CFU/ 100 mL, the diversity of the gram-negative non-fermenting bacteria still existed. The ultraviolet (UV) light and the 0.05 µm filter did not alter the heterotrophic bacteria population or the *Pseudomonas* species found in the preceding points of the Flow Sheet. Both the UV light and the last 0.05 µm filter were found to keep the heterotrophic population as low as 100 CFU/ 100 mL of purified water, although those two apparatus showed no efficiency over the gram-negative non-forming bacteria, such as the species of *Pseudomonas, Flavobacterium*, and *Acinetobacter.*

The dynamic flow of water, passing through the purification stages, carried and spread bioburden throughout every spot of the operational units. Although the heterotrophic enumeration was reduced by at least one log cycle from the 7th to the 8th stage in the storage tank, the diversity of gram-negative bacteria was found to be higher than that at the other points. At point 08, the renewing of the water is slower and the room temperature of the tank favored microorganism adjustment. After exposure to the UV light, the diversity of gram-negative bacteria dropped drastically in the species *P. aeruginosa* and *P. paucimobilis*, (BBL Crystal identification kit) and *P. picketti* (API 20 NE identification kit), the populations of which were kept constant in the water flowing through the 0.05 μm filter. The UV light oxidative process and the 0.05 μm filter at the point – of – use were able to guarantee the maintenance of total enumeration lower than the maximum of 40 CFU detected in 100 mL of water and restrict the distribution of gram-negative bacteria. Maybe a biofilm was established and bacteria might continue to be released from the biofilm into bulk water. Appropriate desinfection of the water treatment system would eliminate the biofilm. This is one of the main reasons why we are studying the best performance of every chemical disinfectant applied at each sampling point.

Procedures should be applied to the system at short-term intervals, in order to control the gram-negative entrance in the system from the municipal source of drinking water (Point 01).

The Minimal Inhibitory Concentration (MIC)

The MIC intervals were expressed in percentage and in mg/L of the chemical agent in contact with the bacteria tested, and are shown in Table 3.

1. Ethyl alcohol (70%, pH= 7.2)

Regarding the exposure to 70% ethanol, *P. aeruginosa* showed the greatest MIC at 17.5%, higher than the MIC of 8.75% which was found for *B. subtilis* and *B. stearothermophilus*, obtained by Vessoni Penna *et al.*[7], emphasizing the importance in the validation of the water purification system for the identification of *P. aeruginosa*.

Trautmann *et al.*[9] did not obtain acceptable long-term results by using chlorination and filtration to eliminate the strains of *P. aeruginosa* from every water tap in the ICU. However, the authors noticed that strains of *P. aeruginosa* were not regularly isolated from hospital personnel, whose hands were disinfected with alcohol (ethanol 70%) before and after contact with patients.

2. Sodium hypochlorite (NaOCl, 0.5%; pH11.9)

All the bacteria studied showed the same level of resistance, i.e. a MIC of 0.25% (2500 mg/L). According to Vessoni Penna et al. [7], *E. coli* presented half a MIC of 0.156% (1560 mg/L) in chlorinated compounds, making it clear that the monitoring of chlorinated waters for *P. aeruginosa* is necessary and mainly critical for warm potable waters, including the cleaning of the water circuits, hydrotherapy, baths, and pools. The *P. aeruginosa›s* resistance to chlorine releasing agents (CRAs) is well reported. Wirtanen *et al.*[10] studied the effects of four commercial disinfectants: (100% alcohol-based) containing isopropanol, a peroxide-base containing hydrogen peroxide (0.5–2.0%) and peracetic acid in formulation, and a chlorine-base containing sodium hypochlorite (0.3–0.8%, pH>9.0). The authors found that the peroxide-based disinfectant was the most efficient on biofilm bacteria, and they confirmed that prolonged treatment with the chlorine-base disinfectant was especially efficient on *Pseudomonas* biofilms.

Therefore, there is an urgent need for seeking the range of CRAs and the adequate time contact that would not induce resistance in bacteria, which inhibits biofilm formation and adherence to the porous apparatus of the system. However, the disinfectants should be chosen according to the operation in process, considering the interference of organic substances in the disinfectant›s activity and effectiveness.

3. Association of peracetic acid (PAA, 0.45%) plus hydrogen peroxide (H_2O_2 2.2%). Solution at 1% (PAA+ H_2O_2; pH = 2.3)

For a solution of peracetic acid (0.45%) + hydrogen peroxide (2.2%), *P. aeruginosa, P. picketti, F. aureum,* and*A. lowffi* presented the highest MIC range from 0.11% to 0.55%, an interval of which was shown to be two, four and eight times greater than that for *P. diminuta* (MIC = 0.056% & 0.275%), *P. alcaligenes* (MIC = 0.028% & 0.137%) and *P. fluorescens* (MIC = 0.014% & 0.068%), respectively.

Considering that the set conditions for MIC were kept constant, independent of the bacteria tested, according to Vessoni Penna *et al.*[7], when vegetative cells such as *Acinetobacter calcoaceticus, Enterobacter cloacae, Escherichia coli, Serratia marcescens*, and *Staphylococcus aureus* were submitted to a H_2O_2 solution at 4.0%,*Escherichia coli* showed the greatest resistance among them with a MIC of 0.25%, which is half that obtained for *P. aeruginosa.* When the same bacteria were tested against peracetic acid, *E. coli* showed a MIC of 0.23%, which is twice as much as that for *P. aeruginosa* in a mixture of peracetic acid + hydrogen peroxide, showing that *P. aeruginosa* and *E. coli* should both be required in the monitoring of treated waters. The solution of PAA + H_2O_2 at 1% is applied to the cleaning of the reverse osmosis membranes and the continuous deionization apparatus for three hours, so as to obtain purified water, and 18 hours for WFI that will be used to prepare parenteral solutions, including peritoneal dialysis solutions.

4. Sodium Hydroxide (0.4%; pH = 12.8)

P. fluorescens and *P. alcaligenes* were the bacteria which were the least resistant to sodium hydroxide, exhibiting a MIC of 0.15%, in relation to the resistance shown by the other bacteria (MIC >0.4%). The principal activity of this chemical compound is the pH adjustment of acidic water in the reverse osmosis (point 06) and the continuous deionization (point 07), with some expected sanitation activity.

5. Citric Acid (0.5%; pH = 2.5)

The bacterium least resistant to citric acid solution was *A. lowffi* (0.06%= 600 mg/L). The most resistant was *P.*

picketti (0.5%= 5000 mg/L); the remainder showed a MIC of 0.25% = 2500 mg/L. This chemical agent is used in the reverse osmosis (point 06) in adjusting the pH of cleaning water before disinfection with a 1% solution of PAA + H_2O_2.

6. Hydrochloric Acid (0.3%; pH = 0.3)

The bacteria least resistant were *A. lowffi* (MIC 0.039% = 390 mg/L) and *P. fluorescens* (MIC 0.078% = 780 mg/L); the remainder had a MIC of 0.156% (1560 mg/L).

It is noteworthy to mention that *P. aeruginosa* was not affected by the chemical adjuvant compounds such as NaOH (0.4%), citric acid (0.5%) and HCl (0.3%) in the concentration applied to the units of the purification system. However, these chemical solutions are expected to show some disinfecting activity, which is not demanding.

7. Sodium Bisulfite (0.5%; pH= 4.0)

All the bacteria showed the same level of resistance (MIC = 0.078%) to sodium bisulfite. This chemical adjuvant is used for the purpose of dechlorinating and preserving multimedia filters; it is further used in water softeners and activated carbon bed filter.

Profile of the identified microorganisms

The gram-negative water bacteria can be significant contaminants in haemodialysis systems. Ferreira *et al.*[13], performed analyses on samples taken from water supplies, after each treatment step and dialysate in haemodialysis units in Rio de Janeiro from 1999 to 2001. They found that the most commonly isolated bacteria were: *Pseudomonas aeruginosa, Stenotrophomonas maltophilia, Acinetobacter anitratus, Acinetobacter lowffi, Pseudomonas (Brevundimonas) diminuta, Pseudomonas fluorescens, Achromobacter xylosoxidans, Moraxella atlantae, Moraxella osloensis, Enterobacter cloacae,* and *Enterobacter aerogenes*. These isolated bacteria pointed out a real risk for haemodialysis patients with developing gram-negative bacteremia, pyrogenic reactions, and peritonitis. The authors emphasized the need for an urgent program for the disinfection of the water systems and dialysis machines [14], in health care centers.

Pseudomonas have been isolated from a great variety of materials, including soils, fresh or sea water, sewage, many types of clinical specimens, and elements commonly handled in clinical laboratories (including distilled water and antiseptic solutions), assorted foods and food industry wastes, flowers, fruits, vegetables, and diseased plants and animals. The *Pseudomonas* species rather than *P. aeruginosa* do not frequently cause infections [4, 8, 11]. *P. aeruginosa* has been found occasionally on the skin, isolated from clinical material [15] removed from the throat (5%), and in feces (3%) of non-hospitalized people. The percentage of patients with gastrointestinal problems from this bacteria increases by 20% after 72 hours of hospitalization. From the frequency with which this opportunistic pathogen is involved in human illnesses, it has been considered the most important human pathogen with respect to the number and types of infections which they cause and their association with morbidity and mortality. *P. aeruginosa* can be also responsible for an infection outbreak in a non-immune endangered community [4].

P. aeruginosa is frequently associated with contaminated potable or even treated water or solutions, such as gastroenteritis through contaminated food, otitis in swimmers, eye infections through contaminated solutions and tap water during lens care, wound infections caused by tap water, and respiratory tract infections caused by contaminated equipment. The most severe community which acquired infections are endocarditis in intravenous drug users, resulting from the use of contaminated parenteral solutions with drugs added, and peritonitis in individuals undergoing contaminated peritoneal dialysis solutions.

The principal reservoirs of *P. aeruginosa* in a hospital environment are [4, 8, 11, 12]: disinfectants, artificial breathing equipment, hydrotherapy equipment and tanks, food, sewers, drains, taps, soap bars, floor cloths, bed frames, air, chairs, cloth towels and tower hangers, mattresses, hoses, and tubes. Dissemination occurs through the hands of personnel (or staff) resulting from direct contact with taps and water reservoirs. The contamination of such water may be the result of clinical, fecal and urine materials.

CONCLUSION

Kawai *et al.*[16] analyzed, by denaturing gradient gel electrophoresis (DGGE), "the bacterial community in partially purified water, which is prepared by an ion exchange from tap water and is used in pharmaceutical manufacturing processes". The authors verified that "the dominant bacterium in purified water" (lower than that of other aquatic environments) could not be detected by soybean casein digest (SCD) and R2A media, although they presented esterase activity. Therefore, the authors emphasized "the importance of culture-independent methods of quality control for pharmaceutical water."

The water used in health center areas and in pharmaceutical industries should be periodically analyzed as a preventive measure against the spreading of microorganisms, allowing measures of improvement to be taken rapidly, as required.

The analysis of treated water for heterotrophic bacteria including *Pseudomonas* species is valuable in the prevention of the formation of biofilms and in the reduction of the amount of pyrogen.

Several solid surfaces can harbor biolfilms in water purification systems, such as stainless steel, nylon materials (deionization apparatus), and polyamide polymers (osmose reverse membranes). Porous surfaces found on the deionization resins and the osmose reverse membranes provide an excellent opportunity to trap inorganic particles and bacteria which begin the formation of biofilms. In general, all cleaning chemical agents perform better in soft water (after water softner apparatus). Chlorine agents have been shown ineffective at removing biofilms. Chlorine agents are used to keep the storage tanks and distribution loop in low microbial densities. Hydrogen peroxide, peroxide containing sanitizers and peracetic acid chemical agents have been found to be highly effective in the prevention and removal of biofilms, in a short period of contact (1–2 minutes) and they are relative non-corrosive.

This work emphasized the urgent need in the continuous monitoring of the water purification performance, including the evaluation of the effectual removal of non-fermenting gram-negative bacteria, with special attention to *Pseudomonas sp*, which showed the most resistance to the chemical agents, at the concentrations used in the system. Maybe the contact time between the sanitary agents (NaOCl and PAA+H_2O_2) and the system points should be reviewed. The 0.5% NaOCl after 18 h contact was shown sufficient to reduce bacteria population over 6-$\log_{10}$. As all the bacteria tested showed similar MIC, maybe the short-time contact of 60 min of this solution (0.5% NaOCl) in the tank with the feeding water should have been greater to guarantee that the gram-negative bacteria that entered the system through the municipal source of drinking water would not survive. Even the contact-time of this sanitary agent (0.5% NaOCl) in the loop of distribution should have been longer to assure 10^{-2} CFU/mL of purified water with less chance of the*Pseudomonas sp* crossing through the consumption point filters. The loop of distribution should have been cleaned first with PAA+H_2O_2 and then with NaOCl to prevent the survival of any microorganism and possible development. Therefore, the contact time of the sanitizers and the different points of application should be reviewed and evaluated in accordance with the compatibility of the equipment and device material.

The effectual application of PAA+H_2O_2 in the osmose reverse membranes required a larger time-contact in the deionization columns, which showed greater heterotrophic enumeration. The double application of first PAA+H_2O_2 followed by NaOCl solutions (except the reverse osmosis membranes) would both reinforce the efficacy over the non-fermenting gram-negative bacteria. The PAA+H_2O_2 sanitary agent is known for its efficiency in preventing biofilm formation and its low pH may enlarge the NaOCl activity favoring microbial effectiveness.

Sodium hydroxide, hydrochloric acid and citric acid were used as chemical adjuvants on the pH adjustment of acid or alkaline water flow in the reverse osmosis membranes, the continuous deionization and the water softeners. At the concentrations used, no sanitation activity was demanded from their application, of which the ability over the tested microbial population was not confirmed, even though after 18 h of contact. Some efficacy shown by those chemical adjuvants was due to the alteration of environmental pH values. Even for extended exposures, *Pseudomonas sp* was shown to be the most resistant to those chemical agents at the tested concentrations.

Therefore the washing of (storage tanks) reservoirs, deionization columns, reverse osmosis membranes, as well as the sanitation of distribution circuits should be carried out by the determination of a schedule established for quality control (bacteriological and chemical) of water systems in the risky areas.

REFERENCES

1. [http://www.sabesp.com.br]
2. Brazilian Ministry of Health Directives: n° 15/MS/ANVS n°211/MS/ANVS 1999, n ° 196/MS/AVNS 1983, n° 36/MS/AVNS 1991, n° 113/MS/AVNS 1993, n° 122/DTN 1993, n° 211/MS/ANVS 1999, and Health Sanitation Code n° 12342,. 1988
3. The United States Pharmacopoeia National Formulary: Water For Pharmaceutical Purposes USP 24/NF. 2000, 19: 2154-2165.
4. Schreckenberger PC, Janda JM, Wong JD, Baron EJ: Algorithms for identification of aerobic gram-negative bacteria. In: Manual of Clinical Microbiology. Edited by: Murray PR. 1999, Washington, American Society For Microbiology, 438-452. 7
5. [http://http//:www.us.filter.com]
6. Palmieri MJ, Caritó SL, Meyer RF: Comparison of rapid NFT and API 20 NE with conventional methods for the identification of gram-negative non-fermentative bacilli from pharmaceutics and cosmetics. Appl Environ Microbiol. 1988, 54: 2838-2841.
7. Vessoni Penna TC, Mazzola PG, Martins AM: The efficacy of chemical agents in cleaning and desinfection programs for hospital nurseries. J BMC Infectious Diseases. 2001, 1: 16-10.1186/1471-2334-1-16.
8. Silva CHPM: Bacteriologia um texto ilustrado (Bacteriology an illustrated text). Rio de Janeiro, Eventos Livraria. 2000
9. Traumann M, Michalsky T, Wiedeck H, Radosavljevic V, Ruhnke M: Tap water colonization with Pseudomonas aeruginosa in a surgical intensive care unit (ICU) in relation to Pseudomonas infections of ICU patients. Infect Control Hosp Epidemiol. 2001, 22: 49-52.
10. Wirtanen G, Salos S, Helander , Mattila Sandholm T: Microbiological methods for testing disinfectant efficiency on Pseudomonas biofilm. Colloids and Surfaces B: Biointerfaces. 2001, 20: 37-50. 10.1016/S0927-7765(00)00173-9.
11. Fernandes AT, Fernandes MOV, Filho NR: Infecção Hospitalar e suas Interfaces na Área da Saúde (Hospital Infection and its Consequences). São Paulo, Atheneu. 2000
12. Arnow PM, Flaherty JP: Non-fermentative gram-negative bacilli. In: Hospital Epidemiology and Infection Control. Edited by: Mayhall CG. 1996, Baltimore, Williams & Wilkins, 162-163.
13. Ferreira JAB, Silva Z, Oliveira RC, Branquinho MR: Bacterial contaminants in water and dialysate used for haemodialysis therapy in Rio de Janeiro (1999–2001). XXI Congresso Brasileiro de Microbiologia (Brazilian Congress of Microbiology). 2001, IH 040: 148-
14. Ferreira JAB, Silva Z, Oliveira RC, Branquinho MR: Microbiological quality of the water used for haemodialysis therapy in Rio de Janeiro (1999–2001). XXI Congresso Brasileiro de Microbiologia (Brazilian Congress of Microbiology). 2001, IH-039: 148-
15. Frota CC, Moreira JLB: Frequency of non-fermentative gram-negative bacilli isolated from clinical materials of patients at the Universidade Federal do Ceará Hospital Complex-Brazil. Brazilian J Microbiol. 1998, 29: 179-182.
16. Kawai M, Matsutera E, Kanda H, Yamaguchi N, Tani K, Nasu M: 16S Ribosomal DNA-Based Analysis of Bacterial Diversity in Purified Water Used in Pharmaceutical Manufacturing Processes by PCR and denaturing Gradient Gel Electrophoresis. Applied and Environmental Microbiology. 2002, 68 (2): 699-704. 10.1128/AEM.68.2.699-704.2002.

Section V - Drinking Water Distribution Systems: Biofilm Microbiology

CHAPTER 18 CHARACTERIZATION, MICROBIAL COMMUNITY STRUCTURE, AND PATHOGEN OCCURRENCE IN URBAN FAUCET BIOFILMS IN SOUTH CHINA

Huirong Lin, Shuting Zhang, Song Gong, Shenghua Zhang, and Xin Yu

INTRODUCTION

During the past decades, scientists have paid great attention to the biofilms formation in drinking water distribution systems (DWDSs) since biofilms formation and its resistance to disinfection were considered to be potential risk in DWDSs [1–3]. Bacteria can grow in bulk water and become attached to pipe walls as biofilms. Biofilms formation in DWDSs can lead to public health issues, such as protecting pathogenic bacterial regrowth, and depletion of disinfectant [4–6]. The microbial quality of tap water is closely related to consumers' health. Thus, knowledge of biofilms behavior in DWDSs and faucet can contribute to the design of effective control strategies.

DWDSs are engineered environments that are subject to frequent, variable disturbances caused by many factors including long residence times, each associated with loss of disinfectant residual and generating higher levels of biofilms growth. These changes are reflected in the new phenotypic characteristics developed by biofilm bacteria and occurred in response to a variety of environmental signals. Phylogenetically diverse bacterial groups can inhabit the biofilms attached to the pipes and the bulk water. The study of bacterial ecology can improve the understanding of the persistence of biofilms and pathogens.

Studies to date suggest that the planktonic-biofilm transition is a complex and highly regulated process. Several methods including culture-based and independent analyses showed that members of the class Proteobacteria, including the Alpha-, Beta-, and Gammaproteobacteria were considered to be the most predominant bacterial groups in water distribution systems [7]. Recent genetic and molecular approaches used to study bacterial and fungal biofilms have identified genes and regulatory circuits important for initial cell-surface interactions, biofilm maturation, and the return of biofilm microorganisms to a planktonic mode of growth [1, 3, 7]. However, the characterization of microbial communities of biofilms in DWDSs is far from being fully understood. To give insight into the characterization and bacterial diversity of biofilms in drinking water distribution systems, microstructure analysis (SEM, CLSM and XRF) was used for the characterization of biofilms on faucet. The bacterial community compositions of real biofilms from 12 urban household taps (Xiamen, southeast of China) were detected using 454 pyrosequencing technique. Potential pathogens and corrosive microorganisms occurrence was also discussed.

Citation: Huirong Lin, Shuting Zhang, Song Gong, Shenghua Zhang, and Xin Yu, "Characterization, Microbial Community Structure, and Pathogen Occurrence in Urban Faucet Biofilms in South China," BioMed Research International, vol. 2015, Article ID 401672, 8 pages, 2015. doi:10.1155/2015/401672

METHODS

Characterization and Visualization of the Biofilms Matrix in DWDSs

In order to observe the biofilms formed on drinking water pipe surface, five 5 × 5 mm removable PVC sheets were inserted in real PVC DWDSs pipe for biofilm attachment. They were sterilized by immersion in 70% ethanol for 2 min individually and air dried. The PVC tap in our lab was twisted off and the sheets were transferred with sterile forceps to the tap; then, the tap was screwed on. After three months, PVC sheets were took out for scanning electron microscopy (SEM) and staining. One PVC sheet was fixed, dried, and viewed with a scanning electron microscope (Hitachi S-4800, Japan). The other four were stained with DAPI (4,6-diamidino-2-phenylindole) for attach cells, FITC (fluorescein isothiocyanate) for extracellular protein, Con-A (concanavalin A) for α-polysaccharide, and CW (calcofluor white) for β-polysaccharide, respectively. Each kind of stain was added on each sheet in the dark. After 15 min the sheets were analyzed using CLSM. CLSM 3D-image was used to visualize the staining biofilms. The CLSM scanned different layers of the staining biofilms. The thickness of the biofilms was got from signals of different layers using the software ZEN 2008 Light Edition.

Biofilms Sampling for 454 Pyrosequencing

It is difficult to get the information on the microbial diversity of biofilms within DWDSs due to limited access and the high cost involved in sampling. In this study, biofilms in taps were collected from a drinking water distribution system of Xiamen, Fujian Province in southeast of China. Different taps with different materials and different ages were studied (Table 1). The taps were removed and ethanol-sterilized cotton swabs were then used to collect biofilms on the taps. Then, the collected biofilms were moved to 2 mL sterilized EP tubes and put into an ice box during transportation. For each sampling point, the physicochemical water quality parameters were measured and listed in Table 1.

Table 1: The biofilms samples used in this study and water quality parameters at each sampling location.

Samples	A1	A2	A3	A4	A5	A6	A7	A8	A9	A10	A11	A12
Tap material	Stainless steel	PVC	Stainless steel	Cast iron	Cast iron	Stainless steel	Stainless steel	Cast iron	Cast iron	Stainless steel	PVC	PVC
Age (year)	2	1	4	9	8	3	5	12	15	6	1	0.5
pH	7.08	6.8	7.01	6.62	7.11	7.03	6.94	7.35	7.26	6.83	6.98	6.7
TSS	40	49	99	75	81	69	39	80	69	73	52	47
Turbidity	0.6	0.3	0.3	0.2	0.4	0.2	0.1	0.3	0.2	0.1	0.4	0.3
Residual chlorine (mg/L Cl_2)	0.55	0.54	0.62	0.46	0.49	0.63	0.65	0.57	0.61	0.62	0.3	0.4
Sulfate (mg/L)	10.4	12.9	31.8	24.2	30.6	13.8	24.2	22	27.8	19.8	17.8	16.5

X-Ray Fluorescence (XRF) Analysis of Biofilms A3 on Corroded Tap

Among the biofilms samples, A3 was red, indicating that the pipe was corroded. In order to analyze the element contents in the biofilms, XRF analysis of the biofilms A3 was performed at the XRF microprobe station (beam line 4W1B) of Beijing Synchrotron Radiation Facility (BSRF), Institute of High Energy Physics of China. The electron energy in the storage ring was 2.2 GeV, with a current range from 78 to 120 mA. The size of the exciting X-ray beam was 10 μm 20 μm. XRF spectra were collected by a PGY Si (Li) solid detector, positioned at 90° to the beam line. The scanning point of the sample was selected and observed by a microscope.

DNA Extraction, 16S rRNA Genes Amplification, and Pyrosequencing

Total DNA of the biofilms samples was extracted and purified using a bead beating method (FastDNATMSPIN Kit for Soil, Bio101 Inc., USA) following the manufacturer's instructions. For pyrosequencing analysis, the V3 and V4 region of 16S rRNA gene was amplified with the primers 338F (5′-ACTCCTACGGGAGGCAGCAG-3′) and 802R (5′-TACNVGGGTATCTAATCC-3′). The PCR was carried out in a total volume of 20 μL: H_2O 13.25 μL, 10xPCR ExTaq Buffer 2.0 μL, DNA template (100 ng/μL) 0.5 μL, primer 338F (10 mmol/L) 1.0 μL, 802R (10 mmol/L) 1.0 μL, dNTP 2.0 μL, and ExTaq (5 U/mL) 0.25 μL. The DNA amplification condition used in this study was initial denaturation at 95°C for 3 min, followed by 35 cycles of 10 s at 95°C, 30 s at 58°C, and 6 sec at 72°C, followed by a final extension at 72°C for 7 min. The PCR products were purified and end-repaired, A-tailed, PE-adapter ligated, and then utilized for pyrosequencing on the 454 Genome Sequencer FLX platform [8].

The unique tags (nonredundant tags) were obtained and aligned against the 16S rRNA database using the BLASTN algorithm. The raw sequences obtained from 454 pyrosequencing were optimized. The redundant tags were deleted by Mothur v.1.11.0 [9]. Sequences with similarities greater than 97% were grouped in one OTU. Using the MEGA5.0 program, the 60 bp representative sequences were aligned by the ClustalW with default settings, and a phylogenetic analysis was performed based on the neighbor-joining method [9].

RESULTS AND DISCUSSION

Characterization and Visualization of the Drinking Water Biofilms Matrix

DWDSs are engineered environments which are subject to frequent, variable disturbances caused by many factors including long residence times due to dead ends and low flow rates, and the latter are associated with loss of disinfectant residual and generating higher levels of biofilm growth. Attachment is required for biofilms formation, and bacteria interact with pipe wall surface through adhesions including polysaccharides and surface proteins, with initial contact often mediated by active motility. The formation of a biofilm is the result of successional development into a mature community, which may require several years before steady state is achieved. For example, it was found that the successional development of a model DW biofilm during a 3-year period was an orderly process resulting in a stable, well-defined community [10]. In order to observe and characterize a real DWDSs biofilm, CLSM and SEM were applied in this study. SEM analysis of the PVC slice revealed different features of biofilms morphology in different fields of the PVC pipe surface. Possible EPS was present on the PVC slice in one field (Figures 1(a) and 1(b)). Sparse populations of mainly rod-shaped bacteria were observed in another field (Figure 1(c)). There were also globular-shaped bacteria (Figure 1(d)). Using Zen 2008 software, the "range" value showed the thickness of the biofilm. CLSM 3D-image results suggested that after three months' growth the biofilm was about 20 μm thick. CLSM results also revealed different distributions of attached cells (Figure 1(e)), extracellular protein (Figure 1(f)), α-polysaccharide (Figure 1(g)), and β-polysaccharide (Figure 1(h)).

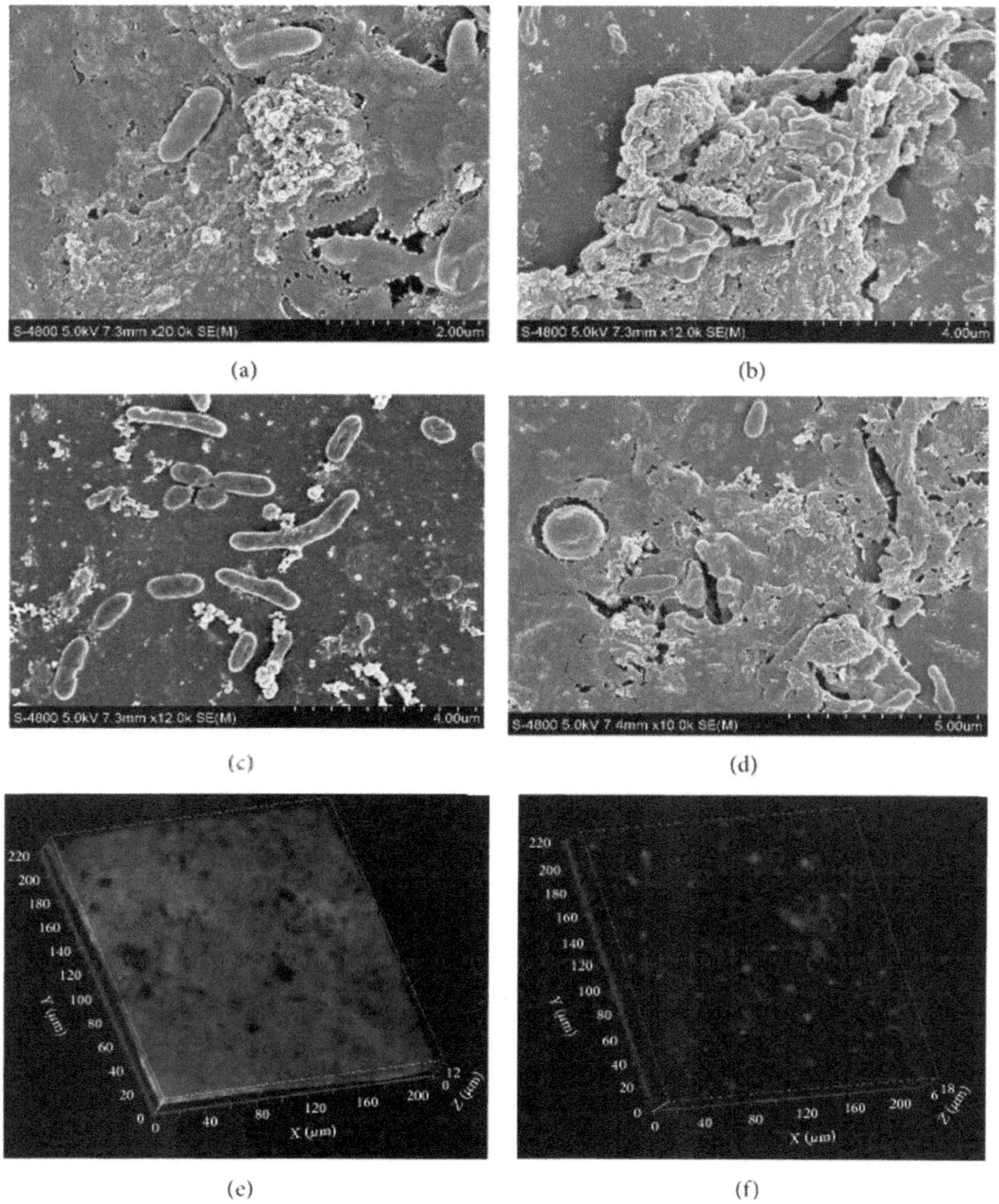

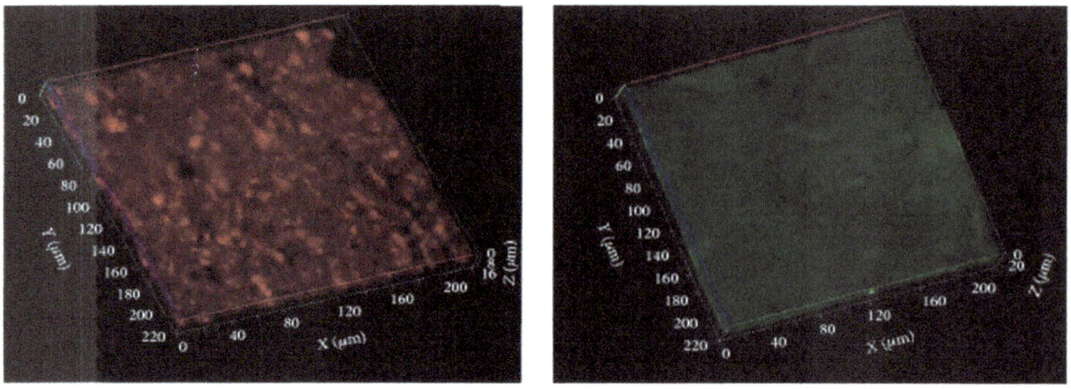

(g) (h)

Figure 1: Visualization of the drinking water biofilms matrix on the PVC slice after three months growth. SEM images of extracellular matrix material (a, b) and rod-shaped bacteria (c) and globular-shaped bacteria (d), CLSM 3D-image of staining biofilms on the surface of the PVC slice after three months growth, DAPI for attach cells (e), FITC for extracellular protein (f), Con-A for α-polysaccharide (g), and CW for β-polysaccharide (h).

During sampling, it was found that there were some yellow iron filings in the biofilm A3, suggesting that the Fe in the stainless steel pipe was released. In order to obtain the elements contents of corroded biofilm A3, XRF was used. Figure 2 was the typical spot scan result of biofilm A3. It showed that K, Ca, Fe, Cu, and Zn were present in biofilm A3, suggesting that this pipe was corroded and metal elements were released which might affect the water quality in the DWDSs. Our study enabled the concurrent visualization and evaluation of the biofilm growth as well as EPS matrix (extracellular proteins and polysaccharide).

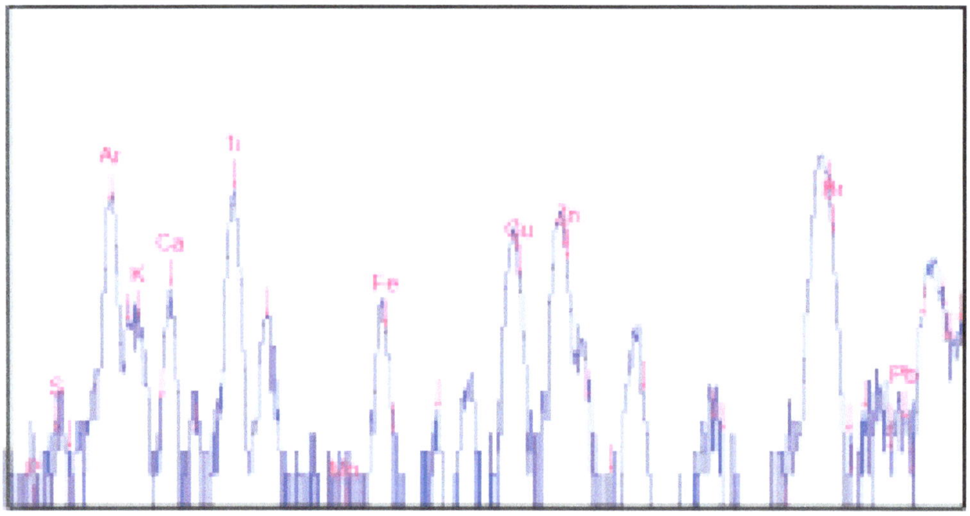

Figure 2: Typical XRF spectra in biofilm A3 from drinking water distribution stainless steel pipe wall. The Ar peak was from the air, while Ti and Br peak were artefact caused by 3M tape for fixing sample.

Diversity of Microbial Communities in the Tap Biofilms

Pyrosequencing is a powerful tool to investigate microorganisms community [11]. Compared with conventional cloning and sequencing methods, pyrosequencing can obtain more sequences and OTUs. In this study, 8 biofilms samples were successfully high-throughput pyrosequenced among the 12 samples. To assess the sequencing depth and the species richness, a rare faction curve was constructed for each sample. Alpha diversity measurement for the samples (Table 2) suggested variations in species richness (Chao1) and species evenness (Simpson index) among different biofilms samples. Higher species richness of a given sample might present a lower species diversity resulting from a lower evenness of the sample. PcoA results indicated that the biofilms samples obtained from different tap materials showed quite different clusters. Biofilms formed on PVC had a more simple microbial community compared with those formed on cast iron and stainless steel. This result was consistent with another study which conducted with 16S rRNA genes sequence [12]. These results indicated that PVC was a more ideal material compared with the other materials as fewer bacteria attached on it.

Table 2: Comparison of phylotype coverage and diversity estimators of the microbial community structure employing the 454 pyrosequencing analysis.

Tap material	Samples	Reads	OTUs	ACE	Chao 1	Shannon	Simpson
PVC	A2	10,804	81	129	139	3.64	0.862
Stainless steel	A1	9,826	186	298	292	4.63	0.898
	A3	36,655	146	294	276	3.93	0.888
	A7	14,605	114	204	202	4.08	0.896
	A10	9,826	351	478	481	5.37	0.927
Cast iron	A5	13,660	106	182	192	3.27	0.831
	A8	13,214	217	383	340	4.43	0.921
	A9	27,517	152	250	256	4.46	0.916

Composition of Microbial Community

DWDSs biofilms are considered to be complex, dynamic microbial assemblages with extensive phenotypic diversity supporting adaptation to different hydraulic and chemical surface conditions [3, 13–15]. As a result, biofilms formed on different materials might have different community structures. The phylogenetic analysis based on the 16S rRNA V3-V4 region sequences from the biofilms samples was showed in Figure 3. The points in this figure were dispersive. It suggested that distinct bacterial communities were detected in these biofilms from different tap materials and different ages. Many factors could influence the bacterial communities including pipe age and pipe material. However, some interesting information could be found. For example, A8 and A9 (biofilms formed on cast iron) located in one region, while A1 and A3 (biofilms formed on stainless steel) located in another region. This result suggested that the pipe materials might be an important factor for the bacterial communities of the biofilms. In addition, biofilms A2 formed on PVC had relatively simple bacterial community structure, with Betaproteobacteria (54.09%) and Alphaproteobacteria (38.24%) as its main community composition (Figure 4). It was quite different from the biofilms formed on cast iron and stainless steel.

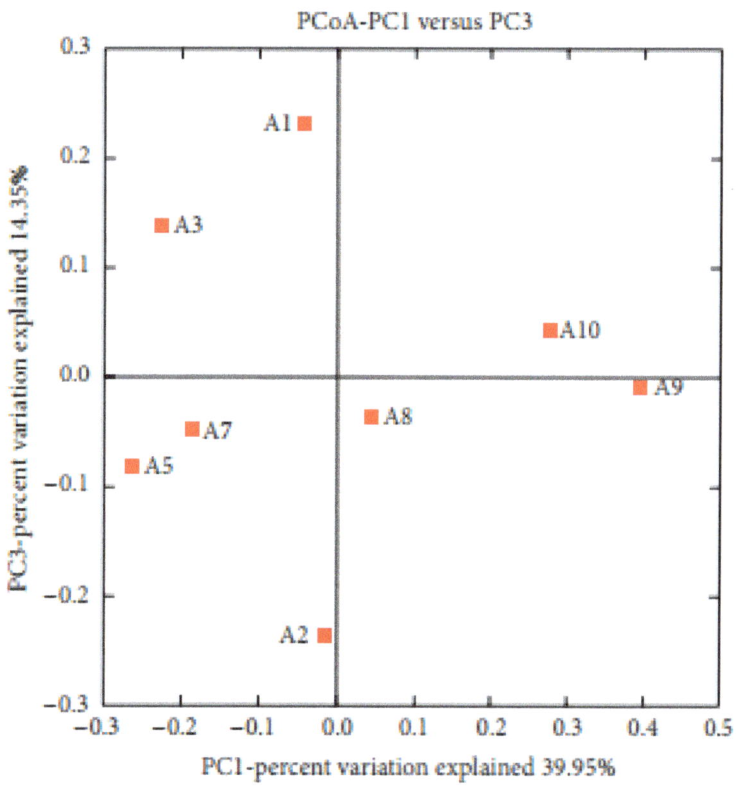

Figure 3: Phylogenetic analysis based on the 16S rRNA V3-V4 region sequences from the biofilms samples.

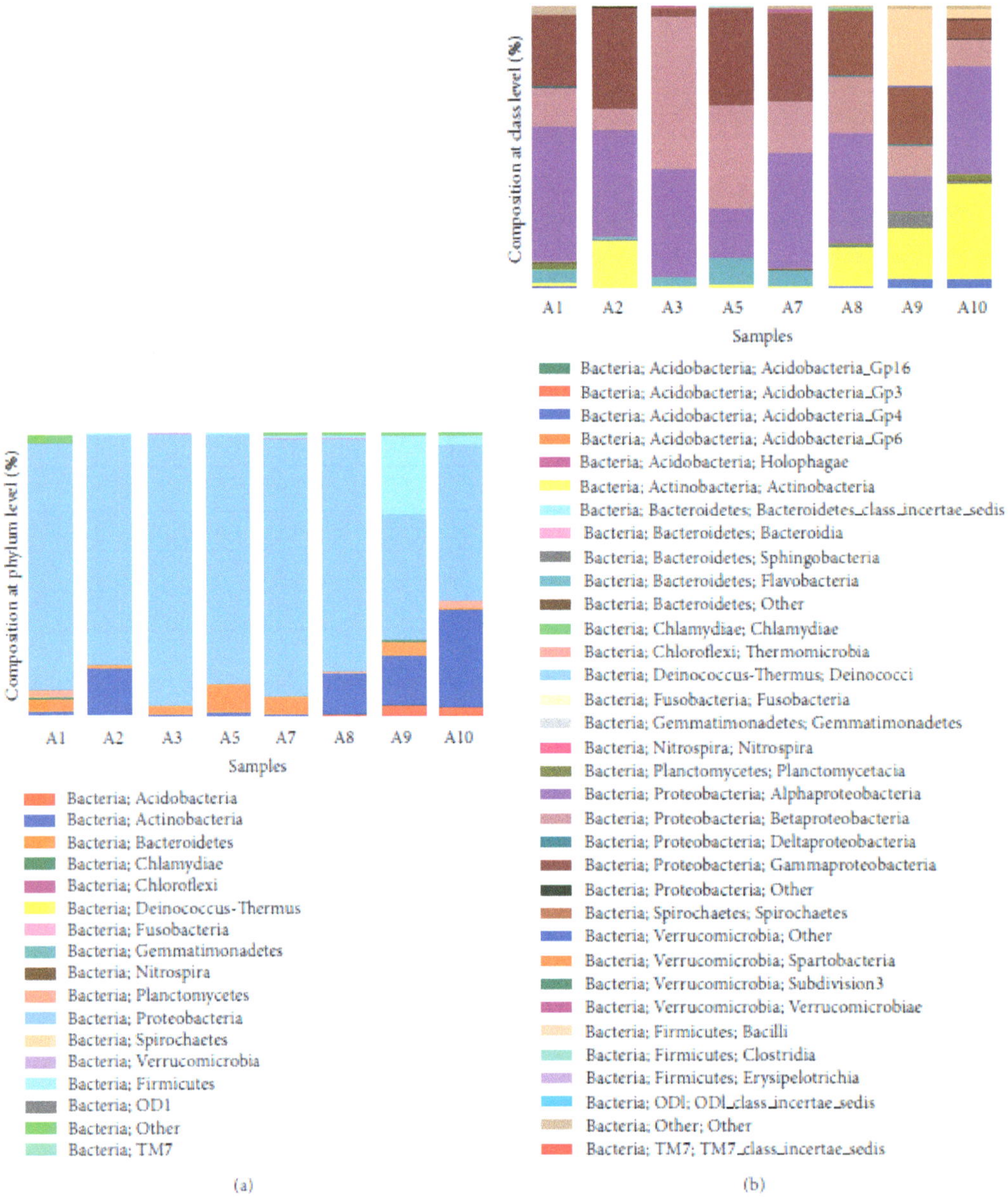

Figure 4: Proportional composition of microbes in the biofilms samples on different materials and from different ages at the phylum level (a) and class level (b). The information of each biofilm sample was listed in Table 1.

The bacterial communities of the samples in both class and species were shown in Figure 4. A high variability of the bacterial core communities of the biofilms was found in this study. Taxonomic analysis revealed that proteobacteria was the common and predominant group in all biofilms samples. These results were broadly consistent with other biofilms studies [7, 16]. Besides, Actinobacteria was the second largest bacterial group in the samples A2, A8, A9, and A10. Some important bacteria should be paid attention to. For example, Sphingomonadales was found in all samples. This might be due to the resistance to chlorine of Sphingomonadales. In addition, Nitrospiraceae and Thiotrichales were found in this study, suggesting the occurrence of nitrification and sulfur transformation.

POTENTIAL PATHOGEN AND CORROSIVE MICROORGANISMS OCCURRENCE

Opportunistic pathogens in drinking water systems are an emerging health concern. In this study, some waterborne opportunistic pathogens were found in the biofilms samples. They were members of species Legionellales, Enterobacteriales, Chromatiales, Pseudomonadales, and Enterococcaceae, although there were differences in numbers for each sample. Abundance of some potential corrosive microorganisms and pathogens in genus level in different bacterial communities was listed in Table 3. These pathogen bacteria in the biofilms may be the risk to human health and lead to clinical and outbreak cases. In previous studies, it was found that environmental pathogens such as Legionella pneumophila and Mycobacterium avium have been shown to be associated with DW biofilms [1, 17, 18]. Our taxonomic findings are consistent with these studies. In some studies, it was found that Mycobacteria were found to be the predominant bacterial genera in DWDSs because they were generally resistant to disinfectants due to their complex cell wall [3, 18–20]. However, it was not detected in this study.

Table 3: Abundance of some potential pathogen and corrosive microorganisms in genus level in different bacterial communities (count).

	genera	PVC	stainless steel				cast iron		
		A2	A1	A3	A7	A10	A5	A8	A9
Pathogen	Escherichia_Shigella	1	0	1	0	0	0	1	0
	Legionella	0	4	5	0	7	0	4	4
	Pseudomonas	13	27	70	39	444	66	429	1219
	Enterococcus	0	0	0	0	0	0	2	0
SOB	Thiobacillus	0	7	0	0	0	0	0	0
IRB	Bacillus	0	0	2	1	382	0	1	1826
	Clostridium_XlVa	0	0	0	0	0	4	0	0
	Clostridium_XVIII	0	0	0	0	0	0	1	0
	Pseudomonas	13	27	70	39	444	66	429	1219

As corrosion was found in the biofilm A3 and XRF result showed high Fe, Cu, and Pb contents in the biofilms sample, the bacterial community information was further analyzed to understand the possible roles of microorganisms in iron corrosion. Thiotrichales (sulfur oxiding bacteria) was found in A1 biofilm sample form on stainless steel. In addition, potential iron reducing bacteria (Bacillales, Clostridiales, and Pseudomonadales) were found in some samples (Table 3). It was reported that IRB might generate Fe_3O_4 in anaerobic conditions, leading to corrosion in DWDSs. As a key impact factor to corrosion, the sulfate contents in the drinking water of this city are usually low according to the general survey. For example, survey data in 2012 showed that there were among the 102 samples, only 2 samples had the sulfate contents higher than 30 mg/L. So in this study, fewer corrosive microorganisms were found in this city, compared with another study which was conducted on the biofilms with higher sulfate contents in water [21]. Nevertheless, these corrosive microorganisms might contribute to corrosion of drinking water pipes. It should be noticed that the genus members within the generaBacillus and Clostridium could affect iron corrosion and modify the corrosion products [22].

CONCLUSIONS

In this study, combined use of SEM, CLSM, and XRF was successfully conducted for the characterization and visualization of the drinking water biofilms matrix. EPS and bacteria (mainly rod-shaped and coccoid) were observed in the biofilms. Four hundred fifty-four pyrosequencing used to characterize bacterial diversity in these biofilms samples suggested that there were differences in the bacterial community composition between different biofilms samples formed on different tap materials with different ages. Proteobacteria (mostly Alpha-, Beta-, and Gamma-) predominated in the biofilms samples. Some potential pathogens (members of species Legionellales, Enterobacteriales, Chromatiales, and Pseudomonadales) were found in the biofilms samples. There were also some potential corrosive microorganisms (Thiotrichales, Bacillales, Clostridiales, and Pseudomonadales). In order to prevent biofilms formation, PVC might be a more ideal material compared with the others. This work might add some new insights into microbial community and its influential factors in DWDSs.

Conflict of Interests

The authors declare that there is no conflict of interests regarding the publication of this paper.

ACKNOWLEDGMENTS

This work was supported by National High-Tech Research and Development (2012AA062607), the National Natural Science Foundation of China (31300109 and 51278482), Fujian Provincial Natural Science Foundation (2013J05087), and the Science and Technology Project of Xiamen (3502Z20132013).

REFERENCES

1. F. Emtiazi, T. Schwartz, S. M. Marten, P. Krolla-Sidenstein, and U. Obst, "Investigation of natural biofilms formed during the production of drinking water from surface water embankment filtration," Water Research, vol. 38, no. 5, pp. 1197–1206, 2004.
2. M. Tachikawa, M. Tezuka, M. Morita, K. Isogai, and S. Okada, "Evaluation of some halogen biocides using a microbial biofilm system," Water Research, vol. 39, no. 17, pp. 4126–4132, 2005.
3. D. Berry, C. W. Xi, and L. Raskin, "Microbial ecology of drinking water distribution systems," Current Opinion in Biotechnology, vol. 17, no. 3, pp. 297–302, 2006.
4. M. W. LeChevallier, W. Schulz, and R. G. Lee, "Bacterial nutrients in drinking water," Applied and Environmental Microbiology, vol. 57, no. 3, pp. 857–862, 1991. View at Google Scholar · View at Scopus
5. C. M. Buswell, Y. M. Herlihy, L. M. Lawrence et al., "Extended survival and persistence

of Campylobacterspp. in water and aquatic biofilms and their detection by immunofluorescent-antibody and -rRNA staining," Applied and Environmental Microbiology, vol. 64, no. 2, pp. 733–741, 1998.

6. J. M. Regan, G. W. Harrington, and D. R. Noguera, "Ammonia- and nitrite-oxidizing bacterial communities in a pilot-scale chloraminated drinking water distribution system," Applied and Environmental Microbiology, vol. 68, no. 1, pp. 73–81, 2002.

7. P.-Y. Hong, C. Hwang, F. Ling, G. L. Andersen, M. W. LeChevallier, and W.-T. Liu, "Pyrosequencing analysis of bacterial biofilm communities in water meters of a drinking water distribution system,"Applied and Environmental Microbiology, vol. 76, no. 16, pp. 5631–5635, 2010.

8. P. D. Schloss, S. L. Westcott, T. Ryabin et al., "Introducing mothur: open-source, platform-independent, community-supported software for describing and comparing microbial communities," Applied and Environmental Microbiology, vol. 75, no. 23, pp. 7537–7541, 2009.

9. K. Tamura, D. Peterson, N. Peterson, G. Stecher, M. Nei, and S. Kumar, "MEGA5: molecular evolutionary genetics analysis using maximum likelihood, evolutionary distance, and maximum parsimony methods,"Molecular Biology and Evolution, vol. 28, no. 10, pp. 2731–2739, 2011.

10. A. C. Martiny, T. M. Jørgensen, H.-J. Albrechtsen, E. Arvin, and S. Molin, "Long-term succession of structure and diversity of a biofilm formed in a model drinking water distribution system," Applied and Environmental Microbiology, vol. 69, no. 11, pp. 6899–6907, 2003.

11. A. Gobet, S. I. Böer, S. M. Huse et al., "Diversity and dynamics of rare and of resident bacterial populations in coastal sands," ISME Journal, vol. 6, no. 3, pp. 542–553, 2012.

12. W. F. Lin, Z. S. Yu, X. Chen, R. Liu, and H. Zhang, "Molecular characterization of natural biofilms from household taps with different materials: PVC, stainless steel, and cast iron in drinking water distribution system," Applied Microbiology and Biotechnology, vol. 97, no. 18, pp. 8393–8401, 2013.

13. P. Deines, R. Sekar, H. S. Jensen, et al., "MUWS (Microbiology in Urban Water Systems)—an interdisciplinary approach to study microbial communities in urban water systems," in Integrating Water Systems, pp. 397–403, 2014.

14. R. S. Roeder, J. Lenz, P. Tarne, J. Gebel, M. Exner, and U. Szewzyk, "Long-term effects of disinfectants on the community composition of drinking water biofilms," International Journal of Hygiene and Environmental Health, vol. 213, no. 3, pp. 183–189, 2010.

15. I. Douterelo, R. L. Sharpe, and J. B. Boxall, "Influence of hydraulic regimes on bacterial community structure and composition in an experimental drinking water distribution system," Water Research, vol. 47, no. 2, pp. 503–516, 2013.

16. R. P. Revetta, V. Gomez-Alvarez, T. L. Gerke, C. Curioso, J. W. S. Domingo, and N. J. Ashbolt, "Establishment and early succession of bacterial communities in monochloramine-treated drinking water biofilms," FEMS Microbiology Ecology, vol. 86, no. 3, pp. 404–414, 2013.

17. J. O. Falkinham, M. D. Iseman, P. de Haas, and D. van Soolingen, "Mycobacterium avium in a shower linked to pulmonary disease," Journal of Water and Health, vol. 6, no. 2, pp. 209–213, 2008.

18. R. Liu, Z. Yu, H. Zhang, M. Yang, B. Shi, and X. Liu, "Diversity of bacteria and mycobacteria in biofilms of two urban drinking water distribution systems," Canadian Journal of Microbiology, vol. 58, no. 3, pp. 261–270, 2012.

19. L. C. Simões, M. Simões, and M. J. Vieira, "Influence of the diversity of bacterial isolates from drinking water on resistance of biofilms to disinfection," Applied and Environmental Microbiology, vol. 76, no. 19, pp. 6673–6679, 2010.

20. R. Liu, Z. Yu, H. Guo, M. Liu, H. Zhang, and M. Yang, "Pyrosequencing analysis of eukaryotic and bacterial communities in faucet biofilms," Science of the Total Environment, vol. 435-436, pp. 124–131, 2012.

21. H. F. Sun, B. Y. Shi, Y. H. Bai, and D. Wang, "Bacterial community of biofilms developed under different water supply conditions in a distribution system," Science of the Total Environment, vol. 472, pp. 99–107, 2014.

22. J. E. Kostka, D. D. Dalton, H. Skelton, S. Dollhopf, and J. W. Stucki, "Growth of iron (III)-reducing bacteria on clay minerals as the sole electron acceptor and comparison of growth yields on a variety of oxidized iron forms," Applied and Environmental Microbiology, vol. 68, no. 12, pp. 6256–6262, 2002.

CHAPTER 19 TEMPORAL VARIATIONS IN THE ABUNDANCE AND COMPOSITION OF BIOFILM COMMUNITIES COLONIZING DRINKING WATER DISTRIBUTION PIPES

John J. Kelly, Nicole Minalt, Alessandro Culotti, Marsha Pryor, Aaron Packman

INTRODUCTION

The pipes that are used to transport drinking water through municipal drinking water distribution systems (DWDS) are challenging habitats for microorganisms. The transported water generally contains chemical disinfectants such as chlorine or chloramine, as well as very low concentrations of organic carbon and inorganic nutrients [1]. Despite these challenges, microbes frequently colonize the interior surfaces of DWDS pipes [1]. Indeed the pipe surfaces may represent the best possible microbial habitats within DWDS, as previous research has shown that surface attachment can enable bacteria to grow in oligotrophic habitats due to the accumulation of nutrients at the solid-liquid interface [2], [3]. In addition, biofilm formation can provide bacteria with protection against chemical disinfectants [4]–[6].

Relatively little is known about the species composition and ecology of biofilms within DWDS. Obtaining samples from below-ground pipes is difficult and expensive [7], and as a result most of the work that has been done on drinking water biofilms has been based on model systems run in the laboratory [6], [8], [9]. While these studies have provided valuable insight into biofilm formation within drinking water, model systems often differ in significant ways from actual DWDS, including duration of biofilm growth, temporal variability, water flow conditions, diversity of pipe materials and the presence or absence of disinfectants. Additionally, most of the work on microbes within DWDS has focused on classical pathogens such as *Vibrio cholerae* and*Salmonella typhi*, emerging pathogens such as *Campylobacter jejuni* and *Legionella pneumophila*, or indicator organisms for fecal contamination, such as coliform bacteria [1]. Many of these studies have used culture-based techniques [1], which are able to assess only a small fraction of natural microbial diversity [10]. In contrast, recent studies using molecular approaches have demonstrated the predominance of nonpathogenic bacterial species within drinking water biofilms [7], [11], [12].

Information regarding the composition and ecology of biofilms within DWDS is valuable for several reasons. First, it improves our general understanding of microbial life in oligotrophic habitats, including built environments. Secondly, biofilms in DWDS are a concern for public health as they have been shown to harbor and protect pathogens from disinfectants and increase pathogen persistence in DWDS [13]. Thirdly, there is evidence that the activities of nitrifying microorganisms

Citation: Kelly JJ, Minalt N, Culotti A, Pryor M, Packman A (2014) Temporal Variations in the Abundance and Composition of Biofilm Communities Colonizing Drinking Water Distribution Pipes. PLoS ONE 9(5): e98542. doi:10.1371/journal.pone.0098542

in DWDS can decrease monochloramine concentrations [14], which could lead to increased microbial growth and possibly increased persistence and transport of pathogens [15]. Finally, the presence of biofilms in DWDS can promote pipe corrosion [16] and cause taste and odor problems in the water [3]. Understanding the microbial composition and development of DWDS biofilms can suggest strategies for management of these problems.

We report here the analysis of biofilms found within pipe samples collected five times over an 18 month period from a DWDS in Pinellas County, FL, USA. Sections of below-ground pipe were cut and transported to the lab, and biofilm communities within the pipes were analyzed by heterotrophic plate counts and tag pyrosequencing of 16S rRNA genes.

MATERIALS AND METHODS

Pipe samples were collected from the municipal drinking water distribution system in Pinellas County, FL, USA, which is operated by Pinellas County Utilities (PCU). The water supply for this system is a blend of groundwater and treated surface water, with desalinated seawater being used periodically as needed. Before entering the DWDS water is treated at the Tampa Bay Water Treatment Plant (TBW) by a four stage process: 1) clarification using the ACTIFLO system (Kruger Inc., Cary, NC), 2) ozone disinfection, 3) biologically active filtration, and 4) disinfection with chlorine. Disinfection within the PCU DWDS is based on maintenance of a chloramine residual, although the utility does switch to chlorine disinfection for a brief period in the summer each year to limit biofilm growth. The switch to chlorine occurred once during our sampling period, specifically between August 1 and September 11, 2011. The average flow through the system during our sampling period was 54.9 million gallons per month.

Sections of six-inch ductile-iron pipe from the main line in the Seminole, FL region within the Pinellas County DWDS were collected periodically over an 18 month period during planned replacement events. Specifically, pipe sections were collected on February 20, 2011, July 20, 2011, December 20, 2011, March 20, 2012, and August 8, 2012. All pipe samples were collected at approximately 9 am, before peak demand which occurs at approximately 10 am. The water main that we sampled was approximately 40–45 years old. The average water age for the main line during our sampling period was 3 to 4 days and the average velocity was 0.4 c.f.s. The sampling location is downstream of a large above ground storage (AGS) tank that was permanently shut down on June 19, 2012 due to concerns about nitrification occurring within the tank. A key outcome of the shut-down of this water tank was decreased water age at our sampling location for the August 2012 sampling date.

On each sampling date the road surface and soil above the water main were excavated using a backhoe. Soil was cleared from around the pipe by hand using a shovel and any soil adhering to the exterior of the pipe was removed using a brush or cloth. The exterior of the pipe was disinfected by pouring a 10% bleach solution over the pipe and simultaneously wiping the pipe exterior with a cloth saturated with 10% bleach solution. A section approximately 1 ft. long was cut from the 6-inch diameter pipe using a scoring-type pipe cutter. The pipe section was capped at one end using a flexible PVC cap with adjustable clamps. The capped pipe section was filled with dechlorinated water from the same water main, which was collected in a plastic carboy and dechlorinated on-site using sodium thiosulfite (~2 g/g Cl_2). The pipe section was filled until overflowing, capped at the other end, placed in a cooler with ice packs, and shipped overnight to Northwestern University, Evanston, IL. Water was also collected from the main for chemical analysis and transported to the Pinellas County Utilities lab in a cooler.

Sample Processing

In the laboratory one cap was removed from the pipe section and the water was carefully poured out. The interior of the pipe section was rinsed gently with filter-sterilized tap water to remove unattached or settled solids. Biofilm samples were collected from three separate, evenly-spaced sections of the pipe interior. For each section, biofilm material was collected by scraping a 5.5 cm wide band around the entire interior circumference of the pipe with a sterile spatula, resulting in a sampling area of 263.3 cm^2. The collected biofilm material was transferred to a sterile 10 ml vial and 3 ml ultrapure water was added. The suspension was homogenized by vortexing and large particles such as corrosion byproducts were allowed to settle out of suspension. From this suspension 100 µl was used for the plate count assay (seemethod below) and 2 ml was used for molecular analysis of the biofilm communities (seemethods below). The 2 ml for molecular analysis was transferred to a 2 ml microcentrifuge tube and centrifuged at 10,000×g for 10 minutes. The supernatant was then discarded and the remaining biofilm pellet was stored at -20°C.

Water Chemistry

All water chemistry analyses were done by Pinellas County Utilities. Analyses were performed based on either EPA Methods or Standard Methods for the Examination of Water and Wastewater [17]. The specific methods used for each assay and results of the water chemistry assays are listed in Table 1. Total haloacetic acid (HAA) concentrations in the source water were determined periodically by EPA method 552.2.

Table 1. Water chemistry.

		Sampling Date				
Analyte	Method	February 2011	July 2011	December 2011	March 2012	August 2012
Temperature (°C)	SM 2550 B	20.6	29.1	20.2	23.8	28.5
pH	SM 4500 H-B	7.6	7.58	7.65	7.52	7.69
Total Organic Carbon (mg L^{-1})	SM 5310-C	1.8	1.9	2.2	2	2.3
Total Phosphorous as P (mg L^{-1})	EPA 365.4	0.35	0.24	0.41	0.34	0.31
Nitrate as N (mg L^{-1})	EPA 300.0	0.08	0.11	0.09	0.08	0.43
Nitrite as N (mg L^{-1})[1]	EPA 300.0	BD	BD	BD	BD	BD
Free Ammonia as N (mg L^{-1})	SM 4500 NH3-F	0.19	0.42	0.21	0.16	0.14
Total chlorine (mg L^{-1})	SM 4500 CL-G	2.9	1.6	2.5	3	3
Monochloramine (mg L^{-1})	SM 4500 CL-G	3.2	1.6	2.2	2.6	2.7
Alkalinity as $CaCO_3$ (mg L^{-1})	SM 2320 B	170	190	170	180	150
Calcium Hardness (mg L^{-1})	SM 2340 B	210	202	202	205	192
Specific Conductance (umhos cm^{-1})	SM 2510 B	549	521	429	447	477
Aluminum (mg L^{-1})[2]	EPA 200.7-DW	BD	BD	BD	BD	BD
Calcium (mg L^{-1})	EPA 200.7-DW	84.3	80.7	81	81.9	65.1
Iron (mg L^{-1})	EPA 200.7-DW	0.055	0.079	0.381	0.046	0.014
Magnesium (mg L^{-1})	EPA 200.7-DW	7.35	6.51	6.39	6.53	7.14

Plate Count Assay

R2A agar was purchased as a dried powder (Fisher Scientific, Pittsburgh, PA) and prepared according to the manufacturer's instructions. Biofilm suspensions were serially diluted from $10^2$to 10^4 in ultrapure water and 100 µl of all dilutions were spread on R2A agar plates. Plates were incubated at 37°C for 36 hours and counted. Counts were normalized based on the surface area of the pipe from which the biofilm had been collected.

Molecular Analysis of Biofilm Communities

DNA was extracted from the frozen biofilm pellets using the Power Biofilm DNA Kit (MoBio Laboratories, Carlsbad, CA) according to the manufacturer's instructions and stored at −20°C. For tag pyrosequencing of bacterial 16S rRNA genes the extracted DNA was sent to Research and Testing Laboratory (Lubbock, TX). Polymerase chain reaction (PCR) amplification was performed using primers 530F and 1100R [18]. The 530F primer was chosen in order to obtain sequences for the V4 hypervariable region, which has been shown to provide species richness estimates comparable to those obtained with the nearly full-length 16S rRNA gene [19]. Sequencing reactions utilized a Roche 454 FLX instrument (Roche, Indianapolis, IN) with Titanium reagents. Sequences were processed using MOTHUR software [20]. Briefly, any sequences containing ambiguities or homopolymers longer than 8 bases were removed. Remaining sequences were individually trimmed to retain only high quality sequence reads and sequences were aligned based on comparison to the SILVA-compatible bacterial alignment database available within MOTHUR. Aligned sequences were trimmed to a uniform length of 127 bases and chimeric sequences were removed using UCHIME [21] run within MOTHUR. Sequences were grouped into phylotypes by comparison to the SILVA-compatible bacterial alignment database and algal chloroplast and mitochondrial sequences were removed from the data set. To avoid any biases associated with different numbers of sequences in each of the samples we randomly subsampled a total of 6,808 sequences from each sample, and used these subsampled sequences for all downstream analyses. Sequences were clustered into operational taxonomic units (OTUs) based on 97% sequence identity using the average neighbor algorithm. Rarefaction curves were produced using MOTHUR. The total OTU richness in each sample was calculated based on the Chao1 richness estimator [22]. The diversity of each sample was assessed based on the Shannon index [23] calculated using the Primer software package (Primer V.5, Primer-E Ltd., Plymouth, UK). The community composition of the individual samples was compared by using MOTHUR to calculate distances between sites based on the theta index [24]. The significance of differences in theta index scores between sites was assessed by analysis of molecular variance (AMOVA) run within MOTHUR. PC-ORD v. 6.08 (MjM Software, Gleneden Beach, Oregon, USA) was used to ordinate the theta index distance matrix via non-metric multidimensional scaling (nMDS) and to determine correlations between the water chemistry data and the axes in the nMDS ordination.

Statistical Analyses

Plate count data and diversity scores were analyzed by one-way analysis of variance (ANOVA) based on sampling date and pairwise comparisons were made by Tukey's post hoc test. Correlations were assessed by determining Pearson product-moment correlation coefficients and Bonferroni-corrected probabilities. Correlation between the relative abundance of *Methylomonas* sequences and the concentration of total HAA in the source water was based on average abundance data for each biofilm sampling date and the HAA concentration from the closest source water sampling date. All statistical analyses were run using Systat 13 (Systat Software, Inc., San Jose, CA) and p values less than 0.05 were considered to be significant.

Data Sharing

All of the sequence data analyzed in this paper can be downloaded from the National Center for Biotechnology

Information (NCBI) Sequence Read Archive (SRA) with accession number SRP038002.

RESULTS

Water Source

The source water for the PCU DWDS is a blend of groundwater, surface water, and desalinated seawater. An approximately equal mix of groundwater and surface water was the most common during the study period, although there were some significant variations in the relative proportions of source waters (Fig. 1). For the four weeks preceding our February 2011 sampling date, the source water averaged 38% Groundwater, 48% surface water and 14% seawater. Between June 20, 2011 and March 20, 2012, a period which included three of our sampling dates, the source water averaged 50% Groundwater, 50% surface water and 0% seawater. Prior to our August 2012 sampling date there were some dramatic shifts in source waters. Between April 14, 2012 and June 9, 2012 the water was predominantly groundwater, with an average of 73% of the water coming from groundwater over that period. Finally, between July 6, 2012 and August 8, 2012 the water was predominantly surface water, with surface water representing an average of 67% of the source water over that period.

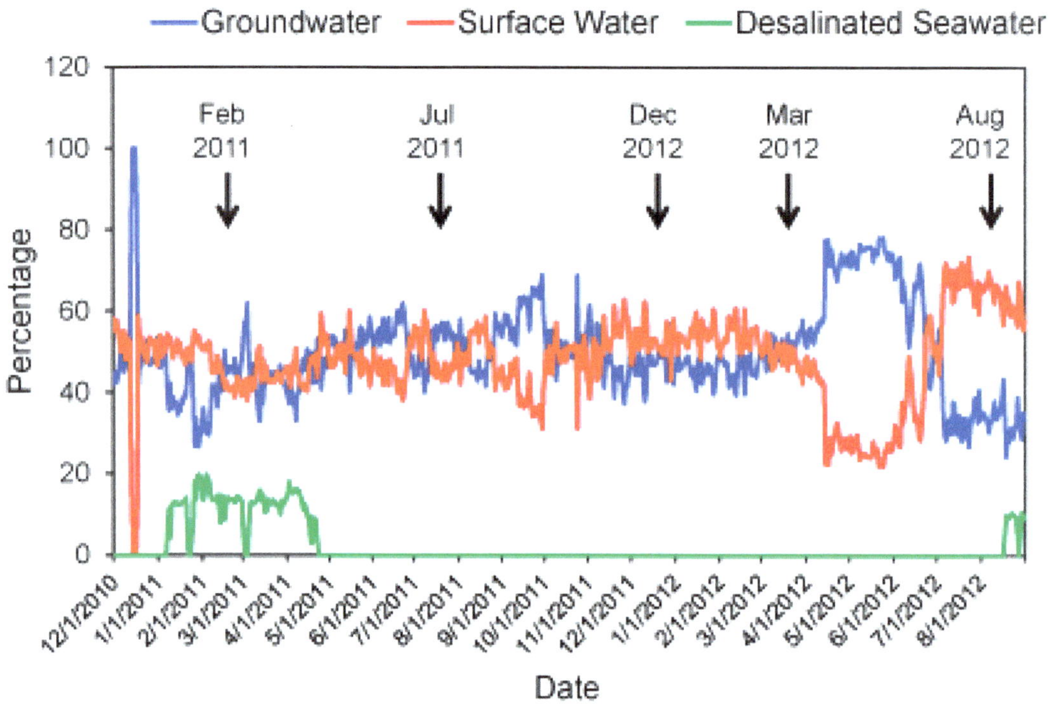

Figure 1. Relative percentages of source waters within PCU drinking water distribution system.

Water Chemistry

Water temperature varied seasonally in the water main from which the biofilm samples were obtained, being higher in summer months (July and August) than winter months (February, March and December) (Table 1). Some water chemistry parameters varied with sampling date but did not show seasonal trends. For example, there were large fluctuations in concentrations of nitrate (from 0.08 to 0.43 mg L^{-1}), free ammonia (from 0.14 to 0.42 mg L^{-1}) and iron (from 0.014 to 0.381 mg L^{-1}). The August 2012 sampling date had the highest concentration of nitrate, approximately four to five times higher than all other sampling dates, and the lowest concentration of free ammonia (Table 1). Analysis of the source water from TBW confirmed a high concentration of nitrate in the source water immediately prior to the August 2012 sampling date (data not shown). Analysis of the source water from TBW also showed fluctuations in concentrations of total haloacetic acids over the course of the study (Fig 2).

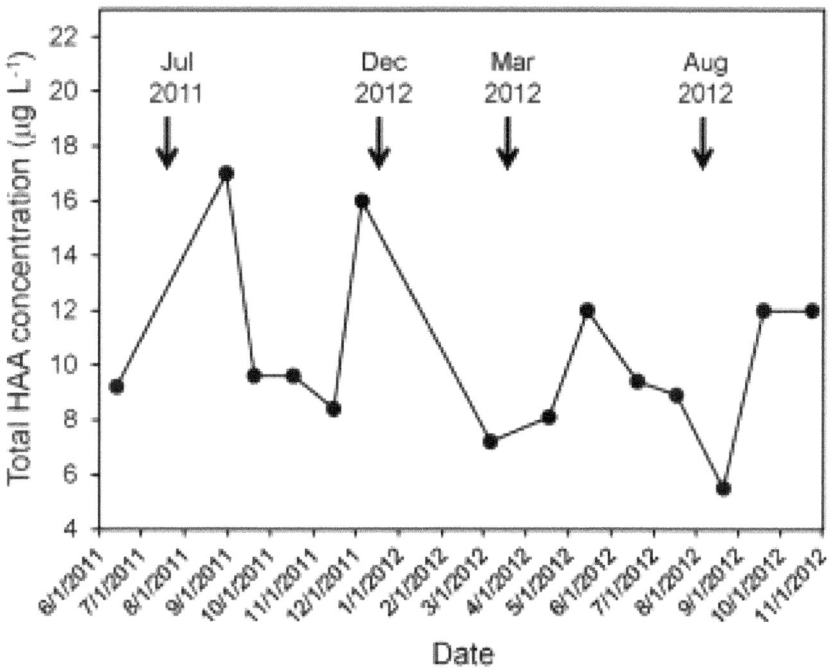

Figure 2. Concentrations of total haloacteic acids (HAA) in the source water for the PCU drinking water distribution system.

Plate Count Assay

Sampling date significantly affected the numbers of bacteria within the pipe biofilms as measured by heterotrophic plate count assay, with bacterial counts varying over 3 orders of magnitude between sampling dates ($p<0.001$; Table 2). Samples from the summer months (July and August) had significantly higher counts than the winter months (February, March and December), and there was a significant positive correlation between plate counts and water temperature ($R^2=0.605$; $p<0.001$). The August 2012 sampling date, which had the highest nitrate concentration, had bacterial counts that were significantly higher than all other sampling dates ($p<0.001$; Table 2), and there was a significant positive correlation between plate counts and nitrate concentration ($R^2=0.799$; $p<0.001$). There was no significant correlation between plate counts and phosphorous concentration ($R^2=0.249$; $p=0.058$). There were also no significant correlations between plate counts and total chlorine ($R^2=0.005$, $p=0.802$) or between plate counts and monochloramine ($R^2=0.285$, $p=0.548$).

Table 2. Numbers of heterotrophic bacteria in pipe biofilms based on plate count assay.

Sampling Date	Number of Bacteria (cfu cm^{-2})[1]	
February 2011	215	a
July 2011	10,887	b
December 2011	2,013	a
March 2012	36	a
August 2012	23,167	c

Bacterial Community Analysis

Tag pyrosequencing of 16S rRNA genes was used to profile the bacteria within the biofilms lining the drinking water pipes. Twelve samples representing three replicate biofilm samples from each of four sampling dates (July 2011, December 2011, March 2012 and August 2012) were sequenced successfully. Despite repeated attempts, DNA from the February 2011 pipe samples could not be amplified with the 530F and 1100R primers, so no sequence data were obtained for this sampling date. After processing, the data set included a total of 159,604 high-quality sequence reads. There was significant variation in the number of sequences obtained for each of the samples, from a low of 8,785 to a high of 35,854. To avoid biases associated with unequal numbers of sequences, 6,808 sequences were randomly selected from each of the twelve samples using the subsample command in MOTHUR, producing a total of 81,696 high quality sequences that were used for all analyses of community composition. With this subsampled data set, rarefaction curves for all samples had reached plateaus (Fig. 3), suggesting that the sequencing depth obtained in this study was adequate to capture most of the diversity within these communities. Similarly, a comparison of the total number of OTUs observed in each sample and the estimated total number of OTUs present in each sample demonstrated that for all samples more than 50% of the estimated total number of OTUs in each sample were detected (Table 3).

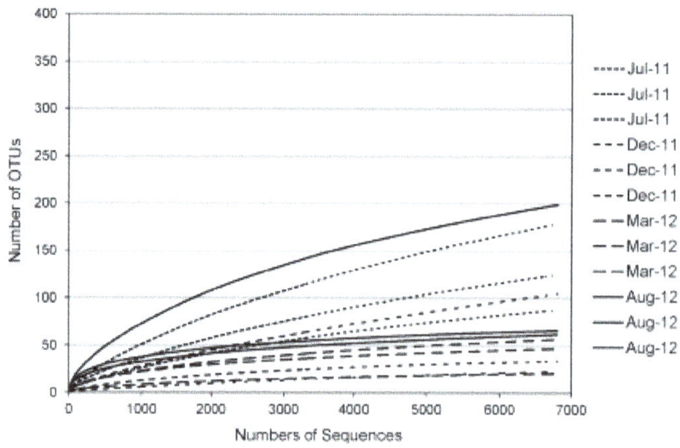

Figure 3. Rarefaction curves for biofilm bacterial communities based on tag pyrosequencing of 16S rRNA genes.

Table 3. Comparison of the number of observed and estimated bacterial OTUs in pipe biofilm communities based on 16S tag pyrosequencing data.

Sampling Date	Observed OTUs[1]	Estimated OTUs[2]	Percent Coverage[3]
July 2011	88	158	55.7%
July 2011	179	332	53.9%
July 2011	125	239	52.3%
December 2011	34	42	81.0%
December 2011	105	194	54.1%
December 2011	23	38	60.5%
March 2012	47	67	70.1%
March 2012	57	87	65.5%
March 2012	21	26	80.8%
August 2012	199	297	67.0%
August 2012	66	74	89.2%
August 2012	62	75	82.7%

nMDS ordination (Fig. 4) and AMOVA analysis (Table 4) indicated that there were significant differences between the biofilm bacterial communities from the different sampling dates. Biofilm bacterial communities from August 2012 were the most distinct and were significantly different from the communities from all other sampling dates (Fig. 4 and Table 4). The nMDS ordination also revealed relationships between the community composition and water chemistry parameters (Fig 4). The composition of the biofilm bacterial community from August 2012 was positively correlated with nitrate concentration and negatively correlated with calcium concentration and hardness. The August 2012 communities were also positively correlated with pH and total organic carbon concentration and negatively correlated with alkalinity. In addition, the separation of the biofilm communities from July 2011 and March 2012 on the nMDS ordination was correlated with the concentrations of free ammonia, total chlorine and monochloramine (Fig. 4). Finally, the nMDS ordination demonstrated that there was variation in bacterial community composition between replicates from both the July and March sampling dates, whereas the December and August samples showed a high degree of similarity between replicates (Fig. 4).

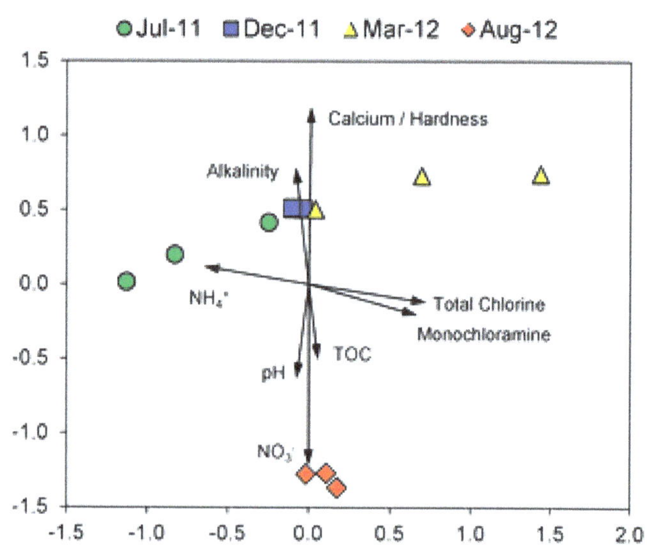

Figure 4. Non-metric multidimensional scaling ordination of biofilm bacterial communities based on tag pyrosequencing of 16S rRNA genes.

Stress value of ordination is 0.129. Vector lines represent correlations between physical and chemical variables and the ordination axes. Variables with correlation values less than 0.5 for both axes are not shown.

Table 4. AMOVA analysis of 16S tag pyrosequencing data from pipe samples.

Comparison	p value
August-July	<0.001
August-December	<0.001
August-March	0.0497
July-December	<0.001
July-March	0.0500
December-March	0.0487

Bacterial Community Diversity

Pyrosequencing analysis identified a total of 677 OTUs (3% distance) within these biofilms, and the number of OTUs per sample ranged from 21 to 199 (Table 3). Despite this large number of OTUs, the diversity of these communities was low, with three of the four sampling dates showing Shannon index scores below 1.2 (Fig 5A). The Shannon diversity index scores for the biofilm bacterial communities varied significantly between sampling dates ($p<0.01$) with the biofilms from August 2012 being the most diverse (Fig. 5A). There was also a significant positive correlation ($p<0.01$) between bacterial abundance in the biofilms and the diversity of the bacterial communities (Fig 5B). As indicated by the low diversity index scores, all of the biofilm communities were dominated by a small number of OTUs, with the ten most abundant OTUs accounting for 93% of the total sequences in the data set.

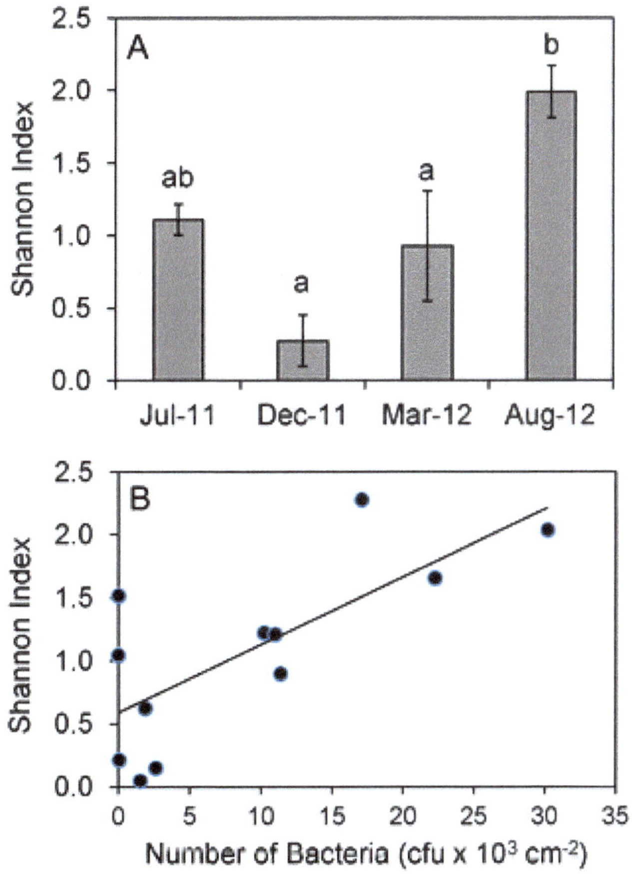

Figure 5. Diversity of bacterial biofilm communities and relationship between bacterial abundance and diversity.

(A) Shannon index scores calculated using 16S rRNA gene tag pyrosequencing data. Each data point represents mean value (n=3) with standard error bars. ANOVA indicated a significant effect of sampling date ($p<0.01$). Data points with different letters are significantly different based on Tukey's posthoc test ($p<0.05$). (B) Relationship between biofilm bacterial abundance and diversity. Linear regression indicated $R2=0.541$. Pearson correlation analysis indicated a significant correlation between numbers of bacteria and diversity concentration and resistance ($p<0.01$).

Bacterial Community Taxonomic Composition

Analysis of pyrosequencing data indicated that sequences corresponding to the genus*Methylomonas* were the most abundant within the biofilm communities, accounting for 41% of the sequences in the total data set (Table 5). Other abundant sequences corresponded to the genera *Acinetobacter*, *Mycobacterium*, *Pseudomonas*, and *Methylobacterium*, as well as an unclassified genus from the family Xanthomonadaceae and an unclassified genus from the class Betaproteobacteria. The relative abundance data indicated some commonalities between the biofilm bacterial communities from the four sampling dates (Table 5). For example,*Methylomonas* accounted for more than 30% of the sequences for three of the four sampling dates. However, the data also illustrate that there were dramatic differences in the composition of biofilm communities from the different sampling dates, as *Methylomonas* accounted for more than 95% of the sequences in December 2011, but less than 2% in August 2012. A previous study suggested a link between *Methylomonas* bacteria in drinking water and HAA[25], and there was a significant positive correlation between the relative abundance of*Methylomonas* sequences in our biofilms and the concentration of total HAA in the source water (R^2=0.962; p=0.027). *Acinetobacter* abundance also varied significantly between sampling dates, accounting for 74% of the sequences in August 2012 biofilms, but representing less than 0.1% of the sequences from the other sampling dates.

Table 5. Relative abundance of most numerically dominant bacterial genera[1].

Bacterial genus	All Samples	July 2011	December 2011	March 2012	August 2012
Methylomonas	40.8	33.6±15.6	95.5±3.5	32.4±21.3	1.3±0.6
Acinetobacter	18.5	0.0±0.0	0.0±0.0	0.0±0.0	74.2±2.4
Mycobacterium	14.7	59.1±14.3	0.2±0.2	0.1±0.1	0.0±0.0
Unclass. Xanthomonadaceae	13.5	0.0±0.0	2.6±2.4	50.4±24.4	0.1±0.1
Pseudomonas	4.7	0.6±0.3	0.3±0.2	15.5±8.6	2.1±0.2
Unclass. Betaproteobacteria	2.3	0.3±0.1	0.1±0.1	0.0±0.0	9.0±1.6
Methylobacterium	1.2	3.7±2.3	0.2±0.2	0.1±0.0	0.8±0.1
Unclass. Bacteria	1.0	0.5±0.2	0.3±0.1	0.0±0.0	3.2±2.8
Massilia	0.9	0.0±0.0	0.0±0.0	0.0±0.0	3.4±0.7
Unclass. Gammaproteobacteria	0.4	0.7±0.4	0.2±0.2	0.7±0.5	0.0±0.0

Sequences corresponding to several genera containing pathogenic species were detected in the biofilms. In a total of 81,696 sequences from all four sampling dates, there were 18*Escherichia* sequences, 5 *Clostridium* sequences and 3 *Streptococcus* sequences detected, so these genera were extremely rare. *Mycobacterium* represented 59% of the sequences in July 2011, while representing less than 1% of sequences from the other sampling dates. The genus *Mycobacterium* includes two well-known human pathogens, *M. tuberculosis* and *M. leprae*, although these species are generally not found in the environment [26]. The genus*Mycobacterium* also includes a large number of non-pathogenic or occasionally pathogenic species [26]. *Acinetobacter* accounted for 74% of the sequences in August 2012 but less than 1% of sequences from the other sampling dates. Bacteria from the genus *Acinetobacter* are a common cause of nosocomial infections among immunocompromised patients, with the most common example being respiratory infections of ventilated patients [27]. Due to the short length of the sequences obtained in our study, we were unable to discriminate any of the sequences down to the species level, so it is unclear whether the sequences from any of these genera represented pathogenic or non-pathogenic organisms.

The pipes analyzed in this study were ductile iron and did not show significant corrosion or the presence of tubercles. The pyrosequencing data identified only a handful of sequences corresponding to genera known to contain iron-oxidizing species: *Acidovorax* (1 sequence),*Aquabacterium* (1 sequence) and *Thiobacillus* (4 sequences) [28], [29]. In addition, no sequences corresponding to any known ammonia oxidizing bacterial genera were detected. However, a few sequences (19 total) from a known nitrite oxidizing genus, *Nitrospira*, were detected, and all of these sequences were found in the July 2011 samples, which also showed the highest concentration of free ammonia (Table 1).

DISCUSSION

Pipe samples were collected from the same region of a drinking water distribution system in Pinellas County, FL on five dates over an 18-month period between February 2011 and August 2012. Water from these pipes showed seasonal variations in temperature and some large fluctuations in concentrations of nitrate, free ammonia and iron. The August 2012 sample had a much higher concentration of nitrate and a lower concentration of free ammonia relative to other sampling dates. The shut-down of the upstream above ground storage tank prior to the August 2012 sampling date probably contributed to the observed differences in water chemistry between the sampling dates. The above ground storage tank was shut down because of nitrification occurring in the tank, which could have contributed to the high nitrate and low free ammonia concentrations in the August 2012 sample. The August 2012 samples also had a different source-water mixture than other dates: for one month prior to our sampling date the source water was composed of a higher percentage of surface water and a lower percentage of groundwater than typical for this system. Analysis of the TBW source water at its point of entry to the DWDS confirmed the high nitrate and low alkalinity of the source water at the time of our August 2012 sampling, indicating that source water was a key driver of the unique aspects of the August 2012 water chemistry.

The abundance of bacteria in the biofilms varied greatly across the sampling dates, with higher bacterial numbers in the summer months and lower bacterial numbers in the winter months. These differences in abundance may have been driven by the strong seasonal differences in temperature in the system, as the water in the summer months was on average 7.3°C warmer than in the winter months, and a temperature change of this magnitude can significantly increase the growth rates of mesophilic bacteria [30]. The connection between bacterial abundance in the biofilms and water temperature was further supported by a statistically significant correlation between these two parameters. The differences in bacterial abundance may also have been related to the availability of inorganic nutrients in the drinking water, specifically nitrogen. We found a significant correlation between bacterial cell numbers and nitrate concentrations. Other studies have indicated that inorganic nutrients can be a limiting factor for bacterial growth in DWDS [1]. In our samples, nitrate concentrations were higher in the summer months than in the winter months and were highest in the August 2012 sample. Finally, our results indicated no significant relationship between the abundance of bacteria within the biofilms and the concentrations of total chlorine or monochloramine. These results suggest that low concentrations of chlorine disinfectants may not be effective at limiting biofilm growth within DWDS, owing to the protection provided by the biofilm matrix, as has been demonstrated by previous studies [4], [5], [6]. It should be noted that we assessed bacterial abundance within the pipe biofilms using heterotrophic plate counts. While the limitations of plate counts as estimates of bacterial abundance are well known [31], this method is commonly used to assess bacterial loads in DWDS, so we chose to use this method to make our results comparable to existing data.

Tag pyrosequencing analysis revealed a total of 677 OTUs in the biofilm bacterial communities, with the estimated total numbers of bacterial OTUs per sample ranging from 26 to 332. A previous pyrosequencing survey of drinking water meters revealed a similarly high number of total bacterial OTUs [7]. Despite the large numbers of OTUs observed in our pipe biofilms, these communities were not very diverse, with all samples having Shannon index scores below 2. These low diversity scores reflect the fact that these communities were dominated by a small number of taxa, specifically *Methylomonas*, *Acinetobacter* and*Mycobacterium*. The diversity of the bacterial biofilm communities was significantly correlated with bacterial abundance, suggesting that when conditions within the DWDS were favorable for bacterial growth (e.g. higher temperature and higher nitrate concentrations) a wider range of bacterial taxa were able to proliferate within the biofilms, whereas when conditions were not favorable (e.g. lower temperature and lower nitrate concentrations), a more limited range of bacteria were able to persist. Within our data set the August 2012 samples seemed to represent the most favorable conditions for biofilm growth, as this sampling date had the highest nitrate concentration, one of the highest water temperatures, and supported biofilms with the highest bacterial numbers and the highest bacterial diversity.

Methylomonas was the most numerically dominant bacterial genus within the biofilms, with its sequences accounting for more than 40% of all of the sequences recovered. *Methylomonas* is a genus of type I methanotrophic bacteria, which obtain their carbon and energy from the oxidation of methane. *Methylomonas* sequences have been detected previously in drinking water [32], and a recent study of water meter biofilms also detected sequences corresponding to the family *Methylococcaceae*, the bacterial family that includes the genus *Methylomonas*,[7]. *Methylobacterium*, a genus of methylotrophic bacteria that oxidize methyl compounds such as methanol but cannot metabolize methane, was also one of the most commonly detected taxa in our biofilms, although it represented just over 1% of the total sequences. Methylotrophic bacteria, specifically *Methylophilus*, were also detected in biofilms within drinking water meters [7]. The factors favoring high abundance of methanotrophic and methylotrophic taxa within drinking water distribution systems are unclear. We would not expect high concentrations of methane or methanol in drinking water, however these compounds could be produced in anoxic sites within DWDS via anaerobic processes such as methanogenesis or fermentation. Another process that might have supported the growth of methylotrophic bacteria is utilization of haloacetic acid, which is a common by-product of chlorination of drinking water. A recent study isolated a *Methylobacterium* strain from a DWDS biofilm that was capable of growth with haloacetic acid as the sole carbon source [25]. The results of our study, which showed a significant correlation between the relative abundance of*Methylomonas* sequences and HAA concentration in the source water, lends further support to this hypothesis.

Our results demonstrated that there was significant variation in the taxonomic composition of the biofilm bacterial communities within the pipe sections across our sampling dates. nMDS and AMOVA analyses indicated that the August 2012 bacterial communities were the most distinct in terms of their composition, and further examination of the composition of these biofilms revealed that the August 2012 communities were dominated by *Acinetobacter*sequences (74% of total sequences), while *Acinetobacter* sequences represented less than 0.1% of sequences from the July, December and March samples. *Acinetobacter* is a genus of Gram-negative, heterotrophic bacteria [33] that is commonly found in soils and groundwater[34]–[36]. *Acinetobacter* is also one of the most common groups of bacteria isolated from drinking water [1], and a number of *Acinetobacter* species have been shown to produce biofilms [37]–[39]. Therefore the presence of *Acinetobacter* in the pipe biofilms was not surprising. However, the dramatic variation in *Acinetobacter* abundance that we observed between August 2012 and the other sampling dates was remarkable. There were several unique features of the August 2012 sampling date that may have contributed to its distinct biofilm composition. First, the source of the drinking water within the distribution system changed prior to August 2012 to predominantly groundwater for a period of three weeks, with this period of groundwater dominance occurring approximately two months prior to the August 2012 sampling. Since *Acinetobacter* are regularly detected in soil and groundwater, this switch to a groundwater dominated system prior to August 2012 might have provided an additional inoculum of *Acinetobacter* that were able to become established within the pipe biofilms. Another related feature of the August 2012 samples was that the drinking water at that sampling time had a much higher nitrate concentration (at least four times higher) than all of the other sampling dates. The nMDS analysis indicated that nitrate concentration was one of the main drivers of the composition of the bacterial communities within the August 2012 biofilms. The August 2012 biofilms also showed bacterial counts that were more than two times higher than any of the other sampling dates, suggesting more biofilm mass which could have generated more anoxic microsites. Although bacteria from the genus *Acinetobacter* are generally aerobes, there are some species within the genus that can utilize nitrate as an electron acceptor when oxygen is not present [40]. In contrast, *Methylomonas*, which was one of the predominant taxa in July 2011, December 2011 and March 2012 but was less than 2% of total sequences in August 2012, is strictly aerobic [41]. Therefore, the higher drinking water nitrate concentration and possibly more anoxic microsites caused by higher bacterial biofilm growth on the August 2012 sampling date may have provided *Acinetobacter* with a competitive advantage over *Methylomonas*.

Mycobacterium abundance also varied considerably over time, as this genus represented 59% of the sequences from the July 2011 sampling date but less than 1% for all other sampling dates. *Mycobacteria* are frequently detected in DWDS and are considered a significant public health issue [42]. The genus *Mycobacterium* consists of approximately 100 species, including a large number of species that are either non-pathogenic or pathogenic under certain situations[26]. For example, nontuberculosis *Mycobacterium* are a major cause of opportunistic infections in immunocompromised hosts [42]. Mycobacteria have been detected previously in this DWDS via culturing, and *M. gordonae* and *M. intracellulare* were the most frequently detected *Mycobacterium* species [43]. *M. gordonae* is among the most frequently reported mycobacteria in drinking water and in DWDS [26] and is generally considered non-pathogenic[44]. *M. intracellulare*, which is part of the *Mycobacterium avium* complex (MAC), has also been detected in DWDS [26]. MAC is the group of non-tuberculosis *Mycobacterium* most commonly associated with human disease, causing primarily pulmonary infections in individuals who are immunocompromised [45]. In this study, we were unable to discriminate the *Mycobacterium* sequences down to the species level, so it is unclear whether the sequences we detected represented potentially pathogenic species.

Several characteristics of *Mycobacteria* enhance their survival in DWDS, including their ability to grow under oligotrophic conditions [46], form biofilms and resist chlorine disinfection [26]. Several recent studies have detected species related to *Mycobacterium* in chlorinated drinking water [47]–[49]. Previous work at the DWDS considered here indicated that the frequency of detection of *Mycobacteria* increased when the disinfectant was switched from chlorine to chloramine in 2002 [43], suggesting that *Mycobacteria* might be less sensitive to chloramine than chlorine. However, a recent study using a model DWDS observed that the relative sensitivity of *Mycobacterium avium* biofilms to chlorine and monochloramine depended on the pipe material [50]. Specifically, *M. avium* was more sensitive to chlorine than chloramine when biofilms were grown on copper pipe, but the reverse was true for *M. avium* biofilms on iron [50], possibly due to corrosion products interfering with free chlorine [51]. Here we found that*Mycobacterium* sequences were most abundant in biofilms from the July 2011 sampling date, which also had the highest level of free ammonia and the lowest levels of total chlorine and monochloramine. These constituents are related, as monochloramine is reductively dehalogenated to ammonia, and monochloramine is the major component of total chlorine in this system. The results of our nMDS analysis indicate that the monochloramine concentration strongly influenced bacterial community composition for the July 2011 sampling date. Therefore, the fact that *Mycobacterium* was predominant only on the sampling date with the lowest level of monochloramine suggests that within the ductile iron pipes in this DWDS monochloramine significantly reduced the abundance of *Mycobacterium* within the biofilms.

An unclassified genus from the family Xanthomonadaceae also varied considerably in abundance across the sampling dates, representing 50% of the sequences from the March 2012 samples but only a small fraction of the sequences from the other sampling dates. The Xanthomonadaceae are obligate aerobic chemoorganotrophs, and this family includes some well-known plant pathogens [33]. Organisms from the family Xanthomonadaceae family were not detected in a previous pyrosequencing survey of drinking water biofilms [7], but these organisms have been isolated from drinking water and from drinking water pipe biofilms using culture based techniques [52]. In this study, we were unable to discriminate the Xanthomonadaceae sequences down to the genus or species level, and the reason for their high abundance in the March 2012 samples is unclear.

The biofilm communities from July 2011 and March 2012 showed much higher variation in composition among replicates than did biofilm communities from December 2011 and August 2012. The December and August samples were each dominated by a single bacterial genus (*Methylomonas* in December and *Acinetobacter* in August) with very little variation between replicates. These data suggest that the environmental conditions in December and August each favored one specific bacterial genus that dominated all of the biofilms on that sampling date. In contrast, biofilms from July and March had several dominant bacterial genera that showed high variations between the replicates, suggesting that conditions during those months produced greater variability by enabling several genera to compete for dominance within the biofilms.

Nitrification is a significant concern in drinking water distribution systems that use chloramine as the secondary residual, as nitrification can lead to a decrease in the chloramine residual, an increase in the growth of heterotrophic bacteria and an increase in concentrations of nitrate and nitrite, which pose risks to human health [53]. Multiple studies have identified nitrifying bacteria in DWDS [53]–[55]. No sequences corresponding to any known ammonia oxidizing bacterial genera were detected in the biofilms analyzed in this study. However, a few sequences from a known nitrite oxidizing genus, *Nitrospira*, were detected in the July 2011 samples, but not in samples from any of the other sampling dates. July 2011 also showed the highest concentration of free ammonia (at least two times higher than all other sampling dates) and unpublished data from PCU confirm that there was a peak in nitrification activity in June and July of 2011. The nMDS analysis indicated that ammonia concentration was a significant driver of the composition of the biofilm bacterial communities from July 2011, and previous studies have indicated that the presence of free ammonia is the principal cause of nitrification in DWDS [56]. Therefore, these data suggest that the high free ammonia concentration combined with the high temperature in July 2011 enabled nitrification to occur within the pipes; however, the lack of detection of ammonia oxidizing bacteria suggests that ammonia oxidation within the biofilms may have been driven by ammonia oxidizing archaea (AOA). Previous studies have detected AOA in drinking water distribution systems [54], but archaea would not have been detected by the bacterial primers used in the current study.

In summary, the results of our study demonstrate that the biofilms within the DWDS pipes were dominated by a few bacterial taxa, specifically *Methylomonas*, *Acientobacter* and*Mycobacterium*, and that the dominant taxa within the biofilms varied dramatically between sampling times. It is likely that these differences in dominant taxa were driven by differences in environmental conditions, and our analysis suggests that nitrate, ammonium, total chlorine, and monochloramine concentrations were key drivers of biofilm bacterial community composition.

Another possibility is that these differences in dominant taxa could have been the result of the founder effect, which stipulates that the founding member of a biofilm will have an advantage over subsequent colonizers and will remain dominant. The founder effect has been suggested as a possible driver of biofilm community composition in a variety of habitats [57]–[60] and could have been a contributing factor to the differences in dominant community members in our biofilms. Further experimental work, which is ongoing in our lab, will be needed to explore the relative contributions of environmental factors and founder effects on the composition of biofilms within drinking water distribution systems.

ACKNOWLEDGMENTS

The authors thank Sharon Waller for coordinating the pipe sampling effort and Kesha Baxi for assistance with data analysis at the start of the project. The authors thank Tim LaPara for his helpful comments on an earlier version of this manuscript.

AUTHOR CONTRIBUTIONS

Conceived and designed the experiments: JK MP AP. Performed the experiments: NM AC. Analyzed the data: JK NM AC. Contributed reagents/materials/analysis tools: JK MP AP. Wrote the paper: JK NM AC MP AP.

REFERENCES

1. Szewzyk U, Szewzyk R, Manz W, Schleifer K (2000) Microbiological safety of drinking water. Annual Reviews in Microbiology 54: 81–127.
2. Marshall KC (1988) Adhesion and growth of bacteria at surfaces in oligotrophic habitats. Can J Microbiol 34: 503–506.
3. Zacheus OM, Lehtola MJ, Korhonen LK, Martikainen PJ (2001) Soft deposits, the key site for microbial growth in drinking water distribution networks. Water Res 35: 1757–1765.
4. LeChevallier MW, Cawthon CD, Lee RG (1988) Factors promoting survival of bacteria in chlorinated water supplies. Appl Environ Microbiol 54: 649–654.
5. Ridgway H, Olson B (1982) Chlorine resistance patterns of bacteria from two drinking water distribution systems. Appl Environ Microbiol 44: 972–987.
6. Liu Y, Zhang W, Sileika T, Warta R, Cianciotto NP, et al. (2011) Disinfection of bacterial biofilms in pilot-scale cooling tower systems. Biofouling 27: 393–402.
7. Hong P, Hwang C, Ling F, Andersen GL, LeChevallier MW, et al. (2010) Pyrosequencing analysis of bacterial biofilm communities in water meters of a drinking water distribution system. Appl Environ Microbiol 76: 5631–5635.
8. Bois FY, Fahmy T, Block J, Gatel D (1997) Dynamic modeling of bacteria in a pilot drinking-water distribution system. Water Res 31: 3146–3156.
9. Martiny AC, Jørgensen TM, Albrechtsen H, Arvin E, Molin S (2003) Long-term succession of structure and diversity of a biofilm formed in a model drinking water distribution system. Appl Environ Microbiol 69: 6899–6907.
10. Amann RI, Ludwig W, Schleifer K (1995) Phylogenetic identification and in situ detection of individual microbial cells without cultivation. Microbiol Rev 59: 143–169.
11. Kalmbach S, Manz W, Szewzyk U (1997) Isolation of new bacterial species from drinking water biofilms and proof of their in situ dominance with highly specific
12. 16S rRNA probes. Appl Environ Microbiol 63: 4164–4170.
13. Schmeisser C, Sto¨ckigt C, Raasch C, Wingender J, Timmis K, et al. (2003) Metagenome survey of biofilms in drinking-water networks. Appl Environ Microbiol 69: 7298–7309.
14. Parsek MR, Singh PK (2003) Bacterial biofilms: An emerging link to disease pathogenesis. Annual Reviews in Microbiology 57: 677–701.
15. Hoefel D, Monis PT, Grooby WL, Andrews S, Saint CP (2005) Cultureindependent techniques for rapid detection of bacteria associated with loss of chloramine residual in a drinking water system. Appl Environ Microbiol 71: 6479–6488.
16. Eichler S, Christen R, Ho¨ltje C, Westphal P, Bo¨tel J, et al. (2006) Composition and dynamics of bacterial communities of a drinking water supply system as assessed by RNA-and DNA-based 16S rRNA gene fingerprinting. Appl Environ Microbiol 72: 1858–1872.
17. Marshall K, Blainey BL (1991) Role of bacterial adhesion in biofilm formation and biocorrosion. In: Anonymous Biofouling and Biocorrosion in Industrial Water Systems: Springer. pp. 29–46.
18. Rice EW, Baird RB, Eaton AD, Clesceri LS, editors (2012) Standard methods for the examination of water and wastewater, 22nd ed. Denver, CO: American Water Works Association.
19. Lane DJ (1991) 16S/23S rRNA sequencing. In: Stackebrandt E, Goodfellow M, editors. Nucleic acid techniques in bacterial systematics. Chichester, UK: John Wiley.
20. Youssef N, Sheik CS, Krumholz LR, Najar FZ, Roe BA, et al. (2009) Comparison of species richness estimates obtained using nearly complete fragments and simulated pyrosequencing-generated fragments in 16S rRNA gene-based environmental surveys. Appl Environ Microbiol 75: 5227–5236.
21. Schloss PD, Westcott SL, Ryabin T, Hall JR, Hartmann M, et al. (2009) ntroducing mothur: Open-source,

platform-independent, community-supported software for describing and comparing microbial communities. Appl Environ Microbiol 75: 7537–7541.

22. Edgar RC, Haas BJ, Clemente JC, Quince C, Knight R (2011) UCHIME improves sensitivity and speed of chimera detection. Bioinformatics 27: 2194– 2200.
23. Chao A (1984) Nonparametric estimation of the number of classes in a population. Scandinavian Journal of Statistics 11: 265–270.
24. Shannon CE (2001) A mathematical theory of communication. ACM SIGMOBILE Mobile Computing and Communications Review 5: 3–55.
25. Yue JC, Clayton MK (2005) A similarity measure based on species proportions.
26. Communications in Statistics-Theory and Methods 34: 2123–2131.
27. Zhang P, LaPara TM, Goslan EH, Xie Y, Parsons SA, et al. (2009)
28. Biodegradation of haloacetic acids by bacterial isolates and enrichment cultures from drinking water systems. Environ Sci Technol 43: 3169–3175.
29. Vaerewijck MJ, Huys G, Palomino JC, Swings J, Portaels F (2005) Mycobacteria in drinking water distribution systems: Ecology and significance for human health. FEMS Microbiol Rev 29: 911–934.
30. Forster D, Daschner F (1998) Acinetobacter species as nosocomial pathogens. European Journal of Clinical Microbiology and Infectious Diseases 17: 73–77.
31. Hedrich S, Schlomann M, Johnson DB (2011) The iron-oxidizing proteobacteria. Microbiology 157: 1551–1564.
32. Emerson D, Fleming EJ, McBeth JM (2010) Iron-oxidizing bacteria: An environmental and genomic perspective. Annu Rev Microbiol 64: 561–583.
33. Madigan MT, Martinko JM, Dunlap PV, Clark DP (2009) Brock biology of microorganisms. San Francisco, CA: Pearson/Benjamin Cummings.
34. Staley JT, Konopka A (1985) Measurement of in situ activities of nonphotosynthetic microorganisms in aquatic and terrestrial habitats. Annual Reviews in Microbiology 39: 321–346.
35. Revetta RP, Pemberton A, Lamendella R, Iker B, Santo Domingo JW (2010)
36. Identification of bacterial populations in drinking water using 16S rRNA-based sequence analyses. Water Res 44: 1353–1360.
37. Brenner D, Krieg N, Staley J, Garrity G, Boone D, et al. (2005) The proteobacteria, part B, the gammaproteobacteria. Bergey's manual of systematic bacteriology 2.
38. McKeon DM, Calabrese JP, Bissonnette GK (1995) Antibiotic resistant gramnegative bacteria in rural groundwater supplies. Water Res 29: 1902–1908.
39. Bifulco JM, Shirey J, Bissonnette G (1989) Detection of acinetobacter spp. In rural drinking water supplies. Appl Environ Microbiol 55: 2214–2219.
40. Shirey JJ, Bissonnette GK (1991) Detection and identification of groundwater bacteria capable of escaping entrapment on 0.45-micron-pore-size membrane filters. Appl Environ Microbiol 57: 2251–2254.
41. Hansen SK, Rainey PB, Haagensen JA, Molin S (2007) Evolution of species interactions in a biofilm community. Nature 445: 533–536.
42. Marin M, Pedregosa A, Laborda F (1996) Emulsifier production and microscopical study of emulsions and biofilms formed by the hydrocarbonutilizing bacteria acinetobacter calcoaceticus MM5. Appl Microbiol Biotechnol 44: 660–667.
43. Tomaras AP, Dorsey CW, Edelmann RE, Actis LA (2003) Attachment to and biofilm formation on abiotic surfaces by acinetobacter baumannii: Involvement of a novel chaperone-usher pili assembly system. Microbiology 149: 3473–3484.
44. Wentzel M, Lo¨tter L, Loewenthal R, Marais G (1986) Metabolic behaviour of acinetobacter spp. in enhanced biological phosphorus removal- a biochemical model. Water S A 12: 209–224.
45. Hanson RS, Hanson TE (1996) Methanotrophic bacteria. Microbiol Rev 60: 439–471.
46. Covert TC, Rodgers MR, Reyes AL, Stelma GN (1999) Occurrence of nontuberculous mycobacteria in environmental samples. Appl Environ Microbiol 65: 2492–2496.
47. Pryor M, Springthorpe S, Riffard S, Brooks T, Huo Y, et al. (2004) Investigation of opportunistic pathogens in municipal drinking water under different supply and treatment regimes. Water Science & Technology 50: 83–90.
48. Weinberger M, Berg SL, Feuerstein IM, Pizzo PA, Witebsky FG (1992)
49. Disseminated infection with mycobacterium gordonae: Report of a case and critical review of the literature. Clinical infectious diseases 14: 1229–1239.
50. Desforges JF, Horsburgh CR Jr (1991) Mycobacterium avium complex infection in the acquired immunodeficiency syndrome. N Engl J Med 324: 1332–1338.
51. Carson LA, Petersen NJ, Favero MS, Aguero SM (1978) Growth characteristics of atypical mycobacteria in water and their comparative resistance to disinfectants. Appl Environ Microbiol 36: 839–846.
52. Beumer A, King D, Donohue M, Mistry J, Covert T, et al. (2010) Detection of mycobacterium avium subsp. paratuberculosis in drinking water and biofilms by quantitative PCR. Appl Environ Microbiol 76: 7367–7370.
53. Falkinham JO, Norton CD, LeChevallier MW (2001) Factors influencing numbers of mycobacterium avium, mycobacterium intracellulare, and other mycobacteria in drinking water distribution systems. Appl Environ Microbiol 67: 1225–1231.

CHAPTER 20 BIOFILM FORMATION BY FILAMENTOUS FUNGI RECOVERED FROM A WATER SYSTEM

Virgínia M. Siqueira and Nelson Lima

INTRODUCTION

Filamentous fungi (ff) have been frequently isolated from aquatic environments such as rivers, streams, lakes, and sea [1]. Water distribution systems (WDS) are nowadays seen as complex aquatic environments in which high diverse microorganisms cohabit, including fungi [2]. Regardless of their importance for human health, little is known about microbial ecology of ff within WDS. Moreover, the focuses of microbial water quality studies still remain on monitoring planktonic microorganisms, despite scientists' awareness that the majority of microorganisms live together as biofilms [3].

Due to its important role in the environment, industry, and medicine, the understanding of mechanisms of biofilm formation has become the focus of biofilm research. Research in ff biofilm in WDS has only recently received attention [4–6]. Most of previous mycological studies are focused on pathogenic fungi, for example,Candida spp. [7–9], Aspergillus fumigatus [10–12], and opportunistic zygomycetes [13]. Increased resistance against antimicrobials is a well-known and still worrying clinical relevant biofilm feature, and wherefore these studies have especially established suitable methods for antimicrobial biofilm susceptibility assay [14–16]. Ramage et al. [17] reported the importance of fungal biofilm phenotype concept in medical and industrial mycological research. These authors described schematically Aspergillus biofilms development and discussed morphological, physiological, and molecular features related to both fungal virulence and enzymatic production.

Studies have reported the detection of ff within biofilms in different habitats such as rocks [18], fuel tanks [19], historic monuments [20–22], acid mine drainage [23], and pipe walls [24]. Nonetheless, the improvement and standardisation of suitable methods for laboratorial studies of ff biofilms are still needed. Characterisation of the development, physiology, architecture, disinfectant resistance, and factors that influence genetic and phenotypic expression of ff biofilms is still unknown for most ff species, even for those frequently recovered from WDS. Additionally, it is important to highlight that ff are eukaryotic organisms which have peculiar characteristics, for example, nutrition by absorption, development of reproductive and vegetative structures (i.e., spore and hyphae), and sexual and asexual reproduction, hence when compared with bacterial and yeast biofilms, different approaches must be employed.

Thus, this work aims to investigate the capability of biofilm formation, and their morphological and physiological characterisation, of fungi isolated from biofilms in a water system.

Citation: Virgínia M. Siqueira and Nelson Lima, "Biofilm Formation by Filamentous Fungi Recovered from a Water System," Journal of Mycology, vol. 2013, Article ID 152941, 9 pages, 2013. doi:10.1155/2013/152941

MATERIAL AND METHODS

Fungal Isolates

Aspergillus sp. 1 (section Nigri), Aspergillus sp. 2 (section Flavi), Alternaria sp., Botrytis sp., Cladosporium sp., and Penicillium sp. were used in this study. These strains were recovered from biofilms developed in a flow chamber reactor setup at the Laboratory of Mycology of the Biological Engineering Centre, University of Minho, Braga, Portugal, during a previous study [25]. The fungi were maintained in MEA (malt extract 20 g, peptone 5 g, agar 20 g, distilled water 1 L) at 25°C.

Biofilm Formation

Spores of each isolate were harvested from 7 days aged pure culture in MEA by adding 2 mL of saline solution (0.85%) into plate. The spore suspension was resuspended and vortexed before quantification. The suspensions were standardized by dilution with saline solution (0.85%) to a final concentration of 10^5 spores/mL and using a Neubauer counter chamber.

For biofilm kinetics, 100 µL of spore suspension and 100 µL of culture broth media MEB (malt extract 20 g, peptone 5 g, distilled water 1 L), R2A broth (proteose peptone 0.5 g; casamino acids 0.5 g; yeast extract 0.5 g; dextrose 0.5 g; soluble starch 0.5 g; dipotassium phosphate 0.3 g, magnesium sulphate $7H_2O$ 0.05 g; sodium pyruvate 0.3 g; distilled water 1 L; pH 7.2), and 100 µL of sterilised tap water were added per well into 96-well, flat-bottomed polystyrene microtiter plates. Media-only blanks were also set up. The plates were incubated at 30°C for 24 h and analyses were made at 4, 8, 12, and 24 h. Time of analysis was chosen based on a protocol for filamentous fungal biofilm in 96-well plates described by [26]. Biofilm visualization and EPS detection were performed using a stereomicroscope (Leica; DFC450 Camera with Leica Application Suite software) and morphological and structural parameters analysis were followed as described by [27].

Biofilm Kinetics

Biofilm Quantification

Biofilm biomass was assessed using a protocol described elsewhere [24]. Briefly, at each time interval (i.e., 4, 8, 12, and 24 h), the spent culture medium was removed from each well and the adherent cells were washed with phosphate-buffered saline (PBS; 0.1 M, pH 7.2). These were airdried and 200 µL of 0.5% (w/v) crystal violet (CV) solution was added for 30 min. The solution was then removed until excess stain was removed. The biofilms were distained by adding 200 µL 95% ethanol to each well. The ethanol was gently pipetted to completely solubilize the CV for 1 min, the ethanol was transferred to a clean 96-well microtitre plate, and the was read using a microtiter plate reader (Model Synergy HT; BIO-TEK). The OD values are proportional to the quantity of biofilm biomass, which comprises hyphae and extracellular polymeric material (the greater the quantity of biological material, the higher the level of staining and absorbance).

MTT Assay

MTT (3-(4,5-dimethyl-2-thiazolyl)-2,5-diphenyl-2H-tetrazolium bromide) is a yellow soluble tetrazolium salt that is converted into an insoluble purple crystal by metabolically active cells. MTT (Sigma) solution was aseptically prepared by dissolving the MTT powder at a concentration of 5 mg/mL in sterile PBS at room temperature and stored at 4°C in a dark, screw-cap container. At each time point, 0.2 mL of MTT solution were added to each well and the 96-well plates were incubated at 37°C during 3 h. After this period, the supernatant was then discarded and 0.2 mL dimethyl sulfoxide (DMSO) was added to each well to solubilize the MTT, which had been cleaved into an insoluble purple formazan through the metabolism of the live cells. Biofilm development was assayed by loading 0.2 mL of the solubilized MTT into a flat-bottom, 96-well polyvinyl chloride microtiter plate, and absorbance measured at OD_{570}nm using a microtiter plate reader (Model Synergy HT; BIO-TEK).

RESULTS

Biofilm formation by Aspergillus sp. 1, Aspergillus sp. 2, Botrytis sp., Alternaria sp., Cladosporium sp., andPenicillium sp. grown in different culture media over 24 h was characterised, as revealed by both crystal violet (CV) biomass estimation and the colorimetric MTT assay. Figure 1 shows the mean value of absorbance level for CV staining (total biofilm), and Figure 2 shows the mean value of absorbance level for MTT staining (viable cells). Additionally, image analysis was used to assess morphological aspects of biofilm formation (Figures 3–8).

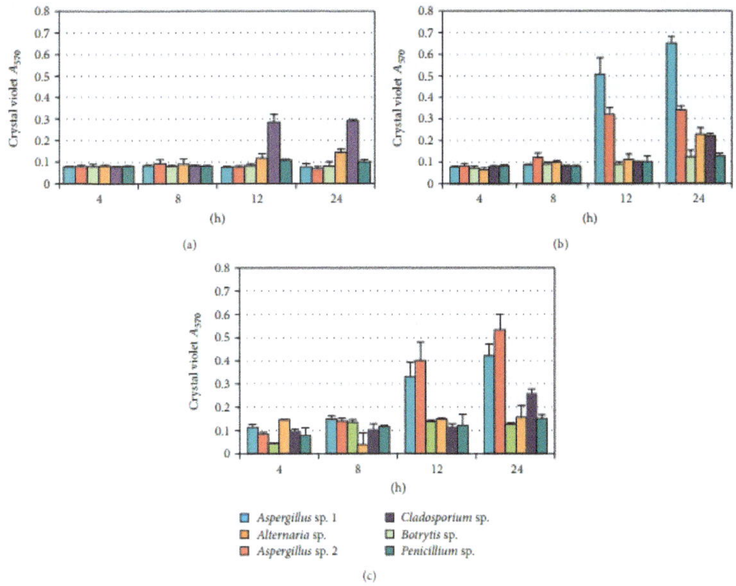

Figure 1: Total biomass (CV) in filamentous fungal biofilm grown in different culture media: (a) water, (b) R2A, and (c) MEB. The means ± standard deviations for at least three replicates are illustrated.

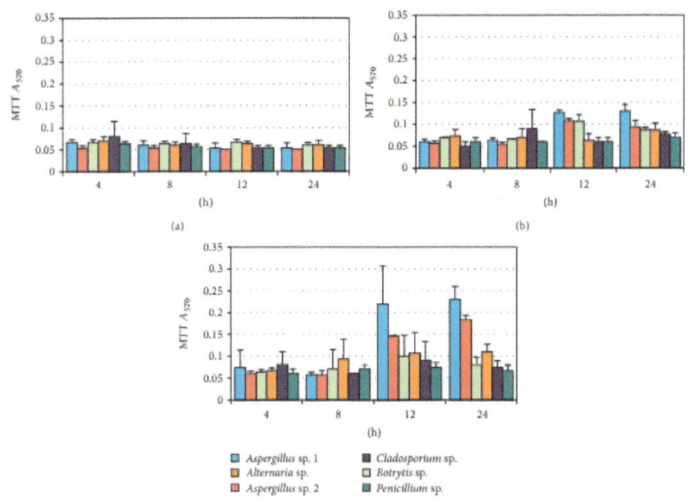

Figure 2: Total viable cells (MTT) in filamentous fungal biofilm grown in different culture media: (a) water, (b) R2A, and (c) MEB. The means ± standard deviations for at least three replicates are illustrated.

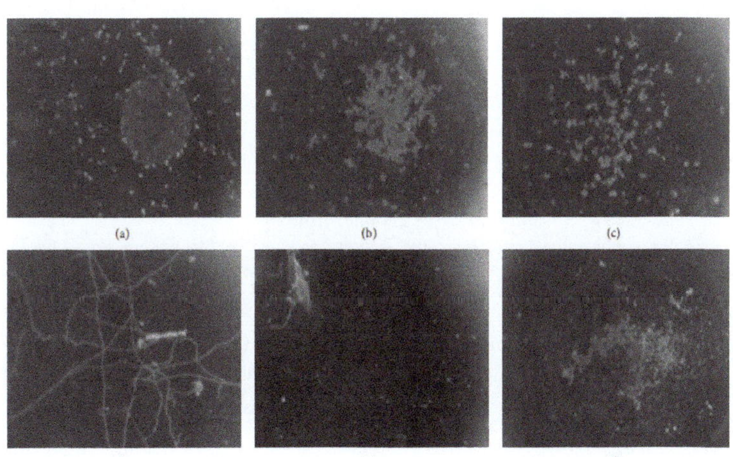

Figure 3: Aspergillus sp. 1 (a), Aspergillus sp. 2 (b), Botrytis sp. (c), Alternaria sp. (d),Cladosporium sp. (e), and Penicillium sp. (f) after 8 h in water.

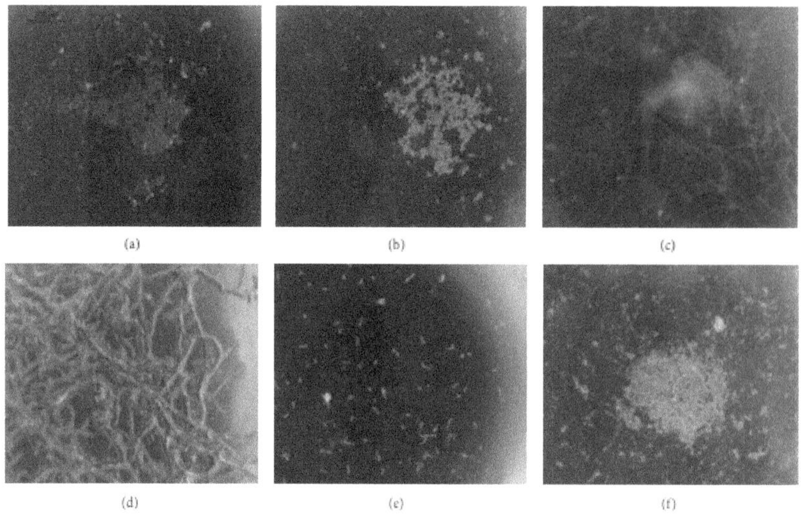

Figure 4: Aspergillus sp. 1 (a), Aspergillus sp. 2 (b), Botrytis sp. (c), Alternaria sp. (d),Cladosporium sp. (e), and Penicillium sp. (f) after 24 h in water.

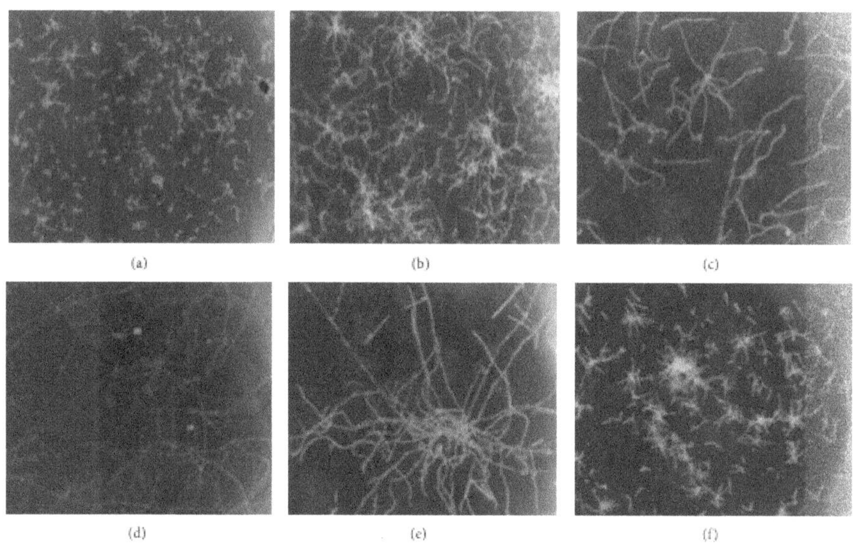

Figure 5: Aspergillus sp. 1 (a), Aspergillus sp. 2 (b), Botrytis sp. (c), Alternaria sp. (d),Cladosporium sp. (e), and Penicillium sp. (f) after 8 h in R2A Broth.

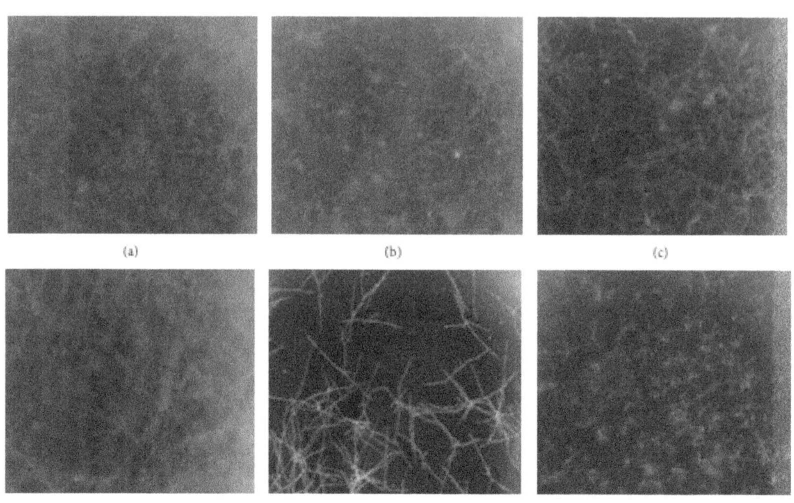

Figure 6: Aspergillus sp. 1 (a), Aspergillus sp. 2 (b), Botrytis sp. (c), Alternaria sp. (d),Cladosporium sp. (e), and Penicillium sp. (f) after 24 h in R2A Broth.

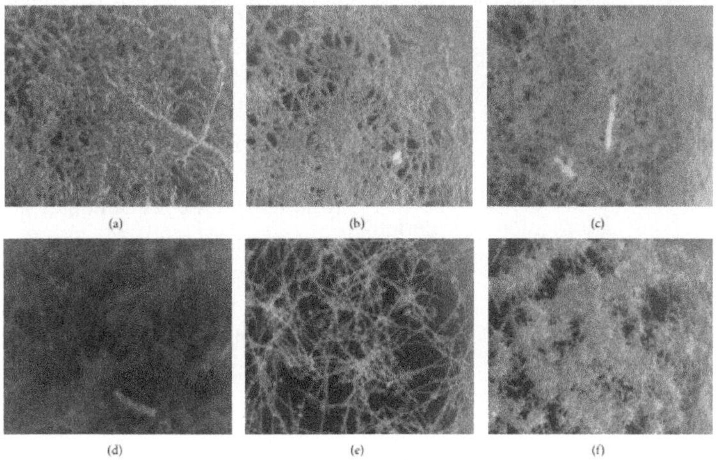

Figure 7: Aspergillus sp. 1 (a), Aspergillus sp. 2 (b), Botrytis sp. (c), Alternaria sp. (d),Cladosporium sp. (e), and Penicillium sp. (f) after 8 h in MEB.

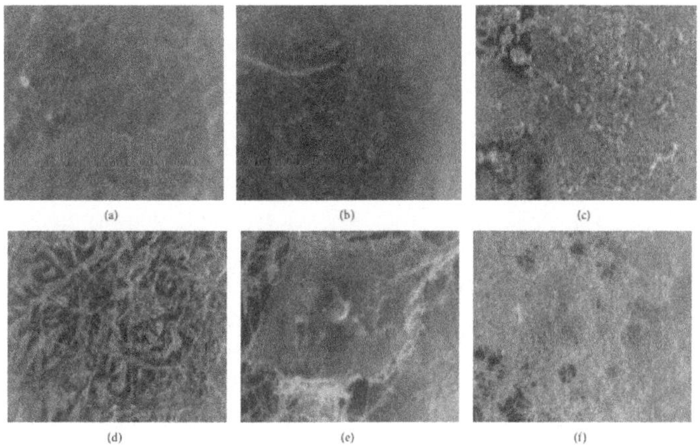

Figure 8: Aspergillus sp. 1 (a), Aspergillus sp. 2 (b), Botrytis sp. (c), Alternaria sp. (d),Cladosporium sp. (e), and Penicillium sp. (f) after 24 h in MEB.

A direct relation was observed between biomass (CV), and biofilm development (Figures 1 and 3–8). On the other hand, a direct relation between biomass (CV) and metabolic active cells (MTT) was not observed (Figures 1 and 2).

In water, biofilms showed a higher cell activity in the first 4 h of analyses (MTT), and lower along time (Figure2(a)). In adverse conditions, spores may enter in a dormancy status with lower cells metabolic activity. In R2A and MEB, biomass and cell activity showed higher values along time. Different patterns of biofilm development were observed between the culture media and fungi. The time when adherence of spores, monolayer development, and EPS production were observed for each fungus biofilm is described in Table 1.

Table 1: Time of detection of spore adherence, monolayer, and EPS production in fungal biofilms grown in different culture media.

Fungus	Water			R2A			MEB		
	Ad	Mn	EPS	Ad	Mn	EPS	Ad	Mn	EPS
Aspergillus sp. 1	—	—	—	4 h	8 h	24 h	<4 h	4 h	8 h
Aspergillus sp. 2	—	—	—	4 h	8 h	24 h	<4 h	4 h	8 h
Alternaria sp.	4 h	8 h	24 h	4 h	8 h	24 h	<4 h	4 h	8 h
Botrytis sp.	—	24 h	—	4 h	8 h	24 h	<4 h	4 h	8 h
Cladosporium sp.	—	—	—	8 h	24 h	*	4 h	8 h	12 h
Penicillium sp.	24 h	—	—	8 h	24 h	*	<4 h	4 h	8 h

Alternaria sp. and Botrytis sp. were the fungi that formed biofilm in water and Penicillium sp. showed spore adherence after 24 h under this same condition (Figures 3 and 4). In R2A, biofilms were formed mainly after 12 h and EPS production was observed in both Aspergillus spp., Alternaria sp., and Botrytis sp. biofilms at 24 h (Figures 5 and 6). In MEB, after 8 h, the fungi formed mature biofilm (Figures 7 and 8), with the exception ofCladosporium sp. that showed mature biofilm in MEB after 12 h. MEB was shown to be the best culture medium for biofilm formation, that is, in which biofilm grew faster and showed a very well-structured shape.

DISCUSSION

The biofilm formation pattern observed in this work resembled the kinetics of biofilm formation of Aspergillus fumigatus recently proposed by Ramage et al. [17].

Previously, Harding et al. 2009 [28] proposed a model for ff biofilms which includes five stages (i) propagule adsorption, (ii) active attachment to a surface, (iii) microcolony formation I, (iv) microcolony formation II, (v) maturation or reproductive development, and (vi) dispersal or planktonic phase.

In this work, reproductive development was not observed, for example, production of conidiophores and spores over 24 h of growth. However, the dispersal phase can be represented by hyphal fragments which also act as dispersal propagules. Because this model has been described based on bacterial and yeast models, it is important to highlight that unique features related to fungal biology distinguish bacterial from fungal biofilm formation: fungi commonly have more than one planktonic form (i.e., sexual and asexual spores, sporangia, and hyphal fragments), and these dispersive forms are not unicellular and often float in water and air as well, and the development of specialised reproductive tissues to produce dispersive forms [28]. Moreover, hyphae growth is characterised by two or three dimensions which is achieved by branching. Branch formation provides a greater regulation growth while bacterial unicellular growth is characterised by colonies which are formed by cells pilling and pushing each other apart [29]. Bacterial and yeast biofilms have been greatly studied in the last twenty years. Consequently, there are well-defined models, criteria, and phenotypes for characterizing bacterial and yeast biofilms. Differently, only in the last few years filamentous fungi had been considered as biofilm-forming microorganisms as bacteria and yeasts, even though many industrial processes and research were actually cell surface adhesion processes (e.g., solid-state fermentation) [30].

Thus, studies in ff biofilm must take in consideration these specific fungal features which leads to a specific way of how biofilm analyses should be carried on. Typical ff biofilm morphology is described as a complex 3D structure with cells usually enclosed within an extracellular matrix consisting of polymeric substances (EPS) [31].

Mainly after 12 h of biofilm growth, a layer of substances was observed surrounding fungal responsible for binding cells and other particulate materials together (cohesion) and to the surface (adhesion), that is, providing the structural support for the biofilm maturation [32]. Further studies in the characterisation of biofilm matrix may provide information about its importance in the biofilm maintenance under adverse conditions such as low nutrient levels and presence of disinfectants in WDS, and its influence in microbial interactions as well. Similar structural characteristics were found in Candida sp. and Aspergillus sp. biofilms, that is, interstitial voids and water channels with cells usually encapsulated within an extracellular matrix which may relate to their pathogenicity [33, 34].

The two isolates of Aspergillus sp. used in this work showed a better capacity in terms of higher biomass and cell metabolism to form biofilm when compared to the other fungal isolates. However, only Alternaria sp. was able to form biofilm in water that may represent a high adaptation to form biofilm under oligotrophic conditions. The genus Alternaria is frequently recovered from WDS ant it is known to provoke allergic responses in susceptible individuals exposed to contaminated water of shower, bath, and saunas or even to water-damaged buildings [35].

In previous studies, the spore density in the initial spore suspension was verified as a key factor in the development of biofilms by ff [24]. Singh et al. [13] found that for Rhizopus oryzae, Lichtheimia corymbifera,and Rhizomucor pusillus, the initial inoculum plays a key role in germination of the adhered spores as well as structural integrity of the biofilms formed, a phenomenon also described in studies in Aspergillus niger [34] andA. fumigatus biofilms [17].

Bonaventura et al. [36] selected an inoculum size of 10^5 CFU/mL, an adhesion time of 1 h, and a biofilm formation time of 72 h as optimal experimental conditions for growing Trichosporon asahii biofilm on polystyrene surfaces. Our findings showed that time for spore adherence and further biofilm development varied for each fungus under different culture conditions. Nonetheless, spore characteristics such as size and surface proprieties may also influence biofilm kinetics.

Airborne fungal spores have a hydrophobic surface which aids dispersal, prevents desiccation, and may provide a barrier to the entry of toxicants [37]. Studies report that fungal spore hydrophobicity influences their capacity to adhere to biological surfaces, and that spore hydrophobicity is significantly influenced by culture conditions [38]. Fungal cultures in culture media which have high nutrient content are likely to produce more hydrophobic spores; the opposite is found when culture media with low nutrient content are used [39]. As spore hydrophobicity may interfere in spore adhesion and consequently in biofilm formation, attention must be paid on previous culture conditions. Siqueira and Lima [40] found that water biofilms of Penicillium spp. have specific hydrophobic hyphae which may be involved in fungal ecological functions. The different levels of hydrophobicity within a biofilm may also interfere in biofilm development and morphology, and consequently when different methods are applied they may significantly interfere in fungal biofilms characteristics. Thus, methods in ff biofilm characterization need urgently to be standardized.

The colorimetric method using the dye MTT has been cogitated as an alternative to traditional methods for in vitro susceptibility testing of fungi [41] and has been applied in fungal biofilms [42, 43]. In this work, MTT values did not correlate with biofilm biomass assessed by CV, mainly in older biofilms, that is, 24 h aged. Older biofilms showed a higher EPS production, which may interfere in the assessment of MTT to cell. Freimoser et al. [44] demonstrated that MTT method serves as a measure for cell densities of the entomopathogenic fungiNeozygites parvispora and Entomophthora thripidum and emphasized that for these two fungi the incubation period had to be longer (i.e., 16 h) because the cell wall might act as an additional barrier for the uptake of MTT. Thus, MTT may not be the best method to evaluate cell viability in mature biofilms.

In conclusion, the results presented here show that the ff studied in this work are able to form biofilms under the applied conditions. Although each fungus presented a different pattern of biofilm development, spore adhesion, monolayer, and EPS production were observed in all fungal species over 24 h of analyses. Moreover, characteristics of spores and culture conditions may play an important role in ff biofilm kinetics and must be taken into consideration for further studies in this area.

REFERENCES

1. C. M. Wurzbacher, F. Bärlocher, and H. P. Grossart, "Fungi in lake ecosystems," Aquatic Microbial Ecology, vol. 59, no. 2, pp. 125–149, 2010.
2. U. Szewzyk, R. Szewzyk, W. Manz, and K. H. Schleifer, "Microbiogical safety of drinking water," Annual Review of Microbiology, vol. 54, pp. 81–127, 2000.
3. J. W. Costerton, K. J. Cheng, G. G. Geesey et al., "Bacterial biofilms in nature and disease," Annual Review of Microbiology, vol. 41, pp. 435–464, 1987.
4. A. B. Gonçalves, I. M. Santos, R. R. M. Paterson, and N. Lima, "FISH and Calcofluor staining techniques to detect in situ filamentous fungal biofilms in water," Revista Iberoamericana de Micologia, vol. 23, no. 3, pp. 194–198, 2006.
5. N. B. Sammon, K. M. Harrower, L. D. Fabbro, and R. H. Reed, "Three potential sources of microfungi in a treated municipal water supply system in sub-tropical Australia," International Journal of Environmental Research and Public Health, vol. 8, no. 3, pp. 713–732, 2011.
6. V. M. Siqueira, H. M. B. Oliveira, C. Santos, R. R. M. Paterson, N. B. Gusmão, and N. Lima, "Filamentous fungi in drinking water, particularly in relation to biofilm formation," International Journal of Environmental Research and Public Health, vol. 8, no. 2, pp. 456–469, 2011.
7. J. Chandra, D. M. Kuhn, P. K. Mukherjee, L. L. Hoyer, T. McCormick, and M. A. Ghannoum, "Biofilm formation by the fungal pathogen Candida albicans: development, architecture, and drug resistance,"Journal of Bacteriology, vol. 183, no. 18, pp. 5385–5394, 2001.
8. L. J. Douglas, "Candida biofilms and their role in infection," TRENDS in Microbiology, vol. 11, pp. 30–36, 2003.
9. D. W. Williams, T. Kuriyama, S. Silva, S. Malic, and M. A. O. Lewis, "Candida biofilms and oral candidosis: treatment and prevention," Periodontology, vol. 55, no. 1, pp. 250–265, 2011.
10. E. Mowat, C. Williams, B. Jones, S. McChlery, and G. Ramage, "The characteristics of Aspergillus fumigatus mycetoma development: is this a biofilm?" Medical Mycology, vol. 47, no. 1, pp. S120–S126, 2009.
11. S. Bruns, M. Seidler, D. Albrecht et al., "Functional genomic profiling of Aspergillus fumigatus biofilm reveals enhanced production of the mycotoxin gliotoxin," Proteomics, vol. 10, no. 17, pp. 3097–3107, 2010.
12. F. M. C. Müller, M. Seidler, and A. Beauvais, "Aspergillus fumigatus biofilms in the clinical setting,"Medical Mycology, vol. 49, no. 1, pp. S96–S100, 2011. View at Publisher · View at Google Scholar · View at Scopus
13. R. Singh, M. R. Shivaprakash, and A. Chakrabarti, "Biofilm formation by zygomycetes: quantification, structure and matrix composition," Microbiology Papers, vol. 157, no. 9, pp. 2611–2618, 2011.
14. D. M. Kuhn, T. George, J. Chandra, P. K. Mukherjee, and M. A. Ghannoum, "Antifungal susceptibility ofCandida biofilms: unique efficacy of amphotericin B lipid formulations and echinocandins,"Antimicrobial Agents and Chemotherapy, vol. 46, no. 6, pp. 1773–1780, 2002.
15. M. M. Harriott and M. C. Noverr, "Candida albicans and Staphylococcus aureus form polymicrobial biofilms: effects on antimicrobial resistance," Antimicrobial Agents and Chemotherapy, vol. 53, no. 9, pp. 3914–3922, 2009.
16. S. Silva, M. Henriques, R. Oliveira, D. Williams, and J. Azeredo, "In vitro biofilm activity of non-Candida albicansCandida species," Current Microbiology, vol. 61, no. 6, pp. 534–540, 2010.
17. G. Ramage, R. Rajendran, M. Gutierrez-Correa, B. Jones, and C. Williams, "Aspergillus biofilms: clinical and industrial significance," FEMS Microbiology Letters, vol. 324, no. 2, pp. 89–97, 2011.
18. A. A. Gorbushina, "Life on the rocks," Environmental Microbiology, vol. 9, pp. 1613–1631, 2007.
19. R. B. Srivastava, M. Awasthi, M. C. Upreti, and G. N. Mathur, "Studies on Aureobasidium pullulansforming biofilm on high strength aluminium alloy, a structural component, in aircraft fuel tanks," Indian Journal of Engineering and Materials Sciences, vol. 13, no. 2, pp. 135–139, 2006.
20. A. A. Gorbushina, J. Heyrman, T. Dornieden et al., "Bacterial and fungal diversity and biodeterioration problems in mural painting environments of St. Martins church (Greene-Kreiensen, Germany),"International Biodeterioration and Biodegradation, vol. 53, no. 1, pp. 13–24, 2004.
21. M. L. Grbić, J. Vukojević, G. S. Simić, J. Krizmanić, and M. Stupar, "Biofilm forming cyanobacteria, algae and fungi on two historic monuments in Belgrade, Serbia," Archives of Biological Sciences, vol. 62, no. 3, pp. 625–631, 2010.
22. W. Lan, H. Li, W. D. Wang, Y. Katayama, and J. D. Gu, "Microbial community analysis of fresh and old microbial biofilms on Bayon Temple sandstone of Angkor Thom, Cambodia," Microbial Ecology, vol. 60, no. 1, pp. 105–115, 2010.
23. B. J. Baker, G. W. Tyson, L. Goosherst, and J. F. Banfield, "Insights into the diversity of eukaryotes in acid mine drainage biofilm communities," Applied and Environmental Microbiology, vol. 75, no. 7, pp. 2192–2199, 2009.
24. E. Mowat, J. Butcher, S. Lang, C. Williams, and G. Ramage, "Development of a simple model for studying the effects of antifungal agents on multicellular communities of Aspergillus fumigatus," Journal of Medical Microbiology, vol. 56, no. 9, pp. 1205–1212, 2007.
25. V. M. Siqueira, Study of filamentous fungal biofilms with molecular and microscopic techniques [Ph.D.

dissertation in Chemical and Biological Engineering], 2011, http://hdl.handle.net/1822/19679.

26. C. G. Pierce, P. Uppuluri, A. R. Tristan et al., "A simple and reproducible 96-well plate-based method for the formation of fungal biofilms and its application to antifungal susceptibility testing," Nature Protocols, vol. 3, no. 9, pp. 1494–1500, 2008. View at Publisher · View at Google Scholar · View at Scopus

27. A. Beauvais, C. Schmidt, S. Guadagnini et al., "An extracellular matrix glues together the aerial-grown hyphae of Aspergillus fumigatus," Cellular Microbiology, vol. 9, no. 6, pp. 1588–1600, 2007.

28. M. W. Harding, L. L. R. Marques, R. J. Howard, and M. E. Olson, "Can filamentous fungi form biofilms?"Trends in Microbiology, vol. 17, no. 11, pp. 475–480, 2009.

29. J. I. Prosser, "Hyphal growth patterns," in Fungal Differentiation: A contemporary Synthesis, J. E. Smith, Ed., pp. 357–396, Marcel-Dekker, New York, NY, USA, 1983.

30. M. Gutiérrez-Correa, Y. Ludeña, G. Ramage, and G. K. Villena, "Recent advances on filamentous fungal biofilms for industrial uses," Applied Biochemistry and Biotechnology, vol. 167, pp. 1235–1253, 2012.

31. P. Stoodley, K. Sauer, D. G. Davies, and J. W. Costerton, "Biofilms as complex differentiated communities," Annual Review of Microbiology, vol. 56, pp. 187–209, 2002.

32. D. G. Allison, "The biofilm matrix," Biofouling, vol. 19, pp. 139–150, 2003.

33. G. K. Villena, T. Fujikawa, S. Tsuyumu, and M. Gutiérrez-Correa, "Structural analysis of biofilms and pellets of Aspergillus niger by confocal laser scanning microscopy and cryo scanning electron microscopy," Bioresource Technology, vol. 101, no. 6, pp. 1920–1926, 2010.

34. G. K. Villena and M. Gutiérrez-Correa, "Production of cellulase by Aspergillus niger biofilms developed on polyester cloth," Letters in Applied Microbiology, vol. 43, no. 3, pp. 262–268, 2006.

35. J. U. Ponikau, "The diagnosis and incidence of allergic fungal sinusitis," Mayo Clinic Proceedings, vol. 74, no. 9, pp. 877–884, 1999.

36. G. Di Bonaventura, A. Pompilio, C. Picciani, M. Iezzi, D. D›Antonio, and R. Piccolomini, "Biofilm formation by the emerging fungal pathogen Trichosporon asahii: development, architecture, and antifungal resistance," Antimicrobial Agents and Chemotherapy, vol. 50, no. 10, pp. 3269–3276, 2006.

37. J. L. Laseter, J. Weete, and D. J. Weber, "Alkanes, fatty acid methyl esters, and free fatty acids in surface wax of Ustilago maydis," Phytochemistry, vol. 7, no. 7, pp. 1177–1181, 1968.

38. T. Singh, R. Saikia, T. Jana, and D. K. Arora, "Hydrophobicity and surface electrostatic charge of conidia of the mycoparasitic Trichoderma species," Mycological Progress, vol. 3, pp. 219–228, 2004.

39. D. J. Holder, B. H. Kirkland, M. W. Lewis, and N. O. Keyhani, "Surface characteristics of the entomopathogenic fungus Beauveria (Cordyceps) bassiana," Microbiology, vol. 153, no. 10, pp. 3448–3457, 2007.

40. V. M. Siqueira and N. Lima, "Surface hydrophobicity of culture and water biofilm of Penicillium spp,"Current Microbiology, vol. 64, pp. 93–99, 2012.

41. J. Meletiadis, J. F. G. M. Meis, J. W. Mouton, J. P. Donnelly, and P. E. Verweij, "Comparison of NCCLS and 3-(4,5-dimethyl-2-thiazyl)-2,5-diphenyl-2H-tetrazolium bromide (MTT) methods of in vitro susceptibility testing of filamentous fungi and development of a new simplified method," Journal of Clinical Microbiology, vol. 38, no. 8, pp. 2949–2954, 2000.

42. B. Jahn, A. Stüben, and S. Bhakdi, "Colorimetric susceptibility testing for Aspergillus fumigatus: comparison of menadione-augmented 3-(4,5-dimethyl-2-thiazolyl)-2,5-diphenyl-2H-tetrazolium bromide and Alamar Blue tests," Journal of Clinical Microbiology, vol. 34, no. 8, pp. 2039–2041, 1996.

43. B. P. Krom, J. B. Cohen, G. E. McElhaney Feser, and R. L. Cihlar, "Optimized candidal biofilm microtiter assay," Journal of Microbiological Methods, vol. 68, no. 2, pp. 421–423, 2007.

44. F. M. Freimoser, C. A. Jakob, M. Aebi, and U. Tuor, "The MTT [3-(4,5-dimethylthiazol-2-yl)-2,5-diphenyltetrazolium bromide] assay is a fast and reliable method for colorimetric determination of fungal cell densities," Applied and Environmental Microbiology, vol. 65, no. 8, pp. 3727–3729, 1999.

CHAPTER 21

FILAMENTOUS FUNGI IN DRINKING WATER, PARTICULARLY IN RELATION TO BIOFILM FORMATION

Virgínia M. Siqueira, Helena M. B. Oliveira, Cledir Santos, R. Russell M. Paterson, Norma B. Gusmão and Nelson Lima

INTRODUCTION

The issue of filamentous fungi in drinking water has matured. This is a direct result of the work of key researchers in the area [1-11], with a crucial issue being identifying isolated strains to the species, rather than simply the genus level. When species identification is undertaken more can be deduced about changes in biodiversity within locations and with time. The properties of the isolates, such as mycotoxin, pigment, and odour formation, are more reliably ascertained because these are associated more with the individual species than the genus. Furthermore, water contaminated with fungi is of importance to hospitals, where immunocompromised patients undergo treatment [12]: *Aspergillus* accounts for the majority of infections, with *Aspergillus fumigatus* accounting for 90% of cases.

Kelley *et al.* [6] provided a comprehensive investigation of filamentous fungi in US distribution systems. In this work,*Alternaria alternata*, *Aspergillus niger*, *Cladosporium* sp., *Epiccocum nigrum* and many *Penicillium* sp. were prevalent. *Penicillium brevicompactum* was rather frequently detected and *P. expansum* considerably less so. However, none of these were detected from the two tap water sites, although *Fusarium* sp. were common together with an unnamed zygomycetes. Very few*Aspergillus* sp. were detected, including *A. fumigatus*. Portuguese tap water was studied more recently [8] and Norwegian and Australian drinking water studies have provided important additional knowledge [9,10,13]. To assist the reader, the most relevant reports from water distribution systems are listed on Table 1.

Citation: Virgínia M. Siqueira, Helena M. B. Oliveira, Cledir Santos, R. Russell M. Paterson, Norma B. Gusmão and Nelson Lima, Filamentous Fungi in Drinking Water, Particularly in Relation to Biofilm Formation, doi:10.3390/ijerph8020456

Table 1. Surveys of fungi in drinking water.

Country, Place, Year	Period of time	Type of water	Main isolation method	Most frequent fungal isolates	Refs.
United Kingdom, 1996	Autumn and Spring	Surface water and network	Membrane filtration, Direct plating and Bating	*Aspergillus*, *Cladosporium*, *Epicoccum*, *Penicillium* and *Trichoderma*	[4]
Greece, Thessaloniki, 1998	One collection (126 samples)	Tap water (hospital and community)	Membrane filtration	*Penicillium*. *Aspergillus* and *Acremonium*	[14]
Greece, 85 haemodialysis units, 1998	One collection (255 samples)	Municipal water supplies of haemodialysis centres	Membrane filtration	*Penicillium* and *Aspergillus*	[15]
Germany, North Rhine-Westphalia, 1998/9	12 months	Drinking water	Pour-plating	*Acremonium*, *Exophiala*, *Penicillium* and *Phialophora*	[5]
Norway, 14 networks, 2002/3	December, June and September	Drinking water (surface and groundwater)	Membrane filtration	*Penicillium*. *Trichoderma* and *Aspergillus*	[9,10]
Portugal, Braga, 2003/4	12 months	Tap water	Membrane filtration	*Penicillium* and *Acremonium*	[8]
Pakistan, Karachi, 2007	One collection (30 samples)	Water (and fruit juice)	Direct plating	*Aspergillus niger* and *A. clavatus*	[16]
Australia, Queensland, 2007/8	18 months	Municipal water	Membrane filtration	*Cladosporium*, *Penicillium*, *Aspergillus and Fusarium*	[13]
Brazil, Recife, 2009/10	5 months	Water treatment plant; tap water	Membrane filtration	*Penicillium*, *Aspergillus* and *Phoma*	[17]
Portugal, Lisbon, 2010	4 months	surface water; spring water; groundwater	Membrane filtration	*Aspergillus*, *Cladosporium*, *Penicillium*	[18,19]

METHODS TO STUDY FUNGI IN TAP WATER

One of the most important problems for studying fungi is the use of isolation media which tends to vary among researchers. Examples of media employed are listed in Table 2. DRBC and DG18 are recommended as general media for the purpose of isolation and enumeration of fungi in foods of high water activity (aw > 0.90), and is therefore suitable for water analyses. DG18 has several advantages; the characteristic colony appearance on agar plates facilitates differentiation of species which makes secondary selection and re-isolation of pure cultures easier to accomplish, compared to DRBC. DG18 also inhibits overgrowth of fast growing fungi such as members of the Mucorales and *Trichoderma*.

Table 2. Principal media used for the analysis of water for fungi.

Medium	Composition	
CMA/2—Corn meal agar half-strength	Corn meal	8.5 g
	Agar	8.5 g
	Distilled water	1000 mL
	pH	5.8–6.2
CZ—Czapek Dox agar	Czapek solution	10 mL
	K_2HPO_4	1 g
	Saccharose	30 g
	Agar	15 g
	Distilled water	1000 mL
	pH	6.0–6.4
	Czapek solution	
	$NaNO_3$	30 g
	KCl	5 g
	$MgSO_4 \cdot 7H_2O$	5 g
	$FeSO_4 \cdot 7H_2O$	0.1 g
	$ZnSO_4 \cdot 7H_2O$	0.1 g
	$CuSO_4 \cdot 5H_2O$	0.05 g
	Distilled water	100 mL

Medium	Composition	
DG18—Dichloran 18 % glycerol agar	Peptone	5 g
	Glucose	10 g
	K_2HPO_4	1 g
	$MgSO_4{\cdot}7H_2O$	0.5 g
	Dichloran (0.2% in ethanol)	1 mL
	Glycerol	220 g
	Chloramphenicol	0.1 g
	Agar	15 g
	Distilled water	1000 mL
	pH	5.4–5.8
DRBC—Dichloran Rose Bengal chloramphenicol agar	Peptone	5 g
	Glucose	10 g
	K_2HPO_4	1 g
	$MgSO_4{\cdot}7H_2O$	0.5 g
	Dichloran	0.002 g
	Rose Bengal	0.025 g
	Chloramphenicol	0.1 g
	Agar	15 g
	Distilled water	1000 mL
	pH	5.4–5.8
MEA—Malt extract agar	Malt extract	20 g
	Peptone	1 g
	Glucose	20 g
	Agar	20 g
	Distilled water	1000 mL
	pH	5.0–5.5
NGRBA—Neopeptone glucose Rose Bengal aureomycin	Neopeptone	5 g
	Glucose	10 g
	0.67 % (w/v) aureomycin solution	5 mL
	1 % (w/v) Rose Bengal solution	3.5 mL
	Agar	15 g
	Distilled water	1000 mL
	pH	6.3–6.7
PDA—Potato dextrose agar	Potato extract	4 g
	Glucose	20 g
	Agar	15 g
	Distilled water	1000 mL
	pH	5.4–5.8
SDA—Sabouraud dextrose agar	Mycological peptone	10 g
	Glucose	40 g
	Agar	15 g
	Distilled water	1000 mL
	pH	5.4–5.8

Finally, DG18 does not readily produce substances toxic to the fungus when exposed to light: Media containing Rose Bengal are light sensitive and produce inhibitory compounds in significant concentrations after two hours exposure to light. The current authors recommend low and high nutritional media to obtain the complete spectrum of fungi present in water.

The phenotypic identification of fungi to species level is based on morphological characteristics such as (a) septation of hypha, and (b) the formation, morphology, patterns and branching frequency of fruiting-bodies and conidiophores. Also, the size, form, colour and ornamentation of the conidia are important characters (Figure 1).

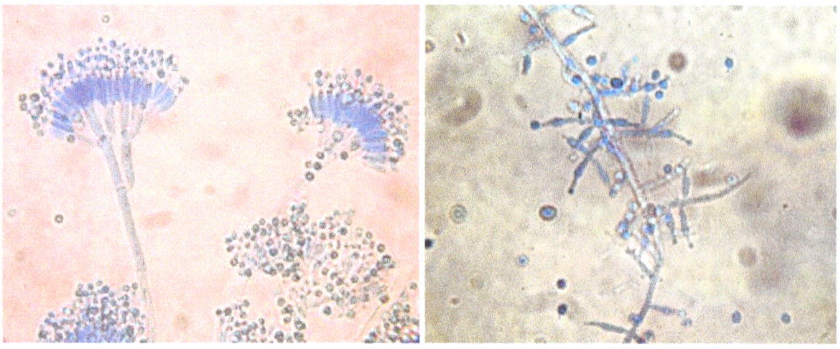

Figure 1. *Penicillium brevicompacum* (left) and *Trichoderma* sp. (right) isolated from tap water grown on MEA and observed under optical microscope after stained with Lactophenol Cotton Blue.

Macroscopic characters such as colony form, structure, size and colour are also important. In addition, appearance and growth on different culture media are used for differentiation (Figure 2). The morphological approach to species identification benefit from supplementation with physiological, biochemical and molecular characters. Ideally, a polyphasic approach to identification is undertaken to cover all aspects of species identity [20,21].

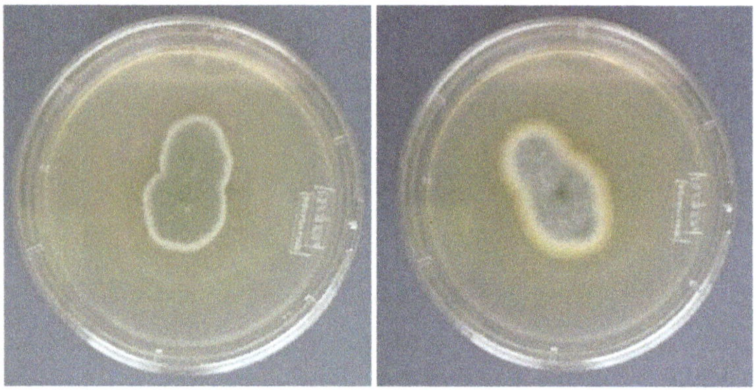

Figure 2. *Penicillium brevicompacum* (left) and *Penicillium aurantiogriseum* (right) isolated from tap water grown on MEA (N.B. it is significant that species names have been obtained rather than simply naming them as *Penicillium* spp.).

FUNGAL WATER BIOFILMS

Biofilms are problematic in water distribution systems as they permit a high concentration of organisms to occur, with the potential to cause some of the problems described above. Their presence can represent a health threat since biofilms play a role in the accumulation, protection and dissemination of pathogens [11,22]. However, the involvement of filamentous fungi in biofilms has not yet been demonstrated satisfactorily due to a dearth of useful techniques.

Sampling Biofilms

The methods which have been employed were often indirect, where the involvement of the fungus cannot be proved categorically. An example of this is placing the inner part of pipelines in contact with fungal growth medium and allowing fungi to emerge (Figure 3).

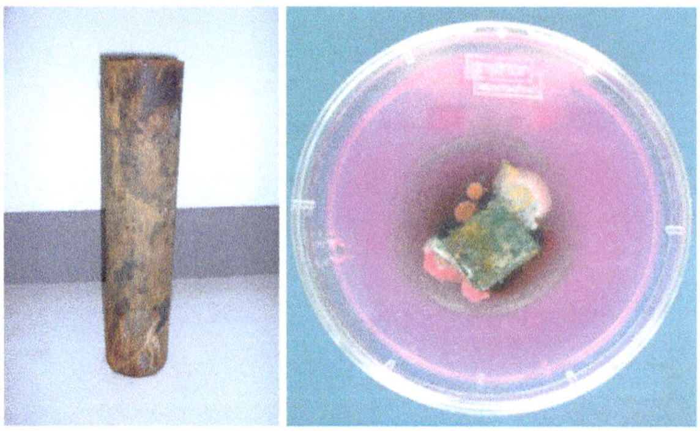

Figure 3. Cast iron tube from a tap water distribution system (left) and a NGRBA plate (see Table 2) with fungi growing from the biofilm present on the internal part of the tube (right).

Nevertheless, many studies on drinking water distributed systems (DWDS) have shown that the major part of the biomass is attached to pipe surfaces in a biofilm [23]. The studies of biofilms in DWDS commonly face problems such as limited access to the biofilm and difficulty in repeating experiments, since pipe sections have to be replaced after sampling. A solution is to construct sampling devices that allow for *in situ* investigation of the attached microbial community.

Examinations of DWDS reveal the complexity that will be required of such a system. Several factors are known to influence biofilm development, including temperature, nutrients, residual disinfectant, the hydraulic regime and the characteristics of the substratum [24]. Moreover, various ecological interactions will occur between microorganisms in biofilms.

Since biofilms are complex, laboratorial systems can be very helpful in determining the influence of individual parameters (e.g., medium composition and biocides application). Various attempts have been made to culture biofilms under controlled but realistic conditions, e.g., the Robbins device [25], the RotoTorque reactor [26], and flow cells [27]. However, do they reflect the *in vivo* situation?

We recommend the use of "sampler devices" to investigate natural filamentous fungi biofilms which were produced for a water distribution system called "*Alto do Céu*" in Recife, Brazil. They were developed largely because of the limitations inherent in the alternative methods mentioned previously. The devices can be employed to: (1) mimic the real conditions of the water network and yet be straight forward to insert and handle, (2) be convenient for transportation and storage, (3) maintain the integrity of biofilms and (4) allow *in situ* analyses of the biofilms. The core of the sampler devices consists of hollow polyvinyl chloride (PVC) pipes within polyethylene or acetate coupons held in place to allow biofilm growth (Figure 4).

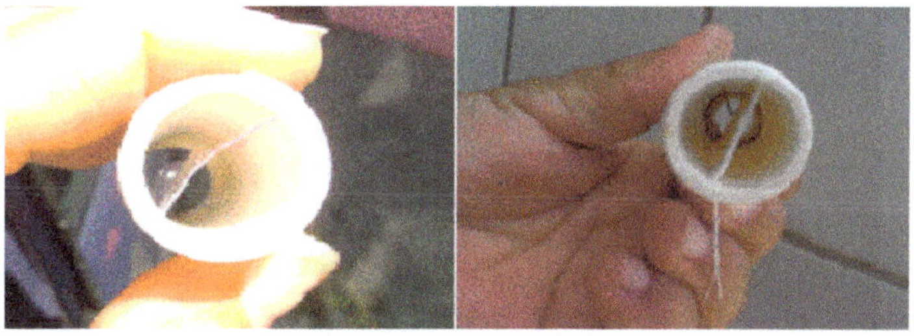

Figure 4. Sampler devices consisted of a hollow PVC pipe within acetate (left) or polyethylene (right) coupons.

The samplers had a diameter of 1.5 cm and a length from 7 to 10 cm. The ends of each sampler form a screw to connect multiple samplers or to close the device with a cap after removal from the water network (Figure 5).

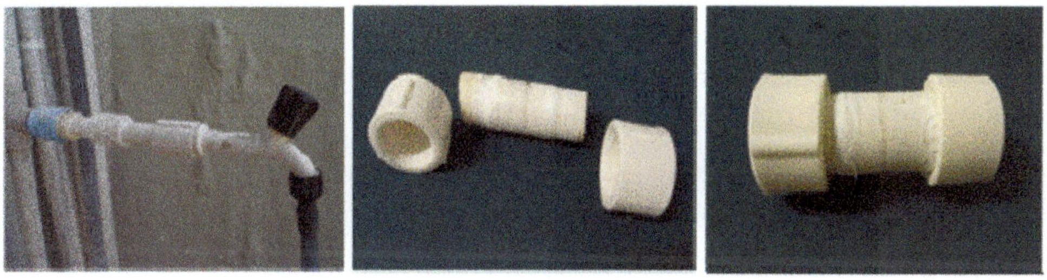

Figure 5. From left to right: (1) sampler device in place, (2) units of samplers to be connected to each other and (3) closed devices for transportation after a suitable period of time.

These features facilitated insertion, handling and removal of each sampler device after collection. The "caps" prevent contact with the external environment during transport. Finally, the pipes can be filled with water, thus maintaining moisture and preserving the integrity of the biofilms formed in the coupons. To study natural biofilms, these sampler devices were installed at different sampling points in the water network: a water treatment plant (reservoir; 20,000 m^3) and three branch connections within the municipal DWDS were selected (Figure 6). The coupons were removed for examination every 15 days for a period of 6 months.

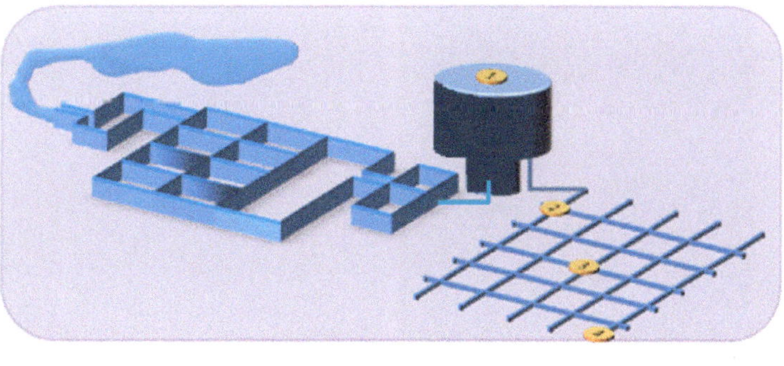

Figure 6. Drinking water distribution system. 1—Water treatment plant and reservoir; 2—beginning; 3—middle and 4—end of the network.

Biofilm Detection

The development of non-invasive and non-destructive techniques such as fluorescence microscopy, enables *in situ*monitoring of microbial biofilm communities. Fluorescence microscopy provides information on cell morphology, metabolism and biodiversity. Simultaneously, data concerning biofilm matrix structure and architecture are provided when used in conjunction with fluorescent molecular probes [28,29]. These data are crucial and are unattainable by conventional approaches (*i.e.*, culturing methods). They contribute to the understanding of "real time" microbial ecology within biofilms. In this context, samplers and real samples collected from *Alto do Céu* were used. The suitability of Calcofluor White MR2 (CW) staining for morphological characterisation, FUN1 for viability and Fluorescent *in situ* Hybridization (FISH) for diversity studies were investigated following the protocol presented in Figure 7.

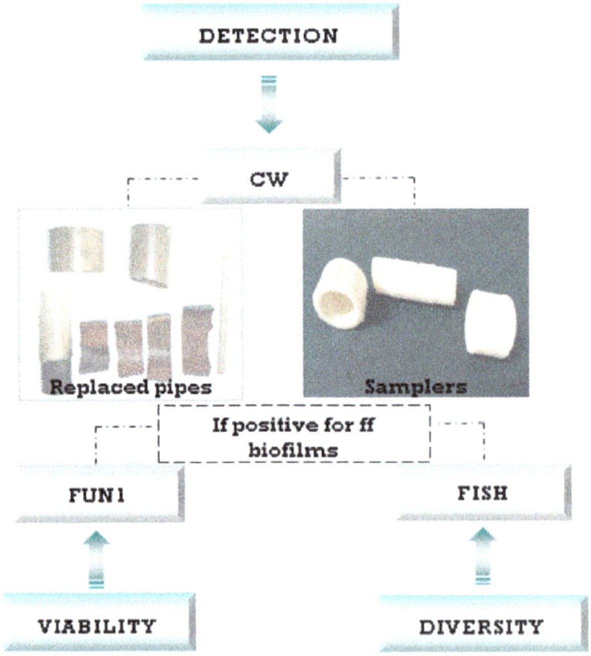

Figure 7. Stepwise approach used to study in situ filamentous fungal (ff) water biofilms using fluorescent staining techniques (for a description of terms see preceding paragraph).

A description of the various stains employed now follows: (1) CW is a fluorescent probe capable of making hydrogen bonds with β-(1→4) and β-(1→3) polysaccharides. CW consists of a symmetrical molecule with two triazole rings and two primary alcohol functions on both sides of an ethylene bridge. The fluorophore shows a high affinity for chitin forming hydrogen bonds with free hydroxyl groups which stains fungal cell walls blue. (2) FUN1 stain discriminates fungal dead cells which have a diffuse yellow-green fluorescence, from the metabolically active cells which have red Cylindrical Intra-Vacuolar Structures (CIVS). (3) For taxon diversity the (a) universal rRNA probe specific for Eukarya EUK516 (5′-ACCAGACTTGCCCTCC-3′, MWG Biotech, Ebersberg, Germany) labelled with the red Cy3 at the 5′ terminal and (b) FUN1429 probe specific for Eumycota (5′-GTGATGTACTCGCTGGCC-3′, MWG Biotech, Ebersberg, Germany) labelled with Oregon-Green at the 5′ terminal for FISH were employed.

A Olympus BX51 epifluorescent microscope, using UV light equipped with 40x/0.30 and 10x/0.65 objectives and a filter set (EX 450-490 nm, EM 520), was used for detecting *in situ* biofilms after fluorescent staining techniques were applied. The images were acquired with a Zeiss AxioCam HRc colour camera using the software CellB®. Storage and handling of reagents were performed as recommended by the supplier. The sampler study shows presumptive filamentous fungi stained with CW colonising the polyethylene coupons after 90 days in contact with water (Figure 8).

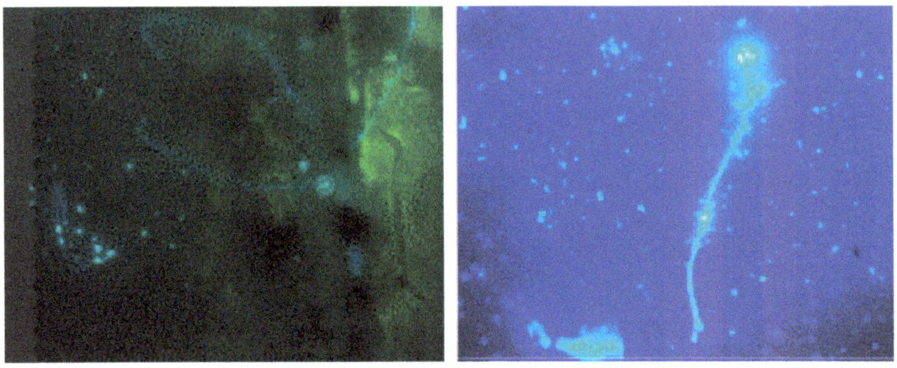

Figure 8. Presumptive filamentous fungal biofilm detection after CW staining on a polyethylene coupon.

Due to the polyethylene PVC autofluorescence, the coupons inside of the samplers were replaced by acetate coupons which have much less autofluorescesce. Notwithstanding this, biofilms were not detected on acetate surfaces of coupons after CW staining.

These unexpected results could occur for several reasons. Acetate coupons are not the optimal surfaces to promote biofilm adhesion, perhaps due to their hydrophobic properties. Alternatively, the length of time of the exposure may be inadequate for the described circumstances. These possibilities require further investigations and other materials also are required to be investigated. In contrast, when CW was applied to the actual replaced pipe samples, filamentous fungi were clearly observed (Figure 9).

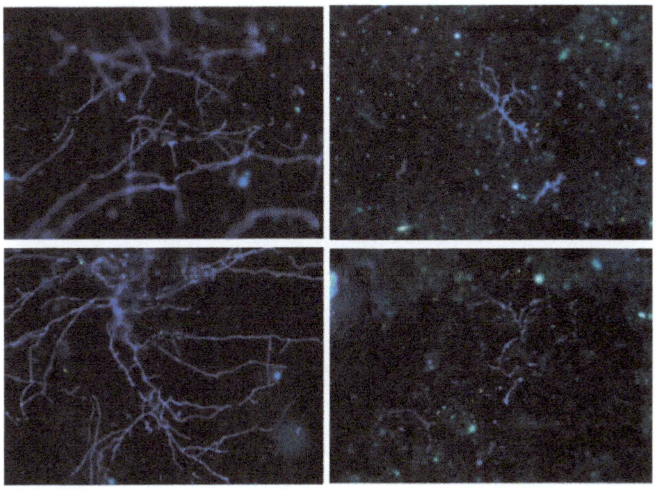

Figure 9. Filamentous fungal biofilms detected after CW staining in replaced pipe samples.

Since the real pipes demonstrated filamentous fungi within the biofilm and deposits, the stepwise approach, as defined inFigure 7, was continued. The FUN1 staining for viability showed red CIVS inside of the fungal vacuoles (Figure 10) demonstrating that the fungi were viable in the biofilm. The CIVS are ATP-dependent which is correlated with fungal viability.

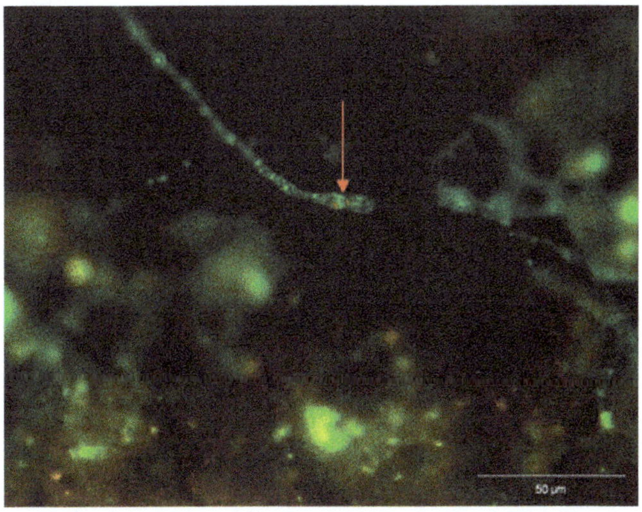

Figure 10. Visualization of red CIVS (arrow) and green diffuse hypha after FUN1 staining biofilm in real replaced pipe samples.

To confirm that the filamentous structures in the pipes were fungi, the samples were submitted to analysis by FISH probes. Figure 11 demonstrates a clear relationship between the CW filamentous structures presented on biofilms with the positive signals for FISH probes. The EUK516 probe shows that eukaryotic microorganisms were present and the FUN1429 probe confirm that these were fungi.

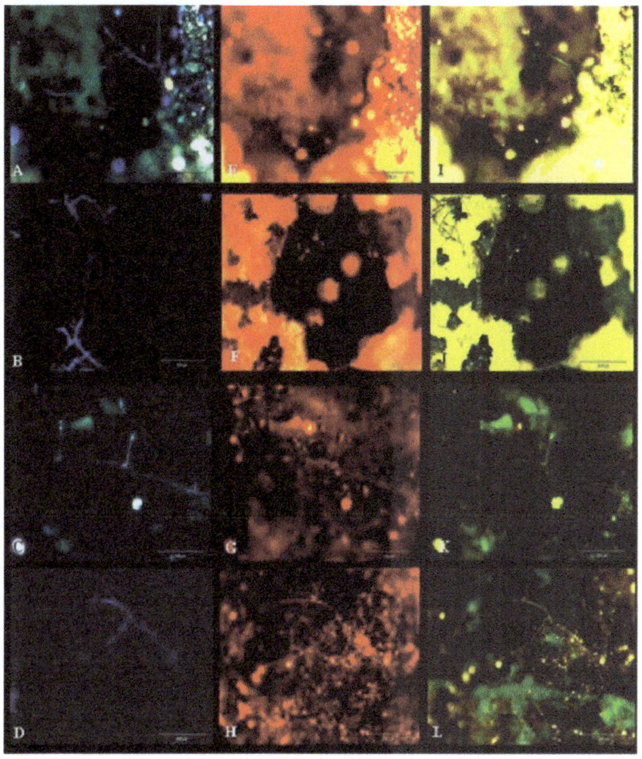

Figure 11. Detection of filamentous fungal water biofilm after CW staining (A-D), and analysis with FISH with EUK516 (E-H) and FUN1429 (I-L) probes in real replaced pipe samples.

Conclusions and Recommended Procedures

The present work is an updated overview concerning the main topics related to filamentous fungal contamination of water distribution system and in relation to biofilm formation. Standard methods are required, equivalent to those that exist for bacterial contamination of drinking water. The methods developed for fungal water biofilms demonstrated the presence of these microorganisms in the biofilms. Recommended standard methods for fungi have been developed and tested by the authors, and may be used to analyse water distribution systems for the occurrence of filamentous fungi [11]. The isolation method is based on NS 4716 [30] with some modifications, consist of filtering 100 mL of the water through a 0.45 µm membrane, placing the membrane on dichloran 18% glycerol agar (DG18) plates and incubating the plates at 20 ºC for 1–2 weeks. The plates must be checked visually every week, the colonies counted and the results presented as colony forming units (CFU)/100 mL water. Isolation of pure single colonies is required for identification, and pure cultures must be subcultured on potato dextrose agar (PDA) or malt extract agar (MEA) (Table 2).

To determine total filamentous fungi, the water sample must be filtered through two 0.45 µm membranes. The membranes are to be placed on neopeptone glucose rose Bengal aureomycin agar (NGRBA) with 5 replicates for filamentous fungal isolation and colony counting. The second filter should be placed onto half strength cornmeal agar (CMA/2) with five replicates for the same purposes. The method devised by us for the analysis of fungal biofilms in Brazil is also recommended more generally. Finally, even more progress will be made only if future studies ensure that strains are identified to species and standard techniques are employed universally.

ACKNOWLEDGMENTS

The authors acknowledge Brazilian Companhia Pernambucana de Saneamento (COMPESA) to support and make possible the work performed by H.M.B. Oliveira. V.M. de Siqueira is supported by the grant SFRH/BD/43719/2008 from Fundação para a Ciência e Tecnologia (FCT), Portugal.

REFERENCES

1. Nagy, LA; Olson, BH. The occurrence of filamentous fungi in drinking water distribution systems. *Can. J. Microbiol***1982**, *28*, 667–671.
2. Nagy, LA; Olson, BH. Occurrence and significance of bacteria, fungi and yeasts associated with distribution pipe surfaces. Proceedings of the American Water Works Association, Water Quality Technology Conference, Houston, TX, December 1985; American Water Works Association: Denver, CO, USA, 1985; pp. 213–238.
3. Franková, E; Horecka, M. Filamentous soil fungi and unidentified bacteria in drinking water from wells and water mains near Bratislava. *Microbiol. Res* **1995**, *150*, 311–313.
4. Kinsey, GC; Paterson, RR; Kelley, J. Methods for the determination of filamentous fungi in treated and untreated waters. *J. Appl. Microbiol. Symp. Suppl* **1999**, *85*, 214S–224S.
5. Göttlich, E; van der Lubbe, W; Lange, B; Fiedler, S; Melchert, I; Reifenrath, M; Flemming, H-C; Hoog, S. Fungal flora in groundwater-derived public drinking water. *Int. J. Hyg. Environ. Health* **2002**, *205*, 269–279.
6. Kelley, J; Kinsey, G; Paterson, R; Brayford, D; Pitchers, R; Rossmore, H. *Identification and Control of Fungi in Distribution Systems*; AWWA Research Foundation and American Water Works Association: Denver, CO, USA, 2003.
7. Paterson, RRM; Lima, N. Fungal Contamination of Drinking Water. In *Water Encyclopedia*; Lehr, J, Keeley, J, Lehr, J, Kingery, TB, III, Eds.; John Wiley & Sons: New York, NY, USA, 2005; pp. 1–7.
8. Gonçalves, AB; Paterson, RRM; Lima, N. Survey and significance of filamentous fungi from tap water. *Int. J. Hyg. Environ. Health* **2006**, *209*, 257–264.
9. Hageskal, G; Knutsen, AK; Gaustad, P; de Hoog, GS; Skaar, I. Diversity and significance of mold species in Norwegian drinking water. *Appl. Environ. Microbiol* **2006**, *72*, 7586–7593.
10. Hageskal, G; Gaustad, P; Heier, BT; Skaar, I. Occurrence of moulds in drinking water. *J. Appl. Microbiol* **2007**, *102*, 774–780.
11. Hageskal, G; Lima, N; Skaar, I. The study of fungi in drinking water. *Mycol. Res* **2009**, *133*, 165–172.
12. Paterson, RRM; Hageskal, G; Skaar, I; Lima, N. Occurrence, problems, analysis and removal of filamentous fungi in drinking water. In *Fungicides: Chemistry, Environmental Impact and Health Effects*; De Costa, P, Bezerra, P, Eds.; Nova Science Publishers: New York, NY, USA, 2009; pp. 379–399.
13. Sammon, NB; Harrower, KM; Fabbro, LD; Reed, RH. Incidence and distribution of microfungi in a treated 9 municipal water supply system in sub-tropical Australia. *Int. J. Environ. Res. Public Health* **2010**, *7*, 1597–1611.
14. Arvanitidou, M; Kanellou, K; Constantinides, TC; Katsouyannopoulos, V. The occurrence of fungi in hospital and community potable waters. *Lett. Appl. Microbiol* **1999**, *29*, 81–84.
15. Arvanitidou, M; Spaia, S; Velegraki, A; Pezarloglou, M; Kanetidis, D; Pangidis, P; Askepidis, N; Katsinas, Ch; Vayonas, G; Katsouyannopoulos, V. High level of recovery of fungi from water and dialysate in haemodialysis units.*J. Hosp. Infect* **2000**, *45*, 225–230.
16. Nazim, S; Dawar, S; Tariq, M; Zaki, MJ. Quantitative estimation of mycoflora in drinking water and fruit juices of Karachi. *Pakistan J. Bot* **2008**, *40*, 1263–1268.
17. Oliveira, HMB. Fungos filamentosos na água e em biofilmes na rede de distribuição de água potável do sistema Alto do Céu, Recife-PE, Masters Thesis, Biologia de Fungos do Departamento de Micologia do Centro de Ciências Biológicas da Universidade Federal de Pernambuco (UFPE), Recife, Brazil. 2010.
18. Pereira, VJ; Basílio, MC; Fernandes, D; Domingues, M; Paiva, JM; Benoliel, MJ; Crespo, MT; San Romão, MV. Occurrence of filamentous fungi and yeasts in three different drinking water sources. *Water Res* **2009**, *43*, 3813–3819.
19. Pereira, VJ; Fernandes, D; Carvalho, G; Benoliel, MJ; San Romão, MV; Barreto Crespo, MT. Assessment of the presence and dynamics of fungi in drinking water sources using cultural and molecular methods. *Water Res* **2010**, *44*, 4850–4859.
20. Santos, C; Paterson, RRM; Venâncio, A; Lima, N. Filamentous fungal characterisations by matrix-assisted laser desorption/ionisation time of flight mass spectrometry. *J. Appl. Microbiol* **2010**, *8*, 375–385.
21. Santos, C; Fraga, ME; Kozakiewicz, Z; Lima, N. Fourier transform infrared as a powerful technique for the identification and characterisation of filamentous fungi and yeasts. *Res. Microbiol* **2010**, *161*, 168–175.
22. Huq, A; Whitehouse, CA; Grim, CJ; Alam, M; Colwell, RR. Biofilms in water, its role and impact in human disease transmission. *Curr. Opin. Biotechnol* **2008**, *19*, 244–247.
23. Boe-Hansen, R; Martiny, AC; Arvin, E; Albrechtsen, H-J. Monitoring biofilm formation and activity in drinking water distribution networks under oligotrophic conditions. *Water Sci. Technol* **2003**, *47*, 91–97.

24. Niquette, P; Servais, P; Savoir, R. Impacts of pipe materials on densities of fixed bacterial biomass in a drinking water distribution system. *Water Res* **2000**, *34*, 1952–1956.

25. McCoy, WF; Bryers, JD; Robbins, J; Costerton, JW. Observations of fouling biofilm formation. *Can. J. Microbiol* **1981**, *27*, 910–917.

26. Van der Wende, E; Characklis, WG; Smith, DB. Biofilms and bacterial drinking water quality. *Water Res* **1989**, *3*, 1313–1322.

27. Caldwell, DE; Lawrence, JR. Study of attached cells in continues-flow slide culture. In *CRC Handbook of Laboratory Systems for Microbial Ecology Research*; Wimpenny, JWT, Ed.; CRC Press: Boca Raton, FL, USA, 1988.

28. Li, Y; Dick, WA; Tuovinen, OH. Fluorescence microscopy for visualization of soil microorganisms—a review. *Biol. Fert. Soils* **2004**, *39*, 301–311.

29. Gonçalves, AB; Santos, IM; Paterson, RRM; Lima, N. FISH and Calcofluor staining techniques to detect *in situ*filamentous fungal biofilms in water. *Revista Iberoamericana de Micologia* **2006**, *23*, 192–196.

30. Determination with the membrane filters method, NS 4716. In *Norwegian Standard for Water Analysis: Microfungi in Water*; The Standardization Organizations in Norway: Oslo, Norway, 1991; pp. 1–12. (In Norwegian)

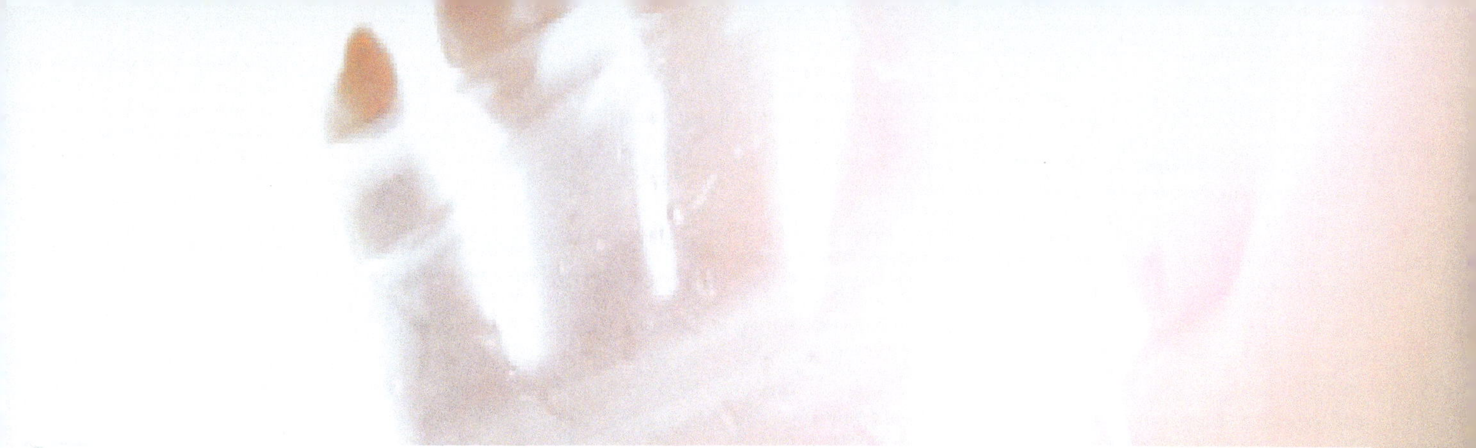

CHAPTER 22

INTERACTION OF LEGIONELLA PNEUMOPHILA AND HELICOBACTER PYLORI WITH BACTERIAL SPECIES ISOLATED FROM DRINKING WATER BIOFILMS

Maria S Gião, Nuno F Azevedo, Sandra A Wilks, Maria J Vieira, Charles W Keevil

BACKGROUND

In natural environments, bacteria can adhere to surfaces forming a complex structure called a biofilm. When embedded in biofilms, microorganisms can be protected from several adverse factors such as temperature, low nutrients and the presence of biocides [1, 2, 3, 4, 5, 6]. Therefore, understanding the ecology of microorganisms in this structure is fundamental in order to obtain a comprehensive knowledge of real systems. In nature, biofilms typically consist of many species of microorganisms that can interact with each other either positively (for instance, the synthesis of a metabolite by one species which can be used in the metabolism of another) or negatively (such as nutrient competition) [7, 8, 9]. One type of biofilm that has been widely studied is that formed in drinking water distribution systems (DWDS) because of its role in the persistence of pathogens in drinking water and the consequent potential for impact on public health [10, 11,12].

Legionella pneumophila is a waterborne pathogen that can cause Legionnaires' disease or Pontiac fever [13,14]. This pathogen is found naturally in fresh water reservoirs and can contaminate drinking water when disinfection is inefficient, being transmitted to man when contaminated aerosols are inhaled [12, 15, 16, 17]. The mode of transmission of *Helicobacter pylori* remains controversial but drinking water as a route of transmission has recently gained recognition [18]. Although no cultivable *H. pylori* have ever been recovered from drinking water systems, molecular techniques such as PCR [19, 20, 21, 22] and peptide nucleic acid (PNA) probes used to target 16 S rRNA in fluorescence *in situ* hybridization (FISH) assays [23, 24], have demonstrated the presence of this pathogen in DWDS. This identification, in addition to epidemiological studies, point to different prevalence of *H. pylori* in the microbial population which is associated with the type of source water. This strongly supports water as a route of transmission [18, 25, 26, 27].

Previous studies have demonstrated that both pathogens can be incorporated into heterotrophic drinking water biofilms and persist for at least 32 days [28, 29]. In the case of *H. pylori*, although no cultivable cells were ever recovered, the presence of a high intracellular rRNA content indicates that cells might be in a viable but non cultivable (VBNC) state

Citation: Maria S Gião, Nuno F Azevedo, Sandra A Wilks , Maria J Vieira , Charles W Keevil, Interaction of legionella pneumophila and helicobacter pylori with bacterial species isolated from drinking water biofilms, DOI: 10.1186/1471-2180-11-57

[28, 30]. It is possible that the loss of *H. pylori* cultivability when associated with heterotrophic biofilms had been due to a negative effect caused by the presence of other microorganisms [31]. Nevertheless, it is also possible that there were other microorganisms present in the biofilm that could have a beneficial effect on *L. pneumophila* or *H. pylori*, as shown by other studies where these pathogens were co-cultured with other microorganisms in liquid media [32, 33]. However, for multi-species biofilms it is technically very challenging to determine which sessile microorganisms could have a positive or negative effect on these pathogens, particularly regarding the intimate associations that occur within biofilms. A particular type of interaction that can facilitate the formation of biofilm is the aggregation of cells, which can occur between cells of the same species (auto-aggregation) or between different species (co-aggregation), and has been well described for isolates of dental plaque species in complex media and aquatic species in potable water [34, 35, 36].

The aim of this work was to study the influence of different autochthonous microorganisms isolated from drinking water biofilms on the incorporation and survival of *L. pneumophila* and *H. pylori* in biofilms. For that, the first part of the work tested all the species used for auto and co-aggregation. Subsequently, dual-species biofilms of *L. pneumophila* and *H. pylori* were formed with the different drinking water bacteria and the results compared with mono-species biofilms formed by *L. pneumophila* and *H. pylori*.

RESULTS

Auto and co-aggregation of *L. pneumophila*and other drinking water bacteria

Initially, the selected biofilm strains were tested for auto- and co-aggregation in test tubes as described by Rickard et al. [35], either alone or with *L. pneumophila*. No co-aggregation was observed for the strains studied, either alone or in pairs with *L. pneumophila* (results not shown).

*L. pneumophila*in biofilms

For the experiments on biofilm formation on uPVC coupons, an inoculum of *L. pneumophila* was prepared containing approximately 3.7×10^7 of total cells ml^{-1} (quantified using SYTO 9 staining). In comparison to total cells, 49% were cultivable on BCYE agar and 50% were detected by PNA-FISH. The inocula of the strains isolated from drinking water biofilms had on average 75% of cultivable cells compared to SYTO 9 stained cells, except in the case of *Mycobacterium chelonae* where the percentage was considerably lower (2.5%).

Figure 1a shows the variation with time of total cells, PNA-cells and cultivable *L. pneumophila* present in a mono-species biofilm. The attachment of *L. pneumophila* cells to the surface occurred in the first 24 hours of the experiment. Moreover, the numbers of total cells (stained by SYTO 9) and PNA stained cells did not change significantly between days 1 and 32 ($P > 0.05$). In contrast, the number of cultivable cells greatly increased in the first two weeks and then decreased significantly in the last weeks of the experiments ($P < 0.05$).

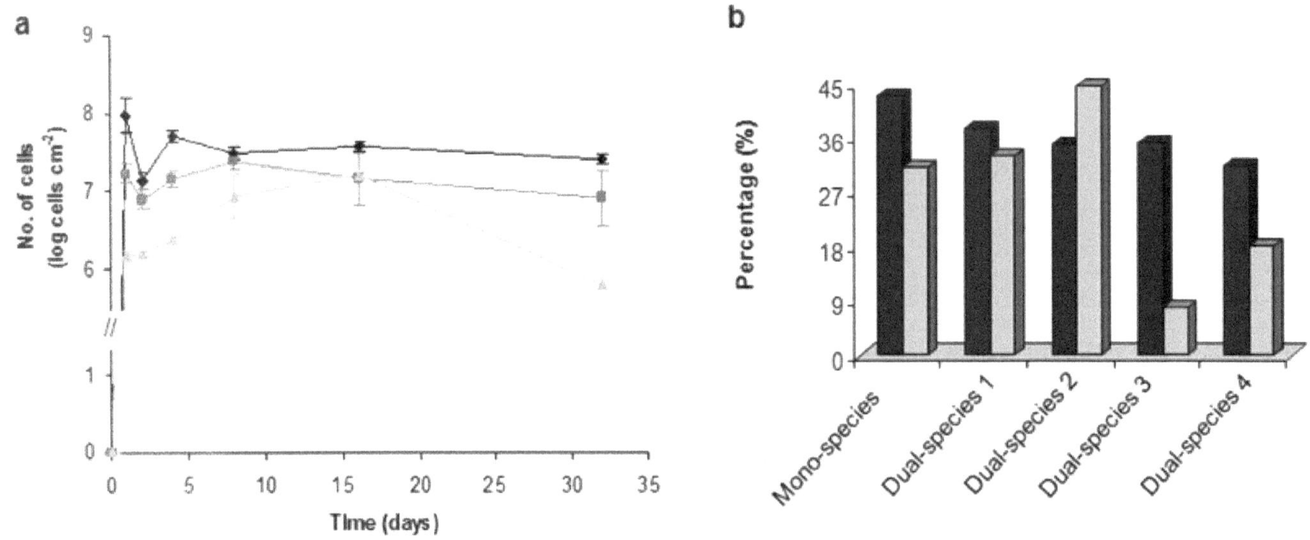

Figure 1 Numbers of *L. pneumophila* cells in mono and dual-species biofilms. Variation of the number of cells of *L. pneumophila* in mono-species biofilm quantified by the three different methods: curves represent the variation of total cell number (black diamond), *L. pneumophila* hybridized with the PNA PLPEN620 probe (dark grey square) and cultivable *L. pneumophila* (light grey triangle); bars represent standard deviation (n = 3) (a). *L. pneumophila* PNA-positive numbers/total cells numbers ratio (dark grey bars) and cultivable *L. pneumophila* numbers/*L. pneumophila*PNA-positive numbers ratio (light grey bars) for the mono-species biofilm and dual-species biofilms of *L. pneumophila* and *V. paradoxus* (Dual-species 1), *L. pneumophila* and *M. chelonae* (Dual-species 2), *L. pneumophila*and *Acidovorax* sp. (Dual-species 3) and *L. pneumophila* and *Sphingomonas* sp. (Dual-species 4); the ratio values were calculated using the average of the values obtained for the six time point samples (b).

For the experiments of *L. pneumophila* in dual-species it was observed that the numbers of *L. pneumophila*PNA-positive cells and cultivable *L. pneumophila* did not change significantly with time after the first day ($P > 0.05$). Table 1 presents the data obtained for the quantification of sessile cells, giving the average values of the samples analyzed at all time points, for mono and dual-species biofilms. The data for the numbers of total cells, total PNA-positive *L. pneumophila* and cultivable *L. pneumophila* in mono and in dual species biofilms were similar ($P > 0.95$), except for the numbers of cultivable *L. pneumophila* when associated with *Acidovorax*sp. which were significantly lower ($P < 0.05$). Figure 1b shows the percentage of PNA-positive *L. pneumophila*in relation to SYTO 9 stained total cells; this was similar for both mono and dual-species biofilms ($P = 1.000$). This indicates that *L. pneumophila* adhere well to uPVC surfaces, either alone or in the presence of *Variovorax paradoxus*, *M. chelonae*, *Acidovorax* sp. And *Sphingomonas* sp., although the morphology of the biofilm appeared to be different for the mono or dual-species (Figure 2a and 2b, respectively). The relationship between cultivable and *L. pneumophila* PNA-positive cells was higher (although not statistically significant, $P > 0.95$) for cells recovered from the *L. pneumophila* - *M. chelonae* biofilm while the numbers of cultivable *L. pneumophila* decreased five-fold when this pathogen was associated with *Acidovorax* sp. and almost four-fold when associated with *Sphingomonas* sp.

Table 1 Average of the total number of cells, *L. pneumophila* PNA-positive, cultivable *L. pneumophila* and cultivable non-*legionellae*cell numbers in mono and dual-species biofilms obtained for all the time points sampled.

Strain in biofilm	Total cells × 10^{-7}(cells cm^{-2})	PNA cells × 10^{-7}(cells cm^{-2})	Cultivable *L. pneumophila*× 10^{-6} (CFU cm^{-2})	Cultivable non -*legionellae*× 10^{-6} (CFU cm^{-2})
L. pneumophila	4.42	1.48	5.25	n.a.
L. pneumophila and*V. paradoxus*	3.51	1.11	4.11	4.49
M. chelonae	4.87	1.05	4.65	0.19
Acidovorax sp.	4.12	1.59	1.05	6.55
Sphingomonas sp.	3.80	0.83	1.45	1.06

n.a. - not applicable.

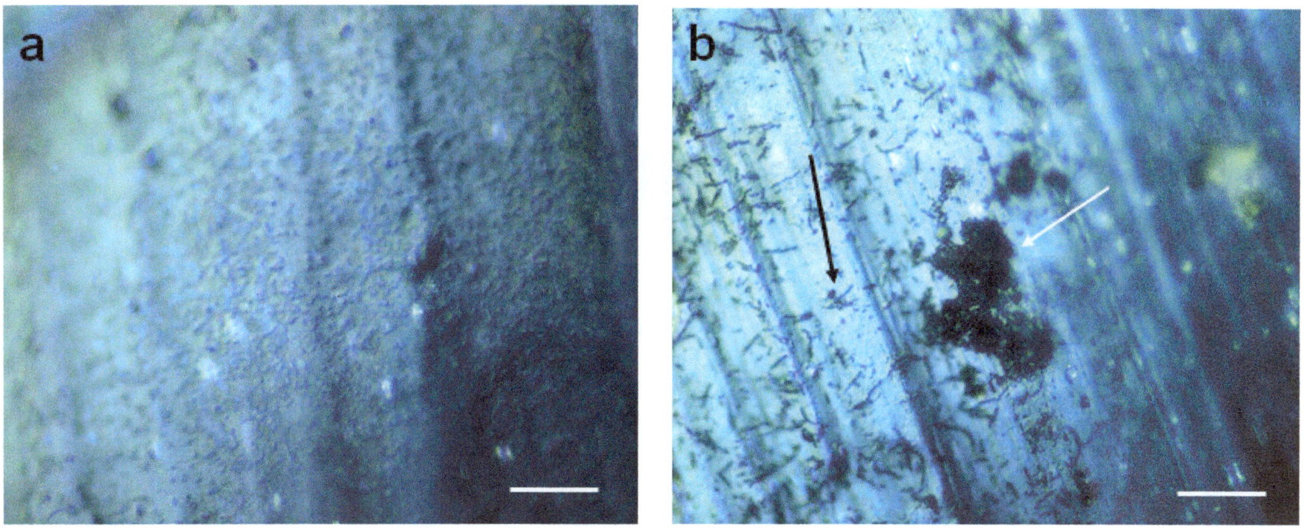

Figure 2 uPVC coupon covered with a mono and dual-species *L. pneumophila* biofilm. Microphotograph of an uPVC coupon visualized under EDIC microscopy covered with a 32 days-old biofilm formed by *L. pneumophila* (a) and *L. pneumophila* and *Sphingomonas* sp. (b). The black arrow indicates individual cells attached to the uPVC surface and white arrow indicates a microcolony. Bars represent 20 µm.

Auto and co-aggregation of *H. pylori*and other drinking water bacteria

The same experiments were repeated using *H. pylori* instead of *L. pneumophila*. For the auto- and co-aggregation of *H. pylori* with drinking water isolates, the same strains were used as selected for the *L. pneumophila* experiments and an additional strain was also included: *Brevundimonas* sp., a bacterium isolated on CBA medium from drinking water biofilms. The results obtained in the test tube assay system showed neither auto nor co-aggregation of *H. pylori* with any of the species investigated.

*H. pylori*in biofilms

The biofilm experiments used the same strains indicated in the previous paragraph. It was observed that for the *H. pylori* inoculum, only 5% of the total cells were cultivable, a value similar to that obtained by Azevedo et al. [37], while 29% were detected by PNA-FISH. Figure 3a and 3b show that *H. pylori* is able to form biofilms, despite the poor cultivability of the cells on agar media. However, while the morphology of *H. pylori* cells from the inoculum was predominantly spiral, after forming biofilms the cells were mainly coccoid shaped.

Figure 3c shows that when in pure culture *H. pylori* adhered to the surface to form the biofilm in the first day followed by a statistically significant decrease ($P < 0.05$) in total cells during day 1 and 4. The same trend was observed for cells quantified using the PNA probe. No cultivable *H. pylori* were recovered on CBA medium. When the biofilm was formed in the presence of *Brevundimonas* sp. the variation with time of total cells and PNA numbers were not statistically significant ($P > 0.05$). Comparing the numbers obtained for pure *H. pylori*biofilms and biofilms grown in the presence of *Brevundimonas* sp. there was no significant difference between the numbers of *H. pylori* detected using the PNA probe (results not shown) nor in the percentage between PNA and total cells numbers ($P > 0.05$).

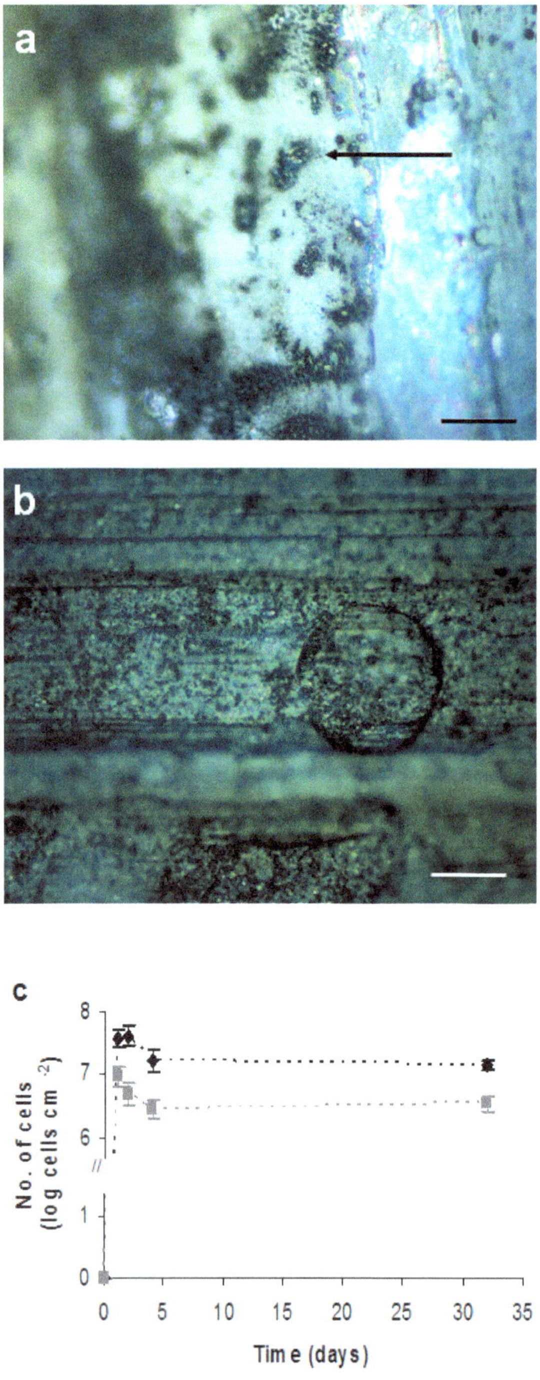

Figure 3 uPVC coupon covered with *H. pylori* biofilm and variation of *H. pylori* numbers in the mono-species biofilm. Microphotograph of an uPVC coupon visualized under EDIC microscopy covered with a mono-species *H. pylori*biofilm after 1 day (a) and 32 days (b) of incubation. Black arrow indicates the presence of a microcolony. Bars represent 20 μm. (c) Variation with time in the total cell number (black diamond) and *H. pylori* PNA-cells (grey square) present in the biofilm. Bars represent standard deviation (n = 3).

In terms of cultivable cells it was observed that no cultivable *H. pylori* were ever recovered from any of the mono or dual-species biofilms at any time point, with the exception of cells recovered from 1 day-old biofilms grown in the presence of *M. chelonae* or *Sphingomonas* sp. (6.67×10^1 and 1.83×10^2 CFU cm^{-2}, respectively).

DISCUSSION

Auto and co-aggregation of *L. pneumophila* and *H. pylori*with drinking water bacteria

In a previous study several bacterial strains were isolated from heterotrophic biofilms formed on uPVC coupons in a two-stage chemostat system [28]. For the present work, the selection of the bacteria used was based on the prevalence of these isolated strains in biofilms, i.e., the strains that were always present in biofilm samples when detected by culture were used rather than those only found intermittently.

In the aggregation studies it was observed that there was no auto-aggregation of any of the bacteria tested in this study, as demonstrated previously for *Brevundimonas vesicularis*, *Acidovorax delafieldii* and *V. paradoxus* [34, 38]. No co-aggregation of *L. pneumophila* or *H. pylori* was observed with any of the bacteria isolated from drinking water biofilms, demonstrating that while all of the bacteria used in this study have the ability to form biofilms they are attaching to the uPVC surfaces without aggregating in the planktonic phase with the other microorganisms [36].

*L. pneumophila*in biofilms

The *L. pneumophila* cells from the inocula prepared for the biofilm experiments were quantified for total, PNA-positive and cultivable cells. Results showed that cultivable and PNA numbers were similar but were only 50% of the numbers obtained by SYTO 9 staining. It is still controversial whether PNA probes detect dead cells or if they just produce a detectable signal with viable cells. PNA probes have been used to detect pathogens in mixed biofilms but it has not been well established if this technique can also detect non-viable cells [23, 29,39]. However the similarity in the cultivable and PNA-positive numbers, and the difference between PNA-labelled and total cells (stained by SYTO 9), strongly indicates that the PNA probe fails to detect dead cells. PNA probes bind specifically to rRNA molecules emitting a signal that can be visualized under microscopy. The intensity of that signal is related to the rRNA content, i.e., the higher the rRNA content the brighter the signal is [40]. A very low content of rRNA would result in insufficient brightness and cells would not be visualized. After cellular death the content of rRNA decreases significantly and therefore some authors have suggested that the emission of a bright signal is a good indication of cell viability [39, 41, 42]. The results obtained in the present study (Figure 1a) support that there is a correlation between the number of viable cells and the number of cells that bind to the PNA probe but further studies should be performed to correlate the PNA-positive cells and their metabolic state.

As it has been demonstrated before by other authors [43, 44], the attachment of *L. pneumophila* cells to the uPVC surface occurred on the first day of biofilm formation and the numbers of total and PNA stained cells, from mono-species biofilms, did not change significantly ($P > 0.05$). Nevertheless, the numbers of cultivable cells increased in the first two weeks and decreased during the rest of the experiment. It has been demonstrated that *L. pneumophila* can survive in tap water for long periods without losing cultivability [45,46], but is not able to replicate in axenic cultures in tap water or in low nutrient media, except when associated with biofilms or parasitizing amoebal species [29, 47, 48]. After two weeks the cultivability decreased but was not completely lost for the 32 days of the experiment which indicates that biofilms are a protective niche for *L. pneumophila*, even in axenic culture. Conversely, PNA-positive numbers with a high fluorescence intensity remained constant and, for the same reason explained before, this suggests that cells are still viable. Moreover, the fact that total *L. pneumophila* and *L. pneumophila* PNA-positive cells remained constant with time indicates that there is no damage to DNA and rRNA, respectively. Conversely, the variation of PNA-positive numbers in dual-species biofilms was used as an indicator of the variation of viable *L. pneumophila* cells inside of those biofilms.

The results of dual-species biofilms showed that when biofilms were formed in the presence of *M. chelonae*the percentage of cultivable *L. pneumophila* in relation to *L. pneumophila* PNA-positive cells was slightly superior compared to mono-species biofilms or dual-species biofilms with the other strains isolated from drinking water. Although the difference is not statistically significant this result indicates that this strain has a small positive effect on *L. pneumophila* cultivability. In contrast, the numbers of cultivable *L. pneumophila*decreased when this pathogen was associated with *Acidovorax* sp. indicating that this species has a negative impact on *L. pneumophila* cultivability. It was also observed that the numbers of cultivable *L. pneumophila*when co-cultivated with *Sphingomonas* sp. decreased and, although the statistical analysis showed that the difference is not significant, the fact that the cultivability was almost four-fold lower appears to reveal an antagonistic effect. Conversely, it appears that both strains affect negatively sessile *L. pneumophila*cultivability, either by competition for nutrients or production of a metabolite toxic to *L. pneumophila*. The fact that these two species were isolated on R2A reveals that they have low nutritional requirements to grow and might even be able to grow in water, contrary to *L. pneumophila* which is not able to grow in absence of, for example, L-cysteine and high iron concentrations [49, 50]. This is corroborated by the values shown in Table 1, where cultivable *Acidovorax* sp. and *Sphingomonas* sp. numbers are 6.55×10^6 and 1.06×10^6 CFU cm^{-2} suggesting that these two microorganisms could be metabolically active in the biofilm despite the poor nutrient concentration of the medium (filtered tap water). Another possible explanation for the lower numbers of cultivable *L. pneumophila* when biofilms were formed in co-culture with *Sphingomonas* sp., can be related to the structure of the biofilm. Figure 2 shows a 32 days-old biofilm formed by *L. pneumophila* and*L. pneumophila* associated with *Sphingomonas* sp. The biofilm formed in the presence of *Sphingomonas* sp. had a different morphology, and although the thickness of the biofilm has not been measured, the presence of microcolonies suggests the presence of thicker structures where anaerobic zones might occur. Wadowsky et al. [33] have demonstrated that in anaerobic conditions *L. pneumophila* loses cultivability and if biofilms formed by *L. pneumophila* and *Sphingomonas* sp. have indeed anaerobic zones, then it is possible that *L. pneumophila* located in those places has become uncultivable. It would therefore be interesting to undertake further research to measure the thickness of different parts of the biofilm and the respective concentration of oxygen and relate those results to the cultivability of cells from those regions. However, the fact that the numbers quantified by the use of a PNA probe remained constant, might indicate that these cells may still be viable and can probably recover cultivability in favorable conditions.

This work clearly demonstrates that *L. pneumophila* can be negatively or positively influenced by other microorganisms present in drinking water. It is important to note that this study was carried out under particular conditions and it will be important to perform more experiments in the future, in particular to study the effect of other drinking water bacteria, the formation of biofilms under dynamic conditions and the incorporation of a disinfectant, such as chlorine.

It is known that other bacteria can influence the growth of *L. pneumophila* either in nutrient-poor environments, such as drinking water, or in rich artificial media. Toze et al. [51] have demonstrated that some bacteria commonly present in heterotrophic biofilms, such as *Pseudomonas* sp. and *Aeromonas* sp., can inhibit the growth of *L. pneumophila* while Wadowsky and Yee [49] demonstrated that *Flavobacterium breve*can support the satellite growth of this pathogen on BCYE agar without L-cysteine. A curious result was obtained by Temmerman et al. [52] who demonstrated that dead cells can also support the growth of this pathogen. Although the mechanisms responsible for the influence of different microorganisms on *L. pneumophila* survival are unknown there is one aspect of *L. pneumophila* microbial ecology that has been already well-established: *L. pneumophila* is not able to grow in drinking water unless associated with biofilms or amoebal species [46, 47, 53]. Hence, the knowledge of how microorganisms affect *L. pneumophila*cultivability is a key factor for the effective control of this pathogen in drinking water and associated biofilms, and requires further investigation.

*H. pylori*in biofilms

In this study the cells recovered from mono-species *H. pylori* biofilms were always uncultivable, for all the time points, which is in contrast to the Azevedo et al. [54] study, where it was demonstrated that after 24 hours sessile *H. pylori* cells were still cultivable. This might be due to the differences in the method of cell removal from the coupons, the quality of water or the type of uPVC substratum. When the biofilm was formed in the presence of *Brevundimonas* sp. no cultivable *H. pylori* cells were ever recovered either. However, for this case, care should be taken in the interpretation of the results. In fact, *Brevundimonas* was able to grow on CBA medium in a faster and more abundant way then *H. pylori*. As such, it is impossible to determine whether *H. pylori* is indeed uncultivable in the presence of this microorganism, or whether it could not be detected because it was overgrown by *Brevundimonas*. We have attempted to solve this issue by using CBA medium supplemented with antibiotics but, as shown by other authors [28], available selective medium for *H. pylori* allows the growth of other species, including *Brevundimonas* sp. The fact that there were no differences in the results for the PNA-positive cell numbers obtained for *H. pylori* in mono-species biofilms and in dual-species biofilms with *Brevundimonas* sp. suggests that this bacterium has little or no effect on the inclusion of *H. pylori* in biofilms.

Cultivable *H. pylori* was never recovered from dual-species biofilms at any time point, independently of the second species used, except when *H. pylori* formed dual-species biofilms in the presence of *M. chelonae* and*Sphingomonas* sp. For these two microorganisms, it was observed that *H. pylori* was able to retain cultivability for a period of between 24 and 48 hours. This suggests that both microorganisms might have a positive effect on the inclusion and survival of this pathogen in drinking water biofilms. The ability of *H. pylori* to adapt to different physico-chemical parameters has been studied by several authors [30, 55, 56, 57, 58], however no studies about the influence of other microorganisms on the survival of this pathogen have been found in the literature except the coculture of *H. pylori* with the protozoan, *Acanthamoeba castellanii* [59]. The interaction of microorganisms in biofilms has been widely studied and in this particular case could be the key for the survival of this microorganism in drinking water systems, even if in a VBNC state. More investigations should therefore be performed concerned with the influence of drinking water microorganisms on *H. pylori*metabolism and survival.

CONCLUSIONS

This work clearly demonstrates that, even in pure culture, both pathogens can adhere to surfaces and form biofilm. *L. pneumophila* can remain cultivable for at least 32 days although less cultivable when associated with *Acidovorax* sp. and *Sphingomonas* sp. The experiments with *H. pylori* demonstrated that this pathogen loses cultivability in less than 24 hours when in mono-species or in dual-species biofilms with *V. paradoxus*,*Acidovorax* sp. and *Brevundimonas* sp., while retaining cultivability for at least 24 hours when biofilms are grown in the presence of *M. chelonae* and *Sphingomonas* sp. Consequently, *M. chelonae* seems to have a positive effect on the cultivability of both pathogens and being a pathogen commonly found in drinking water systems [60, 61], can play an important role in the control of these two pathogens. Control of this mycobacterial opportunistic pathogen and other biofilm species that can have a synergetic effect on *L. pneumophila* and *H. pylori* might provide an important contribution towards the supply of safe drinking water as both *L. pneumophila* and *H. pylori* have been found to be chlorine resistant [62, 63].

METHODS

Culture maintenance

In this work, *L. pneumophila* NCTC 12821 and *H. pylori* NCTC 11637 strains were used. Strains of *V. paradoxus*, *M. chelonae*, *Acidovorax* sp., *Sphingomonas* sp. and *Brevundimonas* sp. were isolated from drinking water biofilms [28, 29]. All strains were maintained in vials frozen at -80°C and recovered by standard plating procedures onto the appropriate media and subcultured once prior to biofilm formation experiments.*L. pneumophila* NCTC 12821, *V. paradoxus* and *M. chelonae* were grown on Buffered Charcoal Yeast Extract (BCYE) agar (Oxoid, UK) for 24 hours at 30°C. *Acidovorax* sp. and *Sphingomonas* sp. were grown on R2A (Oxoid, UK) for 48 hours at 22°C. *H. pylori* NCTC 11637 and *Brevundimonas* sp. were grown on Columbia Agar (Oxoid, UK) supplemented with 5% (v/v) defibrinated horse blood (CBA) (Oxoid, UK) and incubated for 48 hours at 37°C in a microaerophilic atmosphere of 10% CO_2, 7% H_2 and 3% O_2 (the remainder being N_2).

Auto- and co-aggregation in test tubes

Prior to the start of the experiments tap water from Southampton, UK, was collected in a transparent flask and left, loosely closed, overnight for chlorine evaporation. Then the water was sterilized by filtration through a 0.2 μm pore size Nylon filter (Pall Gelman, UK). All bacterial species were suspended in this dechlorinated and filtered tap water, with the following characteristics, provided by the water company (Southern Water, UK): pH 7.3; turbidity 0.10 FTU; conductivity 504 μS cm^{-1}; total organic carbon 0.649 mg l^{-1}; total iron 16 μg Fe l^{-1}; free chlorine 0.21 mg Cl_2 l^{-1}; total chlorine 0.26 mg Cl_2 l^{-1}. The inocula had a final concentration of approximately 2×10^8 cells ml^{-1}. For autoaggregation, 3 ml of each suspension was transferred into a sterile test tube, whereas for co-aggregation experiments 1.5 ml of either *L. pneumophila* or *H. pylori* suspension were mixed with 1.5 ml of each one of the species isolated from drinking water biofilms. At times 0, 1, 2, 4, 6, 8, 24 and 48 hours, tubes were vortexed for 10 seconds and observed for co-aggregation according to the scale described by Rickard et al. [35]. All experiments were performed in duplicate.

Coupon preparation

Unplasticized polyvinylchloride (uPVC) coupons of 1 cm^2 were used as a substratum for biofilm growth as it is a commonly used material in drinking water pipelines. To remove grease and wax from the coupons, prior to biofilm growth, they were immersed in water and detergent for 5 min, washed with a bottle brusher, rinsed twice in distilled water and air-dried. Subsequently, they were washed in 70% (v/v) ethanol to remove any organic compounds and autoclaved at 1 atm and 121°C [64].

Biofilm formation

To form the mono-species biofilms of *L. pneumophila* NCTC 12821 and *H. pylori* NCTC 11637 the inocula were prepared by suspending the cells in 50 ml of dechlorinated and filtered tap water to give a final concentration of approximately 10^7 cells ml^{-1}. The mono-species biofilms were used as a control. The dual-species biofilm inocula were prepared by mixing *L. pneumophila* or *H. pylori* with *V. paradoxus*, *M. chelonae*, *Acidovorax* sp. or *Sphingomonas* sp. in 50 ml of filter-sterilized tap water to a final concentration of 10^7 cells ml^{-1} of each microorganism. For the experiments with *H. pylori* an inoculum was also prepared with this pathogen and*Brevundimonas* sp. All suspensions were homogenized by vortexing and 4 ml of each inoculum were transferred to 6-well microtiter plates containing one uPVC coupon in each well. Plates were incubated in the dark at 22°C and two coupons of each biofilm type were removed after 1, 2, 4, 8, 16 and 32 days, and gently rinsed to remove loosely attached cells on the surface of the biofilm. One coupon was used for direct observation under a Nikon Eclipse E800 episcopic differential interference contrast/epifluorescence (EDIC/EF) microscope (Best Scientific, UK) [65] using the EDIC channel to directly visualise biofilm. The other coupon was scraped to quantify sessile cells.

Quantification of sessile cells

At each time point coupons were removed from the wells and rinsed three times in filtered tap water to remove planktonic cells from the biofilm and coupons surfaces. The coupons were then transferred to a 15 ml centrifuge tube (Greiner Bio-one, UK) containing 2 ml of filter-sterilized tap water and autoclaved glass beads of 2 mm diameter (Merck, UK). To remove the biofilm from the coupon surfaces the tubes were then vortexed for 1 min. The vortexing step also promoted the homogenization of the suspensions prior to the quantification of total cells, PNA-positive cells and cultivable cells, as described below. Preliminary experiments showed that vortexing with glass beads removed the biofilm formed under these conditions, although it was still possible to observe a few dispersed cells on the uPVC surface.

Total cells were quantified using the SYTO 9 staining method (Molecular Probes, Invitrogen, UK). In summary, 1 ml of an appropriate dilution was mixed with 0.5 μl of SYTO 9, incubated in the dark for 15 minutes, filtered through a 0.2 μm pore size polycarbonate black Nucleopore® membrane (Whatman, UK) and allowed to air-dry. Then a drop of non-fluorescent immersion oil (Fluka, UK) and a coverslip were added before observation under the Nikon Eclipse E800 EDIC/EF microscope (Best Scientific, UK) [65]. As the cells were homogenously distributed, 10 fields of view on each membrane were chosen at random and the number of cells counted (×100 objective lens).

L. pneumophila was quantified using the specific PNA probe PLPNE620 (5'-CTG ACC GTC CCA GGT-3') and *H. pylori* by the use of a PNA probe with the following sequence 5'- GAGACTAAGCCCTCC -3'(Eurogentec, Belgium). PNA-FISH was carried out by filtering 1 ml of an appropriate dilution through a 0.2 μm Anodisc membrane (Whatman, UK). This was left to air dry. For the quantification of *L. pneumophila* the membrane was covered with 90% (v/v) ethanol to fix the cells and again air dried. The hybridization, washing and microscopy observation method was performed as described by Wilks and Keevil [42]. For *H. pylori*quantification the membrane was covered with 4% (w/v) paraformaldehyde followed by 50% (v/v) ethanol for 10 minutes each to fix the cells and air dried. The hybridization, washing and microscopy observation method was performed as described by Guimarães et al. [66].

Cultivable numbers of all bacteria were determined by plating 40 μl of an appropriate dilution on the respective agar medium, as described above in the section «Culture maintenance». BCYE plates were incubated aerobically for two days at 30°C, R2A for seven days at 22°C and CBA plates were incubated for seven days at 37°C in a microaerophilic atmosphere. It is recommended that the incubation of BCYE to quantify *L. pneumophila* from environmental samples goes for up to ten days. However it was observed that for these samples if the BCYE plates were incubated for more than two days the colonies would overgrow in diameter and it would be impossible to distinguish individual colonies. Therefore two days was chosen as the incubation time.

Statistical analysis

The homogeneity of variances of total number, PNA and cultivable cells and the relation between *L. pneumophila* of cells and total cells was checked by the Levene test for equality of variances using a statistical package (SPSS Inc., Chicago IL, USA). Results were subsequently compared by a one-way ANOVA followed by a Bonferroni *post hoc* test. Differences were considered relevant if $P < 0.05$.

REFERENCES

1. Camper A, Burr M, Ellis B, Butterfield P, Abernathy C: Development and structure of drinking water biofilms and techniques for their study. J App Microbiol. 1999, 85: 1S-12S. 10.1111/j.1365-2672.1998.tb05277.x.
2. Costerton JW, Lewandowski Z, Caldwell DE, Korber DR, Lappin-Scott HM: Microbial biofilms. Ann Rev Microbiol. 1995, 49 (1): 711-745. 10.1146/annurev.mi.49.100195.003431.
3. Emtiazi F, Schwartz T, Marten SM, Krolla-Sidenstein P, Obst U: Investigation of natural biofilms formed during the production of drinking water from surface water embankment filtration. Water Res. 2004, 38 (5): 1197-1206. 10.1016/j.watres.2003.10.056.
4. Fletcher M, Marshall KC: Are solid surfaces of ecological significance to aquatic bacteria. Advances in microbial ecology. Edited by: Marshall KC. 1982, New York: Plenum Publishing Corp, 6: 199-230.
5. LeChevallier MW, Cawthon CD, Lee RG: Factors promoting survival of bacteria in chlorinated water supplies. App Environ Microbiol. 1988, 54 (3): 649-654.
6. Ellwood DC, Keevil CW, Marsh PD, Brown CM, Wardell JN, Le Roux N: Surface-Associated Growth [and Discussion]. Philos T R Soc B. 1982, 297: 517-532. 10.1098/rstb.1982.0058.
7. Christensen BB, Haagensen JAJ, Heydorn A, Molin S: Metabolic commensalism and competition in a two-species microbial consortium. App Environ Microbiol. 2002, 68 (5): 2495-2502. 10.1128/AEM.68.5.2495-2502.2002.
8. Rao D, Webb JS, Kjelleberg S: Competitive interactions in mixed-species biofilms containing the marine bacterium Pseudoalteromonas tunicata. App Environ Microbiol. 2005, 71 (4): 1729-1736. 10.1128/AEM.71.4.1729-1736.2005.
9. Watnick P, Kolter R: Biofilm, city of microbes. J Bacteriol. 2000, 182 (10): 2675-2679. 10.1128/JB.182.10.2675-2679.2000.
10. Berry D, Xi CW, Raskin L: Microbial ecology of drinking water distribution systems. Curr Opin Biotechnol. 2006, 17 (3): 297-302. 10.1016/j.copbio.2006.05.007.
11. Szewzyk U, Szewzyk R, Manz W, Schleifer KH: Microbiological safety of drinking water. Ann Rev Microbiol. 2000, 54: 81-127. 10.1146/annurev.micro.54.1.81.
12. Keevil CW: Pathogens in environmental biofilms. The Encyclopedia of Environmental Microbiology. Edited by: Bitton G. 2002, New York: Wiley, 2339-2356.
13. McDade JE, Shepard CC, Fraser DW, Tsai TR, Redus MA, Dowdle WR: Legionnaires' disease - isolation of a bacterium and demonstration of its role in other respiratory disease. New Engl J Med. 1977, 297 (22): 1197-1203. 10.1056/NEJM197712012972202.
14. Pasculle W: Update on Legionella. Clin Microbiol Newslett. 2000, 22 (13): 97-101. 10.1016/S0196-4399(00)89124-2.
15. Devos L, Boon N, Verstraete W: Legionella pneumophila in the environment: The occurrence of a fastidious bacterium in oligotrophic conditions. Rev Environ Sci Biotechnol. 2005, 4 (1): 61-74. 10.1007/s11157-004-8174-1.
16. Hsu SC, Martin R, Wentworth BB: Isolation of Legionella species from drinking water. App Environ Microbiol. 1984, 48 (4): 830-832.
17. Steinert M, Hentschel U, Hacker J: Legionella pneumophila: an aquatic microbe goes astray. FEMS Microbiol Rev. 2002, 26 (2): 149-162. 10.1111/j.1574-6976.2002.tb00607.x.
18. Azevedo NF, Guimaraes N, Figueiredo C, Keevil CW, Vieira MJ: A new model for the transmission of Helicobacter pylori: Role of environmental reservoirs as gene pools to increase strain diversity. Crit Rev Microbiol. 2007, 33 (3): 157-169. 10.1080/10408410701451922.
19. Bunn JEG, Mackay WG, Thomas JE, Reid DC, Weaver LT: Detection of Helicobacter pylori DNA in drinking water biofilms: implications for transmission in early life. Lett App Microbiol. 2002, 34 (6): 450-454. 10.1046/j.1472-765X.2002.01122.x.
20. Mackay WG, Gribbon LT, Barer MR, Reid DC: Biofilms in drinking water systems - A possible reservoir for Helicobacter pylori. Water Sci Technol. 1998, 38 (12): 181-185. 10.1016/S0273-1223(98)00822-1.
21. Park SR, Mackay WG, Reid DC: Helicobacter sp recovered from drinking water biofilm sampled from a water distribution system. Water Res. 2001, 35 (6): 1624-1626. 10.1016/S0043-1354(00)00582-0.
22. Voytek MA, Ashen JB, Fogarty LR, Kirshtein JD, Landa ER: Detection of Helicobacter pylori and fecal indicator bacteria in five North American rivers. J Water Health. 2005, 3 (4): 405-422.
23. Bragança SM, Azevedo NF, Simões LC, Keevil CW, Vieira MJ: Use of fluorescent in situ hybridisation for the visualisation of Helicobacter pylori in real drinking water biofilms. Water Sci Technol. 2007, 55 (8): 387-393.
24. Queralt N, Bartolome R, Araujo R: Detection of Helicobacter pylori DNA in human faeces and water with different levels of faecal pollution in the north-

east of Spain. J App Microbiol. 2005, 98 (4): 889-895. 10.1111/j.1365-2672.2004.02523.x.

25. Engstrand L: Helicobacter in water and waterborne routes of transmission. J App Microbiol. 2001, 90: 80S-84S. 10.1046/j.1365-2672.2001.01356.x.
26. Gomes BC, Martinis ECP: The significance of Helicobacter pylori in water, food and environmental samples. Food Control. 2004, 15 (5): 397-403. 10.1016/S0956-7135(03)00106-3.
27. Klein PD, Graham DY, Gaillour A, Opekun AR, Smith EO: Water source as risk factor for Helicobacter pylori infection in Peruvian children. Lancet. 1991, 337 (8756): 1503-1506. 10.1016/0140-6736(91)93196-G.
28. Gião MS, Azevedo NF, Wilks SA, Vieira MJ, Keevil CW: Persistence of Helicobacter pylori in heterotrophic drinking water biofilms. App Environ Microbiol. 2008, 74 (19): 5898-5904.
29. Gião MS, Wilks SA, Azevedo NF, Vieira MJ, Keevil CW: Comparison between standard culture and peptide nucleic acid 16 S rRNA hybridization quantification to study the influence of physico-chemical parameters on Legionella pneumophila survival in drinking water biofilms. Biofouling. 2009, 25 (4): 335-343.
30. Azevedo NF, Pacheco AP, Keevil CW, Vieira MJ: Nutrient shock and incubation atmosphere influence recovery of culturable Helicobacter pylori from water. App Environ Microbiol. 2004, 70 (1): 490-493. 10.1128/AEM.70.1.490-493.2004.
31. Tait K, Sutherland IW: Antagonistic interactions amongst bacteriocin producing enteric bacteria in dual species biofilms. J App Microbiol. 2002, 93 (2): 345-352. 10.1046/j.1365-2672.2002.01692.x.
32. Surman SB, Morton LHG, Keevil CW: The dependence of Legionella pneumophila on other aquatic bacteria for survival on R2A medium. Int Biodeter Biodegr. 1994, 13: 223-236. 10.1016/0964-8305(94)90062-0.
33. Wadowsky RM, Wolford R, McNamara AM, Yee RB: Effect of temperature, pH, and oxygen level on the multiplication of naturally occurring Legionella pneumophila in potable water. App Environ Microbiol. 1985, 49 (5): 1197-1205.
34. Buswell CM, Herlihy YM, Marsh PD, Keevil CW, Leach SA: Coaggregation amongst aquatic biofilm bacteria. J App Microbiol. 1997, 83 (4): 477-484. 10.1046/j.1365-2672.1997.00260.x.
35. Rickard AH, Leach SA, Hall LS, Buswell CM, High NJ, Handley PS: Phylogenetic relationships and coaggregation ability of freshwater biofilm bacteria. App Environ Microbiol. 2002, 68 (7): 3644-3650. 10.1128/AEM.68.7.3644-3650.2002.
36. Kolenbrander PE, London J: Adhere today, here tomorrow: oral bacterial adherence. J Bacteriol. 1993, 175 (11): 3247-3252.
37. Azevedo NF, Almeida C, Fernandes I, Cerqueira L, Dias S, Keevil CW, Vieira MJ: Survival of gastric and enterohepatic Helicobacter spp. in water: Implications for transmission. App Environ Microbiol. 2008, 74 (6): 1805-1811. 10.1128/AEM.02241-07.
38. Rickard AH, McBain AJ, Ledder RG, Handley PS, Gilbert P: Coaggregation between freshwater bacteria within biofilm and planktonic communities. FEMS Microbiol Lett. 2003, 220 (1): 133-140. 10.1016/S0378-1097(03)00094-6.
39. Azevedo NF, Vieira MJ, Keevil CW: Development of peptide nucleic acid probes to detect H. pylori in diverse species potable water biofilms. Biofilm communities: Order from chaos?. Edited by: McBain A, Allison C, Brading M, Rickard A, Verran J, Walker J. 2003, Cardiff: Bioline, 231-239.
40. Pernthaler A, Pernthaler J, Eilers H, Amann R: Growth Patterns of Two Marine Isolates: Adaptations to Substrate Patchiness?. Appl Environ Microbiol. 2001, 67 (9): 4077-4083. 10.1128/AEM.67.9.4077-4083.2001.
41. Lehtola MJ, Torvinen E, Miettinen LT, Keevil CW: Fluorescence in situ hybridization using peptide nucleic acid probes for rapid detection of Mycobacterium avium subsp avium and Mycobacterium avium subsp paratuberculosis in potable water biofilms. App Environ Microbiol. 2006, 72 (1): 848-853. 10.1128/AEM.72.1.848-853.2006.
42. Wilks SA, Keevil CW: Targeting species-specific low-affinity 16 S rRNA binding sites by using peptide nucleic acids for detection of legionellae in biofilms. App Environ Microbiol. 2006, 72 (8): 5453-5462. 10.1128/AEM.02918-05.
43. Rogers J, Dowsett AB, Dennis PJ, Lee JV, Keevil CW: Influence of plumbing materials on biofilm formation and growth of Legionella pneumophila in potable water systems. App Environ Microbiol. 1994, 60 (6): 1842-1851.
44. Rogers J, Dowsett AB, Dennis PJ, Lee JV, Keevil CW: Influence of temperature and plumbing material selection on biofilm formation and growth of Legionella pneumophila in a model potable water system containing complex microbial flora. App Environ Microbiol. 1994, 60 (5): 1585-1592.
45. Ohno A, Kato N, Yamada K, Yamaguchi K: Factors influencing survival of Legionella pneumophila serotype 1 in hot spring water and tap water. App Environ Microbiol. 2003, 69 (5): 2540-2547. 10.1128/AEM.69.5.2540-2547.2003.
46. James BW, Mauchline WS, Dennis PJ, Keevil CW, Wait R: Poly-3-hydroxybutyrate in Legionella pneumophila, an energy source for survival in low nutrient environments. App Environ Microbiol. 1999, 65 (2): 822-827.
47. Murga R, Forster TS, Brown E, Pruckler JM, Fields BS, Donlan RM: Role of biofilms in the survival of Legionella pneumophila in a model potable water system. Microbiology. 2001, 147: 3121-3126.
48. Steinert M, Emody L, Amann R, Hacker J: Resuscitation of viable but nonculturable Legionella pneumophila Philadelphia JR32 by Acanthamoeba castellanii. App Environ Microbiol. 1997, 63 (5): 2047-2053.
49. Wadowsky RM, Yee RB: Satellite growth of Legionella pneumophila with an environmental isolate of Flavobacterium breve. App Environ Microbiol. 1983, 46 (6): 1447-1449.

50. James BW, Mauchline WS, Fitzgeorge RB, Dennis PJ, Keevil CW: Influence of iron limited continuous culture on physiology and virulence of Legionella pneumophila. Infect Immun. 1995, 63 (11): 4224-4230.

51. Toze S, Sly LI, MacRae IC, Fuerst JA: Inhibition of growth of Legionella species by heterotrophic plate count bacteria isolated from chlorinated drinking water. Curr Microbiol. 1990, 21 (2): 139-143. 10.1007/BF02091832.

52. Temmerman R, Vervaeren H, Noseda B, Boon N, Verstraete W: Necrotrophic growth of Legionella pneumophila. App Environ Microbiol. 2006, 72 (6): 4323-4328. 10.1128/AEM.00070-06.

53. Rogers J, Keevil CW: Immunogold and fluorescein immunolabeling of Legionella pneumophila within an aquatic biofilm visualized by using episcopic differential interference contrast microscopy. App Environ Microbiol. 1992, 58 (7): 2326-2330.

54. Azevedo NF, Almeida C, Cerqueira L, Dias S, Keevil CW, Vieira MJ: Coccoid form of Helicobacter pylori as a morphological manifestation of cell adaptation to the environment. App Environ Microbiol. 2007, 73 (10): 3423-3427. 10.1128/AEM.00047-07.

55. Azevedo NF, Pinto AR, Reis NM, Vieira MJ, Keevil CW: Shear stress, temperature, and inoculation concentration influence the adhesion of water-stressed Helicobacter pylori to stainless steel 304 and polypropylene. App Environ Microbiol. 2006, 72 (4): 2936-2941. 10.1128/AEM.72.4.2936-2941.2006.

56. Mouery K, Rader BA, Gaynor EC, Guillemin K: The stringent response is required for Helicobacter pylori survival of stationary phase, exposure to acid, and aerobic shock. J Bacteriol. 2006, 188 (15): 5494-5500. 10.1128/JB.00366-06.

57. Nilsson H-O, Blom J, Al-Soud WA, Ljungh A, Andersen LP, Wadstrom T: Effect of cold starvation, acid stress, and nutrients on metabolic activity of Helicobacter pylori. App Environ Microbiol. 2002, 68 (1): 11-19. 10.1128/AEM.68.1.11-19.2002.

58. West AP, Millar MR, Tompkins DS: Effect of physical environment on survival of Helicobacter pylori. J Clin Pathol. 1992, 45 (3): 228-231. 10.1136/jcp.45.3.228.

59. Winiecka-Krusnell J, Wreiber K, Von Euler A, Engstrand L, Linder E: Free-living amoebae promote growth and survival of Helicobacter pylori. Scand J Infect Dis. 2002, 34 (4): 253-256. 10.1080/00365540110080052.

60. Dailloux M, Laurain C, Weber M, Hartemann P: Water and nontuberculous mycobacteria. Water Res. 1999, 33 (10): 2219-2228. 10.1016/S0043-1354(98)00466-7.

61. Fischeder R, Schulzerobbecke R, Weber A: Occurrence of mycobacteria in drinking water samples. Zentralblatt fur Hygiene und Umweltmedizin. 1991, 192 (2): 154-158.

62. Gião M, Wilks S, Azevedo N, Vieira M, Keevil C: Validation of SYTO 9/Propidium Iodide Uptake for Rapid Detection of Viable but Noncultivable Legionella pneumophila. Microb Ecol. 2009, 58 (1): 56-62.

63. Gião MS, Azevedo NF, Wilks SA, Vieira MJ, Keevil CW: Effect of Chlorine on Incorporation of Helicobacter pylori into Drinking Water Biofilms. App Environ Microbiol. 2010, 76 (5): 1669-1673.

64. Keevil CW: Continuous culture models to study pathogens in biofilms. Method Enzymol. 2001, 337: 104-122. 10.1016/S0076-6879(01)37010-6.

65. Keevil CW: Rapid detection of biofilms and adherent pathogens using scanning confocal laser microscopy and episcopic differential interference contrast microscopy. Water Sci Technol. 2003, 47 (5): 105-116.

66. Guimarães N, Azevedo NF, Figueiredo C, Keevil CW, Vieira MJ: Development and application of a novel peptide nucleic acid probe for the specific detection of Helicobacter pylori in gastric biopsy specimens. J Clin Microbiol. 2007, 45 (9): 3089-3094.

Section VI - Biological Treatment and Biostability of Drinking Water

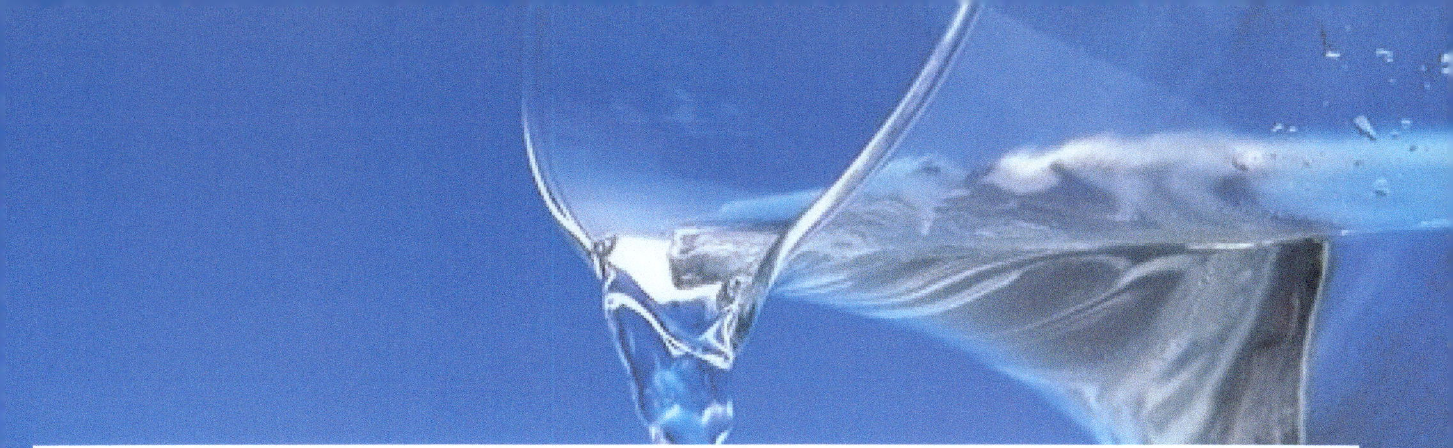

CHAPTER 23

MICROBIOLOGICAL QUALITY OF DRINKING WATER FROM DISPENSERS IN ITALY

Giorgio Liguori, Ivan Cavallotti, Antonio Arnese, Ciro Amiranda, Daniela Anastasi, Italo F Angelillo

BACKGROUND

It is well known that the quality and safety of the drinking water continues to be an important public health issue [1, 2], because its contamination has been frequently described as responsible for the transmission of infectious diseases that have caused serious illnesses and associated mortality worldwide [3, 4, 5, 6]. Clearly, point-of-use water quality is a critical public health indicator [2].

Over the past decade, there has been a markedly increase in the consumption of water derived from different sources in place of tap water for drinking use in many regions of the world. One of these alternative sources is the water from dispensers, which is popular mainly in office buildings and commercial stores, that are often presented as systems that are able to improve some characteristics of water and easy to use and to maintain. However, concerns have been sometimes raised about the quality of this source due to its potential to cause waterborne outbreaks associated with drinking water, particularly in sensitive and immunocompromised populations [2].

International drinking water-quality monitoring programs have been established in order to prevent or to reduce the risk of contracting water related infections. In Italy, the water for human consumption, including the water coming from dispensers, according to the European Community Directive guidelines, is required to be free from any pathogenic microorganism as well as chemical contaminations, which may be hazardous to the human health [7, 8].

To the best of our knowledge, very few studies have been conducted to this end dealing with the quality of drinking water from coolers [9, 10, 11, 12]. Therefore, the objective of the present study was to investigate the chemical and microbiological quality parameters of drinking water plumbed in water coolers in commercial stores in an area of Southern Italy.

METHODS

The data for this study were collected, during the period between February 2005 and September 2007, from the alphabetical list of the commercial stores located in the geographic area of the

Citation: Giorgio Liguori, Ivan Cavallotti, Antonio Arnese, Ciro Amiranda, Daniela Anastasi and Italo F Angelillo, Microbiological quality of drinking water from dispensers in Italy, DOI: 10.1186/1471-2180-10-19

city of Naples, in the South of Italy. From this list, 41 stores were selected using a simple random sampling technique.

Before the study, as part of the process of informed consent, all selected commercial stores received an envelope with a letter informing that a research project was being conducted and describing the study, the voluntary nature of participation, and assurance of privacy and anonymity. For each participant, information about the date of installation, time since last ordinary and extraordinary maintenance of water coolers was collected with a self-administered questionnaire. The water coolers were produced by different companies, all were supplied from tap water and at the sampling time the mechanism of cooling was functioning. At the end of the study, each store received the results of the microbiological and chemical quality parameters investigated.

Collection of water samples

Water samples were collected from coolers (carbonated and non-carbonated water) and tap water corresponding to the tap water used for the cooler from the same store. Each sample was analyzed for total viable count (TVC), qualitative microbial indicators, and chemical parameters of organic contamination. Two litres of each water sample were collected without flushing before sampling and sterilizing the outer surfaces of the faucets. Water samples were collected in sterile sample bottles containing sodium thiosulfate (100 mg/L). Samples were transported on blue ice in an insulated, double-walled container to the laboratory for analysis and primary isolation within 6 hours of sampling.

All analyses were performed according to the current Italian [7] and European regulations on drinking water for human consumption [8]. The reference values for the water in order to be declared potable are the following: *Enterococcus* spp., *Escherichia coli*, and *Pseudomonas aeruginosa* should not be detectable in 100 ml; TVC less than 100 and 20 colony forming units (CFU) per mm at 22°C and 37°C, respectively; nitrite and ammonium less than 0.5 mg/L; free chlorine comprised between the values of 0.2 and 0.8 mg/L; and pH between the values of 6.5 and 9.5.

Microbiological parameters

The microbiological analyses of all water samples were conducted as follows:

1) TVC: 1 mL of each sample was included with 20 mL of Water Plate Count agar (bioMérieux Italia) and two sets of plates were prepared for all samples with each sample diluted until 10^{-3} on three dishes. One set was incubated at 22°C for 72 hours and the other set at 37°C for 48 hours (prEN ISO 6222). All colonies were counted as CFU/mL of the water sample;
2) *Escherichia coli*: 250 mL of each sample was filtered through a 0,45 µm cellulose membrane filter, placed on Tergitol 7 TTC agar (bioMérieux Italia), and plates were incubated at 37°C for 18/24 hours (ISO 9308-1). The species identification was conducted using standardized identification system API 20E (bioMérieux Italia);
3) *Enterococcus* spp.: 250 mL of each sample was filtered through a 0,45 µm cellulose membrane filter, placed on Slanetz-Bartley agar (bioMérieux Italia), and plates were incubated at 37°C for 48 hours. If typical colonies (red/brown/pink) were present, the membrane was transferred on pre-warmed (44°C) plates of Bile Aesculina Azide agar (bioMérieux Italia) and incubated at 44°C for 2 hours (ISO 7899-2). Typical brown/black colonies were identified as *Enterococcus* spp. using standardized identification system API 20 Strep (bioMérieux Italia);
4) *Pseudomonas* spp.: 250 mL of each sample was filtered through a 0,45 µm cellulose membrane filter, placed on Pseudomonas CN agar (Cetrimide-Nalidixic Acid, bioMérieux Italia), and plates were incubated at 37°C for 48 hours, blue/green colonies were isolated on Plate Count agar (bioMérieux Italia) at 37°C for 24 hours, and after the oxydase test (bioMérieux Italia), the species identification was conducted using standardized identification system API 20NE (bioMérieux Italia) (prEN ISO 12780);
5) Other microorganisms: singles colonies growing on Tergitol 7 TTC agar (bioMérieux Italia) were transferred on McConkey agar (bioMérieux Italia), and plates were incubated at 37°C for 24-48 hours; after the oxydase test (bioMérieux Italia), the species identification was conducted using standardized identification systems API 20E/20NE (bioMérieux Italia).

Chemical analyses

pH: The pH was determined electrometrically by using the technique recommended in the Standard Methods [13].

Residual free chlorine

The residual free chlorine content was measured using the *N,N*-diethyl-*p*-phenylenediamine (DPD) colorimetric method at the time of sample collection (colorimetric DPD method; Microquant; Merck, Darmstadt, Germany) [13].

Ammonium

For ammonium ions determination, 50 mL of the water sample and the calibration samples were mixed with 1 mL of a potassium tetraiodiomercurate solution. After 20 minutes reaction time at room temperature in NH_3-free atmosphere, the solution was examined photometrically at a wavelength of 420 nm in cuvettes of appropriate path length (IRSA-CNR, Rome, Italy).

Nitrite

For nitrite ion determination, 50 mL of the water sample and the calibration samples were mixed with 2 mL of a freshly prepared mixture of equal parts of sulphanilic acid solution and 1-naphthylamine solution. After 2 hours at 20°C in darkness the extinction at 530 nm was measured [14].

Statistical analysis

Basic descriptive summaries were used to describe measures of central tendencies and dispersion of water characteristics and microbial concentrations. The χ^2 and the Fisher's exact test were used to determine whether statistically significant differences existed in the contamination levels between coolers water (non-carbonated and carbonated) and tap water. The simple linear regression was used to determine whether statistically significant associations existed between the bacterial counts of the coolers water (non-carbonated and carbonated) and the time since the last filter was substituted. Statistical significance was assessed using two-sided tests with p-values of ≤ 0.05. Analyses were performed using the statistical package Stata [15].

RESULTS

Of the 41 randomly selected commercial stores, 38 agreed to participate for a response rate of 94.7%. The time since the last maintenance of water coolers, comprehensive of filter substitution, in the participating stores ranged between 1 and 24 months.

A description of the data regarding microbiological characteristics of drinking water dispensed by the sampled water from coolers and tap according to the Italian legislation is provided in Additional file 1. It should be noted that *Enterococcus* spp. and *Escherichia coli* were not detected in any of the water samples. In 17% of the samples of tap water after incubation at 22°C and 37°C the number of aerobic bacteria was higher than the stated drinking water limits for TVC of < 100 CFU/mL and < 20 CFU/mL, respectively. *Pseudomonas aeruginosa* was found in only one sample of the tap water and in 28.9% and 23.7% of the non-carbonated and carbonated water samples, respectively. The microbiological results for the water coolers indicated that the total bacteria counts at 22°C and 37°C was higher than the required values in 71% and 81% for the non-carbonated water and in 86% and 88% for the carbonated one, respectively. The overall mean bacteria counts at 22°C and 37°C in the water samples were respectively 102.9 CFU/mL and 86.3 CFU/mL for the tap, 569.7 CFU/mL and 331.8 CFU/mL for the non-carbonated, and 542.1 CFU/mL and 355.9 CFU/mL for the carbonated.

The results of the statistical analysis conducted to determine whether differences exist among the three different types of water with regard to microbial measures showed no significant difference between the number of microorganisms recovered from the non-carbonated and carbonated water from coolers for the bacteria count at 22°C ($\chi^2 = 2.55$, $p = 0.18$) and at 37°C ($\chi^2 = 0.82$, $p = 0.55$), and for *Pseudomonas aeruginosa*($\chi^2 = 0.26$, $p = 0.8$), respectively. The tap water was always of excellent bacteriological quality and it was superior than the water from coolers. Indeed, a statistically significant higher proportion of positive microbial counts has been recorded for both bacterial counts at 22°C and 37°C in the non-carbonated ($\chi^2 = 25.55, p < 0.0001$; $\chi^2 = 34.73, p < 0.0001$) and carbonated ($\chi^2 = 40.07, p < 0.0001$; $\chi^2 = 42.95, p < 0.0001$) waters compared with the tap water. The number of positive samples for *Pseudomonas aeruginosa* was significantly higher in the non-carbonated (Fisher's exact test $p = 0.003$) and carbonated (Fisher's exact test $p = 0.015$) water coolers samples compared with the samples of tap water. Moreover, the results of the simple linear regression analysis showed that no statistically significant associations between bacterial counts at 22°C and 37°C either for the non-carbonated ($p = 0.16$ and $p = 0.15$) or the carbonated water ($p = 0.21$ and $p = 0.14$) from the sampled coolers in relation with the time since the last filter was substituted.

Other microorganisms were isolated from 6 (20%), 25 (65.8%), and 27 (71.1%) samples of the tap, non-carbonated, and carbonated waters. The bacteria were identified mainly to be *Pseudomonas* species, which was recovered respectively from 6 (20%) samples of the tap water and from 19 (50%) samples either of the non-carbonated or the carbonated waters, and the mean concentrations were 48.3 CFU/mL, 241.5 CFU/mL, and 137.2 CFU/mL, respectively. Species of *Stenotrophomonas*, *Pasteurella*, *Enterobacteria*, and*Flavobacterium* were also isolated mainly from the non-carbonated or the carbonated waters.

With regard to the chemical parameters, in all samples the nitrite, ammonium, and free active chlorine residual did not exceed the reference values of the drinking water regulation. The mean average values of the three parameters for the tap water were 0.06 mg/L (range 0.001-0.15) for nitrite and 0.08 mg/L (range 0.01-0.25) for both ammonium and free active chlorine residual; whereas, for the carbonated and non-carbonated waters the average values were 0.076 (range 0-0.025) and 0.06 mg/L (range 0-0.025) for nitrite, 0.08 (range 0-0.3) in both waters for ammonium, and 0.3 (range 0.2-0.4) and 0.29 mg/L (range 0.2-0.4) for free active chlorine residual, respectively. Finally, the pH of the tap and non-carbonated waters did not exceed the reference value and both means were 7.8 ranging from 6.8 and 8.4, whereas for the carbonated the vast majority of the samples (86.8%) had a value lower than the reference limit with an overall mean of 6 and a range of 5.2 and 6.8.

DISCUSSION

This study sought to determine the quality of drinking water dispensed by water coolers from commercial stores in comparison with tap water in the geographic area of Naples, Italy.

In this investigation, the microbiological quality of the drinking water was satisfactory for the chemical indicators of organic contamination in all samples, probably because the values of microbial counts were not high enough to modify them. It should be noted that the same pattern has not been observed for the quantitative and qualitative microbiological parameters. Indeed, should be of concern the finding that a large number of non-carbonated and carbonated water sampled from coolers revealed a bacteria count higher than the limits stated for TVC. Moreover, contamination with *Escherichia coli* and *Enterococcus* spp. were not observed in any of the tap and dispensers water samples. The absence of these microorganisms, considered to represent an indicator of faecal contamination, renders the water satisfactory and safe with no health implications. The finding observed in the present study regarding the presence of a faecal indicator bacterium was in accordance with similar studies recently conducted [10, 12]. Indeed, in water from coolers *Escherichia coli* and *Enterococcus* spp. were absent [10, 12] and *Pseudomonas aeruginosa* has been detected in 24.1% of the water samples [10]. Furthermore, in contrast in a survey conducted in Canada on the microbiological quality of water from coolers located in residences and workplaces with respectively 28% and 36% of the collected samples contaminated by at least one coliform or indicator bacterium and/or one pathogenic bacterium [9].

In addition, we were interested to determine whether the tap water used was responsible for the contamination of the water dispensed by coolers. None of the tap water samples had a bacterial count higher than the water coolers and none of the samples were contaminated with coliforms. Thus, tap water was not directly responsible of water coolers contamination. These findings suggest that the contamination may be caused by the accumulation of small quantity of microorganisms from tap water or from faucet surface which are concentrated at filters.

It was interesting to find out that the results of the statistical analysis indicated that strongly and highly significant differences in quality and quantity of the microbiological parameters between the water coolers samples and the tap water samples. Indeed, the aerobic plate counts were higher in the coolers compared with the tap water and *Pseudomonas aeruginosa* was more frequently detected in the non-carbonated and carbonated water coolers samples than in those of tap water. These findings are in accordance with the two already mentioned studies, since the aerobic plate counts was higher in coolers compared with spring water [10] and a significantly higher proportion of water cooler samples resulted contaminated than tap water [9]. Therefore, a periodic adequate disinfection of water dispensers had to be indicated in order to keep the level of microbiological contamination under control. The validity of this recommendation is supported by the results of a study that showed that the periodic application of hydrogen peroxide (3%) of microfiltered water dispensers led to a reduction in the concentrations of *Pseudomonas aeruginosa* and to obtain water with bacteria counts conforming to Italian regulations for drinking water [12]. Furthermore, the data from this study demonstrated that no significant differences in bacterial counts occur between the non-carbonated and carbonated water in relation with the time since the last filter was substituted.

CONCLUSION

The data presented here raise concern about the microbiological quality of the drinking water plumbed in water coolers and highlights the importance of adopting appropriate monitoring system with changing filters according to their use and the disinfection of the water in order to prevent or to diminish the chances of contamination of this water source.

DECLARATIONS

Acknowledgements

Preliminary results have been presented at the Research Meeting of the Second University of Naples, July 1-4, 2007, Naples, Italy.

REFERENCES

1. Hrudey SE, Hrudey EJ: Published case studies of waterborne disease outbreaks-evidence of a recurrent threat. Water Environ Res. 2007, 79: 233-245. 10.2175/106143006X95483.View ArticlePubMedGoogle Scholar
2. Reynolds KA, Mena KD, Gerba CP: Risk of waterborne illness via drinking water in the United States. Rev Environ Contam Toxicol. 2007, 192: 117-158.
3. Marshall JK, Thabane M, Garg AX, Clark WF, Salvadori M, Collins SM, The Walkerton Health Study Investigators: Incidence and epidemiology of irritable bowel syndrome after a large waterborne outbreak of bacterial dysentery. Gastroenterology. 2006, 131: 445-450. 10.1053/j.gastro.2006.05.053.
4. Jones AQ, Majowicz SE, Edge VL, Thomas MK, MacDougall L, Fyfe M, Atashband S, Kovacs SJ: Drinking water consumption patterns in British Columbia: an investigation of associations with demographic factors and acute gastrointestinal illness. Sci Total Environ. 2007, 388: 54-65. 10.1016/j.scitotenv.2007.08.028.
5. O'Reilly CE, Bowen AB, Perez NE, Sarisky JP, Shepherd CA, Miller MD, Hubbard BC, Herring M, Buchanan SD, Fitzgerald CC, Hill V, Arrowood MJ, Xiao LX, Hoekstra RM, Mintz ED, Lynch MF, The Outbreak Working Group: A waterborne outbreak of gastroenteritis with multiple etiologies among resort island visitors and residents: Ohio, 2004. Clin Infec Dis. 2007, 44: 506-512. 10.1086/511043.
6. Peace T, Mazumder A: Tracking patterns of enteric illnesses in populations and communities. Environ Health Perspect. 2007, 115: 58-64.
7. European Council Directive 98/83/EC of 3 November, 1998. On the quality of water intended for human consumption. Official J Europ Commun. 330: 32-54.
8. Decreto Legislativo 2 febbraio 2001, n. 31. Attuazione della direttiva 98/83/CE relativa alla qualità delle acque destinate al consumo umano. Gazzetta Ufficiale della Repubblica Italiana. Supplemento n. 52 del 3 marzo. 2001
9. Lévesque B, Simard P, Gauvin D, Gingras S, Dewailly E, Letarte R: Comparison of the microbiological quality of water coolers and that of municipal water systems. Appl Environ Microbiol. 1994, 60: 1174-1178.
10. Baumgartner A, Grand M: Bacteriological quality of drinking water from dispenser (coolers) and possible control measures. J Food Prot. 2006, 69: 3043-3046.
11. Sacchetti R, De Luca G, Zanetti F: Control of *Pseudomonas aeruginosa* and *Stenotrophomonas maltophilia*contamination of microfiltered water dispensers with peracetic acid and hydrogen peroxide. Int J Food Microbiol. 2009, 132: 162-166. 10.1016/j.ijfoodmicro.2009.04.017.
12. Zanetti F, De Luca G, Sacchetti R: Control of bacterial contamination in microfiltered water dispensers (MWDs) by disinfection. Int J Food Microbiol. 2009, 128: 446-452. 10.1016/j.ijfoodmicro.2008.10.007.
13. American Public Health Association: Standards methods for the examination of water and wastewater. 2005, American Public Health Association, Washington, D.C, 21
14. German Standard Methods: German Standard Methods for the Examination of Water, Wastewater and Sludge. 1980, DIN 38409-H41-1, ISO 6060-1980.
15. Stata Corporation: Stata reference manual release 10. 2007, TX, USA: College Station

CHAPTER 24

NEW PERSPECTIVES IN MONITORING DRINKING WATER MICROBIAL QUALITY

Ma José Figueras and Juan J. Borrego

Keywords: Indicator microorganisms, drinking water quality, WSP, European drinking water directive, climate change

INTRODUCTION

The evaluation of the microbiological quality of drinking water aims to protect consumers from illness due to consumption of water that may contain pathogens such as bacteria, viruses and protozoa, and thus to prevent drinking-water related illness outbreaks. This has been, and still is nowadays an important challenge. For the past century, this evaluation has been performed through the analysis in finished drinking water of faecal pollution indicators, which are expected to predict the potential presence of pathogenic microorganisms in the water. However, scientists, engineers, public health officials and water pollution control agencies have faced cases in which the quality of water have showed the presence of indicators when water was already served to the consumers. In addition, drinking water outbreaks have occurred both in presence or absence of indicator organisms and involved pathogenic microorganisms that have contaminated the drinking water, and that either were not eliminated during treatment, or the latter failed at the time of the outbreak. The United States Centre for Disease Control has reported 780 disease outbreaks associated with the consumption of contaminated drinking waters from 1971 to 2006, which affected 577,094 persons [1]. A number of outbreaks have also occurred in Europe. For instance in Spain, in the 1999–2006 period, 413 outbreaks were recorded that involved 23,642 cases [2]. These outbreaks occurred despite specific legislations designed to prevent them, and the associated microbial control measures being carried out. The World Health Organization (WHO) has been very active in this field developing important guidelines of universal application and has promoted, in recent years, a more preventive approach than only checking the quality of the finished drinking water [3,4]. This "Water Safety Plans" (WSPs) approach takes into account all factors that endanger the quality of drinking water from the source to the final tap water at the consumer's home. The WHO alone, and in collaboration with the International Water Association (IWA), has developed several guideline documents that are freely accessible through Internet, the most recent of which is the *Water Safety Plan Manual* [4]. Furthermore, both, in collaboration with the Organization for Economic Co-operation and Development (OECD), published a book entitled Assessment of the Safety of Drinking Water, underlying the challenges for the 21st century [5]. Another WHO document is *Emerging Issues in Water and Infectious Diseases* that reviews the problem of emerging pathogens and other aspects that endanger water safety [6]. All of them are

Citation: Ma José Figueras and Juan J. Borrego, New Perspectives in Monitoring Drinking Water Microbial Quality, doi:10.3390/ijerph7124179

important reference manuals associated with water quality [3–6]. In 2003 the European Union initiated an extensive revision of the existing Drinking Water Directive (98/83/EC) [7], and is currently deciding what modifications will be included in the new and updated Directive in order to increase the quality of drinking water and protect public health.

In previous studies, we and other authors, have reviewed the definitions of index and indicator organisms used to evaluate the microbiological quality of water [3,8–12], as well as the relevance of some of them, e.g., faecal streptococci, or their relationship with pathogenic bacteria, e.g., *Salmonella*, also involved in drinking water outbreaks [9,10,13]. The specific methods used for the analysis of indicators have also been reviewed in detail [8,12,14], so in this update we will review recent advances in relation to indicator/index organisms. Furthermore, this overview aims to introduce the current proposed modifications for the new EU Drinking Water Directive, among which the WSPs are a key element. The WSPs principles will be presented in order to raise awareness of water quality professionals so that they can prepare for future developments. The expected influence of climate change in the quality of drinking water will be discussed.

FAECAL INDICATORS IN DRINKING WATER CONTROL

There are hundreds of different enteric microorganisms that are known to infect humans. Enteric microorganisms are excreted in the faeces of infected individuals or animals, and may directly or indirectly contaminate water intended for human consumption [3,8,14–17]. Since the adoption of disinfection practices by drinking water utilities, the incidence of waterborne diseases has decreased drastically. However, the WHO estimates that in developing countries, some two million children die each year of infectious diseases associated with contaminated water [17,18]. According to the American Society for Microbiology "many serious health problems could be eliminated if more countries adopted water quality practices, including the simple steps of source water protection and disinfection to ensure safe water supplies" (http://www.epa-gov./OWOW/watershed/statewidn/table.htm). Therefore, the control of microbial pathogens must be carried out by the use of a multibarrier approach, including source protection, proper treatment and disinfection, and optimal distribution maintenance [19,20]. This approach has been adopted in the form of the WSPs, by the WHO [3,4] and will be discussed in the present review.

The presence of enteric pathogens in drinking waters is of great concern, and thus, legislation either in Europe, USA and other countries requires analysis of indicators to determine the microbiological quality of these waters. Ideally we would like to analyze the waters for the presence and quantification of specific enteric pathogens. However, many waterborne pathogens are still difficult to detect and/or quantify in waters and for most of the newly recognized agents, easy methods to detect them in water samples have still to be developed [12,14]. The introduction of molecular methods has advanced the recognition of these new agents and their benefits were recently reviewed [21]. However, the routine application of these methods for the analysis of pathogens is not a reality yet and is restricted to research studies or to cases of suspected outbreaks. Nowadays, new approaches based on virulence factor-activity relationships to discover and detect emerging waterborne pathogens are being explored [22]. Therefore, the most useful tool to determine the potential presence of pathogenic microorganisms in waters is the analysis of several microorganisms classed as either "indicator, or, index" organisms [8,9]. These indicators must fulfil the requirements indicated in a previous study [8].

To avoid the ambiguity in the term "microbial indicator", the following three groups are now recognized: process microbial indicators, faecal indicators and index and model organisms. Process indicators comprise a group of organisms that demonstrate the efficacy of a process; faecal indicators are those organisms that indicate the presence of faecal contamination, hence, they only infer that pathogens may be present; index and model organisms include a group or species indicative of pathogenic presence and behaviour, respectively [9]. The use of index and indicator organisms to assess the microbiological and sanitary quality of waters is well established and has been practiced for almost a century. The most widely used indicators are coliforms (total coliforms), faecal or thermotolerant coliforms, *Escherichia coli*, enterococci (faecal streptococci or intestinal enterococci) and bacteriophages.

Coliforms

By definition, coliform bacteria are facultative anaerobes, Gram-negative, non-spore-forming, rod-shaped bacteria that ferment lactose with acid production in 24 to 48 h at 36 ºC, and indole-negative. Coliforms belong to the family*Enterobacteriaceae* and include *Escherichia, Enterobacter, Klebsiella* and *Citrobacter, Kluyvera, Leclercia* genera, and some members of the genus *Serratia*. These bacteria were classically used as indicators of faecal contamination of waters because they were considered to be inhabitants of the intestinal tracts of homeothermic animals [8,11]. However, the ability of some coliforms to grow in natural waters, the lack of correlation between the number of coliforms and those of pathogenic microorganisms, and the detection of atypical strains has led them to become unsuitable faecal indicators [8,11,23,24]. Furthermore, several studies have demonstrated presence of coliforms in drinking water distribution systems associated with biofilm growth problems [25,26]. The coliform bacteria, traditionally termed the "total coliform" group, have been the primary standard for potable water in most of the world. However, many regulatory agencies have questioned its utility as an indicator. For these reasons, this is one of the parametres that has been eliminated from the European legislation for the management of the quality of bathing waters (2006/7/EC) and most probably will also disappear in the modification of the current drinking water legislation (98/83/CE) [7]. Nowadays, coliforms are typically associated with treatment effectiveness, and should be absent from adequately treated plant effluents [25,26]. The presence of coliforms in the distribution system, while possibly due to inadequate treatment, could also be due to cross-connections or failure to maintain an adequate disinfectant residual [26,27].

Faecal Coliforms

These bacteria conform to all the criteria used to define total coliforms plus the requirement that they grow and ferment lactose with the production of acid at 44.5 °C. For this reason, "thermotolerant coliform" would be the scientifically accurate term for this group [8,11,23,24]. Bacteria in this coliform subgroup have been found to have a positive correlation with faecal contamination of warm-blooded animals [8,11,15,23,24]. However, some thermotolerant coliform bacteria that conform to this definition also belong to the genus *Klebsiella* and have been isolated from environmental samples in the apparent absence of faecal pollution [8,11,15]. Similarly, other members of the thermotolerant coliform group, including *Escherichia coli*, have been detected in some pristine areas [28], and associated with regowth events in potable water distribution systems [25]. Faecal coliforms display a survival pattern similar to these of bacterial pathogens but their usefulness as indicators of protozoan and viral contamination is limited, therefore, tended to be replaced by *E. coli* in several legislations [26].

Escherichia coli

E. coli is a member of faecal coliform group, being a more specific indicator for the presence of faecal contamination. In addition, *E. coli* conforms to taxonomic as well as functional identification criteria and is enzymatically distinguished by the lack of urease and presence of β-glucuronidase. One disadvantage associated with this organism as an indicator is that it has been consistently found in pristine tropical rain forest aquatic and plant systems as well as soils [28,29]. Additionally, it seems to survive for short periods in aquatic temperate environments [23,30]. *E. coli* is the faecal indicator of choice used in WHO Guidelines for Drinking-water Quality [3,26], and several countries include this organism in their regulations as the primary indicator of faecal pollution (*i.e.*, Europe, USA). Although it has long been known that *E. coli* can cause disease in humans, the bacteria naturally occurs in the lower part of the gut of warm-blooded animals [11,31]. Its role as an enteric pathogen has been reinforced with the discovery of *E. coli* O157:H7 associated with haemorrhagic enteritis and haemolytic uremic syndrome, that was responsible of producing several drinking water outbreaks, and some of them lack of β-glucuronidase activity [32,33].

Faecal Streptococci, Enterococci or Intestinal Enterococci

This group of microorganisms has received widespread acceptance as useful indicators of microbiological water quality, since: (i) they show a high and close relationship with health hazards associated with the water use, mainly for gastrointestinal symptoms; (ii) they are always present in faeces of warm-blooded animals; (iii) they unable to multiply in sewage-contaminated waters; and (iv) their die-off is less rapid than those of coliforms in water, and persistence patterns are similar to those of potential waterborne pathogenic bacteria [8,11,35–39].

Faecal streptococci, enterococci and intestinal enterococci are three synonyms used to refer to species described as members of the genus *Enterococcus*, which also fulfil Sherman's criteria (growth at 10 °C and 45 °C, resistance to 60 °C, growth at pH 9.6 and at 6.5% NaCl, and reduction of 0.1% methylene blue) [8,10,11]. They comprise species of different sanitary significance and survival characteristics and, in addition, the proportions of the species of this group are not the same in animal and human faeces [8,10,11]. *Enterococcus faecalis* and *Ent. faecium,* are the predominant species in human faeces and sewage [10,11,37]. In a European study that investigated enterococcal populations in animals, humans, and the environment the most common species detected were *Ent. faecium* (33%), *Ent. faecalis* (29%), and *Ent. hirae* (24%) [38]. This*Enterococcus* species distribution in human and animal hosts has been recently confirmed using a molecular multiplex PCR technique [39].

Despite the definitions provided above for the indicators (total coliforms, faecal coliforms, *E. coli* and enterococci) in practical terms these are determined on the basis of the biochemical reactions evaluated in culture media that are recognized either by the appearance of characteristic colonies (with a specific colour as response to this reaction in chromogenic substrates) and/or by the emission of fluorescence. Colour and fluorescence are also the responses expected in presence/absence tests in liquid media either in bottles or in a Most Probable Number approach designed as blisters or microplate systems that enable quantification [8,11]. Microbiological methods for indicators are far from perfect because they can produce false positive and negative results [8,11,35–37,40–42].

Molecular methods are useful both to monitor natural communities of bacteria, and to track specific bacterial markers in complex environments. Length-heterogeneity polymerase chain reaction (LHPCR) and terminal restriction fragment length polymorphism (T-RFLP) of 16S rDNAs of anaerobic bacteria have been used by Field *et al.* [43] to develop an alternative indicator that distinguishes the source of faecal pollution in water.

Bacteriophages

Several bacteriophage groups have also been classically used as faecal and viral indicators, and as models to evaluate the efficiency of the chlorination of drinking waters [44–46]. The proposed groups are somatic coliphages, F (male)-specific RNA bacteriophages (FRNA phages) and phages of *Bacteroides fragilis* [47–49].

Somatic coliphages are specific viruses of *E. coli* and have been commonly used as indicators of faecal and/or sewage pollution in several water types and as biotracers to identify pollution sources in surface waters and aquifers [50,51]. In addition, they may also serve as indicators for assessing the removal efficiency during the treatment of water and wastewater treatment plants [52]. On the basis of the differences in origin and ecology between enteric viruses and somatic coliphages, it is doubtful to conclude that this phage group could successfully be used in all situations as enteric viruses indicators [47], and they may not be a useful indicator of a distribution system integrity problem, even when the problem involves the introduction of faecal contamination [53].

The use of FRNA phages was proposed as faecal pollution indicators and as model viruses in water hygiene on the basis of: (i) their similar sizes and shapes to human enteric viruses; (ii) their correlation with the sewage contamination degree; and (iii) their inability to replicate in the water ecosystem [46]. However, the low incidence of this phage group in human faeces and its low specificity for its bacterial host, suggest that they would multiply in the sewerage system [48,54,55]. Hence, the presence of FRNA phages in water should be primarily used as an index of sewage pollution rather than faecal pollution [56].

Bacteroides fragilis is a strict anaerobe found in high concentrations in the human intestinal tract and dies rapidly when discharged into environmental waters. A phage of the strain HSP 40 of *B. fragilis* (isolated from Hospital San Pablo, Barcelona, Spain) has been proposed as a specific index of human faecal pollution of waters [49], because: (i) phages against this bacterial strain are human specific and are not isolated from the faeces of other homoeothermic animals; (ii) *B. fragilis*HSP 40 phages are consistently isolated from sewage, faecally-polluted waters, and their sediments but not from unpolluted samples; (iii) the levels of phages are related to the degree of pollution; (iv) *B. fragilis* phages always outnumber human enteric viruses; and (iv) in model experiments, no replication of these phages has been observed under simulated environmental conditions [57]. The low prevalence of these phages in waters with low and moderate levels of faecal pollution and the complex methodology for their recovery are the main drawbacks for the general use of these viruses as an indicator group [58,59].

DRINKING WATER OUTBREAKS

The total number of drinking water-related illness in the USA has been estimated at 19 million/year; however, this figure depends upon the approach considered [60]. The detected water-borne outbreaks are considered to be just the tip of the iceberg of the total drinking-water-related illness. In fact the actual disease burden in Europe, as in other parts of the world, is difficult to estimate and is, most likely underestimated [61]. Outbreaks have the potential to be rather large as in the case of the Milwaukee (USA) *Cryptosporidium* outbreak that affected over 400,000 people in 1993 [1]. At least 325 drinking water-associated outbreaks of parasitic protozoan diseases have been reported all over the world between 1954 and 2003, over 30% (106) of all outbreaks were documented from Europe, *Giardia duodenalis* and *Cryptosporidium parvum* accounting for the majority of these outbreaks [62]. The first outbreak that provided evidence that *E. coli* O157:H7 was transmitted by drinking water occurred in a small rural town in Missouri (USA) that had an unchlorinated water supply [63]. There were 243 cases, of whom 86 presented bloody stools, 32 were hospitalized, four died and two developed haemolytic ureamic syndrome (HUS). HUS is a severe disease that may cause an acute renal failure, which may require dialysis or kidney transplantation.

Other outbreaks like the one of Walkerton, Ontario (Canada) in 2000, which affected over 2,300 cases, revealed a mixed aetiology by *Cryptosporidium* and *Escherichia coli* 0157:H7 [60]. In our view, outbreaks of mixed aetiologies should be more common than what is detected because the sewage contamination drinking water contains many potential pathogenic microorganisms. A good example of this is the outbreak that occurred in South Bass Island (Ohio, USA) in 2004, which revealed a massive microbiological contamination (with total coliforms, *E. coli*, enterococci, *Campylobacter*, *Arcobacter*, coliphages and adenoviruses) of the ground water used for drinking water producing 1,450 cases of gastroenteritis [64,65]. This outbreak also revealed that the deterioration of the water occurred over years, and that a poorly known microorganism*Arcobacter* was implicated [64]. The latter is often confounded with *Campylobacter* if inappropriate molecular identification methods are applied. Furthermore, *Arcobacter* is frequently present in human sewage showing a good correlation with indicators of faecal pollution [66–68]. However, despite source water showing a high prevalence of *Arcobacter* spp. appropriate treatment can remove these microorganisms as well as noroviruses from the finished drinking water [68].

In Europe, monographic water outbreak reports (e.g., those produced by CDC in the USA [1]) are not available, because drinking water is defined as food, and therefore reporting is included with food-borne outbreaks [69]. In 2007, only 17 water-borne outbreaks were reported by eight countries [69], clearly indicating an under-reporting. They involved 10,912 cases, with 232 hospitalizations. The main microorganisms involved were *Campylobacter*, norovirus, *Giardia* and*Cyptosporidium*. Interestingly, the biggest outbreaks had multiple aetiologies, one involving 453 registered cases in Denmark and a large outbreak with 8,000 cases in Finland of which approximately 1,000 sought medical attention and 200 were hospitalized [69]. In the latter three major (*Campylobacter*, norovirus, *Giardia*), and three minor causative agents (*Salmonella*Enteritidis, *Clostridium difficile* and rotavirus) were isolated from the patients, and all the causative agents were also isolated from water samples [69]. Multiple microorganisms (enteroviruses, *Giardia*, *Cryptosporidium*, *Campylobacter* and *Arcobacter*) were recovered from the patients in an outbreak that occurred in Slovenia in 2008 [70]. This reinforces our idea that multiple microorganism aetiologies maybe more common than as well as the prevalence of *Arcobacter*.

Principal Failures Associated with Outbreaks and Lessons Learned

Prevention and containment of outbreaks requires examination of the causative events responsible for their occurrence. As indicated by Risebro *et al.* [71], retrospective analysis of outbreaks of enteric diseases can be used to inform outbreak investigators, facilitate corrective measures, and further develop multi-barrier approaches. In this sense these authors developed an outbreak fault tree that was applied to 61 enteric outbreaks related to public drinking water supplies in the EU. The approach found that failures in the source water and in water treatment, independently or together, were the cause of more than 50% (34/61) of the outbreaks. Faults at the distribution system occurred less frequently (19/61 outbreaks) but were often solitary events contributing heavily towards the outbreak (a mean % score of 87.42). Livestock and rainfall in the catchment with none or inadequate filtration of the water sources contributed to *Cryptosporidium* outbreaks. Of the 23 protozoan parasite outbreaks that showed one treatment causative event, 90% of these events were filtration deficiencies. However, by contrast, for bacterial, viral, and mixed pathogen outbreaks, disinfection deficiencies were associated with 75% of the outbreaks [71].

Excessive rainfall has been an important contributor to historical waterborne disease outbreaks [72–77]. In fact, most of the bacteriological parametres (heterotrophic bacteria, *E. coli*, total coliforms, faecal streptococci, and *Clostridium perfringens*counts) increased considerably during extreme runoff events as do the concentrations of *Giardia* and *Cryptosporidium* [74–76]. Another important lesson learned, after an outbreak, is that once *Cryptosporidium* has colonized a drinking water system, it can persist for a long time despite vigorous and repetitive flushing of the system [77]. The investigators suggested that oocysts were being trapped in the biofilm in the distribution network and then were being released back into the supply. A very long persistence of norovirus in the water distribution system was also observed in the outbreak that occurred in Finland mentioned previously, and it required advising people in the affected areas to boil the water before use for a ten-week period [69].

Further Health Consequences of Gastroenteritis Outbreaks

Nowadays, it is known that gastroenteritis may have other important health sequels, like reactive arthritis, irritable bowel syndrome, cancer predisposition, to name a few [17]. In this sense, a study carried out with the patients of the Walkerton outbreak in Canada, showed that 15.7% of the asymptomatic patients during the outbreak, and in 17.6 and 21.6% of those who had moderate and severe symptoms of acute gastroenteritis respectively, showed problems with arthritis 4.5 years later [78]. Gastroenteritis is also associated with subsequent post-infectious irritable bowel syndrome [79], and HUS [63].

Such outbreaks can generate high societal alarm, which can result in the introduction of new Drinking Water Regulations. An example is the case of the biggest worldwide outbreak of *Legionella* that motivated the first Spanish legislation in relation to the control of this microorganism. Another example is the incorporation of significant requirements for drinking water providers, following the Walkerton outbreak [80]. In fact the latter outbreak has also influenced the EU legislation for drinking water, which had incorporated the control of *Cryptosporidium* in specific circumstances. These interventions lead to a significant decline in cryptosporidiosis [81].

ROLE OF FAECAL INDICATORS IN THE PROTECTION OF PUBLIC HEALTH: ALTERNATIVE INDICATORS AND RECOVERY OF INJURED BACTERIA

The failure of measurements of single indicator organisms to correlate with pathogens suggests that public health is not adequately protected by simple monitoring schemes based on detection of a single indicator, particularly at the detection limits routinely employed. In addition, the classical microbial indicators proposed present several shortcomings and they cannot be used in all water types. In this instance, other indicators, named "Alternative", should be used to determine the possible threat to public health [3,10,11,26].

Alternative Indicators

The use of the sulphite-reducing members of the genus *Clostridium* (*C. perfringens*) as indicators of faecal pollution is based on: (i) the presence of these microorganisms in the faeces of all warmblooded animals as well as in sewage, (ii) more stability in environmental waters and greater resistance to the disinfection processes than most pathogens, and (iii) successful use in monitoring sewage-contaminated waters [3,10,11,26,82]. Nevertheless, sulphite-reducing clostridia are considered ubiquitous in aquatic sediments and the spore form explains their persistence, although they can be used as indicators of remote or non-point faecal pollution or to evaluate the virus and cyst inactivation in the drinking water disinfection processes [10,11,83]. However, the WHO [3,24,82] does not recommend clostridia for routine distribution system monitoring because, due to their length of survival, they may be detected long after (and far from) the pollution event, leading to possible false alarms.

Heterotrophic plate count (HPC) or total aerobic bacteria were among the first parametres used to monitor the safety of finished drinking water. However, presently they have become an indicator of general water quality within distribution systems [3,26,84]. It provides a good operational monitoring parametre that measures the deterioration of water quality through distribution systems. It is considered that the bacteriological content of drinking-water leaving treatment plants should contain only very low levels of heterotrophic and aerobic spore-forming microorganisms [26]. In fact this parametre, evaluated both at 22° and 37 °C, is included in the EU drinking water legislation, and is also required to be evaluated monthly for the control of *Legionella* in the Spanish and other specific legislations to control the latter microorganism. A series of review papers appeared in 2004, including one by the WHO, which evaluated the role of this parametre in water as a control measure in drinking water safety management [83].

Members of the genus *Pseudomonas* are possibly the microorganisms most often isolated from bodies of water. However, contrary to the previously discussed indicators, their presence does not necessarily indicate a possible risk to public health.*P. aeruginosa* was found to be more resistant than acid-fast bacteria during ozonation processes, demonstrating its resistance to chemical disinfection and thus its usefulness in the analysis of waters that receive chemical disinfection, including drinking waters [10,83,86–88]. Their role and significance in water has recently been reviewed by Mena and Gerba [88]. These authors estimated the health risks associated with the exposure to *P. aeruginosa*, and conclude that the risk derived from drinking water ingestion is low; however, it is slightly higher if the subject is taking an antibiotic to which this microorganism is resistant.

The mycobacteria belong to a group of microorganisms considered emerging pathogens of increasing importance. Their role as aetiological agents of waterborne disease is still not completely understood, although this group includes pathogenic species such as *Mycobacterium tuberculosis* or *M. bovis* and other atypical mycobacteria, for example, *M. intracellulare* and *M. avium* [4,89]. According to the WHO these bacterial species are relatively resistant to treatment and disinfection and have been detected in well operated and maintained drinking-water supplies with HPC less than 500/mL and total chlorine residuals of up to 2.8 mg/L. Furthermore, the growth of these organisms in biofilms reduces the effectiveness of disinfection [4].

Aeromonas are considered autochthonous microorganisms of water and are responsible of producing several diseases in humans, some of them related to water exposure or consumption of contaminated water [4,90–92]. Considerable new knowledge has been accumulated in recent years about the taxonomy, virulence properties and disease presentations of the species included in this genus [92–108]. The introduction of new molecular approaches, e.g., housekeeping genes enabled to recognise new species from freshwater (*A. fluvialis* and *A. rivuli*), tap water (*A. tecta*) and new and/or relevant clinical species like *A. taiwanensis*, *A. saranelli* and *A. aquariorum* [93–96]. The latter three species have been associated mainly with extraintestinal infections [94,95]. These clinical species together with the previously known species associated with different human disease (*A. hydrophila*, *A. caviae*, *A. veronii* bt. sobria, *A. veronii* bt. veronii, *A. jandaei* and *A. schubertii*) [90–92] should nowadays be considered relevant for public health. *Aeromonas* can be readily isolated from drinking water distribution systems, where they appear to survive well, to proliferate at low temperatures and to be associated with pipe biofilms where populations may survive at high chlorine levels [4,91,97]; and therefore, they may be considered as potential indicators of disinfection efficacy and biofilm development [4,91]. In The Netherlands, the public health authorities defined maximum values for *Aeromonas* densities, *i.e.*, 20 CFU/100 mL for finished water, and 200 CFU/100 mL for drinking water in the distribution system [91]. The factors that influence the occurrence and population sizes of *Aeromonas* spp. in water distribution systems include organic content, temperature, the residence time of water in the distribution network, and the presence of residual chlorine [4]. Strains isolated from drinking water contain virulence factors [100–105]. In one of the latter studies, it was demonstrated a clonal relationship between the isolates recovered from patients with diarrhoea and those recovered from the drinking water [103]. However, the role of *Aeromonas* in gastroenteritis has been questioned, but many arguments support its true association with diarrheal disease [98]. Recently a new case of *Aeromonas* HUS has been published and previous described cases were reviewed [66]. Furthermore, shigatoxin genes homologous to those of *E. coli*O157:H7 responsible of HUS were found to be present in some *Aeromonas* strains, reinforcing its role of as an aetiological agent of HUS [105]. The public health importance of this finding, together with that derived from the recent isolation of the clinical relevant species *A. aquariorum* [94], in chironomid egg masses, which may infest drinking water systems, needs to be further clarified [106].

Some groups of human viruses have also been proposed as alternative indicators for the control of drinking water quality, such as adenoviruses and polyomaviruses [107,108], on the basis that adenoviruses have been found to be significantly more stable than faecal indicator bacteria and other enteric viruses during UV treatment. Some researchers have suggested enteroviruses or noroviruses as indicators of other enteric viruses [109,110]. However, these viruses exhibit seasonal fluctuations and epidemic spikes [111]. Griffin *et al.* [112] have proposed Torque teno virus as a more appropriate indicator of viral pathogens in drinking waters due to its characteristics. Torque teno virus is a small, non-enveloped DNA virus that is likely to exhibit similar transport characteristics to pathogenic enteric viruses. Torque teno virus is unique among enteric viral pathogens in that it appears to be ubiquitous in humans, elicits seemingly innocuous infections, and does not exhibit seasonal fluctuations or epidemic spikes. Torque teno virus is transmitted primarily via the faecal-oral route and can be assayed using rapid molecular techniques.

Recovery of Injured Bacteria

Indicator bacteria become injured in water and wastewater following sublethal exposure to a wide variety of chemical and physical agents [15,113–115]. Such bacteria are unable to form colonies on most selective media, and between 10% and 90% of indicator bacteria in treated drinking water may be injured [116]. As a consequence, injured cells are undetected in water, leading to an underestimation of the faecal pollution level of the finished drinking water in distribution networks [117]. The advantages of several methods to detect injured bacteria have been reviewed earlier [3,11,26,114]. The detection of injured bacteria in treated waters may be indicative of the potential regrowth in the distribution system, due especially to the presence of high levels of nutrients, and thus may provide guidance in the diagnosis of problems within water distribution systems [15].

EUROPEAN LEGISLATION ON DRINKING WATER

At present, the Council Directive 98/83/EC [7] is the legislation that is applied for the protection of the quality of water intended for human consumption. Within this Directive it was indicated that the standards included in this legislation were meant to be reviewed by the European Commission (EC) every five years to adapt them to the latest scientific state of the art. In this sense the Commission initiated a process in 2003 in which a wide range of stakeholders participated to discuss the key elements that could be modified in light of current knowledge and advances in technology. The agreed elements were: (i) the inclusion of a more preventative approach for improving the quality of drinking water based on the evaluation of risk to contamination from the source water to the tap through a risk assessment approach in line with the defined WSPs developed by the WHO [3,4,26]; (ii) to update and review both chemical and microbiological parametres and to include standard methods for monitoring, sampling and analysis; (iii) to pay special attention to small water supply systems, which are now known to be those at higher risk globally; and (iv) to introduce criteria for construction products in contact with drinking water. In this process, the EC engaged the WHO to advise on how the WSP concept could be incorporated in the revised drinking water legislation and EC has set up expert groups and employed consultants to address all the key elements mentioned: http://ec.europa.eu/environment/water/water-drink/revision_en.html. In relation to the microbiological parametres it has been agreed that *E. coli* and enterococci have proven to be useful and therefore both parametres will remain in the new proposal; however, it has been recommended that the sampling frequency for enterococci is increased to one similar to that required for monitoring *E. coli*. Another point raised is that the current reference Membrane Filtration method ISO 9308-1 for *E. coli* is not suitable, and the proposal is to change it, and include more than one method. It has also been agreed that the coliform bacteria parametre is not suitable as a faecal indicator but may serve other purposes (*i.e.*, operational monitoring). It is considered that if coliforms are finally kept in the new Directive, there are issues concerning their method of analysis and their definition that should be reviewed. There is also a general agreement about removing *C. perfringens* from the list of parametres for routine compliance monitoring. In relation to total colony count, where the existing Directive the standard is "no abnormal change", it is recognized that this needs to be rephrased, and that more clear guidance value is needed. The report can be consulted at: http://circa.europa.eu/Public/irc/env/drinking_water_rev/library?l=/microbiological/17102007_28022008/_EN_1.0_&a=d.

An important interlink should exist in the new proposed Directive with other already existing or planned EU legislation related to the quality of water, e.g., Water Framework Directive 2000/60/EC (WFD) or the Groundwater Directive 2006/118/EC (GWD). The WFD has a major role in the protection and the prevention of pollution of water resources and therefore may be an important driving force to improve the quality of raw water. An important integration between responsible authorities dealing with the obligations arising from these legislations will be required.

WHO GUIDELINES AND WATER SAFETY PLANS

The WHO recommend WSPs as the most effective approach for consistently ensuring the safety of a drinking-water supply, because this approach manages the risk from the catchment or water sources to the consumer's tap [3,4,26]. The WSP approach is based on the hazard analysis and critical control point (HACCP) system, used classically in the food industry for controlling food quality. The risk assessment of the complete water system (catchment to tap), included in the WSP, should provide a better understanding of the risks of contamination by pathogens at each step along the system. Then preventative strategies should be designed (a multi-barrier protection) in order to correctly manage these risks to efficiently and effectively protect public health. This approach does not rely solely on end point testing, but on the establishment of critical control points that will be subject to on-line monitoring. The parameters that can be measured on-line and in real-time are: free chlorine, water pressure, dissolved oxygen and turbidity, for which safety critical limits are established. Any sudden anomalous changes in any of these parameters may indicate a problem within the system that can be managed before water is supplied to the consumer. The introduction of these early warning or control parametres from source to tap, that can predict or alert of a possible deterioration of the drinking water quality before it is distributed to the population, are the key elements of the WSPs. The microbiological analysis for the identification and enumeration of indicator microorganisms are too slow (require 24–48 h), and therefore are not suitable for this, however, they have an important role as validation tools because they verify that the barriers work properly and that the complete process is under control.

In reality many big water companies have long been adopting the principles of risk assessment and risk management (mostly in the form of operational procedures) for their treatment works and distribution networks, and therefore adaptation to these approaches will not be difficult but beneficial as already reported [118,http://www.gov.ns.ca/nse/water/docs/NSWaterStrategy.pdf]. Within an EU research financed project (Healthy Water) one of the objectives was to train water companies on the principles of the WSPs, the experience demonstrated that the companies acknowledged the benefit of the new approach, and some of the largest water companies have incorporated it already. However, small water companies would only be implementing WSPs when the approach becomes a mandatory requirement under the new EU legislation. In reality legislations are good driving forces for such improvements as outlined previously. For instance the national Spanish legislation on *Legionella* promulgated after the world largest outbreak, requires a part of the microbiological controls two other obligations: (i) to pass a training course for those responsible of handling installations at risk of propagating the legionellosis; and (ii) to implant control plans based on the methodology of HACCP. According to Bartram *et al.* [85] health care facilities should have general water safety plans as part of their infection control strategy. Such plans may be generic (e.g., applicable to health centres in general) or specific for larger buildings (*i.e.*, hospitals and nursing homes) and should address microbial growth in addition to control of external contamination by *P. aeruginosa*, and*Legionella*. The WSPs have to be developed by a team and require:

- Specific measures to protect raw water used to produce drinking water (*i.e.*, fencing).
- Ensuring the appropriate level of treatment at the water company and during storage and distribution pipe networks to customer's tap is maintained to guarantee the water quality.
- Ensuring that customers are aware of their role and responsibility for keeping water wholesome in their properties—it includes public buildings as well as private homes

Protection of the entire catchment areas is the first step of the multiple-barrier protection concept. Modelling can be used for establishing microbial risks in drinking water catchments and can be an excellent management tool [118] in the development of the WSP. There is much evidence that inappropriate water handling is one of the main sources of water contamination at the consumer's homes. Considering this, WHO prepared a specific manual for *Managing Microbial Water Quality in Piped Distribution Systems* [26, http://www.who.int/water_sanitation_health/dwq/924156251X/en/].

Further information, of how to implement a WSP, can be found in the WHO technical guidance documents (http://www.who.int/water_sanitation_health/dwq/wsp170805.pdf), including the WSP Manual [4,http://whqlibdoc.who.int/publications/2009/9789241562638_eng.pdf] in which specific case studies are presented. A dedicated web site on the WSPs has been developed by the International Water Association (IWA) (http://www.wsportal.org/ibis/water-safety-portal/eng/welcome). In addition, WHO and IWA developed guiding documents to initiate such process considering all levels of resources available, so that it can be implemented all over the world, even at poorly developed countries: (www.unwater.org/worldwaterday/.../WSP_RoadMap_Final_3_19_10.pdf).

QUALITY OF DRINKING WATER AND CLIMATE CHANGE

The main impacts of climate change on water availability are flooding and droughts. However, besides these quantitative impacts, the climate change will affect the surface water quality [4,120]. The climate change determinants affecting water quality are mainly the air temperature, the increase of extreme hydrological events, soil drying-rewetting cycles and solar radiation. First of all, temperature is the main factor affecting almost all physico-chemical equilibriums and biological reactions. Consequently, several transformations or effects related to water will be favoured by water temperature increase such as dissolution, solubilisation, complexation, degradation, evaporation, *etc.* This phenomenon globally leads to the concentration increase of dissolved substances in water but also to the concentration decrease of dissolved gasses, such as oxygen. Floods and droughts will also modify water quality by direct effects of dilution or concentration of dissolved substances [4,121]. A positive effect is the concentration decrease of some pollutants due to a low water velocity, which allow the assimilation of nutrients by aquatic plants and the adsorption/complexation of heavy metals on suspended matter and settling [122]. Runoff and solid material transportation are the main consequences of heavy rainfalls; for example, in the temperate zone, climate change will decrease the number of rainy days but increase the average volume of each rainfall event [73,123]. As a consequence, drought–rewetting cycles may impact water quality as it enhances decomposition of organic matter into streams [124]. A study performed by Nichols *et al.* [125] provided evidence that both low rainfall and heavy rain precede many drinking water outbreaks, and therefore both should be considered when assessing the health impacts of climate change. Solar irradiation increase could also alter water quality and especially characteristics of natural organic matter in freshwater systems both by warming and UVB radiation (increasing photolysis) [126].

Waterborne pathogens could be spread within the freshwater after a contamination by animal or human waste due to heavy rainfall discharge in combined sewer systems (CSS) [4,74,127]. When the flow exceeds the CSS capacity, the sewers overflow directly into surface water body [127]. Pednekar *et al.* [128] have studied coliform load in a tidal embayment and shown that storm-water coming from the surrounding watershed is a primary source of coliform. Moreover, higher water temperatures will probably lead to a pathogen survival increase in the environment, although there is still no clear evidence [129]. Floods often led to a contamination of groundwater and additional disease outbreaks [4,130]. Even though the risk of diseases outbreaks linked to drinking waters is low in developed countries, private supplies would be at risk [4,26,129]. In addition, an increase in temperature threats water quality with regard to waterborne diseases especially cholera disease in Asia and South America [129]. It was shown that the increased UV radiation due to ozone layer depletion provokes the breaking down of bioavailable organic compounds, minerals and micronutrients, stimulating the bacterial activity in aquatic ecosystems [126].

Fishes, green algae and diatoms are often used as water quality indicators of climate changes in waters. Daufresne and Böet [131] observed a change in fish communities due to temperature changes. WHO had also prepared a document called Vision 2030:"The resilience of water supply and sanitation in the face of climate change" that aims to increase our understanding of how anticipated climate change may affect drinking water:http://www.who.int/water_sanitation_health/publications/9789241598422_cdrom/en/.

CONCLUSIONS

The microbial contamination of drinking water and its control constitutes a major issue worldwide, because it is still a major source of infection and can cause mortality, especially in the children of developed countries, and threatens the health of the population of developed regions, as illustrated by recorded outbreaks. The latter are however, considered to be underestimated because the major symptom (*i.e.*, diarrhoea) auto limits by itself without treatment in most healthy people. However, the older and immunocompromized people are at higher risk. Today we also know that this type of infection may have important sequels (e.g., reactive arthritis, irritable bowel syndrome, cancer predisposition, *etc.*). The microbiological controls applied to drinking water, have relied on the analysis of faecal pollution indicators in the finished dinking water. The classical indicators have served together with the improvements on the treatment and disinfection to control waterborne outbreaks. However, the use of these indicators may be substituted by the direct detection of pathogenic microorganisms, e.g., in the case of pathogenic viruses [53,132,133]. In addition, the lessons learned from outbreaks are key elements that should guide the proper management of drinking water. The shortcoming of bacterial indicators to predict parasites and viruses, which can be more resistant to disinfection, and the fact that information derived from the microbiological analysis is not immediate (neither is obtained in a continuous manner), have motivated the development of more preventive approaches, like the Water Safety Plans proposed by the WHO. Their application for the management of drinking water, either in big companies, small ones or in undeveloped countries can be foreseen as an important gain for the future. The adoption of this strategy in the EU legislation, which is in process of being modified, is a promising guarantee for the improvement of quality of dinking water in Europe. The recognition in this modification that small water supplies are at the highest risk, and the introduction of measures to control these supplies more efficiently will contribute to the expected improvement. Furthermore, the accumulated knowledge on the impact of climate change allows preparation of strategies to mitigate its impact on the quality of drinking water.

ACKNOWLEDGEMENTS

This work was supported by funds from the European Commission for the EPIBATHE (SSPI-2006-022618) and HEALTHY WATER (FOOD-CT-2006-036306) projects and partially by an AquaFactory International grant. The authors are solely responsible for the content of this publication. It does not represent the opinion of the European Commission. The European Commission is not responsible for any use that might be made of data appearing therein.

REFERENCES

1. Craun, GF; Brunkard, JM; Yoder, JS; Roberts, VA; Carpenter, J; Wade, T; Calderon, RL; Roberts, JM; Beach, MJ; Roy, SL. Causes of outbreaks associated with drinking water in the United States from 1971 to 2006. *Clin. Microbiol. Rev* **2010**,*23*, 507–528.
2. Vigilancia epidemiológica de brotes de transmisión hídrica en España. 1999–2006. *Bol. Epidemiol. Semanal* **2008**, *16*, 25–28.
3. *Guidelines for Drinking-Water Quality*, 3rd ed; World Health Organization: Geneva, Switzerland, 2008; Volume 1, Recommendations.
4. Bartram, J; Corrales, L; Davison, A; Deere, D; Drury, D; Gordon, B; Howard, G; Rinehold, A; Stevens, M. *Water Safety Plan Manual: Step-by-Step Risk Management for Drinking Water Suppliers*; World Health Organization: Geneva, Switzerland, 2009.
5. Dufour, A; Snozzi, M; Koster, W; Bartram, J; Ronchi, E; Fewtrell, L. *Assessing Microbial Safety of Drinking Water Improving Approaches and Methods*; World Health Organization and the Organisation for Economic Co-operation and Development, IWA Publishing Alliance House: London, UK, 2003. Available online: https://www.who.int/water_sanitation_health/dwq/9241546301full.pdf (accessed on 1 December 2010).
6. *Emerging Issues in Water and Infectious Diseases*; World Health Organization: Geneva, Switzerland, 2003. Available online: http://www.who.int/water_sanitation_health/emerging/emergingissues/en/ (accessed on 1 December 2010).
7. European Union. *Council Directive 98/83/EC of 3 November 1998 on the quality of water intended for human consumption (98/83/EC)*, 1998. Available online: (http://ec.europa.eu/environment/water/water-drink/index_en.html (accessed on 1 December 2010).
8. Borrego, JJ; Figueras, MJ. Microbiological quality of natural waters. *Microbiologia SEM (Int. Microbiol.)* **1997**, *13*, 413–426.
9. Ashbolt, NJ; Grabow, WOK; Snozzi, M. Fewtrell, L, Barthram, J, Eds.; Indicators of microbial water quality. In *Water Quality: Guidelines, Standards and Health*; IWA Publishing: London, UK, 2001; pp. 289–316.
10. Borrego, JJ; Castro, D; Figueras, MJ. Bitton, G, Ed.; Fecal streptococci/enterococci in aquatic environments. In*Encyclopedia of Environmental Microbiology*; John Wiley & Sons: New York, NY, USA, 2002; pp. 1264–1278.
11. Figueras, MJ; Robertson, W; Pike, E; Borrego, JJ. Bartram, J, Rees, G, Eds.; Sanitary inspection and microbiological water quality. In *Monitoring Bathing Waters A Practical Guide to the Design and Implementation of Assessments and Monitoring Programmes*; E & FN Spon: London, UK, 2000; pp. 113–167.
12. Köster, W; Egli, T; Ashbolt, N; Botzenhart, K; Burlion, N; Endo, T; Grimont, P; Guillot, E; Mabilat, C; Newport, L; Niemi, M; Payment, P; Prescott, A; Renaud, P; Rust, A. Analytical methods for microbial water quality testing. In*Assessing Microbial Safety of Drinking Water: Improving Approaches and Methods*; OECD, World Health Organization: Geneva, Switzerland, 2003; pp. 237–292.
13. Borrego, JJ; Castro, D; Figueras, MJ. Bitton, G, Ed.; Salmonella *in aquatic environments*. In *Encyclopedia of Environmental Microbiology*; John Wiley & Sons: New York, NY, USA, 2002; pp. 2789–2802.
14. National Research Council, Evaluation of risk: a phased approach to monitoring microbial water quality. In *Indicators for Waterborne Pathogens*; The National Academic Press: Washington, DC, USA, 2004; pp. 13–16.
15. Toranzos, GA; McFeters, GA; Borrego, JJ; Savill, M. Hurst, CJ, Crawford, RL, Garland, JL, Lipson, DA, Mills, AL, Stetzenbach, LD, Eds.; Detection of microorganisms in environmental freshwaters and drinking waters. In *Manual of Environmental Microbiology*, 3rd ed; American Society for Microbiology Press: Washington, DC, USA, 2007; pp. 249–264.
16. Leclerc, H; Schwartzbrod, L; Dei-Cas, E. Microbial agents associated with waterborne diseases. *Crit. Rev. Microbiol***2002**, *28*, 371–409.
17. Asbolt, NJ. Microbial contamination of drinking water and disease outcomes in developing regions. *Toxicology* **2004**,*198*, 229–238.
18. Postel, SL. Water and world population growth. *J. Am. Water Works Assoc* **2000**, *92*, 131–138.
19. Morris, J. Filtered or waivered, you need source protection. *N. Engl. Water Works Assoc* **1994**, *9*, 182–191.
20. *Preventing Waterborne Disease: Is Your System at Risk*; American Water Works Association: Denver, CO, USA, 1994.
21. Brettar, I; Höfle, MG. Molecular assessment of bacterial pathogens—A contribution to drinking water safety. *J. Curr. Opin. Biotech* **2008**, *19*, 274–280.
22. Cangelosi, GA. Prospects for applying virulence factor-activity relationships (VFAR) to emerging waterborne pathogens. *J. Water Health* **2009**, *7*, S64–74.
23. Moriñigo, MA; Cornax, R; Castro, D; Martinez-Manzanares, E; Borrego, JJ. Viability of *Salmonella* spp. and indicator microorganisms in seawater using membrane diffusion chambers. *Antonie van Leeuwenhoek* **1990**, *57*, 109–117.
24. Figueras, MJ; Polo, FL; Inza, I; Guarro, J. Past, present and future perspective of the EU bathing water directive. *Marine Pollut. Bull* **1997**, *34*, 148–156.
25. LeChevallier, MW; Welch, WJ; Smith, DB. Full-scale of factors related to coliform regrowth in drinking water. *Appl. Environ. Microbiol* **1996**, *62*, 2201–2211.
26. World Health Organization, Ainsworth, R, Ed.; *Safe Piped Water: Managing Microbial Water Quality in Piped Distribution Systems*; IWA Publishing: London, UK, 2004.
27. LeChevallier, MW. The case for maintaining a disinfectant residual. *J. Am. Water Works Assoc* **1999**, *91*, 86–94.

28. Hazen, TC; Toranzos, GA. McFeters, GA, Ed.; Tropical source water. In *Drinking Water Microbiology*; Springer-Verlag: New York, NY, USA, 1990; pp. 32–54.

29. Lasalde, C; Rodriguez, R; Smith, HH; Toranzos, GA. Heterogeneity of *uid*A gene in environmental *Escherichia coli*populations. *J. Water Health* **2005**, *3*, 297–304.

30. Borrego, JJ; Arrabal, F; De Vicente, A; Gomez, LF; Romero, P. Study of microbial inactivation in the marine environment. *J. Water Pollut. Control Fed* **1983**, *55*, 297–302.

31. Schierack, P; Walk, N; Reiter, K; Weyranch, KD; Wieler, LH. Composition of intestinal *Enterobacteriaceae* populations of healthy domestic pigs. *Microbiology* **2007**, *153*, 3830–3837.

32. Hunter, PR. Drinking water and diarrhoeal disease due to *Escherichia coli*. *J. Water Health* **2003**, *1*, 65–72.

33. Schets, FM; During, M; Italiaander, R; Heijnen, L; Rutjes, SA; van der Zwaluw, WK; de Roda Husman, AM. *Escherichia coli* O157:H7 in drinking water from private water supplies in the Netherlands. *Water Res* **2005**, *39*, 4485–4493.

34. Kay, D; Fleischer, JM; Salmon, RL; Jones, F; Wyer, MD; Goodfree, AF; Zelenauch-Jacquotte, Z; Shore, R. Predicting likelihood of gastroenteritis from sea bathing: results from randomized exposure. *Lancet* **1994**, *344*, 905–909.

35. Audicana, A; Perales, I; Borrego, JJ. Modification of kanamycin-esculin-azide agar to improve selectivity in the enumeration of fecal streptococci from water samples. *Appl. Environ. Microbiol* **1995**, *61*, 4178–4183.

36. Figueras, MJ; Inza, I; Polo, FL; Feliu, MT; Guarro, J. A fast method for the confirmation of fecal streptococci from M-Enterococcus medium. *Appl. Environ. Microbiol* **1996**, *62*, 2177–2178.

37. Figueras, MJ; Inza, I; Polo, FL; Guarro, J. Evaluation of the oxolinic acid-esculin-azide medium for the isolation and enumeration of faecal streptococci in a routine monitoring programme for bathing waters. *Can. J. Microbiol* **1998**, *44*, 998–1002.

38. Kühn, I; Iversen, A; Burman, LG; Olsson-Liljequist, B; Franklin, A; Finn, M; Aarestrup, F; Seyfarth, AM; Blanch, AR; Vilanova, X; Taylor, H; Caplin, J; Moreno, MA; Dominguez, L; Herrero, IA; Möllby, R. Comparison of enterococcal populations in animals, humans, and the environment—a European study. *Int. J. Food. Microbiol* **2003**, *88*, 133–145.

39. Layton, BA; Walters, SP; Lam, LH; Boehm, AB. *Enterococcus* species distribution among human and animal hosts using multiplex PCR. *J. Appl. Microbiol* **2010**, *109*, 539–547.

40. Fricker, CR; DeSarno, M; Warden, PS; Eldred, BJ. False-negative β-D-glucuronidase reactions in membrane lactose glucuronide agar medium used for the simultaneous detection of coliforms and *Escherichia coli* from water. *Lett. Appl. Microbiol* **2008**, *47*, 539–542.

41. Fricker, CR; Warden, PS; Eldred, BJ. Understanding the cause of false negative β-D-glucuronidase reactions in culture media containing fermentable carbohydrate. *Lett. Appl. Microbiol* **2010**, *50*, 547–551.

42. Schets, FM; Nobel, PJ; Strating, S; Mooijman, KA; Engels, GB; Brouwer, A. EU Drinking Water Directive reference methods for enumeration of total coliforms and *Escherichia coli* compared with alternative methods. *Lett. Appl. Microbiol* **2002**, *34*, 227–231.

43. Field, KG; Bernhard, AE; Brodeur, TJ. Molecular approaches to microbiological monitoring: fecal source detection.*Environ. Monit. Assess* **2003**, *81*, 313–326.

44. Moriñigo, MA; Wheeler, D; Berry, C; Jones, C; Muñoz, MA; Cornax, R; Borrego, JJ. Evaluation of different bacteriophage groups as faecal indicators in contaminated natural waters in Southern England. *Water Res* **1992**, *26*, 267–271.

45. Leclerc, H; Edberg, S; Pierzo, V; Delattre, JM. Bacteriophages as indicators of enteric viruses and public health risk in groundwaters. *J. Appl. Microbiol* **2000**, *88*, 5–21.

46. Borrego, JJ. Rana, BC, Ed.; Coliphages as faecal pollution indicators. In *Pollution and Biomonitoring*; McGraw-Hill: New York, NY, USA, 1995; pp. 117–159.

47. Havelaar, AH; Pot-Hogeboom, WH. F-specific RNA-bacteriophages as model viruses in water hygiene: ecological aspects. *Water Sci. Technol* **1988**, *20*, 399–407.

48. Jofre, J; Bosch, A; Lucena, F; Girones, R; Tartera, C. Evaluation of *Bacteroides fragilis* bacteriophages as indicators of the virological quality of water. *Water Sci. Technol* **1986**, *18*, 167–177.

49. Duran, AE; Muniesa, M; Moce-Llivina, L; Campos, C; Jofre, J; Lucena, F. Usefulness of different groups of bacteriophages as model microorganisms for evaluating chlorination. *J. Appl. Microbiol* **2003**, *95*, 29–37.

50. Borrego, JJ; Moriñigo, MA; de Vicente, A; Cornax, R; Romero, P. Coliphages as an indicator of faecal pollution in water. Its relationship with indicator and pathogenic microorganisms. *Water Res* **1987**, *21*, 1473–1480.

51. Paul, JH; Rose, JB; Brown, J; Shinn, EA; Miller, S; Farrah, SR. Viral tracer studies indicate contamination of marine waters by sewage disposal practices in Key Largo, Florida. *Appl. Environ. Microbiol* **1995**, *61*, 2230–2234.

52. Harvey, RW. Microorganisms as tracers in ground water injection and recovery experiments: A review. *FEMS Microbiol. Rev* **1997**, *20*, 461–472.

53. Hot, D; Legeay, O; Jacques, J; Gantzer, C; Caudrelier, Y; Guyard, K; Lange, M; Andreoletti, L. Detection of somatic phages, infectious enteroviruses and enterovirus genomes as indicators of human enteric viral pollution in surface waters. *Water Res* **2003**, *37*, 4703–4710.

54. Havelaar, AH; Furuse, K; Hogeboom, WH. Bacteriophages and indicator bacteria in human and animal faeces. *J. Appl. Bacteriol* **1986**, *60*, 252–262.

55. Cornax, R; Moriñigo, MA; Gonzalez-Jaen, F; Alonso, MC; Borrego, JJ. Bacteriophages presence in human faeces of healthy subjects and patients with gastrointestinal disturbances. *Zentbl. Bakteriol* **1994**, *281*, 214–224.

56. DeBartolomeis, J; Cabelli, VP. Evaluation of an *Escherichia coli* host strain for enumeration of F male-specific bacteriophages. *Appl. Environ. Microbiol* **1991**, *57*, 1301–1305.

57. Lucena, F; Lasofras, J; McIntosh, D; Forcadell, M; Jofre, J. Effect of distance from the pollution focus on relative concentrations of *Bacteroides fragilis* phages and coliphages in mussels. *Appl. Environ. Microbiol* **1994**, *60*, 2272–2277.

58. Cornax, R; Moriñigo, MA; Paez, IG; Muñoz, MA; Borrego, JJ. Application of direct plaque assay for detection and enumeration of bacteriophages of *Bacteroides fragilis* from contaminated-water samples. *Appl. Environ. Microbiol* **1990**,*56*, 3170–3173.

59. Cornax, R; Moriñigo, MA; Balebona, MC; Castro, D; Borrego, JJ. Significance of several bacteriophage groups as indicators of sewage pollution in marine waters. *Water Res* **1991**, *25*, 673–678.

60. Reynolds, KA; Mena, KD; Gerba, CP. Risk of waterborne illness via drinking water in the United States. *Rev. Environ. Contam. Toxicol* **2008**, *192*, 117–58.

61. Risebro, HL; Hunter, PR. Surveillance of waterborne disease in European member states: a qualitative study. *J. Water Health* **2007**, *5*, 19–38.

62. Karanis, P; Kourenti, C; Smith, H. Waterborne transmission of protozoan parasites: A worldwide review of outbreaks and lessons learnt. *J. Water Health* **2007**, *5*, 1–18.

63. Swerdlow, DL; Woodruff, BA; Brady, RC; Griffin, PM; Tippen, S; Donnell, HD; Geldreich, E; Payne, BJ; Meyer, A; Wells, JG; *et al.* A waterborne outbreak in Missouri of *Escherichia coli* O157:H7 associated with bloody diarrhoea and death. *Ann. Intern. Med* **1992**, *117*, 812–819.

64. Fong, TT; Mansfield, LS; Wilson, DL; Schwab, DJ; Molloy, SL; Rose, JB. Massive microbiological groundwater contamination associated with a waterborne outbreak in Lake Erie, South Bass Island, Ohio. *Environ. Health Perspect***2007**, *115*, 856–864.

65. O'Reilly, CE; Bowen, AB; Perez, NE; Sarisky, JP; Shepherd, CA; Miller, MD; Hubbard, BC; Herring, M; Buchanan, SD; Fitzgerald, CC; Hill, V; Arrowood, MJ; Xiao, LX; Hoekstra, RM; Mintz, ED; Lynch, MF. A waterborne outbreak of gastroenteritis with multiple etiologies among resort island visitors and residents: Ohio, 2004. *Clin. Infect. Dis* **2007**,*44*, 506–512.

66. Figueras, MJ; Collado, L; Guarro, J. A new 16S rDNA-RFLP method for the discrimination of the accepted species of*Arcobacter*. *Diagn. Microbiol. Infect. Dis* **2008**, *62*, 11–15.

67. Collado, L; Inza, I; Guarro, J; Figueras, MJ. Presence of *Arcobacter* spp. in environmental waters correlates with high levels of fecal pollution. *Environ. Microbiol* **2008**, *10*, 1635–1640.

68. Collado, L; Kasimir, G; Perez, U; Bosch, A; Pinto, R; Saucedo, G; Huguet, JM; Figueras, MJ. Occurrence and diversity of *Arcobacter* spp. along the Llobregat River catchment, at sewage effluents and in a drinking water treatment plant.*Water Res* **2010**, *44*, 3696–3702.

69. European Food Safety Authority (EFSA), European Centre for Disease Prevention and Control (ECDC), The Community Summary Report. In *Food-borne Outbreaks in the European Union in 2007*; The EFSA Journal: Brussels, Belgium, 2009; pp. 107–108.

70. Kopilovic, B; Ucakar, V; Koren, N; Krek, M; Kraigher, A. Waterborne outbreak of acute gastroenteritis in a costal area in Slovenia in June and July 2008. *Eurosurveillance* **2008**, *13*, 1–3.

71. Risebro, HL; Doria, MF; Andersson, Y; Medema, G; Osborn, K; Schlosser, O; Hunter, PR. Fault tree analysis of the causes of waterborne outbreaks. *J. Water Health* **2007**, *5*, 1–18.

72. Auld, H; MacIver, D; Klaassen, J. Heavy rainfall and waterborne disease outbreaks: the Walkerton example. *J. Toxicol. Environ. Health. A* **2004**, *67*, 1879–1887.

73. Nichols, G; Lane, C; Asgari, N; Verlander, NQ; Charlett, A. Rainfall and outbreaks of drinking water related disease and in England and Wales. *J. Water Health* **2009**, *7*, 1–8.

74. Kistemann, T; Classen, T; Koch, C; Dangendorf, F; Fischeder, R; Gebel, J; Vacata, V; Exner, M. Microbial load of drinking water reservoir tributaries during extreme rainfall and runoff. *Appl. Environ. Microbiol* **2002**, *68*, 2188–2197.

75. Thurston-Enriquez, JA; Gilley, JE; Eghball, B. Microbial quality of runoff following land application of cattle manure and swine slurry. *J. Water Health* **2005**, *3*, 157–171.

76. Mons, C; Dumetre, A; Gosselin, S; Galliot, C; Moulin, L. Monitoring of *Cryptosporidium* and *Giardia* river contamination in Paris area. *Water Res* **2009**, *43*, 211–217.

77. Howe, AD; Forster, S; Morton, S; Marshall, R; Osborn, KS; Wright, P; Hunter, PR. *Cryptosporidium* oocysts in a water supply associated with a cryptosporidiosis outbreak. *Emerg. Infect. Dis* **2002**, *8*, 619–624.

78. Garg, AX; Pope, JE; Thiessen-Philbrook, H; Clark, WF; Ouimet, J. Walkerton health study investigators. Arthritis risk after acute bacterial gastroenteritis. *Rheumatology* **2008**, *47*, 200–204.

79. Thabane, M; Simunovic, M; Akhtar-Danesh, N; Garg, AX; Clark, WF; Collins, SM; Salvadori, M; Marshall, JK. An outbreak of acute bacterial gastroenteritis is associated with an increased incidence of irritable bowel syndrome in children. *Am. J. Gastroenterol* **2010**, *105*, 933–939.

80. Holme, R. Drinking water contamination in Walkerton, Ontario: positive resolutions from a tragic event. *Water Sci. Technol* **2003**, *47*, 1–6.

81. Lake, IR; Nichols, G; Bentham, G; Harrison, FC; Hunter, PR; Kovats, SR. Cryptosporidiosis decline after regulation, England and Wales, 1989-2005. *Emerg. Infect. Dis* **2007**, *13*, 623–625.

82. Payment, P; Franco, E. *Clostridium perfringens* and somatic coliphages as indicators of the efficiency of drinking water treatment for viruses and protozoan cysts. *Appl. Environ. Microbiol* **1993**, *59*, 2418–2424.

83. Chauret, CP; Radzimiski, CZ; Lepull, M; Creason, R; Andrews, RC. Chlorine dioxide inactivation of *Cryptosporidium parvum* oocyts and bacterial spore indicators. *Appl. Environ. Microbiol* **2001**, *67*, 2993–3001.

84. World Health Organization, Assessing Microbial Safety of Drinking Water. In *Improving Approaches and Methods*; IWA Publishing: London, UK, 2003.

85. Bartram, J; Cotruvo, J; Exnerd, M; Fricker, C; Glasmacher, A. Heterotrophic plate count measurement in drinking water safety management. Report of an Expert Meeting Geneva, 24–25 April 2002. *Int. J. Food Microbiol* **2004**, *92*, 241–247.

86. Ribas, F; Perramon, J; Terradillos, A; Frias, J; Lucena, F. The *Pseudomonas* group as indicator of potential regrowth in water distribution system. *J. Appl. Microbiol* **2000**, *88*, 704–710.

87. Toranzos, GA. Bitton, G, Ed.; Pseudomonas. In *Encyclopedia of Environmental Microbiology*; John Wiley & Sons: New York, NY, USA, 2002; pp. 2632–2639.

88. Mena, KD; Gerba, CP. Risk assessment of *Pseudomonas aeruginosa* in water. *Rev. Environ. Contam. Toxicol* **2009**, *201*, 71–115.

89. Dailloux, M; Laurain, C; Weber, M; Hartemann, PH. Water and nontuberculous mycobacteria. *Water Res* **1999**, *33*, 2219–2228.

90. Figueras, MJ. Clinical relevance of *Aeromonas*. *Rev. Med. Microbiol* **2005**, *16*, 145–153.

91. Martin-Carnahan, A; Joseph, SW. Brenner, DJ, Krieg, NR, Staley, JT, Garrity, GM, Eds.; Order XII. Aeromonadales ord. nov. In *Bergey's Manual of systematic Bacteriology*; Springer: NewYork, NY, USA, 2005; Volume 2, Part B, pp. 556–578.

92. Janda, JM; Abbott, SL. The genus *Aeromonas*. Taxonomy, pathogenicity and infection. *Clin. Microbiol. Rev* **2010**, *23*, 35–73.

93. Demarta, A; Küpfer, M; Riegel, P; Harf-Monteil, C; Tonolla, M; Peduzzi, R; Monera, A; Saavedra, MJ; Martínez-Murcia, A. *Aeromonas tecta* sp. nov., isolated from clinical and environmental sources. *Syst. Appl. Microbiol* **2008**, *31*, 278–286.

94. Figueras, MJ; Alperi, A; Saavedra, MJ; Ko, WC; Gonzalo, N; Navarro, M; Martínez-Murcia, AJ. Clinical relevance of the recently described species *Aeromonas aquariorum*. *J. Clin. Microbiol* **2009**, *47*, 3742–3746.

95. Alperi, A; Martínez-Murcia, AJ; Ko, WC; Monera, A; Saavedra, MJ; Figueras, MJ. *Aeromonas taiwanensis* sp. nov. and*Aeromonas sanarellii* sp. nov., clinical species from Taiwan. *Int. J. Syst. Evol. Microbiol* **2010**, *60*, 2048–2055.

96. Figueras, MJ; Alperi, A; Beaz-Hidalgo, R; Stackebrandt, E; Brambilla, E; Monera, A; Martínez-Murcia, AJ. *Aeromonas rivuli* sp. nov. isolated from the upstream region of a karst water rivulet in Germany. *Int J Syst Evol Microbiol* **2010**.

97. Bomo, AM; Storey, MV; Ashbolt, NJ. Detection, integration and persistence of aeromonads in water distribution pipe biofilms. *J. Water Health* **2004**, *2*, 83–96.

98. Figueras, MJ; Horneman, AJ; Martinez-Murcia, A; Guarro, J. Controversial data on the association of *Aeromonas* with diarrhoea in a recent Hong Kong study. *J. Med. Microbiol* **2007**, *56*, 996–998.

99. Figueras, MJ; Aldea, MJ; Fernández, N; Aspíroz, C; Alperi, A; Guarro, J. *Aeromonas* hemolytic uremic syndrome. A case and a review of the literature. *Diagn. Microbiol. Infect. Dis* **2007**, *58*, 231–234.

100. Chacon, MR; Soler, L; Groisman, EA; Guarro, J; Figueras, MJ. Type III secretion system genes in clinical *Aeromonas*isolates. *J. Clin. Microbiol* **2004**, *42*, 1285–12877.

101. Figueras, MJ; Suarez-Franquet, A; Chacon, MR; Soler, L; Navarro, M; Alejandre, G; Grasa, B; Martinez-Murcia, AJ; Guarro, J. First record of the rare species *Aeromonas culicicola* from a drinking water supply. *Appl. Environ. Microbiol***2005**, *71*, 538–541.

102. Pablos, M; Rodríguez-Calleja, JM; Santos, JA; Otero, A; García-López, ML. Occurrence of motile *Aeromonas* in municipal drinking water and distribution of genes encoding virulence factors. *Int. J. Food. Microbiol* **2009**, *135*, 158–164.

103. Pablos, M; Remacha, MA; Rodríguez-Calleja, JM; Santos, JA; Otero, A; García-López, ML. Identity, virulence genes, and clonal relatedness of *Aeromonas* isolates from patients with diarrhea and drinking water. *Eur. J. Clin. Microbiol. Infect. Dis* **2010**, *29*, 1163–1172.

104. Yu, CP; Chu, KH. Molecular quantification of virulence gene-containing *Aeromonas* in water samples collected from different drinking water treatment processes. *Environ Monit Assess* **2010**.

105. Alperi, A; Figueras, MJ. Human isolates of *Aeromonas* possess Shiga toxin genes (*stx1* and *stx2*) highly similar to the most virulent gene variants of *Escherichia coli*. *Clin. Microbiol. Infect* **2010**, *16*, 1563–1567.

106. Figueras, MJ; Beaz-Hidalgo, R; Senderovich, Y; Laviad, S; Halpern, M. Re-identification of *Aeromonas* isolates from chironomid egg masses as the potential pathogenic bacteria. *Aeromonas aquariorum Environ Microbiol Rep* **2010**.

107. Pina, S; Puig, M; Lucena, F; Jofre, J; Girones, R. Viral pollution in the environment and in shellfish: human adenovirus detection by PCR as an index of human viruses. *Appl. Environ. Microbiol* **1998**, *64*, 3376–3382.

108. Albinana-Gimenez, N; Miagostovich, MP; Calgua, B; Huguet, JM; Matia, L; Girones, R. Analysis of adenoviruses and polyomaviruses quantified by qPCR as indicators of water quality in source and drinking-water treatment plants.*Water Res* **2009**, *43*, 2011–2019.

109. Kopecka, H; Dubrou, S; Prevot, J; Marechal, J; Lopez-Pila, JM. Detection of naturally occurring enteroviruses in waters by reverse transcription, polymerase chain reaction, and hybridization. *Appl. Environ. Microbiol* **1993**, *59*, 1213–1219.

110. Metcalf, TG; Melnick, JL; Estes, MK. Environmental virology: from detection of virus in sewage and water

by isolation to identification by molecular biology—a trip of over 50 years. *Annu. Rev. Microbiol* **1995**, *49*, 461–487.

111. Haramoto, E; Katayama, H; Oguma, K; Yamashita, H; Tajima, A; Nakajima, H; Ohgaki, S. Seasonal profiles of human noroviruses and indicator bacteria in a wastewater treatment plant in Tokyo, Japan. *Water Sci. Technol* **2006**, *54*, 301–308.
112. Griffin, JS; Plummer, JD; Long, SC. Torque teno virus: an improved indicator for viral pathogens in drinking waters.*Virol. J* **2008**, *5*, 112.
113. Huq, A; Rivera, ING; Colwell, RR. Colwell, RR, Grimes, DJ, Eds.; Epidemiological significance of viable but nonculturable microorganisms. In *Nonculturable Microorganisms in the Environment*; ASM Press: Washington, DC, USA, 2000; pp. 301–323.
114. Kahlisch, L; Henne, K; Groebe, L; Draheim, J; Höfle, MG; Brettar, I. Molecular analysis of the bacterial drinking water community with respect to live/dead status. *Water Sci. Technol* **2010**, *61*, 9–14.
115. Cordoba, MA; Del Coco, VF; Minvielle, MC; Basualdo, JA. Influencing factors in the occurrence of injured coliforms in the drinking water distribution system in the city of La Plata, Argentina. *J. Water Health* **2010**, *8*, 205–211.
116. McFeters, GA. McFeters, GA, Ed.; Enumeration, occurrence and significance of injured indicator bacteria in drinking water. In *Drinking Water Microbiology*; Springer-Verlag: New York, NY, USA, 1990; pp. 478–492.
117. Bucklin, KE; McFeters, GA; Amirtharaja, A. Penetration of coliforms through municipal drinking water filters. *Water Res* **1991**, *25*, 1013–1017.
118. Gunnarsdóttir, MJ; Gissurarson, LR. HACCP and water safety plans in Icelandic water supply: preliminary evaluation of experience. *J. Water Health* **2008**, *6*, 377–382.
119. Ferguson, CM; Croke, BF; Beatson, PJ; Ashbolt, NJ; Deere, DA. Development of a process-based model to predict pathogen budgets for the Sydney drinking water catchment. *J. Water Health* **2007**, *5*, 187–208.
120. Delpla, I; Jung, AV; Baures, E; Clement, M; Thomas, O. Impacts of climate change on surface water quality in relation to drinking water production. *Environ. Int* **2009**, *35*, 1225–1233.
121. Van Vliet, MTH; Zwolsman, JJG. Impact of summer droughts on the water quality of the Meuse river. *J. Hydrol* **2008**,*353*, 1–17.
122. Prathumratana, L; Sthiannopkao, S; Kim, KW. The relationship of climatic and hydrological parameters to surface water quality in the lower Mekong River. *Environ. Int* **2008**, *34*, 860–866.
123. Brunetti, M; Maugeri, M; Nanni, T. Changes in total precipitation rainy days and extreme events in northeastern Italy.*Int. J. Climatol* **2001**, *21*, 861–871.
124. Evans, CD; Monteith, DT; Cooper, DM. Long-term increases in surface water dissolved organic carbon: observations, possible causes and environmental impacts. *Environ. Pollut* **2005**, *137*, 55–71.
125. Nichols, G; Lane, C; Asgari, N; Verlander, NQ; Charlett, A. Rainfall and outbreaks of drinking water related disease and in England and Wales. *J. Water Health* **2009**, *7*, 1–8.
126. Soh, YC; Roddick, F; van Leeuwen, J. The future of water in Australia: the potential effects of climate change and ozone depletion on Australian water quality, quantity and treatability. *Environmentalist* **2008**, *28*, 158–165.
127. Charron, DF; Thomas, MK; Waltner-Toews, D; Aramini, JJ; Edge, T; Kent, RA. Vulnerability of waterborne diseases to climate change in Canada: a review. *J. Toxicol. Environ. Health Part A* **2004**, *67*, 1667–1677.
128. Pednekar, AM; Grant, SB; Jeong, Y; Poon, Y; Oancea, C. Influence of climate change, tidal mixing, and watershed urbanization on historical water quality in Newport Bay, a saltwater wetland and tidal embayment in southern California. *Environ. Sci. Technol* **2005**, *39*, 9071–9082.
129. Hunter, PR. Climate change and waterborne and vector-borne disease. *J. Appl. Microbiol* **2003**, *94*, 37S–46S.
130. Curriero, FC; Patz, JA; Rose, JB; Lele, S. The association between extreme precipitation and waterborne disease outbreaks in the United States, 1948–1994. *Am. J. Public Health* **2001**, *91*, 1194–1199.
131. Daufresne, M; Boët, P. Climate change impacts on structure and diversity of fish communities in rivers. *Global Change Biol* **2007**, *13*, 2467–2478.
132. Verheyen, J; Timmen-Wego, M; Laudien, R; Bousaad, I; Sen, S; Koc, A; Uesbeck, A; Mazou, F; Pfister, H. Detection of adenoviruses and rotaviruses in drinking water sources used in rural areas of Benin, West Africa. *Appl. Environ. Microbiol* **2009**, *75*, 2798–2801.
133. Bosch, A; Guix, S; Sano, D; Pinto, RM. New tools for the study and direct surveillance of viral pathogens in water.*Curr. Opin. Biotechnol* **2008**, *19*, 295–301.

Section VII - Bioterrorism and Drinking Water Safety

CHAPTER 25 BIOSURVEILLANCE: A REVIEW AND UPDATE

Nicholas E. Kman and Daniel J. Bachmann

INTRODUCTION

Since the terrorist attacks of September 11, 2001, and the anthrax release in the following month, there has been a heightened interest in bioterrorism surveillance. The years immediately following these attacks were met with increased awareness and funding from the federal government. This paper will focus on the methods that we can use to prepare ourselves and detect these bioagent attacks.

The anthrax attacks of 2001, the SARS outbreak in 2004, and the recent H1N1 Influenza outbreak remind us that an essential component of preparedness for bioterrorism includes surveillance methods that can detect and monitor the course of an outbreak and thus minimize associated morbidity and mortality [1]. Surveillance of a population can be achieved in several ways. Syndromic surveillance occurs by monitoring clinical manifestations of certain illnesses. This type of surveillance occurs when health-related data, like International Classification of Diseases Ninth Revision (ICD-9) codes, are analyzed to signal possibility of an outbreak. Laboratory surveillance occurs by looking for certain markers or laboratory data. The Laboratory Response Network (LRN) is the United States' laboratory system for detecting, confirming, and reporting bioterrorism agents. Within the LRN, sentinel laboratories are tasked with singling out suspicious specimens for further testing in higher-tier labs. Environmental surveillance is the process by which the ambient air or the environment is continually sampled for the presence of biological agents [2].

Unfortunately, the practice of bioterrorism surveillance remains poorly studied. A recent systematic review of 29 biosurveillance systems concluded that there is insufficient evidence to determine which of these systems is best [1]. One thing is known. Whether it is an astute clinician like the one who made the first diagnosis of anthrax in 2001 or the complex chemical lab techniques that are used to detect plague, we must maintain our ability to identify and respond to a biologic terrorist attack.

BACKGROUND

Surveillance is recognized as the single most important public health instrument for identifying public health events of global concern, particularly infectious diseases that are emerging [3]. Not only is the use of surveillance helpful for bioterror attacks, but also the information generated by surveillance systems is also useful in the recognition and response to emerging

Citation: Nicholas E. Kman and Daniel J. Bachmann, "Biosurveillance: A Review and Update," Advances in Preventive Medicine, vol. 2012, Article ID 301408, 9 pages, 2012. doi:10.1155/2012/301408

infectious diseases. These epidemics are not related to traditional bioterror agents but their public health significance can be equally alarming. The recent H1N1 Influenza outbreak is a prime example of this.

The four functions of basic surveillance include (1) detecting cases of disease in specific populations and reporting the information, (2) analyzing and confirming reported case information to detect outbreaks, (3) providing timely and appropriate responses at the local/regional level to allow appropriate national level prevention and control of disease outbreaks, and (4) providing epidemiologic intelligence information to assist in long-term management of public health and health-care policies and programs [3].

Surveillance in health care refers to the continual systematic collection, analysis, interpretation, and dissemination of data [4]. Early methods of public health surveillance have been passive and voluntary. This process occurred when patients were diagnosed with a reportable communicable disease and local health departments were notified by clinicians, hospitals, or laboratories. Time would pass as information meandered through local and state health departments. Although many of the key components of surveillance occur at the local level, it takes many working parts for this to occur in a timely fashion.

Passive surveillance is an important component to global biosurveillance. It has the advantages of being inexpensive, easy to implement, and free of technologic barriers. However, it likely is not rapid and accurate enough to be used alone to respond to a bioterrorist attack. Passive surveillance is used best with other methods to quickly identify the treat and institute public health protection measures such as immunization, prophylaxis, and quarantine.

Active surveillance is the method of tracking emerging infectious disease threats. Active surveillance involves outreach to actively collect disease information from specific groups, such as sentinel medical providers or hospitals. Typically, active surveillance is undertaken to look for a specific disease. Active surveillance is more labor intensive and requires more public health resources than passive surveillance [5].

Systems for bioterrorism surveillance for public health require 3 key features: timeliness, high sensitivity and specificity, and routine analysis of data [1]. Timeliness of diagnosis is vital as the effectiveness of most treatments hinges on early detection. To this end, the electronic collection and reporting of surveillance data has improved detection as compared with manual methods [1]. High sensitivity is necessary as, without this, systems may fail to detect cases of bioterrorism-related illness which could result in delays in detection. On the opposite end, systems with inadequate specificity may have frequent false alarms, which will result in costly public health responses. Using the example of a food-borne illness outbreak, a system with low sensitivity may miss the sentinel cases and not identify the trend until the outbreak is already widespread. This compromises the ability of the surveillance system to adequately mount an effective public health response to the outbreak. Using the same example, a system with low specificity may identify cases which are not truly related to an outbreak resulting in an unnecessary public health response with diversion of resources from other true outbreaks. Sensitivity and specificity are typically inversely related such that optimization of one characteristic is at some expense to the other. Striking the optimal balance between these two characteristics for any given surveillance system is difficult [1].

SYNDROMIC SURVEILLANCE

The first key to identifying a potential bioterrorism event is to maintain a strong index of suspicion. The initial cases of West Nile Fever Virus in 1999 and the deliberate release of anthrax in 2001 were ultimately diagnosed by astute clinicians working hand-in-hand with lab technicians, not by public health surveillance systems. Syndromic data are gathered before laboratory results are reported; therefore, health departments may be able to recognize increases in disease incidence before formal diagnoses are made and to respond to outbreaks early in their course. For this reason, the CDC, state and local public health agencies, and the US Government and military have invested heavily in syndromic surveillance.

Methods of syndromic surveillance include many clues and data points which public health personnel can use to identify patterns. Data sources such as nurse hotline calls, over-the-counter medication purchases, and chief complaints from emergency-department visits can monitor illness clusters [6]. Some other clues to suspicious events include sharp rises in the frequency or severity of communicable diseases, including those in animals. Additional red flags include an unusual cluster or age distribution, occurrence of rare diseases, presence or lack of exposure history, travel to an endemic location, unexplained deaths, or pathogens with unusual antimicrobial resistance [7].

In response to the events of 2001, new types of surveillance systems were developed to detect epidemics through population-based reporting of symptoms tracked by time and region [8]. Many cities and states in the United States use syndromic surveillance, which monitors nonspecific, prediagnostic indicators for disease outbreaks in near real-time to provide an early warning of infectious disease outbreaks in their communities. Syndromic surveillance systems (SSS) monitor descriptive data from clinical diagnoses, chief complaints, and behaviors (e.g., school and work absenteeism, illness-related 911 calls, emergency room admissions for symptoms indicative of infectious disease) to infer patterns suggestive of an outbreak [9]. A comparison of syndromic surveillance with traditional clinical recognition is presented in Table 1 [10].

Table 1: Characteristics of bioterrorism-related epidemics that affect detection through clinical recognition versus syndromic surveillance

Characteristics[a]	Clinical recognition[b]	Syndromic surveillance[c]
Duration and variability of incubation period	Broader distribution of incubation period increases likelihood that patients with short incubation-period disease would be diagnosed before a statistical threshold of syndromic cases is exceeded.	More narrow distribution of incubation period which leads to a steeper epidemic curve in the initial phases increase likelihood that statistical threshold would be exceeded sooner.
Duration of nonspecific prodromal phase	Shorter prodrome increases likelihood of recognition or diagnosis at more severe or fulminant stage.	Longer prodrome increases likelihood that increase in syndromic manifestations would be detectable and that recognition of more severe stage (at which a diagnosis is more apt to be made) would be delayed.
Presence or absence of clinical sign that would heighten suspicion of diagnosis	Presence increases likelihood of earlier clinical recognition and diagnosis (e.g., mediastinal widening on chest X-ray in inhalational anthrax or multiple cases of rare disease presenting at similar time).	Absence decreases likelihood that diagnosis would be considered clinically, increasing opportunity for earlier detection by means of syndromic surveillance.
Likelihood of making diagnosis in the course of routine clinical evaluation	If diagnosis is apt to be made in the course of a routine diagnostic evaluation (not dependent on clinical suspicion of specific bioterrorism infection), early diagnosis through clinical care is likely.	If diagnosis is dependent on the use of a special test that is unlikely to be ordered in the absence of clinical suspicion of diagnosis, then diagnosis in clinical care may be delayed, increasing the opportunity for early detection through syndromic surveillance.

[a]Infection or disease attributes that may affect detection of an epidemic.
[b]Increases likelihood of initial detection through routine clinical care and reporting.
[c]Increases likelihood of initial detection through syndromic surveillance.

The most important determinants of detection for any given SSS were analyzed in a methodological review of 35 evaluations of outbreak detection in automated SSSs [11]. These determinants are key to taking one or more high-volume data feeds and differentiating the outbreak cases or "signal" from the baseline cases or "noise." The determinants were subdivided into characteristics of the system and characteristics of the outbreak being monitored. While evaluations using natural outbreaks were best suited to answer qualitative questions, simulated outbreaks were also useful to allow greater flexibility and increased quantitative results [11].

The influential system characteristics identified included representativeness or sampling approach of the system, the outbreak detection algorithm, and the specificity of the algorithm. For example, systems that monitor a larger proportion of the population have a higher sensitivity for detecting an outbreak. Similarly, systems that only monitor one type of clinical setting—such as ED visits only—were less sensitive. Furthermore, the studies that relied on simulated outbreaks suggested that temporal surveillance was more sensitive when the algorithm considered multiple days of data at each decision point versus data from each day individually. Important determinants related to the outbreak included magnitude and shape of the signal and timing of the outbreak. Intuitively, signals with a rapid rise over a short period of time improved outbreak detection as compared with those that rose more slowly over time. The ideal magnitude of the signal for consistent detection is not clear. The studies indicated magnitudes ranging from 10% up to as much as 60%. Similarly, the influence of the timing of the signal was not consistent, though there was a better detection when the outbreak occurred in context of a lower baseline of activity [11]. Based on these characteristics, one could envision an ideal SSS that monitored a large population at multiple clinical venues over multiple days at a time and flagged signals with rapid rise over a low baseline to at least a magnitude of 10%.

Almost immediately after the terrorist attacks of September 11, 2001, The New York City Department of Health and Mental Hygiene (NYCDOHMH) collaborated with the CDC to initiate an emergency-department-based syndromic surveillance for agents [12]. The system looked for symptoms that could be associated with a bioagent release such as respiratory distress, rash, gastrointestinal symptoms, neurologic impairment, and sepsis. Providers filled out forms with patient data that were analyzed by epidemiologists. This system was up and running in 15 New York City ED's within 2 days of its conception.

Syndromic surveillance systems monitor health care utilization patterns using data collected in real time, usually electronically. One example of a SSS is the Electronic Surveillance System for the Early Notification of Community-Based Epidemics (ESSENCE), which automatically downloads ICD-9 codes from U.S. Department of Defense health care facilities [1]. This novel use of ICD-9 codes is one way to group patient visits into syndromes. There are more than 10,000 ICD-9 codes available [13]. Patient visits are grouped by ESSENCE algorithms into one of eight syndromes based on lists of selected ICD-9 codes. If an increase in number of visits for a syndrome is noted, the clinic can be contacted for more information and an investigation can be launched.

Started in November of 2003, BioSense is a CDC Internet-based syndromic surveillance application designed for the early detection of intentional and natural infectious disease outbreaks [12]. BioSense receives data electronically from several sources. The Department of Veterans Affairs and Department of Defense provide ICD-9 codes for visits to their facilities. Retail pharmacies provide sales information on over-the-counter medications, and Laboratory Corporation of America provides information on laboratory tests ordered.

After examination by CDC analysts, public health officials can access their summary reports.

Current SSSs monitor the average pattern of patients reporting to primary care physicians or emergency-departments and signal an alarm whenever the pattern changes. Reporting sources include emergency-departments, intensive care units, hospital admission and discharge systems, and laboratories [8]. The Rapid Syndrome Validation Project (RSVP) relies on physicians to enter data on patients presenting with a syndrome of interest into a computer that has a touch-screen interface with RSVP [14].

The Emergency Department is the most common clinical source for surveillance data, though other sources of data have been proven to be useful. The Real-time Outbreak and Disease Surveillance Laboratory (RODS) Pennsylvania is the biosurveillance system for the Commonwealth of Pennsylvania. In production since 1999, it monitors 3 million visits to emergency rooms from 137 emergency-departments a year and simultaneously monitors 1262 retail stores in Pennsylvania for disease outbreaks. By utilizing the National Retail Data Monitor (NRDM), they have found a strong correlation that exists between the purchase of over-the-counter (OTC) medications and emergency room visits for constitutional illnesses. This information is useful for predicting coming epidemics as the tracking patterns of influenza and seasonal gastrointestinal illnesses often precede trends in hospital data [15]. One study demonstrated that OTC electrolyte sales preceded hospital visits for gastrointestinal and respiratory illnesses by 2.4 weeks [16].

The Connecticut Department of Public Health has been effectively using an SSS based on unscheduled hospital admissions since 2001. The Hospital Admission Syndromic Surveillance (HASS) system monitors 32 Connecticut-based acute-care hospitals with required reporting for eleven syndromic categories. Daily monitoring of data with weekly comprehensive analysis allows identification of disease clusters and routine public health followup for further action or response [17].

Syndromic surveillance efforts have been expanded to include outpatient monitoring also. This type of system takes advantage of the experience of ambulatory care physicians, who are also likely to be among the first to encounter patients during the prodrome of any potential bioterrorism-related illness. One such system developed with a private large ambulatory multispecialty group practice in Eastern Massachusetts demonstrated that surveillance coverage of 5–10% of a region's population may be adequate to detect significant clusters of interest. Several ideal components of this particular system included the automated collection of information, the use of preexisting data from a standard healthcare database, and the minimal cost for its implementation and continuous administration [18].

Although most systems for syndromic surveillance are continuously collecting, analyzing, and reporting data, some systems are designed for short-term use at mass-gatherings thought to be terrorist targets. These SSSs are referred to as event-based or "drop-in" surveillance [1]. One such "drop-in" surveillance system studied by the Bioterrorism Preparedness and Response Program demonstrated fair-to-good agreement of patient classification into an appropriate syndrome category when comparing use of Emergency Department chief complaints to discharge diagnoses. The findings were suggestive that use of discharge diagnoses may increase surveillance validity for "drop-in" and even possibly automated surveillance systems [19]. It is thought that syndromic surveillance systems are best used synergistically with laboratory surveillance.

ALTERNATIVE SURVEILLANCE SYSTEMS

The Centers for Disease Control and Prevention (CDC) has pioneered surveillance systems for monitoring other indicators of disease beside the traditional symptom- and diagnosis-based data used for clinical and syndromic surveillance. One such system is the Early Aberration Reporting System (EARS). This is a free tool which has been utilized and modified in both cities (Boston, NYC, Los Angeles) and in state public health agencies (Georgia, Florida, Tennessee, North Carolina, and Mississippi). It uses nontraditional public health data sources including school absenteeism rates, over-the-counter medication sales, 911 calls, veterinary data, and ambulance run data [20].

One novel epidemiologic surveillance approach has been developed by Google Inc and the CDC during the influenza season of 2007-08. This system monitored the health-seeking behavior of millions of users per day in the form of queries to online search engines. Ginsberg et al. demonstrated use of their model to estimate influenza-like illness within 85–96% of CDC-reported actual illness prevalence for the mid-Atlantic region of the USA. The advantages of this internet-based system were that illness statistics were available with a reporting lag of only one day, compared to the 7–14 day lag of CDC surveillance reports [21].

Though the Google surveillance system was specifically designed to monitor for influenza-like illness, the concept is more broadly applicable to other infectious pathogens such as bioterror agents. In addition to earlier detection of outbreaks, other advantages include freely available information to both the public and the government officials, automated processing with near real-time dissemination, and relative inexpensiveness for operation. Unfortunately, the specificity of internet-based surveillance remains unclear and could create more issues related to a high false-positive rate. These systems also require large populations with adequate internet access across regions and socioeconomic classes [22].

Another emerging example of a web-based surveillance system is the HealthMap Project. This collaborative undertaking performs extraction, categorization, filtration, and integration of aggregated reports from multistream real-time internet surveillance data [23]. The round-the-clock process involves automated data mining assisted by analyst review and reclassification. This system specifically focuses on identifying the "breaking news" trends to avoid overwhelming public health officials with low-impact problems [23]. The HealthMap system was applied to the H1N1 outbreak of 2009 with impressive results. The time difference between report of suspected cases and confirmed cases of H1N1 influenza was tracked by country with an overall median lag time of 12 days [24]. This time period can and will significantly alter the impact of the subsequent public health response. Further integration of these types of innovative systems with more traditional surveillance offers the greatest promise for future surveillance of emerging diseases [24].

The true utility of the SSSs is the dissemination and integration of its main output: surveillance data. In 2007, the CDC's Office of Critical Information Integration and Exchange created of the CDC created the BioPHusion Center with a mission to provide a CDC-wide resource that facilitates the exchange, integration, and visualization of relevant information from a variety of sources to enhance agency and programmatic situational awareness for decision-making and early event detection. Its goal is to share timely and actionable information to public health programs and leaders at the national, state, local, tribal, and global levels. They use data from a wide variety of governmental, private, and other sources to create an integrated daily report of potential events available through their Public Health Information Integration Portal [25]. Other publicly accessed CDC resources for information exchange include the Epidemic Information Exchange (Epi-X) and the Public Health Information Network (PHIN).

LABORATORY SURVEILLANCE

Clinical laboratories have been the cornerstone of diagnosis in infectious diseases of public health importance. In 1999, the CDC, the Federal Bureau of Investigation (FBI), and the Association of Public Health Laboratories (APHL) established the Laboratory Response Network (LRN) of about 120 laboratories [26]. The RODS laboratory is an example of one such system used for active surveillance as well as research efforts in the field of biosurveillance [15]. The mission of the LRN is to maintain an integrated network of laboratories that are fully equipped to respond to acts of chemical or biological terrorism, emerging infectious diseases, and other public health emergencies [26, 27]. In addition to identifying agents, the LRN is responsible for developing protocols for the handling, identifying, and reporting of potential biological agents to other national security agencies [2].

The LRN includes federal laboratories (CDC), state and local public health labs, military labs (the United States Army Medical Research Institute for Infectious Diseases (USAMRIID)), food testing (FDA), environmental laboratories, veterinary laboratories (United States Department of Agriculture), and international laboratories (Canada, the United Kingdom, and Australia) [2]. The laboratories involved in the LRN are divided into levels A through D, based on capabilities and function. Table 2 describes the levels of labs and their function [2, 26,27].

Table 2: Laboratory divisions within the laboratory response network

Level A Laboratory (Sentinel Labs)	Tier 1	Approximately 2300 hospital and clinic labs were likely first to receive specimens. Role is to rule out and refer to a lab within LRN to confirm a diagnosis.
Level B Laboratory (Reference Labs)	Tier 2	Increased capabilities to confirm diagnoses of biological agent. County public health labs where role is confirmatory testing, initial susceptibility testing, and referral.
Level C Laboratory (Reference Labs)	Tier 2	Much like level B, State public health labs that confirm diagnosis and refer to national laboratory. There are approximately 160 reference labs (B and C).
Level D Laboratory (National Labs)	Tier 3	National laboratories whose primary responsibility is to further characterize the agent (CDC, USAMRIID have biosafety level IV (BSL-4) capabilities).

As described in Table 2 above, sentinel laboratories are the first tier of the LRN and are responsible for sorting through their daily routine clinical tests to find suspicious biothreat specimens. The response to a local outbreak is the first and, perhaps, most important level. Sentinel laboratories must operate using Biosafety Level 2 procedures and possess a class II certified biological safety cabinet [27]. These labs are staffed by workers with only basic sentinel lab training. When a suspected biothreat agent is identified by one of these workers, it is sent to the local and state public health labs that comprise the second tier of the LRN. These reference laboratories then perform rapid confirmatory testing while maintaining biosafety level (BSL-3) facilities [7]. Once this threat is confirmed, it is passed on to the third tier in the LRN, the national laboratories. The national laboratories are equipped with the most secure containment labs (BSL-4) that they can use if necessitated by the agent.

Effective communication between the laboratory divisions is essential to preparedness. Ongoing dialogue between clinicians, sentinel laboratories, and LRN reference laboratories is essential to confirm the diagnosis quickly. In the USA, if a sentinel laboratory cannot rule out a bioterrorism agent, then it must be referred to an LRN reference laboratory [7]. For cases garnering high suspicion, state public officials are typically contacted and specimens are transported under the jurisdiction of law enforcement.

The following principles described by Pien et al. guide clinicians with respect to sentinel laboratory evaluation of potential bioterrorism agents [7]. (1) The initial evaluating physician should obtain optimum specimen collection instructions from the sentinel laboratory and alert them to the possibility of dangerous pathogen. (2) To maximize speed, accuracy, and safety, sentinel laboratories should limit culture manipulation to what is required by LRN reference labs. (3) Labs are to not inoculate highly suspected smallpox, hemorrhagic fever viruses, alphaviruses, or any unknown viral agents of potential bioterrorism into cell culture. Local public health authorities or the CDC should be contacted prior to collection. (4) Labs are directed to not send environmental (e.g., packages, powders, letters, soil, or water), food, animal, or plant specimens to sentinel laboratories for analysis. Instead, these should be referred directly to a LRN reference laboratory. (5) Finally, to reduce the risk of laboratory-acquired infection, restrict manipulation of certain potential agents (e.g.,Francisella tularensis, Brucella species, Coxiella burnetii, Burkholderia mallei, and Burkholderia pseudomallei) to environments under certified class II biological safety cabinet or BSL-3 conditions [7].

The CDC is responsible for monitoring the reference labs via regular proficiency training [30]. It cannot do the entire job alone and delegates some of the work to states local municipalities. This introduces some pitfalls and inconsistencies to the LRN. Many states and locals have different laws regulating this reporting. Additionally, as some of the sentinel labs are privately owned, only moderate oversight can occur at this level. To this end, a recent survey showed that only 73.8% of reporting labs indicated that they had sufficient personnel, equipment, and training to respond to a bioterrorism event [27]. Another study performed exercises with three category A organisms of bioterrorism (Anthrax, Plague, and Tularemia). In this study, sentinel laboratories only correctly identified 84% of bioterrorism agents [26]. This study showed that sentinel lab performance is improving, but still not likely at this optimal goal.

ENVIRONMENTAL SURVEILLANCE

There are two categories of environmental detection systems currently in existence, the remote or standoff detection of aerosol clouds and the point detection systems of the environment [2].

Remote Detection Systems

One way that remote detection systems monitor for potential biothreats from a distance is by the observation of aerosolized masses or clouds. Finding and evaluating the contents of a cloud is referred to as "standoff" detection [28]. On its most basic level, these detectors aim to alert military or civilian public health personnel to the presence of an approaching cloud. After the initial identification of the cloud, a more detailed assessment of the contents, such as water droplets, inert inorganic material, dead biotic particulates, or nonpathogenic microbes, is pursued [28]. Remote or standoff detection surveillance systems include cloud recognition by Doppler radio and radar, the Army's long- and short-range biological standoff detection systems. The Army's standoff detection systems are capable of detecting aerosol clouds from long distances, as well as determining their composition using ultraviolet light reflectance [2].

Point Detection Systems

Point detection systems are those that sample an environmental source, attempting to detect and identify the agent. Specific identification of a biologic agent by rapid diagnostics at the site of the attack can be done using immunologic assays, genetic assays, and mass spectrometry [2]. These systems can further be differentiated by the type and location of sample collected. For example, The Interim Biological Agent Detector is used on US naval ships to monitor the air for an increase in particulate concentrations [1]. Biowatch is an example of an environmental detection system that takes aerosol samples from locations in fixed sites, such as airports or public buildings.

Biowatch

In July of 2003, the Department of Homeland Security (DHS), the Environmental Protection Agency (EPA), and the CDC introduced the Biowatch program—a federal monitoring system intended to speed detection of specific biological agents that could be released in aerosolized form during a biological attack. Biowatch air sampling devices are deployed in 31 major U.S. cities. The air samples typically are tested daily for signs of the particular biological agents being monitored [2, 29].

The core purpose and intent of Biowatch is to hasten the public health response to a covert bioattack. This would allow rapid distribution of medical countermeasures, like antibiotics or vaccinations, thereby saving lives [30]. To this end, there are 500 air filters in these 31 urban areas that work as Biowatch sensors. These sensors have also been deployed to select indoor venues and are used to monitor mass-gathering events, such as the Super Bowl. This nationwide surveillance system uses distributed aerosol collectors to capture airborne ppapers onto removable dry filters that are transported daily to LRN laboratories for analysis [30]. An expanded deployment of the same technology in 2005 was referred to as Generation 2 Biowatch. Generation 2 Biowatch reportedly can sample and report detection from 10 to 36 hours [31]. Biowatch sensors are intended to be integrated into a complex network of environmental monitoring, medical surveillance activities, and public health response. It is thought that this integration of public awareness information, as well as syndromic, laboratory, and environmental surveillance technologies and systems, would be the best defense against a bioterrorist attack [2, 31].

Biowatch is not perfect. As currently operated, Biowatch filters are collected every 24 hours and delivered to local laboratories, where they are analyzed according to prescribed protocols. If this analysis recognizes one of the five biothreat agents that the system is designed to detect, it is termed a Biowatch Actionable Result (BAR). Laboratories report BARs to local public health officials, who must then decide how to respond. This decision is not taken lightly. The decision to treat a BAR as evidence of a bioattack could have huge consequences if it were a false alarm, including destructive impacts on the community's confidence in the public health system. Since 2003, there have been a number of BARs, though none have been the result of a biological attack. In some BAR cases, Biowatch samples contained material that was genetically similar to that found in Biowatch target organisms. These cases turned out to be from microbes that are present in the ambient environment but do not represent a threat to humans. Progress has been made in developing lab tests that distinguish these close relatives of bioweapons and work on more specific lab assays is ongoing [31].

Warnings from Biowatch would only be timelier than current health care systems under specific circumstances. Those are if a large-scale aerosol attack were to use certain biological agents and occur where Biowatch is deployed and if Biowatch successfully detects the biological agent [29].

Generation 3 Biowatch is currently in development. The next evolution of environmental sensor technology has been referred to as a "lab in a box" [31]. Gen 3 Biowatch would be more sophisticated than the current Biowatch sensors, with the ability to automatically collect outdoor air samples, perform molecular analysis of the samples, and report the results electronically to provide near-real time reporting. The target requirements for Generation 3 are reduction of time to diagnosis to 4 hours, increasing targeted biothreat agents monitored, reducing unit procurement costs down to $80,000 per detector unit, and detection sensitivity and false-positive rates remaining consistent with the current system's performance [31].

COST/BENEFIT

A bioterrorist incident is considered a low probability but high-cost event. The costs are high, because many agents go undetected until the onset of symptoms when treatment is less effective and more expensive [32]. That said, it is not economically feasible to the government to undertake a blanket deployment of biosensors [2]. Despite the growing cost, Congress continues to pass legislation intended to strengthen the nation's biological surveillance by increasing funding of federal and state biological surveillance. A 2007 DHS report documented that since the events of 2001, almost $32 billion has been allocated to biodefense and biosurveillance in the USA alone [3].

As of 2005, Biowatch costs per year were approximately $13,672,096. This figure includes labor costs, site upgrades, supplies, travel, training, and other operation and maintenance costs [32]. Most agree that this cost is justified if the probability of a bioterrorism incident remains high as the benefits of Biowatch improve.

As the Biowatch network is presently planned to expand with greater capability, this will increase the costs of the Biowatch Generation 3 system as compared to the currently deployed Generations 1 and 2 systems. Considering the operational complexity of current US biosurveillance systems, it is imperative that the operational advantages and feasibility of the proposed system be carefully evaluated and that actual performance of Generation 3 be tested in field conditions before large technology acquisition investments are made. The Department of Homeland Security will continue to work collaboratively to conduct and oversee developmental and operational tests of Biowatch 3 [31].

LIMITATIONS/CURRENT CHALLENGES

Despite the massive increase in funding and resources that has catapulted our capability for increased biosurveillance over the past decade, both in the US and abroad, there remain several challenges that must be addressed. The value of disease surveillance systems to public health officials is greatest when several systems are used together. The primary limitation of disease surveillance at this point is the limited coordination and lack of interoperability among the various private and federal surveillance systems.

As part of the 9/11 Commission Act, the National Biosurveillance Integration Center (NBIC) was created within the Department of Homeland Security to integrate information and support an interagency biosurveillance community. A 2009 report from the US Government Accountability Office on the state of biosurveillance and resource use concluded that there exists confusion, uncertainty, and skepticism around the value of the interagency community, as well as the mission and purpose of the NBIC within that community. Furthermore, there was a lack of clarity about roles, responsibilities, joint strategies, policies, and procedures for operating across agency borders [33].

Each individual system provides useful information, though no single system is complete [5]. Since it is possible to travel to most places in the world in less time than the incubation period for many infectious diseases, our networks must be expanded to allow for global surveillance [9]. The World Health Organization produced a major overhaul of their International Health Regulations in 2005 with a specific focus on the coordination of the global public health response to natural disasters, accidental release, or deliberate use of biological and chemical agents that can affect global public health [3]. But this cooperation must exist at every level—local, state, federal, and international—to maximize the effects of surveillance.

Another challenge of our current surveillance approach is the consequences of false-positive activation. These systems must be designed for high sensitivity given the overlap of commonplace pathogens with potential bioterror agents. Unfortunately, this may often sacrifice the specificity of the systems. The false alarms may be due to technical malfunctions or to naturally occurring events, such as the detection of anthrax in areas with large concentrations of cattle [32]. The subsequent mobilization of significant resources is not only costly but can be very distracting and generate overwhelming public distress.

The acquisition of data to fuel a surveillance system, especially the syndromic and clinical-based ones, may be challenged by concerns for privacy of protected health information (PHI). Though the Health Insurance Portability and Accountability Act (HIPAA) Privacy Rule allows for essential exchanges of health data during a public health emergency, the flow of PHI may be slowed by misunderstandings of the Privacy Rule's accounting requirement. This obstacle regarding HIPAA exceptions requires education of the necessary parties prior to the event of a bioterrorist attack [34].

CONCLUSIONS

Halting the spread of a bioterrorism attack will take a combination of the surveillance systems described above. It is only through active study, proper funding, and creative invention that we will be able to improve these systems.

REFERENCES

1. D. M. Bravata, K. M. McDonald, W. M. Smith et al., "Systematic review: surveillance systems for early detection of bioterrorism-related diseases," Annals of Internal Medicine, vol. 140, no. 11, pp. 910–922, 2004.·
2. M. Karwa, B. Currie, and V. Kvetan, "Bioterrorism: preparing for the impossible or the improbable,"Critical Care Medicine, vol. 33, no. 1, pp. S75–S95, 2005. ··.
3. C. Castillo-Salgado, "Trends and directions of global public health surveillance," Epidemiologic Reviews, vol. 32, no. 1, pp. 93–109, 2010.
4. D. J. Persell and C. H. Robinson, "Detection and early identification in bioterrorism events," Family and Community Health, vol. 31, no. 1, pp. 4–16, 2008.
5. T. K. Sell, "Understanding infectious disease surveillance: Its uses, sources, and limitations," Biosecurity and Bioterrorism, vol. 8, no. 4, pp. 305–309, 2010.

6. L. Uscher-Pines, C. L. Farrell, S. M. Babin et al., “Framework for the development of response protocols for public health syndromic surveillance systems: case studies of 8 US states,” Disaster Medicine and Public Health Preparedness, vol. 3, no. 1, pp. S29–S36, 2009. ··. ·
7. B. C. Pien, J. R. Saah, S. E. Miller, and C. W. Woods, “Use of sentinel laboratories by clinicians to evaluate potential bioterrorism and emerging infections,” Clinical Infectious Diseases, vol. 42, no. 9, pp. 1311–1324, 2006.
8. V. Sintchenko and B. Gallego, “Laboratory-guided detection of disease outbreaks: three generations of surveillance systems,” Archives of Pathology and Laboratory Medicine, vol. 133, no. 6, pp. 916–925, 2009..·
9. E. R. Choffnes, “Improving infectious disease surveillance,” in Bulletin of Atomic Scientists, University of Chicago, Chicago, Ill, USA, 2008, http://www.thebulletin.org/print/web-edition/op-eds/improving-infectious-disease-surveillance.
10. J. W. Buehler, R. L. Berkelman, D. M. Hartley, and C. J. Peters, “Syndromic surveillance and bioterrorism-related epidemics,” Emerging Infectious Diseases, vol. 9, no. 10, pp. 1197–1204, 2003.·
11. D. L. Buckeridge, “Outbreak detection through automated surveillance: a review of the determinants of detection,” Journal of Biomedical Informatics, vol. 40, no. 4, pp. 370–379, 2007.
12. S. M. Borchardt, K. A. Ritger, and M. S. Dworkin, “Categorization, prioritization, and surveillance of potential bioterrorism agents,” Infectious Disease Clinics of North America, vol. 20, no. 2, pp. 213–225, 2006.
13. J. A. Betancourt, S. Hakre, C. S. Polyak, and J. A. Pavlin, “Evaluation of ICD-9 codes for syndromic surveillance in the electronic surveillance system for the early notification of community-based epidemics,” Military Medicine, vol. 172, no. 4, pp. 346–352, 2007.
14. Zelicoff, J. Brillman, D. W. Forslund, et al., The Rapid Syndrome Validation Project (RSVP), Sandia National Laboratories, Albuquerque, NM, USA, 2001.
15. “Real-time Outbreak and Disease Surveillance Laboratory at the University of Pittsburgh,” 2011, https://www.rods.pitt.edu/site/content/blogsection/9/69/.
16. W. R. Hogan, F.-C. Tsui, O. Ivanov et al., “Detection of pediatric respiratory and diarrheal outbreaks from sales of over-the-counter electrolyte products,” Journal of the American Medical Informatics Association, vol. 10, no. 6, pp. 555–562, 2003.
17. Z. F. Dembek, K. Carley, A. Siniscalchi, and J. Hadler, “Hospital admissions syndromic surveillance—Connecticut, September 2001-November 2003,” Morbidity and Mortality Weekly Report, vol. 53, supplement, pp. 50–52, 2004.
18. R. Lazarus, K. Kleinman, I. Dashevsky et al., “Use of automated ambulatory-care encounter records for detection of acute illness clusters, including potential bioterrorism events,” Emerging Infectious Diseases, vol. 8, no. 8, pp. 753–760, 2002.
19. T. Fleischauer, B. J. Silk, M. Schumacher et al., “The validity of chief complaint and discharge diagnosis in emergency department-Based syndromic surveillance,” Academic Emergency Medicine, vol. 11, no. 12, pp. 1262–1267, 2004.
20. Early Aberration Reporting System (EARS), “Centers for Disease Control and Prevention,” Atlanta, Ga, USA, 2010, http://emergency.cdc.gov/surveillance/ears/.
21. J. Ginsberg, M. H. Mohebbi, R. S. Patel, L. Brammer, M. S. Smolinski, and L. Brilliant, “Detecting influenza epidemics using search engine query data,” Nature, vol. 457, no. 7232, pp. 1012–1014, 2009.
22. K. Wilson and J. S. Brownstein, “Early detection of disease outbreaks using the Internet,” CMAJ, vol. 180, no. 8, pp. 829–831, 2009.
23. J. S. Brownstein, C. C. Freifeld, B. Y. Reis, and K. D. Mandl, “Surveillance sans frontières: internet-based emerging infectious disease intelligence and the HealthMap project,” PLoS Medicine, vol. 5, no. 7, pp. 1019–1024, 2008.
24. J. S. Brownstein, C. C. Freifeld, E. H. Chan et al., “Information technology and global surveillance of cases of 2009 H1N1 influenza,” New England Journal of Medicine, vol. 362, no. 18, pp. 1731–1735, 2010.
25. H. Rolka, D. Walker, and J. Heitgerd, “An Overview of CDC’s OCIIX BioPHusion Program,” Integrated Surveillance Seminar Series, June 2009.
26. E. A. Wagar, M. J. Mitchell, K. C. Carroll et al., “A review of sentinel laboratory performance identification and notification of bioterrorism agents,” Archives of Pathology and Laboratory Medicine, vol. 134, no. 10, pp. 1490–1503, 2010.
27. B. T. Kalish, C. A. Gaydos, Y. H. Hsieh et al., “National survey of Laboratory Response Network sentinel laboratory preparedness,” Disaster Medicine and Public Health Preparedness, vol. 3, no. 1, pp. S17–S23, 2009.
28. M. Kosal, “The basics of chemical and biological weapons detectors: research story of the week,” Centers for Nonproliferation Studies, November 2003, http://cns.miis.edu/stories/031124.htm.
29. Institute of Medicine, “Biowatch and Public Health Surveillance: Evaluating Systems for the Early Detection of Biological Threats,” http://www.iom.edu/Reports/2010/BioWatch-Public-Health-Surveillance-Evaluating-Systems-Early-Detection-Biological-Threats.aspx.
30. J. F. Regan, A. J. Makarewicz, B. J. Hindson et al., “Environmental monitoring for biological threat agents using the autonomous pathogen detection system with multiplexed polymerase chain reaction,”Analytical Chemistry, vol. 80, no. 19, pp. 7422–7429, 2008. ··.
31. Department of Homeland Security (DHS), “Testimony of Tara O›Toole Before the House Subcommittee on Homeland Security Appropriations, on Biosurveillance,” http://www.dhs.gov/ynews/testimony/testimony_1271436311919.shtm.
32. H. Schneider, “Protecting public health in the age of bioterrorism surveillance: is the price right?” Journal of Environmental Health, vol. 68, no. 5, pp. 9–13, 2005.

33. GAO report on biosurveillance, “Developing a collaboration strategy is essential to fostering interagency data and resource sharing,” US General Accounting Office, Washington, DC, USA, 2009, http://www.gao.gov/new.items/d10171.pdf.

34. J. G. Hodge Jr., E. F. Brown, and J. P. O›Connell, “The HIPAA privacy rule and bioterrorism planning, prevention, and response,” Biosecur Bioterror, vol. 2, no. 2, pp. 73–80, 2004.

CHAPTER 26

BIOCRIMES, MICROBIAL FORENSICS, AND THE PHYSICIAN

Steven E. Schutzer, Bruce Budowle, Ronald M. Atlas

Attending physicians, regardless of their specialty or setting of practice, may suspect or learn that their patient has been attacked with a biological agent. In such cases, it is important to be aware of the interactions that may occur with law enforcement, and of the role that the evolving science of microbial forensics [1] may play in the investigation. Physicians are in key positions to preserve critical evidence and, thereby, contribute to the chain of custody, and, at the same time, offer suggestions to help develop the field of microbial forensics.

This article provides guidance to physicians who believe that one of their patients is a victim of an act of bioterrorism or of another biocrime, and who are compelled by law, or with the patient's consent, wish to assist law enforcement in an investigation. In this regard, there are instructive lessons that can be learned from cases of past biocrimes and from analogies to more familiar cases of sexual assault and child abuse (Table 1).

RECENT BIOCRIMES

The most widely publicized bioterrorism event in the United States was the US anthrax mail attacks of 2001. In this case, an astute physician diagnosed the index case of systemic anthrax [2] that set off national panic and a federal investigation that is still ongoing. Highly publicized in Europe was the assassination of a Bulgarian exile in London using ricin, a toxin extracted from castor beans, [3,4] which was delivered to him using an umbrella. Less publicized are other biocrimes such as the case of a laboratory worker in Texas intentionally infecting hospital co-workers with Shigella dysenteriae. Biocrimes are much less likely to occur than many infectious diseases, such as HIV/AIDS. In fact, it is often the abundance of naturally occurring infections that may make the detection of a biocrime difficult. However, there are also documented cases of non-bioterrorism biocrimes (Box 1) [5,6], which far exceed the number of documented bioterrorism acts, as well as many hoaxes where physicians and clinical laboratories were involved in determining if there was a real threat to exposed individuals.

Box 1. Examples of Non-Bioterror Biocrimes

- Intentional Salmonella typhi food contamination in France from 1910 to1918
- A Yersinia pestis attack by injection in 1933 in India
- Deliberate use of HIV-infected blood and secretions to inflict harm

Source: [5,6].

Citation: Schutzer SE, Budowle B, Atlas RM (2005) Biocrimes, microbial forensics, and the physician. PLoS Med 2(12): e337. doi:10.1371/journal.pmed.0020337

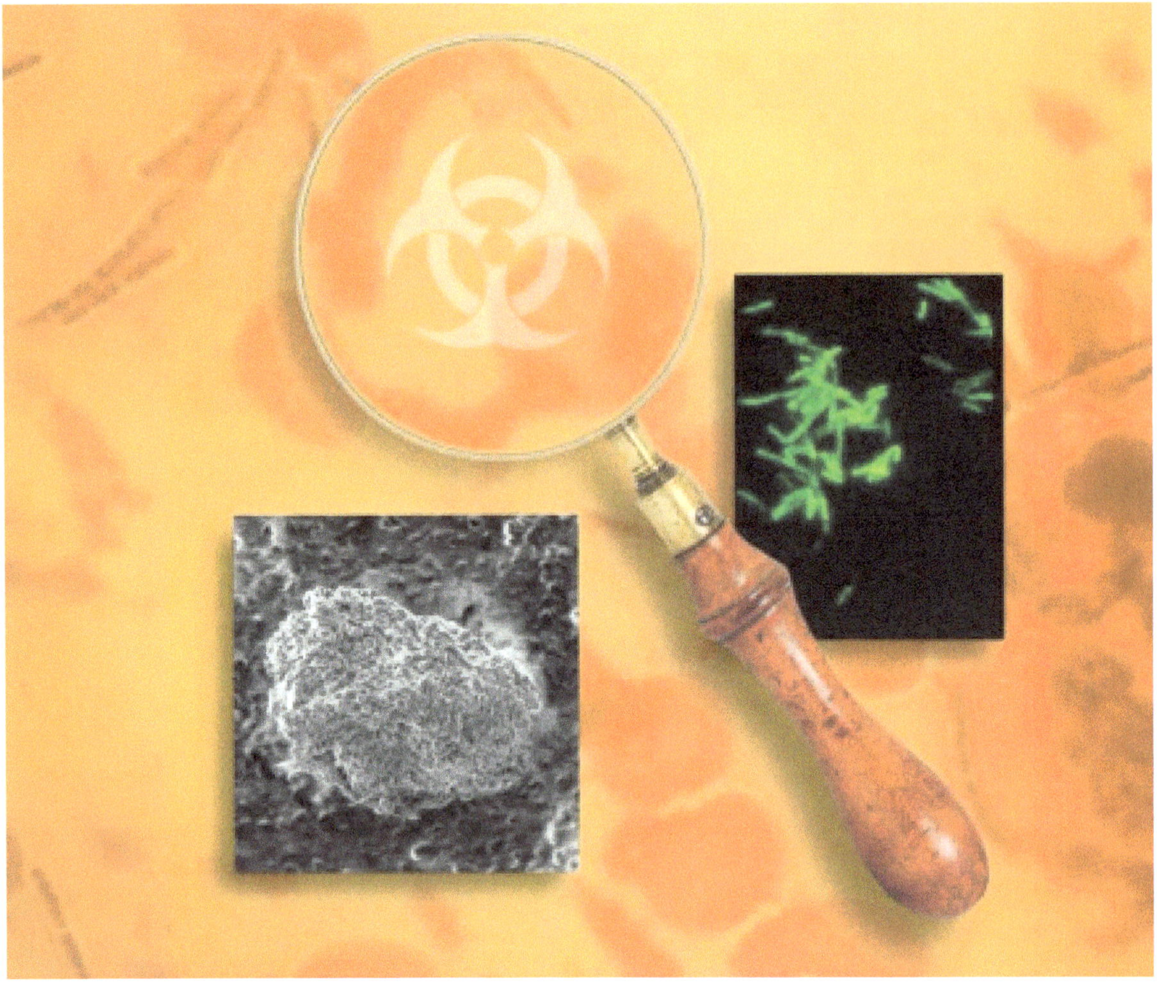

A biocrime is similar to an assault crime, except, instead of a gun or knife, the weapon is a pathogen or a toxin. In the US, acts of bioterrorism are federal crimes that are governed by different responses by law enforcement and public health agencies than those that govern other biocrimes [7]. Most biocrimes and their subset of bioterrorism cases will involve public health agencies because of the nature of a disease threat to the public.

Table 1: Comparison of Biocrimes, Sexual Assaults, and Child Abuse in the US

Variable	Biocrime	Sexual Assault	Child Abuse (Physical, Mental, and/or Sexual)
Validated analysis kit available	None currently available.	Sexual assault analysis kit is available. It is oriented at collection of semen, saliva, sperm, hair, fingernails, and skin, which in general are amenable to the same preservation process of sample evidence. Chain-of-custody protocols are outlined with kit.	Depends on the amount of time since last incident with victim. Documentation by photography and imaging might be considered.
Time factors and limits	Time may be of the essence to recover isolate closest to inoculum, in case of a microbe, or greatest quantity of toxin.	Viable motile sperm may no longer be found after eight hours in general; found up to two to three days in cervical mucosa; nonmotile sperm (very limited in number) found up to 17 days in cervical mucosa.	Delays may occur as victim does not understand, is fearful, or is powerless to report incident. Perpetrators may interfere, as they do not want to have the act discovered.
Established standardized training courses available	None available for medical personnel.	Multiple courses available.	Multiple courses available.
Privacy issues	Yes	Yes	Yes
Primary forensic evidence	Laboratory diagnostics, including identification of etiologic agent: causative microorganisms and their nucleic acids, additives, and stabilizers.	Human DNA, hair, saliva, skin, and spermatozoa.	Photographic documentation and possible physical evidence.
Potential factors for compromise of forensic evidence	Decontamination processes of bathing, personal hygiene, and household cleaning may destroy primary microbial material and extra material evidence (throwing out envelopes, cleaning surfaces, or washing clothes). Contamination with other microbes, and preculture therapy with antibiotics.	Bathing, personal hygiene, and household cleaning may destroy material evidence (semen, hair, skin, and saliva, which contain DNA and other markers, and fibers; destruction of evidence by washing clothes, bedding, or carpet). Other sexual contacts (consensual) in immediate time before or after assault; and spermicides.	Some of the sexual assault factors apply here.

The numerous hoaxes that are biocrimes include white powders found in letters that proclaim the presence of anthrax, and threatening notes claiming ricin contamination of baby food. Ricin currently appears to be a prevalent bioweapon, particularly as a tool for extortion. These potential ricin threats demonstrate the impact of bioterrorism on patient care: physicians had to monitor patients who might have ingested the poisoned food and, hence, were distracted from caring for other patients [8]. Hoaxes can be challenging for the physician, who must distinguish between symptoms and signs that could be toxin-related and those that are just variants of normal health. This challenge was compounded in a recent case where there were trace findings of an inactive form of ricin in baby food. Nonetheless, a hoax is also a crime, and the physician should not discard any evidence simply because material appears innocuous.

As environmental biothreat sensors expand into the workplace and public places, they will be relied on as sentinels for possible biothreat releases. In the US, sensors for anthrax are being placed in some postal offices and at some major public events, and air monitoring is being carried out in major cities and transit systems (the BioWatch system). When such sensors indicate a possible bioterrorist attack, even if the signal is later found to be a false signal, the public will react, and are likely to seek out their own physicians for medical diagnostics, therapy, and advice. Potential cases that follow a public alert warrant evaluation, collection of patient samples, and possibly institution of prophylactic treatments until an alert is deemed a false alarm. Such a situation occurred in March 2005 following a presumptive positive detection of anthrax in a US Department of Defense mailroom. This situation necessitated treating over 900 potential victims with antibiotics as a precaution. However, despite the newer sensor technologies, physicians will likely remain the best and definitive authorities on the presence of an infection.

Reporting a Biocrime

Patients who believe that they have been a victim of a biocrime generally want both a medical and law enforcement response—that is, they want medical treatment, and they want the perpetrator to be found, prosecuted, and punished. Requirements for how physicians and other members of the health-care system initially report a suspected biocrime are governed at the local level (but regulations are ever-changing, and it is important to check current information at the time it is needed). Many US state regulations mandate that diagnostic laboratories report preliminary isolates of certain microbes to public health authorities, who, in turn, are sometimes compelled to notify law enforcement, depending on the isolate. Certain states have laws that mandate that a physician report specific diseases or unusual clinical manifestations to public health authorities and, in some cases, directly to law enforcement. The latter requirement is analogous to what is expected when there is suspicion of child abuse or gunshot wounds, and failure to notify authorities is itself a crime. The regulations of the Health Insurance Portability and Accountability Act (HIPAA) of 1996 established national standards for health-care transactions as well as security and privacy policy for health-related information, and physicians should be aware of these standards. However, even though these regulations severely restrict the release of patient health–related information, they still permit physicians and health-care facilities to release otherwise protected health information in instances of suspected crimes or threats to public health [9,10].

Given the limited guidance for reporting suspected biocrimes, physicians could face several dilemmas. For example, some patients may not want to report a crime or a disease condition, yet may have reasonable concerns. At the beginning of the HIV/AIDS epidemic, when no treatment was available, a diagnosis of HIV infection caused many patients to fear discrimination and loss of employment, a situation that persists in many parts of the world. Similar questions may arise in bioterrorism events about insurance coverage if an event is deemed an act of war. Communication between physician and patient should help the patient understand the pros and cons of notifying law enforcement of a suspected biocrime, including whether withholding notification could place others at risk. At the very least, discussion can strengthen the doctor–patient relationship.

Fear and embarrassment of reporting potential false alarms to law enforcement or public health authorities may also be a concern for physicians and patients. But if a suspicion is not reported, a critical situation may go unrecognized and continue to worsen. Early notification to law enforcement authorities may provide valuable time and direction for investigative leads. It is expected that there will be many more negatives (false alarms) than positives when alerting law enforcement. Early reports to public health authorities may stem an epidemic. It is a misconception that you must wait for a firm diagnosis before reporting a potential case to authorities. There are many other misconceptions about biocrimes (Table 2).

Guidance in the US concerning the reporting of suspicions of biocrimes is provided by the Centers for Disease Control and Prevention (CDC; http://www.cdc.gov), the Federal Bureau of Investigation (FBI; http://www.fbi.gov), and the Department of Homeland Security (DHS;http://www.dhs.gov) (Table 2). A joint statement by the FBI, the CDC, and the DHS advises calling the FBI and public health authorities if a suspicious situation arises [11]. Local public health departments are advised to notify the FBI before notifying the CDC. Specifically, "the FBI must be notified for any case of smallpox or pulmonary anthrax, uncommon agent or disease, an illness caused by a microorganism with markedly atypical features, an illness due to aerosol or food or water sabotage, as opposed to a usual transmission route, one or more clusters of illnesses that remain unexplained after a preliminary investigation; deliberate chemical, industrial, radiation or nuclear release" [12]. Calls or online tips should be directed to the FBI (https://tips.fbi.gov). Interpol and the World Health Organization are also developing response plans to help the public and physicians respond to suspected biocrimes and acts of bioterrorism (http://www.who.int/topics/bioterrorism/en).

Table 2: Possible Misconceptions about Biocrimes in the US

Misconception	Reality
In instances of suspected biocrimes, the FBI and other law enforcement agencies do not want to be notified until the diagnosis is confirmed.	The FBI and CDC Web sites both encourage notification. The FBI provides for online notification, and suggests contacting the local FBI office (information on Web and phone directories). A document issued jointly by the FBI, DHS, and US Department of Health and Human Services/CDC gives guidance on how to respond to a suspicious letter or container with a potential biological threat (see [11]). The CDC requests that all incidents of apparent or threatened bioterrorism be voluntarily reported to the CDC by state public health officials, immediately following notification of the FBI and local law enforcement agencies. State health officials should call the 24-hour notification telephone number ([770] 488-7100) at the CDC Emergency Preparedness and Response Branch. For general inquires to CDC telephone 1-800-CDC-INFO. Even with simple suspicious packages, FBI instructions for biological or chemical materials are to isolate—don't handle. Call 911 (Police).
Physicians must be hypervigilant to make a diagnosis or develop a differential diagnosis in which a biocrime is included, especially during the initial encounter.	Symptoms and signs from a covert attack with a biological agent will often not be distinct during the immediate encounter in a busy emergency room or office visit. Suspicion increases after initial symptoms, when signs evolve with time or a laboratory report raises suspicions of an unusual infection for the geographic area or season. The patient may be admitted to the hospital under a different doctor's care or discharged. News reports may raise suspicions, and reports of isolated cases to the public health authorities may reach a threshold to alert suspicion (advantage of notifying health department early). In the case of an overt attack or threat, e.g, a powder in a letter, or a threat note (also a federal crime with a hoax), the situation is analogous to a sexual assault victim who wishes to report a crime—action is mandatory. Both law enforcement and public health authorities need to be informed simultaneously, according to state law.
It is likely to be the infectious disease specialist or emergency room physician, trained to recognize manifestations caused by microbes, who will make the diagnosis.	Depending on the organ system involved, the patient may present to a different specialist. Some victims from the anthrax mail attacks presented to dermatologists for cutaneous lesions.
If you do not recognize the possibility that the patient is a victim of a biocrime, it is too late to obtain evidence that will be useful to determine who was responsible or that will hold up in court.	Although early recognition and early collection of evidence samples are best, collection of evidence at any stage may still have important probative value. As an infection evolves, the responsible microbe can be recovered.
You could be violating the patient's privacy and confidentiality.	State laws govern when the physician or the laboratory must report a disease, set of symptoms, or microbial isolate irrespective of patient consent. In instances where public health or national security is threatened or a crime is suspected, regulations of the HIPAA permit the release of certain private information to law enforcement or public health authorities when requested. In most cases of a biocrime, informed patients will likely be eager to cooperate. Thus, it will be less likely that a doctor must struggle with confidentiality and public welfare or legal obligations.
A patient who will not consent to notification of public health or law enforcement authorities is concealing criminal activities.	A patient may have legitimate concerns that they could be denied insurance coverage, stigmatized, or denied career advancement. This is analogous to fears about being diagnosed as HIV positive at the beginning of the US epidemic. Occupational accident resulting in exposure to a biological agent may raise issue of carelessness.

HIPAA, Health Insurance Portability and Accountability Act.

The Physician's Role in Collecting Evidence

Although finding the perpetrator of a crime is a law enforcement function, the actions of attending physicians can help with microbial forensics—the scientific discipline dedicated to analyzing evidence from a biocrime or an act of bioterrorism, and that seeks to authenticate a piece of the puzzle for attribution. Implicit in the term attribution is the identification of the responsible party or the exclusion of the innocent [13].

Many physicians are familiar with the treatment of sexual assault victims, and the need to collect and preserve evidence when the patient consents. Sexual assault analysis kits have been validated to preserve semen, saliva, hair, blood, and skin. They also provide instructions on how to maintain a chain of custody to ensure that there has been no tampering with the evidence. Chain of custody is the process that assures integrity of the evidence, and ensures that there is documentation of the time the evidence is handled and each individual handling or examining the evidence. Courses exist in crime scene investigation, evidence collection, and chain of custody of the evidence in suspected sexual assault cases, and there are often well-trained support personnel who can assist patients and physicians. Such evidence collection guidance and support structures are not well-developed for biocrimes.

In contrast to typical human DNA forensic investigations, with microbial forensics, a chain of custody might not be enacted at the initial stage of medical diagnostics. Good diagnostic practices could permit samples to be used as supporting evidence in a criminal investigation. In some cases, samples can be obtained subsequently under a stricter chain-of-custody process. Law enforcement authorities can assist with such documentation processes.

Microbial Forensics

Microbial forensics includes the full scope of forensic evidence, such as analyses of microbes, materials used to prepare, stabilize, and deliver the toxin or pathogen, and fingerprints, hair, fiber, and pollen [1,14]. The laboratory analyses used for microbial forensics may include molecular sequencing, microbiological cultures, biochemistry, electron microscopy, crystallography, and mass spectrometry. These analyses go well beyond those used for medical diagnoses and epidemiologic investigations [15]. They require, however, the same substances used by the physician for diagnostics, for example, body fluid samples and microbial cultures. In this regard, the physician and the clinical laboratory have critical roles in the collection and initial analyses of samples for microbial forensics.

In the 2001 anthrax letter attacks, the preservation of the initial and subsequent isolates enabled microbial forensic methods to identify the strain in the attacks as the Ames strain ofBacillus anthracis. Analyses were based initially on a method to identify variable-number tandem repeat sequences, and later on, whole genome sequencing. Comparisons with existing strains in culture collections narrowed the likely source to a laboratory as opposed to being obtained directly from nature [16–18]. Fortunately, the initial strain from the Florida patient (the index case) and strains isolated from other victims, as well as spores from the letters, were preserved for future analyses. Microbial forensic analysis is now able, with the help of specialized facilities such as the Institute for Genome Research (TIGR), to determine the whole genome sequence of the approximately 5 million bases of B. anthracis to identify the polymorphisms that may be signatures of the bioweapon [17].

A major thrust of microbial forensics will, therefore, be the analysis of nucleic acids that can relate the genome of the pathogen to specific sources. This analysis is analogous to human DNA forensic analysis, which is being widely used to prosecute criminals and to exonerate the innocent [19]. But there are important differences between the analyses of microbial genomes and those used in human DNA forensics. Because of the sheer number of potential pathogens that could be employed as a weapon, identifying genetic markers for microbes is a more daunting task than identifying human DNA. In the case of human identification, only one species is involved, and it is often possible to identify an individual person. Viruses and most bacteria are haploid. Microbes primarily reproduce asexually, but can also evolve by recombination, horizontal gene transfer, and gene duplication. Therefore, statistical methodologies and interpretation will require different tools than are currently used for comparing and estimating the rarity of (diploid) human DNA profiles [20,21]. Nevertheless, obstacles due to genetic complexity can be reduced by obtaining samples as early as possible.

If physicians suspect a biocrime, they should take steps to ensure the preservation of the diagnostic samples so that they are not prematurely destroyed. Physicians may also advise the patient to preserve additional material that may prove useful for a criminal investigation. Just as in sexual assaults, in a suspected biocrime, the patient's personal articles may carry traditional forensic evidence that is of equal value to the information revealed by the microbe itself. Unlike sexual assault evidence, in a suspected biocrime, procedures used to preserve one particular microbe may be deleterious for other microbes and for physical evidence (such as fingerprints, culture media, isotopes, hair, and environmental material). In addition, procedures useful for preserving one microbe may be insufficient to preserve another that may be unknown at the time.

History and Physical Examination

The physician's record of the patient's history and physical examination is evidence that can be expected to be part of any public health and forensic investigation (and subsequent legal proceedings) in either a true attack or a hoax. The physician's ability to interpret the clinical history and physical examination may go beyond differential diagnoses—for example, it can help establish timelines of exposure and of the evolution of disease, which will have forensic and public health implications. The physician is positioned to assist in identification and collection of evidence, and initiate the chain of custody that protects the integrity of the evidence (see supporting online material of [1]; [14,22]) or, at a minimum, maintains good medical practice, akin to that used for transfusion of blood products.

The Case of *Louisiana v. Schmidt*

The case of *Louisiana v. Schmidt*, in which HIV-infected blood was used as the weapon in an attempted murder [23,24], is instructive for the microbial forensics system. A vial of HIV-infected blood was found in the office of a suspect, a gastroenterologist. The challenge for microbial forensics was to provide evidence that this was, or was not, the source of the victim's HIV infection. HIV, an RNA virus, undergoes rapid mutation, so any direct genetic comparison of the donor source and the recipient (the victim) is complicated. In this case, analysis focusing on both rapidly and more slowly mutating genes of HIV proved to be useful. Examination of strains from the vial, the victim, and control samples (known samples from other patients with HIV residing in the same geographic region as the victim) revealed that the viral RNA from the victim was more closely aligned to that from the vial in the suspect's office than to isolates from other patients in the area. The clinical history and clinical laboratory data obtained by attending physicians provided supporting evidence of the uninfected status of the victim prior to this injection. The victim's prior HIV-negative status was documented by blood donation screenings and negative-HIV tests of sexual partners, prior to the injection. The evidence was presented in a US criminal court. Based on the composite epidemiologic and microbial forensic evidence presented, a conviction for attempted murder was obtained.

This case illustrates several points. Sample collection and documentation by the attending physician are paramount to the biocrime investigation. If an attending physician is suspicious about an acquired infection, especially with an organism that is known to mutate rapidly, more frequent sampling and preservation of those samples are important. The sample may contain other clues (the victim, in this case, also was allegedly injected with blood that was hepatitis C positive). These samples could be helpful both epidemiologically, if there were an outbreak, and forensically, if there were an intentional incident. Analysis of these patient samples and other specimens may determine who was the source, and who was the victim. This case showed that even though the earliest isolates were not obtained, when the possibility of a biocrime was considered, there was still sufficient time to obtain valuable specimens, even with this rapidly mutating virus. The case also illustrates that microbial evidence can be informative, but it is rarely the sole deciding evidence. When considered in conjunction with other evidence—in this situation, epidemiological and clinical data—the case was very strong.

Instructive lessons can also be found by reviewing other cases in the literature [5], such as the laboratory technician who poisoned her co-workers with a laboratory stock of Shigella dysenteriae type 2 in muffins [25], and the poisoning of salad bars to skew an election for political gain [26].

Although our primary focus has been on the role that the practicing physician can play, it is important to remember that medical examiners or coroners can also serve as sentinels for discovering acts of bioterrorism and biocrime, as well as collecting pertinent microbial forensic evidence [27]. They have the statutory authority to investigate deaths that are sudden, suspicious, violent, and unattended. Moreover, the medical examiner or coroner may encounter victims that were never examined by practicing physicians. Autopsies can be crucial for diagnosis of unknown infections and for acquiring evidence for subsequent criminal investigations [1,19]. For example, in 1979, in Sverdlovsk, USSR, at least 66 people died during an anthrax outbreak. The official Soviet government position was that the victims were infected by eating contaminated meat. Autopsy data were inconsistent with the proclaimed cause of death, and, instead, supported the proposition that the disease was inhalational anthrax due to an accidental aerosol emission from a secret military weapons facility [28].

Close working relationships should be developed between the medical examiner/coroner and public health and law enforcement entities to alert one another of possible outbreaks (whether natural or intentional) as soon as possible. For suspected attacks, the medical examiner/coroner should immediately collect case-specific death investigation information and establish a chain of custody. Fortunately, medical examiners and coroners have a long-established relationship with law enforcement. However, if there are questions regarding notification or evidence collection, the medical examiner/coroner can contact the proper public health and law enforcement entities (see Table 2).

CONCLUSIONS

Physicians and other health-care providers are positioned to recognize suspicious situations and alert public health and law enforcement officials. This alone may be the most important step physicians can take (Box 2).

Box 2. Measures That a Physician May Take toward Securing Evidence in Cases of Biocrimes

- Maintain primary role in caring for the patient, even at the risk of compromising evidence collection.
- Discuss the situation with the patient, including options for interaction with and disclosure to public health and law enforcement officials.
- If permitted by patient consent, or if required by law, alert as early as possible public health authorities and law enforcement, who can provide the necessary expertise or guidance to collect and preserve evidence.
- Do not assume one agency will notify the other in a time-sensitive period. Ensure that notification has occurred.
- Maintain well-documented medical records because documentation of history, physical examination, and patient course may constitute evidence.
- Obtain samples that may serve as evidence early, frequently, and under a defined chain-of-custody process.
- Once a biocrime is suspected, ensure that the clinical laboratory does not discard microbial isolates, but preserves them for forensic analyses or transfers them under a chain-of-custody procedure (along with accessory material such as the transport tube initially used to transport microbial isolates from patient to laboratory). Law enforcement and public health personnel participating in the Laboratory Response Network can provide this assistance.

In cases of biocrimes, physicians may interact with nonmedical authorities—who often do not fully appreciate that a trusting doctor–patient relationship is crucial for proper care and healing, and that information should be private. It is helpful to inform such officials about the importance of the doctor–patient relationship at the outset so they can be sensitive to the obligations of physicians. Just as with a sexual assault case, once there is recognition of the possibility of a bioterrorism act or other biocrime, the physician should discuss the entire situation with the patient, explaining what can be done with the consent of the patient and what actions physicians are mandated to take to comply with public health and legal requirements. This communication will likely strengthen the patient's relationship with the physician. Within the context of microbial forensics, if the patient consents, or the law requires it, the physician can facilitate preservation of evidence. To the extent possible, earlier, more, and serial sampling of evidence is best.

Physicians can ultimately serve their patients by acting, in the traditional role, as a healer, and by working with public health and law enforcement entities to help prevent further attacks and to achieve justice. As with sexual assaults, identification and conviction of the attacker can bring closure and provide a degree of security to the patient, who can then evolve from being a victim to being a survivor [29]. Physicians and their colleagues are likely to have creative ideas to contribute to the field of microbial forensics. Their input is encouraged and welcomed.

ACKNOWLEDGMENTS

The authors are indebted to Stephen A. Morse (CDC), Barbara K. Richardson (Mt. Sinai Medical Center, New York), and Suzanne Atkin (University of Medicine and Dentistry of New Jersey—New Jersey Medical School) for their suggestions and critical review of the manuscript.

REFERENCES

1. Budowle B, Schutzer SE, Einseln A, Kelley LC, Walsh AC, et al. (2003) Public health. Building microbial forensics as a response to bioterrorism. Science 301: 1852–1853.
2. Bush LM, Abrams BH, Beall A, Johnson CC (2001) Index case of fatal inhalational anthrax due to bioterrorism in the United States. N Engl J Med 345: 1607–1610.
3. Marks JD (2004) Medical aspects of biologic toxins. Anesthesiol Clin North America 22: 509–32. vii.
4. Mayor S (2003) UK doctors warned after ricin poison found in police raid. BMJ 326: 126.
5. Carus WS (1999) Bioterrorism and biocrimes: The illicit use of biological agents in the 20th century center for counterproliferation research. Washington (D.C.): National Defense University. Available: http://www.ndu.edu/centercounter/Full_Doc.pdf. Accessed 18 August 2005.
6. Budowle B, Murch RS, Chakraborty R (2005) Microbial forensics: The next forensic challenge. Int J Legal Med. In press.
7. Department of Health and Human Services (2003) Select agents and toxins. Title 42, Code of Federal Regulations, part 73. Washington (D. C.): Department of Health and Human Services.
8. Health Talk (2004) FDA responds to ricin baby food contamination. San Francisco: Health Talk. Available: http://www.healthtalk.ca/ricin_07282004_8273.php. Accessed 18 August 2005.
9. Annas GJ (2003) HIPAA regulations—A new era of medical-record privacy? N Engl J Med 348: 1486–1490.
10. Hodge JG, Brown EF, O'Connell JP (2004) The HIPAA privacy rule and bioterrorism planning, prevention, and response. Biosecur Bioterror 2: 73–80.
11. Federal Bureau of Investigation, Department of Homeland Security, Centers for Disease Control and Prevention (2004) Guidance on initial responses to a suspicious letter/container with a potential biological threat. Atlanta: Centers for Disease Control and Prevention. Available: http://www.bt.cdc.gov/planning/pdf/suspicious-package-biothreat.pdf. Accessed 18 August 2005.
12. Centers for Disease Control and Prevention (2001) Local health officer is informed of a bioterrorist incident or threat. Atlanta: Centers for Disease Control and Prevention. Available: http://www.bt.cdc.gov/emcontact/determine.asp. Accessed 18 August 2005.
13. Murch RS (2003) Microbial forensics: Building a national capacity to investigate bioterrorism. Biosecur Bioterror 1: 117–122.
14. United States Federal Bureau of Investigation Laboratory Division (1999) Handbook of forensic services. Washington (D.C.): United States Federal Bureau of Investigation Laboratory Division. Available: http://www.fbi.gov/hq/lab/handbook/intro.htm Accessed 18 August 2005.
15. Yeskey K, Morse SA (2003) Physician recognition of bioterrorism-related diseases. In: Roy MJ, editor. Physician's guide to terrorist attack. Totowa: Humana Press. pp. 39–46.
16. Keim P, Smith KL (2002) Bacillus anthracis evolution and epidemiology. Curr Top Microbiol Immunol 271: 21–32.
17. Read TD, Salzberg SL, Pop M, Shumway M, Umayam L, et al. (2002) Comparative genome sequencing for discovery of novel polymorphisms in Bacillus anthracis. Science 296: 2028–2033.
18. Pearson T, Busch JD, Ravel J, Read TD, Rhoton SD, et al. (2004) Phylogenetic discovery bias in Bacillus anthracis using single-nucleotide polymorphisms from whole-genome sequencing. Proc Natl Acad Sci U S A 101: 13536–13541.
19. Lander ES, Budowle B (1994) DNA fingerprinting dispute laid to rest. Nature 371: 735–738.
20. Budowle B (2004) Genetics and attribution issues that confront the microbial forensics field. Forensic Sci Int 146: S185–S188.
21. Budowle B, Chakraborty R (2004) Genetic considerations for interpreting molecular microbial forensic evidence. In: Doutremepuich C, Morling N, editors. Progress in forensic genetics 10. Amsterdam: Elsevier. pp. 56–58.
22. American Society of Crime Laboratory Directors/Laboratory Accreditation Board (2004) Laboratory management and operations. Garner (North Carolina): American Society of Crime Laboratory Directors/Laboratory Accreditation Board. Available: http://www.ascld-lab.org/legacy/aslablegacymanagement.html. Accessed 18 August 2005.
23. Metzker ML, Mindell DP, Liu XM, Ptak RG, Gibbs RA, et al. (2002) Molecular evidence of HIV-1 transmission in a criminal case. Proc Natl Acad Sci U S A 99: 14292–14297.
24. Heitpas J, McMullen LK, Mindell DP, Hanson HL, Rice CM (2005) Keeping track of viruses. In: Breeze RG, Budowle B, Schutzer SE, editors. Microbial forensics. San Diego: Academic Press. pp. 55–97.
25. Kolavic SA, Kimura A, Simons SL, Slutsker L, Barth S, et al. (1997) An outbreak of Shigella dysenteriae type 2 among laboratory workers due to intentional food contamination. JAMA 278: 396–398.
26. Torok TJ, Tauxe RV, Wise RP, Livengood JR, Sokolow R, et al. (1997) A large community outbreak of salmonellosis caused by intentional contamination of restaurant salad bars. JAMA 278: 389–395.
27. Nolte KD, Hanzlick RL, Payne DC, Kroger AT, Oliver WR, et al. (2004) Medical examiners, coroners, and biologic terrorism: A guidebook for surveillance and case management. MMWR Recomm Rep 53: 1–27.

28. Jackson PJ, Hugh-Jones ME, Adair DM, Green G, Hill KK, et al. (1998) PCR analysis of tissue samples from the 1979 Sverdlovsk anthrax victims: The presence of multiple Bacillus anthracis strains in different victims. Proc Natl Acad Sci U S A 95: 1224–1229.

29. Hampton HL (1995) Care of the woman who has been raped. N Engl J Med 332: 234–237.

30. Breeze R, Budowie B, Schutzer S (2005) Microbial Forensics. San Diego: Academic Press. 448 p.

Section VIII - Water Treatment Technologies for Developing Countries

CHAPTER 27

GEOGRAPHICAL INEQUALITIES IN USE OF IMPROVED DRINKING WATER SUPPLY AND SANITATION ACROSS SUB-SAHARAN AFRICA: MAPPING AND SPATIAL ANALYSIS OF CROSS-SECTIONAL SURVEY DATA

Rachel L. Pullan, Matthew C. Freeman, Peter W. Gething, Simon J. Brooker

EDITORS' SUMMARY

Background

Access to a safe drinking-water supply (a water source that is protected from contamination) and to adequate sanitation facilities (toilets, improved latrines, and other facilities that prevent people coming into contact with human urine and feces) is essential for good health. Unimproved drinking-water sources and sanitation are responsible for 85% of deaths from diarrhea and 1% of the global burden of disease. They also increase the transmission of parasitic worms and other neglected tropical diseases. In 2000, world leaders set a target of reducing the proportion of the global population without access to safe drinking water and basic sanitation to half of the 1990 level by 2015 as part of Millennium Development Goal (MDG) 7 ("Ensure environmental sustainability"; the MDGs are designed to improve the social, economic, and health conditions in the world's poorest countries). Between 1990 and 2010, more than 2 billion people gained access to improved drinking-water sources and 1.8 billion gained access to improved sanitation. In 2011, 89% of the world's population had access to an improved drinking-water supply, 1% above the MDG target, and 64% had access to improved sanitation (the MDG target is 75%).

Why Was This Study Done?

Despite these encouraging figures, the WHO/UNICEF Joint Monitoring Programme for Water Supply and Sanitation (JMP) estimates that, globally, 768 million people relied on unimproved drinking-water sources, 2.5 billion people did not use an improved sanitation facility, and more than 1 billion people (15% of the global population) were defecating in the open in 2011. The JMP estimates for 2011 also reveal national and sub-national inequalities in drinking-water supply and sanitation coverage but a better understanding of geographic inequalities is needed to track progress towards universal coverage of access to improved water and sanitation and to identify the populations that need the most help to achieve this goal. Here, the researchers use cross-sectional household survey data and modern statistical approaches to produce a

Citation: Pullan RL, Freeman MC, Gething PW, Brooker SJ (2014) Geographical Inequalities in Use of Improved Drinking Water Supply and Sanitation across Sub-Saharan Africa: Mapping and Spatial Analysis of Cross-Sectional Survey Data, doi:10.1371/journal.pmed.1001626

comprehensive map of the coverage of improved drinking-water supply and improved sanitation at high spatial resolution for sub-Saharan Africa and to investigate geographic inequalities in coverage. Cross-sectional household surveys collect health and other information from households at a single time-point, including data on use of safe water and improved sanitation.

What Did the Researchers Do and Find?

The researchers extracted data on reported household use of an improved drinking-water supply (for example, a piped water supply), improved sanitation facilities (for example, a flushing toilet), and open defecation from 138 national household surveys undertaken between 1991 and 2012 in 41 countries in sub-Saharan Africa. They developed statistical models to fit these data and used the models to estimate coverage at the district (second administrative) level across sub-Saharan Africa for 2012. For ten countries, the estimated coverage of access to improved drinking water at the district level within individual countries ranged from less than 25% to more than 75%. Within-country ranges of a similar magnitude were estimated for coverage of access to improved sanitation (21 countries) and for open defecation (16 countries). Notably, rural households in the districts with the lowest coverage of access to improved water supply and sanitation within a country were 1.5–8 times less likely to access improved drinking water, 2–18 times less likely to access improved sanitation, and 2–80 times more likely to defecate in the open than rural households in districts with the best coverage. Finally, countries with high levels of inequality in improved drinking-water source coverage also experienced high levels of inequality in improved sanitation coverage.

What Do These Findings Mean?

These findings identify important geographic inequalities in the coverage of access to improved water sources and sanitation that were previously hidden within national statistics. The accuracy of these findings depends on the accuracy of the data on water supplies and sanitation provided by household surveys, on the researchers' definitions for improved water supplies and sanitation, and on their statistical methods. Nevertheless, these findings confirm that, to achieve universal coverage of access to improved drinking-water sources and sanitation, strategies that target the areas with the lowest coverage are essential. Moreover, the maps and the analytical approach presented here provide the means for monitoring future reductions in inequalities in the coverage of access to improved water sources and sanitation and thus reflect a major priority of the post-2015 development agenda.

INTRODUCTION

Remarkable gains have been made in the provision of drinking-water supply and sanitation (WSS) globally, with over 2 billion people reportedly gaining access to improved drinking-water sources and 1.8 billion to improved sanitation between 1990 and 2010 [1]. The WHO and UNICEF's Joint Monitoring Programme for Water Supply and Sanitation (JMP) has reported that the world is "on track" to reach the Millennium Development Goal (MDG) target for water supply of reducing by half the proportion of the population without access to sustainable and safe water [1]. In spite of such global gains, important national and sub-national inequalities in WSS coverage remain, including significant rural-urban and regional disparities [1],[2]. National studies also highlight systematic inequalities in WSS access, typically focusing on rural-urban or socio-economic disparities and coverage in hard to reach groups [3]–[9].

Access to safe water and sanitation facilities is a fundamental human right. Unimproved drinking water and sanitation are responsible for an estimated 1% of global disability-adjusted life years (DALYs) [10], and 85% of diarrhoea mortality can be attributed to inadequate water, sanitation, and hygiene practices (WaSH) [11]. Populations with inadequate WSS are also disproportionately affected by the neglected tropical diseases (NTDs) [12]–[17], and as such WSS are increasingly recognised as critical for sustaining the impact of control and elimination strategies that rely on geographically targeted mass drug administration [18],[19]. There have however been few published analyses of sub-national geographical distribution of WSS at policy relevant scales. To better understand inequality within and between countries, and provide a bench mark for tracking progress and help prioritize resource allocation, there is a clear need to develop policy-relevant data on sub-national inequality and a standardised approach to mapping sub-national coverage in WSS.

Population-based national household cluster-sample surveys, such as demographic and health surveys (DHS) and multiple indicator cluster surveys (MICS), provide a wealth of information on the coverage of health and development indicators, including WSS, within countries and have been used previously to map geographical variation in bed net coverage [20], anaemia[21], and under-nutrition [22]. There is an important trade-off however between accuracy and spatial resolution that must be resolved before these data can be truly useful for stakeholders. Surveys are usually powered to provide accurate data for provinces or regions, although access to WSS is likely to vary markedly at this spatial scale. Modern statistical approaches, including small area estimation (SAE) and Bayesian spatial conditional autoregressive (CAR) models, can help tackle the problem of providing feasible estimates for smaller geographical areas, whilst explicitly acknowledging uncertainty associated with data powered to be representative at larger spatial scales [23]–[27].

In this paper we combine multiple national surveys using spatial statistical methods to investigate differences in use of improved drinking water, improved sanitation, and open defecation at small spatial scales; identify specific geographical areas where coverage is substantially worse than national averages; and explore relative geographic inequalities within countries. Our analysis focuses on sub-Saharan Africa (SSA), as this is the region where progression towards the MDGs for water and sanitation is often least successful [1],[28]. Our intent is not to replace existing national JMP estimates, but instead to develop robust maps of contemporary WSS coverage at a finer spatial resolution useful for service delivery providers, policy makers and those planning investment within governments, implementers, and donors. The work is conducted within the context of the Global Atlas of Helminth Infections project (www.thiswormyworld.org), which aims to develop a suite of geographical resources and tools for NTD control.

METHODS

Data Sources

WSS data were sourced from national household cluster-sample surveys undertaken as part of multiple indicator cluster surveys (MICS) (http://www.childinfo.org/mics4_surveys.html; implementation supported by UNICEF), Demographic and Health Surveys (DHS) and national malaria and AIDS indicator surveys (MIS/AIS) (http://www.measuredhs.com/data/available-datasets.cfm; USAID) and living standard measurement studies (LSMS) (http://iresearch.worldbank.org/lsms/lsmssurveyFinder.htm; World Bank). These surveys are based on probability sampling using existing sampling frames (primarily population censuses) and are conducted by trained enumerators conducting household visits, typically achieving household response rates >97%. Informed consent is obtained from all respondents before participating. All available DHS, MICs, and LSMS surveys conducted in SSA since 1990 that contained the necessary modules to calculate household WSS use were included (n=138 surveys) [29]. Access to improved drinking water and sanitation were defined using the criteria outlined by the JMP, and are measured by reported use. For MDG monitoring, improved sanitation facilities are defined as those that "hygienically separate human excreta from human contact," whilst improved ("safe") drinking-water sources are defined as those that are "protected from outside contamination (especially faecal contamination)." The proportion of households reporting open defecation was also recorded. In the subset of data where distance to water source was available (n=121 surveys), we also recorded the proportion of households using an accessible, improved drinking water source. This was defined as one within 1 km (or 15 minutes) of the household, which has been suggested as an appropriate distance for meeting the MDG targets [30]. Further details of indicator definition, and comparison with those used by the JMP, are provided in Box 1.

Box 1. Comparison with the Joint Monitoring Programme Methodology

Data availability

- Available data sources included in this analysis (DHS, MICs, and LSMS) make up 42% percent of all data in the JMP database for the same countries. Data available to the JMP that could not be used in this analysis include national surveys implemented by governments and census data.

Common definitions

- ***Improved sources of drinking water***: piped water into the dwelling; piped water to yard/plot/compound; public tap or standpipe; tubewell or borewell; protected dug well; protected spring; rainwater.
- ***Improved sanitation***: flush toilet; piper sewer system; septic tank; ventilated improved pit latrine (VIP); pit latrine with slab; composting toilet.
- ***Open defecation***: No facilities or bush or field.
- *Not used by JMP.* ***Accessible, improved drinking water source***: an improved drinking water source (see above) within either 1 km or 15 minutes from the household. Data available in 85% of surveys.

Shared sanitation facilities

- **Contrasting variable definition:** JMP classify all shared facilities as unimproved, and make assumptions on the proportion shared when this data are not available. In our main analysis, we do not distinguish between shared and private sanitation facilities as this information is only available for 70% of included surveys, although a sub-analysis is performed to compare differences
- ***Implications***: Coverage of improved sanitation in countries with large numbers of households reporting shared facilities will be considerably higher than current JMP estimates

Unimproved pit latrines

- **Contrasting variable definition:** older DHS surveys (phase 4 and earlier, pre-2003) do not distinguish the type of pit latrine (improved or unimproved), and for these surveys JMP make assumptions on the proportion that can be classified as improved. In our analysis all such latrines were included as unimproved to avoid misclassification.
- ***Implications***: In our analysis past coverage may have been underestimated for countries with older DHS surveys, and temporal gains in coverage overestimated.

Methodology

- ***JMP methods***: for each country, separate linear time trends are fit through national (rural and urban) coverage estimates from survey and census data. Regression slopes are extrapolated two years outside available data; beyond this slope is assumed to be zero for a maximum of six years after which estimates are not made. Population averaging (to determine rural, urban, and overall coverage rates) relies on reported survey weights.
- ***Spatial multilevel methods***: data from all countries are treated as a continuous time series and fit within a single multilevel model. This considers the hierarchical structure of the whole dataset: surveys sites are nested within administrative areas, within countries. The model estimates an average intercept and an average slope with residual variances across countries, whilst accounting for sub-national spatial correlation and non-spatial variation. In practice, this means that when there is reliable information for a specific country, predictions will closely follow the country survey points, whereas when there is little temporal (or sub-nationally disaggregated) information trends will tend to follow the regional (or national) mean. Population averaging uses high resolution population data generated by the WorldPop project [33],[34].

For each of the available surveys, the proportion of households reporting use of an improved drinking water source, improved sanitation, and open defecation were calculated at the cluster (survey site) level. Each survey site was then coded as either urban or rural as defined by the survey and was allocated, where possible, to sub-national administrative areas. For reference, in most SSA countries first administrative areas (admin1, often referred to as provinces, regions, or states) represent on average approximately 1.35

million people and are typically further subdivided into three to 15 second administrative areas (admin2, often referred to as districts, representing approximately 200,000 people). Admin1 and admin2 digital boundaries were derived for each country using the United Nations Second Administrative Level Boundaries dataset project (SALB) ([31]): for 74 surveys, survey sites were linked to admin2 using the provided survey site co-ordinates, whilst admin1 and admin2 names were used to match survey site locations for the remaining 64 surveys. For 45 of these, survey sites could be matched to admin1 only as admin2 names were not available. In these instances, survey site data were attributed to all admin2 contained within the admin1 boundaries, and weighted to reflect the number of admin2 this represented. No survey data were available for Botswana and Eritrea, and these countries (representing 0.85% of the population of SSA) are therefore not included in the analysis.

Digital gridded population surfaces and urban/peri-urban extent maps for 2012 at 100 m^2resolution were provided by the WorldPop project [32],[33]. These population surfaces were used in combination with the admin2 boundaries to extract total admin2 populations and the proportion of each admin2 population living in urban/peri-urban and rural areas. As a confirmation step, the overall proportion urban generated for each country for 2012 was compared with estimates produced by the UN Urbanization Prospects [34] suggesting good agreement (r=0.72, p<0.001).

Data Analysis

For each country, available data for urban and rural populations were plotted on a timescale from 1990 to 2012, and the temporal trend examined visually. Access to improved water and sanitation has risen approximately linearly over time, although there are large differences in this trend between countries, and between urban and rural populations within countries. It was therefore decided that a random coefficient model incorporating a linear time trend would be appropriate, allowing different intercepts and temporal slopes for urban and rural populations by country, but assuming an overall linear change over time.

For each water and sanitation indicator, spatially explicit logistic regression models were therefore developed and fitted using a Bayesian framework. These consider the hierarchical structure of the whole dataset: survey sites are nested within admin2, which are themselves nested within admin1 and countries. Instead of calculating an intercept and slope separately for each country, as is currently done by the JMP, the multilevel model estimates an average intercept and an average slope for rural and urban populations, with residual variances across countries. Neighbouring admin2 within countries are also expected to have similar coverage rates, and so an admin2-level random effect is included that explicitly models the correlation between neighbouring admin2 within countries using a conditional autoregressive (CAR) covariance structure. Models also include a regional random effect with an unstructured covariance structure, to account for non-spatial dependence between survey sites within admin1 areas.

Essentially, the model structure implies that countries with scarce data over time will follow the mean trend, whilst countries with reliable information (i.e., several surveys across multiple time points) will closely follow the survey points with less associated uncertainty. Similarly, when there is little spatially disaggregated data available, sub-national predictions will follow national means, whilst if data coverage at the admin2 level is good (i.e., survey sites located in admin2 throughout the country) admin2-level coverage will be smoothed towards local, neighbouring values thus counteracting the problem of outliers resulting from under-sampling. Models were fit within a Bayesian framework in the software package WinBUGS [35] using a Markov chain Monte Carlo (MCMC) algorithm, a robust and flexible inference platform that is well suited to complex problems and that explicitly incorporates uncertainty in input data and model parameters. MCMC is an iterative, stochastic simulation technique, and as such this uncertainty is represented in the form of predictive posterior distributions.

Predicting Coverage in 2012 and Analysis of Geographical Inequality

The developed models were used to predict coverage of water and sanitation for the urban and rural populations of each admin2 for 2012; overall population estimates are population-weighted averages of the urban and rural numbers. Following the approach of Gomez-Rubio and colleagues [24] and Banerjee and colleagues [36], we used information provided by the spatially correlated admin2-level random effect to make predictions for admin 2 without data. To prevent unstable predictions for areas with few or no neighbours with data, this was only performed for countries with available data at admin2 level. For seven countries with no data located at admin2 level, predictions were made for admin1 only. At each MCMC iteration, multiplying the urban and rural population surfaces with the admin2 predicted coverage (or where applicable, coverage by admin1) and then aggregating allowed estimation for the overall population locally and nationally. Resulting uncertainty in predicted coverage will vary geographically, depending on observed sampling variation and the density of survey sites within countries and across time.

Sub-national inequality was evaluated using absolute and relative measures. As absolute measures, the range of coverage (by admin2) for each country was computed, and admin2 with coverage significantly lower (and higher) than the national mean identified based on 95% Bayesian credible intervals (BCIs). As a country-level measure of sub-national relative inequality, we developed a relative geographic inequality (RGI) index, derived from a geographical GINI coefficient. The GINI coefficient is a measure commonly used to describe the distribution of income within a country [37], and has been widely adapted for other applications, including measuring inequalities in health care provision [38],[39] and disease incidence across population groups [40]. It is based on the Lorenz curve, an accumulated frequency curve that compares the distribution of a specific variable with a uniform distribution that represents equality. To calculate Lorenz curves for each country, administrative areas were ranked smallest to largest by their share of national use; the cumulative proportion of use was then calculated and plotted against the cumulative percentage of population. The greater the deviation of the Lorenz curve from the diagonal line of equal distribution, the greater the inequality. The GINI coefficient was then calculated as twice the area between the diagonal and the Lorenz curve. Values can range from 0 to 1, with 0 representing perfect equality and 1 total inequality [37]. Importantly however, in contrast to income, WSS coverage is a bounded variable (i.e., it cannot fall outside the range 0%–100%) and as such the GINI coefficient would be expected to decline as coverage increases. Outlier countries (with higher or lower levels of inequality given their level of coverage) were thus identified using linear regression of GINI coefficient against national coverage for overall, rural, and urban populations and the RGI score generated as the difference between the observed and expected GINI coefficient given national coverage based upon this modelled relationship. This approach draws upon the methods of Jamison and colleagues in their investigation of the relationship between survival curves and life expectancy, which follow similar rules [41]. In the case of open defecation, the GINI and RGI scores were calculated on the basis of households not reporting open defecation (i.e., with access to any type of sanitation) to enable better comparison with the other variables.

Ethics Statement

This is a secondary analysis of previously collected and published household survey data, and as such ethical approval was not required for this work. Ethical clearance for each included survey was provided by review boards in each country, with informed consent provided by all participants. Details of this process vary by survey.

RESULTS

National WSS Coverage

In total, we obtained 138 DHS, MICs, and LSMS surveys conducted between 1991 and 2012, representing over 1.15 million households from over 50,000 survey sites located in 2,751 admin2 across SSA (outlined in Table 1). Figure 1 emphasises the high data coverage in West and East Africa, and low coverage in much of Southern and Central Africa. Data coverage was very limited for South Sudan and much of Angola.

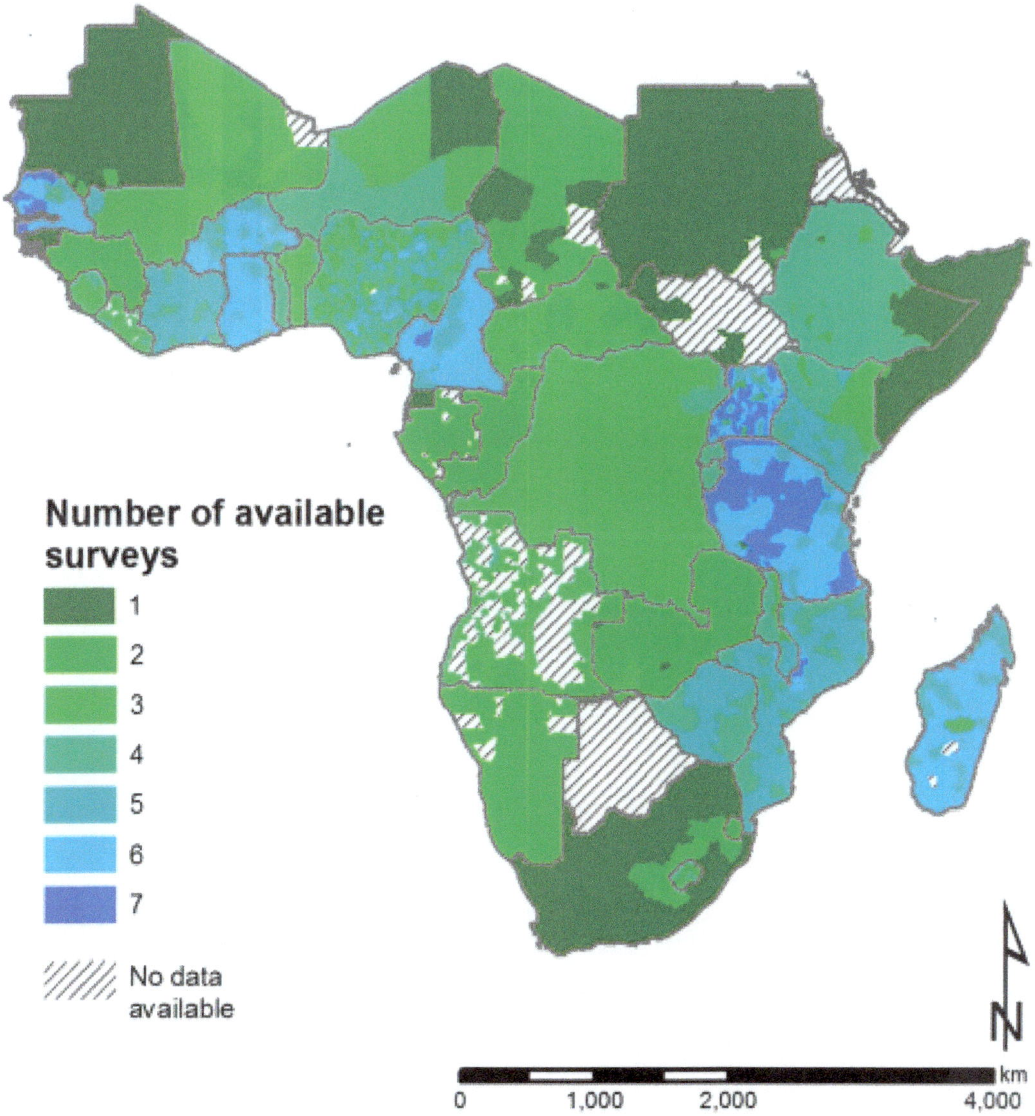

Figure 1: Availability of nationally representative, cluster survey data on improved drinking water and sanitation across sub-Saharan Africa for the period 1990–2012.

Data are linked to second administrative areas where possible and if not to the first administrative level; administrative boundaries are provided by the United Nations Second Administrative Level Boundaries (SALB) project.

Table 1: Regional summary of water and sanitation coverage data sources and quantity for 41 sub-Saharan African countries

	Available surveys		Data Located to Second Administrative Area		Data Located to First Administrative Area	
Region	**Number by type**	**Date range**	**Households**	**Survey Sites**	**Households**	**Survey Sites**
Central	11 DHS/MIS/AIS; 6 MICs	1994–2012	38,800	1,434	25,238	993
East	41 DHS/MIS/AIS; 7 MICs; 3 LSMS	1992–2012	294,468	12,573	164,519	8,127
Southern	9 DHS; 1 MICs	1998-2010	61,815	2,796	32,925	1,878
West	43 DHS; 17 MICs	1991–2011	311,136	13,047	254,395	11,036

Sub-Saharan countries not included are Botswana, Cape Verde, Comoros, Djibouti, Eritrea, Reunion, Sao Tome and Principe, and Seychelles, representing <1.0% of the population of SSA in 2012.

AIS, AIDs Indicator Surveys; DHS, Demographic and Health Surveys; LSMS, Living Standard Measurement Studies; MICs, Multiple Indicator Cluster Surveys; MIS, Malaria Indicator Surveys.

Table 2 provides predicted national estimates and admin2 summaries; the predicted distributions of coverage in urban and rural populations are shown in Figure 2. Model results suggest that in 2010, 62.0% (95% BCI: 61.5%–62.4%) of the population of SSA (excluding Eritrea and Botswana) reported using an improved drinking-water source, although at national levels this was seen to vary between as low as 11.5% (95% BCI: 6.2%–19.0%) in rural Somali populations to 99.0% (95% BCI: 98.7%–99.3%) in urban populations in Namibia. Similar ranges are seen for reported use of improved sanitation: whilst we predict that 42.5% (95% BCI: 42.1%–43.1%) of the total population in SSA used improved sanitation, this ranges from less than 30% for 11 countries to greater than 70% in the top four performing countries (Ghana, Guinea-Bissau, Equatorial Guinea, and Rwanda). For six countries, it is estimated that at least 50% of households habitually defecate in the open (Niger, Namibia, Benin, Burkina Faso, South Sudan, and Chad).

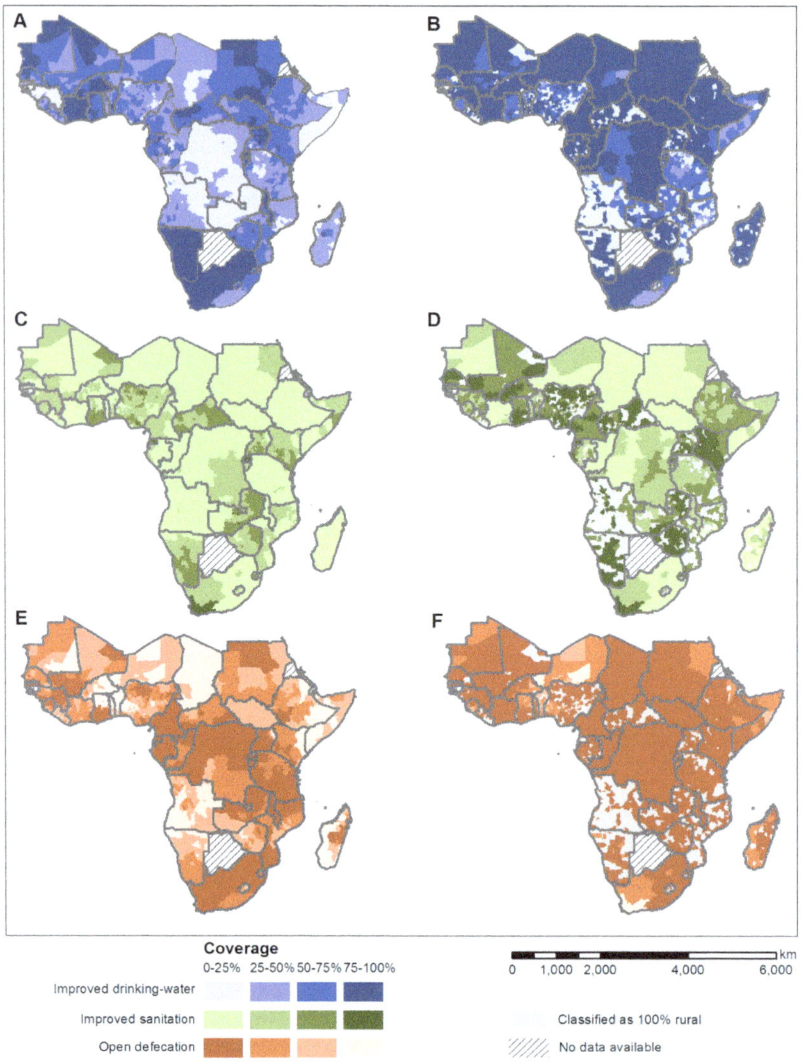

Figure 2: Predicted population coverage in 2012 for rural and urban populations, by second administrative area.

Access to improved drinking-water supply in (A) rural and (B) urban populations; access to improved sanitation in (C) rural and (D) urban populations; and open defecation in (E) rural and (F) urban populations. Model results showing posterior median predicted coverage (i.e. most likely value) for each second administrative area. No data was available for Botswana and Eritrea (hatched). Each indicator was modelled independently.

Table 2: National estimates and admin2 summaries resulting from the modelling procedures for access to improved drinking-water source, improved sanitation, and open defecation

Country	Indicator	MDG-7C Achieved[a] (Yes/No)	National Coverage		Admin2-Level Statistics		RGI Score[c]
			JMP Estimate	Model Estimate (95% BCI)	Range in Predicted Admin2 Coverage	Median Admin2 SD[b]	
Central SSA							
Angola	Drinking water	No	53.4%	44.5% (43.4%–45.7%)	9.0%–84.7%	3.2%	↑ 0.04
	Improved sanitation	No	58.7%	37.5% (36.7%–38.3%)	2.4%–92.8%	2.0%	↑ 0.18
	Open defecation	—	25.7%	44.3% (43.3%–45.2%)	0.8%–92.8%	3.4%	↑ 0.08
CAR	Drinking water	No	67.1%	75.1% (73.8%–76.3%)	37.6%–98.6%	2.0%	0.00
	Improved sanitation	No	33.8%	58.3% (56.6%–60.0%)	28.7%–91.8%	2.7%	↓ −0.06
	Open defecation	—	19.9%	25.2% (23.9%–26.5%)	1.7%–59.1%	1.9%	↓ −0.02
Congo	Drinking water	Yes	72.4%	74.6% (73.3%–76.0%)	21.2%–87.3%	1.6%	0.01
	Improved sanitation	No	17.8%	16.6% (14.9%–18.5%)	1.4%–22.5%	1.3%	↓ −0.21
	Open defecation	—	8.3%	10.6% (9.6%–11.8%)	4.4%–40.3%	1.5%	0.00
DR Congo	Drinking water	No	46.2%	43.2% (42.4%–44.1%)	9.0%–93.8%	1.5%	↑ 0.07
	Improved sanitation	No	30.7%	27.6% (26.7%–28.5%)	7.7%–65.6%	1.7%	−0.01
	Open defecation	—	9.0%	17.3% (16.6%–18.1%)	2.5%–48.2%	1.5%	−0.02
Equatorial Guinea	Drinking water	No		72.8% (39.9%–96.8%)	55.6%–85.1%	18.3%	↓ −0.03
	Improved sanitation	No		82.0% (34.6%–98.2%)	63.7%–86.6%	16.6%	−0.04
	Open defecation	—		0.2% (0.0%–1.4%)	0.1%–0.2%	0.6%	0.00
Gabon	Drinking water	Yes	87.9%	88.2% (87.8%–88.5%)	30.3%–99.0%	2.1%	↑ 0.05
	Improved sanitation	No	32.9%	53.7% (53.0%–54.5%)	3.9%–79.3%	1.2%	↑ 0.06
	Open defecation	—	1.2%	3.0% (2.7%–3.2%)	0.2%–49.3%	0.4%	0.00
East SSA							
Burundi	Drinking water	No	74.4%	78.8% (78.2%–79.4%)	47.9%–95.4%	1.7%	0.00
	Improved sanitation	No	50.1%	40.1% (39.2%–40.9%)	7.9%–88.1%	2.2%	−0.03
	Open defecation	—	2.6%	3.1% (2.8%–3.5%)	0.2%–12.5%	0.5%	0.00
Ethiopia	Drinking water	No	49.0%	62.8% (62.2%–63.4%)	31.5%–99.8%	1.5%	↓ −0.03
	Improved sanitation	No	20.7%	18.4% (17.9%–19.0%)	4.2%–84.0%	1.1%	↑ 0.05
	Open defecation	—	45.0%	37.9% (37.2%–38.5%)	3.6%–90.2%	1.6%	↓ −0.03
Kenya	Drinking water	No	60.9%	71.8% (71.1%–72.6%)	44.7%–95.3%	1.8%	−0.01
	Improved sanitation	No	29.4%	64.5% (63.6%–65.3%)	23.5%–95.4%	2.2%	−0.01
	Open defecation	—	13.7%	16.0% (15.5%–16.7%)	0.1%–83.2%	1.3%	0.02
Madagascar	Drinking water	No	48.1%	44.9% (44.2%–45.5%)	16.4%–96.7%	2.1%	0.01
	Improved sanitation	No	13.7%	7.7% (7.4%–7.9%)	0.7%–44.3%	0.4%	↑ 0.12
	Open defecation	—	39.4%	43.2% (42.7%–43.8%)	1.3%–88.8%	1.6%	↑ 0.07
Malawi	Drinking water	Yes	83.7%	82.6% (82.2%–82.9%)	72.8%–96.5%	0.5%	↓ −0.02
	Improved sanitation	No	52.9%	21.7% (21.1%–22.2%)	9.3%–41.8%	0.8%	↓ −0.17
	Open defecation	—	6.5%	10.2% (9.9%–10.5%)	3.3%–16.7%	0.5%	↓ −0.03
Mozambique	Drinking water	No	47.2%	49.5% (48.7%–50.1%)	20.5%–89.7%	2.7%	−0.02
	Improved sanitation	No	19.1%	35.8% (34.9%–36.8%)	5.4%–91.5%	2.9%	0.03
	Open defecation	—	41.8%	24.3% (23.7%–24.9%)	0.5%–55.3%	1.9%	−0.01
Rwanda	Drinking water	No	68.9%	85.6% (85.1%–86.1%)	65.3%–95.8%	1.5%	0.00
	Improved sanitation	No	61.3%	85.0% (84.3%–85.7%)	71.0%–91.3%	1.4%	−0.02
	Open defecation	—	2.2%	1.1% (0.9%–1.3%)	0.1%–4.3%	0.3%	0.00
Somalia	Drinking water	No	29.5%	33.2% (18.5%–45.8%)	3.2%–59.5%	4.4%	0.03
	Improved sanitation	No	23.6%	56.6% (39.5%–78.9%)	6.0%–93.0%	13.5%	↑ 0.10
	Open defecation	—	52.5%	41.1% (32.9%–47.0%)	7.7%–96.1%	5.2%	↑ 0.07
South Sudan	Drinking water	No	56.4%	65.5% (24.6%–95.4%)	39.9%–84.7%	23.0%	↓ −0.08
	Improved sanitation	No	8.9%	27.2% (0.9%–97.6%)	19.0%–28.3%	30.4%	↓ −0.30
	Open defecation	—	77.0%	61.8% (9.2%–91.7%)	60.6%–68.1%	26.6%	↓ −0.28
Sudan	Drinking water	No	55.4%	81.0% (55.7%–91.5%)	64.7%–96.7%	13.1%	0.01
	Improved sanitation	No	23.5%	23.2% (1.8%–92.4%)	8.3%–44.9%	32.8%	↓ −0.13

	Improved sanitation	No	23.5%	23.2% (1.8%–92.4%)	8.3%–44.9%	32.8%	↓ −0.13
	Open defecation	—	46.3%	34.3% (17.8%–49.2%)	7.8%–59.5%	9.7%	↓ −0.04
UR Tanzania	Drinking water	No	74.8%	74.2% (73.8%–74.6%)	18.2%–97.8%	1.8%	↓ −0.05
	Improved sanitation	No	35.0%	50.9% (50.4%–51.4%)	5.7%–95.1%	2.2%	↑ 0.06
	Open defecation	—	8.6%	8.6% (8.4%–8.8%)	0.1%–88.0%	0.8%	0.01
Uganda	Drinking water	Yes	53.3%	48.1% (47.5%–48.7%)	18.7%–86.9%	2.2%	↑ 0.03
	Improved sanitation	No	11.9%	25.7% (25.2%–26.3%)	5.3%–92.4%	1.9%	−0.03
	Open defecation	—	12.4%	13.0% (12.7%–13.4%)	0.1%–50.5%	1.2%	0.01
Zambia	Drinking water	No	64.1%	35.8% (34.6%–37.0%)	4.7%–90.7%	1.5%	↑ 0.15
	Improved sanitation	No	42.1%	62.9% (60.2%–65.4%)	15.6%–92.0%	3.6%	0.04
	Open defecation	—	17.0%	22.8% (21.6%–24.2%)	0.6%–72.6%	2.2%	0.03
Southern SSA							
Lesotho	Drinking water	No	77.7%	76.8% (75.8%–77.8%)	44.9%–94.0%	2.9%	−0.01
	Improved sanitation	No	26.3%	37.7% (36.5%–38.9%)	12.6%–72.6%	3.4%	↓ 0.09
	Open defecation	—	36.8%	37.4% (36.5%–38.6%)	2.2%–88.1%	3.3%	0.01
Namibia	Drinking water	Yes	93.4%	91.6% (90.8%–92.4%)	66.6%–99.7%	1.0%	↑ 0.04
	Improved sanitation	No	32.3%	47.5% (46.3%–48.8%)	3.7%–94.2%	2.4%	↑ 0.13
	Open defecation	—	51.6%	51.8% (50.4%–53.1%)	5.5%–94.7%	2.5%	↑ 0.09
South Africa	Drinking water	No	91.5%	79.4% (49.8%–95.7%)	43.1%–96.4%	11.2%	↑ 0.03
	Improved sanitation	No	74.0%	45.9% (18.1%–76.3%)	24.3%–92.8%	16.8%	−0.01
	Open defecation	—	7.0%	8.6% (1.9%–25.8%)	1.2%–25.6%	4.9%	−0.01
Swaziland	Drinking water	Yes	72.2%	80.4% (78.3%–82.3%)	55.9%–98.6%	2.5%	0.00
	Improved sanitation	No	57.0%	64.9% (61.4%–68.4%)	43.8%–84.8%	3.7%	↓ −0.08
	Open defecation	—	14.7%	14.1% (12.7%–15.8%)	0.3%–36.4%	2.3%	0.01
Zimbabwe	Drinking water	No	80.0%	72.5% (71.8%–73.1%)	48.8%–93.9%	1.7%	0.00
	Improved sanitation	No	40.2%	63.7% (63.0%–64.4%)	21.2%–93.7%	1.6%	0.01
	Open defecation	—	25.6%	27.9% (27.3%–28.5%)	1.2%–77.9%	1.6%	0.01
West SSA							
Benin	Drinking water	No	76.0%	72.9% (71.8%–74.0%)	33.0%–99.5%	2.0%	0.01
	Improved sanitation	No	14.2%	51.3% (50.0%–52.8%)	9.4%–95.2%	2.7%	↑ 0.05
	Open defecation	—	54.3%	55.2% (54.1%–56.4%)	8.8%–94.2%	2.3%	↑ 0.04
Burkina Faso	Drinking water	Yes	80.0%	84.4% (84.0%–84.8%)	61.4%–94.8%	1.0%	0.01
	Improved sanitation	No	18.0%	39.4% (38.8%–40.0%)	4.7%–80.2%	1.4%	0.01
	Open defecation	—	57.9%	59.1% (58.6%–59.6%)	13.8%–93.6%	1.2%	0.02
Cameroon	Drinking water	No	74.4%	70.5% (70.0%–71.0%)	40.3%–94.8%	1.3%	0.00
	Improved sanitation	No	47.8%	50.7% (50.2%–51.2%)	19.8%–79.2%	1.4%	−0.06
	Open defecation	—	6.2%	6.0% (5.7%–6.2%)	0.4%–26.7%	0.4%	0.01
Chad	Drinking water	No	50.0%	37.3% (33.5%–42.4%)	10.0%–65.4%	2.8%	↓ −0.09
	Improved sanitation	No	11.7%	1.6% (1.2%–2.2%)	0.2%–8.5%	0.4%	↑ 0.10
	Open defecation	—	65.4%	82.1% (79.5%–84.1%)	34.6%–97.5%	1.5%	↑ 0.07
Côte d'Ivoire	Drinking water	No	79.9%	88.3% (87.7%–88.9%)	73.9%–98.0%	1.0%	0.01
	Improved sanitation	No	23.9%	12.4% (11.8%–13.1%)	1.3%–35.4%	0.5%	↑ 0.06
	Open defecation	—	27.1%	28.6% (27.5%–29.6%)	6.1%–54.5%	2.0%	↓ −0.03
Gambia	Drinking water	Yes	89.3%	64.8% (52.1%–75.4%)	45.0%–94.6%	2.5%	↓ −0.03
	Improved sanitation	Yes	67.7%	99.9% (99.8%–99.9%)	99.0%–100.0%	0.0%	0.03
	Open defecation	—	2.3%	1.3% (1.0%–1.7%)	0.0%–8.2%	0.5%	0.00

Ghana	Drinking water	Yes	86.3%	72.0% (71.4%–72.6%)	53.1%–92.4%	1.8%	↓ −0.03
	Improved sanitation	No	13.5%	71.8% (71.3%–72.3%)	11.6%–94.1%	1.7%	0.03
	Open defecation	—	18.4%	19.2% (18.8%–19.5%)	1.4%–92.3%	1.1%	↑ 0.04
Guinea	Drinking water	No	73.6%	42.0% (40.1%–44.0%)	12.1%–97.8%	1.9%	↑ 0.03
	Improved sanitation	No	18.5%	48.0% (44.8%–51.4%)	27.1%–86.2%	3.3%	↓ −0.06
	Open defecation	—	19.7%	20.5% (19.1%–22.0%)	0.1%–57.5%	1.4%	0.01
Guinea-Bissau	Drinking water	Yes	71.7%	82.6% (42.2%–96.8%)	40.4%–91.3%	20.8%	↑ 0.03
	Improved sanitation	No	19.0%	80.4% (28.7%–95.1%)	26.9%–90.0%	23.2%	0.03
	Open defecation	—	24.9%	8.3% (2.5%–21.5%)	3.6%–47.9%	12.0%	0.00
Liberia	Drinking water	No	74.4%	67.3% (66.1%–68.6%)	28.4%–88.5%	2.7%	↓ −0.03
	Improved sanitation	No	18.2%	32.3% (31.2%–33.5%)	5.8%–62.1%	1.9%	−0.03
	Open defecation	—	44.3%	46.3% (45.2%–47.5%)	20.7%–83.9%	2.7%	↓ −0.05
Mali	Drinking water	No	65.4%	67.9% (67.0%–68.8%)	46.0%–95.4%	1.5%	−0.02
	Improved sanitation	No	21.6%	42.5% (41.4%–43.7%)	13.2%–93.8%	1.8%	0.00
	Open defecation	—	13.5%	16.4% (15.8%–17.1%)	0.1%–61.2%	0.8%	0.01
Mauritania	Drinking water	No	49.6%	59.7% (52.4%–67.9%)	42.0%–87.1%	4.5%	↓ −0.05
	Improved sanitation	No	26.6%	31.2% (12.4%–44.6%)	2.0%–48.4%	6.8%	0.01
	Open defecation	—	50.9%	39.6% (25.6%–46.5%)	22.9%–85.9%	6.9%	↓ −0.03
Niger	Drinking water	No	50.3%	57.7% (55.8%–59.5%)	39.2%–90.7%	2.2%	↓ −0.06
	Improved sanitation	No	9.6%	4.1% (3.7%–4.5%)	0.4%–30.6%	0.3%	↑ 0.13
	Open defecation	—	78.1%	80.1% (79.1%–81.0%)	0.1–99.8%	1.8%	↑ 0.06
Nigeria	Drinking water	No	61.1%	59.7% (59.1%–60.2%)	16.5%–91.4%	2.7%	−0.01
	Improved sanitation	No	30.6%	60.7% (60.2%–61.2%)	10.0%–95.9%	2.6%	0.03
	Open defecation	—	22.7%	29.2% (28.7%–29.5%)	0.6%–88.4%	2.4%	↑ 0.02
Senegal	Drinking water	No	73.4%	78.6% (78.2%–79.1%)	28.1%–98.2%	0.8%	↑ 0.06
	Improved sanitation	No	51.4%	57.9% (57.4%–58.4%)	17.5%–94.7%	0.9%	↑ 0.06
	Open defecation	—	16.8%	19.2% (18.8%–19.7%)	0.6%–48.9%	0.9%	0.00
Sierra Leone	Drinking water	No	57.5%	61.5% (60.5%–62.5%)	43.1%–84.5%	1.0%	↓ −0.03
	Improved sanitation	No	12.9%	48.2% (47.1%–49.4%)	31.9%–72.1%	1.1%	↓ −0.10
	Open defecation	—	27.1%	28.7% (27.6%–29.7%)	11.4%–59.5%	0.9%	↓ −0.03
Togo	Drinking water	No	59.0%	58.5% (57.4%–59.7%)	26.1%–79.3%	1.5%	−0.01
	Improved sanitation	No	11.4%	37.4% (36.3%–38.4%)	2.9%–76.7%	1.0%	↑ 0.10
	Open defecation	—	53.7%	41.2% (40.4%–42.0%)	7.0%–87.5%	1.1%	↑ 0.03

[a]MDG 7C is to halve the proportion of the population without sustainable access to safe drinking water and basic sanitation. Classification of achievement is based on figures produced by the WHO/UNICEF JMP for Water Supply and Sanitation for 1990 and 2010 [91].

[b]Median administrative area standard deviation (SD) is taken from the posterior distribution for standard deviation at the administrative area level generated by the hierarchical model, and reflects the amount of data available for a given country (increased SD reflects greater uncertainty and less data).

[c]RGI is a measure of relative inequality in access sub-nationally within a given country given national coverage levels. Negative values indicate a lower than expected inequality, whilst positive values indicate greater than expected inequality. ↓ Indicates a score significantly lower than 0 and ↑ Indicates a score significantly higher than 0.

doi:10.1371/journal.pmed.1001626.t002

For 30 countries there was sufficient survey data to investigate reported use of accessible, improved drinking water supply (defined as those <1 km away or within a 30-minute round trip) and for 27 countries sufficient survey data to investigate reported use of private, improved sanitation (defined as one that is not shared with another household). Overall, in 2012 only 55.0% of rural populations using an improved drinking water source used an "accessible" source (range: 23.9% in Uganda to 98.8% in Namibia) rising to 74.1% of urban populations (range: 39.9% in Cote d'Ivoire to 95.8% in Mali). As can be seen in Figure 3, this urban-rural disparity is less apparent for private and shared improved sanitation: overall, 54.1% of rural populations using improved sanitation had access to a private latrine (range: 12.6% in Liberia to 88.7% in Mozambique) compared with 48.9% of urban populations (range: 14.4% in Togo to 82.7% in Mozambique). Given the reduced data availability for these additional indicators, further analysis at a sub-national level is restricted to use of any improved drinking-water source and any improved sanitation facility.

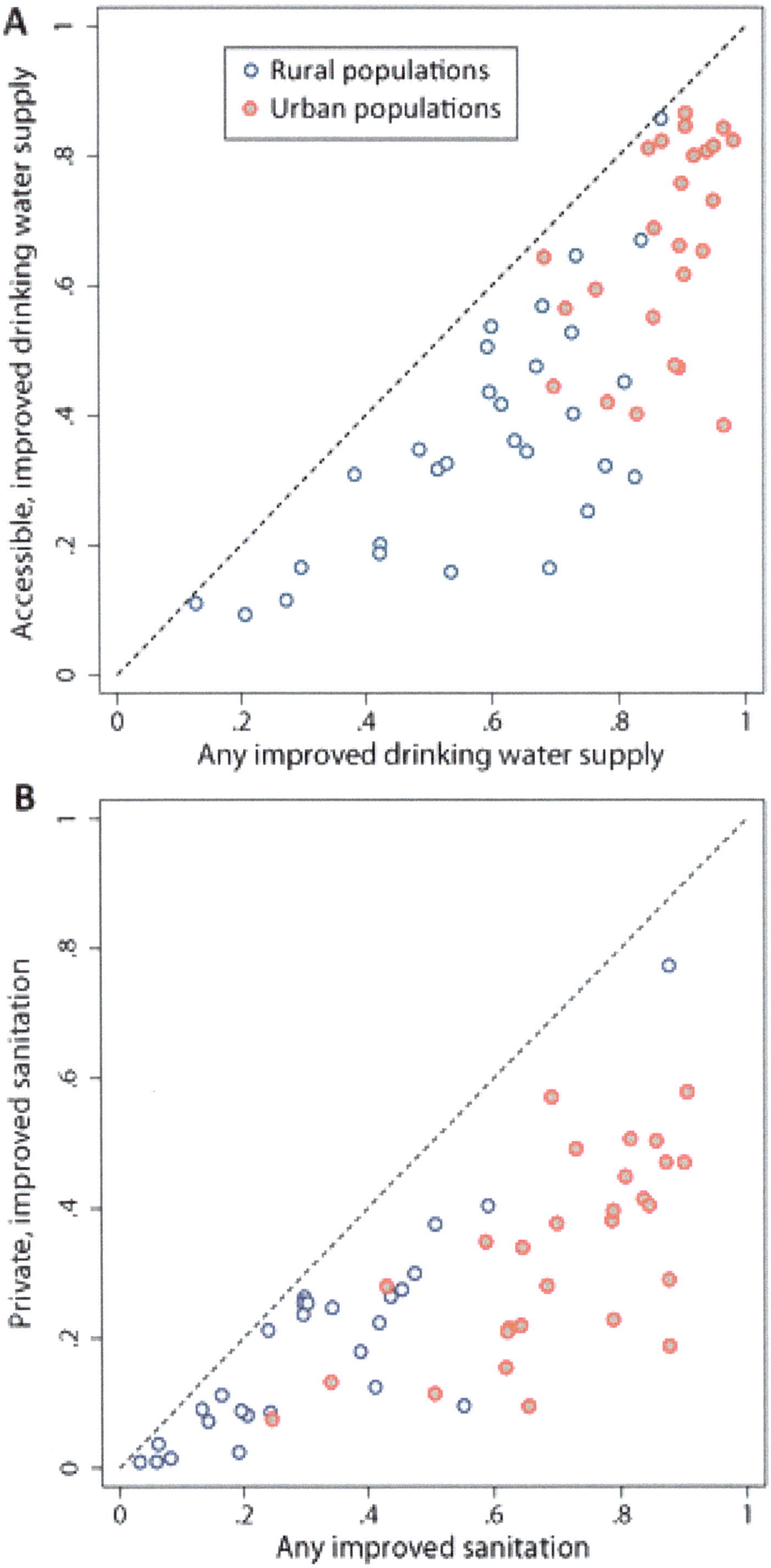

Figure 3: Comparison of (A) any improved drinking water source against accessible, improved drinking water source and (B) any improved sanitation against private improved sanitation.

Dots show national comparisons for urban (red) and rural (blue) populations. An accessible drinking water supply is defined as one within 15 minutes of the household; private sanitation is defined as a facility used by only one household.

Sub-national Variation in WSS Coverage

Substantial within-country variation in use is predicted for all three main outcomes, with some of the greatest differences in coverage between admin2 seen for the highly populous countries of Nigeria, Ethiopia, and DR Congo. For ten countries, admin2-level coverage in use of an improved drinking-water source ranged within country from less than 25% to over 75%; within-country ranges of the same magnitude were seen for improved sanitation for 21 countries and open defecation for 16 countries. Figure 4 highlights those admin2 where coverage of both improved drinking water and improved sanitation were significantly lower than national averages based on 95% BCIs (shown in red), highlighting the considerable overlap in admin2 with low access to improved drinking water and improved sanitation. Similarly, when considering all countries in SSA, correlation between admin2-level coverage of improved drinking water and improved sanitation is reasonably high (r=0.44, p<0.001) although as can be seen in Figure 5 this varies considerably by country.

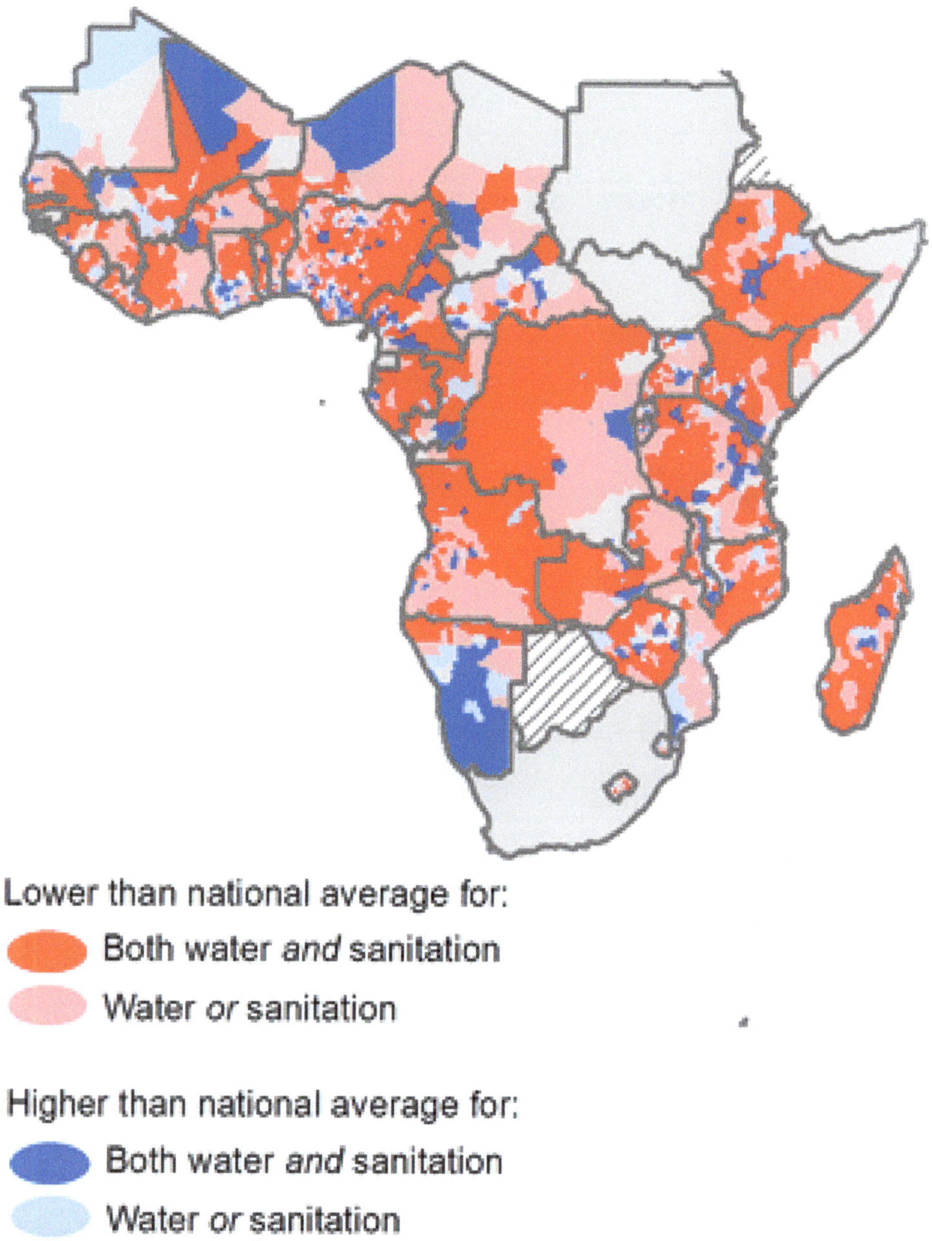

Figure 4: Predicted second administrative areas that differ significantly from national mean coverage.

Second administrative areas shaded red have significantly lower coverage than the national average, based on 95% BCI, for either both (dark red) or one (light red) of improved drinking water and improved sanitation; administrative areas shaded blue have significantly higher coverage rates than the national average. Administrative areas shaded grey are not significantly different from the national mean.

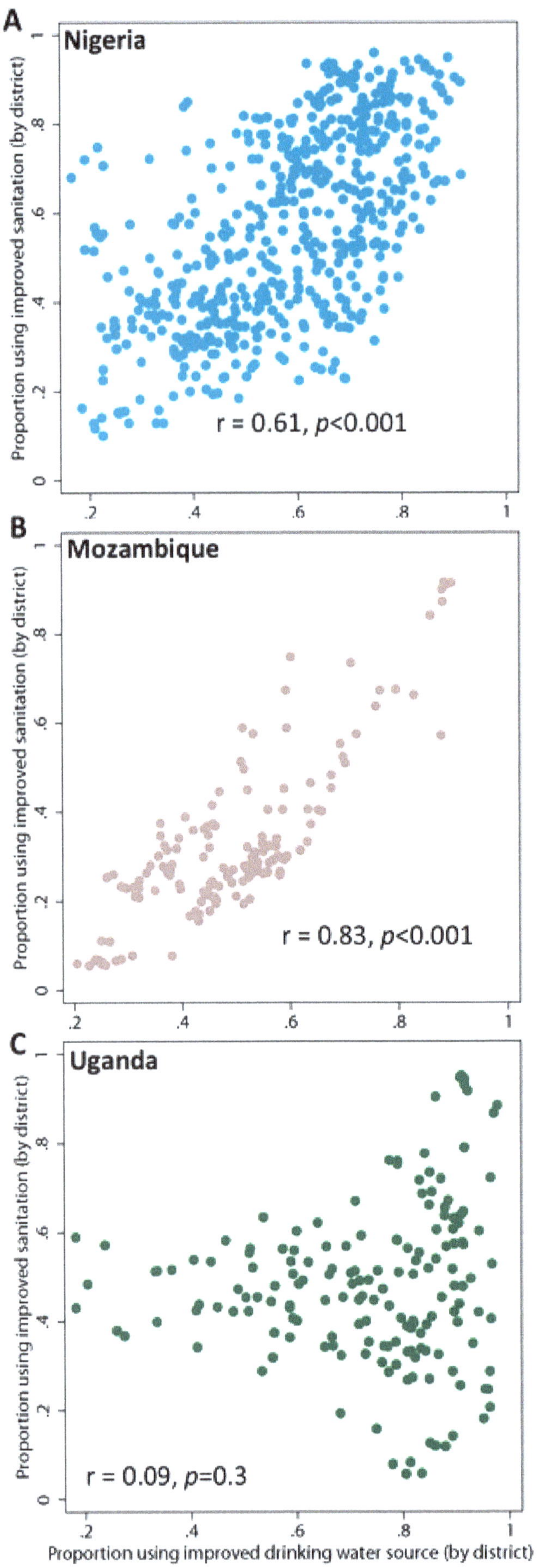

Figure 5: Comparison of coverage in improved drinking water against improved sanitation in overall population, by second administrative area.

Comparisons are made for (**A**) Nigeria, (**B**) Mozambique and (**C**) Uganda. *r* is the Pearson pairwise correlation coefficient. Each dot represents one administrative area.

http://dx.doi.org/10.1371/journal.pmed.1001626.g005

The distribution of coverage across all admin2 within each country is shown in Figure 6. For use of an improved drinking-water source, the majority of countries display a linear pattern of coverage, where the distance between quintiles is similar from lowest to highest levels of access. For access to improved sanitation (where coverage is generally lower) a larger number of countries display a top inequality pattern, meaning that coverage in the top quintile is substantially higher than the rest (e.g., Ethiopia, Gabon, Mali, Mozambique, Madagascar, South Africa). In contrast, several countries (including Malawi, Chad, and Sudan) display very little variation between quintiles, especially for access to improved sanitation.

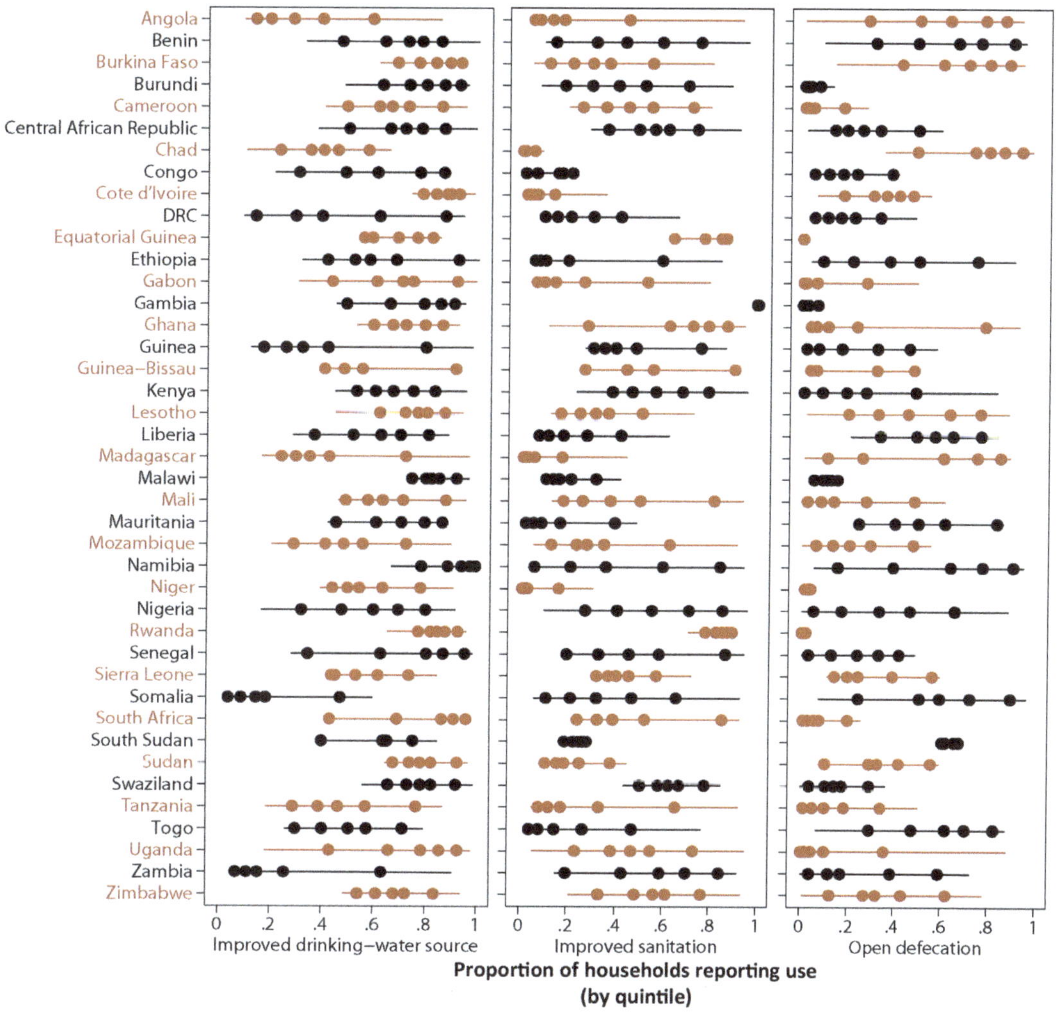

Figure 6: Distribution of modelled WSS coverage across administrative areas by country. Second administrative areas were stratified into quintiles based on coverage of each indicator. Dots show median proportion of households with access for each quintile; lines show the full range in coverage.

Inequality Analysis

The degree of geographic inequality in access to WSS varies substantially across SSA. As would be expected, plotting GINI against national coverage reveals strong negative relationships for all three indicators (range in r −0.78 to −0.88, $p<0.001$); the linear regression of GINI scores against national coverage thus provides a straightforward mechanism for identifying outlier countries with lower, or higher, levels of inequality than would be expected given their level of coverage. As shown in Figure 7, there are a number of outliers for all three WSS indicators, most noticeably for use of improved sanitation. For use of improved drinking water and any type of sanitation, higher than expected levels of inequality are seen across the full range of national coverage. In contrast, for use of improved sanitation countries with low national coverage are more likely to have higher than expected levels of geographical inequality. There are also some notable outliers with lower than expected geographical inequality given their national coverage rates, especially in access to improved sanitation. The first of these is South Sudan, although this is most likely due to a paucity of data to adequately describe sub-national variation. However Congo, Malawi, and Sierra Leone all display low geographical inequality in access to both improved drinking water and sanitation.

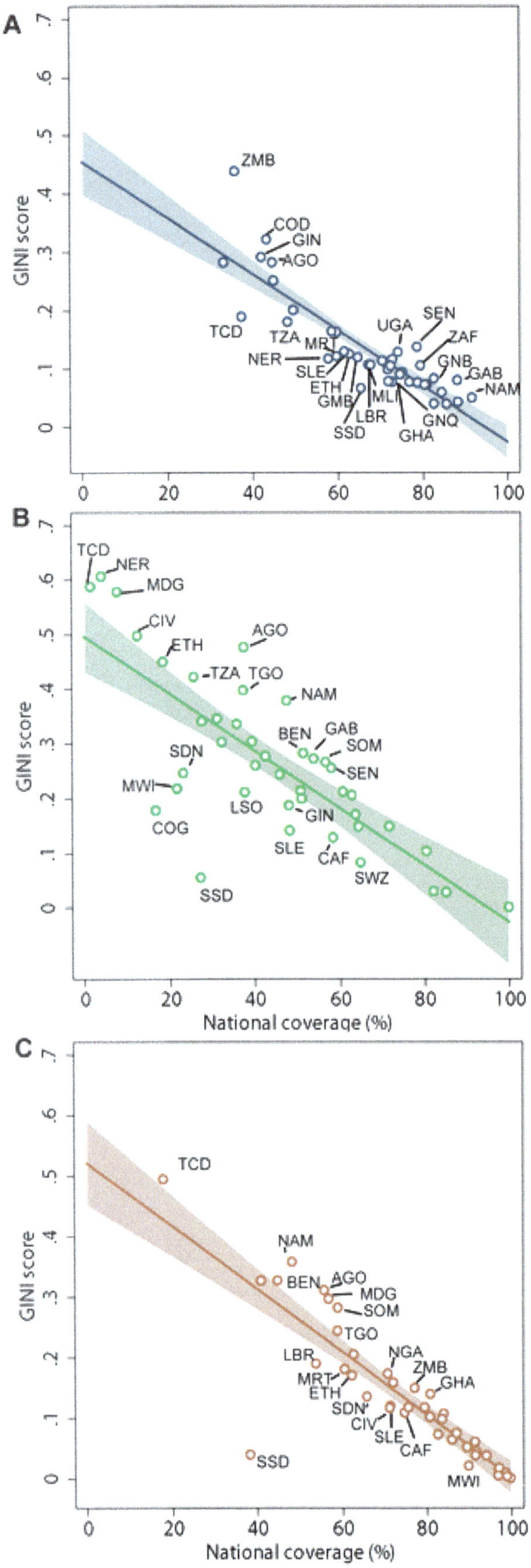

Figure 7: Empirical relationship between inequality (GINI score) as a function of national coverage.

Plots are shown for (**A**) use of improved drinking water, (**B**) use of improved sanitation facilities, and (**C**) use of any type of sanitation. All plots show the linear regression prediction (solid line) with 95% confidence interval (shaded area). Labelled countries (by 3-letter ISO codes) are those with GINI scores significantly higher or lower than would be expected, given national coverage.

For reference, the difference between the observed and expected GINI scores for the overall population (RGI) are provided in Table 2. There was no evidence for correlation between RGI scores for improved drinking water and sanitation for the overall population at national levels (r=0.18, p>0.2). When considering rural and urban populations separately however, there was some evidence that countries with higher levels of relative inequality for use of an improved drinking-water source (i.e., positive RGI scores) also experienced higher levels of inequality for use of improved sanitation (shown in Figure 8; rural populations r=0.47, p=0.002; urban populations r=0.39, p=0.01).

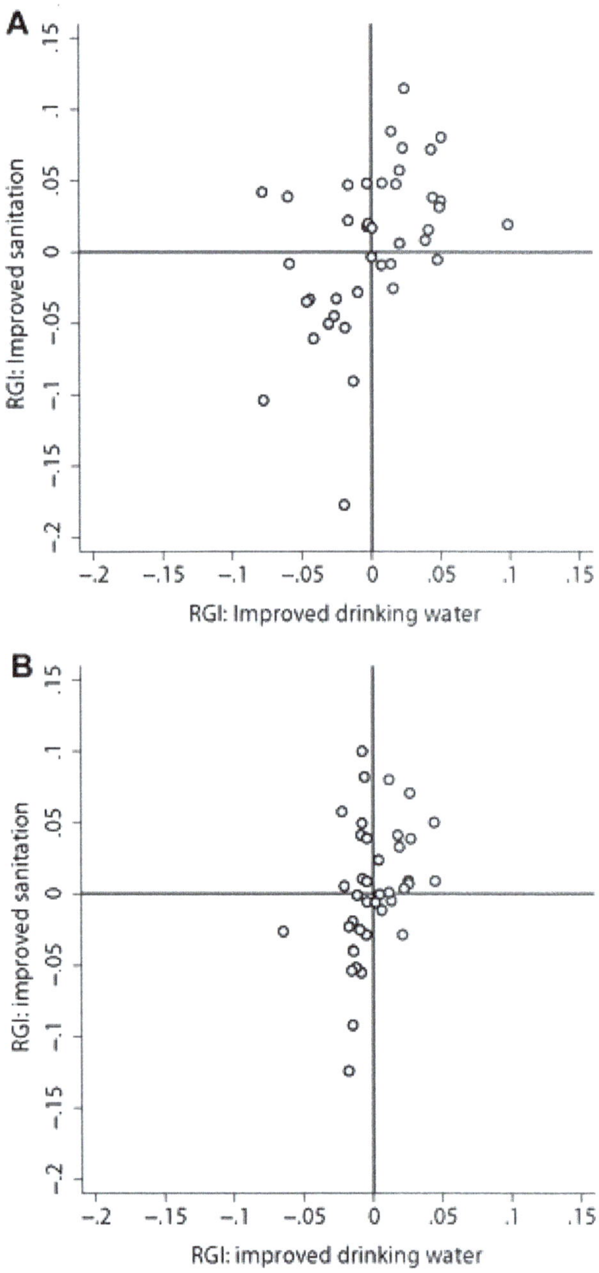

Figure 8: Relationship between relative geographical inequality for use of improved drinking water and RGI for use of improved sanitation for (**A**) rural populations (correlation (r)=0.47, p=0.002) and (**B**) urban populations (r=0.39, p=0.01).

DISCUSSION

It is already known that coverage of WSS varies considerably between countries in SSA [1]. To our knowledge, this is the first attempt to systematically map and analyse geographical inequalities in WSS within countries. We selected three widely reported WSS coverage indicators (access to improved drinking water, access to improved sanitation, and open defecation) and have generated a new resource for researchers and country stakeholders. Our robust analytical approaches, which explore geographical heterogeneity and geographic inequalities, showed consistent findings for all the three indicators that have important implications not only for the WSS sector, but also the international public health and development communities, policy makers, and donors.

In October 2013, key sector stakeholders issued a joint statement calling on the international community to ensure that the post-2015 development agenda be framed around the principle of equality, emphasising that future goals, targets, and indicators must rely on disaggregated data to allow inequalities to be effectively measured [42]. Our results suggest that there are substantial geographical inequalities in use of WSS across SSA that surpass simple urban-rural disparities and are of similar magnitude to the large socio-economic inequalities highlighted in a number of national studies [2]–[8],[12],[43]–[45]. In almost all countries, rural households in lowest coverage admin2 were 1.5 to 8 times less likely to access improved drinking water, 2 to 18 times less likely to access improved sanitation, and 2 to 80 times more likely to defecate in the open than rural households with the best access. Even across urban populations, coverage in use of improved drinking water varies on average by 30% nationally, use of improved sanitation by 5%, and dependence upon open defecation by 26% nationally, emphasising how population growth and rapid urbanisation may frequently outstrip service provision in poorer urban areas [46]. Considerable overlap in those admin2 with significantly worse coverage than the national average for both improved drinking water and improved sanitation suggests that vulnerable and marginalized populations often suffer the compounded effects of unsafe water and poor sanitation.

Perhaps our most striking finding, however, is that high geographical inequality (i.e., higher than would be expected given absolute national coverage) was seen across all levels of national coverage, although for improved sanitation this was typically greater than would be expected for those countries with lowest national coverage. In addition, those countries with high levels of relative inequality in improved drinking water also experienced higher levels of relative inequality in use of improved sanitation. This finding suggests that countries struggling to increase coverage often also struggle with issues of poor targeting of resources, or patchy implementation of government or NGO delivered interventions, and need to develop strategies and investment plans with reduction of inequalities in mind. Notably, evidence from across the public health and development arena suggests that, unless governments and stakeholders deliberately adopt strategies aimed at reaching lowest coverage areas and population groups, it is unlikely that countries will achieve universal coverage [43],[47]–[49]. It is likely that areas without access to improved water supply and basic sanitation are not only the poorest, but most challenging in terms of environmental conditions and demand for resilient infrastructure, such as low or inconsistent rainfall, poor soils and deep water tables, and a paucity of available markets and materials. The publication in 2012 of the first quantitative aquifer storage maps for the African continent reveals that most countries have the potential for sufficient well-placed hand-pumps to support rural populations; however, the potential for higher yielding boreholes (required to support rapid urbanisation) is limited and will require careful planning[50]. We hope that the information presented here, when combined with quantitative mapping of groundwater resources at sub-national levels, can provide a useful tool to accurately target WSS investments geographically to those admin2 where they are most needed. These data also demonstrate the promise of using geographically disaggregated data as a complementary tool to stratifying on the basis of socio-economic status [51] when systematically measuring progress in reducing inequalities within low coverage countries.

It is commonly recognised that there is no ideal measure for expressing the magnitude of inequalities [37],[52],[53], and as such we have presented both absolute (within country ranges and quintiles) and relative measures. Despite an increasing focus on spatial analysis within health research, there are relatively few publications concerning measurement methods for geographic inequality in health outcomes [54]–[57], and very few papers have explored the use of absolute or relative metrics for sub-national coverage data such as this [58],[59]. We have developed an RGI score, an adaptation of the GINI coefficient. This widely applied coefficient is a ratio analysis measure, and is independent of scale and population size, making it helpful for comparing diverse countries and groups within countries, such as urban and rural populations. It also takes full account of the entire distribution, rather than simply comparing those at the extremes. However, it is a relative measure, and as such countries with vastly differing national coverage can share the same score if the relative distribution is similar. Its use is also limited with bounded variables, such as coverage. Generating an RGI however, which compares the observed GINI coefficient with that expected given existing levels of coverage, helps circumvent the negative correlation that is to be expected when generating GINI coefficients using bounded indicators and allows identification of countries with large relative inequality in provision of WSS. Metrics such as these that accurately align with development principles, including equality of access, are urgently needed. However we would like to add a note of caution: measures of inequality should not be used in isolation, but rather in combination with indicators of absolute coverage. Importantly, equality of use does not reflect how much overall coverage has been achieved, which means that a country could achieve significantly low levels of relative inequality whilst still having very low overall coverage, as clearly demonstrated by Sierra Leone and Ethiopia.

Crucially, our results are entirely dependent upon the quality of the water and sanitation data available. DHS, MICs, and LSMS surveys contain a standardised water and sanitation module in which survey questions and response categories are fully harmonised [29], helping to ensure internal consistency and comparability between countries. However, the only available proxy for sustainable access to improved drinking water source and sanitation is self-reported use and ownership. These reports are rarely if ever supplemented by visual inspection to confirm the functionality and whereabouts of the water source, or correct use of the latrine. Whilst many epidemiological studies rely on self-report to measure type and use of such facilities, to our knowledge, very few studies have examined the reliability of self-reported water and sanitation practices and few have used objective measures to assess actual use [60].

Our analysis is based on those indicators currently recommended for monitoring the MDG targets, but these indicators are subject to considerable debate [61]. For example, not all water sources that are classified as improved will provide water that is safe to drink [62]. Correction with data on drinking water quality would likely have a large impact on estimates, calling into question whether the MDG water target would actually be met by 2015 [61]. For example, the only nationally representative water quality data that exists for SSA, to our knowledge, suggests that only 72% of improved drinking-water sources in Ethiopia were in compliance with WHO and national guidelines on drinking water quality [63]. There is currently massive lack of capacity in many countries to generate reliable and representative information on drinking water quality [64], and so we are unable to make appropriate corrections. Neither have we been able to account for the reliability of water supplies—studies across SSA have suggested that up to around one-third of hand-pumps may be non-functioning—nor of household treatment of water [65]. Our preliminary analysis of distance to water source however has suggested that only a little over half of all rural households reported as using an improved water source were using one within 30-minutes round trip of their household. This is especially striking given that a recent systematic review and meta-analysis has suggested significant increase in illness risk in people living further away from their water source [66]. Distance is an important determinant of the quantity of water brought to the household and used for drinking, cooking, and hygiene behaviours [65], and given our results we would suggest that future monitoring activities should always attempt to quantify the time taken for households to obtain their water.

Similarly controversial is the definition of improved sanitation. For example, a latrine that is correctly used by all members of the household and well maintained can provide an effective barrier to the transmission of faecal-oral disease, whereas a latrine that is not used correctly or that is poorly maintained can actually become a focus for the transmission of disease. Our reported use measures that rely upon the potentially subjective definition of an "improved" pit latrine. Again to avoid approximation, we only include latrines classified as having a slab as improved, which is supported by evidence from Tanzania suggesting the presence of helminth eggs in a large majority (71%) of soil samples collected from more simple pit latrines [67]. Nevertheless, it is not possible to correct for the appropriate use and cleanliness of all latrines classified as improved. Finally, as detailed in Box 1, we have classified all sanitation facilities that use improved technology as improved, regardless of whether they are shared, primarily to avoid approximating when data were unavailable (e.g., all pre-2004 surveys). Although this will include some public sanitation facilities (which may often be used by too many people, be poorly maintained, expensive or distant, and even post a risk for interpersonal violence[68]–[70]), in most instances these represent private facilities shared between five households or fewer. There is little evidence surrounding the health impacts of such facilities [71],[72], but it has been argued that a high proportion are probably safe for health [73]. This will however mean that our estimates of coverage for improved sanitation will be substantially higher than those produced by the JMP, who define all shared facilities as unimproved, preventing them from being directly comparable.

The multilevel modelling approach used here is transparent and flexible, offering advantages over the traditional country-specific linear regression models currently employed by the JMP. As demonstrated in a recent similar exploration of multilevel modelling for estimating national access to drinking water and sanitation [46], using a single model for all countries provides a continuous time series, providing additional information for those countries with scarce data. Our spatial Bayesian approach also enables use of information from neighbouring areas to improve predictions in unsampled areas [24], and quantifies the uncertainty resulting from modelling sub-national estimates at small spatial scales. Some limitations should be noted: we still relied on fitting linear time trends whereas a more flexible curve fitting approach might be more appropriate [46]; we did not include further covariates in the model; we did not consider uncertainties associated with the survey estimates arising from non-sampling errors; and we modelled household, not population, use (i.e., for each survey site we modelled the proportion of households reporting use). The latter may systematically bias our estimates upwards, as it is likely that individuals living in smaller (and potentially richer) households with greater access to WSS will be overrepresented. It will also be important to validate this work as new data become available.

Although not intended to reproduce the JMP, and not directly comparable for improved sanitation, in most instances our estimates are in keeping with official estimates (improved drinking water: r=0.77, root mean squared error [rmse] 11.3%; open defecation: r=0.92, rmse=7.7%) although they do differ substantially for individual countries, up to 33% for access to an improved drinking-water source and 19% for access to improved sanitation. These differences may be attributed in part to our hierarchical modelling approach—which combines both a more robust handling of temporal trends, and potentially more accurate population weighting as a result of disaggregation. However, the reduced amount of data available for this analysis for many countries (we did not have access to national census data or government surveys) will have played a very important role. For example, the largest difference we see for access to improved drinking water is for Guinea (32%), for which we relied on only two DHS surveys in contrast to the nine surveys and censuses available to JMP.

The implications of geographical inequalities in WSS coverage go beyond logistic considerations of effective service provision. Being able to target where inequalities exist will bring us closer to the goal of universal water and sanitation, which will fulfil human rights obligations. Understanding geographic inequities in WSS coverage can also provide insight into the epidemiology and control of many infectious diseases. Access to safe water and improved sanitation significantly reduce not only diarrhoeal disease [74],[75], but also water-borne diseases such as cholera, typhoid, and cryptosporidiosis [12], and NTDs such as soil-transmitted helminthiasis (STH), schistosomiasis, and trachoma [13]–[16],[76]. In recent years, there has been a concerted effort to develop geographical resources on a range of infectious and tropical diseases [77]–[79] and combining these data with knowledge of local WSS provision and other externalities [18],[80] can provide an empirical basis on which to interpret and predict the impact of public health programmes. For example, there is a large body of evidence that the risk of cholera transmission is greatly reduced as water and sanitation coverage improves [81]–[83], with recent mathematical models suggesting that WSS interventions may be as effective as the oral cholera vaccine in averting cholera cases, with the greatest impacts seen when the two are implemented in combination [84]. Similarly, it is unlikely that drug-based interventions alone will eliminate or control STH, schistosomiasis, and trachoma where WSS coverage remains low [18],[85]–[88]. Whilst growing resolve for collaboration and coordination between the WSS and health sectors is encouraging, there is still a pressing need to build a strong evidence base for collaborative programming [89]. For instance, it has been argued that an historic dissociation of WSS and health sectors has led to major problems in developing and maintaining essential infrastructure [65], and as such the public health community must play a role in setting health-based targets and indicators [51]. An essential component of this is the development of more effective and cross-sectoral coverage and impact indicators [89]. These should be able to effectively measure improvements in both WSS and infection control, and should consider in detail all potential factors that contribute to infection risk.

Although reducing inequalities was not a key element of the original health-related MDGs, there is a growing consensus that monitoring indicators solely at national levels fails to incentivise the targeting of areas of greatest need and potential greatest impact [7],[42],[90]. Here, we have revealed substantial levels of inequality in contemporary access to both improved drinking-water supplies and sanitation and open defecation within countries, and have shown how mapping the geographical distribution of WSS at policy relevant scales can help to make visible those deprived subgroups that were previously hidden within national statistics. As a consequence, we would urge the JMP to consider providing sub-national estimates of coverage within its country profiles, coupled with summary statistics of relative inequality. Mapping a phenomenon, however, does not explain it, and reasons for inequalities are likely to vary substantially between countries. Detailed investigation of the influence of contextual and programmatic factors on contemporary coverage, as well as changing patterns of inequality over time, are outside the scope of the present analysis, but are issues we intend to address in future work at both national and regional scales. More generally, this work has highlighted those countries that are struggling to target resources to areas of greatest need, and it is the responsibility of the international community to provide these countries with assistance in the development of strategies and investment plans to reduce inequality and marginalization. Finally, our intention was to create a vital resource for both researchers and planners, and to this end sub-national maps can be viewed on www.ntdmap.org and the data are available onwww.thiswormyworld.org.

ACKNOWLEDGMENTS

We would like to thank Professor Dean Jamison for comments and advice on an earlier version of this manuscript.

REFERENCES

1. UNICEF WHO (2013) Progress on sanitation and drinking-water - 2013 update. Geneva: World Health Organization and UNICEF.
2. WHO/UNICEF Joint Monitoring Programme for Water Supply and Sanitation (JMP) (2011) Drinking Water Equity, Safety and Sustainability: Thematic Report on Drinking Water 2011. Geneva: World Health Organization and UNICEF.
3. Parkar A, Gosling L (2011) Equity and Inclusion in Sanitation and Hygiene in Africa. 3rd AfricanSan Conference. Kigali, Rwanda.
4. Khan MM, Kramer A, Khandoker A, Prufer-Kramer L, Islam A (2011) Trends in sociodemographic and health-related indicators in Bangladesh, 1993–2007: will inequities persist? Bull World Health Organ 89: 583–593. doi: 10.2471/blt.11.087429
5. Bailie RS, carson BE, McDonals EL (2004) Water supply and sanitation in remote indigenous communities–priorities for health development. Aust N Z J Public Health 28: 409–414. doi: 10.1111/j.1467-842x.2004.tb00021.x
6. Soares LC, Griesinger MO, Dachs JN, Bittner MA, Tavares S (2002) Inequities in access to and use of drinking water services in Latin America and the Caribbean. Rev Panam Salud Publica 11: 386–396. doi: 10.1590/s1020-49892002000500013
7. Rheingans R, Cumming O, Andrewson J, Showalter J (2012) Exploring inequalities in sanitation-related disease burden and estimating the potential impacts of pro-poor targeting. Sanitation and Hygiene Applied Reserach for Equity Available: http://www.shareresearch.org/LocalResources/EquityResearchReport.pdf.
8. Fehr A, Sahin M, Freeman MC (2013) Sub-national Inequities in Philippine Water Access Associated with Poverty and Water Potential. J WASH Dev 3: 638–645. doi: 10.2166/washdev.2013.115
9. UNICEF WHO (2012) Progress on drinking-water and sanitation: 2012 update. Geneva: World Health Organization and UNICEF.
10. Lim SS, Vos T, Flaxman AD, Danaei G, Shibuya K, et al. (2012) A comparative risk assessment of burden of disease and injury attributable to 67 risk factors and risk factor clusters in 21 regions, 1990–2010: a systematic analysis for the Global Burden of Disease Study 2010. Lancet 380: 2224–2260.
11. Lopez AD, Mathers CD, Ezzati M, Jamison DT, Murray CJL (2006) Global Burden of Disease and Risk Factors. Washington DC: World Bank and Oxford University Press.
12. Bartram J, Cairncross S (2010) Hygiene, sanitation and water: forgotten foundations of health. PLoS Med 7: e1000367. doi: 10.1371/journal.pmed.1000367

13. Ziegelbauer K, Speich B, Mausezahl M, Bos R, Keiser J, et al. (2012) Effect of sanitation on soil-transmitted helminth infection: systematic review and meta-analysis. PLoS Med 9: e1001162. doi: 10.1371/journal.pmed.1001162

14. Esrey SA, Potash JB, Roberts L, Shiff C (1991) Effects of improved water supply and sanitation on ascariasis, diarrhoea, dracunculiasis, hookworm infection, schistosomiasis, and trachoma. Bull World Health Organ 69: 609–621.

15. Emerson PM, Lindsay SW, Alexander N, Bah M, Dibba SM, et al. (2004) Role of flies and provision of latrines in trachoma control: cluster-randomised controlled trial. Lancet 363: 1093–1098. doi: 10.1016/s0140-6736(04)15891-1

16. Stocks ME, Ogden S, Hadded D, Addiss DG, McGuire C, Freeman MC (2014) Effect of water, sanitation and hygiene on the prevention of trachoma: a systematic review and meta-analysis. PLoS Med 11: e1001605. doi: 10.1371/journal.pmed.1001605

17. Singer B, de Castro MC (2007) Bridges to sustainable tropical health. Proc Natl Acad Sci USA 104: 16038–16043. doi: 10.1073/pnas.0700900104

18. Spiegel JM, Dharamsi S, Wasan KMYA, Singer B, Hotez PJ, et al. (2010) Which new approaches to tackling neglected tropical diseases show promise? PLoS Med 7: e1000255. doi: 10.1371/journal.pmed.1000255

19. World Health Organization (2013) Sustaining the drive to overcome the global impact of neglected tropical diseases. Second WHO report on neglected tropical diseases. Geneva, Switzerland: World Heath Organization.

20. Noor AM, Amin AA, Akhwale WS, Snow RW (2007) Increasing coverage and decreasing inequity in insecticide-treated bed net use among rural Kenyan children. PLoS Med 4: e255. doi: 10.1371/journal.pmed.0040255

21. Soares Magalhães RJ, Clements ACA (2011) Mapping the Risk of Anaemia in Preschool-Age Children: The Contribution of Malnutrition, Malaria, and Helminth Infections in West Africa. PLoS Med 8: e1000438. doi: 10.1371/journal.pmed.1000438

22. Nube M, Sonneveld BG (2005) The geographical distribution of underweight children in Africa. Bull World Health Organ 83: 764–770.

23. Rao JNK (2003) Small area estimation Hoboken, New Jersey: John Wiley & Sons, Inc.

24. Gomez-Rubio V, Best N, Li G, Clarke P (2010) Bayesian statistics for small area estimation. Imperial College London Available: http://eprints.ncrm.ac.uk/1686/1/BayesianSAE.pdf.

25. Jiang J, Lahiri P (2006) Mixed model prediction and small area estimation. Test 15: 1–96. doi: 10.1007/bf02595419

26. Singh BB, Shukla GK, Kundu D (2005) Spatio-temporal models in small area estimation. Survey Method 31: 183–196.

27. Best N, Richardson S, Thomson A (2005) A comparison of Bayesian spatial models for disease mappping. Stat Method Med Res 14: 35–59. doi: 10.1191/0962280205sm388oa

28. Luh J, Baum R, Bartram J (2013) Equity in water and sanitation: developing an index to measure progressive realisation of the human right. Int J Hyg Environ Health 216: 662–671. doi: 10.1016/j.ijheh.2012.12.007

29. World Health Organization, UNICEF (2006) Core questions on drinking water and sanitation for household interviews. Geneva: World Health Organization and UNICEF.

30. United Nations (2003) Indicators for monitoring the Millennium Development Goals: definitions, rationale, concepts and sources. ST/ESA/STAT/SER.F/95. United Nations Publication

31. SALB The Second Administrative Level Boundaries data set project. Available: http://salbgeonetwork.grid.unep.ch/geonetwork/srv/en/main.home

32. The WorldPop Project. High resolution, contemporary data on human population distributions. Available: http://www.worldpop.org.uk

33. Linard C, Gilbert M, Snow RW, Noor AM, Tatem AJ (2012) Population Distribution, Settlement Patterns and Accessibility across Africa in 2010. PLoS ONE 7: e31743. doi: 10.1371/journal.pone.0031743

34. UNPD World urbanization prospects: the 2011 revision population database. New York: United Nations Population Division (UNDP).

35. Lunn DJ, Thomas A, Best N, Spiegelhalter D (2000) WinBUGS - a Bayesian modelling framework: concepts, structure, and extensibility. Stat Comp 10: 325–337.

36. Banerjee S, Carlin BP, Gelfand AE (2004) Hierarchical Modeling and Analysis for Spatial Data. Boca Raton, Florida: Chapman & Hall/CRC.

37. Wagstaff A, Paci P, van Doorslaer E (1991) On the measurement of inequalities in health. Soc Sci Med 33: 545–557. doi: 10.1016/0277-9536(91)90212-u

38. Pan American Health Organisation (2001) Measuring Health Inequalities: Gini Coefficient and Concentration Index. Epidemiol Bull PAHO 22: 3–4.

39. Schwanke Khilji SU, Rudge JW, Drake T, Chavez I, Borin K, et al. (2013) Distribution of selected healthcare resources for influenza pandemic response in Cambodia. Int J Equity Health 12: 82. doi: 10.1186/1475-9276-12-82

40. Green G, Yu BN, Marrie RA (2013) Exploring the implications of small-area variation in the incidence of multiple sclerosis. Am J Epidemiol 178: 1059–1066. doi: 10.1093/aje/kwt092

41. Jamison DT, Jha P, Malhotra V, Verguet S (2013) Human health: the twentieth century transformation of health - its magnitude and value. In: Lomborg B, editor. How much have global problems cost the world. New York: Cambridge University Press. pp. 214.

42. The Friends of Water, UNICEF, Special Rapportueur on the human right to safe drinking water and sanitation

OHCHR (2013) Joint statement. Measuring inequalities in the post-2015 development agenda. New York

43. Taylor-Robinson DC, Maayan N, Soares-Weiser K, Donegan S, Garner P (2012) Deworming drugs for soil-transmitted intestinal worms in children: effects on nutritional indicators, haemoglobin and school performance. Cochrane Database Syst Rev 7: CD000371. doi: 10.1002/14651858.cd000371.pub5

44. Hutton G, Haller L (2004) Evaluation of the costs and benefits of water and sanitation improvements at the global level Geneva: World Health Organization.

45. Hutton G, Haller L, Bartram J (2007) Global Cost-Benefit Analysis of Water Supply and Sanitation Interventions. J Water Health 5: 481–502. doi: 10.2166/wh.2007.009

46. Wolf J, Bonjour S, Pruss-Ustun A (2013) An exploration of multilevel modelling for estimating access to drinking-water and sanitation. J Water Health 11: 64–77. doi: 10.2166/wh.2012.107

47. Bundy DA, Kremer M, Bleakley H, Jukes MC, Miguel E (2009) Deworming and development: asking the right questions, asking the questions right. PLoS Negl Trop Dis 3: e362. doi: 10.1371/journal.pntd.0000362

48. Murray CJL, Vos T, Lozano R, Naghavi M, Flaxman AD, et al. (2012) Disability-adjusted life years (DALYs) for 291 diseases and injuries in 21 regions, 1990—2010: a systematic analysis for the Global Burden of Disease Study 2010. Lancet 380: 2197–2223.

49. Laxminarayan R, Chow J, Shahid–Salles SA (2006) Intervention Cost–Effectiveness: Overview of Main Messages. In: Jamison DT, Breman JG, Measham AR, Alleyne C, Claeson M et al.., editors. Disease Control Priorities in Developing Countries, 2nd Edition. New York Oxford University Press. pp. 35–86.

50. MacDonald AM, Bonsor HC, Dochartaigh BEO, Taylor RG (2012) Quantitative maps of groundwater resources in Africa. Environ Res Let 7: 024009. doi: 10.1088/1748-9326/7/2/024009

51. Special Rapportueur on the human right to safe drinking water and sanitation (2012) Integrating non-discrimination and equality into the post-2015 development agenda for water, sanitation and hygiene. Available: http://www.ohchr.org/en/Issues/WaterAndSanitation/SRWater/Pages/AnnualReports.aspx

52. Barros AJD, Victora CG (2013) Measuring coverage in MNCH: determining and interpreting inequalities in coverage of maternal, newborn and child health interventions. PLoS Med 10: e1001390. doi: 10.1371/journal.pmed.1001390

53. Harper S, King NB, Meersman SC, Reichman ME, Breen N, et al. (2010) Implicit value judgements in the measurement of health inequality. Milbank Q 88: 4–29. doi: 10.1111/j.1468-0009.2010.00587.x

54. Congdon P, Campos RM, Curtis SE, Southall HR, Gregory IN, et al. (2001) Quantifying and explaining changes in geographical inequality of infant mortality in England and Wales since the 1890s. Int J Popn Geog 7: 35–51. doi: 10.1002/ijpg.203

55. Anselmi L, Fernandes QF, Hanson K, Lagarde M (2013) Accounting for geograhpical inequalities in the assessment of equity in health care: a benefit incidence analysis. Lancet 381: S9. doi: 10.1016/s0140-6736(13)61263-5

56. Low A, Low A (2004) Measuring the gap: quantifying and comparing local health inequalities. J Public Health 26: 388–395. doi: 10.1093/pubmed/fdh175

57. Brown (1994) Using Gini-style indices to evaluate the spatial patterns of health practitioners: theoretical considerations and an application based on Alberta data. Soc Sci Med 38: 1243–1256. doi: 10.1016/0277-9536(94)90189-9

58. Reading R, Colver A, Openshaw S, Jarvis S (1994) Do interventions that improve immunisation uptake also reduce social inequalities in uptake? BMJ 308: 1142–1144. doi: 10.1136/bmj.308.6937.1142

59. Barr LM, Pressey RL, Fuller RA, Segan DB, McDonald-Madden E, et al. (2011) A new way to measure the World's protected area coverage. PLoS ONE 6: e24707. doi: 10.1371/journal.pone.0024707

60. Clasen T, Fabini D, Boisson S, Taneja J, Song J, et al. (2012) Making sanitation count: developing and testing a device for assessing latrine use in low-income settings. Environ Sci Technol 46: 3295–3303. doi: 10.1021/es2036702

61. Clasen TF (2012) Millennium Development Goals water target claim exaggerates achievement. Trop Med Int Health 17: 1178–1180. doi: 10.1111/j.1365-3156.2012.03052.x

62. Wright J, Gundry S, Conroy R (2004) Household drinking water in developing countries: a systematic review of microbiological contamination between source and point-of-use. Trop Med Int Health 9: 106–117. doi: 10.1046/j.1365-3156.2003.01160.x

63. Tadesse D, Desta A, Geyid A, Girma W, Fisseha S, et al.. (2010) Rapid assessment of drinking-water quality in the Federal Democratic Republic of Ethiopia. Country report of the pilot project implementation in 2004–2005. Geneva: World Health Organization and UNICEF.

64. WHO/UNICEF Joint Monitoring Programme for Water Supply and Sanitation (JMP) (2010) JMP Technical Task Force Meeting on Monitoring Drinking-water Quality. Geneva: World Health Organization and UNICEF.

65. Hunter PR, MacDonald AM, Carter RC (2010) Water Supply and Health. PLoS Med 7: e1000361. doi: 10.1371/journal.pmed.1000361

66. Wang X, Hunter PR (2010) Short report: systematic review and meta-analysis of the association between self-reported diarrhoeal disease and distance from home water source. Am J Trop Med Hyg 83: 582–584. doi: 10.4269/ajtmh.2010.10-0215

67. Baker SM, Ensink JH (2012) Helminth transmission in simple pit latrines. Trans R Soc Trop Med Hyg 106: 709–710. doi: 10.1016/j.trstmh.2012.08.002

68. Bapat M, Agarwal I (2003) Our needs, our priorities; women and men from the slums in Mumbai and Pune

talk about their needs for water and sanitation. Environ Urban 15: 71–86. doi: 10.1630/095624703101286754

69. World Bank (2006) Kenya, Inside Informality: Poverty, Jobs, Housing and Services in Nairobi's Slums (No. 36347-KE). Water and Urban Unit 1, Africa Region, World Bank

70. Amnesty International (2010) Kenya: Insecurity and Indignity: Women's Experiences in the Slums of Nairobi, Kenya. London, UK: Amnesty International Publications.

71. Montgomery MA, Desai MM, Elimelech M (2010) Comparing the effectiveness of shared versus private latrines in preventing trachoma in rural Tanzania. Am J Trop Med Hyg 82: 693–695. doi: 10.4269/ajtmh.2010.09-0540

72. Tumwebaze IK, Orach CG, Niwagaba C, Luthi C, Mosler HJ (2013) Sanitation facilities in Kampala slums, Uganda: users' satisfaction and determinant factors. Int J Environ Health Res 23: 191–204. doi: 10.1080/09603123.2012.713095

73. WHO/UNICEF Joint Monitoring Programme for Water Supply and Sanitation (JMP) (2010) JMP Technical Task Force Meeting on Sanitation and Methods for Estimating Progress. Geneva: World Health Organization and UNICEF.

74. Waddington H, Snilstveit B, White H, Fewtrell L (2009) Water, sanitation and hygiene interventions to combat diarrhoea in developing countries. New Delhi, India: International Initiative for Impact Evaluation. p 119.

75. Pullan RL, Sturrock HJW, Soares Magalhaes RJ, Clements ACA, Brooker SJ (2012) Spatial parasite ecology and epidemiology: a review of methods and applications. Parasitol 139: 1870–1887. doi: 10.1017/s0031182012000698

76. Schmidlin T, Hurlimann E, Silue KD, Yapi RB, Houngbedji C, et al. (2013) Effects of Hygiene and Defecation Behavior on Helminths and Intestinal Protozoa Infections in Taabo, Côte d'Ivoire. PLoS ONE 8: e65722. doi: 10.1371/journal.pone.0065722

77. Pullan RL, Brooker SJ (2012) The global limits and population at risk of soil-transmitted helminth infections in 2010. Parasit Vectors 5: 81. doi: 10.1186/1756-3305-5-81

78. Smith JL, Flueckiger RM, Hooper PJ, Polack S, Cromwell EA, et al. (2013) The geographical distribution and burden of trachoma in Africa. PLoS Negl Trop Dis 7: e2359. doi: 10.1371/journal.pntd.0002359

79. Schur N, Hurlimann E, Garba A, Traore MS, Ndir O, et al. (2011) Geostatistical Model-Based Estimates of Schistosomiasis Prevalence among Individuals Aged</=20 Years in West Africa. PLoS Negl Trop Dis 5: e1194. doi: 10.1371/journal.pntd.0001194

80. Manderson L, Aagaard-Hansen J, Allotey P, Gyapong M, Sommerfeld J (2009) Social research on neglected diseases of poverty: Continuing and emerging themes. PLoS Negl Trop Dis 3: e332. doi: 10.1371/journal.pntd.0000332

81. Waldman RJ, Mintz ED, Papowitz HE (2013) The cure for cholera: improving access to safe water and sanitation. N Engl J Med 368: 592–594. doi: 10.1056/nejmp1214179

82. Sepulveda J, Valdespino JL, Garcia-Garcia L (2006) Cholera in Mexico: the paradoxical benefits of the last pandemic. Int J Infect Dis 10: 4–13. doi: 10.1016/j.ijid.2005.05.005

83. Roses Periago M, Frieden TR, Tappero JW, De Cock KM, Aasen B, et al. (2012) Elimination of cholera transmission in Haiti and the Dominican Republic. Lancet 379: e12–e13. doi: 10.1016/s0140-6736(12)60031-2

84. Dangour AD, Watson L, Cumming O, Boisson S, Che Y, et al. (2013) Interventions to improve water quality and supply, sanitation and hygiene practices, and their effects on the nutritional status of children. Cochrane Database Syst Rev 8: CD009382. doi: 10.1002/14651858.cd009382.pub2

85. Ngondi JM, Gebre T, Shargie EB, Adamu L, Teferi T, et al. (2010) Estimation of effects of community intervention with antibiotics, facial cleanliness, and environmental improvement (A,F,E) in five districts of Ethiopia hyperendemic for trachoma. Br J Ophthalmol 94: 278–281. doi: 10.1136/bjo.2009.168260

86. Ngondi JM, Matthews F, Reacher M, Baba S, Brayne C, et al. (2008) Associations between active trachoma and community intervention with Antibiotics, Facial cleanliness, and Environmental improvement (A,F,E). PLoS Negl Trop Dis 2: e229. doi: 10.1371/journal.pntd.0000229

87. World Health Organization (2012) Eliminating soil-transmited helminthiases as a public health problem in children. Progress report 2001–2010 and strategic plan 2010–2020. Geneva: World Health Organization.

88. Utzinger J, Raso G, Brooker SJ, de Savigny D, Tanner M, et al. (2009) Schistosomiasis and neglected tropical diseases: towards integrated and sustainable control and a work of caution. Parasitol 136: 1859–1874. doi: 10.1017/s0031182009991600

89. Freeman MC, Ogden S, Jacobson J, Abbott D, Addiss DG, et al. (2013) Integration of water, sanitation and hygiene for the prevention and control of neglected tropical diseases: a rationale for inter-sectoral collaboration. PLoS Negl Trop Dis 7: e2439. doi: 10.1371/journal.pntd.0002439.

90. UN System Task Team on the post-2015 UN development agenda (2013) Statistics and indicators for the post-2015 development agenda. New York: United Nations.

91. WHO/UNICEF Joint Monitoring Programme (JMP) for Water Supply and Sanitation. Available: http://www.wssinfo.org/fileadmin/user_upload/resources/1268174016-JMP_Core_Questions.pdf.

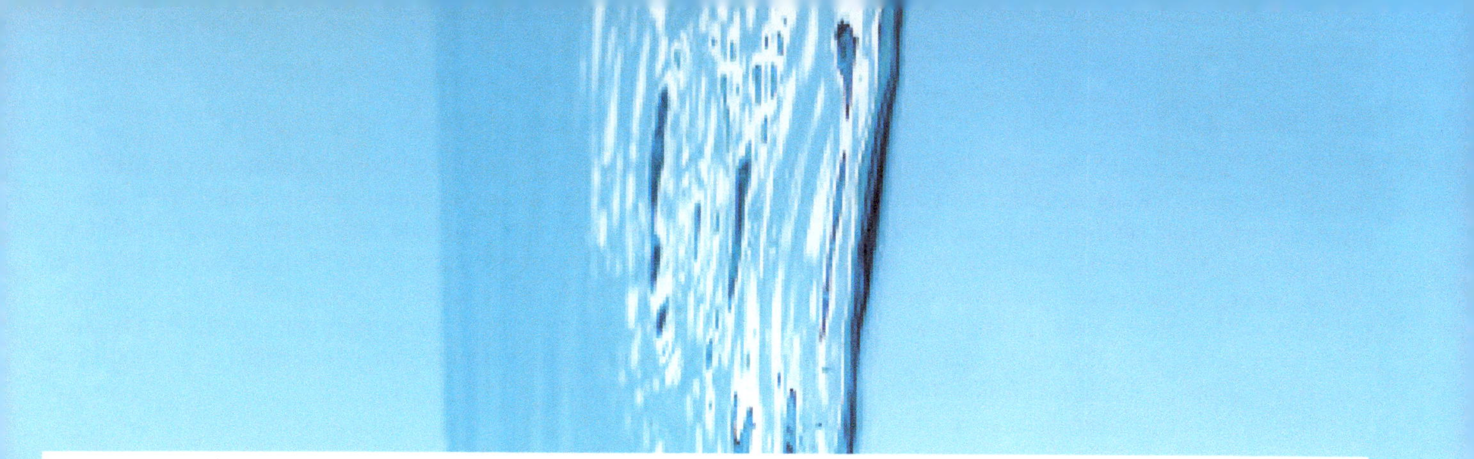

CHAPTER 28

FECAL CONTAMINATION OF DRINKING-WATER IN LOW- AND MIDDLE-INCOME COUNTRIES: A SYSTEMATIC REVIEW AND META-ANALYSIS

Robert Bain, Ryan Cronk, Jim Wright, Hong Yang, Tom Slaymaker, Jamie Bartram

EDITORS' SUMMARY

Background

Access to clean water is fundamental to human health. The importance of water to human health and wellbeing is encapsulated in the Human Right to Water, reaffirmed by the United Nations in 2010, which entitles everyone to "sufficient, safe, acceptable and physically accessible and affordable water for personal and domestic uses." A step towards such universal access to water is Millennium Development Goal (MDG) target 7c that aims to halve the proportion of the population without sustainable access to safe drinking-water. One of the indicators to help monitor progress towards this target used by the Joint Monitoring Project (JMP—an initiative of the World Health Organization and UNICEF) is "use of an improved source." Improved sources include piped water into a dwelling, yard, or plot, or a standpipe, borehole, and protected dug well. Unimproved sources are those that do not protect water from outside contamination, such as unprotected wells, unprotected springs, and surface waters.

Why Was This Study Done?

While this simple categorization may reflect established principles of sanitary protection, this indicator has been criticized for not adequately reflecting safety, suggesting that reported access to safe water might be overestimated by billions of people by not accounting for microbial water safety or more fully accounting for sanitary status. So the researchers conducted a systematic review and meta-analysis to investigate whether water from improved sources is less likely to exceed health-based guidelines for microbial water quality than water from unimproved sources and to what extent microbial contamination varies between source types, between countries, and between rural and urban areas.

What Did the Researchers Do and Find?

The researchers comprehensively searched the literature to find appropriate studies that investigated fecal contamination of all types of drinking-water in low and middle-income countries. The researchers included studies that contained extractable data on *Escherichia coli*or thermotolerant coliform (the WHO recommended indicators of fecal contamination) collected by appropriate techniques. The authors also assessed studies for bias and quality and used a statistical method (random effects meta-regression) to investigate risk factors and settings where fecal contamination of water sources was most common.

Citation: Bain R, Cronk R, Wright J, Yang H, Slaymaker T, Bartram J (2014) Fecal Contamination of Drinking-Water in Low- and Middle-Income Countries: A Systematic Review and Meta-Analysis. PLoS Med 11(5): e1001644. doi:10.1371/journal.pmed.1001644

Using these methods, the authors included 319 studies reporting on 96,737 water samples. Most studies were from sub-Saharan Africa, southern Asia, or Latin America and the Caribbean. They found that overall, the odds (chance) of contamination within a given study were considerably lower for "improved" sources than "unimproved" sources (odds ratio=0.15). However, in 38% of 191 studies, over a quarter of samples from improved sources contained fecal contamination. In particular, protected dug wells were rarely free of fecal contamination. The researchers also found that water sources in low-income countries, and rural areas were more likely to be contaminated (both had odds ratios of 2.37).

What Do These Findings Mean?

These findings show that while water from improved sources is less likely to contain fecal contamination than unimproved sources, they are not consistently safe. This study also provides evidence that by equating "improved" with "safe," the number of people with access to a safe water source has been greatly overstated, and suggests that a large number and proportion of the world's population use unsafe water. As studies rarely reported stored water quality or sanitary risks, the accuracy of these findings may be limited. Nevertheless, the findings from this study suggest that the Global Burden of Disease 2010 may greatly underestimate diarrheal disease burden by assuming zero risk from improved water sources and that new indicators are needed to assess access to safe drinking water. Therefore, greater use should be made of other measures, such as sanitary inspections, to provide a complementary means of assessing safety and to help identify corrective actions to prevent water contamination.

INTRODUCTION

The importance of water to human health and wellbeing is encapsulated in the Human Right to Water and Sanitation, which entitles everyone to "sufficient, safe, acceptable physically accessible and affordable water for personal and domestic uses" [1], as reaffirmed by the United Nations General Assembly and Human Rights Council in 2010 [2]. Millennium Development Goals (MDGs) Target 7c aims "to halve the proportion of the population without sustainable access to safe drinking-water ..." [3], a step towards universal access. "Use of an improved source" was adopted as an indicator for monitoring access to safe drinking-water globally (Table 1) and relies on national censuses and nationally representative household surveys as the primary sources of data.

Table 1: Types of improved source and the estimated proportion of the global population using these as their primary source of drinking-water

Source Category[a]	Description	Global Population Using Water Source in 2010[b] (%)		
		Urban	Rural	Total
Household or yard connection	Piped water into dwelling, also called a household connection, is defined as a water service pipe connected with in-house plumbing to one or more taps. Piped water to yard/plot, also called a yard connection, is defined as a piped water connection to a tap placed in the yard or plot outside the dwelling.	80	29	54
Standpipe	Public tap or standpipe is a public water point from which people can collect water. A standpipe is also known as a public fountain or public tap. Public standpipes can have one or more taps and are typically made of brickwork, masonry, or concrete.	6	8	7
Borehole	Tubewell or borehole is a deep hole that has been driven, bored, or drilled, with the purpose of reaching groundwater supplies. Boreholes/tubewells are constructed with casing, or pipes, which prevent the small diameter hole from caving in and protects the water source from infiltration by runoff water.	8	30	18
Protected dug well	Protected dug well is a dug well that is protected from runoff water by a well lining or casing that is raised above ground level and a platform that diverts spilled water away from the well. A protected dug well is also covered, so that bird droppings and animals cannot fall into the well.	2[c]	10[c]	6[c]
Protected spring	The spring is typically protected from runoff, bird droppings, and animals by a "spring box," which is constructed of brick, masonry, or concrete and is built around the spring so that water flows directly out of the box into a pipe or cistern, without being exposed to outside pollution.	<1[c]	3[c]	2[c]
Rainwater	Rainwater refers to rain that is collected or harvested from surfaces (by roof or ground catchment) and stored in a container, tank, or cistern until used.	<1	2	1

Source: UNICEF/WHO [5].
[a]Households using bottled water as their primary source of water for drinking are generally considered to use an improved source if one of the above sources is used for washing and cooking. The JMP estimates that 6% of the urban population and 1% of the rural population primarily use bottled water as their source of drinking-water.
[b]An estimated 3% of the global population use surface waters and a further 8% use "other unimproved sources" such as tanker trucks, unprotected dug wells, and unprotected springs.
[c]Published estimates include do not distinguish protected and unprotected. In the absence of data we assumed half are protected.

The Joint Monitoring Programme for Water Supply and Sanitation (JMP) of WHO/UNICEF categorizes a drinking-water source type as improved if "by nature of its construction or through active intervention, [it] is protected from outside contamination, in particular from contamination with faecal matter" [4]. Improved source types include piped water into dwelling, yard, or plot, standpipe, borehole, protected dug well or spring, and rainwater. Unimproved source types are those that do not protect water from outside contamination (unprotected wells, unprotected springs, surface waters, and tanker trucks). While the categorization reflects well-established principles of sanitary protection, on announcing that the target had been met in 2010, the JMP cautioned that the MDG indicator does not take water quality measurements into account [5]. The indicator has been criticized for not adequately reflecting safety [6]–[8], with some estimates suggesting that reported access to safe water might be overestimated by billions of people [9],[10], by not accounting for microbial water safety [8] or more fully accounting for sanitary status [9].

Diseases related to contamination of drinking-water constitute a major burden on public health. The principal risk to health is from ingestion of water contaminated with feces containing pathogens that cause infectious diseases such as cholera and other diarrheal diseases, dysenteries, and enteric fevers [11],[12]. The burden of water-related disease varies according to context and is highest in low-income settings where diarrhea remains a leading cause of child deaths [13]. Systematic reviews of epidemiological evidence from intervention studies[14]–[18], and especially outbreak investigations [19],[20], suggest drinking-water quality plays an important role in fecal-oral transmission, though the magnitude of the effect has been contested owing to a limited number of blinded trials [21]. It is difficult to isolate the effects of one component of the multiple and interrelated fecal-oral pathways, which are highly context-specific.

WHO publishes widely recognized Guidelines for Drinking-water Quality (GDWQ) (4th edition) that include criteria for assessing health risks and setting targets for improving water safety[12]. Direct measurement of pathogens is complex but techniques for assessing fecal contamination using fecal indicator bacteria (FIB) are well established and widely applied. The WHO GDWQ recommend using *E. coli*, or alternatively thermotolerant coliform (TTC), and new enzymatic methods have made quantification simpler, cheaper, and more robust [22],[23]. The WHO GDWQ recommend that *E. coli*, or alternatively TTC, be used in assessing fecal contamination of drinking-water [12]. The WHO guideline value for *E. coli* ("none detected in any 100-ml sample") [12] is reflected in the standards of most OECD member states and low- and middle-income countries (LMICs). The WHO GDWQ further suggest the use of a risk classification to prioritize interventions as higher levels of indicator organisms are generally indicative of greater levels of fecal contamination. A commonly used risk classification is based on the number of indicator organisms in a 100 ml sample, which includes: <1, "very low risk"; 1–10, "low risk"; 10–100, "medium risk"; >100, "high risk" or "very high risk" [24],[25]. However FIB are imperfect and their level does not necessarily equate to risk [26]; since quality varies both temporally and spatially, occasional sampling may not accurately reflect actual exposure.

A complementary approach in safety assessment is the identification of hazards and preventative risk management measures through "sanitary inspection" of a water source and its surroundings [24],[27]. The improved source indicator is in effect a very simplified form of sanitary inspection. Like FIB, sanitary inspections have long been a tool in assessing drinking-water safety. In 1904, Prescott and Winslow stated, "[t]he first attempt of the expert called in to pronounce upon the character of a potable water should be to make a thorough sanitary inspection…" [28]. Standardized forms can be used to assess sanitary risk and derive a summary measure, the sanitary risk score. These forms typically include questions about the integrity of protective elements, such as fencing or well covers, and the proximity of hazards such as latrines; forms are available for different types of water source. Like water quality, some sanitary risk factors may vary spatially and temporally. The approach can be combined with microbiological analysis, either to yield a risk cross-tabulation [24],[25] or as a part of a more detailed Water Safety Plan [29].

In January 2012, WHO and UNICEF established working groups to develop targets and indicators for enhanced global monitoring of drinking-water, sanitation, and hygiene post-2015. The water working group proposed to continue using the improved water source classification as part of a revised set of indicators for assessing progressive improvements in service [30]. This review was commissioned to assist the group in evaluating the evidence linking improved source types and health-related indicators of water quality. The following specific questions were considered in order to determine the potential and limits of classification by source type in assessing safety in future global reporting: (i) Is water from improved sources less likely to exceed health-based guidelines for microbial water quality than water from unimproved sources? (ii) To what extent does microbial contamination vary between source types, between countries, and between rural and urban areas? (iii) Are some types of water source associated with higher risk scores as assessed by sanitary inspection?

METHODS

We conducted a systematic review of studies of fecal contamination of drinking-water in LMICs in adherence with PRISMA guidelines[31].

Search Strategy

Studies were identified from both peer-reviewed and grey literature. To identify peer-reviewed literature, the topic "water quality" was combined with terms to restrict the search to drinking-water and either a measure of microbial water quality (e.g., "coli") or sanitary risk (e.g., "sanitary inspection"). We further restricted the search to LMICs using a list of country names based on the MDG regions [32]. Online databases were searched including PubMed, Web of Science, and the Global Health Library. Grey literature was sourced from a variety of sites including those used in previous drinking-water–related reviews [33]–[35]. Translated search terms (Chinese, French, Portuguese, and Spanish) were used to identify additional studies. An email requesting submissions of relevant studies was distributed to water sector professional networks. We searched bibliographies of included studies and contacted authors where full texts could not be obtained through other means. Searches were conducted between 7th January and 1st August 2013.

Eligibility and Selection

Studies were included in the review provided they: reported on water quality, at either the point of collection or consumption, from sources used for drinking that would not be classified as surface waters by the JMP; contained extractable data on TTC or *E. coli* with sample volumes not less than 10 ml; were published between January 1990 (the baseline year for MDG targets) and August 2013; included results from at least ten separate water samples from different water sources of a given type or, in the case of piped systems, individual taps, and in the case of packaged waters, brands; reported data from LMICs as defined by the MDG regions [32](thereby excluding 55 high-income countries, comprising 18.1% of the global population in 2010 [36]); were published in languages spoken by at least one author (Chinese, English, French, Portuguese, or Spanish); and included sufficient detail about the water sources and associated results with a water source with sufficient detail to be categorized (refer to Figure 1for details). Other indicators such as coliphage and direct pathogen detection are not as widely used and are not included in this review [37]. We did not include studies that only assessed surface waters as these are generally considered unfit for drinking. We included bottled water and sachet water that do not form part of the JMP improved source classification (which is concerned with the household's primary source of water for drinking, cooking, and personal hygiene [38]) but are nonetheless important sources of drinking-water in many countries.

Independent primary screening of English language titles and abstracts for studies was conducted by two authors (RB and RC). If any reviewer selected a study, we referred to the full text. Data from eligible studies were extracted into a standardized spreadsheet and 10% of the English language texts were subjected to independent quality control by a second author (RB and RC). Screening and extraction of data in other languages was conducted by one author (RB or HY).

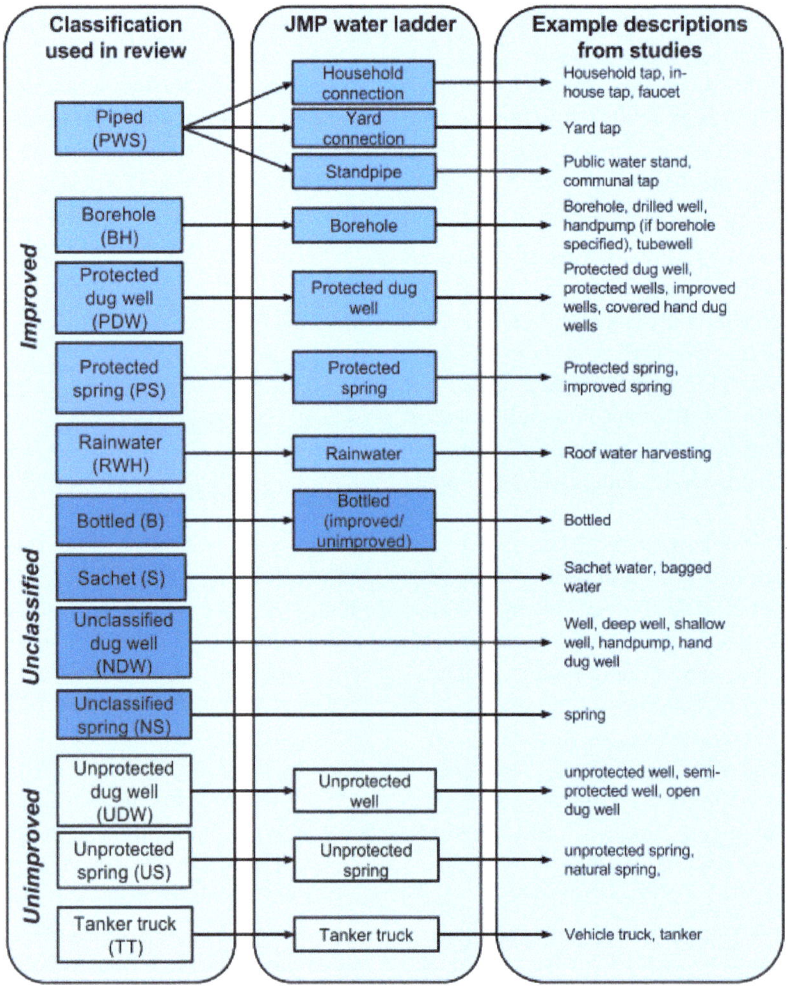

Figure 1: Matching drinking-water source types to the classification used by the Joint Monitoring Programme.

Data Extraction and Matching

Where possible we extracted or calculated the following information for each type of water source in the studies: (i) total number of samples and proportion containing *E. coli* or TTC; (ii) proportion of samples within microbial risk categories (<1 or not detected, 1–10, 10–100, and >100 *E. coli* or TTC per 100 ml); (iii) geometric mean, mean, or median levels of *E. coli* or TTC; and (iv) risk categories according to the sanitary inspection ("low," "medium," "high," and "very high" risk) as reported in the studies. For intervention studies (other than the provision of an improved source, for example the protection of unprotected springs), estimates could be based on either the baseline or control group; when both were available we used whichever had the largest sample size. For studies reporting both *E. coli* and TTC, we used only the *E. coli*results. Where repeated measures were taken at the same source and data permitted we extracted the lowest compliance level (e.g., wet season data) with WHO Guideline values as well as the overall proportion of samples containing FIB. We identified countries as "low," "lower middle," "upper middle," and "high" income using the 2013 World Bank classification[39]. We recorded whether studies took place during or shortly after emergencies or natural disasters and if they were in non-household settings such as schools and health facilities. We identified additional study characteristics expected to influence water quality, including the setting (urban/rural), season (wet/dry or period of sampling), and study design [34].

Each type of water source in a given study was classified as improved or unimproved and matched to a specific water source type following the classification used in household surveys including the Demographic and Health Surveys [38]. We recorded whether samples had been taken directly from the water source or after storage, for example in the home. Where the appropriate match could not be determined, our approach differed depending on the type of source. We grouped groundwater sources from studies that did not distinguish between protected and unprotected (unclassified dug well, unclassified spring) and we created groups for studies of other sources such as bottled and sachet water. Further information about the matching is available in Figure 1.

Study Quality and Bias

Studies were rated for quality on the basis of the criteria summarized in Table 2. A quality score between 0 and 13 for each study was determined on the basis of the number of affirmative responses. We also categorized studies on the basis of anticipated susceptibility to bias in estimating the compliance to health-based guidelines and the extent of microbial contamination; our categories were: case-control or cohort, intervention, diagnostic study, cross-sectional survey, and longitudinal survey. Any study of at least 6 months duration and more than two samples at each water point was categorized as longitudinal. We identified studies where authors indicated whether selection was intended to be representative or selection had been randomized.

Table 2: Quality criteria used to assess studies of microbial water quality

Criterion	Question
Selection described	Do the authors describe how the water samples were chosen, including how either the types of water source or their users were selected?
Selection representative	Did the authors detail an approach designed to provide representative picture water quality in a given area?
Selection randomized	Was sampling randomized over a given study area or population?
Region described	Does the study report the geographic region within the country where it was conducted?
Season reported	Were the seasons or months of sampling reported?
Quality control	Were quality control procedures specified or referred to?
Method described	Are well-defined and appropriate methods of microbial analysis described or referenced?
Point of sampling	Was the point at which water was sampled well defined? (For example whether the water was collected from within a household storage container or directly from a water source)
Handling described	Are sample handling procedures described, including sample collection, transport method, and duration?
Handling minimum criteria	Does sample handling and processing meet the following criteria: transport on ice or between 2–8°C, analysis within 6 hours of collection, and specified incubation temperature?
Accredited laboratory	Was the microbial analysis conducted in an accredited laboratory setting?
Trained technician	Do the authors state whether trained technicians conducted the water quality assessments or the analyses were undertaken by laboratory technicians?
External review	Was the study subject to peer review or external review prior to publication?

Analysis

Because of the extent of heterogeneity between studies, we chose to plot cumulative density functions (CDFs) of the proportion of samples with detected (>1 per 100 ml) and high (>100 per 100 ml) FIB in each study to compare water source types between studies. This approach has been used in a systematic review of prevalence of schizophrenia [40]. CDFs are used to qualitatively assess the proportion of studies reporting frequent and high levels of microbial contamination. Measures of central tendency from studies were not included in the meta-analysis because of limited reporting of measures of dispersion, inadequate explanation of the handling of censored data, and the difficulty in reconciling diverse reported measures of central tendency (e.g., geometric versus arithmetic mean) [41].

Random effects meta-regression was used to investigate risk factors and settings where fecal contamination is most common and other possible explanations for the observed heterogeneity between studies [42]. A logit transformation is recommended for the analysis of proportions[43] and was applied to both the proportion of samples with detectable (>1 per 100 ml) and high (>100 per 100 ml) levels of FIB. The *metareg* function in Stata was used after a continuity correction of ±0.5 where the proportion of samples positive was zero or one [44], and we estimated the within study variance for each proportion as the reciprocal of the binomial variance [45]. Subgroup analysis included variables defined *a priori* (including water source type, rural versus urban, and income-level) and defined *a posteriori* (for example if piped water had been treated prior to distribution). We separately evaluated piped and other improved sources for those variables reaching significance at the 5% level in bivariate analysis for all source types.

Studies that included both improved and unimproved sources were then combined using meta-analysis with the odds ratio (OR) as the effect measure. We calculated a pooled estimate of the protective effect of an improved source and corresponding confidence intervals using the*metan* function in Stata. We then assessed the influence of small study bias by the funnel plot method and performed an Egger's test using a normal likelihood approximation. The extent of heterogeneity in protective effect was determined using Higgins I^2 and corresponding confidence intervals were calculated [42]. Calculations were performed in Stata 13SE.

RESULTS

Search Results

As shown in Figure 2, in total, 6,586 reports were identified through database searches. A further 1,274 reports were identified from grey literature and correspondence with experts. Most studies were excluded because they did not test water that was clearly used for drinking, did not associate results with a water source type, or did not include enough different water sources or in the case of packaged water, brands. Studies often did not provide an adequate description of the water sources to allow them to be matched to the JMP source categories; this limitation was particularly the case for ground water sources. For example, several studies reported results for "hand pumps" (a description of the technology above ground) but did not provide details about well construction. Although these may often be boreholes, hand pump conversions are also applied to dug wells. Other studies simply described water sources as "wells" or "springs." Some studies provided details that are not captured in the JMP classification, such as whether water from a piped supply had been treated. Full texts could not be obtained for 99 potentially relevant reports, many of which were conference presentations and most of which were identified from bibliographies. The remaining 310 reports[6],[24],[46]–[353] were incorporated in our review and provide information on 96,737 water samples. The total number of studies is higher (319) due to a small number of multi-country reports. On average each study provides information on 1.7 water source types, resulting in a database with 555 datasets.

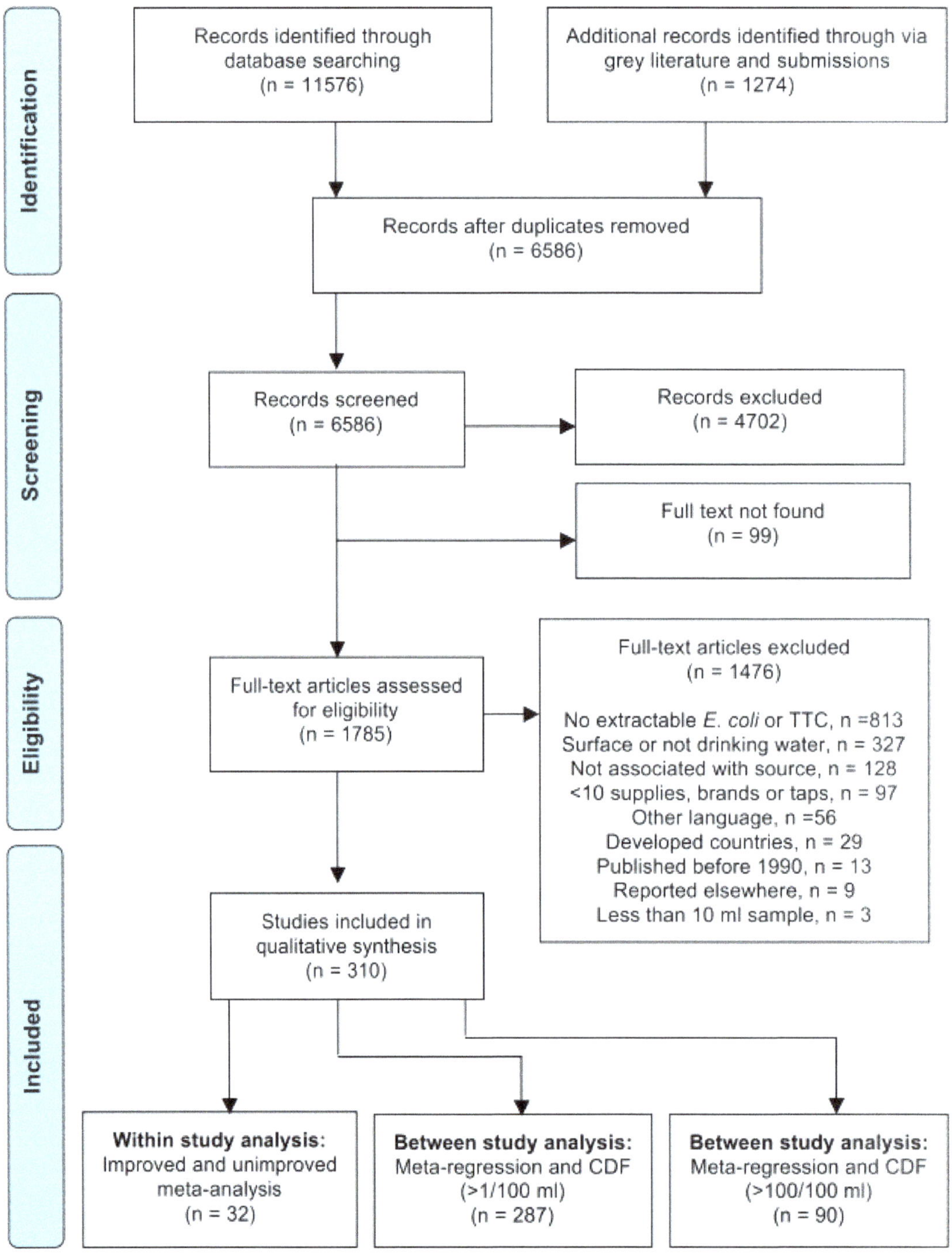

Figure 2: Flowchart for a review of safety of sources of drinking-water.

Study Characteristics

Characteristics of included studies are summarized in Table 3. The review is dominated by cross-sectional studies (n=241, 75%) with fewer longitudinal surveys (n=39, 12%). Authors report selecting sources or households at random in a minority of studies (n=68, 21%); most of these studies selected sources randomly within a region or community rather than at national level. The main exceptions were the Rapid Assessment of Drinking-Water Quality (RADWQ) studies commissioned by WHO and UNICEF, of which five have been published[64],[65],[164],[281],[322] and a repeated cross-sectional study in Peru for which only the total coliform results have previously been reported but for which we were able to secure *E. coli*data [227].

Table 3: Characteristics of included studies

Characteristic	Studies	Datasets	Samples
	Number (%)	Number (%)	Number (%)
Setting			
Urban	146 (46)	227 (41)	30,038 (31)
Rural	130 (41)	243 (44)	34,850 (36)
Both urban and rural	41 (13)	83 (15)	31,767 (33)
Unclassified setting	2 (1)	2 (0)	82 (0)
Emergencies	13 (4)	26 (5)	2,897 (3)
Non-household	17 (5)	21 (4)	2,121 (2)
Point of sampling			
Stored water	50 (15)	74 (13)	19,965 (21)
Directly from source	293 (92)	481 (87)	76,772 (79)
Water supply			
Improved	209 (65)	273 (49)	56, 268 (58)
Piped	118 (37)	119 (21)	32,348 (33)
Borehole	83 (26)	83 (15)	11,452 (12)
Protected dug well	36 (11)	36 (6)	8,697 (9)
Protected spring	11 (3)	11 (2)	978 (1)
Rainwater	25 (8)	25 (5)	2,793 (3)
Unimproved	62 (19)	71 (13)	5,594 (6)
Unprotected dug well	49 (15)	49 (9)	4,577 (5)
Unprotected spring	16 (5)	16 (3)	810 (1)
Tanker truck	6 (2)	6 (1)	207 (0)
Unclassified	167 (53)	213 (38)	35,087 (36)
Sachet	15 (5)	15 (3)	1,305 (1)
Bottled	35 (11)	35 (6)	2,339 (2)
Dug well	49 (15)	49 (9)	4,577 (5)
Spring	16 (5)	16 (3)	810 (1)
Design			
Randomized	68 (21)	131 (24)	31,210 (32)
Representative	74 (23)	148 (27)	37,614 (39)
Cohort or case control	5 (2)	15 (3)	4,114 (4)
Intervention	22 (7)	47 (8)	9,799 (10)
Cross-sectional survey	241 (75)	404 (73)	48,559 (50)
Longitudinal survey	39 (12)	66 (12)	32,302 (33)
Diagnostic	12 (4)	23 (4)	1,963 (2)
Parameter			
E. coli	152 (48)	270 (49)	32,298 (33)
TTC only	167 (52)	285 (51)	64,439 (67)
Language			
English	276 (86)	502 (90)	81,349 (84)
Spanish	6 (2)	8 (1)	3,024 (3)
Portuguese	24 (8)	29 (5)	9,146 (9)
French	4 (1)	5 (1)	187 (0)
Chinese	9 (3)	11 (2)	3,031 (3)
Reporting			
Presence/absence of FIB	287 (90)	499 (90)	90,056 (93)
Microbial risk classification	90 (28)	165 (30)	23,953 (25)
Mean FIB	80 (25)	136 (25)	15,530 (16)
Geometric mean FIB	34 (11)	68 (12)	11,797 (12)

Range of FIB	74 (23)	108 (19)	9,407 (10)
Standard deviation of FIB	21 (7)	38 (7)	4,417 (5)
Sanitary risk	44 (14)	82 (15)	15,808 (16)
WHO sanitary risk	12 (4)	31 (6)	9,160 (9)
Sanitary risk classification	17 (5)	44 (8)	10,667 (11)
Sample Size[a]			
Small ($n = 10–30$)	NA	192 (35)	3,711 (4)
Medium ($n = 31–100$)	NA	187 (34)	11,615 (12)
Large ($n = 101–6{,}021$)	NA	176 (32)	81,411 (84)
Quality[b]			
Low (1–5)	113 (36)	199 (36)	27,892 (29)
Medium (6–7)	94 (29)	142 (26)	16,980 (17)
High (8–13)	112 (35)	214 (39)	51,865 (54)
Total	319 (100)	555 (100)	96,737 (100)

[a]Terciles by datasets.
[b]Terciles by study.

Study quality varied greatly spanning from a quality score of 1 to 12 and with an interquartile range of 5 to 8 (Figure S1). Whereas most studies described the analytical method used to detect *E. coli* or TTC (80%), how water sources were selected (67%), and the setting in which the study took place (86%), fewer specified quality control procedures (15%), met the basic sample handling criteria (25%), used trained technicians to conduct the water quality tests (15%), or arranged testing in an accredited laboratory (12%) (Figure S2). Most studies were from sub-Saharan Africa, southern Asia, or Latin America and the Caribbean (Figure 3). The majority of included studies investigated water quality at the source. Studies reporting on the quality of water stored in households by provenance were less common (n=49), and few of these compare quality of stored water with that of the associated source (n=26). Several studies took place during or after emergencies [97],[201] and natural hazards, including cyclones [235], floods [78],[208], droughts [341], and tsunamis [130],[147],[331]. Non-household settings such as schools and health facilities were addressed in a small number of studies (n=17). Few studies separately report water quality information from slum or peri-urban settings (n=7).

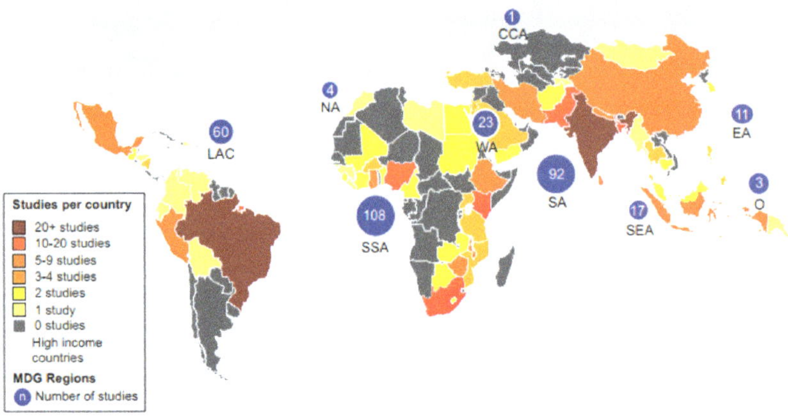

Figure 3: Map of study locations.

Qualitative Synthesis

In Figures S3 and S4 levels of microbial contamination are shown using the FIB level classification (<1, 1–10, 10–100, and >100 FIB per 100 ml), grouped by type of improved water source. These results are broadly in agreement with a comparison using measures of central tendency (Figure S5) and show great variability in the likelihood and extent of contamination between studies and source types.

Large studies with random sampling demonstrate marked differences in water quality between countries; for example less than 0.01% of samples from utility piped supplies in Jordan [281]were found to contain TTC compared with 9% to 23% of utility piped supplies in the other four RADWQ countries [64],[65],[164],[322]. Only one national randomized study differentiated between rural and urban areas; the proportion of samples from piped supplies containing *E. coli*was found to be substantially higher in rural (61%, n=101) than urban (37%, n=1470) areas in Peru [227].

In comparison to microbial testing, sanitary inspections are less widely practiced or data are rarely published. Sanitary inspection procedures vary considerably and are usually adapted to the local context; of the 44 studies reporting sanitary inspections only 12 used standardized WHO forms. In Figure S6 the sanitary risk levels as reported in nine studies are compared with the proportion of samples containing FIB and suggest that there is no strong association between these two measures.

Between Studies Analysis: CDF and Meta-regression

The number of studies reporting high proportions of samples contaminated or high levels of FIB is lower for improved sources as can be seen in Figure S7. Yet, in 38% of 191 studies reporting the quality of improved sources, at least a quarter of samples exceeded recommended levels of FIB. Figure S8 shows CDFs by source type with similar patterns to those from the FIB level classification.

Results of the meta-regression are shown in Table 4. We find that country income-level is a significant determinant of water quality and the odds of contamination are 2.37 times (95% CI 1.52–3.72 [p=0.001]; Table 4) higher in low-income countries compared with wealthier countries. However this result is not significant when separately considering piped and other improved sources (Tables S1 and S2).

Table 4: Between studies meta-regression

Variables	Proportion of Samples >1 FIB per 100 ml			Proportion of Samples >100 FIB per 100 ml		
	Obs.	OR [95% CI]	p-Value	Obs.	OR [95% CI]	p-Value
Source type						
Improved vs. unimproved	291	0.14 [0.08–0.25]	<0.001	87	0.13 [0.05–0.33]	<0.001
Piped vs. other improved	239	0.53 [0.32–0.89]	0.017	68	0.47 [0.18–1.20]	0.11
Protected vs. unprotected groundwater	90	0.26 [0.11–0.60]	0.002	31	0.37 [0.09–1.52]	0.16
Treated piped vs. untreated piped[a]	69	0.07 [0.02–0.27]	<0.001	18	0.10 [0.01–0.72]	0.025
Stored vs. source	474	2.09 [1.16–3.78]	0.015	140	1.85 [0.68–5.04]	0.23
Setting						
Low-income vs. other	414	2.37 [1.52–3.71]	<0.001	122	1.30 [0.59–2.86]	0.52
Rural vs. urban	344	2.37 [1.47–3.81]	<0.001	96	1.18 [0.49–2.83]	0.71
Study characteristics						
Thermotolerant vs. *E. coli*	417	1.08 [0.70–1.67]	0.72	122	0.99 [0.45–2.19]	0.98
Publication year	415	0.96 [0.93–1.00]	0.029	122	0.96 [0.91–1.02]	0.16
Random vs. non-random selection	417	0.92 [0.53–1.57]	0.75	122	0.60 [0.25–1.44]	0.25
High quality vs. lower quality[b]	417	0.90 [0.54–1.49]	0.68	122	0.51 [0.21–1.23]	0.13
Longitudinal vs. cross-sectional	372	1.00 [0.58–1.73]	0.99	116	1.05 [0.40–2.73]	0.93
Wet vs. dry	51	0.99 [0.32–3.10]	0.99	22	0.93 [0.09–9.34]	0.95
Reporting format						
Measure of central tendency	417	2.31 [1.45–3.69]	<0.001	122	1.22 [0.55–2.68]	0.62
Microbial risk classification	417	2.30 [1.45–3.67]	<0.001	–	–	–

With the exception of stored versus source, we restricted the analysis to source water samples. We excluded emergencies from the meta-regression.
[a]Post hoc analysis.
[b]Top tercile of studies versus bottom two terciles.
Obs.,number of observations.

Meta-regression showed a substantial difference in the proportion of samples containing FIB between urban and rural areas (OR=2.37 [95% CI 1.47–3.81], p<0.001). There is weak evidence to suggest that piped supplies are more likely to be contaminated in rural areas (OR=2.4 [95% CI 0.98–5.92], p=0.054; Table S1), but no evidence of differences for all other improved source types (OR=1.19 [95% CI 0.52–2.72], p=0.67; Table S2).

Protection of groundwater (OR=0.26 [95% CI 0.11–0.60]; p=0.002; Table 3) and treatment of piped water (OR=0.07 [95% CI 0.02–0.27], p<0.001; Table 3) were both strongly related to better water quality. Contamination of stored water was more likely than water at the source (OR=2.09 [95% CI 1.16–3.78], p=0.015; Table 3), including for piped supplies (OR=2.35 [95% CI 1.08–5.12], p=0.032; Table S1).

Within Studies Analysis: Meta-analysis

Figure 4 is a forest plot showing the ORs of contamination for improved sources compared to unimproved sources from eligible studies. Meta-analysis of randomized and non-randomized studies showed that improved sources are less likely to be contaminated (pooled OR=0.15 [0.10–0.21]; Figure 4); the protective effect was found to be greater in randomized studies and none of the randomized studies found contamination to be more frequent in improved sources. Heterogeneity was relatively high (I^2=80.3% [95% CI 72.9–85.6) indicating that the protective effect varies considerably between settings. The OR for a small number of studies was greater than one, suggesting that in some settings improved sources may not always be less likely to contain FIB than the unimproved alternative.

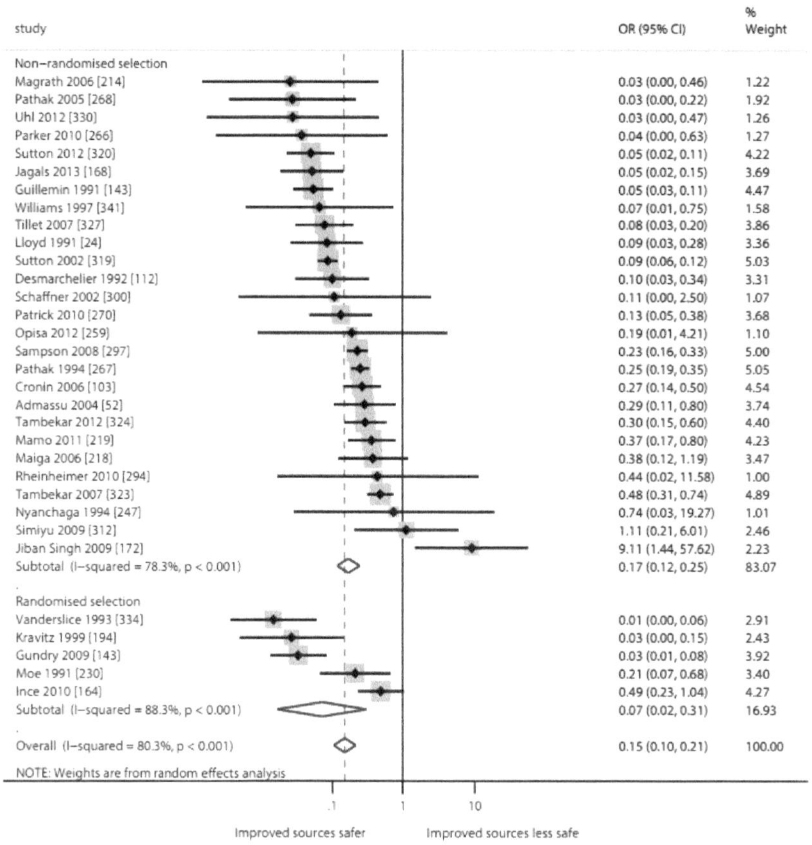

Figure 4: Forest plot of the odds of fecal contamination for improved and unimproved sources.

Assessment of Bias

Egger's test found no evidence of small study effects for the meta-analysis of improved versus unimproved sources (p=0.64; Figure S9). Meta-regression did not detect evidence of bias due to study design, lack of randomization, study quality, or season ("wet" or "dry"). As expected, we find that levels of FIB (risk classification or measures of central tendency) were more likely to be reported in studies where water was more contaminated with FIB (Tables 4, S1, and S2) and therefore may constrain comparisons between studies. Publication year was also related to the proportion of samples containing FIB (OR=0.96 [95% CI 0.93–1.00]; p=0.029; Table 4), but this was not significant when separately considering piped and other improved sources (Tables S1 and S2). Studies testing the same sources in different seasons report considerable variation in microbial water quality (Table S3).

DISCUSSION

Safety of Improved Sources

We demonstrate that water from improved sources is less likely to contain FIB than unimproved sources. Using meta-analysis we compare water from unimproved sources and improved sources within the same study and found improved sources were less likely to contain FIB (OR=0.15 [95% CI 0.10–0.21]) (Figure 4), with the greatest protective effects in studies where selection of water sources was randomized. Comparison between studies of improved versus unimproved sources yielded an OR of 0.14 (95% CI 0.08–0.25) for the presence of FIB and 0.13 (95% CI 0.05–0.34) for samples exceeding 100 FIB per 100 ml (Table 4). Cumulative density plots show markedly lower contamination for improved sources relative to unimproved sources.

While improved sources clearly offer a greater degree of protection compared to unimproved sources, they are not all nor consistently safe [354]. In particular, protected dug wells were rarely free of fecal contamination and it is not uncommon for these sources to contain high levels of FIB. High levels of contamination were occasionally reported for boreholes and piped water, which are typically perceived as high quality and lower risk. Risk factors for microbial contamination of piped supplies include intermittency [198] and inadequate chlorination [20],[355]. For boreholes and dug wells, the reasons for fecal contamination can be more difficult to ascertain owing to the possibility of aquifer contamination and/or inadequate sanitary completion. In many cases contamination is associated with poor hygiene and inadequate sanitation, but specific risks can be readily identified through sanitary inspection of the water source and its surroundings and may in part explain the heterogeneity in FIB concentrations observed for a given source type.

Studies included in this review show great variability in water quality for water sources of a given type, suggesting considerable scope for reducing exposure to fecal contamination through systematic management of water safety. Microbial water quality displays substantial heterogeneity between studies and we found a high I^2 when comparing improved and unimproved sources from the same study (I^2=80.3% [72.9–85.6]; Figure 4).

Across all studies, we find a higher risk of contamination in rural areas compared to urban areas (OR=2.37 [1.47–

3.81], $p<0.001$; Table 4) and a higher risk of contamination in low-income countries (OR=2.37 [1.52–3.71], $p<0.001$; Table 4). Higher risk in rural areas is consistent with a recent multi-country study of over 25,000 hand pumps, which found greater risk of non-functioning water sources in areas distant from district centers [356].

There is some evidence to suggest that overall water quality has gradually improved over time. Publication year was associated with the proportion of samples containing FIB (OR 0.96 [0.93–1.00], $p=0.023$; Table 4); this may reflect a progressive trend towards greater use of source types associated with less contamination and potentially a lessening of population-level exposure.

We found only limited published data on sanitary risk, suggesting that sanitary risk inspection techniques are not widely used and/or reported, despite being well-established in national and international drinking-water guidelines [12]. Studies report sanitary risks in some sources that are not found to contain detectable FIB, indicating that infrequent monitoring does not provide assurance of water safety. Conversely, low sanitary risk scores are reported for water sources that contain FIB, indicating that sanitary inspections alone do not capture water safety fully. Although individual studies report different levels of sanitary risk, the data are too few to draw general conclusions. Moreover, the use of different questions for each source type and their equal weighting limits the comparability of sanitary risk scores. While a simple prevalence of factor scoring is unlikely to be appropriate in more complex systems, there has been progress in developing scoring for quality of water safety plans and higher scoring utilities have been linked to both improved water quality and health outcomes [357].

LIMITATIONS OF THE REVIEW

At the outcome level, our principal analyses are based on the proportion of samples detecting FIB rather than compliance to health guidelines over the course of a year. These prevalence measures overestimate compliance to health guidelines and national standards that require minimum sampling frequencies. Furthermore, where the sample volume is less than the recommended 100 ml [12], contamination is less likely to be detected and will be detected less frequently.

At the study level, our review was limited by infrequent reporting of a consistent measure of central tendency (or of individual sample data), sanitary risk inspections, and stored water quality. In analysis, we combine studies that used diverse sample handling and microbiological analytical methods and these factors may account for some of the variability in reported water quality. The majority of the included studies were cross-sectional and do not provide information on temporal variability in water quality. Few studies achieved robust random selection of water sources, and few received high scores for study quality (14% with >9 out of 13; Figure S1) with description of quality control procedures, meeting handling criteria, and statement of season(s) of sampling most frequently omitted quality factors. Many studies, particularly of groundwaters, were excluded as we could not match water source types or determine whether they were "improved."

At the review level, we may not have identified all studies that meet the inclusion criteria. To capture additional studies would have required the screening of tens of thousands of records, as we were unable to identify more specific search terms. Two sources of water quality information that could be used in future studies and monitoring: regulatory surveillance and utility quality control data are likely to be extensive and not well represented as they may not be published and publicly available. Publicly available data from these sources rarely matched our inclusion criteria, usually because of failure to report sample sizes or associate water quality with source type. We identified few studies in languages other than English despite conducting searches in four other languages, and several regions are underrepresented (Figure 3) including Caucasus and Central Asia and Oceania for which studies may be available in other languages. Since few studies separately report water quality in slums, we combined studies of slum and peri-urban populations with those taking place in formal urban areas and we were therefore unable to investigate intra-urban disparities [7]. There may be a small number of errors in the database; in the 10% of English language studies independent extraction <0.5% errors were identified.

There are two sources of bias that will tend to cause overestimation of the safety of improved sources. Firstly, for many source types, water (including unreliable piped systems, public standpipes, and wells) is collected at the source, carried to, and stored in the household—affording multiple opportunities for contamination—such that final water quality is often worse than in the associated source [34]. Few studies identified the source type of origin of stored water; those that did supported the suggestion that stored water is more frequently contaminated and contaminated at higher levels. Data interpretation is confounded because samples are not paired [204] or are temporally displaced. Despite the potential to improve matters, evidence for the impact of household water treatment on stored water quality is inconsistent [295],[358]. Secondly, most studies were cross-sectional. One-off or infrequent sampling overestimates safety by missing seasonal [359],[360] or sporadic contamination and longitudinal studies suggest seasonal effects can be substantial (Table S3).

Monitoring Implications: Improving on Improved

There is a widely perceived hierarchy for water source desirability, typically with piped-to-household-tap as the ideal. Such general hierarchies combine many aspects of water service and their value, including quantity, affordability, accessibility, and reliability or continuity of service as well as safety [361]. It has been argued that a more graduated approach to monitoring than the "improved"/"unimproved" dichotomy is required [24],[362]. These concepts were reflected in initial JMP working group proposals that called for the monitoring of "basic" and "intermediate" service levels representing improvements in quality, continuity of supply, and accessibility [30]. Such monitoring could contribute to assessing progressive realization of the Human Right to Water and Sanitation [363] and to encourage improvements in service delivery, including improvements within water source-type categories.

A variety of indicators could be used to improve global estimates of water safety. Account of safety could be enhanced by ranking or scoring improved source types according to the proportion of sources showing fecal contamination (in descending order): piped (treated), boreholes, protected springs, rainwater, piped (untreated), and protected dug wells. Although based on *a posteriori* analysis, we find that untreated piped supplies are much more likely to be contaminated than treated piped supplies (OR=0.07 [95% CI 0.02–0.27], $p<0.001$; Table 4) and may usefully be considered separately. This ranking could be refined as more data become available but may have unintended consequences such as disincentivising improvements within each source type and could discourage the use of some source types in regions where these may in fact provide comparatively safe water. We find that bottled and sachet water are typically high quality, but their environmental sustainability has been questioned [318]. They are excluded from global progress efforts not for reasons of quality but because they provide insufficient quantity for domestic uses, such as cooking and hygiene, other than direct consumption.

Adjustment of the improved source indicator for country-specific source-type compliance with microbial water quality guidelines would capture the substantial heterogeneity both between and within countries, highlighting disparities in the use of safe drinking-water [364]. It would provide a more robust and consistent means of assessing safety than the use of source-type classifications alone and would enable improvements in quality to be reflected in monitoring. This review indicates that such an adjustment would have a large (downward) impact on international estimates of the number of people using safe drinking-water, in agreement with previous estimates based on five studies [8],[9].

Prevalent FIB are imperfect indices of water safety [26] and health risk. They are known to be more sensitive to chlorine than some important waterborne pathogens such as cryptosporidium[12], and can persist or multiply in some tropical waters [365]. These factors and the fact of temporal variability in water quality suggest that it would be strongly preferable to combine periodic measurement of water quality with assessment of sanitary status through sanitary inspection or water safety plans [29] in assessing water safety. These approaches can serve to highlight the condition, operation, and maintenance of water sources and provide a more complete picture than access to infrastructure alone. Such approaches would benefit from standardization to ensure comparability. However, adjustment for the proportion of samples containing FIB would not account for all hazards to health, including the two major chemical hazards: fluoride [366],[367] and arsenic [368]. The JMP has outlined an approach that could be taken that combines a hierarchy of measures of water quality with sanitary risk or management data [22].

Cost-effectiveness arguments suggest basing future monitoring efforts on both regulatory and utility data and building monitoring capacity especially in peri-urban and rural areas of low-income countries. Special initiatives such as dedicated water safety surveys (e.g., RADWQ) and integration of water quality testing in household surveys are likely to play an interim role and will assist in filling gaps in the available data [22].

Implications for Public Health Policy

Our review provides strong evidence that by equating "improved" with "safe," the number of people with access to a safe water source has been greatly overstated, and suggests that a large number and proportion of the world's population use unsafe water. We analyze the implications following a framework of health sector functions in environmental health [369] and highlight key implications for policy in Box 1.

Box 1. Key Implications for Policy

- Fecal contamination of drinking-water is widespread globally, especially in low-income countries and rural areas, and affects many improved sources.
- The Global Burden of Disease 2010 study may greatly underestimate diarrheal disease burden by assuming zero risk from improved sources.
- Adjustment of safe drinking-water coverage estimates for water quality and ideally sanitary risk would highlight disparities and enable improvements in quality to be reflected in monitoring.
- Piped water is not a panacea: high levels of contamination have been reported in a range of settings and water stored in the household, often motivated by an intermittent or distant source, is more likely to be contaminated, especially in rural areas.
- Quality and sanitary risks are heterogeneous indicating that it is possible to substantially enhance safety and reduce exposure through incremental improvements in service.
- Greater use should be made of sanitary inspections as these provide a complementary means of assessing safety and are able to identify corrective actions to prevent contamination.
- Studies of microbial contamination and sanitary risk could be improved by adhering to higher standards, including those outlined in our quality criteria.

Policy Makers

Health policy makers framing post-2015 goals for achieving universal health coverage and reducing the global burden of disease need to ensure that targets and indicators go beyond health care services and address underlying determinants of health including progressive improvements in access to safe drinking-water, sanitation, and hygiene services. Adequate quantities of safe water at home are essential for good health [370] and, together with improvements in sanitation and hygiene, considered one of the more cost-effective interventions to protect and improve public health [371].

Through the process of implementing the MDGs, great strides have been made in increasing coverage of improved water sources although an estimated 783 million people still lack an improved source, most living in sub-Saharan Africa and Southeast Asia [5]. This review confirms that there are also pronounced disparities in access to "safe" water between and within countries. The Human Right to Water and Sanitation calls for progressive reduction in inequalities, and public health considerations suggest that reducing exposure among the most vulnerable, including the poor, undernourished, and immunocompromised, is a key public health concern. Health policy makers therefore have an important role to play in advocating for health protecting policies in other sectors, including those by actors concerned with water supply services.

Standard Setting for Water Quality

The health sector plays a substantive role in drinking-water quality standard setting in many countries; this is logical since the underlying rationale is based on health concerns. The evidence presented here suggests that failures in water safety are frequent in LMICs, and that effective standard setting would combine outcomes measures (such as the measurement of FIB) with the verification of preventive or protective measures through sanitary inspection, water safety plans, or similar approaches. Future standard setting should take account of inequalities in access so as to direct efforts to those most affected and be informed by the availability of effective interventions.

Surveillance

The health sector also plays an important role in environmental health including drinking-water quality surveillance in many countries. However in many LMICs surveillance of water quality is limited outside large consolidated urban centers and enforcement of guidelines can be weak[372]. Surveillance is typically weakest in rural areas where levels of access to "improved sources" are lowest and the likelihood of contamination is greatest. The lack of disaggregated data for peri-urban and urban slum areas is also a problem in many countries. To-date the focus of public health policy, targets, and monitoring has been a household's primary source of drinking-water [373], but there is growing concern over inadequate water, sanitation, and hygiene services in non-household settings, such as schools and health care facilities [30],[373]. A public health perspective suggests that health sector surveillance should focus particularly on these settings where the risk of exposure is high.

Health Care Settings

Ensuring adequate environmental health in health care settings is a key responsibility of the health profession [374], but in many countries access to water, sanitation, and hygiene in health care facilities remains inadequate. While data were few, those studies that addressed water safety in health care settings documented water safety deficiencies [142],[235]. Furthermore, nationally representative surveys of health care facilities in Uganda and Rwanda found two-thirds of health care facilities to lack an improved water source [375],[376]. There are also opportunities to better protect and improve health through incorporation of water safety components in health programs such as those focused on specific diseases [377] or sensitive life stages such as maternal and child health [378]. Given the vulnerability of the populations using them and the potential for health facilities to serve as models for the wider community, the health sector has both a duty of care and an opportunity to advance health through better management of water safety in its own facilities.

Outbreak Investigation

Despite clear evidence of its prevalence, the global burden of disease attributable to fecal contamination of drinking-water remains poorly understood. Data on outbreaks of waterborne disease [19] and of the impact of interventions to improve water quality on endemic disease[15],[16] provide evidence of the importance of fecal contamination. The associated studies frequently provide only weak insight into causal factors that might otherwise contribute to improved preventive action. There is an opportunity to enhance current outbreak investigation, advancing its role from one of curtailment to general prevention, thereby improving the ability to retrieve information that can be generalized for future prevention.

Waterborne Disease Burden

Study of the national and global burden of disease provides an opportunity to enhance public health protection and increase cost-effective action by focusing efforts on disease burdens and risk factors of greatest significance. The recent Global Burden of Disease study [379] based its estimates on the assumption of zero risk for those supplied by improved drinking-water sources and no additional benefit of a piped supply on premises [354]. The findings of this review indicate that fecal contamination of drinking-water is widespread, particularly in rural areas and low-income countries. Some improved source types, especially protected dug wells and protected springs, are frequently and sometimes highly contaminated. Contamination reported in piped supplies in especially rural but also urban areas is concerning given that these serve the majority (63%; Table 1) of the world's population and the use of this source type is expanding rapidly in many countries, especially in China [380]. In assuming improved sources are safe [379], current estimates may greatly underestimate waterborne disease burden and this gap would be expected to grow as improved source coverage increases.

Implications for Research

We have applied analytical tools usually associated with the medical sciences (meta-analysis of prevalence) to the study of environmental contaminants. There are differences in the underlying data that merit highlighting and limit the transferability of these techniques. Firstly, robust sampling frames are usually available for the selection of households (e.g., from national statistical offices), but random selection of water sources is more challenging. Studies such as RADWQ that adapted stratified cluster sampling techniques used in household surveys to address this problem have been subject to methodological criticism [381]. Future directions to achieve representative samples could include the use of water point mapping[382] or satellite imagery to create lists of water points but both are relatively complex and approaches based on population may be more feasible. Secondly, whereas epidemiological studies seek to measure outcome variables such as the number of events at a given point in time ("point prevalence") or the rate at which they occur ("incidence"), environmental studies often seek to determine whether a threshold condition of safety has been exceeded ("compliance"). Even brief failures in safety may negate much of the potential health benefits of otherwise safe water [383],[384]. As a consequence, assessments of safety can be particularly susceptible to the frequency of monitoring and temporal representativity can be as important as spatial or population-based randomization.

In order to assess levels of compliance with regulatory standards or international guidelines based on infrequent surveys or limited data, research is required to understand the effect of repeated sampling of both source and stored water that results from variability over time (e.g., seasonality) and replicate sampling (sequential testing). Given their potential to inform assessments of safety, there is also a need to improve understanding of the relative importance of sanitary risks and their temporal variation for which very limited data are currently available. There is unlikely to be a simple correlation between sanitary risks and microbial contamination [103],[209] but the predictive value of sanitary risks may be much higher when accounting for compliance over time because some infrastructure failures will only lead to contamination in the presence of a co-factor such as rainfall. There is also a need to better understand the role of water collection and storage on microbial contamination and the associated risk to health.

We find strong evidence of differences in water quality between rural and urban areas, including for piped supplies. Further work is needed to characterize intra-urban differences; we encourage randomized water quality surveys to include slum or peri-urban populations as part of the sampling frame. Finally, the approaches taken in this review could be extended to other drinking-water contaminants, such as arsenic and fluoride.

CONCLUSIONS

Fecal contamination of drinking-water in LMICs is widespread. We demonstrate that improved sources are in general safer than unimproved sources of drinking-water, but they are not universally nor consistently free of fecal contamination. In 38% of 191 studies at least a quarter of samples from improved sources exceeded WHO recommended levels of FIB. By equating "use of an improved source" with "safe," international estimates greatly overstate access to safe drinking-water. Substantial differences are observed in the presence and levels of contamination between countries, between urban and rural regions, and between water source types. Infrequent measurements of water quality alone tend to overestimate safety and so an improved future strategy would combine sanitary status with water quality measurements.

Supporting Information

Figure S1 Study quality rating for 319 studies.

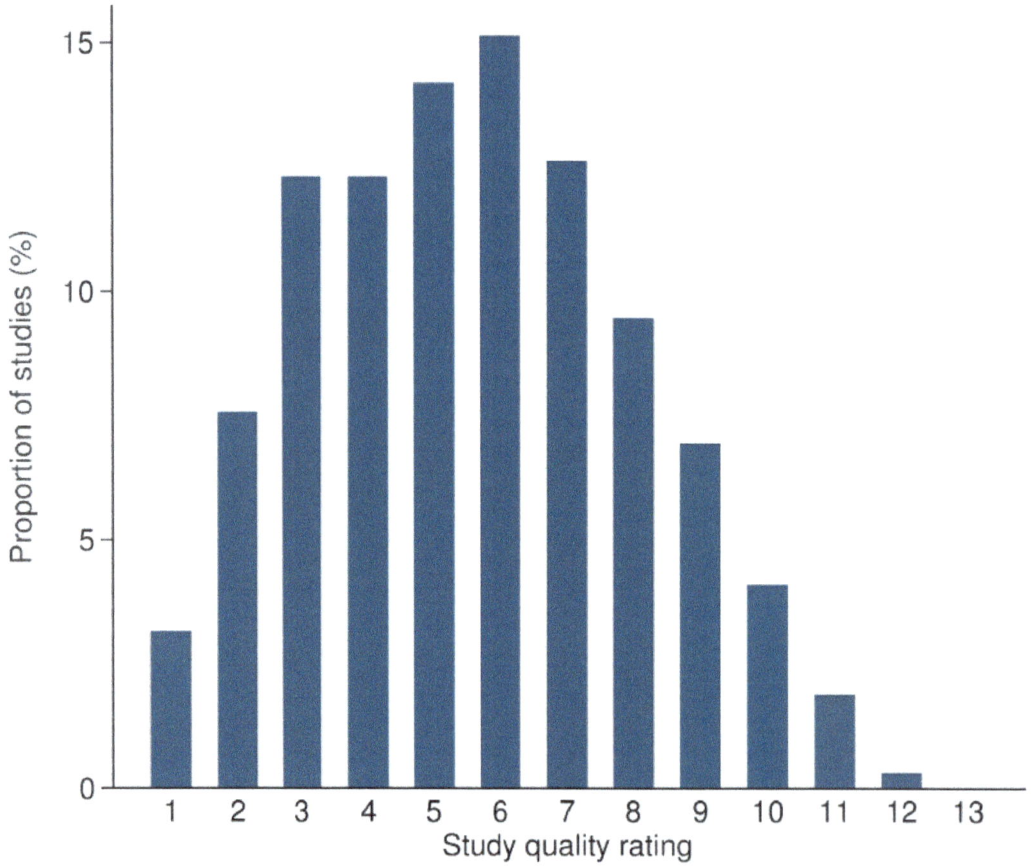

Figure S2 Frequency of 13 quality criteria being met by 319 included studies.

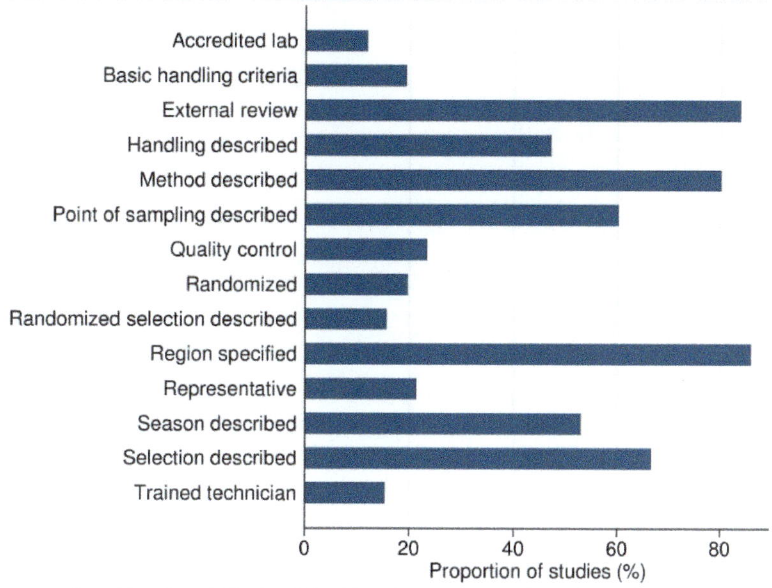

Figure S3 Fecal indicator bacteria level classification for improved sources by source type and study. Included studies are those for which a FIB level classification was reported or could be calculated. Sample size in curved parentheses. Reference number in square brackets. Source types are: BH, borehole; PWS, piped water supply; PS, protected spring; PDW, protected dug well; RWH, rainwater harvesting.

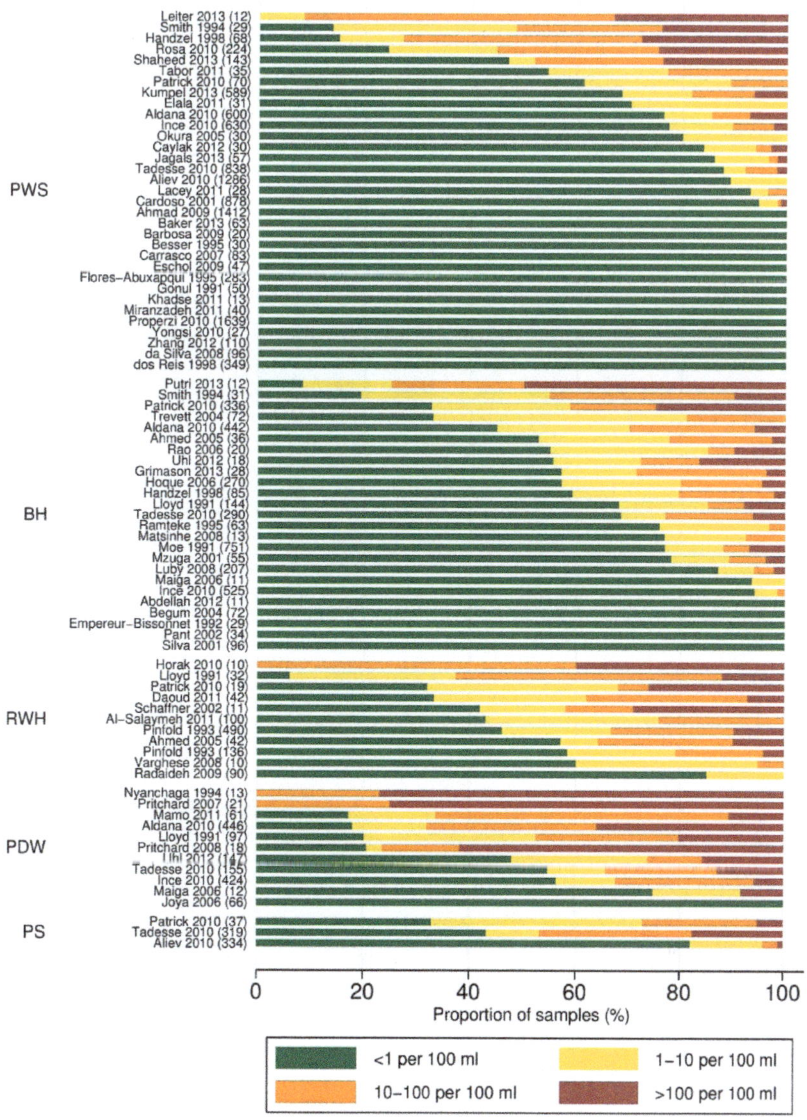

Figure S4 Fecal indicator bacteria level classification for unimproved and unclassified sources by source type and study. Included studies are those for which a FIB level classification was reported or could be calculated. Sample size in curved parentheses. Reference number in square brackets. Source types are: UDW, unprotected dug well; US, unprotected spring; TT, tanker truck; B, bottled; S, sachet. NDW, dug well (unclassified); NS, springs (unclassified).

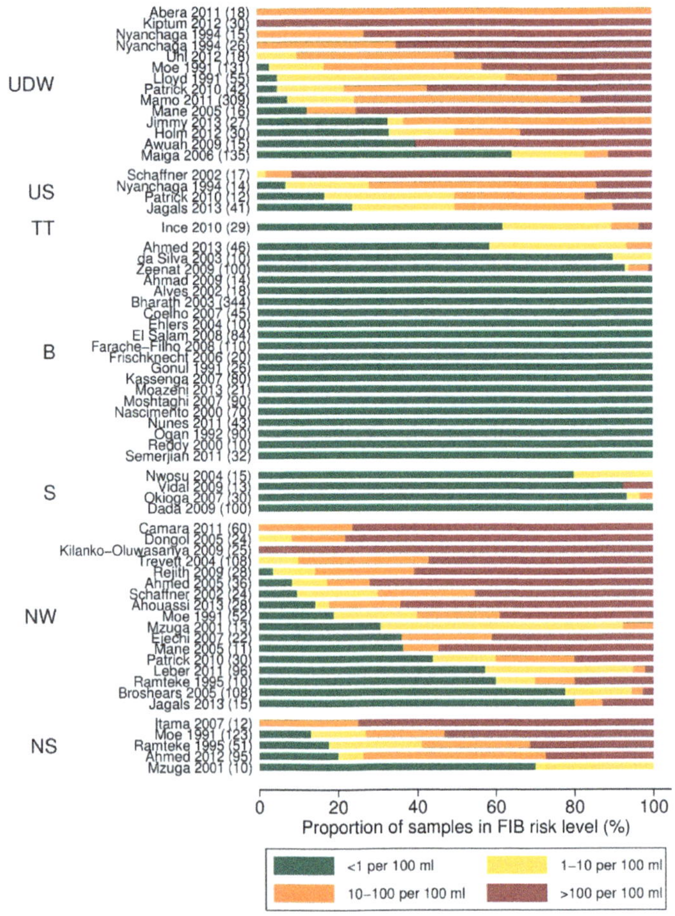

Figure S5 Measures of central tendency reported by included studies, by source type. Size of circles proportional to number of water samples evaluated.

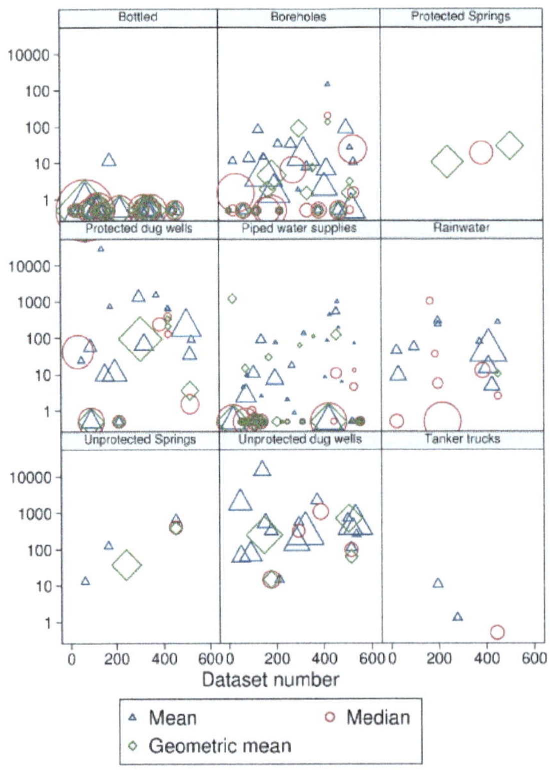

Figure S6 Comparison of sanitary risk levels and proportion of samples containing fecal indicator bacteria in studies using WHO standardized inspection forms.

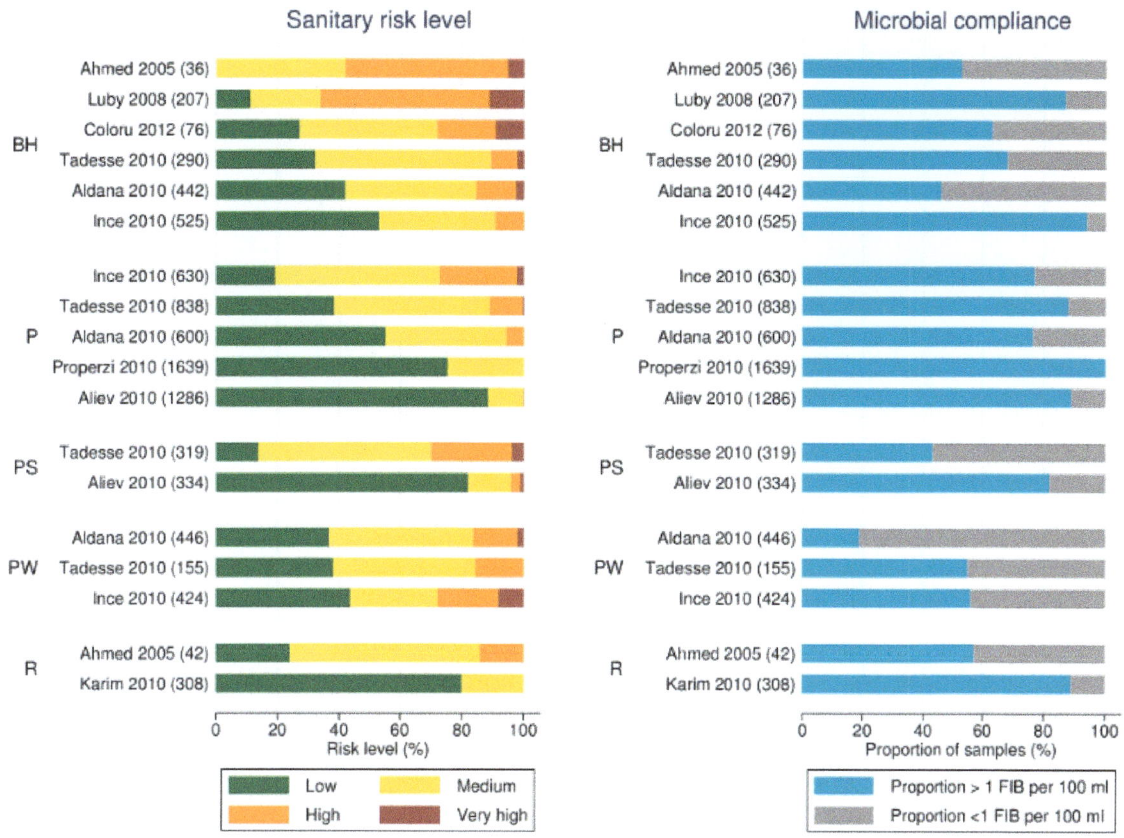

Figure S7 Cumulative density functions for the proportion of samples in each study with detectable (.1 per 100 ml) and high (.100 per 100 ml) E. coli or TTC, by improved and unimproved source.

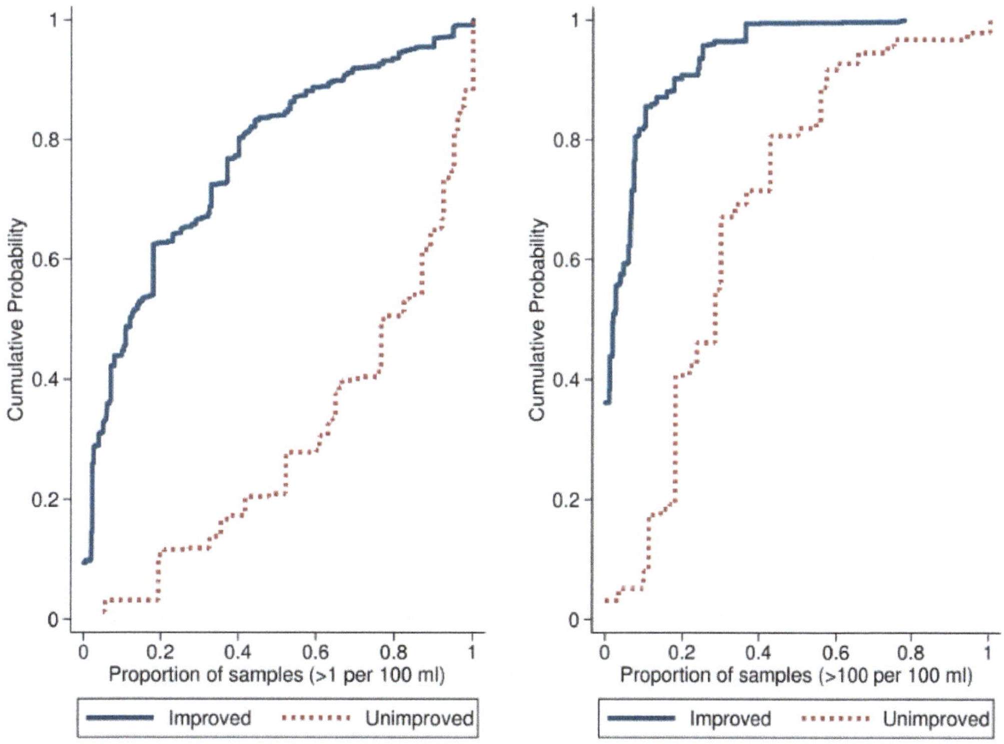

Figure S8 Cumulative density function of the proportion of samples containing fecal indicator bacteria in each study for improved (left) and unimproved (right) sources by type.

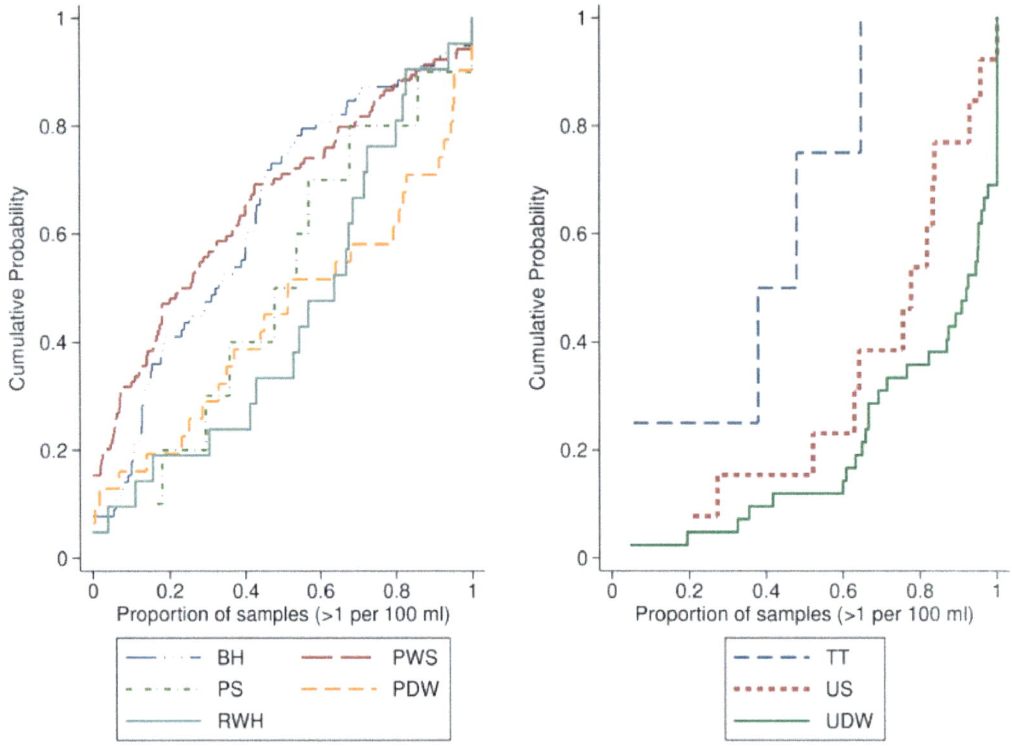

Figure S9 Funnel plot for the odds ratio comparing the safety of improved and unimproved sources in a given study.

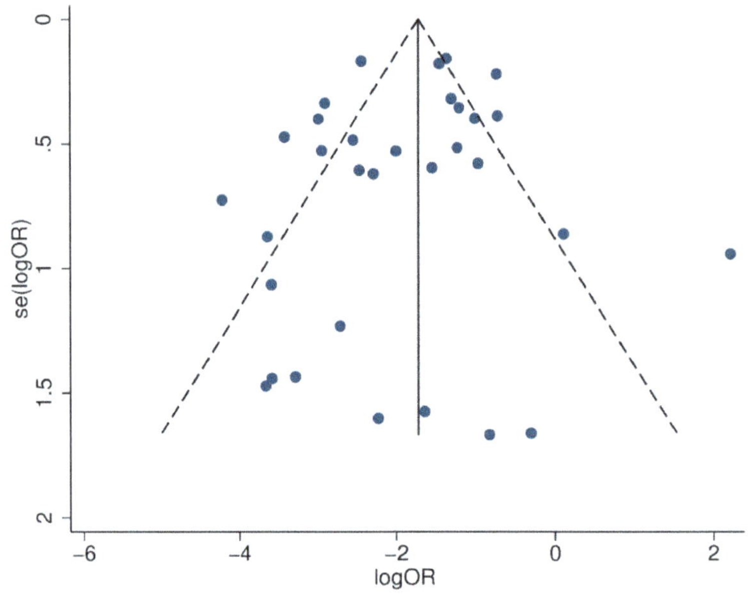

Table S1 Between studies meta-regression for piped supplies

Variables	Proportion of samples >1 FIB per 100 ml			Proportion of samples >100 FIB per 100 ml		
	Obs.	**OR** [95% CI]	**p-value**	**Obs.**	**OR** [95% CI]	**p-value**
Setting						
Low-income vs other	100	1.21 [0.44-3.31]	0.706	22	1.12 [0.17-7.55]	0.903
Rural vs urban	85	2.41 [0.98-5.92]	0.054	16	2.64 [0.46-15.13]	0.253
Sampling location						
Stored vs source	129	2.35 [1.08-5.12]	**0.032**	30	3.99 [0.96-16.58]	0.056
Study characteristics						
Publication Year	100	0.98 [0.92-1.05]	0.640	22	0.93 [0.79-1.09]	0.346
Reporting format						
Measure of central tendency	101	1.98 [0.78-5.05]	0.151	22	1.73 [0.23-13.02]	0.576
Microbial risk classification	101	1.08 [0.41-2.83]	0.875	-	-	-

Table S2 Between studies meta-regression for other improved sources

Variables	Proportion of samples >1 per 100 ml			Proportion of samples >100 per 100 ml		
	Obs.	OR [95% CI]	p-value	Obs.	OR [95% CI]	p-value
Setting						
Low-income vs other	138	1.21 [0.63-2.33]	0.565	46	1.71 [0.61-4.78]	0.300
Rural vs urban	111	1.19 [0.52-2.72]	0.671	33	0.70 [0.18-2.7]	0.597
Sampling location						
Stored vs source	183	2.29 [1.11-4.72]	**0.026**	53	0.80 [0.26-2.46]	0.694
Study characteristics						
Publication Year	137	0.99 [0.94-1.04]	0.710	46	0.99 [0.92-1.07]	0.837
Reporting format						
Measure of central tendency	138	1.50 [0.77-2.93]	0.228	46	0.67 [0.23-1.9]	0.440
Microbial risk classification	138	1.73 [0.87-3.44]	0.114	-	-	-

Other improved includes boreholes, rainwater, protected dug wells and protected springs.

Table S3 Variation in microbial safety during the year, findings of included studies for selected source types

Source	Findings
Rain-water	One study in costal Bangladesh found more frequent contamination during the rainy season than the dry season [166]. Another study in India found that all rainwater systems were free of contamination in winter and many were contaminated in summer [335].
Piped	Three studies of municipal water supplies in India show no clear trend for between seasons: in Gangtok, 13 taps were contaminated with TTC during the monsoon whereas none were positive in the summer or winter [182]; in Shillong, non-compliance at consumer endpoints was highest in the summer (100%, n=26) and somewhat lower post-monsoon and in winter (88% and respectively 85%) [184]; in Guwahati, noncompliance was lower post-monsoon (17/49) than summer or winter (31/49, 36/49) [183]. In urban and rural Brazil, Nogueira et al. [243] found treated waters to range from 0% to 6% (taps and reservoirs), with the lowest levels in the summer. Yassin et al. [347] report data from Palestine that show compliance rates ranging from 0-17% for piped water throughout the year, but it is not clear whether these data represent surveillance in the same areas.
Boreholes	Hoque et al. [159] found that the proportion of contaminated boreholes varied considerably lower in the dry season (37%) than during or after the monsoon (61%, and 57% respectively). However, the geometric mean and median values of TTC counts at the tubewell sites did not vary significantly between the surveys. Mertens et al [224] note that boreholes with handpumps provided the best quality water throughout the year and were therefore not as susceptible to seasonal trends as other groundwater sources. Potgieiter et al. [277] found significant differences between the dry and rainy seasons TTC compliance for private but not communal boreholes in Limpopo province, South Africa. Ranges were similar, but the levels were somewhat higher (respectively 7.8 to 19.0 and 4.4 to 7.5 per 100 ml).
Dug wells	Ejechi et al. [118] found wet season counts to be significantly higher than dry season counts for TTC for both covered and uncovered hand dug wells combined (GM of 3 vs. 14 per 100 ml). Knappett et al. [191] noted substantial variation in the proportion of samples positive for *E. coli* based on weekly grab samples from 35 private wells (30-80% noncompliance). However, the level of contamination was generally low, with the 75th percentile ranging from 1-5 CFU/100 ml. In an area of Sri Lanka where nearly all protected dug wells were contaminated, Mertens et al. [224] found little variation in the proportion of samples with TTC (from 93% to >99%). The geometric mean contamination for the contaminated wells ranged from 50 to 150 per 100 ml and was highest during the monsoon in December. Yassin et al. [347] found that noncompliance levels for wells ranged from 0 to 10%, peaking in August. Razzolini et al. [291] collected water samples from 177 wells every other month. The authors concluded that there was no obvious relationship between rainfall and contamination levels. Geometric mean contamination level for the wells ranged from to around 2 to 50. Karnchanawong et al. [179] found rainy season and dry season ranges to differ (53-55000 vs. 13-5700) for shallow wells near a waste disposal site.

Note: Several studies did not disaggregate temporal trends by supply type or provided information about the susceptibility of more specific types of wells. For example, Nogueira et al. [243] found greater variation for springs and private wells than for treated piped supplies. The proportion of samples positive for TTC increased with both rainfall and water temperature. A study of various pump designs in Malawi found that the level of contamination was approximately twice as high in the wet season as the dry season [325].

ACKNOWLEDGMENTS

We thank Mellanye Lackey for assistance in selecting the search terms and electronic databases and the librarians at UNC for locating many of the full texts. We are grateful for the support provided by Benjamin Mann in generating the map of studies by country and to Manpreet Singh and Wilson Cheng for discussions regarding implications for health policy. The authors wish to thank Joe Brown for insightful comments on earlier versions of the manuscript. This paper was shared with the JMP Water Working Group and valuable comments were received from the group. In particular, we would like to thank David Bradley, Rick Johnston and Philip Peters. We would also like to thank Annette Prüss-Üstün and Sophie Bonjour at WHO for sharing a list of water quality studies.

REFERENCES

1. Committee on Economic Socal and Community Rights (2002) General comment no. 15, The right to water. UN Doc. E/C.12/2010. New York: United Nations.
2. United Nations (2010) Resolution on Human Right to Water and Sanitation. United Nations General Assembly. A/64/292. New York: United Nations.
3. United Nations (2013) UN Goal 7: Ensure Environmental Sustainability. Available: http://www.un.org/millenniumgoals/environ.shtml. Accessed 1 December 2013.
4. UNICEF/WHO Joint Monitoring Programme: definitions and methods. Available: http://www.wssinfo.org/definitions-methods/. Accessed 1 December 2013.
5. WHO/UNICEF (2013) Progress on Sanitation and Drinking-Water: 2012 Update WHO/UNICEF. Available: http://www.wssinfo.org/definitions-methods. Accessed 1 December 2013.
6. Godfrey S, Labhasetwar P, Wate S, Pimpalkar S (2011) How safe are the global water coverage figures? Case study from Madhya Pradesh, India. Environ Monit Asses 176: 561–574.
7. Schäfer D, Werchota R, Dölle K (2007) MDG monitoring for urban water supply and sanitation: catching up with reality in Sub-Saharan Africa. Eschborn: Deutsche Gesellschaft für Technische Zusammenarbeit (GTZ).
8. Bain RES, Gundry SW, Wright JA, Yang H, Pedley S, et al. (2012) Accounting for water quality in monitoring access to safe drinking-water as part of the Millennium Development Goals: lessons from five countries. Bull World Health Organ 90: 228–235A.
9. Onda K, LoBuglio J, Bartram J (2012) Global access to safe water: accounting for water quality and the resulting impact on MDG progress. Int J Environ Res Public Health 9: 880–894.
10. Payen G (2011) Worldwide needs for safe drinking water are underestimated: billions of people are impacted. Paris: AquaFed.
11. White GF, Bradley DJ, White AU (1972) Drawers of water: domestic water use in East Africa. Chicago: The University of Chicago Press.
12. WHO (2011) Guidelines for drinking-water quality. Geneva: World Health Organization.
13. Liu L, Johnson HL, Cousens S, Perin J, Scott S, et al. (2012) Global, regional, and national causes of child mortality: an updated systematic analysis for 2010 with time trends since 2000. Lancet 379: 2151–2161.
14. Esrey SA, Feachem RG, Hughes JM (1985) Interventions for the control of diarrhoeal diseases among young children: improving water supplies and excreta disposal facilities. Bull World Health Organ 63: 757–772.
15. Fewtrell L, Colford JM Jr (2005) Water, sanitation and hygiene in developing countries: interventions and diarrhoea—a review. Water Sci Technol 52: 133–142.
16. Clasen T, Roberts I, Rabie T, Schmidt W, Cairncross S (2006) Interventions to improve water quality for preventing diarrhoea. Cochrane Database Syst Rev 19: CD004794.
17. Waddington H, Snilstveit B (2009) Effectiveness and sustainability of water, sanitation, and hygiene interventions in combating diarrhoea. Journal of Development Effectiveness 1: 295–335.
18. Cairncross S, Hunt C, Boisson S, Bostoen K, Curtis V, et al. (2010) Water, sanitation and hygiene for the prevention of diarrhoea. Int J Epidemiol 39: i193–i205.
19. Ligon GC (2012) Waterborne disease outbreaks: a systematic review of the health effects of drinking water system failures, Chapel Hill (North Carolina): University of North Carolina at Chapel Hill.
20. Craun GF, Brunkard JM, Yoder JS, Roberts VA, et al. (2010) Causes of Outbreaks Associated with Drinking Water in the United States from 1971 to 2006. Clin Microbiol Rev 23(3): 507–528.
21. Schmidt WP, Cairncross S (2009) Household water treatment in poor populations: is there enough evidence for scaling up now? Environ Sci Technol 43: 986–992.
22. WHO/UNICEF (2010) JMP Technical Task Force Meeting on Monitoring Drinking-water Quality. Geneva and New York: WHO/UNICEF Joint Monitoring Programme. Available: http://www.wssinfo.org/fileadmin/user_upload/resources/JMP-Task-Force-Meeting-on-Monitoring-Drinking-water-Quality.pdf. Accessed 1 December 2013.
23. Bain RES, Bartram JK, Elliott M, Matthews RL, McMahan L, et al. (2012) A summary catalogue of microbial drinking water tests for low and medium resource settings. Int J Environ Res Public Health 9: 1609–1625.
24. Lloyd BJ, Bartram JK (1991) Surveillance solutions to microbiological problems in water-quality control in developing-countries. Water Sci Technol 24: 61–75.
25. WHO (1997) Guidelines for Drinking-Water Quality. Geneva: World Health Organization
26. Gleeson C, Gray NF (1997) The coliform index and waterborne disease: problems of microbial drinking water assessment. London; New York: E & FN SPON. xii, 194 p. p.
27. WHO (2012) Water safety planning for small community water supplies: Step-by-step risk management guidance for drinking-water supplies in small communities. Geneva: World Health Organization.

28. Prescott SC, Winslow CEA (1904) Elements of Water Bacteriology, with special reference to sanitary water analysis (1904). New York: John Wiley & Sons.

29. Davison A, Howard G, Stevens M, Callan P, Fewtrell L, et al.. (2005) Water safety plans: Managing drinking-water quality from catchment to consumer. Geneva: World Health Organization.

30. WHO/UNICEF (2013) Proposal for consolidated drinking water, sanitation and hygiene targets, indicators and definitions. Geneva and New York: WHO/UNICEF. Available: http://www.wssinfo.org/fileadmin/user_upload/resources/A-proposal-for-consolidated-WASH-goal-targets-definitions-and-indicators_version7_Nov22_final.pdf. Accessed 1 December 2013.

31. Moher D, Liberati A, Tetzlaff J, Altman DG (2009) Group P (2009) Preferred reporting items for systematic reviews and meta-analyses: the PRISMA statement. PLoS Med 6: e1000097.

32. UN (2013) Millennium Development Indicators: World and regional groupings. Available: http://mdgs.un.org/unsd/mdg/Host.aspx?Content=Data/RegionalGroupings. Accessed 1 December 2013.

33. Opryszko MC, Huang H, Soderlund K, Schwab KJ (2009) Data gaps in evidence-based research on small water enterprises in developing countries. J Water Health 7: 609–622.

34. Wright J, Gundry S, Conroy R (2004) Household drinking water in developing countries: a systematic review of microbiological contamination between source and point-of-use. Trop Med Int Health 9: 106–117.

35. Wright JA, Yang H, Walker K, Pedley S, Elliott J, et al. (2012) The H(2)S test versus standard indicator bacteria tests for faecal contamination of water: systematic review and meta-analysis. Trop Med Int Health 17: 94–105.

36. United Nations Department of Economic and Social Affairs (2012) World Population Prospects: The 2012 Revision. Available: http://esa.un.org/wpp/. Accessed 1 December 2013.

37. Rompre A, Servais P, Baudart J, de-Roubin MR, Laurent P (2002) Detection and enumeration of coliforms in drinking water: current methods and emerging approaches. J Microbiol Methods 49: 31–54.

38. WHO/UNICEF (2006) Core questions on drinking-water and sanitation for household surveys WHO/UNICEF. Geneva and New York: WHO/UNICEF. Available: http://www.who.int/water_sanitation_health/monitoring/oms_brochure_core_questionsfinal24608.pdf. Accessed 1 December 2013.

39. World Bank (2013) Country and lending groups. Available: http://data.worldbank.org/about/country-classifications/country-and-lending-groups. Accessed 1 December 2013.

40. Saha S, Chant D, Welham J, McGrath J (2005) A systematic review of the prevalence of schizophrenia. PLoS Med 2: e141.

41. McBride GB (2005) Using statistical methods for water quality management: issues, problems, and solutions. Hoboken (New Jersey): Wiley-Interscience. xxiv, 313 p. p.

42. Borenstein M (2009) Introduction to meta-analysis. Chichester, U.K.: John Wiley & Sons. xxviii, 421 p. p.

43. Warton DI, Hui FK (2011) The arcsine is asinine: the analysis of proportions in ecology. Ecology 92: 3–10.

44. Sweeting M, J. Sutton A, C. Lambert P (2004) What to add to nothing? Use and avoidance of continuity corrections in meta-analysis of sparse data. Statistics in Medicine 23: 1351–1375.

45. Rodríguez G. (2007) Logit Models for Binary Data. Lecture Notes on Generalized Linear Models. Available: http://data.princeton.edu/wws509/notes/c3.pdf. Accessed 30 December 2013.

46. Abdellah AM, Abdel-Magid HM, Yahia NA (2012) Assessment of Drinking Water Microbial Contamination in Al-Butana Region of Sudan. Journal of Applied Sciences (Faisalabad) 12: 856–862.

47. Abera S, Zeyinudin A, Kebede B, Deribew A, Ali S, et al. (2011) Bacteriological analysis of drinking water sources. Afr J Microbiol Res 5: 2638–2641.

48. Abu Amr SS, Yassin MM (2008) Microbial contamination of the drinking water distribution system and its impact on human health in Khan Yunis Governorate, Gaza Strip: seven years of monitoring (2000–2006). Public Health 122: 1275–1283.

49. Abu Mayla YS, Abu Amr SS (2010) Chemical and microbiological quality of drinking water in Gaza Strip, Palestine. Science Vision 10: 80–88.

50. Abusafa A, Arafat HA, Abu-Baker M, Khalili KN (2012) Utilisation of drinking water from rainwater-harvesting cisterns in the Palestinian territories: assessment of contamination risk. Int J Environment and Waste Management 9: 358–371.

51. Addo KK, Mensah GI, Bekoe M, Bonsu C, Akyeh ML (2009) Bacteriological quality of sachet water produced and sold in Teshie-Nungua suburbs of Accra, Ghana. African Journal of Food Agriculture Nutrition and Development 9.

52. Admassu M, Wubshet M, Gelaw B (2004) A survey of bacteriological quality of drinking water in North Gondar. Ethiop J Health Dev 18: 112–115.

53. Aga T, Beka NC, Eziashi AC (2010) Spatial variation in groundwater quality of Jos metropolis and environs, north-central Nigeria. Continental J Environmental Sciences 4: 1–11.

54. Agard L, Alexander C, Green S, Jackson M, Patel S, et al. (2002) Microbial quality of water supply to an urban community in Trinidad. Journal of Food Protection 65: 1297–1303.

55. Ahmad M, Bajahlan AS (2009) Quality comparison of tap water vs. bottled water in the industrial city of Yanbu (Saudi Arabia). Environ Monit Asses 159: 1–14.

56. Ahmed K, Ahmed M, Ahmed J, Khan A (2012) Risk Assessment by Bacteriological Evaluation of Drinking Water of Gilgit-Baltistan. Pakistan Journal of Zoology 44: 427–432.

57. Ahmed MF, Shamsuddin SAJ, Mahmud SG, Rashid HU, Deere D, et al.. (2005) Risk Assessment of Arsenic

Mitigation Options (RAAMO). Dhaka, Bangladesh: APSU.

58. Ahmed SA, Hoque BA, Mahmud A (1998) Water management practices in rural and urban homes: a case study from Bangladesh on ingestion of polluted water. Public Health 112: 317–321.

59. Ahoussi KE, Koffi YB, Kouassi AM, Osemwegie I, Biémi J (2013) Influence of Anthropogenic Activities of Groundwater from Hand Dug Wells within the Precarious Settlements of Southern Abidjan, Côte d'Ivoire: Case of the Slums of Anoumabo (Marcory) and Adjouffou (Port-Bouët). Journal of Water Resource and Protection 5: 427–439.

60. Akoachere JF, Omam LA, Massalla TN (2013) Assessment of the relationship between bacteriological quality of dug-wells, hygiene behaviour and well characteristics in two cholera endemic localities in Douala, Cameroon. BMC Public Health 13: 692.

61. Al-Khatib IA, Arafat HA (2009) Chemical and microbiological quality of desalinated water, groundwater and rain-fed cisterns in the Gaza strip, Palestine. Desalination 249: 1165–1170.

62. Al-Salaymeh A, Al-Khatib IA, Arafat HA (2011) Towards Sustainable Water Quality: Management of Rainwater Harvesting Cisterns in Southern Palestine. Water Resour Manag 25: 1721–1736.

63. Albert J, Luoto J, Levine D (2010) End-User Preferences for and Performance of Competing POU Water Treatment Technologies among the Rural Poor of Kenya. Environ Sci Technol 44: 4426–4432.

64. Aldana JM (2010) Rapid assessment of drinking-water quality in the Republic of Nicaragua: country report of the pilot project implementation in 2004–2005. Geneva: WHO/UNICEF. Available: http://www.wssinfo.org/fileadmin/user_upload/resources/RADWQ_Nicaragua.pdf. Accessed 1 December 2013.

65. Aliev S, Shodmonov P, Babakhanova N, Schmoll O (2010) Rapid assessment of drinking-water quality in the Republic of Tajikistan: country report of the pilot project implementation in 2004–2005. Geneva: WHO/UNICEF. Available: http://www.wssinfo.org/fileadmin/user_upload/resources/RADWQ_Tajikistan.pdf. Accessed 1 December 2013.

66. Alves NC, Odorizzi AC, Goulart FC (2002) Microbiological analysis of mineral water and drinking water of reservoir supplies, Brazil. Rev Saúde Pública 36: 749–751.

67. Amouei A, Miranzadeh MB, Shahandeh Z, Taheri T, Asgharnia HA, et al. (2012) A Study on the Microbial Quality of Drinking Water in Rural Areas of Mazandaran Province in North of Iran (2011). Journal of Environmental Protection 3: 605–605.

68. Ampofo JA, Andoh A, Tetteh W, Bello M (2007) Microbiological quality and health risks of packaged water produced in southern Ghana. Journal of Applied Science and Technology 12: 88–97.

69. An Y-J, Breindenbach GP (2005) Monitoring E-coli and total coliforms in natural spring water as related to recreational mountain areas. Environ Monit Asses 102: 131–137.

70. Andrade CdS, Leite CC, da Silva MD, de Assis PN, Guimaraes AG (2004) Quality of water used in the beach huts of Salvador-Bahia seashore microbiological aspects. Revista do Instituto Adolfo Lutz 63: 215–219.

71. Arnold M, Vanderslice JA, Taylor B, Benson S, Allen S, et al. (2013) Drinking water quality and source reliability in rural Ashanti region, Ghana. J Water Health 11: 161–172.

72. Augoustinos MT, Venter SN, Kfir R (1995) Assessment of water quality problems due to microbial growth in drinking water distribution systems. Environ Toxic Water 10: 295–299.

73. Austin CJ (1993) Chlorinating household water in The Gambia. 19th WEDC conference. Accra, Ghana: WEDC.

74. Awuah E, Nyarko KB, Owusu PA, Osei-Bonsu K (2009) Small town water quality. Desalination 248: 453–459.

75. Ayantobo OO, Oluwasanya GO, Idowu OA, Eruola AO (2012) (2012) Water Quality Evaluation of Hand-dug Wells in Ibadan, Oyo State, Nigeria. Special Publication of the Nigerian Association of Hydrological Sciences 231–239.

76. Aydin A (2007) The microbiological and physico-chemical quality of groundwater in West Thrace, Turkey. Pol J Environ Stud 16: 377–383.

77. Bahram A, Hamid B, Akram N (2012) Estimation of coliform contamination rate and impact of environmental factor on bacterial quality of tube well water supplies in Khorramdarreh County, Iran. Afr J Biotechnol 11: 7912–7915.

78. Baig SA, Xu X, Navedullah, Muhammad N, Khan ZU, et al. (2012) Pakistan's drinking water and environmental sanitation status in post 2010 flood scenario: humanitarian response and community needs. Journal of Applied Sciences in Environmental Sanitation 7: 49–54.

79. Baker KK, Sow S, Kotloff KL, Nataro JP, Farag TH, et al. (2013) Quality of piped and stored water in households with children under five years of age enrolled in the Mali Site of the Global Enteric Multi-Center Study (GEMS). Am J Trop Med Hyg 89(2): 214–22.

80. Barbosa DA, Lage MM, Badaro ACL (2009) Microbiological quality of water drinking fountains of a university campus in Ipatinga, Minas Gerais. Nutrir Gerais 3: 505–517.

81. Barthiban S, Lloyd BJ (2012) Validity of the application of open dug well sanitary survey methodology in the development of a water safety plan in the Maldives islands. Brebbia CA, Zubir SS, editors. Management of natural resources, sustainable development and ecological hazards III. Ashurst, UK: Wit Press. pp. 369–380.

82. Barthiban S, Lloyd BJ, Maier M (2012) Sanitary hazards and microbial quality of open dug wells in the Maldives Islands. Journal of Water Resource and Protection 4: 474–474.

83. Bartram JK (1996) Optimizing the monitoring and assessment of rural water supplies. Guildford, UK: University of Surrey.

84. Begum J, Ahmed K, Bora KN (2004) Isolation and identification of coliform bacteria from different sources of drinking water. Nature, Environment and Pollution Technology 3: 51–53.

85. Besser RE (1995) Prevención de la transmisión del cólera: evaluación rápida de la calidad del agua municipal en Trujillo, Perú. Boletin - Oficina Sanitaria Panamericana 119: 189–194.

86. Bharath J, Mosodeen M, Motilal S, Sandy S, Sharma S, et al. (2003) Microbial quality of domestic and imported brands of bottled water in Trinidad. Int J Food Microbiol 81: 53–62.

87. Bordalo AA, Savva-Bordalo J (2007) The quest for safe drinking water: an example from Guinea-Bissau (West Africa). Water Research 41: 2978–2986.

88. Bracho MG, Moron V, Luzardo M, Montiel M, Botero L (2008) Hepatitis A virus, adenovirus 40–41 and bacteriophages in water for human consumption. Ciencia (Maracaibo) 16: 271–278.

89. Broshears RE, Amin Akbari M, Chornack MP, Mueller DK, Ruddy BC (2005) Inventory of ground-water resources in the Kabul Basin, Afghanistan. Reston (Virginia): U.S. Geological Survey (USGS).

90. Camara O (2011) Pollution microbiologique des eaux souterraines dans le quartier Tanghin de Ouagadougou: etats des lieux et perspectives. Ouagadougou, Burkina Faso: Foundation 2iE.

91. Cardoso ALSP, Tessari ENC, de Castro AGM, Kanashiro AMI (2001) Avaliação da qualidade microbiológica de águas minerais comercializadas em supermercados da cidade de Alfenas. Higiene Alimentar 15.

92. Carrasco L, Mena KD, Mota LC, Ortiz M, Behravesh CB, et al. (2008) Occurrence of faecal contamination in households along the US-Mexico border. Lett Appl Microbiol 46: 682–687.

93. Caylak E, Tokar M (2012) Metallic and microbial contaminants in drinking water of Cankiri, Turkey. E-Journal of Chemistry 9: 608–614.

94. Chaidez C, Soto M, Martinez C, Keswick B (2008) Drinking water microbiological survey of the Northwestern State of Sinaloa, Mexico. J Water Health 6: 125–130.

95. Chemuliti JK, Gathura PB, Kyule MM, Njeruh FM (2002) Bacteriological qualities of indoor and out-door drinking water in Kibera sub-location of Nairobi, Kenya. East Afr Med J 79: 271–273.

96. Chen D, Lan L (2011) The analysis of drinking-water sanitation in cities and towns of Nanping in the year 2008-2010. Chinese Journal of Health Laboratory Technology 21: 2519–2521.

97. Chung C (2011) The challenges of a water system management handover in Eastern Ethiopia - from the United Nations Refugee Agency to a local community. Cambridge: Massachusetts Institute of Technology.

98. Clasen T, Saeed TF, Boisson S, Edmondson P, Shipin O (2007) Household water treatment using sodium dichloroisocyanurate (NaDCC) tablets: a randomized, controlled trial to assess microbiological effectiveness in Bangladesh. Am J Trop Med Hyg 76: 187–192.

99. Coelho DA, Faria e Silva PM, Veiga SMOM, Fiorini JE (2007) Avaliação da qualidade microbiológica de águas minerais comercializadas em supermercados da cidade de Alfenas. Higiene Alimentar 21: 88–92.

100. Coloru B, Mgaya S, Taubert R (2012) Appropriate technologies for rural water supply: a comparative study between "rope pumps" and conventional piston-pumps on water quality and other sustainability parameters. Milan: ACRA/SHIPO.

101. Copeland CC, Beers BB, Thompson MR, Fitzgerald RP, Barrett LJ, et al. (2009) Faecal contamination of drinking water in a Brazilian shanty town: importance of household storage and new human faecal marker testing. J Water Health 7: 324–331.

102. Costello DH (2013) An Evaluation of a Water, Sanitation, and Hygiene (WASH) Program for rural communities in Northern Afghanistan. Corvallis: Oregon State University.

103. Cronin AA, Breslin N, Gibson J, Pedley S (2006) Monitoring source and domestic water quality in parallel with sanitary risk identification in northern Mozambique to prioritise protection interventions. J Water Health 4: 333–345.

104. Cui B, Deng H, Jing Y, Yue F, Yu Q, et al. (2012) Investigation on the quality of groundwater in Rural Fengtai District. Journal of Environmental Hygiene 2: 280–284.

105. da Silva MEZ, Santana RG, Guilhermetti M, Filho IC, Endo EH, et al. (2008) Comparison of the bacteriological quality of tap water and bottled mineral water. Int J Hyg Environ Health 211: 504–509.

106. Dada AC (2009) Sachet water phenomenon in Nigeria: Assessment of the potential health impacts. Afr J Microbiol Res 3: 15–21.

107. Daoud AK, Swaileh KM, Hussein RM, Matani M (2011) Quality assessment of roof-harvested rainwater in the West Bank, Palestinian Authority. J Water Health 9: 525–533.

108. de Sa LLC, de Jesus IM, Santos ECO, Vale ER, Loureiro ECB, et al. (2005) Microbiological quality of drinking water in two areas following sanitation interventions – Belém, Pará State, Brazil. Epidemiol Serv Saude 14: 171–180.

109. de Siqueira LP, Shinohara NK, de Lima RM, de Paiva Jdo E, de Lima Filho JL, et al. (2010) Microbiological evaluation of drinking water used in feeding units. Cien Saude Colet 15: 63–66.

110. Degbey C, Makoutode M, Fayomi B, Brouwer CD (2010) La qualité de l ′ eau de boisson en milieu professionnel à Godomey en 2009 au Bénin Afrique de l'Ouest. J Int Santé Trav 1: 15–22.

111. Degbey C, Makoutode M, Ouendo EM, Fayomi B, De Brouwer C (2008) The quality of well water in the municipality of Abomey-Calavi in Benin. Environnement Risques & Sante 7: 279–283.

112. Desmarchelier P, Lew A, Caique W, Knight S, Toodayan W, et al. (1992) An evaluation of the hydrogen sulphide water screening test and coliform counts for water

quality assessment in rural Malaysia. Trans R Soc Trop Med Hyg 86: 448–450.

113. Devoto F, Duflo E, Dupas P, Pariente W, Pons V (2011) Happiness on tap: piped water adoption in Urban Morocco. Cambridge (Massachusetts): National Bureau of Economic Research.

114. Dongol B, Merz J, Schaffner M, Nakarmi G, Shah P, et al. (2005) Shallow groundwater in a middle mountain catchment of Nepal: quantity and quality issues. Environ Geol 49: 219–229.

115. dos Reis JAP (1998) Perfil Higienico sanitario das aguas de consumo do distrito federal. Revista de Saude do Distrito 9: 32–35.

116. Egwari L, Aboaba OO (2002) Environmental impact on the bacteriological quality of domestic water supplies in Lagos, Nigeria. Revista de Saude Publica 36: 513–520.

117. Ehlers MM, van Zyl WB, Pavlov DN, Muller EE (2004) Random survey of the microbial quality of bottled water in South Africa. Water SA 30: 203–210.

118. Ejechi BO, Olobaniyi SB, Ogban FE, Ugbe FC (2007) Physical and sanitary quality of hand-dug well water from oil-producing area of Nigeria. Environ Monit Assess 128: 495–501.

119. Ejechi EO, Ejechi BO (2008) Safe drinking water and satisfaction with environmental quality of life in some oil and gas industry impacted cities of Nigeria. Soc Indic Res 85: 211–222.

120. El-Salam MM, Al-Ghitany EM, Kassem MM (2008) Quality of bottled water brands in Egypt Part II: biological water examination. J Egypt Public Health Assoc 83: 468–486.

121. Eldin MNA, Madany IM, Al-Tayaran A, Al-Jubair AH, Gomaa A (1993) Quality of water from some wells in Saudi Arabia. Water Air Soil Poll 66: 135–143.

122. Empereur-Bissonnet P, Salzman V, Monjour L (1992) Evaluation of a new type of transport and storage recipient for improving the quality of drinking-water in rural african areas. Bulletin De La Societe De Pathologie Exotique 85: 390–394.

123. Escamilla V, Knappett PSK, Yunus M, Streatfield PK, Emch M (2013) Influence of latrine proximity and type on tubewell water quality and diarrheal disease in Bangladesh. Ann Assoc Am Geogr 103: 299–308.

124. Eshcol J, Mahapatra P, Keshapagu S (2009) Is fecal contamination of drinking water after collection associated with household water handling and hygiene practices? A study of urban slum households in Hyderabad, India. J Water Health 7: 145–154.

125. Esterhuizen L, Fossey A, Lues JFR (2012) Dairy farm borehole water quality in the greater Mangaung region of the Free State Province, South Africa. Water SA 38: 803–806.

126. Falcone-Dias MF, Emerick GL, Farache-Filho A (2012) Mineral water: a microbiological approach. Wa Sci Technol 12: 556–562.

127. Farache Filho A, Dias MFF (2008) Qualidade microbiológica de águas minerais em galões de 20 litros. Alim Nutr 19: 243–248.

128. Farache Filho A, Dias MFF, Taromaru RH, Cerqueira CS, Duque JD (2008) Qualidade microbiológica de águas minerais não carbonatadas em embalagens de 1,5 litros, comercializadas em araraquara-sp. Alim Nutr 19: 421–425.

129. Ferguson AS, Layton AC, Mailloux BJ, Culligan PJ, Williams DE, et al. (2012) Comparison of fecal indicators with pathogenic bacteria and rotavirus in groundwater. Sci Total Environ 431: 314–322.

130. Ferretti E, Bonadonna L, Lucentini L, Della Libera S, Semproni M, et al. (2010) A case study of sanitary survey on community drinking water supplies after a severe (post-Tsunami) flooding event. Annali Dell Istituto Superiore Di Sanita 46: 236–241.

131. Fiore MM, Minnings K, Fiore LD (2010) Assessment of biosand filter performance in rural communities in southern coastal Nicaragua: an evaluation of 199 households. Rural Remote Health 10: 1483.

132. Flores-Abuxapqui JJ, Suarez-Hoil GJ, Puc-Franco MA, Heredia-Navarrete MR, Vivas-Rosel MD, et al. (1995) Bacteriological quality of drinking water in the City of Merida, Mexico. Salud Publica Mex 37: 236–239.

133. Frischknecht DS (2006) Análise de NMP de coliformes em águas minerais comercializadas no Distrito Federal. Brasília: Universidade de Brasília.

134. Garode AM, Nanoty VD, Bodhankar MG (1998) Bacteriological status of drinking water in and around Chikhli town of Buldana district, Maharashtra. Pollution Research 17: 293–294.

135. Gaytan M, Castro T, Bonilla P, Lugo A, Vilaclara G (1997) Preliminary study of selected drinking water samples in Mexico City. Revista Internacional de Contaminacion Ambiental 13: 73–78.

136. Gelinas Y, Randall H, Robidoux L, Schmit J-P (1996) Well water survey in two districts of Conakry (Republic of Guinea), and comparison with the piped city water. Water Research 30: 2017–2026.

137. Genthe B, Seager J, Vundule C, Strauss N, Maforah F (1996) The effect of water supply, handling and usage on water quality in relation to health indicies in developing communities Stellenbosch, South Africa: Water Research Council.

138. Ghenghesh KS, Belhaj K, Algaui A, Alturki E, Eltomi A (2004) Blessed water: bacteriological quality of drinking water obtained from mosques in Tripoli-Libya. Int J Antimicrob Ag 24: S150.

139. Gonul SA, Karapinar M (1991) The microbiological quality of drinking water supplies of Izmir City: the incidence of Yersinia enterocolitica. Int J Food Microbiol 13: 69–73.

140. Gorter AC, Alberts JH, Gago JF, Sandiford P (1995) A randomized trial of the impact of rope-pumps on water quality. J Trop Med Hyg 98: 247–255.

141. Grimason AM, Beattie TK, Morse TD, Masangwi SJ, Jabu GC, et al.. (2013) Classication and quality of groundwater supplies in the Lower Shire Valley, Malawi – part 2: classification of borehole water supplies in Chikhwawa, Malawi. Water SA 39.

142. Guedes ZBL, Oriá HF, Britto NPB, Neto JWS, Silveira; JWd, et al. (2004) Controle sanitario da agua consumida nas unidades de saude do municipio de Fortaleza, CE. Higiene Alimentar 18: 28–31.

143. Guillemin F, Henry P, Uwechue N, Monjour L (1991) Faecal contamination of rural water supply in the Sahelian area. Water Research 25: 923–927.

144. Guimaraes APRC, Serafini AB (2005) Avaliação da qualidade microbiológica de amostras de água mineral natural, envasada, comercializadas em Goiânia, Goiás. Goiânia, Brazil: UFG.

145. Gundry SW, Wright JA, Conroy RM, Du Preez M, Genthe B, et al. (2009) Child dysentery in the Limpopo Valley: a cohort study of water, sanitation and hygiene risk factors. J Water Health 7: 259–266.

146. Gupta SK, Sheikh MA, Islam MS, Rahman KS, Jahan N, et al. (2008) Usefulness of the hydrogen sulfide test for assessment of water quality in Bangladesh. J Appl Microbiol 104: 388–395.

147. Gupta SK, Suantio A, Gray A, Widyastuti E, Jain N, et al. (2007) Factors associated with E. coli contamination of household drinking water among tsunami and earthquake survivors, Indonesia. Am J Trop Med Hyg 76: 1158–1162.

148. Gwimbi P (2011) The microbial quality of drinking water in Manonyane community: Maseru District (Lesotho). African Health Sciences 11: 474–480.

149. Handzel T (1998) The effect of improved drinking water quality on the risk of diarrheal disease in an urban slum of Dhaka, Bangladesh: A home chlorination intervention trial. Chapel Hill: University of North Carolina at Chapel Hill.

150. Haruna R, Ejobi F, Kabagambe EK (2005) The quality of water from protected springs in Katwe and Kisenyi parishes, Kampala city, Uganda. Afr Health Sci 5: 14–20.

151. Hashmi I, Farooq S, Qaiser S (2009) Chlorination and water quality monitoring within a public drinking water supply in Rawalpindi Cantt (Westridge and Tench) area, Pakistan. Environ Monit Asses 158: 393–403.

152. Hashmi I, Farooq S, Qaiser S (2009) Incidence of fecal contamination within a public drinking water supply in Ratta Amral, Rawalpindi. Desalination and Water Treatment 11: 124–131.

153. Henry FJ, Rahim Z (1990) Transmission of diarrhoea in two crowded areas with different sanitary facilities in Dhaka, Bangladesh. J Trop Med Hyg 93: 121–126.

154. Herath AT, Abayasekara CL, Chandrajith R, Adikaram NKB (2012) Temporal variation of microbiological and chemical quality of noncarbonated bottled drinking water sold in Sri Lanka. J Food Sci 77: M160–M164.

155. Hira-Smith MM, Yuan Y, Savarimuthu X, Liaw J, Hira A, et al. (2007) Arsenic concentrations and bacterial contamination in a pilot shallow dugwell program in West Bengal, India. J Environ Sci Health A Tox Hazard Subst Environ Eng 42: 89–95.

156. Holm RH (2012) Survey of water quality in northern Malawi, Africa, and communication of human health risk. Pullman: Washington State University.

157. Holt S (2009) A survey of water storage practices and beliefs in households in Bonao, Dominican Republic in 2005. Atlanta: Georgia State University.

158. Hoque BA, Sack RB, Siddiqi M, Jahangir AM, Hazera N, et al. (1993) Environmental health and the 1991 Bangladesh cyclone. Disasters 17: 143–152.

159. Hoque BA, Yamaura S, Sakai A, Khanam S, Karim M, et al. (2006) Arsenic mitigation for water supply in bangladesh: appropriate technological and policy perspectives. Water Qual Res J Can 41: 226–234.

160. Horak HM, Chynoweth JS, Myers WP, Davis J, Fendorf S, et al. (2010) Microbial and metal water quality in rain catchments compared with traditional drinking water sources in the East Sepik Province, Papua New Guinea. J Water Health 8: 126–138.

161. Howard G, Luyima P (1999) Urban and peri-urban water supply surveillance programme report on water supply surveillance in ten selected urban areas of Uganda. Loughborough, UK: Loughborough University.

162. Hubbard B, Gelting R, Baffigo V, Sarisky J (2005) Community environmental health assessment strengthens environmental public health services in the Peruvian Amazon. Int J Hyg Envir Heal 208: 101–107.

163. Hussain J, Hussain I, Ojha KG, Sharma KC (2001) Bacteriological pollution of ground water in textile city, Bhilwara, Rajasthan (India). Asian J Chem 13: 1123–1126.

164. Ince M, Bashire D, Oni OOO, Awe EO, Ogbechie V, et al.. (2010) Rapid assessment of drinking-water quality in the Federal Republic of Nigeria: country report of the pilot project implementation in 2004–2005. Geneva: WHO/UNICEF. Available: http://www.wssinfo.org/fileadmin/user_upload/resources/RADWQ_Nigeria.pdf. Accessed 1 December 2013.

165. Isaac-Marquez AP, Lezama-Davila CM, Ku-Pech PP, Tamay-Segovia P (1994) Sanitary quality of water supply for human consumption in Campeche. Salud Publica Mex 36: 655–661.

166. Islam MA, Sakakibara H, Karim MR, Sekine M, Mahmud ZH (2011) Bacteriological assessment of drinking water supply options in coastal areas of Bangladesh. J Water Health 9: 415–428.

167. Itama E, Olaseha IO, Sridhar MKC (2006) Springs as supplementary potable water supplies for inner city populations: a study from Ibadan, Nigeria. Urban Water J 3: 215–223.

168. Jagals P, Barnard TG, Mokoena MM, Ashbolt N, Roser DJ (2013) Pathogenic Escherichia coli in rural household container waters. Water Sci Technol 67: 1230–1237.

169. Jagals P, Bokako TC, Grabow WOK (1999) Changing consumer water-use patterns and their effect on microbiological water quality as a result of an engineering intervention. Water SA 25: 297–297.

170. Jagals P, Grabow WOK, Williams E (1997) The effects of supplied water quality on human health in an urban development with limited basic subsistence facilities. Water SA 23: 373–378.

171. Jayatilleke SK, Gunawardana SS (2009) Microbiological quality of well water in Kalutara District. Ceylon Med J 54: 44–45.

172. Jiban Singh M, Somashekar RK, Prakash KL, Shivanna K (2009) Bacteriological assessment of groundwater in Arkavathi and Vrishabhavathi basins, Bangalore, Karnataka. Journal of Ecology and The Natural Environment 1: 156–159.

173. Jimmy DH, Sundufu AJ, Malanoski AP, Jacobsen KH, Ansumana R, et al. (2013) Water quality associated public health risk in Bo, Sierra Leone. Environ Monit Assess 185: 241–251.

174. Joya SA, Mostofa G, Yousuf J, Islam A, Elahi A, et al. (2006) One solution to the arsenic problem: a return to surface (improved dug) wells. J Health Popul Nutr 24: 363–375.

175. Junaidu AU, Adamu RYA, Ibrahim MTO (2001) Bacteriological assessment of pipe borne water in Sokoto metropolis, Nigeria. Tropical Veterinarian 19: 160–162.

176. Jung JH, Yoo CH, Koo ES, Kim HM, Na Y, et al. (2011) Occurrence of norovirus and other enteric viruses in untreated groundwaters of Korea. J Water Health 9: 544–555.

177. Kanyerere T, Levy J, Xu Y, Saka J (2012) Assessment of microbial contamination of groundwater in upper Limphasa River catchment, located in a rural area of northern Malawi. Water SA (Pretoria) 38: 581–595.

178. Karim MR (2010) Microbial contamination and associated health burden of rainwater harvesting in Bangladesh. Water Sci Technol 61: 2129–2135.

179. Karnchanawong S, Koottatep S, Ikeguchi T (1993) Monitoring and evaluation of shallow well water quality near a waste disposal site. Environ Int 19: 579–587.

180. Kassenga GR (2007) The health-related microbiological quality of bottled drinking water solid in Dar es Salaam, Tanzania. J Water Health 5: 179–185.

181. Kerala State Pollution Control Board (1991) The bacterial quality of water in selected wells in Kerala: an investigation. Trivandrum SEU-Kerala.

182. Khadse GK, Kalita M, Pimpalkar SN, Labhsetwar PK (2011) Drinking water quality monitoring and surveillance for safe water supply in Gangtok, India. Environ Monit Asses 178: 401–414.

183. Khadse GK, Kalita MD, Labhsetwar PK (2012) Change in drinking water quality from source to point-of-use and storage: a case study from Guwahati, India. Environ Monit Asses 184: 5343–5361.

184. Khadse GK, Kalita MD, Pimpalkar SN, Labhasetwar PK (2011) Surveillance of drinking water quality for safe water supply-a case study from Shillong, India. Water Resour Manag 25: 3321–3342.

185. Khan SU, Bangash FK (2001) Drinking water quality forecast of Peshawar valley on the basis of sample data. J Chem Soc Pakistan 23: 243–252.

186. Khush R, London A, Arnold J, Ramaswamy K (2009) Evaluating the sustainability and impacts of water, sanitation & hygiene interventions. San Francisco: Aquaya.

187. Kilanko-Oluwasanya GO (2009) Better safe than sorry: towards appropriate water safety plans for urban self supply systems. Cranfield, UK: Cranfield University.

188. Kimani-Murage EW, Ngindu AM (2007) Quality of water the slum dwellers use: the case of a Kenyan slum. J Urban Health 84: 829–838.

189. Kiptum CK, Ndambuki JM (2012) Well water contamination by pit latrines: a case study of Langas. International Journal of Water Resources and Environmental Engineering 4: 35–43.

190. Klasen S, Lechtenfeld T, Meier K, Rieckmann J (2012) Benefits trickling away: the health impact of extending access to piped water and sanitation in urban Yemen. Journal of Development Effectiveness 4: 537–565.

191. Knappett PSK, McKay LD, Layton A, Williams DE, Alam MJ, et al. (2012) Unsealed tubewells lead to increased fecal contamination of drinking water. J Water Health 10: 565–578.

192. Knight SM, Toodayan W, Caique WC, Kyi W, Barnes A, et al. (1992) Risk factors for the transmission of diarrhoea in children: a case-control study in rural Malaysia. Int J Epidemiol 21: 812–818.

193. Korfali SI, Jurdi M (2009) Provision of safe domestic water for the promotion and protection of public health: a case study of the city of Beirut, Lebanon. Environ Geochem Hlth 31: 283–295.

194. Kravitz JD, Nyaphisi M, Mandel R, Petersen E (1999) Quantitative bacterial examination of domestic water supplies in the Lesotho Highlands: Water quality, sanitation, and village health. Bull World Health Organ 77: 829–836.

195. Kremer M, Leino J, Miguel E, Zwane AP (2011) Spring cleaning: rural water impacts, valuation, and property rights institutions. Q J Econ 126: 145–205.

196. Kremer M, Null C, Zwane AP (2008) Trickle down: diffusion of chlorine for drinking water treatment in Kenya. Available: http://www.stanford.edu/group/SITE/archive/SITE_2008/segment_1/papers/de_giorgi_SITE_Trickle_Down.pdf. Accessed 1 December 2013.

197. Kromoredjo P, Fujioka RS (1991) Evaluating three simple methods to assess the microbial quality of drinking water in indonesia. Environ Toxic Water 6: 259–270.

198. Kumpel E, Nelson KL (2013) Comparing microbial water quality in an intermittent and continuous piped water supply. Water Res 47: 5176–5188.

199. Lacey S, Lopez R, Frangos C, Khodadoust A (2011) Water quality degradation after water storage at household level in a piped water system in rural Guatemala. International Journal for Service Learning in Engineering 6: 118–129.

200. Lal M, Kaur H (2006) A microbiological study of bottled mineral water marketed in Ludhiana. Indian J Public Health 50: 31–32.

201. Lantagne D, Clasen T (2011) Assessing the Implementation of Selected Household Water Treatment and Safe Storage (HWTS) Methods in Emergency Settings. London: London School of Hygiene and Tropical Medicine.

202. Leber J, Rahman MM, Ahmed KM, Mailloux B, van Geen A (2011) Contrasting influence of geology on E. coli and arsenic in aquifers of Bangladesh. Ground Water 49: 111–123.

203. Leiter M, Levy J, Mutiti S, Boardman M, Wojnar A, et al. (2013) Drinking water quality in the Mount Kasigau region of Kenya: a source to point-of-use assessment. Environmental Earth Sciences 68: 1–12.

204. Levy K, Nelson KL, Hubbard A, Eisenberg JNS (2008) Following the water: a controlled study of drinking water storage in northern coastal Ecuador. Environ Health Persp 116: 1533–1540.

205. Lin M-Y, Chen X-L, You T-Z (2009) Hygienic quality of bottled drinking water in Longyan from 2006 to 2008. Chinese Journal of Public Health Engineering/Zhongguo Weisheng Gongcheng Xue 8: 168–169.

206. Lira I, Matamoros A, Jacob F, Baltodano F, Harvey R (2008) Measurement of the quality of potable water by detection of H2S-producing bacteria; comparison with detection of fecal coliforms by the Millipore method in rural Nicaragua. Matagalpa/Jinotega, Nicaragua: ENACAL/DAR. Available: http://tascanica.org/images/Pathoscreenreport.pdf. Accessed 1 December 2013.

207. Long WF, Yang JJ, Xu QJ (2012) Microbial contamination of household drinking water in five Haikou residential groups. Huanjing yu Zhiye Yixue 29: 566–568.

208. Luby S, Islam MS, Johnston R (2006) Chlorine spot treatment of flooded tube wells, an efficacy trial. J Appl Microbiol 100: 1154–1158.

209. Luby SP, Gupta SK, Sheikh MA, Johnston RB, Ram PK, et al. (2008) Tubewell water quality and predictors of contamination in three flood-prone areas in Bangladesh. J Appl Microbiol 105: 1002–1008.

210. Lukacs H (2002) From design to implementation: innovative slow sand filtration for use in developing countries. Cambridge: Massachusetts Institute of Technology.

211. Machdar E, van der Steen NP, Raschid-Sally L, Lens PN (2013) Application of Quantitative Microbial Risk Assessment to analyze the public health risk from poor drinking water quality in a low income area in Accra, Ghana. Sci Total Environ 449: 134–142.

212. Macy JT, Dunne EF, Angoran-Benie YH, Kamelan-Tano Y, Kouadio L, et al. (2005) Comparison of two methods for evaluating the quality of stored drinking water in Abidjan, Cote d'lvoire, and review of other comparisons in the literature. J Water Health 3: 221–228.

213. Magan M, Bile KM, Kazi BM, Gardezi Z (2010) Safe water supply in emergencies and the need for an exit strategy to sustain health gains: lessons learned from the 2005 earthquake in Pakistan. East Mediterr Health J 16 Suppl: S91–97

214. Magrath J (2006) Towards sustainable water-supply solutions in Rural Sierra Leone: a pragmatic approach, using comparisons with Mozambique. Oxford: Oxfam/WaterAid.

215. Mah U, Haq T, Perveen F, Sultana L (2007) Bacteriological analysis of groundwater of Karachi. Pakistan J Sci Ind R 50: 273–277.

216. Mahasneh IA (1992) Isolation and characterization of faecal indicator bacteria from urban and rural natural drinking water sources. Biomedical Letters 47: 347–354.

217. Mahfouz AA, Abdel-Moneim M, al-Erian RA, al-Amari OM (1995) Impact of chlorination of water in domestic storage tanks on childhood diarrhoea: a community trial in the rural areas of Saudi Arabia. J Trop Med Hyg 98: 126–130.

218. Maiga H, Maiga B, Sutton S (2006) Self-supply in Mali. Waterlines 25: 13–14.

219. Mamo E, Mekonta L, Butterworth J, Sutton S (2011) New insights on the oldest approach: family wells in Ethiopia. 6th Rural Water Supply Network Forum 2011. Uganda. Kampala, Uganda: RWSN.

220. Mane VR, Chandorkar AA, Kumar R (2005) Prevalence of pollution in surface and ground water sources in the rural areas of Satara Region, Maharashtra, India. Asian Journal of Water, Environment and Pollution 2: 81–87.

221. Massoud MA, Al-Abady A, Jurdi M, Nuwayhid I (2010) The challenges of sustainable access to safe drinking water in rural areas of developing countries: case of Zawtar El-Charkieh, Southern Lebanon. J Environ Health 72: 24–30.

222. Matsinhe NP, Juizo D, Rietveld LC, Persson KM (2008) Water services with independent providers in peri-urban Maputo: challenges and opportunities for long-term development. Water SA 34: 411–420.

223. Mazengia E, Chidavaenzi MT, Bradley M, Jere M, Nhandara C, et al. (2002) Effective and culturally acceptable water storage in Zimbabwe: maintaining the quality of water abstracted from upgraded family wells. J Environ Health 64: 15–18.

224. Mertens TE, Fernando MA, Marshall TFDC, Kirkwood BR, Cairncross S, et al. (1990) Determinants of water quality availability and use in Kurunegala Sri Lanka. Trop Med Parasitol 41: 89–97.

225. Metwali RM (2003) Water quality of some wells in Taiz City (Yemen Republic) and its surroundings. Folia Microbiol (Praha) 48: 90–94.

226. Michelina AF, Bronharoa TM, Dareb F, Ponsanoc EHG (2006) Qualidade microbiológica de águas de sistemas de abastecimento público da região de Araçatuba, SP. Higiene Alimentar 20: 90–95.

227. Miranda M, Aramburu A, Junco J, Campos M (2010) State of the quality of drinking water in households in children under five years in Peru, 2007–2010. Rev Peru Med Exp Salud Publica 27: 506–511.

228. Miranzadeh MB, Heidari M, Mesdaghinia AR, Younesian M (2011) Survey of microbial quality of drinking water in rural areas of Kashan-Iran in second half of 2008. Pak J Biol Sci 14. 59–63.

229. Moazeni M, Atefi M, Ebrahimi A, Razmjoo P, Vahid Dastjerdi M (2013) Evaluation of chemical and microbiological quality in 21 brands of Iranian bottled drinking waters in 2012: a comparison study on label and real contents. Journal of Environmental and Public Health 2013: 4.

230. Moe CL, Sobsey MD, Samsa GP, Mesolo V (1991) Bacterial indicators of risk of diarrhoeal disease from drinking-water in the Philippines. Bull World Health Organ 69: 305–317.

231. Momba MNB, Obi CL, Thompson P (2009) Survey of disinfection efficiency of small drinking water treatment plants: challenges facing small water treatment plants in South Africa. Water Sa 35: 485–493.

232. Mora-Bueno D, Sanchez-Pena LC, del Razo LM, Gonzalez-Arias CA, Medina-Diaz IM, et al. (2012) Presencia de arsénico y coliformes en agua potable del municipio de Tecuala, Nayarit, México. Rev Int Contam Ambie 28: 127–135.

233. Morgan P (1990) Rural water supplies and sanitation. London: Macmillan. 358 p.

234. Moshtaghi H, Boniadian M (2007) Microbial quality of drinking water in Shahrekord (Iran). Research Journal of Microbiology 2: 299–302.

235. Mosley LM, Sharp DS, Singh S (2004) Effects of a tropical cyclone on the drinking-water quality of a remote Pacific Island. Disasters 28: 405–417.

236. Moyo S, Wright J, Ndamba J, Gundry S (2004) Realising the maximum health benefits from water quality improvements in the home: a case from Zaka district, Zimbabwe. Phys Chem Earth 29: 1295–1299.

237. Mpenyana-Monyatsi L, Onyango MS, Momba MNB (2012) Groundwater quality in a South African rural community: a possible threat to public health. Pol J Environ Stud 21: 1349–1358.

238. Muzuga JM, Tole MP, Ucakuwun EK (1998) The impact of geology and pit latrines on ground water quality in Kwale District. Dunes, groundwater, mangroves and birdlife in costal Kenya. Nairobi, Kenya: Acts Press.

239. Mzuga JM, Tole MP, Ucakuwun EK (2001) Contamination of groundwater resources by pit latrines in Kwale District, Kenya. Discov Innovat 13: 203–212.

240. Nascimento AR, Azevedo TKL, Filho NEM, Rojas MOAI (2000) Qualidade microbiológica das águas minerais consumidas na cidade de São Luís - MA. Higiene Alimentar 14: 69–72.

241. NEERI (2005) Study on surveillance of drinking water quality in selected cities/towns in India. Volume I: surveillance volume II: city appraisal. Nehru Marg, Nagpur: NEERI.

242. Nikaeen M, Pejhan A, Jalali M (2009) Rapid monitoring of indicator coliforms in drinking water by an enzymatic assay. Iranian Journal of Environmental Health Science and Engineering 6: 7–10.

243. Nogueira G, Nakamura CV, Tognim MCB, Abreu Filho BA, Dias Filho BP (2003) Microbiological quality of drinking water of urban and rural communities, Brazil. Revista de Saude Publica 37: 232–236.

244. Norton DM, Rahman M, Shane AL, Hossain Z, Kulick RM, et al. (2009) Flocculant-disinfectant point-of-use water treatment for reducing arsenic exposure in rural Bangladesh. Int J Environ Heal R 19: 17–29.

245. Nunes SM, Fuzihara TO (2011) Availacao microbiologica das aguas envadas e commercializadas na regiao do ABC, SP. Higiene Alimentar 25: 195–199.

246. Nwosu JN, Ogueke CC (2004) Evaluation of sachet water samples in Owerri metropolis. Nigerian Food Journal 22: 164–170.

247. Nyanchaga EN (1994) Rehabilitation of hand-dug wells and springs. Aqua (Oxford) 43: 233–237.

248. Obiri-Danso K, Okore-Hanson A, Jones K (2003) The microbiological quality of drinking water sold on the streets in Kumasi, Ghana. Lett Appl Microbiol 37: 334–339.

249. Ogan MT (1992) Microbiological quality of bottled water sold in retail outlets in Nigeria. J Appl Bacteriol 73: 175–181.

250. Okagbue RN, Dlamini NR, Siwela M, Mpofu F (2002) Microbiological quality of water processed and bottled in Zimbabwe. Afr J Health Sci 9: 99–103.

251. Okioga T (2007) Water quality and business aspects of sachet-vended water in Tamale, Ghana [Master of Engineering dissertation]. Cambridge: Massachusetts Institute of Technology.

252. Okura MH, Siqueira KB (2005) Enumeracao de coliformes totais e coliformes termotolerantes em agua de abastecimento e de minas. Higiene Alimentar 19: 86–91.

253. Olaoye OA, Onilude AA (2009) Assessment of microbiological quality of sachet-packaged drinking water in Western Nigeria and its public health significance. Public Health 123: 729–734.

254. Oloruntoba EO, Sridhar MK (2007) Bacteriological quality of drinking water from source to household in Ibadan, Nigeria. Afr J Med Med Sci 36: 169–175.

255. Omar KB, Potgieter N, Barnard TG (2010) Development of a rapid screening method for the detection of pathogenic Escherichia coli using a combination of Colilert Quanti-Trays/2000 and PCR. Wa Sci Technol 10: 7–13.

256. Omer EOM, Sallam T (2012) Bacteriogical investigation of drinking water in Shendi Locality, River Nile State, Sudan. Open Access Scientific Reports 1: 329.

257. Onifade AK, M LR (2008) Microbiological analysis of sachet water vended in Ondo State, Nigeria. Environmental Research Journal 2: 107–110.

258. Oo KN, Han AM, Hlaing T, Aye T (1991) Bacteriologic studies of food and water consumed by children in Myanmar: 1. The nature of contamination. J Diarrhoeal Dis Res 9: 87–90.

259. Opisa S, Odiere MR, Jura WG, Karanja DM, Mwinzi PN (2012) Faecal contamination of public water sources in informal settlements of Kisumu City, western Kenya. Water Sci Technol 66: 2674–2681.

260. Opryszko MC, Guo Y, Macdonald L, Macdonald L, Kiihl S, et al. (2013) Impact of water-vending kiosks and hygiene education on household drinking water quality in rural Ghana. Am J Trop Med Hyg 88: 651–660.

261. Ordinioha B (2011) A survey of the community water supply of some rural riverine communities in the Niger

Delta region, Nigeria: Health implications and literature search for suitable interventions. Niger Med J 52: 13–18.

262. Oswald WE, Lescano AG, Bern C, Calderon MM, Cabrera L, et al. (2007) Fecal contamination of drinking water within peri-urban households, Lima, Peru. Am J Trop Med Hyg 77: 699–704.

263. Oyedeji O, Olutiola PO, Moninuola MA (2010) Microbiological quality of packaged drinking water brands marketed in Ibadan metropolis and Ile-Ife city in South Western Nigeria. Afr J Microbiol Res 4: 96–102.

264. Pan F, Meng K, Zhao H (2011) Hygienic investigation on rural drinking water in Daxing District of Beijing. Occupation and Health 27: 2140–2141.

265. Pant H, Sarfaraz S, Iyengar L (2002) Evaluation of L-Cystine-amended H2S test for the detection of faecal pollution in potable water: comparison with standard multiple tube fermentation method. World Journal of Microbiology and Biotechnology 18: 317–320.

266. Parker AH, Youlten R, Dillon M, Nussbaumer T, Carter RC, et al. (2010) An assessment of microbiological water quality of six water source categories in north-east Uganda. J Water Health 8: 550–560.

267. Pathak SP, Gopal K (1994) Antibiotic resistance and metal tolerance among Coliform sp. from drinking water in a hilly area. J Environ Biol 15: 139–139.

268. Pathak SP, Gopal K (2005) Efficiency of modified H2S test for detection of faecal contamination in water. Environ Monit Assess 108: 59–65.

269. Pathak SP, Gopal K (2008) Prevalence of bacterial contamination with antibiotic resistant and enterotoxigenic fecal coliforms in treated drinking water. J Toxicol Env Heal A 71: 427–433.

270. Patrick JM (2010) Recommendations for at-risk water supplies in Capiz Province, Philippines: using water source and community assessments. Cambridge: Massachusetts Institute of Technology.

271. Phattharangrong N, Chantratong N, Jitsurong S (1998) Bacteriological quality of holywater from Thai temples in Songkhla Province, southern Thailand. J Med Assoc Thailand 81: 547–550.

272. Pickering AJ, Davis J, Walters SP, Horak HM, Keymer DP, et al. (2010) Hands, water, and health: fecal contamination in Tanzanian communities with improved, non-networked water supplies. Environ Sci Technol 44: 3267–3272.

273. Pinfold JV, Horan NJ, Wirojanagud W, Mara D (1993) The bacteriological quality of rainjar water in rural northeast Thailand. Water Research 27: 297–302.

274. Piranha JM, Pacheco A, Gamba RC, Mehnert DU, Garrafa P, et al. (2006) Faecal contamination (viral and bacteria) detection in groundwater used for drinking purposes in Sao Paulo, Brazil. Geomicrobiol J 23: 279–283.

275. Platenburg RJPM, Zaki M (1993) Patterns of water quality in rural areas of Assyut Governorate, Egypt. Water Sci Technol 27: 55–55.

276. Potgieter N, Becker PJ, Ehlers MM (2009) Evaluation of the CDC safe water-storage intervention to improve the microbiological quality of point-of-use drinking water in rural communities in South Africa. Water SA 35: 505–516.

277. Potgieter N, Mudau LS, Maluleke FRS (2006) Microbiological quality of groundwater sources used by rural communities in Limpopo Province, South Africa. Kroiss H, Potgieter N, editors. London: IWA Publishing. 371–377 p.

278. Pradhan B (2004) Rural communities› perception on water quality and water borne disease: the case of Bungamati Village Development Committee in Kathmandu Valley, Nepal. Journal of Nepal Health Research Council 2.

279. Pritchard M, Mkandawire T, O'Neill JG (2007) Biological, chemical and physical drinking water quality from shallow wells in Malawi: case study of Blantyre, Chiradzulu and Mulanje. Phys Chem Earth 32: 1167–1177.

280. Pritchard M, Mkandawire T, O'Neill JG (2008) Assessment of groundwater quality in shallow wells within the southern districts of Malawi. Phys Chem Earth 33: 812–823.

281. Properzi F (2010) Rapid assessment of drinking-water quality in the Hashemite Kingdom of Jordan: country report of the pilot project implementation in 2004–2005. Geneva: WHO/UNICEF. Available: http://www.wssinfo.org/fileadmin/user_upload/resources/RADWQ_Jordan.pdf. Accessed 1 December 2013.

282. Putri LSE, Kustanti N, Yunita E (2013) Effect of land use on ground water quality (a casestudy from Ciracas Sub District, East Jakarta, Indonesia). International Journal of Bioscience, Biochemistry and Bioinformatics 3: 33–36.

283. Qi W (2007) Analysis of water quality of 186 samples from homemade wells of Yuci District of Jinzhong City in 2005. Preventive Medicine Tribune 13: 536–537.

284. Quick RE, Kimura A, Thevos A, Tembo M, Shamputa I, et al. (2002) Diarrhea prevention through household-level water disinfection and safe storage in Zambia. Am J Trop Med Hyg 66: 584–589.

285. Radaideh J, Al-Zboon K, Al-Harahsheh A, Al-Adamat R (2009) Quality assessment of harvested rainwater for domestic uses. Jordan Journal of Earth and Environmental Sciences 2: 26–31.

286. Rainey RC, Harding AK (2005) Drinking water quality and solar disinfection: effectiveness in peri-urbanhouseholds in Nepal. J Water Health 3: 239–248.

287. Ramteke PW (1995) Comparison of standard most probable number method with three alternate tests for detection of bacteriological water quality indicators. Environ Toxic Water 10: 173–178.

288. Ramteke PW, Bhattacharjee JW, Pathak SP, Kalra N (1992) Evaluation of coliforms as indicators of water-quality in India. J Appl Bacteriol 72: 352–356.

289. Rao PNS, Suresh K, Basavarjappa KG (2006) Bacteriological quallity of potable water in Davangere city. Journal of Communicable Diseases 38: 381–384.

290. Rawat V, Jha SK, Bag A, Singhai M, Rawat CMS (2012) The bacteriological quality of drinking water in Haldwani Block of Nainital District, Uttarakhand, India. J Water Health 10: 465–470.

291. Razzolini MTP, Santos TFD, Bastos VK (2010) Detection of Giardia and Cryptosporidium cysts/oocysts in watersheds and drinking water sources in Brazil urban areas. J Water Health 8: 399–404.

292. Reddy PS, Ata-Ur-Rasheed MD, Sharma S (2000) Microbiological analysis of bottled water. Indian J Med Microbiol 18: 72–76.

293. Rejith PG, Jeeva SP, Vijith H, Sowmya M, Hatha AAM (2009) Determination of groundwater quality index of a Highland Village of Kerala (India) using geographical information system. J Environ Health 71: 51–58.

294. Rheinheimer DS, GonÇalves CS, Bortoluzzi EC, Pellegrin JBR, da Silva JLS, et al. (2010) Qualidade de águas subterrâneas captadas em fontes em função da presença de proteção física e de sua posição na paisagem. Eng Agríc, Jaboticabal 30: 948–957.

295. Rosa G, Miller L, Clasen T (2010) Microbiological effectiveness of disinfecting water by boiling in rural Guatemala. Am J Trop Med Hyg 82: 473–477.

296. Rufener S, Maeusezahl D, Mosler H-J, Weingartner R (2010) Quality of drinking-water at source and point-of-consumption-drinking cup as a high potential recontamination risk: a field study in Bolivia. J Health Popul Nutr 28: 34–41.

297. Sampson M (2008) Water quality assessment of hand dug wells. Phnom Penh, Cambodia: Resource Development International.

298. Santana A, Silva SCFL, Farani IO, Amaral CHR, Macedo VF (2003) Qualidade microbiológica de águas minerais. Ciênc Tecnol Aliment 23: 190–194.

299. Sasikaran S, Sritharan K, Balakumar S, Arasaratnam V (2012) Physical, chemical and microbial analysis of bottled drinking water. Ceylon Med J 57: 111–116.

300. Schaffner M (2002) Drinking water quality assessment and improvement in the Jhikhu Khola watershed. Bern: Universität Bern.

301. Semerjian LA (2011) Quality assessment of various bottled waters marketed in Lebanon. Environ Monit Asses 172: 275–285.

302. Shaheed A, Orgill J, Ratana C, Montgomery MA, Jeuland A, et al. (2013) Water quality risks of "improved " water sources: evidence from Cambodia. Trop Med Int Health 19: 186–194.

303. Shaikh SA, Gul N, Sultana L (2008) Surveillance of drinking water of Karachi City: microbiological quality. Pakistan J Sci Ind R 51: 272–275.

304. Shakya (2004) An integrated drinking water quality assessment of rainwater harvesting jars and the related socio-economic conditions and gender issues of Chappani VDC-1, Palpa District. Available: http://www.aehms.org/pdf/Shakya%20&%20Sharma%20Proceedings%20FE.pdf. Accessed 30 December 2014.

305. Shar AH, Kazi YF, Kanhar NA, Soomro IH, Zia SM, et al. (2010) Drinking water quality in Rohri City, Sindh, Pakistan. Afr J Biotechnol 9: 7102–7107.

306. Shar AH, Kazi YF, Soomro IH (2009) Antibiotic susceptibility of thermo-tolerant escherichia coli 2 isolated from drinking water of Khairpur City, Sindh, Pakistan. Pakistan Journal of Biological Sciences 12: 648–652.

307. Shar AH, Kazi YF, Zardari M, Soomro IH (2007) Enumeration of total and faecal coliform bacteria in drinking water of Khairpur City, Sindh, Pakistan. Bangladesh J Microbiol 24: 163–165.

308. Sharma BK, Sharma LL, Durve VS (2008) Assessment of hand pump waters in three tribal dominated districts of southern Rajasthan, India. Journal of Environmental Science & Engineering 50: 133–136.

309. Shrestha S, Malla SS, Aihara Y, Kondo N, Nishida K (2013) Water Quality at Supply Source and Point of Use in the Kathmandu Valley. Journal of Water and Environment Technology 11: 331–340.

310. Silva EF, Salgueiro AA (2001) Avaliacao da qualidade bacteriologica de agua de pocos na regio metropolitana de Recife - P. Higiene Alimentar. 15: 73–78.

311. Simango C, Dindiwe J, Rukure G (1992) Bacterial-contamination of food and household stored drinking-water in a farmworker community in Zimbabwe. Cent Afr J Med 38: 143–149.

312. Simiyu S (2009) Assessment of spring water quality and quantity, and health implications in Tongaren division, Nzoia River catchment, Kenya. Afr J Ecol 47: 99–104.

313. Singh KP, Gaur A, Chandra H, Ray PK (1993) Drinking water quality status of Bankura District (West Bengal) as indicator of environmental pollution. Agrawal VP, Abidi SAH, Verma GP, Singh KP, editors. Muzaffarnagar (India): Society of Bioscience. 71–78 p.

314. Smith PG, Sabone TG (1994) Drinking water quality in the Gantsi District of Botswana. Int J Environ Heal R 4: 141–147.

315. Soto FRM, Fonseca YSK, Antunes DV, Risseto MR, Amaku M, et al. (2005) Microbiological evaluation of public water supply in schools of Ibiuna-SP: comparative study of water quality in conduit and post conduit. Revista do Instituto Adolfo Lutz 64: 128–131.

316. Soto FRM, Fonseca YSK, Risseto MR, de Azevedo SS, Arini MdLB, et al. (2006) Monitoring the quality of water from groundwater located in the rural public schools of the Ibiuna municipality/SP: microbiological and physicochemical parameters, and factors for environmental risk. Revista do Instituto Adolfo Lutz 65: 106–111.

317. Souza Coelho MI, Mendes ES, Soares Cruz MC, Bezerra SS, Pinheiro e Silva RP (2010) Evaluation of the microbiological quality of mineral water consumed in the metropolitan region of Recife, Pernambuco State. Acta Scientiarum Health Science 32: 1–8.

318. Stoler J, Fink G, Weeks JR, Otoo RA, Ampofo JA, et al. (2012) When urban taps run dry: sachet

water consumption and health effects in low income neighborhoods of Accra, Ghana. Health Place 18: 250–262.

319. Sutton S (2002) Community led improvements of rural drinking water supplies. Shewsbury, UK: SWL Consultants. Available: http://www.rural-water-supply.net/en/resources/details/249. Accessed 1 December 2013.

320. Sutton S, Butterworth J, Mekonta L (2012) A hidden resource: household-led rural water supply in Ethiopia. The Netherlands: IRC International Water and Sanitation Centre. Available: http://www.irc.nl/content/download/185695/850981/file/A%20hidden%20resource%20web_2012.pdf. Accessed 1 December 2013.

321. Tabor M, Kibret M, Abera B (2011) Bacteriological and physicochemical quality of drinking water and hygiene-sanitation practices of the consumers in bahir dar city, ethiopia. Ethiop J Health Sci 21: 19–26.

322. Tadesse D, Desta A, Geyid A, Girma W, Fisseha S, et al.. (2010) Rapid assessment of drinking-water quality in the Federal Democratic Republic of Ethiopia: country report of the pilot project implementation in 2004–2005. Geneva: WHO/UNICEF.

323. Tambekar DH, Hirulkar NB, Werulkar SA (2007) Multidrug resistance in Salmonella typhi isolated from drinking water in Amravati. Nature, Environment and Pollution Technology 6: 285–288.

324. Tambekar DH, Neware BB (2012) Water quality index and multivariate analysis for groundwater quality assessment of villages of rural india. Science Research Reporter 2: 229–235.

325. Taylor H, Ebdon J, Phillips R, Chavula G, Kapudzama O (2012) Assessment of drinking water quality for low-cost water options in rural Malawi. Brighton, UK: University of Brighton.

326. Tewari S, Ramteke PW, Garg SK (2003) Evaluation of simple microbiol tests for detection of fecal coliforms directly at 44.5 C. Environ Monit Asses 85: 191–198.

327. Tillet W (2007) An investigation into the impacts and challenges of implementing self supply in Eastern Uganda. Cranfield, UK: Cranfield University.

328. Tole MP (1997) Pollution of groundwater in the coastal Kwale District, Kenya. IAHS Publications Series of the Proceedings and Reports - Intern Assoc Hydrological Sciences 240: 287–310.

329. Trevett AF, Carter RC, Tyrrel SF (2004) Water quality deterioration: a study of household drinking water quality in rural Honduras. Int J Environ Heal R 14: 273–283.

330. Uhl VW, Daw A, Baron JA (2012) How a city gets its drinking water a case study - capital city of Monrovia, Liberia. Lambertville (New Jersey): UHL & Associates, Inc.

331. Vaccari M, Collivignarelli C, Tharnpoophasiam P, Vitali F (2010) Wells sanitary inspection and water quality monitoring in Ban Nam Khem (Thailand) 30 months after 2004 Indian Ocean tsunami. Environ Monit Assess 161: 123–133.

332. Valente JPS, Lopes CAM, Caminhas AMT, Horacio A (1999) Bacteriological evaluation of the water supplies of the Eldorado municipality — Vale do Ribeira (SP). Revista do Instituto Adolfo Lutz 58: 9–13.

333. van Geen A, Ahmed KM, Akita Y, Alam MJ, Culligan PJ, et al. (2011) Fecal contamination of shallow tubewells in Bangladesh inversely related to arsenic. Environ Sci Technol 45: 1199–1205.

334. Vanderslice J, Briscoe J (1993) All coliforms are not created equal - a comparison of the effects of water source and in-house water contamination on infantile diarrheal disease. Water Resour Res 29: 1983–1995.

335. Varghese J, Jaya DS (2008) Drinking water quality assessment of rain water harvested in ferrocement tanks in Alappuzha District, Kerala (India). Journal of Environmental Science & Engineering 50: 115–120.

336. Vidal J, Consuegra A, Gomescaseres L, Marrugo J (2009) Assessment of the microbiological quality of water packed in bags manufactured in Sincelejo-Colombia. Revista Mvz Cordoba 14: 1736–1744.

337. Vollaard AM, Ali S, Smet J, van Asten H, Widjaja S, et al. (2005) A survey of the supply and bacteriologic quality of drinking water and sanitation in Jakarta, Indonesia. SE Asian J Trop Med 36: 1552–1561.

338. Warner NR, Levy J, Harpp K, Farruggia F (2008) Drinking water quality in Nepal›s Kathmandu Valley: a survey and assessment of selected controlling site characteristics. Hydrogeology Journal 16: 321–334.

339. Watts R (1992) Low-cost water supplies and their contribution to health. Afr Health 15: 10–11.

340. Welch P, David J, Clarke W, Trinidade A, Penner D, et al. (2000) Microbial quality of water in rural communities of Trinidad. Rev Panam Salud Publica 8: 172–180.

341. Williams MA, Moran P, Nhandara C, Hove I, Charimari L, et al. (1997) Contamination of traditional drinking water sources during a period of extreme drought in the Zvimba communal lands, Zimbabwe. Cent Afr J Med 43: 316–321.

342. Wright J, Cronin A, Okotto-Okotto J, Yang H, Pedley S, et al. (2013) A spatial analysis of pit latrine density and groundwater source contamination. Environ Monit Asses 185: 4261–4272.

343. Wu Z (2008) Drinking water test of rural centralized water supply in Dianjiang. Prev Med 46: 123–125.

344. Xavier RP, Siqueira LP, Chaves Vital FA, Soares Rocha FJ, Irmao JI, et al. (2011) Microbiological quality of drinking rainwater in the inland region of Pajeu, Pernambuco, northeast Brazil. Revista do Instituto de Medicina Tropical de Sao Paulo 53: 121–124.

345. Yao Y-B, Li Q-J, Liu X-Z (2012) Health survey of rural drinking water in Jiangxi Province from 2008 to 2010. Journal of Environment and Health 29: 340–342.

346. Yasin N, Shah N, Khan J, Saba NU, ul Islam Z (2012) Bacteriological status of drinking water in the peri-urban areas of Rawalpindi and Islamabad-Pakistan. Afr J Microbiol Res 6: 169–175.

347. Yassin MM, Abu Amr SS, Al-Najar HM (2006) Assessment of microbiological water quality and its

relation to human health in Gaza Governorate, Gaza Strip. Public Health 120: 1177–1187.

348. Yongsi HBN (2010) Suffering for water, suffering from water: access to drinking-water and associated health risks in Cameroon. J Health Popul Nutr 28: 424–435.

349. Zeenat A, Hatha AAM, Viola L, Vipra K (2009) Bacteriological quality and risk assessment of the imported and domestic bottled mineral water sold in Fiji. J Water Health 7: 642–649.

350. Zeilhofer P, Zeilhofer LV, Hardoim EL, Lima ZM, Oliveira CS (2007) GIS applications for mapping and spatial modeling of urban-use water quality: a case study in District of Cuiaba, Mato Grosso, Brazil. Cad Saude Publica 23: 875–884.

351. Zhang XD, Ge ZX, Xiang JH, Zhu MF, Zhang JF, et al. (2012) Water quality in automatic water vending machines in Pudong New Area: an evaluation using drinking water quality index. Huanjing yu Zhiye Yixue 29: 231–233.

352. Zuin V, Ortolano L, Alvarinho M, Russel K, Thebo A, et al. (2011) Water supply services for Africa›s urban poor: the role of resale. J Water Health 9: 773–784.

353. Elala D, Labhasetwar P, Tyrrel SF (2011) Deterioration in water quality from supply chain to household and appropriate storage in the context of intermittent water supplies. Wa Sci Technol 11.

354. Shaheed A, Orgill J, Montgomery MA, Jeuland MA, Brown J (2014) Why "improved" water sources are not always safe. Bull World Health Organ 92: 283–289.

355. Hrudey SE, Hrudey EJ, Pollard SJT (2006) Risk management for assuring safe drinking water. Environ Int 32(8): 948–957.

356. Foster T (2013) Predictors of Sustainability for Community-Managed Handpumps in Sub-Saharan Africa: Evidence from Liberia, Sierra Leone, and Uganda. Environ Sci Technol 47: 12037–12046.

357. Gunnarsdottir MJ, Gardarsson SM, Elliott M, Sigmundsdottir G, Bartram J (2012) Benefits of water safety plans: microbiology, compliance, and public health. Environ Sci Technol 46: 7782–7789.

358. Psutka R, Peletz R, Michelo S, Kelly P, Clasen T (2011) Assessing the microbiological performance and potential cost of boiling drinking water in urban Zambia. Environ Sci Technol 45: 6095–6101.

359. Blum D, Huttly SRA, Okoro JI, Akujobi C, Kirkwood BR (1987) Bacteriological quality of traditional water sources in North-Eastern Imo State, Nigeria. Epidemiol Infect 99: 429–437.

360. Wright RC (1986) Seasonality of bacterial quality of water in a tropical developing country (Sierra Leone). J Hygiene-Camb 96: 75–82.

361. Kayser GL, Moriarty P, Fonseca C, Bartram J (2013) Drinking water service delivery frameworks for monitoring evaluation, policy and planning: a review. Int J Environ Res Public Health Vol 10, No 10, p 4812–4835.

362. Bartram J (2008) Improving on haves and have-nots. Nature 452: 283–284.

363. de Albuquerque C, Roaf V (2012) On the right track: good practices in realising the rights to water and sanitation. Geneva: OHCRH.

364. Yang H, Bain R, Bartram J, Gundry S, Pedley S, et al. (2013) Water safety and inequality in access to drinking-water between rich and poor households. Environ Sci Technol 47: 1222–1230.

365. Rivera SC, Hazen TC, Toranzos GA (1988) Isolation of fecal coliforms from pristine sites in a tropical rain forest. Appl Environ Microbiol 54: 513–517.

366. Ayoob S, Gupta AK (2006) Fluoride in drinking water: a review on the status and stress effects. Crit Rev Env Sci Tec 36: 433–487.

367. Amini M, Mueller K, Abbaspour KC, Rosenberg T, Afyuni M, et al. (2008) Statistical modeling of global geogenic fluoride contamination in groundwaters. Environ Sci Technol 42: 3662–3668.

368. Amini M, Abbaspour KC, Berg M, Winkel L, Hug SJ, et al. (2008) Statistical modeling of global geogenic arsenic contamination in groundwater. Environ Sci Technol 42: 3669–3675.

369. Rehfuess EA, Bruce N, Bartram JK (2009) More health for your buck: health sector functions to secure environmental health. Bull World Health Organ 87: 880–882.

370. Hunter PR, MacDonald AM, Carter RC (2010) Water Supply and Health. PLoS Med 7: e1000361.

371. Hutton G (2012) Global costs and benefits of drinking-water supply and sanitation interventions to reach the MDG target and universal coverage. Geneva: WHO. WHO/HSE/WSH/12.01.

372. Howard G, Bartram J (2005) Effective water supply surveillance in urban areas of developing countries. J Water Health 3: 31–43.

373. Bradley DJ, Bartram JK (2013) Domestic water and sanitation as water security: monitoring, concepts and strategy. Philos T Roy Soc A 371: 20120420.

374. WHO (2008) Essential environmental health standards in health care. Geneva: WHO.

375. Ministry of Health Uganda, Macro International Inc. (2008) Uganda Service Provision Assessment Survey 2007. Uganda; Calverton (Maryland): Ministry of Health; Macro International Inc.

376. Ministry of Health Rwanda, Macro International Inc. (2008) Rwanda Service Provision Assessment Survey 2007. Rwanda; Calverton (Maryland): NIS, MOH; Macro International Inc.

377. Peletz R, Mahin T, Elliott M, Harris MS, Chan KS, et al. (2013) Water, sanitation, and hygiene interventions to improve health among people living with HIV/AIDS: a systematic review. AIDS 27: 2593–2601.

378. Benova L, Cumming O, Campbell OM (2014) Systematic review and meta-analysis: association between water and sanitation environment and maternal mortality. Trop Med Int Health 19: 368–387.

379. Lim SS, Vos T, Flaxman AD, Danaei G, Shibuya K, et al. (2012) A comparative risk assessment of burden of disease and injury attributable to 67 risk factors and risk

factor clusters in 21 regions, 1990–2010: a systematic analysis for the Global Burden of Disease Study 2010. Lancet 380: 2224–2260.

380. Yang H, Wright JA, Gundry SW (2012) Water accessibility: boost water safety in rural China. Nature 484: 318.

381. WHO/UNICEF (2012) Review: sampling issues in tools for water quality monitoring. Geneva and New York: WHO/UNICEF Joint Monitoring Programme. Available: http://www.wssinfo.org/fileadmin/user_upload/resources/Report-of-Review-WQ-sampling-Geneva-18-19June2012_final.pdf. Accessed 1 December 2013.

382. Jimenez A, Perez-Foguet A (2008) Improving water access indicators in developing countries: a proposal using water point mapping methodology. Wa Sci Technol 8: 279–287.

383. Brown J, Clasen T (2012) High adherence is necessary to realize health gains from water quality interventions. PLoS One 7: e36735.

384. Hunter PR, Zmirou-Navier D, Hartemann P (2009) Estimating the impact on health of poor reliability of drinking water interventions in developing countries. Sci Total Environ 407: 2621–2624.

CHAPTER 29

QUALITY OF SOURCE WATER AND DRINKING WATER IN URBAN AREAS OF MYANMAR

Hiroshi Sakai, Yatsuka Kataoka, and Kensuke Fukushi

INTRODUCTION

Access to safe water is a significant issue in developing countries. According to a WHO report, around 780 million people globally do not have access to adequate water supply sources [1]. Additionally, 2.5 billion people do not have access to suitable sanitation facilities. Furthermore, about 2 million people die every year due to diarrheal diseases. Therefore, access to safe water is a crucial requirement in developing countries, where infrastructure is not always provided and often needs to be expanded. Due to the mismatches of urban planning and actual residential area, some areas must rely on inadequate private water supplies. This is a distinct issue in slum area and periurban areas.

Myanmar is a developing country in Southeast Asia. Even among the developing countries, Myanmar falls into the category of least developed countries by United Nations criteria [2]. The water infrastructure needs to be developed for the country's further economic development. However, very few water-quality data are currently available. To the best of our knowledge, only a single water-quality survey has been conducted [3]. That study reported water quality in Lake Inle in the northeastern part of the country. Some water-quality data for dams are provided on the webpage of the Water Environment Partnership in Asia [4] but only for limited quality parameters. To our knowledge, no other information has been reported, and the current water-quality and sanitation situations in Myanmar therefore remain unclear. We visited the country to perform a water-quality survey and assess the current situation with respect to water infrastructure. The survey was conducted in two urban areas, Yangon and Nay Pyi Taw, the former and current capital, respectively. This article reveals the water-quality and sanitation situations in Myanmar for the first time.

MATERIALS AND METHODS

Study Area

We surveyed two urban areas, Nay Pyi Taw and Yangon. The city of Nay Pyi Taw became the capital in 2008; the city of Yangon was the previous capital. A survey of drinking-water sources and quality was undertaken. The details of the location of source waters are shown in Figure 1. In Nay Pyi Taw, source waters from a deep well and two dams were examined. In the Yangon area, environmental waters in lakes and rivers were examined.

Citation: Hiroshi Sakai, Yatsuka Kataoka, and Kensuke Fukushi, "Quality of Source Water and Drinking Water in Urban Areas of Myanmar," The Scientific World Journal, vol. 2013, Article ID 854261, 5 pages, 2013. doi:10.1155/2013/854261

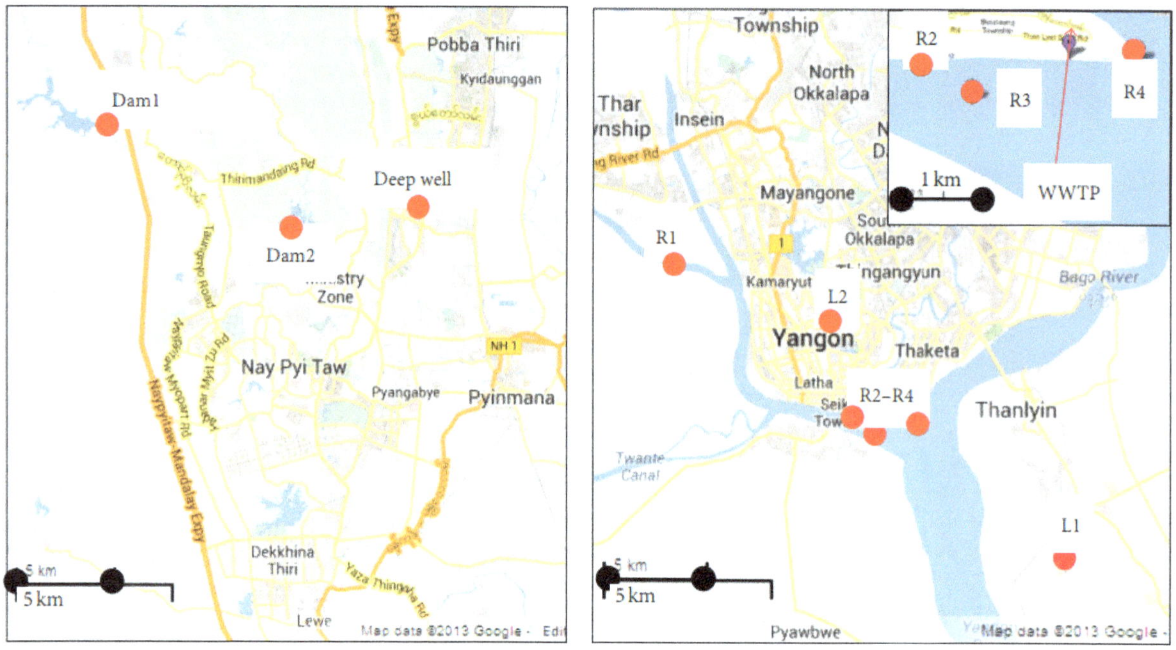

(a) (b)

Figure 1: Maps of sampling locations in (a) Nay Pyi Taw and (b) Yangon.

Drinking water was collected from various sources including public pots, nonpiped taps, piped taps, and bottled waters, as shown in Table 1. Three bottled waters (500 mL) from different companies were obtained commercially. Piped tap water was collected from three taps in Yangon and Nay Pyi Taw. Nonpiped tap water was collected at a pagoda and at another building (building D). In Myanmar, a pagoda is a meeting place for Buddhists, and complementary drinking water is provided. The tap water at the pagoda was provided by a nonpiped supply source. It was treated by a point-of-use (POU) facility and then stored. Building D was situated outside Yangon city in an area that is not served by the Yangon City Development Committee (YCDC) tap-water service. The water supply system to the building was privately operated by a POU facility that used a combination of a reverse osmosis (RO) membrane treatment and an ultraviolet (UV) disinfection system.

Table 1: Summary of drinking-water samples

Type	Description	Location
Pot	Pot A	Yangon
	Pot B	Suburban Yangon
Nonpiped	Pagoda A	Yangon
	Building D	Suburban Yangon
	After treatment	
	Before treatment	
Piped	Building A	Yangon
	Building B	Nay Pyi Taw
	Building C	Nay Pyi Taw
Bottle	Manufacturer A	—
	Manufacturer B	—
	Manufacturer C	—

Drinking water was also collected from public pots located on the roadside (Figure 2). In the Yangon area, pots are filled with water, covered, and placed along the roadside for public drinking purposes. We collected samples from various pots and examined the water quality, with a focus on bacterial analysis.

Figure 2: Roadside pots containing water for drinking.

Measured Parameters

Bacteria

The numbers of E. coli and total coliform bacteria and the heterotrophic plate counts (HPC) were determined using a commercial kit (Petrifilm, 3M, USA) at each site. E. coli and total coliform bacteria were incubated at 37°C for 24 hours, and HPC was incubated at 37°C for 48 hours. The validity of this kit has been confirmed [5,6], and it showed high correlations for E. coli and total coliform. High correlations were also reported for HPC, although small differences can arise with different incubation conditions [7].

Chemical Parameters

Dissolved organic carbon (DOC), dissolved total nitrogen (DTN), and anions were measured with a TOC analyzer (TOC-L, Shimadzu). Anion concentrations (F^-, Cl^-, Br^-, NO_2^-, NO_3^-, PO_4^-, and SO_4^-) were determined by ion chromatography (IC-861, Metrohm) after filtration through a 0.45 μm polytetrafluoroethylene (PTFE) membrane.

Heavy Metals

Dissolved metals were analyzed by ICP-MS (7500 Series, Agilent) after filtration through a 0.45 μm PTFE membrane.

RESULTS AND DISCUSSIONS

Source Water Quality in Nay Pyi Taw

Water quality parameters in two dams and a deep well in Nay Pyi Taw were examined and compared. A summary of the results, shown in Figure 3, indicates generally good water quality. No E. coli was detected in 1 mL samples at all locations, indicating good bacterial water quality. Total coliform levels were 14 and 3 CFU/mL at the two dams. These values are close to 10 CFU/mL, which is the "class A" Japanese environmental standard for lake water [8]. The DOC was 3.5 mg/L at Dam 1 and 3.0 mg/L at Dam 2, and it was 0.8 mg/L in the deep well. Values of DOC at the two dams were acceptable for water sources, considering that 3 mg/L of TOC in finished water has been adopted as a drinking-water standard in Japan [9]. A previous study had reported about 30–50 mg C/L of TOC in Lake Inle [3]. Compared with the results for Lake Inle, these two dams have a much better water quality.

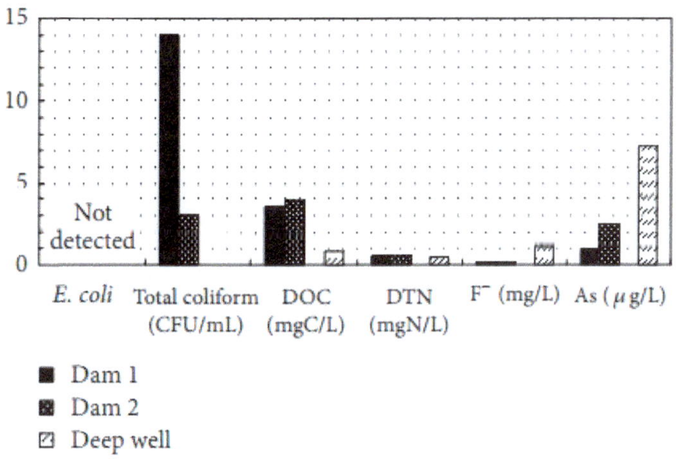

Figure 3: Comparison of water quality in dams and a deep well water at Nay Pyi Taw.

Two important observations were made with regard to the deep well water quality. Fluoride ion levels were 1 mg/L, which is close to the 1.5 mg/L WHO guideline value [10]. The As level was 7.9 µg/L, which is also close to the WHO guideline value of 10 µg/L. Therefore, adequate water treatment must be provided before the well can be used as a drinking-water source. Overall, the water quality in the two dams and the deep well could be considered fair, and the water could be used for drinking with appropriate treatment.

Environmental Water Quality in Yangon

Environmental water quality was also surveyed in the Yangon region. The results of bacterial parameters are shown in Table 2. The Japanese standard for total coliform levels in river water is 0.5 CFU/mL for class AA, 10 CFU/mL for class A, and 50 CFU/mL for class B [8]. River waters in Yangon were found to be close to the class B standard, indicating that they can be used for drinking after advanced treatment. Among the sampling points on the Yangon River, R2 was located on the left riverbank, whereas R3 and R4 were located in the center of the river. The total coliform value was highest at R2 on the left riverbank, which was closest to urban activity. There was also a discharge from a wastewater treatment plant on the left riverbank, which would have contributed to the larger number of total coliform recorded at R2.

Table 2: Bacterial water quality in Yangon

Location	R1	R2	R3	R4	L1	L2
Total coliform (CFU/mL)	52	820	32	38	10	110
E. coli (CFU/mL)	20	510	15	16	<1	2

L1 and L2 are recreational lakes in the Yangon region. At L1, E. coli was not detected in 1 mL samples, whereas the level was 2 CFU/mL at L2. Total coliform was also high, with 110 CFU/mL at L2. A previous report indicated total coliform levels of 18–137 CFU/mL in Lake Inle [3]. Considering these numbers, L2 was not suitable as a drinking-water source.

The chemical water-quality parameters are summarized in Figure 4. In river water samples, the DOC was less than 3 mg C/L, which satisfies the Japanese drinking-water-quality standard [9]. Levels of the Cl^- ion tended to increase downstream. The levels of Br ion and As displayed a similar trend, although the decrease was not as marked. In contrast to the elemental measurements, DOC, DTN, and nitrate were stable along the river flow. It was therefore assumed that the sources of Cl^-, Br^-, and As were different from those of carbon and nitrogen.

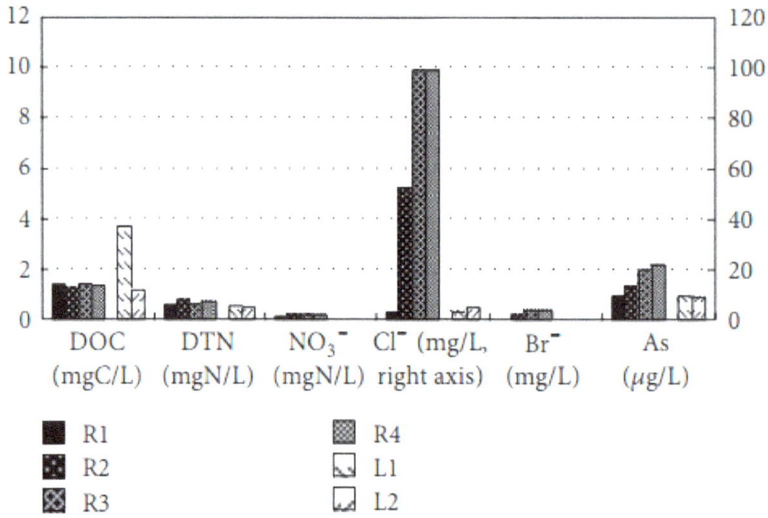

Figure 4: Comparison of water quality in rivers and lakes at Yangon.

Samples from L1 and L2 had good water quality with regard to chemical parameters, except for high DOC in L1. Overall, lake and river waters were good in terms of their chemical parameters, but the levels of bacterial contamination needed improvement.

Drinking-Water Quality

Potable drinking-water quality was surveyed for various water sources in Myanmar including (i) public pots, (ii) piped water supply in Yangon and Nay Pyi Taw, (iii) nonpiped water supply, and (iv) bottled water.

For bacterial water quality, E. coli was not detected in 1 mL water samples from any water source. However, there was a clear trend in the HPC, as shown in Figure 5. Pot water had HPC levels of 3200 and 1500 CFU/mL. In nonpiped tap water, the HPC level was 1700 and 1170 CFU/mL, and, in piped tap waters, the HPC was 140, 600, and 1200 CFU/mL. Of the three bottled waters examined, HPC was detected in two bottles at 760 and 12 CFU/mL, but HPC was not detected in the other bottled water sample. HPCs have various incubation conditions [11], and different incubation conditions produce different values. The Japanese government has adopted a value of 2000 CFU/mL as a water-quality standard for HPC by incubation at 20°C for 7 days [9]. This is equal to about 740 CFU/mL by incubation at 37°C for 2 days in Petrifilm, which we used in our preliminary investigation (unpublished data). From these results, we concluded that water from all pots, all nonpiped taps, one piped tap, and one bottled water exceeded the converted value of the Japanese drinking-water-quality standard and may not be suitable for drinking.

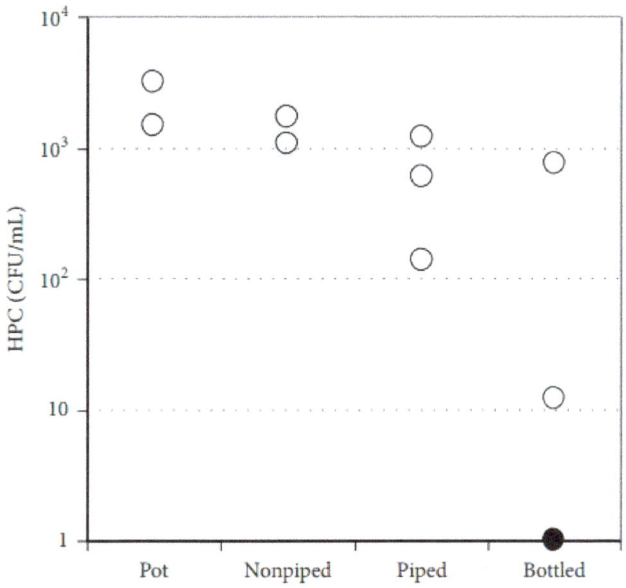

Figure 5: Heterotrophic plate count for potable water.

As shown in Figure 5, HPC was highest in the pots, followed by nonpiped taps, piped taps, and bottled waters. Nonpiped water was surveyed at two taps, at a pagoda and at a building outside the YCDC tap water service area. At those taps, water was supplied after treatment by POU facilities.

Piped water was surveyed at two buildings in Nay Pyi Taw and at one building in Yangon. Piped water supplies contained some residual chlorine: 0.01 mg/L at two taps in Nay Pyi Taw and 0.03 mg/L at a tap in Yangon. At a tap in Yangon, chlorine was present entirely as free chlorine. This residual chlorine may have contributed to the upkeep of water quality in the piped water supply.

It is interesting to note the water-quality distribution in bottled waters, with one bottled water supply having worse water quality than two piped taps. This probably results from the source water quality and the treatment efficiency of different bottled water companies. Other water-quality parameters were also monitored, including As and F^-. All measured items satisfied the Japanese drinking-water-quality standards [9].

Treatment Efficiency of a POU Facility

We investigated the performance of a POU facility at a building in Yangon, which was situated outside the YCDC piped water supply area. The water source was ground water, which was treated by an RO membrane followed by UV disinfection. The removal ratios of bacteria (HPC), carbon and nitrogen, anions, and metals are shown in Figure 6. The listed heavy metals were removed with high efficiency. Anions were also removed with high efficiency except for nitrate, for which only a 45% removal was achieved. Because UV treatment does not remove anions and metals, these elements were removed by the RO membrane.

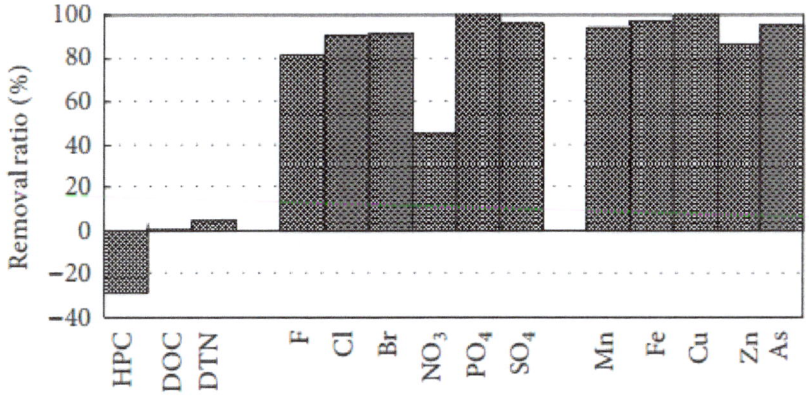

Figure 6: Removal ratio for a point-of-use (POU) facility.

In contrast, DOC and DTN removal was very low: 1% DOC and 5% DTN. The DOC and DTN contents of raw water were 0.55 mg C/L and 0.23 mg N/L, respectively. Considering the removal of anions and metals, the RO membrane would have worked well. A possible explanation is that most organic matter and nitrogen in raw water have a very small molecular weight and can pass through the RO membrane. Further investigation of the molecular weight distribution would confirm this assumption.

It is noteworthy that the bacterial removal ratio was negative. Considering the removal of anions and metals, bacteria could be removed by the RO membrane treatment. UV treatment also contributes to the suppression of bacterial activity, but UV treatment has no residual effect. Therefore, bacterial regrowth may occur in a storage tank after UV treatment. UV treatment has a high potential to be installed in POU facilities because of its ease of handling and maintenance. Maintaining bacterial water quality after UV treatment is an important issue outside the piped water supply area.

CONCLUSION

This study investigated the water quality in urban areas of Myanmar and produced an overview of the current situation. River, dam, lake, and well water samples were examined and found to be of generally good quality. As and F^- were present at relatively high concentrations and must be removed before deep wells can be used. Heterotrophic plate counts in drinking water were highest in pots, followed by nonpiped tap water, piped tap water, and bottled water samples. Measures need to be taken to improve the poor water quality in pots and nonpiped taps.

ACKNOWLEDGMENTS

The authors would like to express appreciations to Dr. Mu Mu Than and Dr. Kyi Kyi Lwin for their kind supports for water sampling. This paper was prepared with a partial support by MEXT through Green Network of Excellence (Eco-Health Project).

REFERENCES

1. WHO, Progress on Drinking Water and Sanitation, World Health Organization, Geneva, Switzerland, 2012.
2. United Nations, "Least developed countries," 2012, http://www.unohrlls.org/en/ldc/25/.
3. F. Akaishi, M. Satake, M. Otaki, and N. Tominaga, "Surface water quality and information about the environment surrounding Inle Lake in Myanmar," Limnology, vol. 7, no. 1, pp. 57–62, 2006. · ·
4. Water Environment Partnership in Asia, "State of water environmental issues, Myanmar," 2003, http://www.wepa-db.net/policies/state/myanmar/myanmar.htm.
5. V. Beloti, J. A. de Souza, M. D. A. F. Barros et al., "Evaluation of Petrifilm EC and HS for total coliforms and Escherichia coli enumeration in water," Brazilian Journal of Microbiology, vol. 34, no. 4, pp. 301–304, 2003. ·
6. H. Schraft and L. A. Watterworth, "Enumeration of heterotrophs, fecal coliforms and Escherichia coli in water: comparison of 3M Petrifilm plates with standard plating procedures," Journal of Microbiological Methods, vol. 60, no. 3, pp. 335–342, 2005. · ·
7. S. A. Mueller, J. E. Anderson, B. R. Kim, and J. C. Ball, "Comparison of plate counts, Petrifilm, dipslides, and adenosine triphosphate bioluminescence for monitoring bacteria in cooling-tower waters," Water Environment Research, vol. 81, no. 4, pp. 401–406, 2009. · ·
8. Ministry of Environment, "Environmental quality standards for water pollution," 2013, http://www.env.go.jp/en/water/wq/wp.pdf.
9. Ministry of Health, Labour, and Welfare, "Water quality standards," 2013, http://www.mhlw.go.jp/english/policy/health/water_supply/4.html.
10. WHO, Guidelines for Drinking-Water Quality, World Health Organization, Geneva, Switzerland, 4th edition, 2011.
11. J. Bartram, J. Cotruvo, M. Exner, C. Fricker, and A. Glasmacher, Heterotrophic Plate Counts and Drinking-Water Safety, International Water Association, London, UK, 2003.

CHAPTER 30

GLOBAL ACCESS TO SAFE WATER: ACCOUNTING FOR WATER QUALITY AND THE RESULTING IMPACT ON MDG PROGRESS

Kyle Onda, Joe Lobuglio And Jamie Bartram

Keywords: Water quality, drinking water, water safety, Millennium Development Goals (MDGs), Joint Monitoring Programme (JMP)

INTRODUCTION

The United Nations Millennium Development Goals (MDGs) include Target 7c, to halve the "proportion of the population without sustainable access to safe drinking-water" between 1990 and 2015 [1]. The Joint Monitoring Programme for Water Supply and Sanitation (JMP) of the World Health Organization (WHO) and the United Nations Children's Fund (UNICEF) reports progress towards meeting this goal [2]. The corresponding MDG indicator is the "proportion of households using water from an improved source," and is reported on a country-by-country basis [1]. Sources are classified as improved or unimproved as shown in Table 1, according to whether they are "protected from outside contamination" [2]. The MDG indicator thereby conflates access to certain water sources with use of safe water. However, data and monitoring mechanisms regarding the safety of water sources at a national scale when the MDG targets were cast were, and remain, scant. As such, at the time the MDGs were developed, there was no credible alternative approach to an indicator that allows for the calculation of a percentage figure easily aggregated to the country and global scales and amenable for use as a target given the types of data available.

Citation: Onda K, LoBuglio J, Bartram, (2012) Global Access to Safe Water: Accounting for Water Quality and the Resulting Impact on MDG Progress. Int. J. Environ. Res. Public Health, 9(3), 880-894; doi:10.3390/ijerph9030880

Table 1: JMP Classification of drinking-water source types as improved or unimproved [2]

Source class	Type of source
Unimproved drinking-water source	Unprotected dug well, unprotected spring, cart with small tank or drum, surface water (e.g., river, dam, lake, pond, stream, canal or irrigation channel) and bottled water
Improved drinking-water source (piped to dwelling, plot or yard)	Piped water connection located inside the user's dwelling, plot or yard
Improved drinking-water source (other sources)	Public taps or standpipes, tube wells or boreholes, protected dug wells, protected springs and rainwater collection

Using this approach, WHO and UNICEF estimate that 5.8 billion people used improved sources in 2010, with 783 million using unimproved water sources [3]. Treating use of an improved source as an indicator for use of safe water is likely to overestimate the population using safe water, since some improved sources may provide water that is microbiologically or chemically contaminated whether at source or by the time it reaches the home and is consumed [4,5]. On the other hand, most unimproved sources do not provide safe drinking water, so under-accounting of safe water coverage due to unimproved sources providing safe water is likely to be small [6,7].

In 2010, WHO and UNICEF released data on water quality and sanitary risk (*i.e.*, risk of contamination) associated with improved sources from five countries as part of the Rapid Assessment of Drinking-Water Quality (RADWQ) study which had been undertaken between October 2004 and April 2005 [8,9,10,11,12]. The RADWQ study is the only source of nationally-representative drinking water quality data amenable to analysis for microbial contamination disaggregated by water source type. A recent analysis of these data found that accounting for water source compliance with WHO water quality guidelines significantly reduced the estimates of safe water access relative to the JMP figures, as well as national-level progress towards the MDG target, in four of the five countries [13]. However, this study did not account for sanitary risk, and did not extrapolate beyond the five countries for which data were available. We use both the water quality and sanitary risk data from the RADWQ studies to extrapolate to other countries, and estimate global figures. We believe our study is the first to use the available nationally-representative water safety data to estimate the proportion of the global population with access to safe drinking water and the concomitant impact on estimates of global progress towards MDG Target 7c with a method that accounts for differences between countries in factors affecting water source safety.

METHODS

Drinking Water Source Type Classification

Here we considered "piped" sources as the subset of MDG-classified improved sources that are piped connections to a user's dwelling, plot, or yard, and "other-improved" sources as all other improved drinking water sources as described in Table 1.

Drinking Water Quality Data

The RADWQ data was collected following the methodology described in Howard *et al.* [14]. The RADWQ data includes a nationally-representative sample of approximately 1,600 improved water sources in each of Ethiopia, Jordan, Nicaragua, Nigeria, and Tajikistan [8,9,10,11,12]. A population-weighted sample was taken of each source type that provided water to at least 5% of the national population in four of the five countries [8,9,10,11,12]. In Nicaragua, no single improved source type other than utility piped supplies covered more than 5% of the population, so several other-improved source types, each covering less than 5% of the population, were sampled [11]. Each sampled source was tested for thermotolerant coliform bacteria (TTC), fluoride, arsenic, and nitrate, and subjected to a standardized sanitary inspection [14]. Sanitary inspections in the RADWQ surveys identified risk factors for faecal contamination for a water source from a standardized list for each water source type of 10 common risks [8,9,10,11,12]. Examples of sanitary risks on these lists included pipe breaks, supply discontinuities, poor drainage, and proximity to latrines and animal waste. The RADWQ studies reported sanitary risk as a categorical variable with four levels: 0–, 3–5, 6–8, and 9–10 sanitary risks [8,9,10,11,12]. The RADWQ studies also provided cross-tabulations of sanitary risk levels and TTC contamination levels, except in the case of Nicaragua, which only reported the percentage of contaminated water sources in each sanitary risk level [8,9,10,11,12].

For each of the five countries, we used the RADWQ data to compute the percentages of piped water sources and other-improved water sources that were "safe" by virtue of not testing positive for TCC [15]. We computed the percentages of safe sampled piped and other-improved sources that also had greater than two sanitary risks on the sanitary risk inspection that accompanied the water quality tests for each sampled water source. We used this threshold to identify systems with all but the lowest aggregated level of sanitary risk reported in the RADWQ studies. We considered any one sanitary risk to represent an elevated risk of water contamination or re-contamination that is not captured by water quality testing at a single point in space and time. Where multiple piped or other-improved water source types were tested within a country, the percentages were computed as an average weighted by the population using each source type. The population receiving drinking water from each source in 2004 to 2005 was estimated from RADWQ project reports.

The RADWQ study made no attempt to assess the safety of water from unimproved sources, other than the case of tanker trucks in Nigeria [9]. In the absence of such data, we assumed that 100% of unimproved sources do not provide safe water.

Synthetic Covariates from Principal Components Analysis

In order to estimate global access to safe water, empirical statistical models capturing the relationship between faecal contamination or sanitary risk proportions and country-level economic, governance, health, social and environmental characteristics were built. Covariates were chosen based on the availability of data for countries with 2010 JMP estimates and on their relation to drinking water quality. The following country-level indicators were explored: gross domestic product per capita (GDP) [16], the World Bank's Government Effectiveness (GE) score [17], the Human Development Index [18], the Water Quality Index (WQI) from Yale's Environmental Performance Index [19], annual aggregate precipitation [16], percent of population attaining tertiary education [20], and under-5 diarrheal morbidity rates [16]. Due to many of these variables being highly correlated, and the small size of our dataset, the models were limited to one or two covariates. Principal Components Analysis (PCA) was used to create uncorrelated synthetic variables that captured the most variance in those national characteristics. This analysis was performed with Stata 10.1.

Predicting Drinking Water Safety Proportions for Piped and Other-Improved Sources

Four fractional logit models were built based on the data points listed in Table 2, one for each of the proportions of: safe piped sources, safe other-improved sources, safe piped sources with elevated sanitary risk, and safe other-improved sources with elevated sanitary risk. We modeled the sanitary risk proportions of safe sources, since faecally-contaminated sources were assumed to be unsafe regardless of their level of sanitary risk. The fractional logit model as developed by Papke and Wooldridge [21], rather than minimizing the sum-of-squares error, is a quasi-maximum likelihood method. The fractional logit is superior to an ordinary least squares (OLS) regression because the predicted values of such models are not guaranteed to be restricted to values between 0 and 1. This method is also superior to the common alternative of performing an OLS regression on the logit transformation of the proportion (log-odds regression), in that the predicted values of the proportions from such a model are not recoverable without making significant assumptions.

Candidate covariates used in the models were the first three PCA components from the analysis described in section 2.3. For each of the four dependent variables, models using combinations of these three components were run and the models having the greatest log-likelihoods while maintaining a significant difference from the null model were chosen.

The resulting models were used to extrapolate the proportions calculated from the RADWQ countries to the remaining countries with JMP data for 2010. The predicted proportion of piped or other-improved sources uncontaminated with TTC for each country was multiplied by the corresponding JMP estimate of the population using piped or other-improved water, respectively, to estimate the population with access to microbiologically safe piped or other-improved water in each country. Similarly, the predicted proportion of uncontaminated piped or other-improved sources with elevated sanitary risk was multiplied by the estimated population with access to uncontaminated piped and other-improved water sources. 95%-confidence interval upper and lower bounds for each figure were calculated by using the standard errors of the model coefficients. This method of calculating confidence intervals assumes the RADWQ studies are perfectly accurate in their reported percentages. This assumption is necessary as the complete RADWQ data sets were not available and we were therefore unable to calculate standard errors for the RADWQ country percentages.

Table 2: Cross-tabulation of TTC contamination and sanitary risk for piped and other-improved water sources [8,9,10,11,12]

Piped *					
	% Sanitary risk category safe	**TTC Count (cfu/100 mL)**			
		<1	**1–10**	**10–100**	**>100**
Ethiopia					
0–2 (Very Low San. Risk)	88%	78	1	8	2
3–5 (Low San. Risk)	89%	220	9	14	5
6–8 (Med. San. Risk)	70%	40	5	11	1
9–10 (High San. Risk)	0%	0	1	0	0
Jordan					
0–2 (Very Low San. Risk)	100%	1233	1	0	0
3–5 (Low San. Risk)	100%	404	0	0	0
6–8 (Med. San. Risk)	NA	0	0	0	0
9–10 (High San. Risk)	NA	0	0	0	0
Nigeria					
0–2 (Very Low San. Risk)	89%	108	10	3	1
3–5 (Low San. Risk)	79%	263	48	23	1
6–8 (Med. San. Risk)	72%	115	14	26	5
9–10 (High San. Risk)	23%	3	3	5	2
Tajikistan					
0–2 (Very Low San. Risk)	91%	1038	100	0	0
3–5 (Low San. Risk)	72%	91	35	1	0
6–8 (Med. San. Risk)	48%	10	11	0	0
9–10 (High San. Risk)	NA	0	0	0	0
Other-Improved **					
	% Sanitary risk category safe	**TTC Count (cfu/100 mL)**			
		<1	**1–10**	**10–100**	**>100**
Ethiopia					
0–2 (Very Low San. Risk)	94%	192	2	3	8
3–5 (Low San. Risk)	72%	567	76	94	46
6–8 (Med. San. Risk)	41%	122	46	76	53
9–10 (High San. Risk)	0%	0	2	0	3
Nigeria					
0–2 (Very Low San. Risk)	78%	364	71	31	0
3–5 (Low San. Risk)	80%	256	31	33	1
6–8 (Med. San. Risk)	81%	105	9	15	1
9–10 (High San. Risk)	38%	12	9	10	1
Tajikistan					
0–2 (Very Low San. Risk)	77%	121	28	6	2
3–5 (Low San. Risk)	89%	146	15	3	0
6–8 (Med. San. Risk)	54%	7	4	1	1
9–10 (High San. Risk)	NA	0	0	0	0

* All crosstabulations from RADWQ project reports; The Nicaragua report did not include them.

** No other-improved sources tested in Jordan.

Adjusting MDG Target 7c Progress Estimates

We re-estimated MDG progress towards global safe water access by accounting for faecal contamination of piped or other-improved water sources, as well as sanitary risks of piped or other-improved sources for which faecal contamination was not detected. The 1990 baseline estimates obtained from JMP were multiplied by the estimated proportions of people with piped and other improved water sources for which such sources were estimated to be unsafe by virtue of having tested positive for TTC. This was repeated to account for piped and other-improved sources with sanitary risks. In the absence of any trend data, proportions estimated based on the 2004–2005 RADWQ data were used to estimate both 2010 and 1990 water source safety. Projections of the global proportion of people without access to safe water in 2015 according to JMP data, and accounting for water quality, were based on linear extrapolations of the 1990–2010 JMP and water quality-adjusted data, respectively. This is the same method used by JMP to make its projections [22]. The 2015 MDG targets were calculated by halving the original and recalculated 1990 baselines.

RESULTS

The results of the RADWQ studies in the form of cross-tabulations of detected TTC contamination levels and sanitary risk scores, aggregated into piped and other-improved sources, are shown in Table 2. Within each country, TTC contamination rates for each water source type generally increase with increasing sanitary risk, indicating the expected association between water contamination and the presence of sanitary risk factors.

The aggregated results of the RADWQ studies are presented in Table 3. The relationship between sanitary risk and TTC contamination differs between countries. For example, even though all Jordan's piped water supplies are uncontaminated, almost a quarter have more than two sanitary risks; while less than 9% of Tajikistan's uncontaminated piped supplies have more than two sanitary risks.

Data were available such that the PCA could be performed for 150 countries, accounting for 92.5% of the global population (Table 4). The first three components explained 84% of the variance among these countries. The first component shows that GDP, GE, HDI, and Tertiary Education rates all covary and contribute similar information. The second component is dominated by annual aggregate precipitation, and the third component by the WQI. The spread of the five RADWQ countries with respect to the dominant variables of the first three components are shown in Table 5.

Table 3: Contamination of piped and other-improved sampled sources with TTC, and sanitary risk of safe sources [8,9,10,11,12]

Source type, by country	Thermotolerant coliform (TTC) contamination *		Sanitary risk of safe sources
	TTC-Uncontaminated (Safe) (%)	Sources sampled (*n*)	>2 Sanitary Risks (%)
Ethiopia			
Piped	87.6	838	60.7
Other-Improved	55.0	764	67.1
Jordan			
Piped	99.9	1639	24.7
Other-Improved	NA	0	NA
Nicaragua			
Piped	85.5	600	47.3
Other-Improved	33.5	888	56.5
Nigeria			
Piped	77.0	630	77.9
Other-Improved	76.0	949	50.6
Tajikistan			
Piped	88.6	1286	8.9
Other-Improved **	82.0	334	55.8

* Data aggregated from RADWQ reports. TTC contamination judged against guideline value of <1 TTC per 100 mL as indicated in the WHO Guidelines for Drinking Water Quality [15]; ** Only 44% of the protected springs in Tajikistan were found to be adequately protected for designation as an improved source.

Table 4: Factor loadings, Eigenvalues, and Variances accounted for from PCA

Variable	Comp 1	Comp 2	Comp 3	Comp 4	Comp 5	Comp 6	Comp 7
Eigenvalue	4.19	1.06	0.623	0.476	0.326	0.180	0.138
pc GDP	0.422	0.121	0.353	0.204	−0.429	0.647	0.199
GE	0.422	0.151	0.336	0.348	−0.0785	−0.715	0.219
HDI	0.441	0.0920	0.258	−0.101	0.399	0.0799	−0.745
Precipitation	−0.102	0.913	0.0051	−0.380	0.0029	−0.0178	0.106
WQI	0.331	0.237	−0.793	0.433	−0.0363	0.0497	−0.123
U5 Diarrhea	−0.384	0.214	0.227	0.642	0.537	0.210	0.100
Tert. Educ.	0.427	−0.137	−0.127	−0.280	0.601	0.131	0.567
Proportion	0.599	0.152	0.0885	0.0679	0.0466	0.0258	0.0197
Cumulative	0.599	0.751	0.839	0.908	0.955	0.980	1.00

Table 5: Spread of countries on dominant PCA Covariates

	GE	Precipitation (mm)	WQI
Ethiopia	−0.35	848	55.3
Jordan	0.08	111	11.9
Nicaragua	−0.96	2391	57.1
Nigeria	−1.2	1150	20.1
Tajikistan	−0.91	691	42.8

We used the fractional logit models shown in Table 6 to predict proportions of piped and other-improved sources that provide safe water in every country for which 2010 JMP estimates were available. A comparison of the RADWQ countries' measured and predicted proportions is shown in Table 7. We estimate that 1 billion (confidence interval 0.75 to 1.6 billion) of the 5.8 billion using piped or other-improved sources receive faecally-contaminated water. This lowers the number of people estimated to use safe water from 5.8 billion (the 2010 JMP figure) to 4.8 billion, and increases the number or people with unsafe water from 0.78 billion to 1.8 billion as of 2010.

Table 6: Fractional Logit Models of Water Source Safety Proportions

	Model			
Variable	% Piped Sources Safe	% Other-Improved Sources Safe	% Safe Piped Sources w/Elevated Sanitary Risk	% Safe Oth. Imp. Sources w/Elevated Sanitary Risk
Comp 1	0.879 **		−1.131 ***	
	(0.286)		(0.285)	
Comp 2	−0.436 *	−0.792 ***		
	(0.209)	(0.159)		
Comp 3				−0.350 ***
				(0.099)
Constant	3.188	0.370	−1.819	0.345
	(0.566)	(0.236)	(0.707)	(0.075)
N	5	4	5	4
Residual df	2	2	3	2
Log pseudo-likelihood	−1.2305	−1.65	−2.119	−1.773

* $p < 0.05$; ** $p < 0.01$; *** $p < 0.001$; one-tailed tests for hypothesized effects, robust standard errors in parentheses.

Of these 4.8 billion using safe water, approximately 1.2 billion people (confidence interval 0.75 to 2.1 billion) receive water from sources that are at risk of faecal contamination by virtue of having greater than two of the common sanitary risks for that source type as defined by RADWQ [14]. If a more stringent definition of safety (requiring both no faecal contamination and low sanitary risk) is used, then the estimate of the number of people with unsafe water is 3 billion, (confidence interval 1.5 billion to 3.9 billion). 302 million people reside in countries for which JMP estimates do not exist. An additional 370 million people reside in countries for which JMP estimates exist, but for which the data necessary for principal components analysis was unavailable. The safety of the water received by these 370 million is not estimated.

Table 7: Comparison of Measured and Predicted Proportions

	% Piped Sources Safe		% Other-Improved Sources Safe		% Safe Piped Sources w/Elevated Sanitary Risk		% Safe Oth. Imp. Sources w/Elevated Sanitary Risk	
Country	*RADWQ*	*Predicted*	*RADWQ*	*Predicted*	*RADWQ*	*Predicted*	*RADWQ*	*Predicted*
Ethiopia	88%	85%	55%	55%	61%	59%	67%	57%
Jordan	100%	98%	NA	87%	25%	18%	NA	48%
Nicaragua	86%	85%	34%	33%	47%	33%	57%	56%
Nigeria	77%	77%	76%	68%	78%	73%	51%	51%
Tajikistan	88%	93%	82%	76%	9%	18%	56%	55%

The full categorization of the world population of piped and other-improved water source users into categories based on faecal contamination and sanitary risk using our estimation methods is shown in Figure 1, where upper and lower bounds are shown in parentheses where available. The world's population is divided into those using improved sources, unimproved sources, and sources of unknown type. The improved source population is divided into those using piped and other-improved sources. Each of these two populations is divided into those using sources that are safe, unsafe, or unestimated safety. The populations using safe piped and safe other-improved sources are divided into those using sources with low and elevated levels of sanitary risks.

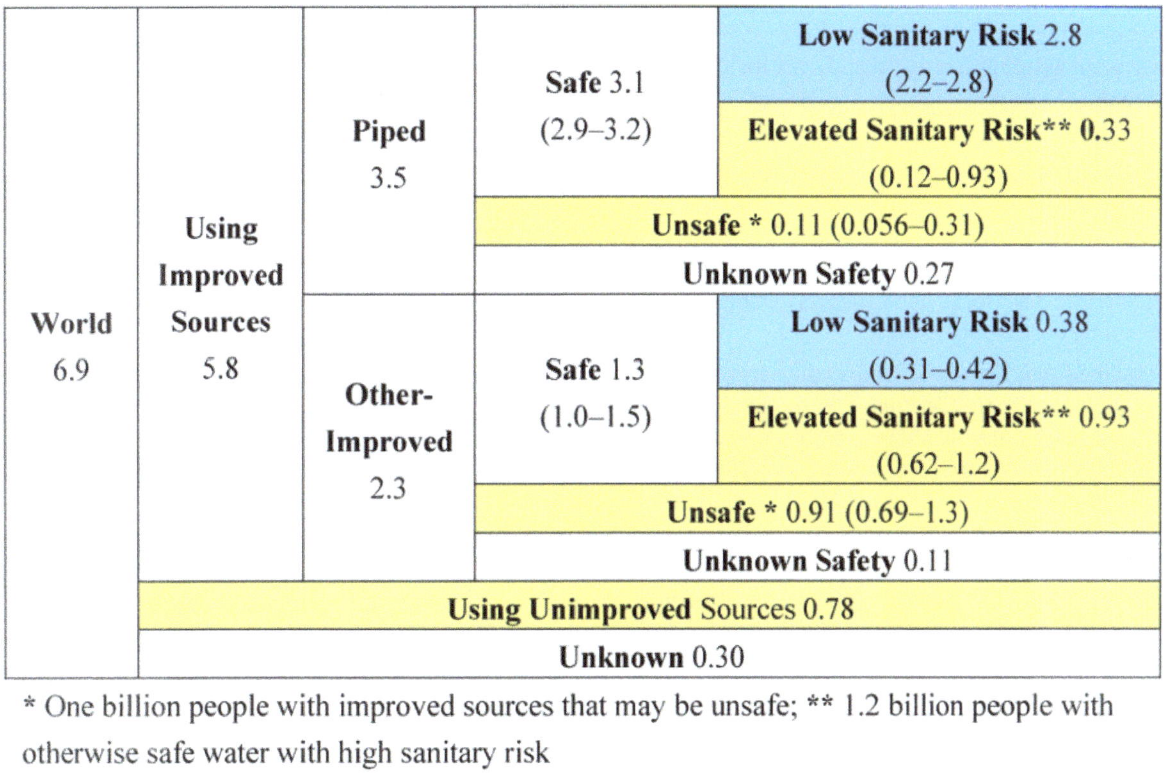

Figure 1: World population by water contamination status and sanitary risk for 2010 (billions).

Percentages of the world population for the purposes of adjusting the MDG Target 7c progress are calculated by assuming the populations with unknown water source types and water source safety are distributed similarly to the rest of the world. Accounting for faecal contamination based on our models increases the 1990 baseline estimate of the population without access to a safe water source from 23% to 37% (see Figure 2). In order to meet MDG Target 7c, the proportion without access to safe water would need to be reduced to 18% by 2015, while trends indicate this figure will be 26%. For 2010, the shortfall is 10 percentage points (680 million people), and we project an 8-percentage point shortfall in 2015, while JMP estimates, based on the indicator of use of an improved source, indicate that the MDG target has already been met.

Using a more stringent definition of water safety that requires water sources be both uncontaminated and of low sanitary risk would result in a 1990 baseline of 53% of the population, a target of 26%, and a 2015 projection of 46%. This would result in a 21-percentage point shortfall in 2010 and a 20-percentage point shortfall in 2015.

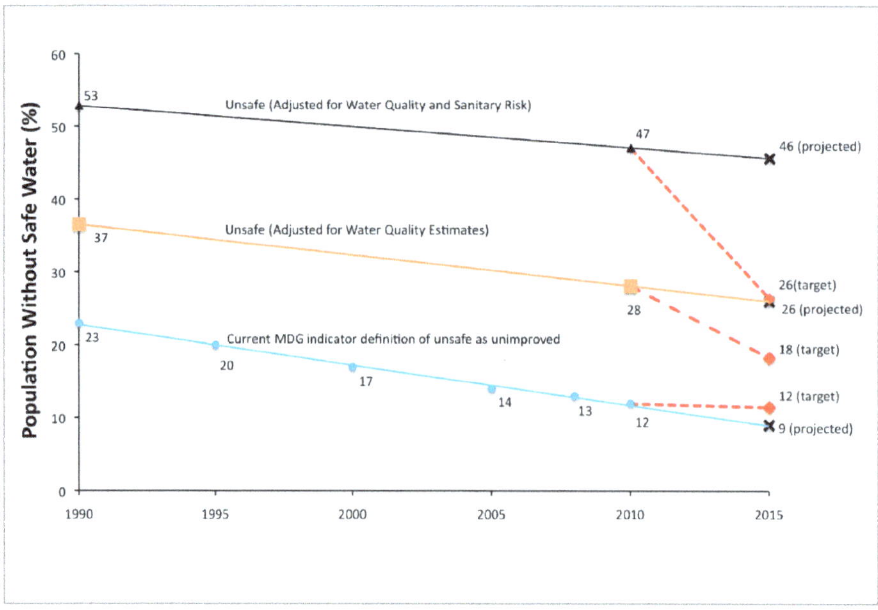

Figure 2: Comparison of MDG Target 7c baseline and target when including and excluding faecal contamination and sanitary risk in water safety.

DISCUSSION

Our estimates of the population with access to safe water were based on several assumptions that could contribute to over- or under-estimation.

We used the >1 cfu/100 mL TTC level as a threshold for considering water microbiologically safe, as did the RADWQ studies. *E. coli* and *Enterrococci* have been shown to be better indicators of waterborne disease risks than total TTC [23]. This could result in an overestimate of the population actually at risk of waterborne disease, as some sources testing positive for TTC may not be positive for *E. coli*.

We assumed that the relationships between water quality and national characteristics described by the models in Table 3 hold for countries other than the five countries assessed by RADWQ. The causal factors determining water quality for a given source type may differ from country to country and the causal relationships determining water quality for a given source type may differ from country to country. In addition, countries differ substantially regarding the mixture of predominant source types. For instance, protected springs were the only non-piped improved water source assessed by RADWQ in Tajikistan, while boreholes and protected wells were assessed in Ethiopia, Nicaragua, and Nigeria [8,9,10,11,12]. All non-piped "improved" sources were aggregated and treated as "other-improved". Given these country-to-country differences, the direction of error will differ from country to country although the overall effect on the global estimates may be relatively small as the errors are aggregated.

The sanitary risk estimates are based on the proportion of safe sources assessed as having greater than two of the sanitary risks listed as possible on the RADWQ survey forms. Moreover, the RADWQ project reports stated that the lists of sanitary risks used in the survey could have been better tailored to country conditions [8,9,10,11,12]. Each of the possible risks on these forms probably contributes to a different level of actual risk for water contamination, but are weighted equally in the RADWQ methodology [8,9,10,11,12,14]. The relationship between the number of sanitary risks and the prevalence and degree of water contamination differs between countries (Table 2). Since any single sanitary risk is significant, use of this threshold (three or more risk factors) is likely to underestimate the population using unsafe water.

We assumed that no contamination occurs between the water point and the point and time of use. However, such contamination is known to occur [5]. Water that is safe in a distribution system at one point may become contaminated at another point before it is received by the user, due to deficiencies in the distribution system [24]. Additionally, water that must be transported manually from the source to the home, and any water stored in the home, as is common with other-improved sources, can become contaminated due to unsanitary storage conditions [25]. We did not account for this due to lack of representative data. The effect may be greater for the estimates for "other-improved" sources than for piped sources. However, the effect is likely to be significant in the case of piped sources with discontinuous service that encourage users to store water in the home. These assumptions would tend to lead to an underestimate of the population using unsafe water.

We assumed that the proportion of piped and other-improved sources that provide safe water found in the RADWQ reports represent the safety of those sources over an entire year. This is likely incorrect, due to the transient nature of many contamination events [26]. One round of water quality testing is unlikely to capture the true extent of microbial contamination that might occur over a long period of time at a given source. As such, the RADWQ data likely substantively underestimate the proportion of water sources that are contaminated over an extended time period. The effect would be to underestimate the population using unsafe water.

We did not consider contamination from nitrates, arsenic, fluoride, or any of the chemical contaminants not tested for by the RADWQ project. This is because waterborne pathogens from faecal sources cause more disease than any other waterborne contaminant, and thermotolerant coliform bacteria measurement serves as an indicator for the presence of faecal contamination [15]. There is likely not complete overlap in terms of the types of contamination a given water source might face, so more water sources are probably unsafe than reliance on microbiological indicators alone would indicate. The effect of these assumptions is to underestimate the population using unsafe water.

We assumed that the proportion of piped and other-improved sources that are faecally contaminated, and the proportion of these with elevated sanitary risk, remained constant across time, at least backwards to 1990 for the purposes of recalculating the MDG baseline, and forwards to 2015 for projected proportions. This assumption was necessary due to a lack of equivalent data for the relevant time periods. The effect of this assumption will vary from country to country, and the overall effect on the global estimates may be relatively small as the errors are aggregated.

Since the RADWQ data was not disaggregated by rural and urban settings, we assumed that the proportions of sampled water sources complying with WHO guidelines and having significant levels of sanitary risk were the same in urban and rural settings. The effect of this assumption will vary from country to country.

We assumed that 100% of unimproved sources do not provide safe water. This is based on a lack of water quality data and a consequent reliance on the same methodology as JMP, treating all such sources as unsafe due to the lack of sustainability of access and lack of protection against contamination characteristic of such sources. This assumption could result in an overestimate of the number of people receiving faecally contaminated water. However, this assumption does not overestimate those receiving water from sources with elevated sanitary risks, since unimproved sources by definition lack sanitary protection. In addition, the MDG indicator accounts for both access and water quality [1]. Since unimproved sources may not meet the access criterion of the indicator, the potential error of the assumption is further reduced.

We did not account for household water treatment (HWT). Use of adequate HWT strategies in households can reduce the health risks of water from contaminated sources (both improved and unimproved sources) and also the risk arising from water contamination during transport and household storage [25,27]. Among a sample of 67 low and middle-income countries, using data from the nationally representative Demographic and Health Surveys (DHS), 1.1 billion people use HWT [27]. However, the DHS surveys used did not confirm responses indicating HWT use with physical indicators in the home of such use. Additionally, it is unknown how these people might be distributed over the actual quality of their source water, how many of these people only use water that they have treated, or to what degree these people sustain the HWT strategy for long periods of time. There also exists the risk of recontamination of water post-treatment [28]. The number of people relying on HWT receiving safe water is likely much smaller than 1.1 billion. As such, the effect of the assumption is likely to be a relatively small overestimate of the population using unsafe water.

The effects of the above assumptions will vary between countries. As such, the country-level estimates for water source compliance and sanitary risk are not presented, as these are likely to be imprecise. However, the overall effect on the global estimates may be relatively small as the errors are aggregated.

The confidence intervals were calculated using the standard errors of the model parameter estimates, and did not account for the standard error of the dependent variable. The standard errors of the reported RADWQ figures were unavailable. This results in an underestimate of the upper bounds and an overestimate of the lower bounds.

CONCLUSIONS

We show that the MDG indicator (proportion of the population using an improved water source) used in assessing progress towards MDG Target 7c results in a substantive underestimate of the proportion of the population using unsafe water. The 2010 JMP data shows that 780 million people (11% of the population) use unimproved sources. We estimate that 1 billion (between a lower bound of 750 million and an upper bound of 1.6 billion) people using piped or other-improved water sources receive unsafe water, meaning 1.8 billion people did not have access to safe water in 2010.

Using a definition for safe water that includes the absence of faecal contamination in a one-off sample from a piped or other-improved source shows that the current indicator, based on the definition of an improved source alone, underestimated the progress required to meet the drinking-water component of MDG Target 7c by 10% of the global population, whereas the 2010 JMP progress estimate suggests that the MDG drinking-water target has already been achieved.

We estimate that an additional 1.2 billion (between 750 million and 2.1 billion) people using safe piped or other-improved sources are using sources that are at elevated risk of contamination unlikely to be detected by one-off, or perhaps even routine monitoring.

While these estimates are imprecise, their magnitude and health and development implications suggest that greater attention is needed to better understand and manage the problem of contamination of improved water sources.

This study highlights the substantive differences between population estimates using water from an improved source and estimates of populations using water free of faecal contamination and water with low risk of faecal contamination. As increasing populations use piped and other-improved water sources it is likely that increasing attention will be paid to water safety. The demand for evidence to inform effective policy making will increase in parallel. Here we use data from large scale dedicated surveys to gain preliminary insights. While the RADWQ methodology has provided valuable information regarding national water safety in terms of contamination and sanitary risk factors, further work is required to identify the most appropriate means to secure greater and more targeted evidence to inform decision-making.

ACKNOWLEDGMENTS

We thank Rolf Luyendijk at UNICEF and Robert Bain for valuable feedback. This work was supported by the International Association of Plumbing and Mechanical Officials (IAPMO). The University of North Carolina at Chapel Hill's Libraries provided support for open access funding.

REFERENCES

1. United Nations Development Group. *Indicators for Monitoring the Millenium Development Goals*; United Nations: New York, NY, USA, 2003.
2. WHO/UNICEF Joint Monitoring Programme for Water Supply and Sanitation. *Progress on Sanitation and Drinking-Water: 2010 Update*; World Health Organization: Geneva, Switzerland, 2010.
3. Data Resources and Estimates of the WHO/UNICEF Joint Monitoring Programme for Water Supply and Sanitation for Water. Available online: http://www.wssinfo.org/data-estimates/introduction/ (accessed on 6 March 2012).
4. Godfrey, S.; Labhasetwar, P.; Wate, S.; Pimpalkar, S. How safe are the global water coverage figures? Case study from Madhya Pradesh, India. *Environ. Monit. Assess.* 2011, *176*, 561–574.
5. Wright, J.; Gundry, S.; Conroy, R. Household drinking water in developing countries: A systematic review of microbiological contamination between source and point-of-use. *Trop. Med. Int. Health* 2004, *9*, 106–117.
6. Kravitz, J.D.; Nyaphisi, M.; Mandel, R.; Petersen, E. Quantitative bacterial examination of domestic water supplies in the Lesotho Highlands: Water quality, sanitation, and village health. *Bull. World Health Organ.* 1999,*77*, 829–836.
7. Guillemin, F.; Henry, P.; Uwechue, N.; Monjour, L. Faecal contamination of rural water supply in the Sahelian area. *Water Res.* 1991, *25*, 923–927.
8. Tadesse, D.; Desta, A.; Geyid, A.; Girma, W.; Fisseha, S.; Schmoll, O. *Rapid Assessment of Drinking-Water Quality in the Federal Democratic Republic of Ethiopia: Country Report of the Pilot Project Implementation in 2004–2005*; World Health Organization: Geneva, Switzerland, 2010.
9. Ince, M.; Bashire, D.; Oni, O.; Awe, E.O.; Ogbechie, V.; Korve, K. *Rapid Assessment of Drinking-Water Quality in the Federal Republic of Nigeria: Country Report of the Pilot Project Implementation in 2004–2005*; World Health Organization: Geneva, Switzerland, 2010.
10. Properzi, F. *Rapid Assessment of Drinking-Water Quality in the Hashemite Kingdom of Jordan: Country Report of the Pilot Project Implementation in 2004–2005*; World Health Organization: Geneva, Switzerland, 2010.
11. Aldana, J.M. *Rapid Assessment of Drinking-Water Quality in the Republic of Nicaragua: Country Report of the Pilot Project Implementation in 2004–2005*; World Health Organization: Geneva, Switzerland, 2010.
12. Aliev, S.; Shodmonov, P.; Babakhanova, N.; Schmoll, O. *Rapid Assessment of Drinking-Water Quality in the Republic of Tajikistan: Country Report Project Implementation in 2004–2005*; World Health Organization: Geneva, Switzerland, 2010.
13. Bain, R.; Gundry, S.; Wright, J.; Yang, H.; Pedley, J.; Bartram, J. Accounting for water quality in MDG monitoring: Lessons from five countries. *Bull. World Health Organ.* Available online: http://www.who.int/bulletin/online_first/11-094284.pdf (accessed on 9 February 2012). in press..
14. Howard, D.; Ince, M.; Smith, M. *Rapid Assessment of Drinking Water Quality: A Handbook for Implementation*; World Health Organization & United Nations Children's Fund: Geneva, Switzerland, 2003.
15. World Health Organization. *Guidelines for Drinking-Water Quality*, 4th; World Health Organization: Geneva, Switzerland, 2011.
16. World Development Indicators. Available online: http://data.worldbank.org/data-catalog/world-development-indicators (accessed on 30 November 2011).
17. Worldwide Governance Indicators. Available online: http://info.worldbank.org/governance/wgi/sc_country.asp(accessed on 30 November 2011).
18. Country Profiles and International Human Development Indicators. Available online: http://info.worldbank.org/governance/wgi/sc_country.asp (accessed on 30 November 2011).
19. Water Quality Index. Available online: http://www.epi2010.yale.edu/Metrics/WaterQualityIndex (accessed on 30 November 2011).
20. EdStats-Education Statistics. Available online: http://web.worldbank.org/WBSITE/EXTERNAL/TOPICS/EXTEDUCATION/EXTDATASTATISTICS/EXTEDSTATS/0,,menuPK:3232818~pagePK:64168427~piPK:64168435~theSitePK:3232764,00.html (accessed on 30 November 2011).
21. Papke, L.E.; Wooldridge, J.M. Econometric methods for fractional response variables with an application to 401(k) plan participation rates. *J. Appl. Econ.* 1996, *11*, 619–632.
22. JMP's method-deriving progress estimates. Available online: http://www.wssinfo.org/definitions-methods/method/ (accessed on 3 February 2012).
23. Moe, C.L.; Sobsey, M.D.; Samsa, G.P.; Mesolo, V. Bacterial indicators of risk of diarrhoeal disease from drinking-water in the Philippines. *Bull. World Health Organ.* 1991, *69*, 305–317.
24. Lee, E.J.; Schwab, K.J. Deficiencies in drinking water distribution systems in developing countries. *J. Water Health* 2005, 109–127.
25. Sobsey, M. *Managing Water in the Home: Accelerated Health Gains from Improved Water Supply*; World Health Organization: Geneva, Switzerland, 2002.
26. Nwachuku, N.; Craun, G.F.; Calderon, R.L. How effective is the TCR in assessing outbreak vulnerability? *J. Am. Water Works Assoc.* 2002, *94*, 88–96.
27. Rosa, G.; Clasen, T. Estimating the scope of household water treatment in low- and medium-income countries. *Am. J. Trop. Med. Hyg.* 2010, *82*, 289–300.
28. McLaughlin, L.A.; Levy, K.; Beck, N.K.; Shin, G.-A.; Meschke, J.S.; Eisenberg, J.N. An observational study on the effectiveness of point-of-use chlorination. *J. Environ. Health* 2009, *71*, 48–53.

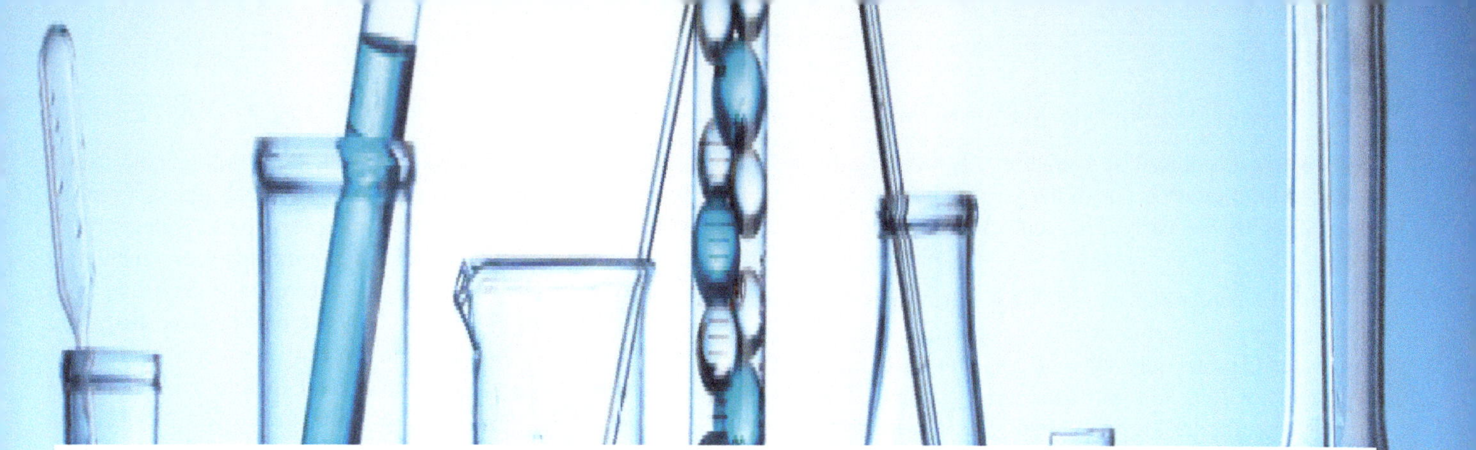

CHAPTER 31

BIOLOGICAL APPROACHES FOR ADDRESSING THE GRAND CHALLENGE OF PROVIDING ACCESS TO CLEAN DRINKING WATER

Mark R Riley, Charles P Gerba and Menachem Elimelech

INTRODUCTION

Water scarcity is a fact of life in arid and semi-arid regions where agricultural, domestic and industrial demands compete for limited resources. Access to clean drinking water presents a monumental challenge that is well documented for the developing world but is a rising problem for more established regions [1]. The problems for both locations are often presented in simplified form as being either a lack of water quantity or a lack of water quality; however, the reality is infrequently so straight-forward. The NAE Grand Challenge document http://www.engineeringchallenges.org/cms/8996/9142.aspx states:

"Lack of clean water is responsible for more deaths in the world than war. About 1 out of every 6 people living today do not have adequate access to water, and more than double that number lack basic sanitation, for which water is needed. In some countries, half the population does not have access to safe drinking water, and hence, is afflicted with poor health. By some estimates, each day nearly 5,000 children worldwide die from diarrhea-related diseases, a toll that would drop dramatically if sufficient water for sanitation was available."

In the so-called developed world, an aging infrastructure plays a large role in problems of providing clean drinking water. Along much of the U.S. east coast, conveyance systems (pipes, pumps, valves, etc.) were designed for a 100-year life span, but were constructed in the middle 1800's. Much of the water infrastructure is of poor quality, which has led to substantial leakage of water and unaccounted-for water totaling one half of that initially introduced. In many cases drinking water pipes are located in close proximity to black water (wastewater) pipes. In a highly connected piping network, pressure can at times be either positive or negative relative to the surrounding water table. In a leaky system, drinking water can mix with a variety of water sources, thus providing one common route for introduction of microbial contaminants into the drinking water supply.

Certainly, the world does possess sufficient water for all of man's endeavors. Globally, water is available in abundance. Unfortunately, water abundance does not match locations and the quality to meet mankind's need. Even within close geographic locations, some regions may be awash in fresh water while other regions, perhaps separated by only a few hundred miles, are afflicted by drought. In many instances, political and economic barriers prevent access to water even in

Citation: Riley MR, Gerba CP, Elimelech M (2011) Biological approaches for addressing the grand challenge of providing access to clean drinking water. Journal of Biological Engineering 5:2 DOI: 10.1186/1754-1611-5-2

areas where it is otherwise available. "Overcoming the crisis in water and sanitation is one of the greatest human development challenges of the early 21st century," a recent U.N. report warns [2].

Public perception of drinking water and its value is complex. Nearly half of the U.S. population avoids drinking tap water [3] and only 40% of respondents to a survey are reportedly willing to contribute some amount of money for improvement of drinking water quality [3]. The majority of these refusals are characterized as protest responses. Surprisingly, the frequency with which an individual drinks tap water does not influence their willingness to support drinking water quality improvements. There are reports [4] that a substantial proportion of the deficiencies in drinking water systems contributing to consumer complaints and waterborne-disease outbreaks are associated not with the conveyance systems, but rather are due to problems in the premise plumbing (that is, plumbing within the home or industrial facility).

Water for drinking and personal use is only a small part of society's total water needs with household water accounting for generally less than 5 percent of total water use. Most water is utilized for agriculture and industry, not accounting for water needed for natural ecological processes. The oceans contain an enormous reserve of water which could be made potable through desalination processes such as those in use in many arid regions, but cost currently limits their use in areas of moderate need. Substantial progress has been made in the past 10 years in improving desalination methods and developing the technology to function on a large scale [5]; however, improvements are still needed and uses for the brackish effluent must be further developed so as not to contaminate naturally available drinking water supplies. In some developing countries, water supplies are contaminated by human wastes (fecal matter and endocrine disrupting compounds), industrial and agricultural wastes (pesticides, solvents), naturally occurring contaminants (arsenic, lead, and other metals), and increasingly due to the results of natural disasters and industrial accidents (typhoons in Southeast Asia; oil drilling in the Gulf of Mexico).

This report will focus on challenges and a few biologically-based solutions to address water quality and water quantity. Specifically we shall address microbial contamination as a water quality problem and the use of desalination as a means to address water quantity limitations.

WATERBORNE PATHOGENS

Meeting the need of providing adequate drinking water has separate tasks based on location and degree of living conditions. The developing world has high childhood mortality with high endemic disease and a need for low cost sustainable treatment [6, 7]. In the developed world, outbreaks and special populations create endemic and chronic diseases. Every time we change our environment, we create new opportunities for new types of organisms to thrive. When we create a new source of water for human use, we open the environment to new types of organisms that might have negative effects on human health and on the natural environment. There is a need for real time monitoring and management of water treatment and water distribution systems that are currently monitored through infrequent batch measurement procedures, which unfortunately, can miss transient, but problematic water quality events. The EPA assessed how much illness from tap water occurs in the U.S. from multiple studies conducted over a similar time frame. Studies reported that from municipal water systems alone there were between 12 - 19.5 million cases of diarrhea/yr [8, 9, 10]. In 2009 Gerba estimated that there were 7,000-20,000 deaths/yr from water-based organisms (unpublished). To place this in context, there were more people who died from water-based diseases then died from AIDS in 2008.

Recent studies on over 500 wells around the U.S. showed that 10-42% of groundwater supply wells are contaminated with enteric viruses [10]; 95% of all individual and small groundwater systems violate one or more primary and secondary drinking water standards in Arizona [11]; and 52% of all point-of-use carbon filters contain an enteric organism after 3 months of use in a municipal system [12]. Approximately 80% of consumer complaints about tap water quality originate in the consumer's home distribution system [4], indicating that management of water quality problems cannot easily be addressed from a single central location.

A broad range of treatment options are available for decentralized sewage systems, but high rates of failure of traditional septic tanks [13] discourages innovation in favor of centralized sewerage installation. Oftentimes, indicator organisms or surrogates are applied for monitoring water quality and developing water treatment methodologies; however, the scientific community does not have consensus on how well these indicators truly reflect the behavior of target organisms, which may be pathogenic, teratogenic, endocrine-disrupting, or problematic in other ways.

Even though potable water distribution systems present a hostile environment for growth of microorganisms due to low nutrient levels as well as the presence of disinfection residuals [14], biofilms form ubiquitously and may provide a near constant source of HPC (heterotrophic plate count) bacteria even when the water has been subjected to disinfection processes including chlorination, chloramination and ultra-violet (UV)-treatment [15, 16]. Heterotrophic bacteria are those that require organic carbon rather than carbon dioxide as a carbon source; all human pathogenic bacteria are heterotrophic. While disinfection residuals within distribution systems reduce the potential for growth of pathogens and regrowth of heterotrophic bacteria, they do not eliminate them [17]. While the presence of heterotrophs does not in itself constitute a human health hazard, they can serve as a means to assess the microbiological quality of the water and the efficacy of water treatment [18].

The U.S. Environmental Protection Agency (U.S. EPA) has recommended that heterotrophic bacterial numbers in drinking water should not exceed 500 colony-forming units (CFU/mL), primarily because of the interference of coliform detection [19]. This number is based on the effect of HPC populations on recovery of total coliforms rather than on a health-based action level [20]. Higher numbers of HPC are often the result of bacterial regrowth, particularly in the distribution system [21] and in the water treatment devices mounted at the household tap [22]. It should be noted that in an evaluation of the relative contribution of heterotrophic bacteria from various sources in the normal diet of an average person in the United States, less than 5% of the average consumer's total heterotrophic bacteria intake was found to be derived from drinking water [23]. However, individuals in high-risk categories, such as the immunocompromised, young children, or the elderly could be at risk from the consumption of drinking water containing opportunistic pathogens that are members of the HPC group [24].

According to Allen and coworkers [20], the number of HPC bacteria in drinking water varies widely, depending on many factors. These include the initial quality of the source water, the types of treatment, and the type and concentration of disinfection residuals. Additionally, water age and ambient temperature of the raw and finished water influence HPC levels.

Based on these points, HPC are often used as an indicator to assess the microbial quality of drinking water even though the clinical and epidemiological evidence that HPC bacteria pose a health risk is lacking.

METHODS TO ADDRESS WATERBORNE PATHOGENS

A number of methods are in development for removing or inactivating waterborne pathogens. Here we focus on a few instructive examples.

In 1993 the largest waterborne outbreak of disease ever documented occurred in Milwaukee, Wisconsin. Over 400,000 individuals developed gastroenteritis and perhaps 100 individuals died in response to consuming drinking water contaminated with the protozoan parasite *Cryptosporidium parvum* [25]. The organism appeared in high concentrations in the city's water source after a period of heavy rains, allowing some of the oocysts of the organism to penetrate the filtration barrier. Infections ranged from acute conditions, such as diarrhea, to chronic conditions with symptoms that remained present even 10 years after initial infection. Chlorine disinfection had little effect on the viability of the oocysts, and in fact, this organism was later detected in 60% of the treated drinking water supplies in the United States. Since that time, new rules for protecting surface water supplies and required water quality measurements have been put into place.

Cryptosporidium is highly chlorine resistant and thus presents a challenge for its elimination from water supplies. Recent research on microbial resistance to treatment and disinfection demonstrates that the microbial surface structure and composition are key to determining the potential for waterborne transmission of emerging pathogens [25].

Commonly used methods for water disinfection in pipes rely on chlorination as the predominant method in part due to chlorine's wide range of activity and to its role as a lingering disinfectant, which unfortunately also leads to harmful residuals. Ultraviolet (UV) light has been shown in some cases to be equivalent to chlorine, but generates no disinfection residuals; the merit of such residuals is still in debate. Hammes and coworkers [26] showed that drinking water that had undergone ozonation and biological filtration, but which was distributed without any disinfectant residuals, had a high level of biological stability with total cell concentrations hardly changing throughout the distribution network. Their work highlights the benefit of utilizing multiple chemical and microbiological parameters to evaluate water quality since correlations could be found between organic carbon and cell counts, but not with HPC.

Treatment technologies need to increasingly address changes in water quality and quantity requirements along with secondary effects of usage of the treatments. Increased use of chlorine and chloramines has resulted in biofilm populations being supplemented with *Mycobacterium*. The increased usage of UV light has resulted in increased discharge of adenovirus in surface waters [27]; the super-surviving fraction has caused the disinfection microbe inactivation curves not to be linear [28]. Rotavirus has no significant inactivation with UV light from 100-350 mJ cm^2. Changes in treatment technology will result in changes in the pathogens to which we are exposed, which will then lead to a problem of detection.

Straub and Chandler [29] addressed some of the fundamental challenges for detection of pathogens in water systems. Some of the key questions they raise include several that are still not fully answered: Is the detection objective enumeration, presence/absence, viability/infectivity; what is an appropriate level of risk (is it one viron per 1000 L of water); what volume of water needs to be or should be collected; is it necessary to detect the full range of likely pathogens or merely a few target organisms; and can the methods be automated.

Additionally, it will be necessary to take better account of historic weather patterns since the majority of all surface and ground water outbreaks are associated with heavy rainfall, much above normal levels [30]. As climate change leads to an increase in the frequency of storms and intensity of rainfall events, as well as an increase in the temperatures of surface and water distribution systems, an increase in the types and numbers of water-based pathogens will result. This will also likely cause an increase in taste and odor problems as a result of the expected increase in algal blooms.

To better prepare for changes in microbial water quality, treatment technologies with real time event monitoring where sensors are located throughout the distribution systems will be essential [31]. The primary challenge for such a system is that measurements must be performed in a continual, always on, manner and be capable of detecting low numbers of indicators and/or pathogens in large volumes of water (100-1000 L per minute). Commercial sensors are available for measurement of chemical water quality parameters including pH, chlorine, electrical conductivity, total organic or dissolved organic carbon, and turbidity. Few devices can detect microorganisms in a continual manner [32] and the ones that are available typically do not perform species identification (pathogenic vs. non-pathogenic strains).

In the past decade, PCR (polymerase chain reaction) based methods have become the most commonly used method for detection of waterborne viruses. PCR and RT-PCR methods have high specificity and sensitivity, can be used for nonculturable or noncytopathic viruses, and require a fairly short time for detection (generally 2-4 hours) [33]. Unfortunately, PCR methods can be limited for environmental analyses in that very small sample volumes are applied (which may lead to non-representative testing), the transcriptase enzymes can be inhibited by waterborne substances including humic acids, and the method cannot discriminate between infective and non-infective viruses. Persistence of infectivity can vary substantially across type of microbe and the environment [34]. Cell culture methods [35], when integrated with PCR may circumvent some of the problems [36, 37] but require comparatively long times (4-30 days) for measurement.

Microchip based devices using antibodies can perform the required levels of detection, but their operation is typically very expensive, especially for a continual monitoring system, although some have promise for reusability [38]. Detection limits for some of these automated systems [39] are reported to be on the order of 10^4 cfu/mL; lower detection limits (more sensitive measurement) will be needed especially for measurement of viral contaminants having low requirements for infectivity. Biosensors have been developed that can perform capture of pathogens followed by real-time PCR using integrated waveguides [40]. Such optical methods have high promise for continual monitoring, however, without an enrichment or PCR step, detection limits tend to be higher than needed for municipal water monitoring. Nanoscale components using nanorods or nanoparticles can enhance signal strength and hence, reduce detection limits [41, 42]; however stability and cost remain problematic. Antibody arrays [43] provide detection of multiple targets at one time; however, the broad-based microbial population of drinking water makes their implementation difficult.

Genomics, metagenomics, and functional genomics will provide information on exposure to pathogens to the level of being able to determine fairly precise locations of origin. Such source tracking has shown that there is more growing in the tap than one would think. Often such detailed analyses are not performed unless a contaminant made the water taste unacceptable or made someone ill. Such problems are likely to increase due to our over-reliance on an aging water infrastructure, from climate effects which increase surface water temperature, and due to unintended consequences of well-meaning applications of water treatment technologies.

For example, the water based pathogen *Naegleria fowleri* occurs in drinking water supply wells where the well temperature is between 20-40°C and feeds on HPC bacteria in the well. The use a more environmentally friendly biodegradable lubricant oil for well pump motors results in increased levels of HPC bacteria, increasing the numbers of *N. fowleri* [44]. *Naegleria* is one of the most resistant water-based pathogens to chlorine disinfectants and UV light; with 98% human mortality if infection occurs. Such a perfect microbial storm impacts how we have to consider unintended consequences of our desire to make a friendlier environment can affect our exposure to waterborne and water-based pathogens. Again, as we change the environmental conditions either through direct or indirect human action, the resultant microbial populations will find within their microenvironments new opportunities to thrive. We need to continuously develop new sources of viable drinking water.

Once pathogens have been detected in a water source, the water may be either discarded or the water itself and the conveyance systems must be treated to prevent recurrence. This is often accomplished by either removing the pathogens using a filtration step or by inactivating the pathogens using radiation or chemical, thermal, or other methods. Multiple-barrier or "hurdle" techniques are most efficient since no single technique can provide the necessary breadth of coverage.

Removal processes often involve passing the water through a coarse filtration system such as gravel or sand filters, which reduce gross turbidity. Such methods are especially effective for removing large diameter contaminants such as algal species and protozoa. Gravity-based settlement within a storage tank is often used but does require substantial holding times since pathogen-settling follows a terminal velocity based on particle diameter. A large, but typical protozoan measuring 250 um (Amoebic dysentery) will settle at a rate on the order of tens of mm per second. Such a simple process is often not sufficient by itself to meet the high pathogen removal level required.

Disinfection or inactivation methods provide a broader means to reduce pathogen load and typically are applied during a second stage. Heat and ultraviolet light treatments disrupt cellular proteins or genetic material in an irreversible manner but do not provide a lasting barrier against pathogen regrowth. Oxidative disinfection methods are effective at inactivating a broad range of pathogens. The effectiveness of such oxidative disinfection is species-specific and usually increases with the time and concentration applied. Chlorination or chloramination deliver a lasting effect but raise concerns for human health impact. Ozonation is highly effective but does not provide a lasting residual. Disinfection efficiency can vary substantially by the water chemistry (pH, salt, turbidity, etc.) and contact time. Unfortunately, disinfectants can react with naturally occurring materials in drinking water to form disinfection by-products (DBPs) which may lead to adverse health effects.

Filtration at the point of use (POU) is highly effective at removing pathogens and addresses concerns about water age. Water age refers to the time that an individual volume element exists in a water conveyance system and plays a major role in water quality deterioration within a distribution system [13]. As water travels through the distribution system it undergoes a variety of chemical and physical changes dependent on the flow rate, age and quality of the pipes and their deposited materials, finished water quality, and ambient temperature. Water age changes over the course of a day due in part to varying consumer demand changes. In general, a shorter water age correlates with few negative alterations in water quality. Water age has been determined to vary across cities based in part on the size and design of the conveyance system and the size of the population served with larger populations having shorter water ages. Regions with local populations over 500,000 generally have water ages on the order of 3-7 days while regions with smaller populations (less than 50,000) are estimated to have water ages of 12-24 days within the water conveyance system [13]. Water quality problems associated with water age include decay of disinfectants, formation of disinfection by-products, microbial regrowth (after decay of disinfectants), sediment deposition, and decreases in quality of taste and odor [13].

Removal of pathogens at the necessary high flow volumes is challenging. Commonly applied filtration suffers from membrane fouling thus motivating the search for new surfaces and coatings. One approach in development utilizes functionalized self-assembled monolayers [45] in which pathogens can be removed based on their electrostatic properties. In general, viruses will adsorb to negatively charged surfaces at pH values below the isoelectric point of the virus, which favors a positive surface charge [46]. If the water pH is above the pI (isoelectric point) of the microbe, then deposition is driven to oppositely charged surfaces. Reversal of the surface charge can in some cases provide a reversible means to remove the deposited microbe and regenerate the surface for future use [32]. Similarly, water pH and surface characteristics can impact transport of microbes through soils. This approach has been utilized to remove microbes from water by use of columns containing sand coated with ferric and aluminum hydroxides [47]. The amount of phosphate in the water (also a challenge for desalination) can influence bacterial adhesion by either reducing attachment (phosphate levels below 0.5 mM) or increasing attachment (phosphate levels above 0.5 mM) [48].

Many types of microorganisms have a natural tendency to attach to wet surfaces, where they then multiply and embed themselves in a slimy matrix forming a biofilm [49]. Biofilm formation on pipes and water conveyance systems contributes to the presence of HPCs in water systems and may play a role in hygiene problems. Since the attachment of microbes to surfaces and the development of biofilm phenotypes occur quickly, it is almost impossible to prevent biofilm formation [50]. The removal and killing of established biofilms require harsh treatments, mostly using oxidizing biocides which may have variable effectiveness depending on the nature of the biofilm. The emergence of bacteria resistant to conventional antimicrobials necessitates that new control strategies be developed. Some new methods employ the use of biological solutions (enzymes, phages, interspecies interactions, and antimicrobial molecules of microbial origin).

Generally, disinfectants do not penetrate the typical biofilm matrix remaining on a surface, such as a pipe, after an ineffective cleaning procedure [51]. Cleaning must be the first step applied and must be done to high levels of cleanliness; this is tremendously challenging for water distribution systems that are not designed for in-place cleaning, rinsing, or intermediate usage.

Enzyme-based detergents have been used as "green chemicals" to remove biofilms [49], especially in the food industry. Proteolytic enzymes have been successful, but are inhibited by the presence of other proteinaceous material. Introduction of surfactants improves the penetration into the biofilm.

Bacteriophages are viruses that infect bacteria and may provide a species-specific means to eliminate biofilms from pipes [52]. The biofilm bacteria would need to be susceptible to the phage. Most biofilms are comprised of a complex mixture of species and so application of this method has many challenges. A related approach utilizes introduction of microbial metabolites, which can interfere with the communal interactions and communications within a complex biofilm. This approach relies upon competition for substrates that can alter the balance of species within a biofilm [53].

DESALINATION

There is a tremendous need for methods to substantially increase water supply, not only to meet current needs but to provide for population growth and increased water usage. Only 3% of water on our planet is fresh, but most of this is either in snow or ice or in aquifers that are challenging to reach. Extraction from groundwater often exceeds the rate of replenishment. An obvious solution for meeting global freshwater demand lies in the world's oceans, but this water is too salty to be used without extensive treatment such as that provided by desalination.

In Israel, where total demand for fresh water exceeds the average natural supply, the freshwater deficit is replaced to a great extent by treated wastewater used for agriculture [54, 6]. However, the treatment of effluents prior to use in irrigation is not adequate for sustaining long-term agricultural production, due to salinization, which can deteriorate soil structure.

Sustainable production of adequate drinking water can only be achieved by desalination of sea and brackish waters [5, 55]. Desalination is a process for extracting salt from seawater and returning water of a higher quality. Desalination is not a new idea and is already used in many regions, particularly in the Middle East. Saudi Arabia alone accounts for about a tenth of global desalination while Israel uses desalination technology to provide about a fourth of its domestic water needs. Such technologies must have lower impact on the environment, which means less energy use and reduced production of CO_2 while utilizing the high salt effluent in ecologically sustainable ways.

Modern desalination plants employ reverse osmosis, which uses a membrane to separate out the salt. More than 12,000 desalination plants are now operational across the world. It has been estimated that up to 25 million m^3 of desalinated water is produced daily around the world [56]. Desalination facilities are expensive to build and require substantial amounts of energy to operate, making desalination suitable mainly for seaside cities in rich countries. It therefore has limited value for impoverished countries, where water supply problems are most serious.

Growing populations and political concerns are prompting governments and investors to look closely at desalination [55]. In a public sector partnership, Long Beach Water, along with the Los Angeles Department of Water & Power and the United States Bureau of Reclamation, has constructed a 300,000 gallon-per-day prototype desalination facility, the largest seawater desalination research and development facility of its kind in the United States. This facility was developed to evaluate new seawater desalination methods which could reduce cost and energy requirements.

Desalination by reverse osmosis has seen major improvements in the past 10 years. Further improvements, however, are likely to be incremental in part because of thermodynamic limitations [5, 55]. A 10-20% reduction in energy use is potentially feasible [55]. Recoveries of seawater desalination are up to 50%, resulting in a brine discharge of environmental concern. Thermodynamics dictate a minimum energy of desalination, where the minimum energy needed to desalt water is independent of the technology or mechanism used for desalination [5].

New technologies that would decrease energy use, and therefore costs, might help desalination's contribution to drinking water supplies. Most of the research to reduce the energy use in reverse osmosis desalination has focused on creating high permeability (flux) membranes and low fouling membranes. The rationale for both improvements is to reduce the required hydraulic pressure and hence, the electric energy needed for the process. Membrane fouling is a sizeable problem, especially when the source water has high solids content or a substantial organic matter load. Most bacteria are removed by pre-treatment, but since the system is not sterile and because we cannot apply chlorine to RO membranes, biofilms will always grow. Such biofouling decreases the water flux, increases energy costs, and requires frequent membrane cleaning or replacement.*Bacillus* sp. and *Pseudomonas* sp. are among the most prevalent bacteria that form biofilms on RO and nanofiltration (NF) membranes used for water treatment [57]. Chlorination cannot be applied to reverse osmosis membranes because oxidants degrade the membrane structure [58] Inorganic scaling [54] can also arise due to phosphate and other inorganic ions present in wastewater effluent.

Nanotechnology has significant promise for use in reverse osmosis. Studies have shown that water flow through the inside of carbon nanotubes (so called nano-osmosis) yields flow rates that are 1,000 times to 10,000 times greater than expected based on the simple fluid mechanics calculations of flow in the pipe [5,59]. The surface inside such structures is smooth and hydrophobic and as a result, water flows as a string of molecules that readily slips through the membrane. Multi-carbon nanotubes have been used, but the diameter is too large for removing salt. Smaller single-walled carbon nanotubes (SWNTs) with a pore size of about 0.7 nm are critical for salt rejection, while a high density of nanotubes is needed for sufficient water productivity through such a membrane [60].

The nanotechnology approach has a biological inspiration for high flux membranes based on the aquaporin proteins extracted from living organisms, which can be incorporated into a lipid bilayer membrane or a synthetic polymer matrix [5]. While an intriguing approach, energy is needed even for membranes with infinite permeability since osmotic pressure of sea water, which is about 25 atm or 400 psi, must be overcome to induce permeation of fresh water through the reverse osmosis membranes. To get any meaningful water flux through the membrane one must exceed the pressure significantly. Additionally, the buildup of salt from the sea water next to the membrane leads to concentration polarization, requiring further energy to drive water through the membrane. The minimum theoretical energy cost for desalination of 50% recovery is 1 kWh/m^3 of water treated; however, considering practical limitations, it is likely not to be less than 1.5 kWh/m^3. The long term goal is to achieve between 1.5-2 kWh/m^3 with a more likely energy cost of around 2 kWh/m^3.

It is necessary to consider other approaches to desalinate salt water. In a process recently developed at Yale University, instead of applying hydraulic pressure, an osmotic pressure difference is used [61, 62]. Ammonia and CO_2 dissolved in water results in ammonium salts of very high concentration and hence, large osmotic pressure driving forces. This draw solution is used to induce natural osmotic flux of water across a semi-permeable membrane from a relatively dilute saline feedwater. The salts of this solution are thermolytic: when low-grade or waste heat is supplied, the salts decompose into ammonia and carbon dioxide gases for simple stripping from solution. These gases, when introduced to water again at a lower temperature, readily reconstitute the desired draw solutes. The major energy source for this "ammonia-carbon dioxide forward osmosis desalination process" is in the form of waste heat, with electric energy use of about 20% of that used in reverse osmosis desalination [63].

The environmental impact of desalination cannot be overlooked as there is a need to develop desalination technologies that consume less prime energy, typically electricity [55]. Methods for using waste heat (or recovering this heat) to desalinate sea water are needed to improve overall plant energy efficiency. Selection of the desalination facility is critical for minimizing ecological impact and can be hindered by contrasting goals. It is desirable to locate a facility far from any other discharge sources so that the intake will be minimally disrupted. Conversely, the plant discharge has lower ecological impact when it can be diluted with other waste streams having lower salinity [64].

The use and discharge of the concentrated brine must also be addressed in better ways so that the environmental impacts of larger and more numerous desalination plants are minimized. Desalination plant discharge can adversely affect aquatic communities due to substantial increases in salinity and the accumulation of metals, hydrocarbons and toxic anti-fouling compounds in receiving waters [65]. The ultimate consequence of this release can be shifts in the populations of plant and animal species due to the altered physical and chemical characteristics of the water [66]. Greater detail in monitoring is needed to ensure that ecological impacts lie within desired limits. However, many published monitoring studies lack sufficient detail with respect to study design and statistical analyses, making conclusive interpretation of results difficult [65].

Even with such advances it seems unlikely that desalination alone will be able to solve the world's water problems. Other approaches will be needed to reduce water use, to recycle what can be further used, and to capture new sources of water. These could include use of biological systems. Two such methods are discussed below.

Many halophyte plants and green algae have been used to desalinate water. Algae have been found in some of the harshest saline conditions on the planet and many are able to complete all of their life cycle processes in highly saline environments. Microalgae have been used in tertiary sewage treatment to eliminate nitrogen and phosphorous compounds [67]. Some microalgae, including *Scenedesmus obliquus* are able to use organic compounds and mixotrophic conditions [68]. Both fresh water and marine algal species can accumulate organic compounds as osmo-regulatory solutes in response to salt stress [69]. Salt-tolerant algal species are also being explored as producers of oil to generate biodiesel transportation fuels.

Microbial desalination of wastewater has been developed by the Bruce Logan Laboratory at Penn State University [70]. Typically desalination and wastewater treatment require substantial power input; however, Logan's approach replaces the power source with a microbial fuel cell. As the bacteria break down nutrients in the wastewater they also draw salt out of the water. Initial devices remove 88 to 94 percent of the salt, which is very good, but not yet as high as traditional desalination. Advances in electrodes and membrane materials along with stabilization of the microbial communities for use in a variety of environments will be necessary for full scale implementation.

OTHER TECHNOLOGIES TO PROVIDE CLEAN WATER

Technologies are being developed to improve recycling of wastewater and sewage treatment so that water can be applied for non-personal uses, such as irrigation, toilet flushing, or specific kinds of industrial purposes. Recycled water could resupply aquifers; however, very effective and consistent treatment methods along with rigorous safeguards are necessary to preserve the safety of recycled water. Advances in membrane technologies may lead to approaches, which can remove specific chemical and biological contaminants while passing along desired nutrients (N, P, and K in particular).

A different approach to the water problem involves developing strategies for reducing water use. Agricultural irrigation consumes enormous quantities of water; in developing countries, irrigation often exceeds 80 percent of total water use. Improved technologies that would more efficiently provide crops with water, such as sub-surface drip irrigation, (which is already commonplace in semi-arid regions of the U.S. and in the Middle East), can substantially reduce agricultural water demand. Despite its high efficiency, drip irrigation has been implemented only slowly in some areas due in part to the high cost of installation, to early problems (now solved) of fouling, and to general skepticism of a technology in which the critical input water is not seen by the farmer.

Methodologies are needed for small scale purification of water, especially in the developing world. Wide-spread use of inexpensive gravity-fed filtration units, solar distillation, and low-power treatment systems could greatly reduce waterborne disease [6, 7]. The ultimate cost to the consumer and ease of use and maintenance, especially in low-income rural areas, are key factors at least as important as technological design requirements. Cheap and efficient techniques using "homemade" filter media, such as slow sand filtration in rural areas, are being tested for their ability to remove chemical and biological contaminants [71]. It is necessary that methods not only be effective, but that they be cost effective and maintained by personnel with only minimal training [71].

SUMMARY

To meet the NAE's grand challenge of providing safe drinking water requires advances in technology to provide safe and substantial quantities of water in addition to addressing societal problems tied to socioeconomics. Removal or inactivation of pathogens are key steps to prevent widespread disease. However, outbreaks of waterborne and water-based pathogens remain problematic even in the developed world. As reserves of accessible fresh water continue to be depleted, desalination techniques will become increasingly important. While thermodynamic limits have nearly been met using advanced nanotechnologies, further refinement to reduce biofouling can decrease total energy costs while increasing sustainability.

We need new indicator microorganisms that are water based pathogens and real time sensors that are capable of organism identification. Detection sensitivity is needed to be able to find 1 organism/1-10,000 mL, along with some assessment of viability. These are great challenges, but we also need technologies for treatment of biofilms in the distribution system. We need technologies for treatment at the tap and for premise plumbing, as well as new disinfectants, fewer by-products, and disinfectants designed to work in conjunction with existing disinfectants.

To this end, the major questions addressed include the following:

- How do multi-scale, natural and engineered systems interact and subsequently affect water quality and quantity?
- How can the fate of chemical and biological water contaminants be monitored, modeled, and controlled? Similarly, should emphasis be placed on detection and monitoring methods of high specificity or broad applicability?
- How can the energy and CO_2 burden of purification methods, including desalination, be provided for in a sustainable manner?
- How can socioeconomic burdens related to access to safe drinking water in the developing world be reduced?

Fully addressing this grand challenge will continue to be a task for engineers and scientists developing and applying a multitude of methods. These challenges are met by advances made at the interface of what are commonly considered scientific and engineering disciplines and can benefit greatly by fully integrating biological components and accounting for biological impacts of engineering design decisions.

ACKNOWLEDGEMENTS

We thank the National Science Foundation for financial support of this publication and for support of the IBE 2010 Conference Session on Addressing the NAE's Grand Challenges through Biological Engineering

REFERENCES

1. Yoon JY, Riley MR: "Grand challenges for biological engineering". Journal of Biological Engineering 2009, 3: 16. 10.1186/1754-1611-3-16
2. United Nations Development Programme: Human Development Report 2006: Beyond Scarcity: Power, Poverty and the Global Water Crisis. New York: Palgrave Macmillan; 2006.
3. Polyzou E, Jones N, Evangelinos KI, Halvadakis CP: "Willingness to pay for drinking water quality improvement and the influence of social capital". J Socio-Econ 2010, 40: 74-80. 10.1016/j.socec.2010.06.010
4. Derby , Mary P, Quintanar DR, Gerba CP, Pepper IL: Implementation of a Novel Intervention to Enhance Public Waterborne Disease Surveillance: Initial Findings. International Society for Disease Surveillance, 2009 Conference presentation
5. Shannon MA, Bohn PW, Elimelech M, Georgiadis JG, Mariñas BJ, Mayes AM: "Science and Technology for Water Purification in the Coming Decades". Nature 2008, 452: 301-310. 10.1038/nature06599
6. Elimelech M: The Global Challenge for Adequate and Safe Water. Journal of Water Supply: Research and Technology - AQUA 2006, 55: 3-10.

7. Montgomery MA, Elimelech M: "Water and Sanitation in Developing Countries: Including Health in the Equation". Environmental Science and Technology 2007, 41: 17-24. 10.1021/es072435t

8. Colford JM, Roy SL, Beach MJ, Hightower A, Shaw SE, Wade TJ: A review of household drinking water intervention trials and an approach to the estimation of endemic waterborne gastroenteritis in the United States. Journal of Water and Health 2006,4(2):71-88. 10.2166/wh.2006.018

9. Messner M, Shaw S, Regli S, Rotert K, Blank V, Soller J: An approach for developing a national estimate of waterborne disease due to drinking water and a national estimate model application. Journal of Water and Health 2006,4(Suppl 2):201-240. 10.2166/wh.2006.024

10. Reynolds , Kelly A, Mena KristinaD, Gerba CharlesP: Risk of Waterborne Illness Via Drinking Water in the United States. Reviews of Environmental Contamination and Toxicology 2008, 192: 117-158. full_text

11. Marrero-Ortiz R, Riley KR, Karpiscak MK, Gerba CP: Groundwater quality of individual wells in small systems in Arizona. JAWWA 2009, 101: 89-100.

12. Miles , Syreeta L, Gerba CharlesP, Pepper IanL, Reynolds KellyA: Point-of-Use Drinking Water Devices for Assessing Microbial Contamination in Finished Water and Distribution Systems. Environ Sci Technol 2009, 43: 1425-1429. 10.1021/es801482p

13. U.S. E.P.A., Anonymous: Effects of water age on distribution system water quality.2002. [http://www.epa.gov/ogwdw/disinfection/tcr/pdfs/whitepaper_tcr_waterdistribution.pdf]

14. Långmarka , Jonas , Storeyb MichaelV, Ashboltc NicholasJ, Stenströma Thor-Axel: The effects of UV disinfection on distribution pipe biofilm growth and pathogen incidence within the greater Stockholm area, Sweden. Water Research 2007, 41: 3327-3336.

15. Mnb Momba, Te Cloete, Sn Venter, Kfir R: Evaluation of the impact of disinfection processes on the formation of biofilms in potable surface water distribution systems. Water Sci Technol 1998,38(8-9):283-289. 10.1016/S0273-1223(98)00703-3

16. Kulakov LA, McAlister MB, Ogden KL, Larkin MJ, O'Hanlon JF: "Analysis of bacteria contaminating ultrapure water in industrial systems". Applied and Environmental Microbiology 2002, 68: 1548-1555. 10.1128/AEM.68.4.1548-1555.2002

17. Payment P, Trudel M: Methods and techniques in virology. New York, NY: Marcel Dekker, Inc; 1993.

18. Camper AK, LeChevallier ML, Broadaway SC, McFeters GA: Growth and persistence of pathogens on granular activated carbon. Appl Environ Microbiol 1985, 50: 1178-1382.

19. USEPA (2001): Drinking Water Infrastructure Needs Survey. In EPA 816-F-01-001. Office of Water, U.S. Environmental Protection Agency, Washington, DC;

20. Allen MJ, Edberg SC, Reasoner DJ: Heterotrophic plate count bacteria-what is their significance in drinking water? Presented at the NSF International/World Health Organization symposium on HPC Bacteria in Drinking Water, April 22-24, 2002, Geneva, Switzerland. International Journal of Food Microbiology 2004, 92: 265-274.

21. Geldreich EE, Taylor RH, Blannon JC, Reasonner DJ: Bacterial colonization study of point-of-use devices. J Am Water Works Assoc 1985, 77: 72-80.

22. Reasoner DJ, Blannon JC, Geldreich EE: Microbiological characteristics of third-faucet point-of-use devices. J Am Water Works Assoc 1987, 79: 60-66.

23. Stine SW, Pepper IL, Gerba CP: "Contribution of drinking water to the weekly intake of heterotrophic bacteria from diet in the United States". Water Research 2005, 39: 257-263. 10.1016/j.watres.2004.09.010

24. Rusin PA, Rose JB, Haas CN, Gerba CP: Risk assessment of opportunistic bacterial pathogens in drinking water. Rev Environ Contam Toxicol 1997, 152: 57-83.

25. Nwachcuku N, Gerba CP: "Emerging waterborne pathogens: can we kill them all?". Current Opinion in Biotechnology 2004, 15: 175-180. 10.1016/j.copbio.2004.04.010

26. Hammes F, Berger C, Köster O, Egli T: "Assessing biological stability of drinking water without disinfectant residuals in a full-scale water supply system". Journal of Water Supply: Research and Technology-AQUA 2010, 59: 31-40. 10.2166/aqua.2010.052

27. Rodriguez RA, Gundy PM, Gerba CP: Composition of BGM and PLC/PRC/5 cell lines for total culturable viral assay of treated sewage. Applied and Environmental Microbiology 2008, 74: 2583-2587. 10.1128/AEM.00626-07

28. Li D, Gu AZ, Shi HC, Yang W: UV inactivation and resistance of rotavirus evaluated by integrated cell culture and real-time RT-PCR assay. Water Research 2009, 43: 3261-3269. 10.1016/j.watres.2009.03.044

29. Straub TM, Chandler DP: "Towards a unified system for detecting waterborne pathogens". Journal of Microbiological Methods 2003, 53: 185-197. 10.1016/S0167-7012(03)00023-X

30. Curriero FC, Patz JA, Rose JB, Lee S: The association between extreme precipitation and waterborne disease outbreaks in the United States. American Journal of Public Health 2001, 91: 1149-1199. 10.2105/AJPH.91.8.1194

31. Sinclair RG, Choi CY, Riley MR, Gerba CP: "Pathogen Surveillance Through Monitoring of Sewer Systems". Advances in Applied Microbiology 2008, 65: 249-269. full_text

32. Vargas CA, Wilhelm AA, Williams J, Lucas P, Reynolds KA, Riley MR: Integrated capture and spectroscopic detection of viruses. Applied Environmental Microbiology 2009, 75: 6431-6440. 10.1128/AEM.02036-08

33. Rodríguez RA, Pepper IL, Gerba CP: "Application of PCR-Based Methods To Assess the Infectivity of Enteric Viruses in Environmental Samples". Applied and Environmental Microbiology 2009, 75: 297-307.

34. Sinclair R, Boone SA, Greenberg D, Keim P, Gerba CP: "Persistence of Category A Select Agents in the Environment". Applied and Environmental Microbiology 2008, 74: 555-563. 10.1128/AEM.02167-07

35. Straub TM, Pepper IL, Gerba CP: Comparison of PCR and cell culture for detection of enteroviruses in sludge-amended field soils and determination of their transport. Appl Environ Microbiol 1995, 61: 2066-2068.

36. Reynolds KA: Integrated cell culture/PCR for detection of enteric viruses in environmental samples. Methods Mol Biol 2004, 268: 69-78.

37. Reynolds KA, Gerba CP, Pepper IL: Detection of infectious enteroviruses by an integrated cell culture-PCR procedure. Appl Environ Microbiol 1996, 62: 1424-1427.

38. Han JH, Yoon JY: "Reusable, polyethylene glycol-structured microfluidic channel for particle immunoassays". Journal of Biological Engineering 2009, 3: 6. 10.1186/1754-1611-3-6

39. Yacoub-George E, Hell W, Meixner L, Wenninger F, Bock K, Lindner P, Wolf H, Kloth T, Feller KA: "Automated 10-channel capillary chip immunodetector for biological agents detection". Biosensors and Bioelectronics 2007, 22: 1368-1375. 10.1016/j.bios.2006.06.003

40. Zhu P, Shelton DR, Karns JS, Sundaram A, Li S, Amstutz P, Tang C-M: "Detection of water-borne E. coli O157 using the integrating waveguide biosensor". Biosensors and Bioelectronics 2005, 21: 678-683. 10.1016/j.bios.2005.01.005

41. Euma NS, Yeom SH, Kwon DH, Kim HR, Kang SW: Enhancement of sensitivity using gold nanorods-Antibody conjugator for detection of E. coli O157:H7. Sensors and Actuators B 2010, 143: 784-788. 10.1016/j.snb.2009.09.054

42. Goeller LJ, Riley MR: Discrimination of bacteria and bacteriophages by Raman spectroscopy and surface enhanced Raman spectroscopy. Applied Spectroscopy 2007, 6: 1-7.

43. Karoonuthaisiri N, Charlermroj R, Uawisetwathana U, Luxananil P, Kirtikara K, Gajanandana O: "Development of antibody array for simultaneous detection of foodborne pathogens". Biosensors and Bioelectronics 2009, 24: 1641-1648. 10.1016/j.bios.2008.08.026

44. Bright KR, Marciano-Cabral F, Gerba CP: Occurrence of Naegleria fowleri in Arizona drinking water supply wells. J Amer Water Works Assoc 2009, 101: 43-50.

45. de los Reyes M, Keegan A, Monis P, Majewski P: Removal of pathogens by functionalised self-assembled monolayers. Journal of Water Supply: Research and Technology-AQUA 2008, 57: 2.

46. Zerda KS, Gerba CP, Hou KC, Goyal SM: "Adsorption of viruses to charged-modified silica". Appl Environ Microbiol 1985, 49: 91-95.

47. Lukasik J, Cheng Y-F, Lu F, Tamplin M, Farrah SR: "Removal of microorganisms from water by columns containing sand coated with ferric and aluminum hydroxides". Water Research 1999, 33: 769-777. 10.1016/S0043-1354(98)00279-6

48. Park SJ, Chang-Gu Lee, Song-Bae Kim: "The role of phosphate in bacterial interaction with iron-coated surfaces". Colloids and Surfaces B: Biointerfaces 2009, 68: 79-82. 10.1016/j.colsurfb.2008.09.015

49. Simoes M, Simoes LC, Vieira MJ: "A review of current and emergent biofilm control strategies". LWT Food Science and Technology 2010, 32: 573-583. 10.1016/j.lwt.2009.12.008

50. Meyer B: "Approaches to prevention, removal and killing of biofilms". International Biodeterioration & Biodegradation 2003, 51: 249-253.

51. Simoes M, Simoes LC, Machado I, Pereira MO, Vieira MJ: Control of flow-generated biofilms using surfactants - evidence of resistance and recovery. Food and Bioproducts Processing 2006, 84: 338-345. 10.1205/fbp06022

52. Hughes KA, Sutherland IW, Jones MV: "Biofilm susceptibility to bacteriophage attack: the role of phage-borne polysaccharide depolymerase". Microbiology 1998, 144: 3039-3047. 10.1099/00221287-144-11-3039

53. Zhao T, Doyle MP, Zhao P: "Control of Listeria monocytogenes in a bio-film by competitive-exclusion microorganisms". Applied and Environmental Microbiology 2004, 70: 3996-4003. 10.1128/AEM.70.7.3996-4003.2004

54. Katz I, Dosoretz CG: "Desalination of domestic wastewater effluents: phosphate removal as pretreatment". Desalination 222(2008):230-242.

55. National Research Council of the National Academies: "Desalination: A National Perspective". Water Science and Technology Board (WSTB), National Academy of Sciences 2008. ISBN-13: 978-0-309-11923-8

56. Lattemann S, Hopner T: "Environmental impact and impact assessment of seawater desalination". Desalination 2008, 220: 1-15. 10.1016/j.desal.2007.03.009

57. Lee J, Oh Byung, Kim Sungyoun, Kim Sung-Jo, Hong Soon, Kim InS: "Fate of Bacillus sp. and Pseudomonas sp. isolated from seawater during chlorination and microfiltration as pretreatments of a desalination plant". Journal of Membrane Science 2010, 349: 208-216. 10.1016/j.memsci.2009.11.045

58. Glater J, Hong S-K, Elimelech M: "The Search for a Chlorine Resistant Reverse Osmosis Membrane". Desalination 1994,95(3):325-345. 10.1016/0011-9164(94)00068-9

59. Hinds B, Chopra N, Rantell T, Andrews R, Gavalas V, Bachas L: Aligned multiwalled carbon nanotube membranes. Science 2004, 303: 62-65. 10.1126/science.1092048

60. Mauter MS, Elimelech M: "Environmental Applications of Carbon Based Nanomaterials". Environmental Science & Technology 2008, 42: 5843-5859.

61. McGinnis RL, Elimelech M: "Global Challenges in Energy and Water Supply: The Promise of Engineered Osmosis". Environmental Science & Technology 2008, 42: 8625-8629.

62. McCutcheon JR, McGinnis RL, Elimelech M: "A Novel Ammonia-Carbon Dioxide Forward (Direct) Osmosis

Desalination Process". Desalination 2005, 174: 1-11. 10.1016/j.desal.2004.11.002

63. McGinnis RL, Elimelech M: "Energy Requirements of Ammonia-Carbon Dioxide Forward Osmosis Desalination". Desalination 2007, 207: 370-382. 10.1016/j.desal.2006.08.012

64. Maugin G, Corsin P: "Concentrate and other waste disposals from SWRO plants: characterization and reduction of their environmental impact". Desalination 2005, 182: 355-364. 10.1016/j.desal.2005.02.033

65. Roberts DA, Johnston EL, Knott NA: "Impacts of desalination plant discharges on the marine environment: A critical review of published studies". Water Research 2010, 44: 5117-5128. 10.1016/j.watres.2010.04.036

66. Miri R, Chouikhi A: "Ecotoxicological marine impacts from seawater desalination plants". Desalination 2005, 182: 403-410. 10.1016/j.desal.2005.02.034

67. El-Sayed AB, El-Fouly MM, Abou El-Nour EAA: Immobilized microalga Scenedesmus sp . for biological desalination of Red Sea water: I. Effect on Growth. Nature and Science 2010,8(9):69-76.

68. Martinez ME, Sanchez S, Jimenez JM, El Yousfi F, Munoz L: Nitrogen and phosphorus removal from urban wastewater by the microalga Scenedesmus obliquus . Bioresource Technol 2000, 73: 263-272. 10.1016/S0960-8524(99)00121-2

69. Mackay MA, Norton RS, Borowitzka LJ: Organic osmoregulatory solutes in cyanobacteria. Journal of General Microbiology 1984, 130: 2177-2191.

70. Logan BE: Microbial Fuel Cells. Hoboken, NJ, USA, Wiley-Interscience; 2008.

71. Devi R, Alemayehu E, Singh V, Kumar A, Mengistie E: Removal of fluoride, arsenic and coliform bacteria by modified homemade filter media from drinking water. Bioresource Technology 2008,99(7):2269-2274. 10.1016/j.biortech.2007.05.002

CHAPTER 32

FIELD ASSESSMENT OF A NOVEL HOUSEHOLD-BASED WATER FILTRATION DEVICE: A RANDOMISED, PLACEBO-CONTROLLED TRIAL IN THE DEMOCRATIC REPUBLIC OF CONGO

Sophie Boisson, Mbela Kiyombo, Larry Sthreshley, Saturnin Tumba, Jacques Makambo, Thomas Clasen

INTRODUCTION

Diarrhoea is responsible for 1.8 million deaths annually, mostly among children under five in developing countries[1]. Much of this disease burden is attributable to unsafe water, poor hygiene and sanitation[2]. An estimated 884 million people worldwide lack access to improved water sources[3]; hundreds of millions more rely on improved sources that are not consistently safe for drinking. Even water that is safe at the point of distribution often becomes contaminated during collection, transport and storage within the home due to poor hygiene conditions and practices[4].

While safe, reliable, piped-in water is an essential goal, treating water at the household or other point of consumption provides a means by which vulnerable populations can improve the quality of their own drinking water[5]. The practice is widespread, with hundreds of millions reporting that they usually treat their water at home before drinking it [6].

There is also evidence that household water treatment is protective against diarrhoea [7], [8],[9] though research suggests that placebo effect and reporting bias play a role in the estimate of effect reported in open trials[10], [11], [12], [13]. Placebo-controlled trials of chlorine-based interventions have been conducted [14], [15], but apart from a recent study in Ghana[16], none have assessed the neutrality of the placebo or the effectiveness of the blinding, and other issues have been raised about their methodological quality. Filters are more difficult to blind among populations relying on unimproved water. If the water is turbid, a placebo that contains no filter medium is readily identified by comparing its effluent with the effective filter. However, a placebo that removes turbidity to ensure blinding will probably also remove pathogens that tend to adhere to the suspended solids; it may also create adsorption sites or promote biofilm adhesion that will also render the "placebo" at least somewhat effective in removing pathogens. To date, the only placebo-controlled trials of household-based filters have been conducted in the United States with municipally treated water that is low in turbidity but also met WHO water quality standards[16], [17], [18], [19]. Thus, these results cannot be generalised to settings with turbid and contaminated water.

Several water treatment methods have been promoted in low-income settings, including disinfection, disinfection/

Citation: Boisson S, Kiyombo M, Sthreshley L, Tumba S, Makambo J, Clasen T (2010) Field Assessment of a Novel Household-Based Water Filtration Device: A Randomised, Placebo-Controlled Trial in the Democratic Republic of Congo. PLoS ONE 5(9): e12613. doi:10.1371/journal.pone.0012613

flocculation, ceramic filtration, solar disinfection and boiling[5].Each has limitations in terms of microbiological effectiveness, cost, acceptability, environmental impact, and sustainability among target populations[20]. Moreover, except for boiling, none of these interventions have achieved scale except in limited settings [6]. This has led to calls for alternative technologies that are effective against the full array of microbial pathogens, that can be deployed and used at a large scale with minimum programmatic support, and that will be embraced by the target population[21].

The Lifestraw Family® is a newly developed household-based gravity filter that employs hollow-fibre membranes to remove waterborne pathogens by ultrafiltration. Independent laboratory testing has shown the device to meet the US Environmental Protection Agency (USEPA) standards for bacteria, viruses and protozoan cysts[22]. The device is designed to treat a minimum of 18,000 L of water and assumed to last for about three years. The manufacturer, Vestergaard-Frandsen SA of Lausanne, Switzerland, plans to sell the filter in large volumes for about US$20.

METHODS

Study design; sample size calculation

The study was designed as a randomised, double-blinded, placebo-controlled trial. Our primary outcome was longitudinal prevalence of diarrhea defined as the number of weeks with diarrhea divided by the total number of weeks under observation. The study was powered to detect a 30% reduction in the mean longitudinal diarrhoea between the two groups. This was a conservative estimate in comparison with the pooled risk reduction of 63% calculated from six previous studies of household filters[8]. The calculation assumed 80% power, $\alpha=0.05$, a baseline longitudinal prevalence of diarrhoea of 5%, and a coefficient of variation of 2. In order to account for potential lost to follow-up (10%) as well as clustering of diarrhoea within household and intermittent surveillance (7-day period prevalence measured repeatedly once a month over the 12-month follow-up period) (10%), we estimated that we needed at least 600 individuals in each arm. Assuming a mean of 5 persons per household, the number of households to be recruited was approximately 120 per arm, or 240 households in total.

Setting and Participant Eligibility

The study was conducted from April 2008 to July 2009 in the rural health zone of Bibanga, 80 km from the city of Mbuji-Mayi in the eastern province of Kasai, the Democratic Republic of Congo (DRC). Despite abundant water resources, more than three quarters of the population in rural areas in the DRC rely on unimproved water sources for drinking, mainly surface water and unprotected springs[23]. With the assistance of the Presbyterian Church of Kinshasa, which has been supporting community health programmes in this area for many years, and staff at the health zone level, we identified possible study sites. Selected communities relied on unimproved water sources that tested over 1000 thermotolerant coliforms (TTC)/100 ml, reported low use of household water treatment, were easily accessible all year round from the reference hospital of Bibanga where the field team was established, and were motivated to take part in the project. In order to meet the sample size requirements, the study was conducted in two neighbouring villages.

Intervention

Each intervention household received a Lifestraw Family filter and each control household received a placebo. The Lifestraw Family is a gravity-fed microbiological water purifier. Water is poured into a 2.5 L plastic vessel, passes through a 27-µm pre-filter, and flows down a 1 m long plastic pipe before passing through the filtration cartridge comprised of hollow-fibres with a 20-nm pore size. The top vessel contains a slow eluding chlorine tablet designed to prevent biofilm formation and increase the life of the cartridge. Treated water is accessed from the side of the cartridge via a tap. The device is cleaned daily by rinsing the pre-filter and backwashing the cartridge using a squeeze-pump and outlet valve mounted on the bottom of the cartridge. The device is designed to treat at least 18,000 L of water with a flow rate of approximately 150 ml per minute or 9 L per hour. In the laboratory, the filter was found to meet the USEPA standards for microbiological water purifiers by reducing bacteria by 6.9 logs, viruses by 4.7 logs and protozoan cysts by 3.6 logs [22].

Placebo

The placebo had the same configuration, appearance and external components as the Lifestraw Family except that (i) the chlorine tablet was removed from the upper vessel to prevent possible microbicidal action, (ii) the filtration membranes were replaced by some extra piping to imitate the weight and effluent flow rate of the real cartridge, and (iii) the 27-µm screen on the pre-filter was removed to minimise retention of microbes adhering to suspended solids. Three weeks of testing in the laboratory confirmed that the placebo removed no bacteria, viruses and protozoan cysts from test water. Despite the challenge in blinding household filters, we determined after piloting that blinding the intervention would be feasible in our study area because the water had low turbidity, ranging from <5 nephelometric turbidity units (NTU) for most of the year to 10 NTU during heavy rains.

Enrolment, Baseline Survey, Randomization and Filter Deployment

After discussing the proposed study with community leaders and obtaining consent from the heads of households, a baseline survey was undertaken in April 2008 to collect information on demographics, socio-economic characteristics, and water, hygiene and sanitation practices. Data collection tools were translated in Tshiluba, the local language, and piloted before use. Following the baseline survey, households were randomly assigned to one of the two groups using a random number generator. Randomisation was stratified by village and was conducted by the trial manager who played no part in the collection of the data. Both the intervention and the placebo were distributed door-to-door by five trained field workers who were unaware of whether the device was an active filter or a placebo. Householders were trained on use and maintenance of the device according to the manufacturer's instructions. They were advised to drink filtered water directly from the tap and not to store filtered water in order to prevent recontamination.

The start of follow-up period was delayed by two months due to initial technical problems with the filters.

Blinding

The allocation sequence was concealed from both field investigators and the study population. In order to blind the intervention among assessors, field workers were divided into two teams. The team responsible for assessing health outcomes was neither involved in the distribution of the filters at the commencement of the trial nor in the assessment of the filter performance and use during follow-up. Any questions from the householders that were related to the filter were referred to and dealt with by the filter assessment team.

Outcome Assessment

Diarrhoea

Investigators interviewed the female head of household or primary care giver of young children once each month over a 12-month period. They recorded any diarrhoea cases in the preceding seven days. Diarrhoea was defined as three or more loose stools passed within a 24-hour period. In an effort to further obscure the outcome of interest from the target population, field assessors also inquired about and recorded presence of fever and cough within the past seven days. Children with diarrhoea were given oral rehydration sachets and instructions on how to use them. When necessary, they were referred to the closest community health post to receive medical care free of charge. Fever and cough were also treated among young children.

Filter Monitoring

Each month, a random sample of 30 filters and 30 placebos (25% of the total number distributed) was monitored. At each household visit, field workers noted the location and condition of the filter and recorded if the respondent was able to use and clean the filter correctly. Filter components found to be damaged were replaced. Flow rate was monitored by filling the top container with 2.5 L of water, opening the tap and measuring the time necessary to fill a 125 ml container with water. The flow rate was expressed in ml per minute.

Water Quality

Influent and effluent paired water samples were collected for each of the selected devices. If the respondent mentioned storing the water once filtered, a third sample was collected from the container designated as the treated water storage vessel. All samples were collected in sterile 125 ml Nalgene sampling bottles and assessed for thermotolerant coliforms (TTC) within 4 h after collection. Microbiological assessment was performed using the membrane filtration technique (APHA *Standard Methods*) on membrane lauryl sulphate medium (Oxoid Limited, Basingstoke, Hampshire, UK) using a DelAgua field incubator (Robens Institute, University of Surrey, Guilford, Surrey, UK). Microbiological performance of the filters was expressed in terms of log reduction value (LRV) calculated as the log of the influent concentration divided by the effluent concentration ($\log_{10}$ influent/effluent).

Compliance

Cross-sectional surveys were conducted among each household eight and fourteen months after distribution. Participants were classified as current users if they reported using the filter 'today or yesterday' and if the field investigator found the filter hung for use with water in the top vessel of the device. Consistency of use was estimated by asking the respondent if he/she had drunk unfiltered water within the previous day. The survey covered further aspects on use and acceptability.

Blinding Assessment

Immediately following the conclusion of the follow-up period, we assessed the effectiveness of blinding among participants. Blinding indices were calculated using methods developed by James and colleagues [24]and Bang and colleagues[25] Female heads of household or primary care giver were asked to identify which device they had received. Surveys were targeted at the respondents to the health surveys because they would be most likely to be influenced by their belief in treatment assignment.

Data Analysis

The analysis of the primary outcome was on an intention-to-treat basis. We used Poisson regression with robust standard errors to estimate the effect of the intervention on the longitudinal prevalence of diarrhea and other health outcomes[26]. We used generalized estimating equations (GEE) to account for clustering at the household level. Categorical data were compared using a Chi square or a Fisher's exact test where appropriate. Continuous variables were compared with a Student t test. Statistical analyses of microbiological data were conducted after $\log_{10}$ transformation of TTC counts to normalize the distribution. Data analysis was conducted in Stata (Stata Corporation, College Station, Texas, US).

Ethics

The study was reviewed and approved by the ethics committee at the London School of Hygiene and Tropical Medicine and the ethics committee at the School of Public Health in Kinshasa. Written consent to participate in the research was obtained from community leaders and the head of each participating household. Investigators explained that half of the study population would be receiving effective microbiological purifiers while the others would receive placebos and that householders should continue their existing water management practices since their device may not be protective against microbial contamination. At the conclusion of the follow-up period, all placebo filters were replaced by effective filters. Following the completion of the study, the results were communicated to all study participants.

RESULTS

Participant Flow

259 households initially volunteered to participate in the study. Nineteen households were excluded because they did not reside in the selected villages; they relied primarily on spring water for drinking, or subsequently elected not to participate. A total of 240 households were enrolled; 120 were assigned to

receive the Lifestraw Family filter and 120 the placebo. Over the 12-month follow-up period, data were collected for 11,236 (81.8%) possible person-weeks of observation. Data were missing for 2492 weeks (18.2%) due primarily to participants leaving the study area or being absent at the time of visit (Figure 1). Over the study period, twenty participants died, six of them were children under the age of five. The number of deaths was 12 in the intervention group and 8 in the control group (p=0.27).

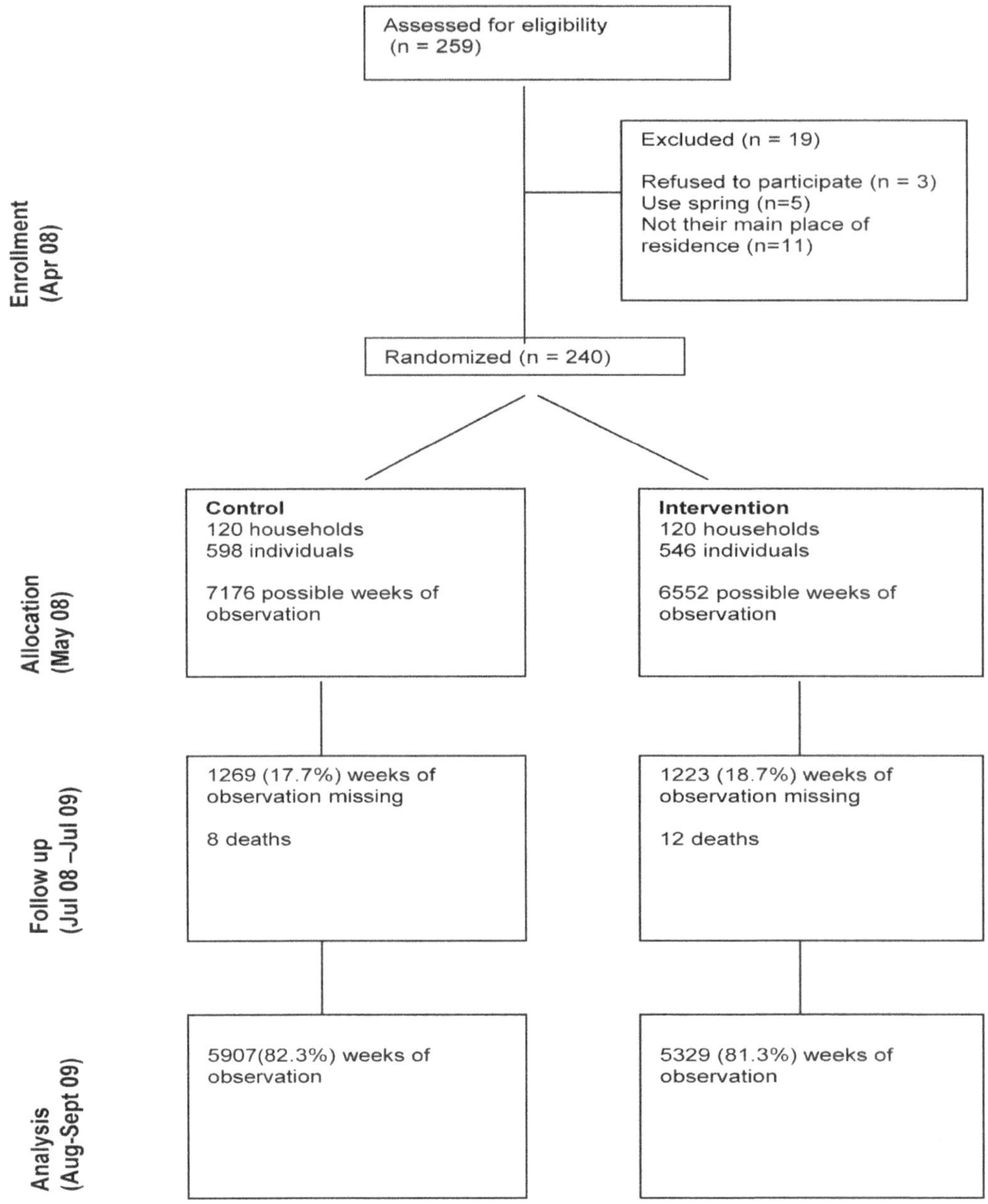

Figure 1: CONSORT diagram showing the flow of participants through the trial.

Baseline Surveys

Intervention and control groups were similar in terms of demographic and socio-economic characteristics and hygiene and sanitation practices (Table 1). Almost all households primarily used river water for drinking. However, intervention households were more likely to store their water in clay pots and access it by dipping a cup into the container compared with control households who often used jerrycans. Only four households reported treating their water sometimes or rarely by boiling or adding bleach. Only 37% of households had a latrine and 51% had soap present in the house at the time of visit (Table 1).

Table 1: Baseline characteristics of participating households

	Control		Intervention		Total	
	N	%	N	%	N	%
Demographic and socio-economic						
Number of households	120	(50)	120	(50)	240	(100)
Number of persons	598	(52.3)	546	(47.7)	1144	(100)
Number of households with children <5	66	(55)	57	(47.5)	123	(51.2)
Number of children <5	105	(17.6)	85	(15.8)	190	(16.6)
Mean number of persons per household	5.0		4.5		4.8	
Mean number of rooms in the house	2.2		2.3		2.3	
Respondent is female	76	(63.3)	76	(63.3)	152	(63.3)
Mean age of respondent	37.5		40.8		39.1	
Level of education						
No formal education	47	(39.2)	38	(31.7)	85	(35.4)
Primary	44	(60.3)	45	(54.9)	89	(57.4)
Secondary	29	(39.7)	36	(43.9)	65	(41.9)
Higher	0	(0)	1	(1.2)	1	(0.6)
Owns						
House	113	(94.2)	116	(96.7)	229	(95.4)
Land	115	(95.8)	117	(97.5)	232	(96.7)
Livestock	59	(49.2)	64	(53.8)	123	(51.5)
Radio	27	(22.7)	34	(28.3)	61	(25.5)
Phone	10	(8.3)	16	(13.3)	26	(10.8)
Bicycle	18	(15)	16	(13.3)	34	(14.2)
Hygiene and sanitation						
Use soap to wash hands	54	(45)	54	(45)	108	(45)
Presence of soap at the time of visit	65	(54.2)	59	(49.2)	124	(51.7)
Received hygiene advice in past 6 months	4	(3.4)	10	(8.4)	14	(5.9)
Presence of latrine	47	(39.2)	41	(34.2)	88	(36.7)
Water handling practices						
Primary source of drinking water						
River	120	(100)	117	(97.5)	237	(98.7)
Rainwater	44	(36.7)	46	(38.3)	90	(37.5)
Spring	15	(12.5)	19	(15.8)	34	(14.2)
Type of drinking water container						
Clay pot	68	(56.7)	83	(69.2)	151	(62.9)
Jerry can	50	(41.7)	30	(25)	80	(33.3)
Other	2	(1.7)	7	(5.8)	9	(3.7)
Vessel opening						
Wide mouth	71	(59.2)	92	(76.7)	163	(67.9)
Narrow mouth	49	(40.8)	28	(23.3)	77	(32.1)
Storage vessels covered	111	(93.3)	113	(95.0)	224	(94.1)
Means of obtaining water						
Pour	48	(41.0)	27	(23.3)	75	(32.2)
Dip	69	(59.0)	89	(76.7)	158	(67.8)
Treat water*	3	(2.5)	1	(0.8)	4	(1.7)

*Treat water sometimes (n = 1) or rarely (n = 3). Treatment methods boil (n = 2), bleach (n = 1), water settle (n = 1).

doi:10.1371/journal.pone.0012613.t001

Diarrhoea Surveillance

At baseline, the prevalence of diarrhoea was similar in both groups (12.6% versus 10.6% for control and intervention groups, respectively). Over the 12-month follow-up period, participants of all ages who received the active filter experienced 15% fewer weeks with diarrhoea compared to those who received a placebo (mean, 2.66 versus 3.15, respectively). However, the confidence interval of the longitudinal prevalence ratio (LPR) adjusted for clustering within the household (LPR 0.85, 95% CI 0.61; 1.20) was wide and included 1. The longitudinal prevalence ratio among children under five was 0.85 (95%CI 0.56; 1.28). Figure 2 shows the prevalence of diarrhoea between intervention and control over time. We observed no difference in the mean longitudinal prevalence of fever (LPR 0.99; 95% CI 0.80; 1.22) or cough (LPR 0.99; 95% CI 0.81; 1.22) between the two groups. Health outcome data are presented in Table 2.

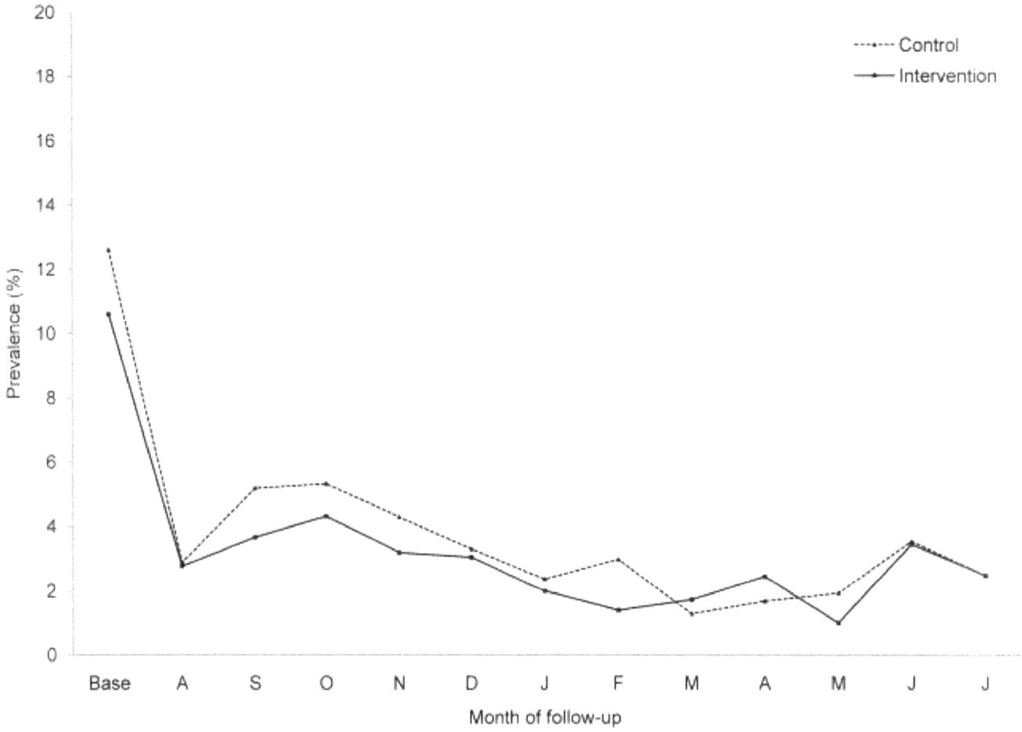

Figure 2: Prevalence of diarrhoea over the course of the study among participants of all ages.

Table 2: Longitudinal prevalence of diarrhoea and other health conditions by age and treatment group

	Mean longitudinal prevalence						LPR (95% CI)	LPR* (95%CI)
	Control			Intervention				
	Weeks of illness	Person-weeks of observation	% Weeks ill	Weeks of illness	Person-weeks of observation	% Weeks ill		
Diarrhoea								
<5	96	1072	**8.96**	60	801	**7.49**	0.84 (0.61; 1.14)	0.85 (0.56; 1.28)
5–15	31	1880	**1.65**	29	1765	**1.64**	1.00 (0.60; 1.65)	0.91 (0.49; 1.67)
>15	59	2945	**2.00**	52	2752	**1.89**	0.94 (0.65; 1.36)	0.95 (0.61; 1.57)
All ages**	186	5907	**3.15**	142	5329	**2.66**	0.85 (0.68; 1.05)	0.85 (0.61; 1.20)
Fever								
<5	249	1072	**23.23**	187	801	**23.35**	1.00 (0.85; 1.19)	1.02 (0.79; 1.30)
5–15	99	1880	**5.27**	123	1765	**6.97**	1.32 (1.02; 1.71)	1.28 (0.89; 1.85)
>15	226	2945	**7.67**	188	2752	**6.83**	0.89 (0.74; 1.07)	0.91 (0.68; 1.22)
All ages**	576	5907	**9.75**	500	5329	**9.38**	0.96 (0.86; 1.08)	0.99 (0.80; 1.22)
Cough								
<5	196	1072	**18.30**	162	801	**20.20**	1.11 (0.92; 1.33)	1.11 (0.85; 1.43)
5–15	163	1880	**8.70**	142	1765	**8.00**	0.93 (0.75; 1.50)	0.89 (0.63; 1.27)
>15	192	2945	**6.50**	201	2752	**7.30**	1.12 (0.93; 1.35)	1.07 (0.82; 1.39)
All ages**	551	5907	**9.33**	505	5329	**9.46**	1.01 (0.90; 1.14)	0.99 (0.81; 1.22)

*Adjusted for clustering within household.
**Age missing for 3 participants.
doi:10.1371/journal.pone.0012613.t002

Water Quality

Each device was tested on average 3 times during follow-up. 580 (81%) of the total possible paired water samples were collected. Missing samples are due to householders being absent or not being in possession of their filter at the time of visit. Source drinking water was highly contaminated, with 75% of household samples showing contamination levels above 1000 TTC/100 ml (Figure 3). The active filter achieved a LRV of 2.98 (95% CI 2.88, 3.08), removing about 99.8% of the indicator bacteria. Overall, 64% of water samples treated with filter were free of TTC and 27% had TTC levels between 1–10 TTC/100 ml. None of the filters produced water with >100 TTC/100 ml consistently over the three visits. Samples from placebos were also contaminated, with 73% of the water samples containing between 100–1000 TTC/100 ml. However, unlike the results from laboratory testing that showed the placebo to be microbiologically ineffective, results from the field showed that the placebo actually removed more than 90% of the TTC from source water (LRV 1.05, 95% CI: 0.93, 1.16).

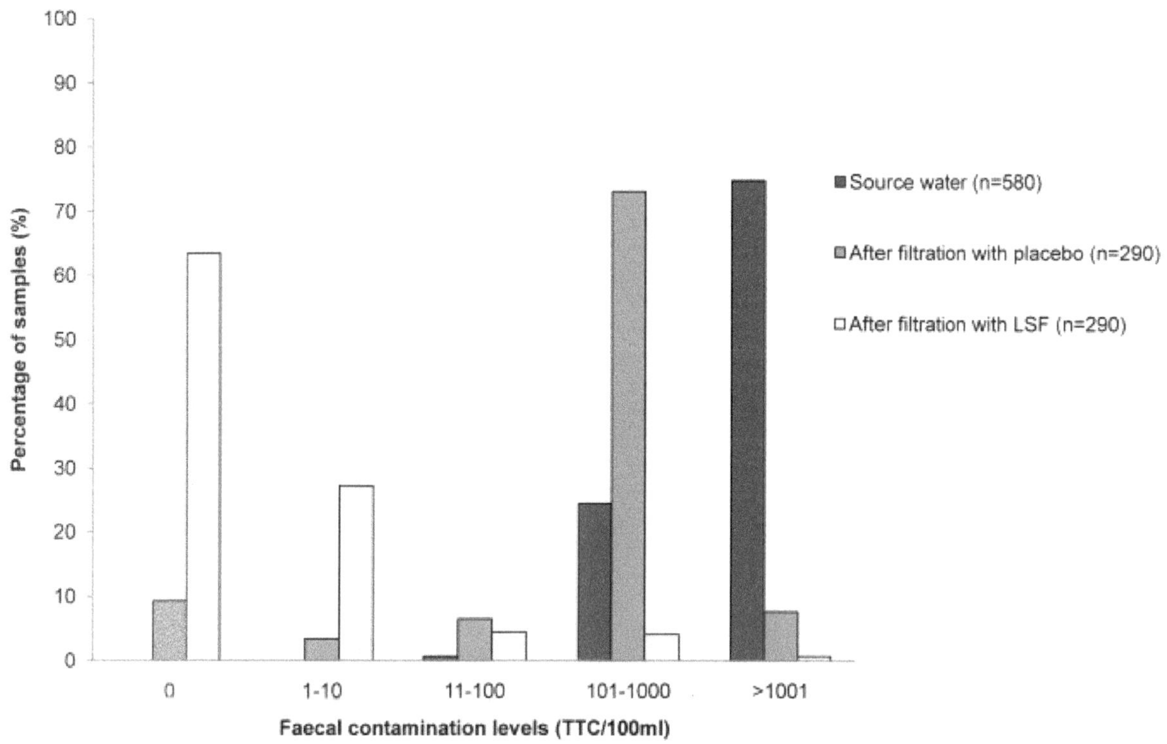

Figure 3: Percentage of water samples by level of contamination (TTC/100 ml).

Flow Rate

The mean flow rate of the filters over the study period was 202 ml/min (95% CI 198, 206) or 12 L/hour. It declined slightly over time (−1.5 ml/min per month, $p<0.002$).

Operation and Maintenance; Acceptability

Over half of the respondents (56%) correctly demonstrated how to clean the filters. The pre-filter was cleaned at each use (40%) or once a day (41%), wheareas the cartrige was generally backwashed once a day (67%). Overall, 36 (12.4%) of the 290 active filters tested were found damaged during visits, mainly due to rodents chewing on the soft hoses (n=35). Intervention households reported liking the filter due to improved aesthetics (88%), taste (92%), odour (56%) and health (35%). Reasons for dissatisfaction were slow flow rate (87%), small size of the top container (85%) and problems with rats (44%).

Compliance

Eight months after distribution, 183 (76%) of the households were present at the time of visit and were still in possession of their filter. 68% of respondents in the intervention group could be defined as current users against 48% in the placebo group ($p<0.001$). However, nearly all adults (83%) and young children (95%) also reported drinking untreated water in the previous day. Fourteen months after distribution, the proportion of current users was slightly higher in both groups (76% versus 69% among intervention and control groups, respectively). Additional details about use are included in Table 3. Subgroup analysis showed no evidence of an association between use and diarrhoea morbidity (Table 4).

Table 3: Description of use among study participants

	Control		Intervention		Total	
	n	%	n	%	n	%
MONTH 8						
Last use (n = 183)*						
Previous day	44	(48.3)	63	(68.5)	107	(58.5)
Previous week	30	(33.0)	14	(15.2)	44	(24.0)
>1 week ago	17	(18.7)	15	(16.3)	32	(17.5)
Consistency of use on previous day (n = 107)						
Respondent drank unfiltered water	43	(97.7)	46	(73.0)	89	(83.2)
Children (<5) drank unfiltered water	31	(93.9)	39	(95.1)	70	(94.6)
Filter accessible to young children	1	(2.3)	6	(9.5)	7	(6.5)
Store filtered water for young children	4	(12.9)	8	(19.5)	12	(16.7)
Additional details on use in previous day (n = 107)						
Respondent drank unfiltered water when						
In the field	33	(76.7)	39	(78.3)	72	(77.9)
In a hurry to drink	30	(69.8)	33	(71.7)	63	(70.8)
Away from village	16	(37.2)	15	(32.6)	31	(34.8)
Other	3	(7.0)	12	(26)	15	(16.8)
Children drank unfiltered water when						
Person operating the filter not present	21	(67.7)	31	(79.5)	52	(74.3)
In a hurry to drink	11	(35.5)	23	(59.0)	34	(48.6)
Away from home	10	(32.3)	13	(33.3)	23	(32.9)
Other	5	(16.1)	7	(17.9)	12	(17.1)
Did not store filtered water for children:						
No container	17	(68)	28	(87.1)	45	(78.9)
Lock the door	6	(24)	3	(9.3)	9	(15.8)
Don't want to always filter, too slow	2	(8)	0	(0)	2	(3.5)
Told not to store water	0	(0)	1	(3.1)	1	(1.7)
MONTH 14						
Last use (n = 190)**						
Previous day	63	(69.2)	75	(75.8)	138	(72.6)
Previous week	14	(15.4)	11	(11.1)	25	(13.2)
>1 week ago	14	(15.4)	13	(13.3)	27	(14.2)

*197 (82%) households present at the time of visit; 183 (93%) of them were still in possession of the filter and ever used it.
**203 (85%) households present at the time of visit; 192 (94%) of them were still in possession of the filter and ever used it + answer missing for 2 households.

Table 4: Longitudinal prevalence of diarrhoea stratified by reported last time of use

	Mean longitudinal prevalence of diarrhoea						LPR (95% CI)
	Control			Intervention			
	Weeks of illness	Person-weeks of observation	% weeks ill	Weeks of illness	person-weeks of observation	% weeks ill	
8 months							
User	71	2475	**2.87**	74	3155	**2.35**	0.82 (0.59; 1.13)
Non-user	68	2420	**2.81**	41	1319	**3.11**	1.11 (0.75; 1.62)
14 months							
User	102	3463	**2.95**	99	3894	**2.54**	0.86 (0.66; 1.10)
Non-user	49	1642	**2.98**	27	1025	**2.63**	0.88 (0.55; 1.40)

Blinding Status

Table 5 shows respondent guesses for each treatment assignment groups. James' method, similar to the kappa statistics, produced a blinding index (BI) score of 0.42 (95%CI 0.38; 0.46). A score of 0 means that all respondents guessed correctly, 1 indicates that all respondents guessed incorrectly and 0.5 indicates random guessing. Bang's method calculates the proportion of correct guesses beyond chance in each treatment group. Bang's BI was 0.96 (95%CI 0.90; 0.99) for the intervention group and −0.63 (95%CI −0.73; −0.53) for the placebo-controlled group. Bang's blinding index varies from −1 to 1. 1 indicates complete lack of blinding, −1 opposite guess about treatment assignment and 0 random guessing. Subgroup analysis showed no evidence of an association between diarrhoea and respondents' guesses.

Table 5: Blinding status of respondents by group assignment at the end of the study

	Group assignment					
Guess	Placebo*		Lifestraw Family*		Total*	
Placebo	17	(18.3)	2	(2.0)	19	(9.9)
Lifestraw Family	74	(79.6)	97	(98.0)	171	(89.1)
Don't know	2	(2.1)	0	(0)	2	(1.0)
Total**	93	(100.0)	99	(100.0)	192	(100.0)

*N (%) - number of respondents and percentage in each group.
**192 (80%) households present at the time of interview and still in possession of the filter.

DISCUSSION

We undertook the first double-blinded, placebo-controlled trial of household-based water filters in a low-income setting with water known to be contaminated with faecal pathogens. This design sought to assess the impact of the intervention in the absence of respondents' bias that is common in open trials. Due to challenges of developing a placebo in such settings and to successfully blinding the intervention, we monitored placebo performance and conducted a post-intervention assessment of blinding among study participants. Filter performance and health impact were monitored for a full year to account for seasonal variations and minimise the potential for exaggerated health impact often associated with shorter-term trials [9].

After adjusting for clustering, members of intervention households had 15% fewer weeks of diarrhoea than those of control households, but the confidence intervals indicated little statistical support (longitudinal prevalence ratio 0.85, 95%CI: 0.61 to 1.20). With the exception of a recent study in the United States among an elderly population[17], this finding is consistent with other placebo-controlled trials of household water treatment interventions [14], [15], [16] which found no protective effect against diarrhoea. However, as we have observed elsewhere, those studies may have had insufficient power to identify a statistically significant impact on diarrhoea[27]. Our sample size also was not sufficiently large to detect a statistically significant difference in diarrhoea of 15%. Moreover, the baseline prevalence of diarrhoea was lower than anticipated and the clustering effect due to repeated measurement and household randomisation was higher. Pos-hoc sample size calculations indicated that we would have needed a study approximately ten times larger to achieve statistical significance.

Moreover, the placebo was not microbiologically neutral, as it removed about 90% of faecal bacteria from the source water used by control households. The reasons for this apparent effectiveness are not clear. Field staff responsible for water quality testing were extensively trained and supervised throughout the study, thereby minimizing the risk of measurement errors. One of the most plausible explanations is the formation of a biofilm resulting from adhesion of suspended solid particles and bacteria on the inner surface of the plastic pipe forming the placebo cartridge. The effectiveness of the placebo rendered our trial a comparison between a 1-log filter and a 3-log filter. Studies have reported an association between 1 log removal of faecal bacteria from drinking water and a reduction in diarrhoeal disease [28], [29]. Our results may therefore understate the effectiveness of the active filter if it were compared to a true placebo. These results suggest that in this setting with relatively high levels of microbial contamination in source water, a filter of superior microbiological performance may be more effective at preventing diarrhoea than one that removes only 90% of waterborne pathogens. This finding, if validated in future studies, would support the need for high performance standards in water treatment devices in order to optimize health benefits.

The blinding of the intervention was not successful. In both treatment groups, the vast majority of survey respondents believed that they had received the active filter, although this proportion was significantly lower in the placebo group. Unsuccessful blinding means that we cannot rule out the possibility that the observed effect on diarrhoea is unbiased. However, the interpretation of blinding indices is not always clear[30]. The fact that a large proportion of control households remained blinded throughout the trial suggests that respondents' bias may have at least been partly reduced. The smaller effect size we observed here may be indicative of a less biased estimate compared with open trials. Our estimate is similar to the pooled estimate of effect of open trials of ceramic filters after adjustment for lack of blinding[13]. The fact that 'control' health conditions (fever and cough) remained unchanged by the intervention also suggests that blinding may have been effective, although the usefulness of this approach to detect the presence of respondents' bias has not been validated. Including a third arm with no intervention would have provided a better understanding of the role of bias in this study.

Under field conditions, the Lifestraw Family filters were effective in removing faecal bacteria from source water. Two-thirds of filtered water samples were free of faecal coliforms while most of the remaining samples had low levels of contamination. The fact that specific filters did not consistently produce contaminated water suggests that contamination may have occurred during collection of the sample, perhaps from the tap. The flow rate was higher than that observed in laboratory conditions, possibly due to lower water turbidity at the study site (compared to lab testing at 15 NTU) or inconsistent use by householders. The damage rate was high although the most common problems were due to rats eating the soft plastic components.

Eight months after distribution, two-thirds of the respondents met the study's definition of current users, although almost none of them drank filtered water exclusively. This pattern of use was seen among both adults and children under five. Participants drank unfiltered water when spending time outside their home, but also when they felt eager to drink and did not want to wait for filtration. Young children did not have access to the filter when their parents were away from home. In accordance with the manufacturer's instructions, householders were advised to use water directly from the filter and not to store treated water due to the risk of recontamination. Consistent with these instructions, almost none of the households stored filtered water for their children, though many lacked a storage container even if they had chosen to do so. The manufacturer has advised that in future deployment of the filters, it will consider changes in instructions to encourage safe storage of treated water or provide a storage vessel for the filtered water to help increase exclusive consumption of treated water, especially by this vulnerable group of young children. However, there is also evidence that even occasional consumption of untreated water may eliminate the protective effect of water treatment[31] and changes to the configuration of the filter may not be sufficient to increase exclusive use unless accompanied by fundamental changes in behaviour to increase compliance.

Our study had certain additional limitations. The study sites were not randomly selected, but were chosen based on eligibility criteria that included high levels of faecal contamination in source water and high prevalence of diarrhoea at baseline. Accordingly, these results are not necessarily generalizable to other populations in the Congo or beyond. Second, the use of a seven-day recall period is known to produce less precise estimates compared with a 48-hour recall period[32].

Our study provides little evidence of a protective effect of the filter against diarrhoea. Nevertheless, an effect of 15%, which we observed but could not confirm here, would represent a substantial impact on diarrhoea, a major killer of young children. Future studies with sufficient power to detect this effect size will be necessary to determine the magnitude of any effect against a neutral placebo and to confirm that the effect is not attributable to chance. Our study also demonstrates the need to monitor placebo performance and the challenge of blinding household-based water treatment interventions under adverse conditions.

ACKNOWLEDGMENTS

We are grateful to the communities of Bakua Tshitongo and Bakua Masela for their participation in the study. We thank the field team: Emery Lukusa, Gaston Kalonji, Joseph Lukonga, Charly Ntanga, Urbain Somue, Jean Mbikayi, Justin Kabongo, Rockie Tshimanga, Jean-Michel Mutombo. We thank the staff from the AXxes project in Kinshasa: Albert Kalonji, Joaquim Lubiba, Ben Munongo and the AXxes coordination staff at Mbuji-Mayi: Bernard Ngoyi, Berry Biangwa, Teddy Mutebwa. We wish to thank the Presbyterian Church of Kinshasa (CPK), in particular Mbuyi Mbiayi, Esperant Kabalosh, Rossi Kajingu. We also thank Wolf-Peter Schmidt for his statistical support. Finally, we wish to thank John M. Colford of the University of California-Berkeley School of Public Health for his helpful comments on the manuscript.

REFERENCES

1. WHO The World Health Report (2005) Make every mother and child count. Geneva: World Health Organisation.
2. Pruss-Unstun A, Bos R, Gore F, Bartram J (2008) Safer water, better health: costs, benefitsandsustainability of interventions to protect and promote health. Geneva: World Health Organization.
3. WHO/UNICEF (2008) Joint Monitoring Programme for water supply and sanitation (JMP). Progress on Drinking-water and Sanitation: special focus on sanitation. Geneva: The World Health Organization and the United Nations Children's Fund.
4. Wright J, Gundry S, Conroy R (2004) Household drinking water in developing countries: a systematic review of microbiological contamination between source and point-of-use. Trop Med Int Health 9: 106–117.
5. Sobsey M (2002) Managing water in the home: accelerating health gains from improved water supply. Geneva: World Health Organization.
6. Rosa G, Clasen T (2010) Estimating the scope of household water treatment in low- and medium-income countries. Am J Trop Med Hyg 82: 289–300.
7. Fewtrell L, Kaufmann RB, Kay D, Enanoria W, Haller L, et al. (2005) Water, sanitation, and hygiene interventions to reduce diarrhoea in less developed countries: a systematic review and meta-analysis. Lancet Infect Dis 5: 42–52.
8. Clasen T, Roberts I, Rabie T, Schmidt W, Cairncross S (2006) Interventions to improve water quality for preventing diarrhoea. Cochrane Database Syst Rev 3: CD004794.
9. Arnold BF, Colford JM Jr (2007) Treating water with chlorine at point-of-use to improve water quality and reduce child diarrhea in developing countries: a systematic review and meta-analysis. Am J Trop Med Hyg 76: 354–364.
10. Schmidt WP, Cairncross S (2009) Household water treatment in poor populations: is there enough evidence for scaling up now? Environ Sci Technol 43: 986–992.
11. Boisson S, Schmidt WP, Berhanu T, Gezahegn H, Clasen T (2009) Randomized controlled trial in rural Ethiopia to assess a portable water treatment device. Environ Sci Technol 43: 5934–5939.
12. Waddington HS, White BH, Fewtrell L (2009) Water, Sanitation and Hygiene Interventions to Combat Childhood Diarrhoea in Developing Countries. (3ie). International Initiative for Impact Evaluation.
13. Hunter PR (2009) Household Water Treatment in Developing Countries: Comparing Different Intervention Types Using Meta-Regression. Environ Sci Technol 43: 8991–8997.

14. Kirchhoff LV, McClelland KE, Do Carmo Pinho M, Araujo JG, De Sousa MA, et al. (1985) Feasibility and efficacy of in-home water chlorination in rural North-eastern Brazil. J Hyg (Lond) 94: 173–180.
15. Austin (1993) Investigations of in-house water chlorination and its effectiveness for rural areas of the Gambia [Dissertation]. New Orleans, LA: Tulane University School of Public Health and Tropical Medicine.
16. Jain S, Sahanoon OK, Blanton E, Schmitz A, Wannemuehler KA, et al. (2010) Sodium dichloroisocyanurate tablets for routine treatment of household drinking water in periurban Ghana: a randomized controlled trial. Am J Trop Med Hyg 82: 16–22.
17. Colford JM Jr, Hilton JF, Wright CC, Arnold BF, Saha S, et al. (2009) The Sonoma Water Evaluation Trial: A Randomized Drinking Water Intervention Trial to Reduce Gastrointestinal Illness in Older Adults. Am J Public Health 11: 1988–1995.
18. Colford JM Jr, Rees JR, Wade TJ, Khalakdina A, Hilton JF, et al. (2002) Participant blinding and gastrointestinal illness in a randomized, controlled trial of an in-home drinking water intervention. Emerg Infect Dis 8: 29–36.
19. Colford JM Jr, Wade TJ, Sandhu SK, Wright CC, Lee S, et al. (2005) A randomized, controlled trial of in-home drinking water intervention to reduce gastrointestinal illness. Am J Epidemiol 161: 472–482.
20. Sobsey MD, Stauber CE, Casanova LM, Brown JM, Elliott MA (2008) Point of use household drinking water filtration: A practical, effective solution for providing sustained access to safe drinking water in the developing world. Environ Sci Technol 42: 4261–4267.
21. Clasen T (2008) Scaling up Household Water Treatment: Looking Back, Seeing forward. Geneva: World Health Organization.
22. Clasen T, Naranjo J, Frauchiger D, Gerba C (2009) Laboratory assessment of a gravity-fed ultrafiltration water treatment device designed for household use in low-income settings. Am J Trop Med Hyg 80: 819–823.
23. International MdPeM (2008) Enquête Démographique et de Santé, République Démocratique du Congo 2007. Calverton, MD: Ministère du Plan et Macro International.
24. James KE, Bloch DA, Lee KK, Kraemer HC, Fuller RK (1996) An index for assessing blindness in a multi-centre clinical trial: disulfiram for alcohol cessation—a VA cooperative study. Stat Med 15: 1421–1434.
25. Bang H, Ni L, Davis CE (2004) Assessment of blinding in clinical trials. Control Clin Trials 25: 143–156.
26. Zou G (2004) A modified poisson regression approach to prospective studies with binary data. Am J Epidemiol 159: 702–706.
27. Clasen T, Bartram J, Colford J, Luby S, Quick R, et al. (2009) Comment on "Household water treatment in poor populations: is there enough evidence for scaling up now?". Environ Sci Technol 43: 5542–5544; author reply 5545–5546.
28. Stauber CE, Elliott MA, Koksal F, Ortiz GM, DiGiano FA, et al. (2006) Characterisation of the biosand filter for E. coli reductions from household drinking water under controlled laboratory and field use conditions. Water Sci Technol 54: 1–7.
29. Tiwari SS, Schmidt WP, Darby J, Kariuki ZG, Jenkins MW (2009) Intermittent slow sand filtration for preventing diarrhoea among children in Kenyan households using unimproved water sources: randomized controlled trial. Trop Med Int Health 14: 1374–1382.
30. Park J, Bang H, Canette I (2008) Blinding in clinical trials, time to do it better. Complement Ther Med 16: 121–123.
31. Hunter PR, Zmirou-Navier D, Hartemann P (2009) Estimating the impact on health of poor reliability of drinking water interventions in developing countries. Sci Total Environ 407: 2621–2624.
32. Zafar SN, Luby SP, Mendoza C (2009) Recall errors in a weekly survey of diarrhoea in Guatemala: determining the optimal length of recall. Epidemiol Infect 1–6.

CHAPTER 33

SOLAR DRINKING WATER DISINFECTION (SODIS) TO REDUCE CHILDHOOD DIARRHOEA IN RURAL BOLIVIA: A CLUSTER-RANDOMIZED, CONTROLLED TRIAL

Daniel Mäusezahl, Andri Christen, Gonzalo Duran Pacheco, Fidel Alvarez Tellez, Mercedes Iriarte, Maria E. Zapata, Myriam Cevallos, Jan Hattendorf, Monica Daigl Cattaneo, Benjamin Arnold, Thomas A. Smith, John M. Colford Jr

EDITORS' SUMMARY

Background

Thirsty? Well, turn on the tap and have a drink of refreshing, clean, safe water. Unfortunately, more than one billion people around the world don't have this option. Instead of the endless supply of safe drinking water that people living in affluent, developed countries take for granted, more than a third of people living in developing countries only have contaminated water from rivers, lakes, or wells to drink. Because of limited access to safe drinking water, poor sanitation, and poor personal hygiene, 1.8 million people (mainly children under 5 years old) die every year from diarrheal diseases. This death toll could be greatly reduced by lowering the numbers of disease-causing microbes in household drinking water. One promising simple, low-cost, point-of-use water purification method is solar drinking water disinfection (SODIS). In SODIS, recycled transparent plastic drinks bottles containing contaminated water are exposed to full sunlight for 6 hours. During this exposure, ultraviolet radiation from the sun, together with an increase in temperature, inactivates the disease-causing organisms in the water.

Why Was This Study Done?

SODIS has been promoted as an effective method to purify household water since 1999, and about 2 million people now use the approach (www.SODIS.ch). However, although SODIS works well under laboratory conditions, very few studies have investigated its ability to reduce the number of cases of diarrhea occurring in a population over a specific time period (the incidence of diarrhea) in the real world. Before any more resources are used to promote SODIS—its effective implementation requires intensive and on-going education—it is important to be sure that SODIS really does reduce the burden of diarrhea in communities in the developing world.

Citation: Mäusezahl D., Christen A., Duran Pacheco G, Tellez FA, Iriarte M, et al. (2009) Solar Drinking Water Disinfection (SODIS) to Reduce Childhood Diarrhoea in Rural Bolivia: A Cluster-Randomized, Controlled Trial. PLoS Med 6(8): e1000125. doi:10.1371/journal.pmed.1000125

In this study, therefore, the researchers undertake a cluster-randomized controlled trial (a study in which groups of people are randomly assigned to receive an intervention or to act as controls) in 22 rural communities in Bolivia to evaluate the ability of SODIS to reduce diarrhea in children under 5 years old.

What Did the Researchers Do and Find?

For their trial, the researchers enrolled 22 rural Bolivian communities that included at least 30 children under 5 years old and that relied on drinking water resources that were contaminated with disease-causing organisms. They randomly assigned 11 communities (225 households, 376 children) to receive the intervention—a standardized, interactive SODIS promotion campaign conducted by Project Concern International (a nongovernmental organization)—and 11 communities (200 households, 349 children) to act as controls. Households in the intervention arm were trained to expose water-filled plastic bottles for at least 6 hours to sunlight using demonstrations, role play, and videos. Mothers in both arms of the trial completed a daily child health diary for a year. Almost 80% of the households self-reported using SODIS at the beginning and end of the study. However, community-based field workers estimated that only 32.1% of households on average used SODIS. Data collected in the child health diaries, which were completed on more than three-quarters of days in both arms of the trial, indicated that the children in the intervention arm had 3.6 episodes of diarrhea per year whereas the children in the control arm had 4.3 episodes of diarrhea per year. The difference in episode numbers was not statistically significant, however. That is, the small difference in the incidence of diarrhea between the arms of the trial may have occurred by chance and may not be related to the intervention.

What Do These Findings Mean?

These findings indicate that, despite an intensive campaign to promote SODIS, less than a third of households in the trial routinely treated their water in the recommended manner. Moreover, these findings fail to provide strong evidence of a marked reduction of the incidence of diarrhea among children following implementation of SODIS although some aspects of the study design may have resulted in the efficacy of SODIS being underestimated. Thus, until additional studies of the effectiveness of SODIS in various real world settings have been completed, it may be unwise to extend the global promotion of SODIS for general use any further.

INTRODUCTION

Globally, 1.8 million people die every year from diarrhoeal diseases the vast majority of whom are children under the age of 5 y living in developing countries [1]. Unsafe water, sanitation, and hygiene are considered to be the most important global risk factors for diarrhoeal illnesses[2].

Recent systematic reviews concluded that interventions to improve the microbial quality of drinking water in households are effective at reducing diarrhoea, which is a principal source of morbidity and mortality among young children in developing countries [3]–[5]. One widely promoted water disinfection method with encouraging evidence of efficacy in laboratory settings is solar drinking water disinfection (SODIS) [6]. Global efforts are underway to promote SODIS as a simple, environmentally sustainable, low-cost solution for household drinking water treatment and safe storage (www.who.int/household_water,www.sodisafricanet.org). SODIS is currently promoted in more than 30 countries worldwide (www.sodis.ch) and in at least seven Latin American countries through the SODIS Foundation including in Bolivia.

Despite this widespread promotion, evidence of the effectiveness of SODIS from field studies is limited. The three reported SODIS trials to date implemented the intervention at the household level, two of them in highly controlled settings that ensured very high compliance[7]–[9]. The highest reduction in incidence (36%) was recorded in a trial carried out among 200 children in an urban slum in Vellore, India [9].

Because SODIS is a behavioural intervention designed to reduce infectious diarrhoea, disease transmission and its interruption likely have community level dynamics [10]. In addition, because SODIS is typically rolled out in practice through community rather than household level promotion, there is an urgent need for effectiveness data from such settings. We conducted a community-randomized intervention trial to evaluate the effectiveness of SODIS in decreasing diarrhoea in children <5 y in rural communities in Bolivia.

METHODS

Ethics Statement

The study was approved by the three human subjects review boards of the University of Basel, Switzerland, the University of California, Berkeley, and the University of San Simon, Cochabamba, Bolivia. The Cochabamba and Totora municipal authorities also approved the study and informed consent was obtained from community leaders and male and female household heads prior to implementation of the study. Informed consent was obtained before randomisation to the treatment arms (Figure 1). Mildly ill children from households participating in the study were provided with and instructed to use oral rehydration salts, or they were referred by field staff to the local health system where clinical services were provided free of charge. The project provided transport and treatment costs for those patients. All project staff completed training on research ethics (www.fhi.org/training/sp/Retc/). Project staff comprised all project personnel of all project partners. Field staff comprised all personnel working in our laboratories and at our Totora field station including data enumerators and data- and project-management staff, supervisors, and community-based field workers living in the study communities.

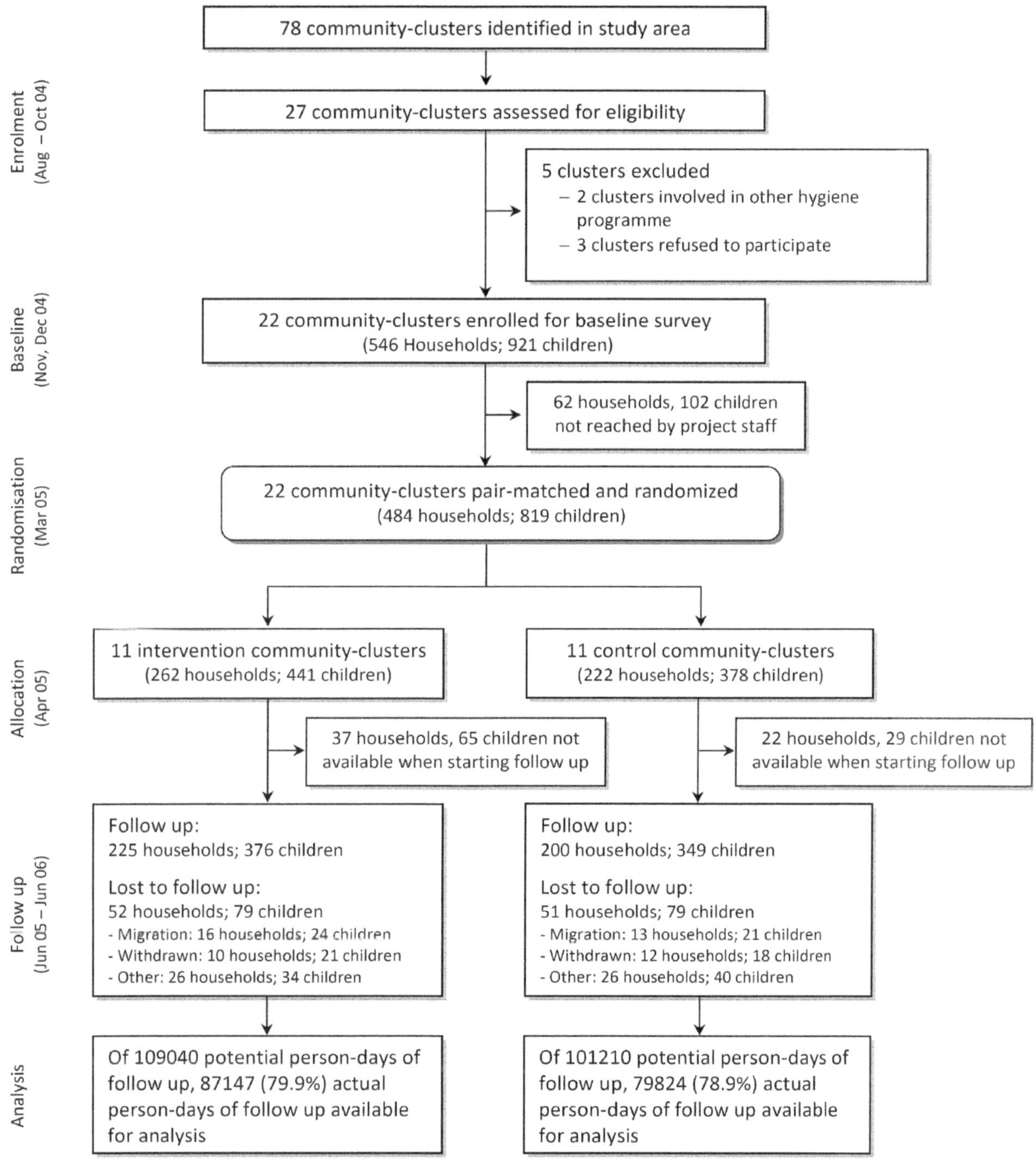

Figure 1: Community-randomized trial flow diagram on point-of-use SODIS in Totora District, Bolivia.

Site and Population

Our trial, the Bolivia Water Evaluation Trial (BoliviaWET), was conducted in an ethnically homogeneous Quechua setting in rural Totora District, Cochabamba Department, Bolivia. Our study was part of a comprehensive SODIS roll-out programme in collaboration with Project Concern International, a nongovernmental organisation (NGO). Most of the local residents are farmers, typically living in small compounds of three buildings with mud floors, with five or more persons sleeping in the same room. Our own surveys showed that 15% of homes have a latrine or other sanitary facilities and that most residents defecate in the nearby environment.

Drinking water is typically stored in 10-l plastic buckets or open jerry cans of 5–20 l in the household. Baseline assessments of the drinking water quality in the home indicated a median contamination of thermotolerant coliforms (TTC) of 32 TTC/100 ml (interquartile range (IQR)=3–344; *n*=223). Samples of at least one water source per community were tested for *Giardia lamblia* and *Cryptosporidium parvum*. The two parasites were detected in 18/24 and 11/23 water samples, respectively.

Parasites were detected by using immunomagnetic separation and PCR techniques [11]. Piped water, when available, is not chlorinated.

Design

Twenty-seven of 78 communities in the study area fulfilled the selection criteria (geographically accessible all year round; at least 30 children <5 y; reliance on contaminated drinking water sources). Two communities were excluded because of other ongoing health and hygiene campaigns, and three communities withdrew participation before baseline activities because of a change in political leadership. Community health workers undertook a census and identified households with at least one child <5 y. All children <5 y were enrolled in the participating villages.

We pair-matched communities on the incidence of child diarrhoea as measured in an 8-wk baseline survey [12]. The intervention was then assigned randomly to one community within each of the 11 consecutive pairs. This assignment was done during a public event because key political stakeholders were worried about possible backlash, public outcry, or a drop-off in group participation, which would result from providing some members with a new benefit while others got "nothing." It was agreed that a public drawing event was necessary to increase perceived fairness among the participating district and municipal authorities. Three authorities, the district head (Alcalde), representatives of the Ministries of Health and Education, and the deputy of the farmers union (Central Campesina), each drew one of two balls (with community codes inscribed that were randomly assigned beforehand) representing paired communities from a concealed box. It was agreed that the first draw assigned the community to the intervention arm. The group allocation was immediately recorded in a protocol by an independent witness. Subsequently, the witness disclosed the sequence, informed the community members and the authorities present in the town hall, and all drawers signed the protocol.

We explicitly chose community-level randomization because important components of the intervention (i.e., community efforts to encourage adoption of the SODIS-method) would occur at the community level. Randomization below the community level would not reflect the reality of scale-up programme implementation, and we would not have captured the potential community-level reinforcement of the behaviour change. Furthermore, community-level randomization is considered ethically optimal, because participants expect to equally benefit from interventions within their community [13]–[15]. Additionally, we believed cross-contamination (of the intervention) between the intervention and control communities was minimised by vast geographical dispersion of the communities. Control communities knew from the beginning of the study that they would receive the intervention as part of the NGO's development plans after study completion. It was not possible for the NGO to carry out the intervention in all the communities at the same time, thus making randomization feasible and acceptable to the three ethical review boards overseeing the study.

Sample size was calculated according to methods outlined by Hayes and Bennett [16], assuming an incidence rate (IR) in the control villages of five episodes/child/year [17], and accounting for clustering, the number of episodes, and the expected effect. We assumed a coefficient of between-cluster variation (k) of similar studies, between 0.1–0.25 (as cited by Hayes and Bennett) and a minimum of 10 child-years of observation per cluster [16]. We calculated that nine pairs of clusters were required to detect a difference of at least 33% in the IR between the control and intervention arms with 80% power, k=0.20 and an alpha level of 0.05. Anticipating a drop-out of at least one cluster per arm and a loss of follow-up of individuals, the final sample size was adjusted to 11 pairs with 30 children per community cluster. We powered the study to detect a 33% reduction in diarrhoea incidence after reviewing the evidence base for point-of-use water treatment at the time of the study's inception in 2002[18].

Implementation of the Intervention

The SODIS intervention was designed according to the published guidelines for national SODIS dissemination (http://www.sodis.ch/files/TrainingManual_sm.pdf). Promotion activities were targeted at primary caregivers and all household members (biweekly), whole communities (monthly), and primary schools (three times) by the NGO as part of its regional community development programme. Eleven communities (262 households and 441 children) were randomized to the intervention; 11 communities (222 households, 378 children) served as a control group (Figure 1). The implementation scheme and detailed description of the intervention in the intervention arm (and the control arms after study end) are described inFigure S1. For a period of 15 mo an intensive, standardised, and repeated interactive promotion of the SODIS method was implemented in the intervention communities beginning 3 mo before the start of follow-up.

Within the intervention arm, participating households were supplied regularly with clean, recycled polyethyleneteraphtalate (PET) bottles. The households were taught through demonstrations, role plays, video, and other approaches to expose the water-filled bottles for at least 6 h to the sun. NGO staff emphasized the importance and benefits of drinking only treated water (especially for children), explained the germ–disease concept, and promoted hygiene measures such as safe drinking water storage and hand washing as they relate to the understanding of drinking water and the faecal–oral route of transmission of pathogens (Figure S1). During household visits the NGO staff encouraged all household members to apply the method, answered questions, and assisted mothers and primary caregivers to integrate the water treatment into daily life. The same intervention (in terms of contents and messages) was supplied to the communities in the control arm by the NGO-staff at the end of the study (Figure S1).

Outcome

The primary outcome was the IR of diarrhoea among children <5 y, defined as number of diarrhoea episodes per child per year obtained from daily assessment of individual diarrhoea occurrence. We applied the WHO definition for diarrhoea of three or more watery bowel movements or at least one mucoid/bloody stool within 24 h [19],[20]. We defined a new episode of diarrhoea as the occurrence of diarrhoea after a period of 3 d symptom-free[20]–[22]. An episode of diarrhoea was labelled "dysentery" if signs of blood or mucus in the stool were recorded at any time. We also calculated the longitudinal prevalence (number of days a child suffered diarrhoea divided by the number of days of observation) because of its closer relation to severity, growth faltering, and mortality than diarrhoea incidence [19],[23]. Severe diarrhoea was defined as the occurrence of diarrhoea on more than 10% of the observed days [24].

Data Collection and Field Staff

The primary outcome was measured by community-based field workers who were recruited nearby and who lived one per community during data collection periods. The field workers were extensively trained in interviewing and epidemiological observation techniques, data checking, recording, and in general approaches to community motivation. Community-based field workers were randomly rotated between communities every 3 mo. Child morbidity was reported by the closest caregiver using the vernacular term "K'echalera," which had been established previously to correspond to the WHO definition of diarrhoea [25]. Mothers or closest caretakers kept a 7-d morbidity diary recording daily any occurrence of diarrhoea, fever, cough, and eye irritations in study participants [25]. Community-based field workers visited households weekly to collect the health diaries, and supervisors revisited an average 7% of homes. Discrepancies between supervisors and community-based field workers' records were clarified during a joint home revisit. Child exposure risks were also assessed by community-based staff interviewing mothers once during baseline and twice during the 1-y follow-up.

Compliance with the SODIS method was measured using four different subjective and objective indicators. Three of the indicators were assessed by field staff independent from the implementing NGO: (i) the number of SODIS-bottles exposed to sunlight and, (ii) the number of bottles ready-to-drink in the living space, and (iii) the personal judgment about families' user-status was provided by community-based field workers living among the families in the intervention arm. Judgement criteria for this main compliance indicator study included observing regular SODIS practice and bottles exposed to sun or ready to drink in the kitchen and being offered SODIS-treated water upon request. The fourth SODIS-use indicator was based on self-reporting and caregivers' knowledge of and attitudes toward the intervention that was assessed at the beginning (i.e., 3 mo after start of the intervention) and at the end of the 12-mo follow-up period.

Statistical Analysis

An intention-to-treat analysis was applied comparing the IR of diarrhoea between children <5 y in intervention and control communities. Diarrhoea prevalence (PR) and severe diarrhoea (SD) were additionally analysed. Generalized linear mixed models (GLMM) were fitted to allow for the hierarchical structure of the study design (pair-matched clusters). In contrast to our original trial protocol we selected the GLMM approach rather than generalized estimating equations (GEE) because recent publications indicated that the latter method requires a larger number of clusters to produce consistent estimates [26].

The crude (unadjusted) model included only the design factors and the intervention effect [12],[27]. Further models included potential confounders (selected a priori: child's age, sex, child hand-washing behaviour, and water treatment at baseline). Following an evaluation of the best fit, the GLMM included the log link function for negative binomial data (IR) and logit for binomial data (PR and SD). Denoting the link function of the outcome Y by $g(E(Y))$, the crude and adjusted models were: $g(E(Y_{ijk}))=\mu+B_i+\tau_j+\xi_{ij}$, and $g(E(Y_{ijk}))=\mu+B_i+\tau_j+\xi_{ij}+x'b$ where Y_{ijk} denotes the observed outcome value for the kth individual from a community allocated to the jth intervention, in the ith pair, μ is the general mean, B_i is the random effect of the ith pair $\approx N(0, \sigma^2_p)$, τ_j is the fixed effect of the SODIS intervention, and ξ_{ij} is the random effect of the interaction of the ith pair with the jth intervention applied to the community $\approx N(0, \sigma^2_{pt})$ (signifying the within-pair cluster variance and used as error term for τ_j), x is the vector of potential confounding factors, and b the vector of the corresponding regression coefficients.

The intracluster correlation coefficient (ICC) and the coefficient of between-cluster variation (k) were calculated after data collection to validate the degree of clustering and our assumptions for the sample size. ICC and k were estimated from the unscaled variance of the IR's GLMM. To estimate the uncertainty of ICC and k, we obtained the 95% credible region (Bayesian equivalent of 95% confidence interval [CI]) through an analogous Bayesian hierarchical regression [28]. Noninformative priors were used. The statistical analyses were performed using SAS software v9.1 (PROC GLIMMIX, SAS Institute Inc.) and WinBUGS v1.4 (Imperial College and MRC).

RESULTS

Participant Flow and Recruitment

Among the 1,187 households in the 22 communities there were 546 that met the inclusion criteria (Figure 1). The median number of participating households with children <5 y per community was 22. Because of political unrest and national election campaigns in 2005 a period of 6 mo passed between the baseline and the start of follow-up. Subsequently, 62 households (102 children) were no longer traceable before randomisation, and 59 households (37 intervention, 22 control) were lost before data collection had started. The loss to follow-up was balanced in intervention and control arms. Data were obtained from 376 children (225 households) in the intervention and 349 children (200 households) in the control arm, thus reaching our originally planned sample size.

Follow-up started in June 2005 and ended in June 2006. During the 51 wk of the study, information on the occurrence of diarrhoea was collected for 166,971 person-days representing 79.9% and 78.9% of the total possible person-days of child observation in intervention and control arms. We excluded from the potential observation time the experience of 94 children who dropped out before the start of follow-up. National festivities, holidays, and political unrest over the entire year amounted to further 9 wk during which outcome surveillance needed to be suspended. The main reasons for incomplete data collection were migration (28%) and withdrawal (67%). Supervisors reevaluated the outcome during 984 unannounced random home visits, and discrepancies between community-based field workers' and supervisors' records were found for five (0.5%) of all visits.

Baseline Characteristics

At baseline the households in the different study arms were well balanced on multiple other factors suggesting successful randomisation (Table 1). The main types of water sources for household chores and drinking were similar in both arms as was the distance to the source (median distance 50 m and 30 m in the control and intervention arms, respectively). Storing water for longer than 2 d was more common among the intervention (26.8%) than the control arm (13.9%). Nearly 30% of all households reported treating water regularly before drinking. Boiling was the most common water treatment before the trial (20.2% in both arms).

Table 1: Baseline community and household characteristics of a community-randomized trial of SODIS

Category	Description	*n* Children or Households	Control 11 Clusters	*n* Children or Households	Intervention 11 Clusters
Demography	Community size: *n* of households [mean (SD)]	—	50 (20)	—	58 (20)
	Household size: *n* of household members [mean (SD)]	N = 222	6.2 (2.1)	N = 262	6.3 (2.6)
	n of children <5 y per household [mean (SD)]	—	1.8 (0.7)	—	1.7 (0.8)
	n of children <5 y per community [mean (SD)]	—	35.3 (6.6)	—	41.4 (9.9)
	Female household head [*n* (%)]	—	20 (9.0)	—	14 (5.4)
	Closest child caregiver (female)	—	223 (99.5)	—	266 (99.6)
	Age of closest child caregiver (y) [mean (SD)]	—	31(9)	—	30 (10)
	n of children <1 y	—	65 (4.7)	—	67 (4.1)
	n of children <5 y	—	369 (26.6)	—	426 (25.9)
Education	Household chief: reported years of education [mean (SD)]	N = 167	4.1 (2.6)	N = 178	4.2 (2.4)
	Closest child caregiver: reported years of education [mean SD)]	N = 179	2.5 (1.9)	N = 198	2.7 (1.8)
Socio-economic	Main occupation of the household chief as farmer	N = 208	180 (86.5)	N = 228	207 (90.8)
variables	Ownership of truck, car, or motorbike	—	12 (5.8)	—	14 (6.2)
	Ownership of radio	—	129 (86.1)		194 (85.1)
	Ownership of bicycle	—	109 (52.4)	—	121 (53.1)
	Ownership of television	—	24 (11.5)		15 (6.6)
	n of rooms in the house [mean (SD)]	—	2.9 (1.4)	—	2.8 (1.2)
Water management	Spring as source of drinking water	N = 208	100 (48.1)	N = 228	136 (59.6)
and consumption	Tap as source of drinking water	—	108 (51.9)	—	129 (56.6)
	River as source of drinking water	—	46 (22.1)	—	29 (12.7)
	Rain as source of drinking water	—	31 (14.9)	—	71 (31.1)
	Dug well as source of drinking water	—	31 (14.9)	—	37 (16.2)
	Distance to water source (m) [median (Q1, Q3)]	—	50 (7.5, 100)	—	30 (6, 150)
	Container for water collection: plastic bucket	—	189 (90.9)	—	205 (89.9)
	Container for water collection: jerry can	—	165 (79.3)	—	156 (68.4)
	Container for water collection: bottles	—	32 (15.4)	—	36 (15.8)
	Container for water collection: jar/pitcher	—	13 (6.3)	—	20 (8.8)
	Container for water collection: barrel	—	10 (4.8)	—	25 (10.9)
	Child's consumption of untreated water (glasses/day) [mean (SD)]	M = 318	1.2 (1.2)	M = 359	1.2 (1.4)
	Treat water before drinking	N = 208	59 (28.4)	N = 228	67 (29.4)
	Store water for >2 d	—	29 (13.9)	—	61 (26.8)
	Water storage container: jerry can	—	23 (11.1)	—	49 (21.5)
	Water storage container: plastic bucket	—	17 (8.2)	—	37 (16.2)
	Water turbidity in water storage container >30 NTU	—	13 (11.2)	—	24 (18.8)
Sanitation	Reported *n* of interviewee's hand washing per day [mean (SD)]	N = 177	3.8 (1.7)	N = 200	4.1 (1.8)
	Reported *n* of child hand washing per day [mean (SD)]	M = 348	2.5 (1.2)	M = 376	2.6 (1.4)
	Child washes hands: before eating	—	228 (65.5)	—	270 (71.8)
	Child washes hands: when hands are dirty	—	62 (17.8)	—	56 (14.9)
	Child washes hands: other occasions	—	58 (16.7)	—	50 (13.3)
	Latrine present	N = 208	27 (13.0)	N = 228	38 (16.7)
	Use of latrine by the interviewee (day or night)	—	15 (7.2)	—	20 (8.8)
	Feces visible in yard	N = 202	121 (59.9)	N = 219	124 (56.6)

Data shows numbers and percentages unless otherwise specified. Baseline data from December 2004.
Abbreviations: 30NTU, threshold for efficacious pathogen-inactivation of the SODIS method; M, number of children; N, number of households; NTU, nephelometric units; SD, standard deviation.
doi:10.1371/journal.pmed.1000125.t001

Intervention and Attendance

The NGO conducted 210 community events and 4,385 motivational household visits in intervention communities; 3,060 visits occurred in the households with children <5 y followed up and analysed for the study, and 1,325 household visits took place in homes that were not taking part in the study. Study households attended a median of nine community events (IQR=5–12) and were visited by the SODIS-programme team a median 11 times at home (IQR=7–18). To ensure a sufficient number of PET bottles, the NGO provided as many SODIS-bottles as required by participants (mean 955 bottles/community).

Diarrhoeal Illness in the Control and Intervention Arm

Children in the SODIS-intervention arm reported a total of 808 episodes or a mean of 3.6 per child per year-at-risk (Table 2). In the control arm there were 887 episodes and an annual mean of 4.3 per child per year. In both arms median length of episodes was 3 d. The unadjusted relative rate (RR) estimate (0.81, 95% CI 0.59–1.12) suggested no statistically significant difference in the number of diarrhoea episodes between the SODIS and control arms of the study (Table 3). In an analysis of the longitudinal prevalence of diarrhoea we found no significant treatment effect (odds ratio [OR]=0.92, 95% CI 0.66–1.29). Furthermore, no strong evidence was detected for the reduction of odds of severe diarrhoea cases (OR=0.91, 95% CI 0.51–1.63) and dysentery (OR=0.80, 95% CI 0.55–1.17).

Table 2: Diarrhoea episodes, length of illness, and days ill with diarrhoea

Health Condition	Class or Parameter	*n*	Control	*n*	Intervention
Diarrhoea illness overview		**Children**		**Children**	
Days under observation	Median (Q1, Q3)	349	263 (213, 274)	376	263 (222, 273)
Days at risk	Median (Q1, Q3)	349	246 (192, 265)	376	247 (202, 265)
n Episodes	Median (Q1, Q3)	349	1 (0, 3)	376	1 (0, 3)
n Dysentery episodes	Median (Q1, Q3)	349	1 (0, 2)	376	1 (0, 2)
Days spent ill	Median (Q1, Q3)	349	4 (0, 11)	376	4 (0, 12)
Episode length (d)	Median (Q1, Q3)	349	3 (1, 5)	376	3 (2, 5)
Days under observation	Total		79,829		87,140
Days at risk	Total		75,077		82,682
n Episodes	Total		887		808
n Dysentery episodes	Total		460		431
Days spent ill	Total		3,111		3,038
Diarrhoea incidence	**Age class**	**Children**	**IR**	**Children**	**IR**
n Episodes/(child × year at risk)	<1	16	7.8	15	11.1
	1–2	67	7.1	70	5.5
	2–3	67	4.3	82	3.8
	3–4	77	3.2	75	2.8
	4–5	71	3.4	80	2.1
	5–6	50	2.7	53	2.5
	Total[a]	349	4.3	376	3.6
Diarrhoea prevalence	**Age class**	**Children**	**Mean (SD)**	**Children**	**Mean (SD)**
n Days ill/(child × year)	<1	16	27.4 (28.3)	15	42.3 (40.7)
	1–2	67	31.4 (42.2)	70	23.0 (26.1)
	2–3	67	19.0 (47.5)	82	16.4 (28.4)
	3–4	77	11.7 (24.5)	75	7.3 (9.7)
	4–5	71	9.5 (15.1)	80	6.2 (12.4)
	5–6	50	6.9 (11.8)	53	7.7 (10.4)
	Total[a]	349	16.5 (32.8)	376	13.5 (22.4)
Diarrhoea illness	**Days spent ill**	**Children**	**Percent**	**Children**	**Percent**
	0 d	97	27.8	126	33.5
	1–2 d	50	14.3	42	11.2
	3–7 d	91	26.1	80	21.3
	8–14 d	49	14.0	59	15.7
	15–21 d	27	7.7	33	8.8
	22–40 d	18	5.2	21	5.6
	>40 d	17	4.9	15	4.0
	Total	349	100	376	100
Diarrhoea illness duration	**Episode duration**	**Episodes**	**Percent**	**Episodes**	**Percent**
	1 day	250	28.2	191	23.6
	2–3 d	303	34.2	292	36.1
	4–7 d	258	29.1	250	30.9
	8–13 d	54	6.1	59	7.3
	>13 d	22	2.5	16	1.9
	Total	887	100	808	100
Prevalence of other symptoms (d/[child × year])		**Children**	**Mean (SD)**	**Children**	**Mean (SD)**
Vomit		349	5.5 (13.2)	376	4.0 (8.9)
Fever		349	21.0 (33.0)	376	15.1 (19.8)
Cough		349	41.9 (48.3)	376	30.9 (39.4)
Eyes irritation		349	12.8 (29.8)	376	8.3 (19.5)

[a]Includes one child per treatment arm with unknown age. SD, standard deviation.
doi:10.1371/journal.pmed.1000125.t002

Table 3: Effect of SODIS on diarrhoea episodes, longitudinal prevalence, severe diarrhoea, and dysentery episodes

Outcome	Model	*n* Children	Parameter	RR/OR	95% CI	*p*-Value
***n* Episodes (RR)**	Unadjusted	725	Intervention	0.81	(0.59–1.12)	0.19
	Adjusted	644	Intervention	0.74	(0.50–1.11)	0.14
			Age	0.75	(0.70–0.81)	<0.001
			Sex	1.03	(0.84–1.26)	0.80
			Water treatment	1.05	(0.81–1.36)	0.69
			Hand washing	0.93	(0.85–1.02)	0.13
Prevalence (OR)	Unadjusted	725	Intervention	0.92	(0.66–1.29)	0.62
	Adjusted	644	Intervention	0.91	(0.64– 1.30)	0.60
			Age	0.67	(0.61–0.73)	<0.001
			Sex	1.05	(0.84–1.31)	0.68
			Water treatment	1.00	(0.76–1.33)	0.97
			Hand washing	0.94	(0.84–1.04)	0.23
Severe diarrhoea (OR)	Unadjusted	643	Intervention	0.91	(0.51–1.63)	0.75
	Adjusted	589	Intervention	1.02	(0.52–2.01)	0.95
			Age	0.52	(0.40–0.67)	<0.001
			Sex	1.12	(0.63–2.01)	0.69
			Water treatment	1.59	(0.81–3.12)	0.18
			Hand washing	0.94	(0.75–1.19)	0.62
Dysentery (OR)	Unadjusted	725	Intervention	0.80	(0.55–1.17)	0.23
	Adjusted	644	Intervention	0.75	(0.47–1.18)	0.20
			Age	0.73	(0.67–0.80)	<0.001
			Sex	1.00	(0.80–1.26)	0.97
			Water treatment	1.15	(0.87–1.53)	0.33
			Hand washing	0.91	(0.82–1.01)	0.06

Number of episodes, *n* of episodes per days at risk; prevalence, *n* of days ill per days under observation; severe diarrhoea, diarrhoea during >10% of all days (only children with more than 100 d of observation are included); unadjusted, general linear mixed models, only design factors and treatment are included; adjusted, effects of treatment and covariates; sex: 0, female; 1, male; water treatment: water treatment at baseline, 0, no treatment; 1, treatment (chlorination or boiling or SODIS); hand washing, reported number of child's hand washing per day at baseline.
doi:10.1371/journal.pmed.1000125.t003

A multivariable model adjusting for age, sex, baseline-existing water treatment practises, and child hand washing was consistent in its estimate of effect (RR=0.74, 95% CI 0.50–1.11). We repeated the analysis by including confounding covariates in the order of occurrence of the variables in Table 3 to confirm that the conclusions were not sensitive to the choice of covariates. None of the models yielded significant results for the effect of SODIS (all p-values>0.1) or resulted in meaningful changes in estimates of ORs. Figure 2 shows the relationship between study time and diarrhoea in the control and intervention arm. We found no statistically significant effect of the interaction of time and intervention in a time-dependent model.

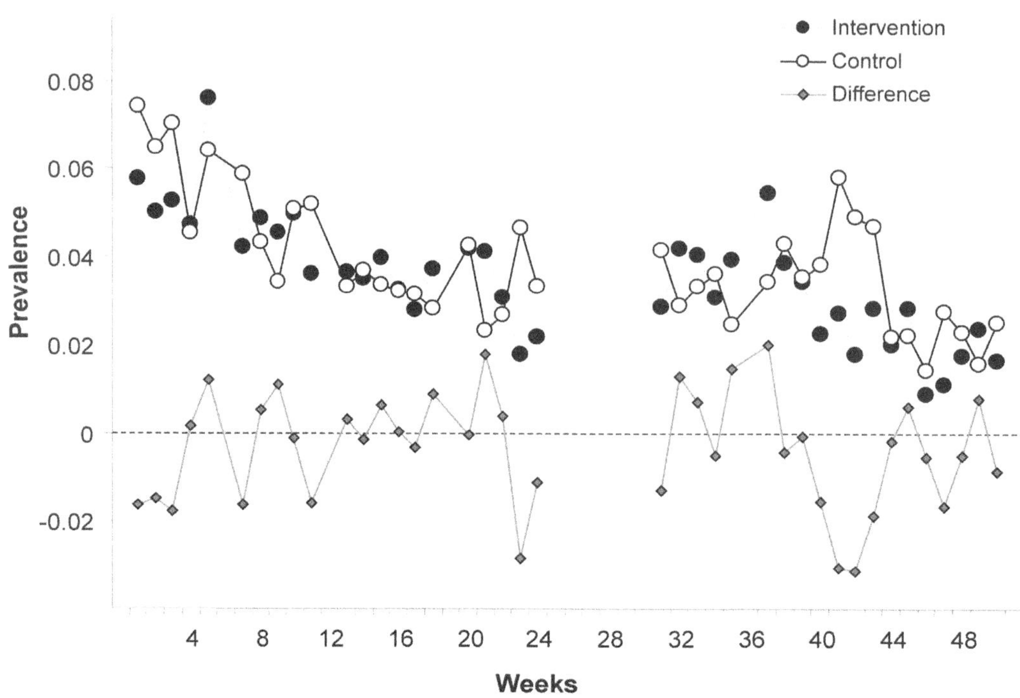

Figure 2: Weekly prevalence of child diarrheal illness.

Weekly points are derived from daily prevalence data of each participating child.

The ICC was estimated as 0.0009 with a 95% posterior credible region between (0.0001, 0.0025); k was estimated to be 0.27 with a 95% confidence region of (0.11, 0.46).

Compliance

Community-based field workers who were living in the communities throughout the study observed a mean SODIS-user rate of 32.1% in the intervention arm (minimum 13.5%, maximum 46.8%, based on their personal judgement) (Figure 3). The mean proportion of households with SODIS-bottles exposed to the sun was 5 percentage points higher than the assessment by community-based field workers. In contrast, almost 80% of the households reported using SODIS at the beginning and end of the follow-up. About 14% of the households used the method more than two-thirds (>66%) of the weeks during observation, and 43% of the households applied SODIS in more than 33% of the observed weeks (Table 4).

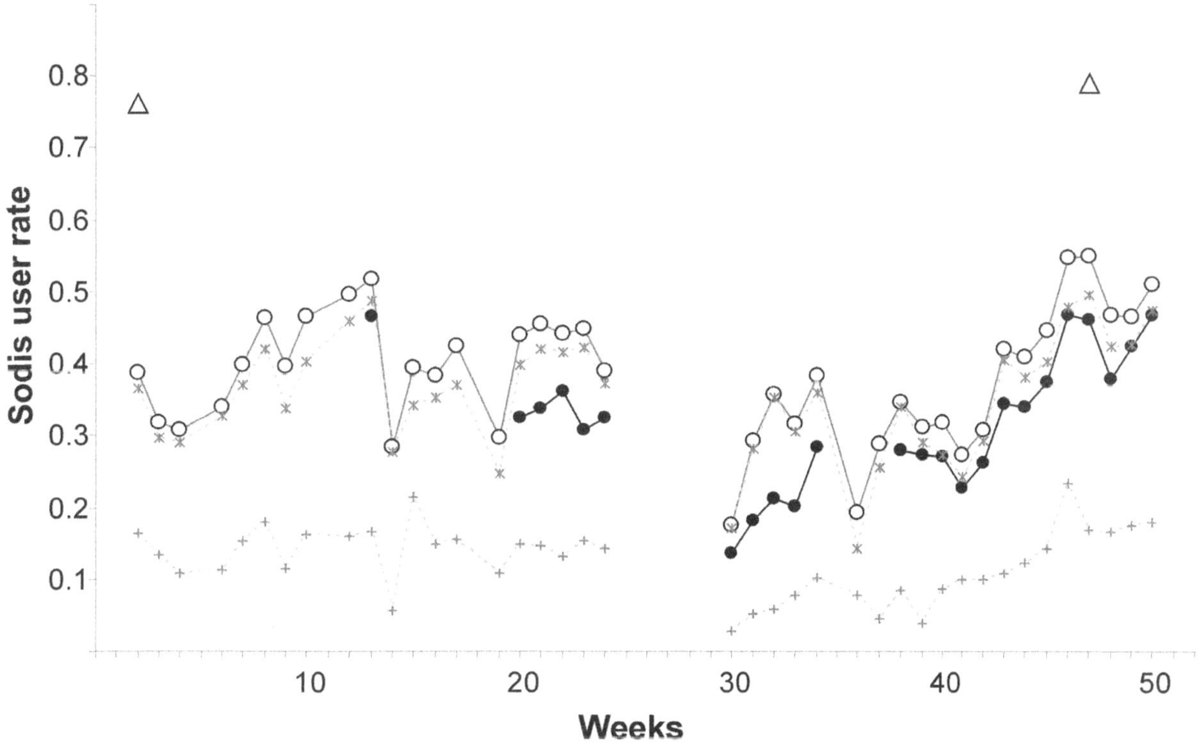

Figure 3: Weekly observed proportion of households using SODIS as point-of-use drinking water purification method.

Open triangles, self-reported SODIS use at the beginning (after 3 mo of initial SODIS promotion) and at the end of follow-up; filled dots, SODIS use observed by project staff living in the community (see Methods for definition); open circles, SODIS bottles observed on the roof and/or in the kitchen; stars, SODIS-bottles on the roof; crosses, SODIS-bottles in the kitchen.

Table 4: Climatic conditions and SODIS use of a cluster-randomized trial involving 22 rural communities of Totora District, Bolivia

Category	Description	Control (n=11 Clusters)	Intervention (n=11 Clusters)
Climate	Percentage of sunny days (>6 h sunshine) [median of clusters (min, max)]	70 (57, 78)	67 (44, 77)
	Average duration of sunshine [median of clusters (min, max)]	7.0 (6.3, 8.0)	7.1 (4.5, 8.3)
SODIS-use	**Observed level of SODIS use[a]**	**Percentage of households**	**Percentage of households**
	0.66–1	0%	14%
	0.33–0.66	0.5%	29%
	0–0.33	99.5%	57%

[a]Proportion of weeks in which SODIS was used, as estimated by community-based project staff at the end of study. Households with <10 wk of observation are excluded.

doi:10.1371/journal.pmed.1000125.t004

Diarrhoeal Illness by Compliance

No positive effect of compliance (proportion of weeks of observed SODIS use) on the IRs in the intervention arm was observed. The incidence did not decline with the increase of weeks using SODIS (Figure 4). Seasonal variation in compliance was observed. The proportion of SODIS-practising households was consistently below average during weeks 4–16 (January 2005–April 2006), which corresponded to the labour intensive cultivating period from November to May.

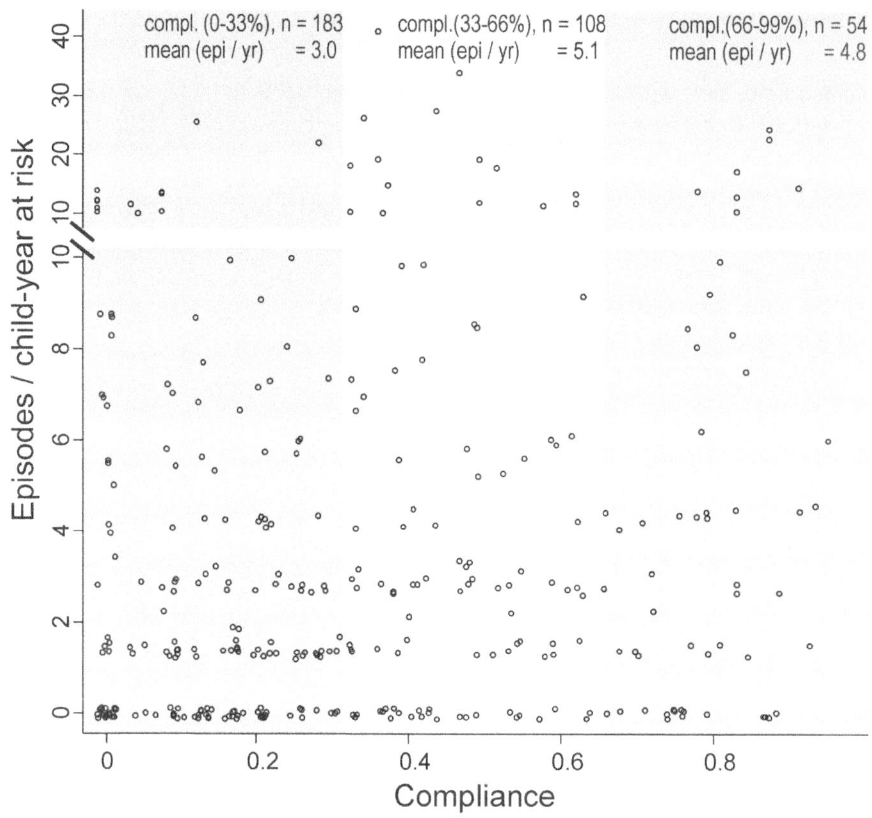

Figure 4: Compliance of using SODIS and child diarrhoea in rural Bolivia.

Compliance of SODIS use is estimated as the proportion of weeks a family has been classified as a SODIS user by community-based project staff. Dots, number of episodes per child-year at risk. Small random noise was added to the dots to avoid over plotting. Only children with at least 110 d under observation are included.

The median proportion of sunny days with more than 6 h of sunshine was 70.2% and 67.2% in intervention and control communities, respectively, consistent with the technical and climatic conditions necessary for the proper functioning of the ultraviolet SODIS purification process[29] during the study (Table 4).

DISCUSSION

We conducted a community-randomized trial within the operations of an ongoing national SODIS-dissemination programme, which provided an intensive training and repeated reinforcement of the SODIS intervention throughout the study period. In this context of a "natural experiment" we found a RR of 0.81 for the IR of diarrhoea episodes among children assigned to SODIS compared to controls. However, the CI was broad and included unity (RR=0.81, 95% CI 0.59–1.12) and, therefore, we conclude that there is no strong evidence for a substantive reduction in diarrhoea among children in this setting. Subsequently, we discuss the primary outcome in the context of other study findings, and explain why we hypothesize that the true effect—if there is any—might be smaller.

First, the estimate for the longitudinal prevalence of diarrhoea was substantially smaller (OR=0.92, 95% CI 0.66–1.29) than the estimate for incidence and there is some evidence that prevalence is a better predictor in terms of mortality and weight gain than incidence [23]. The absence of a time-intervention interaction in our time-dependent analysis suggested no increased health benefits with the ongoing intervention. Furthermore, within the intervention arm, there was no evidence that increased compliance was associated with a lower incidence of diarrhoea (Figure 4). However, we interpret this post hoc subgroup analysis cautiously because compliant SODIS users might differ in important ways from noncompliant users. A compliant SODIS user might be more accurately keeping morbidity diaries, whereas less compliant families may tend to underreport diarrhoeal illness. Or, households with a high burden of morbidity might be more likely to be compliant with the intervention. Both of these scenarios could lead to an underestimation of the effectiveness of SODIS.

Further, analysing the laboratory results from 197 randomly selected stool specimens also did not provide convincing evidence for an intervention effect: the proportion of *C. parvum* was lower in the intervention children (5/94 versus 2/103), but other pathogens were found at similar proportions in intervention and control children (*G. lamblia*, 39/94 versus 40/103; *Salmonella*sp., 2/94 versus 3/104; *Shigella* sp., 3/94 versus 3/104). In further exploring the occurrence of other illness symptoms we found the prevalence of eye irritations and cough to be lower in the intervention group compared to the control group. This difference could be the result of the hygiene component in the intervention that increased hygiene awareness among the treatment communities. An alternative explanation is that the lack of blinding led to biased (increased) health outcome reporting in the intervention group.

Due to the nature of the intervention neither participants nor personnel were blinded to treatment assignment. Ideally, blinding to the intervention allocation should apply to the

NGO staff administering the SODIS intervention and our enumerators assessing outcomes [30]. Although the former could not be blinded in our study (for obvious reasons), the latter would inevitably be able to identify the intervention status of the cluster through the visible display of bottles to sunlight in the village or directly at the study home during home visits. These problems are consistent with nearly all household water treatment interventions [5] and other public health cluster randomized trials [31],[32]. Schmidt and Cairncross [33] recently argued that reporting bias may have been the dominant problem in unblinded studies included in a meta-analysis reporting a pooled estimate of a 49% reduction of diarrhoea in trials investigating the effects of drinking water quality interventions [5]. However, their review of only four available blinded trials showing no effect demonstrates weak support for contrast. In addition, all of the blinded trials exhibited analytical shortcomings or had very broad CIs suggesting very low power. In the absence of blinding—unavoidable in many behavioural change interventions or household water treatment studies—we believe that data collection independent from the implementation is a crucial factor. Future reviews should include reporting on such additional quality parameters.

In our study the lack of blinding may have reduced motivation in the control communities. However, the number of households lost during follow-up and the number of days under observation were almost identical in both arms. Additionally, the control communities knew that they would receive the intervention after study end. Finally, a reduction of diarrhoea frequency of 20% might be insufficient to be well perceived, i.e., have a noticeable impact in a population with a high burden of child diarrhoea and will, thus, not result in a sustainable behavioural change. Faecal contamination in about 60% of the yards indicates a highly contaminated environment with presumably a large potential for transmission pathways other than consuming contaminated water. This simultaneous exposure to a multiplicity of transmission pathways may explain why we found no significant diarrhoea reduction due to SODIS.

On the other hand, our result of a 19% reduction in diarrhoeal episodes appears to be roughly consistent with results of the two other SODIS trials both from Maasai cultural settings conducted by Conroy and colleagues among children <6 y and 5–16 y of age. They report a 16% reduction (in <6 y olds, 2-wk prevalence of 48.8% in intervention, and 58.1% in control group) [8] and a 10.3% reduction in the 2-wk diarrhoea prevalence (in 5–16 y olds) [7]. However, these randomized controlled trials were undertaken in a socio-cultural setting assuring a 100% compliance (as stated by the authors) in water treatment behaviour through social control by Maasai elders who promoted the method [7],[8]. In the results presented in these studies adjusted models with post hoc selected covariates were presented (i.e., no unadjusted models were provided). These trials were carried out in conditions of heavily contaminated drinking water and very high diarrhoea rates—important considerations when attempting to generalize these results. The only other—quasi-randomized—trial to estimate the effect of solar water disinfection was carried out in the urban slum in Vellore and resulted in a remarkable reduction of diarrhoea among children <5 y (IR ratio, 0.64; 95% CI 0.48–0.86) despite 86% of SODIS users also drinking untreated water [9].

To our knowledge this is the first community-randomized trial and the largest study so far to assess the effectiveness of the SODIS method under typical social and environmental conditions in a general rural population setting where children drink untreated water.

Our study was sufficiently powered to detect a 33% reduction in the effectiveness of the SODIS intervention, and we accounted for clustered design in our analysis. On the basis of a post hoc sample size calculations using the model-based estimate for the between-cluster variability (k=0.27), we would have needed a study 2.5 times larger for a 20% difference to be significant.

The implementing NGO, which had global experience in disseminating SODIS, adapted a campaign to the local and cultural needs and also involved the public health and educational system in the roll-out. This comprehensive SODIS campaign resulted in a mean SODIS usage of 32% on any given study day. In using the SODIS-use indicator on the basis of the personal judgement of community-based staff, we intended to measure actual use in combining objective, visible signs of use (e.g., bottles exposed to sunlight) with proxies more responsive to actual treatment behaviour (e.g., SODIS water can be offered to drink upon request). We consider this a restrictive, more conservative definition of SODIS use compared to that in other studies, which recorded reported use [9] or the number of bottles exposed to sunlight[34]. Both are indicators that can easily and reliably be measured, but which are prone to over-reporting due to low specificity for actual use. Further studies will need to validate different compliance indicators and formally assess the dimension of reporting bias.

It is possible that respondents would like to please field staff and over-report use out of courtesy. Also, observing exposed bottles on the roof may overestimate use (Figure 3), because some households were noted anecdotally to have placed bottles on the roof to avoid discussions with the SODIS-implementing NGO staff. Figure 3 is indicative of this phenomenon, as reported use at the beginning and reported use and satisfaction with the method at the end of the study reached the 80% mark—a usage figure consistent with other studies relying on reported compliance [9] and evaluation reports from grey literature. We conclude that self-reported SODIS use may overestimate compliance and a combination of reported and objectively measurable indicators provides more accurate SODIS-compliance data.

There are limitations to our study. As in other studies [24],[35], we observed a decline in the reporting of child diarrhoea during the observational period in both arms (Figure 2). If true, seasonal variation of diarrhoea could be one possible cause; increased awareness leading to more attention to basic hygiene and hence to illness reduction may be another reason. Alternatively, the pattern could be due to survey fatigue.

Despite a comprehensive and intensive intervention promotion campaign, we detected no strong evidence for a significant reduction in the IR of diarrhoea in children <5 y in families using SODIS in our trial in a typical setting in rural Bolivia. We believe that clearer understandings of the discrepancy between laboratory and field results (obtained under typical environmental and cultural conditions), the role of compliance in effectiveness, and a direct comparison of SODIS to alternate drinking water treatment methods are needed before further global promotion of SODIS.

ACKNOWLEDGMENTS

The authors greatly acknowledge the families who participated in the study, the dedication of our study communities, and the support of the authorities of Cochabamba and Totora District that made this project possible. Project Concern International (PCI) authorized embedding this evaluation trial in their local community development plans—we specifically thank the PCI implementation teams of Carlos Morante and Luciano Cespedes and their field staff for their relentless efforts to intertwine research and development approaches. We are grateful to the study team: Roy Cordova (administrator Bolivia site), Freddy Arauco (data entry, supervisor), Abrahan Cuevas, Fernando Salvatierra, David Villaroel, Dora Claros, Elmer Garvizu, Alfonso Claure (field supervisors and sample collectors), Claudia Lazarte (medical practitioner), and the entire MMS team (Morbidity Monitoring Staff). We thank Sonia Peredo, Gabriela Almanza, and Gonzalo Fillips (laboratory staff); Jenny Rojas, Edgar Sejas, Ana Maria Romero, and Mirjam Mäusezahl for their personal and institutional support. Lee Riley (UCB), Alan Hubbard (UCB), and Joseph Eisenberg, University of Michigan and formerly at UCB, contributed to the development of the research plan for this trial. We greatly appreciate the manifold administrative support of Catherine Wright (UCB) and Ulrich Wasser (STI, Basel). Marcel Tanner (STI) kindly reviewed the manuscript. Markus Niggli, Tim Haley, Michael Hobbins, and Stephan Indergand contributed to the study implementation and initial analyses. The contents of the paper are solely the responsibility of the authors and do not necessarily represent the official view of the National Institutes of Health (NIH). Portions of this manuscript were presented at the 12th International Congress on Infectious Diseases in Lisbon, Portugal, on June 16, 2006.

REFERENCES

1. WHO (2005) The World Health Report 2005 - make every mother and child count. Geneva: World Health Organization.
2. Pruss A, Kay D, Fewtrell L, Bartram J (2002) Estimating the burden of disease from water, sanitation, and hygiene at a global level. Environ Health Perspect 110: 537–542.
3. Fewtrell L, Kaufmann RB, Kay D, Enanoria W, Haller L, Colford JM Jr (2005) Water, sanitation, and hygiene interventions to reduce diarrhoea in less developed countries: a systematic review and meta-analysis. Lancet Infect Dis 5: 42–52.
4. Zwane AP, Kremer M (2007) What works in fighting diarrheal diseases in developing countries? A critical review. National Bureau of Economic Research Working Paper Series No. 12987. March 2007. Available: http://www.nber.org/papers/w12987.
5. Clasen T, Schmidt WP, Rabie T, Roberts I, Cairncross S (2007) Interventions to improve water quality for preventing diarrhoea: systematic review and meta-analysis. BMJ 334: 782.
6. Sobsey M (2002) Managing water in the home: Accelerated health gains from improved water. WHO/SDE/WSH/02.07. Geneva: World Health Organization.
7. Conroy RM, Elmore-Meegan M, Joyce T, McGuigan KG, Barnes J (1996) Solar disinfection of drinking water and diarrhoea in Maasai children: a controlled field trial. Lancet 348: 1695–1697.
8. Conroy RM, Meegan ME, Joyce T, McGuigan K, Barnes J (1999) Solar disinfection of water reduces diarrhoeal disease: an update. Arch Dis Child 81: 337–338.
9. Rose A, Roy S, Abraham V, Holmgren G, George K, et al. (2006) Solar disinfection of water for diarrhoeal prevention in southern India. Arch Dis Child 91: 139–141.
10. Eisenberg JN, Scott JC, Porco T (2007) Integrating disease control strategies: balancing water sanitation and hygiene interventions to reduce diarrheal disease burden. Am J Public Health 97: 846–852.
11. McCuin RM, Bukhari Z, Sobrinho J, Clancy JL (2001) Recovery of Cryptosporidium oocysts and Giardia cysts from source water concentrates using immunomagnetic separation. J Microbiol Methods 45: 69–76.
12. Murray DM (1998) Design and analysis of group-randomized trials. New York, Oxford: Oxford University Press.
13. Edwards SJ, Braunholtz DA, Lilford RJ, Stevens AJ (1999) Ethical issues in the design and conduct of cluster randomised controlled trials. BMJ 318: 1407–1409.
14. Chingono A, Lane T, Chitumba A, Kulich M, Morin S (2008) Balancing science and community concerns in resource-limited settings: Project Accept in rural Zimbabwe. Clin Trials 5: 273–276.
15. Ranson MK, Sinha T, Morris SS, Mills AJ (2006) CRTs–cluster randomized trials or "courting real troubles": challenges of running a CRT in rural Gujarat, India. Can J Public Health 97: 72–75.
16. Hayes RJ, Bennett S (1999) Simple sample size calculation for cluster-randomized trials. Int J Epidemiol 28: 319–326.
17. Ministry of Health, Bolivia. Situación de salud Bolivia (2004) La Paz, Bolivia: Ministry of Health. Available: http://www.sns.gov.bo/snis/.
18. Clasen T, Roberts I, Rabie T, Schmidt W, Cairncross S (2006) Intervention to improve water quality for preventing diarrhoea. Cochrane Database Syst Rev 3: CD004794.
19. WHO (1988) Persistent diarrhoea in children in developing countries: memorandum from a WHO meeting. Bull World Health Organ 66: 709–717.
20. Baqui AH, Black RE, Yunus M, Hoque AR, Chowdhury HR, et al. (1991) Methodological issues in diarrhoeal diseases epidemiology: definition of diarrhoeal episodes. Int J Epidemiol 20: 1057–1063.
21. Morris SS, Cousens SN, Lanata CF, Kirkwood BR (1994) Diarrhoea–defining the episode. Int J Epidemiol 23: 617–623.
22. Wright JA, Gundry SW, Conroy R, Wood D, Du PM, et al. (2006) Defining episodes of diarrhoea: results from a three-country study in Sub-Saharan Africa. J Health Popul Nutr 24: 8–16.

23. Morris SS, Cousens SN, Kirkwood BR, Arthur P, Ross DA (1996) Is prevalence of diarrhea a better predictor of subsequent mortality and weight gain than diarrhea incidence? Am J Epidemiol 144: 582–588.
24. Luby SP, Agboatwalla M, Painter J, Altaf A, Billhimer W, et al. (2006) Combining drinking water treatment and hand washing for diarrhoea prevention, a cluster randomised controlled trial. Trop Med Int Health 11: 479–489.
25. Hobbins MA (2004) Home-based drinking water purification through sunlight: from promotion to health effectiveness [PhD dissertation]. Basel, Switzerland: Swiss Tropical Institute, University Basel. 214 p.
26. Young ML, Preisser JS, Qaqish BF, Wolfson M (2007) Comparison of subject-specific and population averaged models for count data from cluster-unit intervention trials. Stat Methods Med Res 16: 167–184.
27. Twisk JWR (2006) Applied multilevel analysis: a practical guide for medical researchers. Cambridge: Cambridge University Press.
28. Turner RM, Omar RZ, Thompson SG (2006) Constructing intervals for the intracluster correlation coefficient using Bayesian modelling, and application in cluster randomized trials. Stat Med 25: 1443–1456.
29. Sommer B, Mariño A, Solarte Y, Salas ML, Dierolf C, et al. (1997) SODIS - an emerging water treatment process. J Water SRT 46: 127–137.
30. Campbell MK, Elbourne DR, Altman DG (2004) CONSORT statement: extension to cluster randomised trials. BMJ 328: 702–708.
31. Kumar V, Mohanty S, Kumar A, Misra RP, Santosham M, et al. (2008) Effect of community-based behaviour change management on neonatal mortality in Shivgarh, Uttar Pradesh, India: a cluster-randomised controlled trial. Lancet 372: 1151–1162.
32. Campbell R, Starkey F, Holliday J, Audrey S, Bloor M, et al. (2008) An informal school-based peer-led intervention for smoking prevention in adolescence (ASSIST): a cluster randomised trial. Lancet 371: 1595–1602.
33. Schmidt WP, Cairncross S (2009) Household water treatment in poor populations: is there enough evidence for scaling up now? Environ Sci Technol 43: 986–992.
34. SODIS Web site. Available: http://www.sodis.ch.
35. Colford JM Jr, Wade TJ, Sandhu SK, Wright CC, Lee S, et al. (2005) A randomized, controlled trial of in-home drinking water intervention to reduce gastrointestinal illness. Am J Epidemiol 161: 472–482.
36. Perera R, Heneghan C, Yudkin P (2007) Graphical method for depicting randomised trials of complex interventions [review]. BMJ 334: 127–129.

Section IX - Bottled Water Microbiology

CHAPTER 34

BACTERIOLOGICAL AND PHYSICAL QUALITY OF LOCALLY PACKAGED DRINKING WATER IN KAMPALA, UGANDA

Abdullah Ali Halage, Charles Ssemugabo, David K. Ssemwanga, David Musoke, Richard K. Mugambe, David Guwatudde, and John C. Ssempebwa

INTRODUCTION

Packaged water is any potable water that is manufactured or processed for sale which is sealed in food-grade bottles, sachets, or other containers and intended for human consumption [1]. Sale of packaged water has exploded all over the world in recent years, largely as a result of public perception that it is safe, tastes better, and has a better quality compared to raw tap water [2–5]. The increment in consumption globally could also be due to a result of an increase in per capita use as well as population growth [6, 7]. Even in countries where tap water quality is considered excellent, demand was so high, making packaged water the fastest growing product of the nonalcoholic beverage market worldwide [8, 9]. The above situation is not any different in the study area, Kampala, and all over Uganda where water packaged in bottles and polythene bags has become a very common consumer product most especially in the urban centres. The high demand may attract substandard products and counterfeits into the market.

With this increase in consumption of packaged water, there is a possibility of producing products that are not fit for human consumption because of monetary interests [10]. Access to safe drinking water is still one of the major challenges of the 21st century [11]. Unsafe water is a global public health threat, placing persons at risk for a host of diarrhoeal diseases as well as chemical intoxication [12]. Although disease outbreaks due to contaminated packaged water are not common, any possible contamination may lead to widespread epidemic because of the high demand and coverage [13]. Therefore access to an adequate and safe water supply is very important and can result in significant benefits to health [14]. Packaged water has ever been implicated as a source of outbreaks of cholera and typhoid fever as well as traveler's disease in countries such as Portugal and Spain [1, 15–17]. Several studies have shown that packaged water can be contaminated with bacteria at various stages of production [18–23]. Under improper or prolonged storage of bottled water, bacteria can grow to levels that may be harmful to human health [1, 4, 5].

In Kampala, production and consumption of packaged water are increasing at a very high rate. However, continuous

Citation: Abdullah Ali Halage, Charles Ssemugabo, David K. Ssemwanga, et al., "Bacteriological and Physical Quality of Locally Packaged Drinking Water in Kampala, Uganda," Journal of Environmental and Public Health, vol. 2015, Article ID 942928, 6 pages, 2015. doi:10.1155/2015/942928

surveillance of its quality at retail premises is not being carried out. This may lead to consumption of poor quality packaged water. There has also been a growing concern about the microbiological quality of the products. Therefore, this study assessed the physical and bacteriological quality of locally packaged water sold for public consumption in Kampala, Uganda.

MATERIALS AND METHODS

Study Setting and Design

This was a cross-sectional study carried out in Kampala, the capital city of Uganda. Kampala is divided administratively into 5 divisions, namely, Central division, Kawempe division, Makindye division, Nakawa division, and Lubaga division. It covers a total area of 195 square kilometers with a population of 1,516,210 [24]. Available sources of drinking water in the city include spring water, piped water, wells, boreholes, and lake. The three major causes of morbidity and mortality in Kampala are malaria, respiratory infections, and diarrhoeal diseases. There are over 20 bottled water brands in Uganda and around 14 of them are in Kampala. However, the common brands on market are 10. The number of sachet water manufacturing premises is not known since most of them operate illegally.

Data Collection and Laboratory Analysis

Selection of the Divisions and Parishes

Three of the five municipalities in the city, namely, Lubaga, Makindye, and Kawempe, were randomly selected for involvement in the study. Balloting method with replacement was used. One parish from each municipality was chosen at random using the same method. Main trading centres of the selected parishes were purposively selected because it is where most of the packaged water brands sold were likely to be found. The selected trading centres in each municipality were Bwaise in Kawempe, Kasubi in Lubaga, and Kabalagala in Makindye municipality.

Sampling of Packaged Water

A total of 90 samples (60 bottled and 30 sachets) of all the common brands of packaged water available on market during the month of January were collected for this study. A total of 10 brands of bottled water and 15 brands of sachet water were sampled. All samples were purchased randomly from several retail outlets and supermarkets in the selected trading centres of Kampala. At the retail outlets, duplicate samples (two) bottles of water of each brand and of the same batch were procured from each of the three selected trading centres, thus 20 samples from each municipality giving a total of 60 bottled samples. All the fifteen sachet water brands available on market were sampled. On average, five different brands of sachet water were sampled in duplicate from each of the three trading centres giving a total of 15 brands and 30 samples of sachet water.

All the samples were retained in their original sealed containers and clearly marked for identification with letters from A to J for bottled water and from K to Y for sachet water samples. They were transported to School of Public Health Laboratory, Makerere University, at Kasangati in Kampala in a sample carrier containing ice packs and analysed within 2 to 4 hours of collection. Then 25% of the samples, a blank sample (a negative control) and spiked sample (a positive control), were sent to National Water and Sewerage Corporation (NWSC) laboratory in Kampala for analysis so as to compare with the analysis findings.

Water Sample Analysis in the Laboratory

The membrane filtration method for bacteriological analysis with M-Lauryl Sulphate Broth as culture medium was used. 100 mL of water from each sample was filtrated using a membrane filter and then incubated at 44.5°C and 37.5°C for 18 hours for faecal and total coliforms, respectively. The samples were analysed in duplicate and the average was recorded. These temperatures are known to be optimal for the growth of these bacteria. Turbidity was measured using turbidity tubes and reported in NTU and pH using Wagtech pH meter. The equipment was calibrated before use. The analysis for total coliform and faecal coliform organisms per 100 mL was reported in terms of cfu/100 mL. Glassware or materials used in this experiment were washed with distilled water and then autoclaved at 121°C for 15 minutes to ensure sterility. The bacteriological and physical quality of the packaged water was determined by comparing it with the national and WHO standards for packaged water which requires water samples of good quality to be with 0 coliforms per 100 mL of water and turbidity of <5 NTU and pH of 6.5–8.5, respectively.

Data Analysis and Management

Data was entered using SPSS version 17.0 computer software for analysis. Data was analysed using both descriptive and inferential statistics. For the descriptive statistics, frequencies and cross tabulations were generated. The main outcome was bacteriological quality of the packaged water (presence or absence of either total coliform or faecal coliform). Other outcomes were turbidity level (within the acceptable limits or not) and pH (within the acceptable limits or not). Mann-Whitney U test was used to compare median total coliform count of bottled water to that of sachet water. Wilcoxon signed-rank test was used to test whether the mean of the packaged water brands was statistically different from the required standard of zero total coliform forming units per 100 mL. Fisher's exact test was used to compare the proportion of bottled water with total coliform to that of sachet water. The strength of association was determined using odds ratio and P values at 95% level of confidence. The test for normality of the data used was Shapiro-Wilk test.

Quality Control

To ensure quality, water samples were transported and analysed within 2–4 hours. Additionally, samples were analysed in duplicate using standard methods. Equipment was calibrated before use, a blank sample and a spiked sample were included in the analysis, and a reference laboratory (NWSC) was used for validation of the findings.

Ethical Considerations

Approval to conduct the study was obtained from Makerere University School of Public Health Higher Degree Research and Ethical Committee. Permission was also obtained from Kampala Capital City Authority.

RESULTS

Bacteriological and Physical Quality of Packaged Drinking Water on Sale

The median total coliform count of sachet and bottled water samples was 3.0 and 0.0 per 100 mL of water, respectively. The median pH and turbidity of both bottled water and sachet water were within the acceptable range. All the samples (both bottled water and sachet water) showed negative growth for faecal coliform as shown in Table 1.

Table 1: Descriptive statistics of the total coliform count, pH, and turbidity in packaged water brands

Parameters	Median	Std. deviation	Range	Max.	Min.	Standards	
Bottled water						WHO	National
Total coliform	0.0	3.44	11	11	0.0	0	0
Faecal coliform	0.0						
pH	7.2	0.32	1.1	7.8	6.8	<8.0	6.5–8.5
Turbidity	0.0	0.0	0.0	0.0	0.0	<5	<5
Sachet water							
Total coliform	3.0	18.9	65.0	65.0	0.0	0	0
Faecal coliform	0.0	0.0	0.0	0.0	0.0	0	0
pH	7.0	0.33	1.0	7.6	6.7	<8.5	6.5–8.5
Turbidity	0.0	0.0	0.0	0.0	0.0	<5	<5

Association between Total Coliform and Type of Packaged Water

Using Wilcoxon signed-rank test, the median total coliform count in sachet water brands was found to be significantly above the acceptable level of zero ($Z = -2.94$, p value = 0.003) but that of bottled water brands was not ($Z = -1.633$, p value = 0.102). Therefore, sachet water is contaminated with total coliform. Mann-Whitney U test showed a statistically significant difference between the median count of total coliform in both sachet water and bottled water brands ((24) = 37.0, $p = 0.027$). Majority (70%, 21/30) of the sachet water samples and a few (15%, 9/60) of bottled water samples analysed showed positive growth for total coliform above the acceptable levels of 0 cfu/100 mL of water set by WHO and national standards. Most (73.3%, 11/15) of sachet water and a few (30%, 3/10) of bottled water brands had total coliform count above the acceptable limit of 0 cfu/100 mL as shown in Table 2.

Table 2: Proportion of packaged water samples and brands exceeding permitted total coliform, faecal coliform, pH, and turbidity limits

Packaged water parameters (cfu/100 mL)	Bottled (n = 60) Number (%)	Sachet (n = 30) Number (%)	WHO standards per 100 mL	National standards per 100 mL
Total coliform			0	0
Exceeding	09 (15.0)	21 (70.0)		
Not exceeding	51 (85.0)	09 (30.0)		
Brands exceeding permitted total coliform counts				
Exceeding	DEG	LOUV		
Not exceeding	ABCFHIJ	KMNPQRSTWXY		
Faecal coliform			0	0
Exceeding	00 (00)	00 (00)		
Not exceeding	60 (100)	30 (100)		
Turbidity				
Exceeding	00 (00)	00 (00)	<5	<5
Not exceeding	60 (100)	30 (100)		
pH			<8.5	6.5–8.5
Exceeding	00 (00)	00 (00)		
Not exceeding	60 (100)	30 (100)		

The odds of sachet water being likely to be contaminated with total coliform were 13 times higher when compared to bottled water (OR = 13.2, CI: 4.12–43.58) as shown in Table 3.

Table 3: Association between packaging materials for drinking water and total coliform occurrence in packaged water

Variable	Bottled water (n = 60) Number (%)	Sachet water (n = 30) Number (%)	Odds ratio (Fisher's exact test)	95% CI
Total coliform				
Present	09 (15.0)	21 (70.0)	13.2	4.12–43.58
Absent	51 (85.0)	09 (30.0)		
Total	60 (100)	30 (100)		

DISCUSSION

The physical and bacteriological state of packaged water is a very important aspect that should be observed by all the packaging companies. Bottled water is generally of good quality for drinking, but if it is not properly protected during bottling and transit, it could be contaminated [16]. As per the national and WHO standards, packaged water should have turbidity of less than 5 NTU, pH range of 6.5–8.5, and total coliform and faecal coliform of 0.00 per 100 mL. As a measure for safety, indicator organisms of coliforms are investigated and their presence generally indicates that the water may contain disease causing agents. Coliforms are in two categories: total (all) coliforms count which indicates general contamination and faecal coliform count which indicates faecal contamination and is a significant indicator of pollution.

The physical properties of packaged water showed that pH and turbidity were within the permissible limits which indicate that packaged water was suitable for consumption. These results are similar to studies conducted in Delhi, India; Alexandria, Egypt; and Kano state, Nigeria [25–27]. However, these results contradict results from studies carried out in Ghana and Nigeria which showed pH values for sachet water ranging from 5.4 to 7.6 [28, 29] and in Amhara, Ethiopia, from 5.3 to 5.5 [30] which implies that some of the packaged water was not suitable for human consumption. Another study carried out in Georgia, United States, showed that sachet ice water had 95 and 3 samples with pH and turbidity outside the acceptable levels [31].

The study findings showed that some bottled water samples and most of the sachet water samples had total coliform above the acceptable levels for human consumption and the deference was statistically significant between the 2 categories ($p = 0.027$). These findings to a greater extent concur with the findings of other studies [9, 17–21, 32–35]. However, they are similar to findings from a study carried out in Lahore, Pakistan [36], where 3 out of the 10 bottled and 4 out of the 15 sachet water brands tested did not meet WHO and national standards of 0 cfu per 100 m, and other studies in sub-Saharan Africa, where detectable total coliforms were found in packaged water [5, 37]. Correspondingly, the proportion of sachet water contaminated with total coliforms at 73.3% was higher than that in a study conducted in Accra, Ghana, where only 18% of the samples exceeded this level [9]. Unlike these studies, in Nigeria, all sachet water samples had total coliforms [38]. The rate of contamination varied according to the type of packaged water. The rate of total coliform contamination in sachet water was higher than that in bottled water. This finding concurs with those of studies conducted in Delhi and Jaipur, India; Dar es Salaam, Tanzania; and Kumasi, Ghana [18, 20, 25, 32]. Furthermore, none of the samples of both bottled water and sachet water were positive for faecal coliform and exceeded the acceptable turbidity level and pH range. This finding is consistent with the finding of other studies conducted by [9, 13, 32,39] who found out no microbial indicators of faecal contamination in packaged water.

The presence of total coliform in packaged water can be linked to a number of factors such as the raw water source used, hygienic practices observed during production, improper storage in unhygienic and high temperature conditions, and lack of protective measures due to the common treatment methods used (ozonation and ultraviolet light) against bacterial regrowth [40, 41]. In view of this, the possibility of regrowth of microorganisms is greatly increased considering that the temperature in Kampala may be as high as 28°C. Studies conducted in the United Arab Emirates and United States of America demonstrated that organisms multiply more easily between 25°C and 37°C [21, 42]. Another study carried out in Freetown, Sierra Leone, showed that increase in concentration of total coliforms may be due to growth of microorganisms already present within the packaged water [5]. The findings suggest need for the government and other stakeholders to intensify surveillance activities and enforce strict hygienic measures in this rapidly expanding industry to improve water quality. The source and treatment process of sachet water therefore need further investigation.

Some of the possible reasons why sachet water is more likely to be contaminated than bottled water could be attributed to better hygienic practices observed in the bottled water industry compared to sachet water industry. These include use of protective sealed caps on bottles, improved and hygienic filling system, source of water, time between production and sale, and use of nonreturnable plastic containers [5, 18, 20]. In Kampala the most common source of water used in sachet water is springs which are most likely to be contaminated [43]. There may also be a possibility that sachet water was contaminated before packaging as observed by [23,44]. Other reasons could be due to plastic sachets that can easily be punctured and leak compared to plastic bottles; sachet water is most likely to be hand filled with water of suspect quality under unhygienic conditions and sold cheaply; it is much easier to reseal a plastic sachet than to mend the broken seal of water bottle for producing counterfeits. Lastly, sachet water producers are operating illegally; therefore, they are not registered and being monitored.

Despite the presence of total coliforms in some packaged water samples, it may be safer than other alternative drinking water sources. Presence of faecal coliform is the most important indicator for water contamination. During the typhoid outbreak in Kampala, majority of the water sources tested were found to be contaminated with faecal coliforms [45]. In addition, spring water sources in Kampala were found to be contaminated with faecal coliform [43]. With only 85% of the population in Kampala having access to improved sources of drinking water [46], many consumers may resort to alternative sources, including sachet water, wells, and springs, among which sachet water may be the safest. Therefore, there is need to balance protection of public health and access to drinking water as emphasized in some studies conducted in Sierra Leone and Ghana [5,47]. These findings suggest the need for the government and other stakeholders to intensify surveillance activities and enforce strict hygienic measures to packaged water industries to improve water quality, The source, treatment process, and stage (retail outlet or manufacturing premises) of packaged water contamination should be investigated.

Study Limitations. In this study we only analysed for turbidity and pH as indicators for physical quality.

CONCLUSIONS AND RECOMMENDATIONS

It can be concluded from the results of this study that both bottled water and sachet water sold at retail outlets were free from faecal contamination. Majority of sachet water samples were contaminated with total coliform above acceptable limits for human consumption. Findings suggest need for the government and other stakeholders to intensify surveillance activities and enforce strict hygienic measures to packaged water industries and retail outlets to improve water quality. The source, treatment process, and stage (retail outlet or manufacturing premises) of packaged water contamination should be investigated.

ACKNOWLEDGMENTS

The authors would like to thank all of the participants of this research. The authors are grateful to Mr. Rudolf Buga for his support throughout the study.

REFERENCES

1. D. W. Warburton, "The microbiological safety of bottled waters," in Safe Handling of Foods, J. M. Farber and E. D. Ewen, Eds., Marcel Dekker, New York, NY, USA, 2000.
2. M. D. F. Doria, N. Pidgeon, and P. R. Hunter, "Perceptions of drinking water quality and risk and its effect on behaviour: a cross-national study," Science of the Total Environment, vol. 407, no. 21, pp. 5455–5464, 2009.
3. L. A Ward, O. L. Cain, R. A. Mullally et al., "Health beliefs about bottled water: a qualitative study," BMC Public Health, vol. 9, no. 1, article 196, 2009.
4. D. Venieri, A. Vantarakis, G. Komninou, and M. Papapetropoulou, "Microbiological evaluation of bottled non-carbonated ('still') water from domestic brands in Greece," International Journal of Food Microbiology, vol. 107, no. 1, pp. 68–72, 2006.
5. M. B. Fisher, A. R. Williams, M. F. Jalloh et al., "Microbiological and chemical quality of packaged sachet water and household stored drinking water in Freetown, Sierra Leone," PLoS ONE, vol. 10, no. 7, Article ID e0131772, 2015.
6. O. M. Sallam, "Political border effects on sustainable management of the arab shared aquifers resources," in Proceedings of the 3rd International Conference on Water Resources and Arid Environments and the 1st Arab water Forum, Environment Agency, Riyadh, Saudi Arabia, November 2008.
7. C. N. Umeh, O. I. Okorie, and G. A. Emesiani, "Towards the provision of safe drinking water: the bacteriological quality and safety of sachet water in Awka, Anambra State," in Proceedings of the 29th Annual Conference and Genral Meeting on Microbes as Agents of Sustainable Development, p. 22, Nigerian Society for Microbiology (NSM), Federal University of Agriculture, Abeokuta, Abeokuta, Nigeria, November 2005.
8. M. F. Doria, "Bottled water versus tap water: understanding consumer›s preferences," Journal of Water and Health, vol. 4, no. 2, pp. 271–276, 2006.
9. J. Stoler, R. A. Tutu, H. Ahmed, L. A. Frimpong, and M. Bello, "Sachet water quality and brand reputation in two low-income urban communities in Greater Accra, Ghana," American Journal of Tropical Medicine and Hygiene, vol. 90, no. 2, pp. 272–278, 2014.
10. WHO, Fact Sheet No.256—Bottled Drinking Water, 2000.
11. J. P. S. Cabral, "Water microbiology. Bacterial pathogens and water," International Journal of Environmental Research and Public Health, vol. 7, no. 10, pp. 3657–3703, 2010.
12. J. M. Hughes and J. P. Koplan, "Saving lives through global safe water," Emerging Infectious Diseases, vol. 11, no. 10, pp. 1636–1637, 2005.
13. K. K. Addo, G. Mensah, B. Donkor, C. Bonsu, and M. Akyeh, "Bacteriological quality of bottled water sold on the Ghanaian market," African Journal of Food, Agriculture, Nutrition and Development, vol. 9, no. 6, 2009.
14. WHO, Guidelines for Drinking-Water Quality: Recommendations, vol. 1 of Incorporating First and Second Addenda, World health organization, Geneva, Switzerland, 3rd edition, 2008.
15. P. A. Blake, M. L. Rosenberg, J. Florencia, J. B. Costa, L. do Prado Quintino, and E. J. Gangarosa, "Cholera in Portugal, 1974. II. Transmission by bottled mineral water," American Journal of Epidemiology, vol. 105, no. 4, pp. 344–348, 1977.
16. A. Mavridou, "Study of the bacterial flora of a non-carbonated natural mineral water," Journal of Applied Bacteriology, vol. 73, no. 4, pp. 355–361, 1992.
17. A. A. Bordalo and A. Machado, "Water bags as a potential vehicle for transmitting disease in a West African capital, Bissau," International Health, vol. 7, no. 1, pp. 42–48, 2014.

18. R. Gangil, R. Tripathi, A. Patyal, P. Dutta, and K. N. Mathur, “Bacteriological evaluation of packaged bottled water sold at Jaipur city and its public health significance,” Veterinary World, vol. 6, no. 1, pp. 27–30, 2013.
19. L. A. Semerjian, “Quality assessment of various bottled waters marketed in Lebanon,” Environmental Monitoring and Assessment, vol. 172, no. 1–4, pp. 275–285, 2011.
20. G. R. Kassenga, “The health-related microbiological quality of bottled drinking water sold in Dar es Salaam, Tanzania,” Journal of Water and Health, vol. 5, no. 1, pp. 179–180, 2007.
21. H. Nsanze, Z. Babarinde, and H. Al Kohaly, “Microbiological quality of bottled drinking water in the UAE and the effect of storage at different temperatures,” Environment International, vol. 25, no. 1, pp. 53–57, 1999.
22. A. C. Dada, “Packaged water: optimizing local processes for sustainable water delivery in developing nations,” Globalization and Health, vol. 7, article 24, 2011.
23. D. K. Dodoo, E. K. Quagraine, F. Okai-Sam, D. J. Kambo, and J. V. Headley, “Quality of ‘sachet’ waters in the Cape Coast municipality of Ghana,” Journal of Environmental Science and Health—Part A: Toxic/ Hazardous Substances and Environmental Engineering, vol. 41, no. 3, pp. 329–342, 2006.
24. UBOS, National Population and Housing Census 2014: Provisional Results, Uganda Bureau of Statistics, Kampala, Uganda, 2014.
25. A. Singla, H. Kundu, P. Basavaraj, S. Singh, K. Singh, and S. Jain, “Physico-chemical and bacterial evaluation of packaged drinking water marketed in Delhi—potential public health implications,” Journal of Clinical and Diagnostic Research, vol. 8, no. 3, pp. 246–250, 2014.
26. H. Z. Ibrahim, H. A. G. Mohammed, and A. M. Hafez, “Physicochemical properties of some bottled water brands in Alexandria Governorate, Egypt,” Journal of the Egyptian Public Health Association, vol. 89, no. 2, pp. 60–65, 2014.
27. M. Sheshe and A. Magashi, “Assessment of physicochemical quality of sachet water produced in selected local government areas of Kano Metropolis, Kano State—Nigeria,” Bayero Journal of Pure and Applied Sciences, vol. 7, no. 2, pp. 31–35, 2015.
28. B. S. N. Cheabu and J. H. Ephraim, “Sachet water quality in Obuasi, Ashanti Region, Ghana,” Journal of Biology, Agriculture and Healthcare, vol. 4, no. 5, pp. 37–42, 2014.
29. O. S. Ayodele and A. S. Ajayi, “Water quality assessment of Otun and Ayetoro Area, Ekiti State, Southwestern Nigeria,” Advancement in Science and Technology Research, vol. 2, no. 1, pp. 8–18, 2015.
30. F. Biadglegne, B. Tessema, M. Kibret et al., “Physicochemical and bacteriological quality of bottled drinking water in three sites of Amhara regional state, Ethiopia,” Ethiopian Medical Journal, vol. 47, no. 4, pp. 277–284, 2009.
31. S. L. Mako, M. A. Harrison, V. Sharma, and F. Kong, “Microbiological quality of packaged ice from various sources in Georgia,” Journal of Food Protection, vol. 77, no. 9, pp. 1546–1553, 2014.
32. K. Obiri-Danso, A. Okore-Hanson, and K. Jones, “The microbiological quality of drinking water sold on the streets in Kumasi, Ghana,” Letters in Applied Microbiology, vol. 37, no. 4, pp. 334–339, 2003.
33. J. Bharath, M. Mosodeen, S. Motilal et al., “Microbial quality of domestic and imported brands of bottled water in Trinidad,” International Journal of Food Microbiology, vol. 81, no. 1, pp. 53–62, 2003.
34. Y. B. Ngwai, A. A. Sounyo, S. M. Fiabema, G. A. Agadah, and T. O. Ibeakuzie, “Bacteriological safety of plastic-bagged sachet drinking water sold in Amassoma, Nigeria,” Asian Pacific Journal of Tropical Medicine, vol. 3, no. 7, pp. 555–559, 2010.
35. O. A. Igbeneghu and A. Lamikanra, “The bacteriological quality of different brands of bottled water available to consumers in Ile-Ife, south-western Nigeria,” BMC Research Notes, vol. 7, article 859, 2014. ·
36. S. Yousaf and M. A. Chaudhry, “Microbiological quality of bottled water available in Lahore city,” Journal of Pioneering Medical Sciences, vol. 3, no. 2, pp. 110–112, 2013.
37. I. A. Thliza, A. U. Khan, D. B. Dangora, and A. Yahaya, “Study of bacteriological load of some brands of sachet water sold in Ahmadu Bello University (Main Campus), Zaria, Nigeria,” International Journal of Current Science, vol. 14, pp. 91–97, 2015.
38. O. Oyedeji, P. O. Olutiola, and M. A. Moninuola, “Microbiological quality of packaged drinking water brands marketed in Ibadan metropolis and Ile-Ife city in South Western Nigeria,” African Journal of Microbiology Research, vol. 4, no. 1, pp. 096–102, 2010.
39. S. Ugochukwu, F. J. Giwa, and A. Giwa, “Bacteriological evaluation of sampled sachet water sold in Samaru-Zaria, Kaduna State, Nigeria,” Nigerian Journal of Basic and Clinical Sciences, vol. 12, no. 1, pp. 6–12, 2015.
40. E. E. Geldreich, Microbial Quality of Water Supply in Distribution Systems, Water Quality-Management, edited by B. Raton, CRC Press, New York, NY, USA, 1st edition, 1996.
41. D. W. Warburton, P. I. Peterkin, K. F. Weiss, and M. A. Johnston, “Microbiological quality of bottled water sold in Canada,” Canadian Journal of Microbiology, vol. 32, no. 11, pp. 891–893, 1986.
42. S. D. Raj, “Bottled water: how safe is it?” Water Environment Research, vol. 77, no. 7, pp. 3013–3018, 2005.
43. R. Haruna, F. Ejobi, and E. K. Kabagambe, “The quality of water from protected springs in Katwe and Kisenyi parishes, Kampala city, Uganda,” African Health Sciences, vol. 5, no. 1, pp. 14–20, 2005.
44. J. Stoler, J. R. Weeks, and G. Fink, “Sachet drinking water in Ghana›s Accra-Tema metropolitan area: past, present, and future,” Journal of Water Sanitation and Hygiene for Development, vol. 2, no. 4, pp. 223–240, 2012.

45. WHO, Typhoid Fever—Uganda, World Health Organization, Geneva, Switzerland, 2015.

46. MoWE, Water Atlas: Kampala District Information, Ministry of Water and Environment, Kampala, Uganda, 2010.

47. N. Yidana, I. Osman, Z. Yidana, J. D. Nyeadi, and C. A. Atogenzoya, "Assessing the compliance of sachet water producers to regulatory standards in Ghana: a study of the tamale metropolis," International Journal of Innovative Research and Development, vol. 3, no. 6, pp. 426–436, 2014.

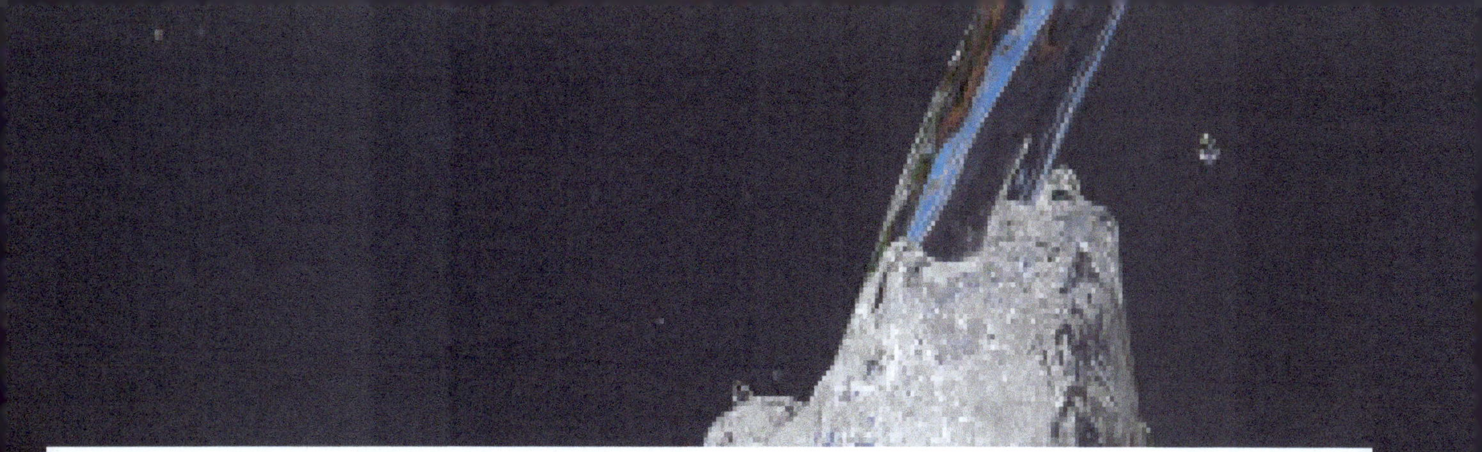

CHAPTER 35

BOTTLED WATER: UNITED STATES CONSUMERS AND THEIR PERCEPTIONS OF WATER QUALITY

Zhihua Hu, Lois Wright Morton and Robert L. Mahler

Keywords: Bottled water, water quality perceptions, ground water quality

INTRODUCTION

Consumption of bottled water is increasing by ten percent every year worldwide, with the fastest growth seen in the developing countries of Asia and South America [1]. The United States (U.S.) is the largest consumer market for bottled water in the world. The U.S. consumption of bottled water in 2008 was estimated to be 8.6 billion gallons, or 27.6 gallons per person [2]. Despite the common belief that bottled water is safer to drink and has better taste than tap water, scientific studies have shown that the belief is not necessarily true [3,4]. Research also shows that the sales and consumption of bottled water can have environmental and social impacts whose consequences are yet to be fully understood [5–7]. After years of substantial growth in sales, the U.S. bottled water market is recently slowing down. The current economic downturn may have played a part in the drop; however, environmental concern is also an important factor. Some research has found that environmental awareness campaigns may have curbed consumer demand [8–10].

Previous studies about bottled water have focused on its production, regulation, sales and consumption, and criticism and concerns. However, few researchers have examined the relationship between consumer use of bottled water and perceptions of drinking water quality. In this article, the authors explore the demographic and social factors associated with bottled water users in the U.S. and the relationship between bottled water use and perceptions of the quality of local water supply. A brief discussion of bottled water and tap water and bottled water consumers is used to develop several hypotheses. These hypotheses are tested using a national dataset representing twenty-one U.S. states. Results and discussion are followed by implications directed toward educators and public policy makers as they fund and develop programs that promote knowledge about health and local drinking water.

Bottled Water vs. Tap Water

Bottled water has been used in place of tap water for its convenience, better taste, and perceived purity [1,3,11]. Perceptions of bottled water being of higher quality, however,

Citation: Hu Z, Morton LW, Mahler RL (2011) Bottled Water: United States Consumers and Their Perceptions of Water Quality. Int. J. Environ. Res. Public Health 2011, 8(2), 565-578 doi:10.3390/ijerph8020565

are challenged by the increasing number of water quality incidents with bottled water [12]. A study showed that only five percent of the bottled water purchased in Cleveland, Ohio had the required fluoride recommended by the state, whereas the sampled tap water 100% met this requirement [3]. The same experiment also conducted bacteria count on both bottled water and tap water samples. The result showed that all of the tap water samples had a bacterial content under 3 CFUs/mL (colony-forming unit, a measure of viable bacterial or fungal numbers) and the bottled water samples' bacterial content ranged from 0.01–4,900 CFUs/mL. Although most of the water bottle samples were under 1 CFU/mL, there were 15 water bottle samples containing 6–4,900 CFUs/mL [3]. Another study focusing on the temperature and duration of storage for bottle water found that the bacterial growth in bottled water was markedly higher than that in tap water, especially at higher temperatures [4].

Many scientific reports on bottled water urge increased public awareness and development of guidelines/regulations on the industry of bottled water [1]. Incidents with bottled water quality are largely reported as associated with lenient regulations on bottled water. Bottled water plants are subject to the U.S. Food and Drug Administration (FDA) monitoring and inspection. Despite specific inspection requirements, bottled water plants are given low priority for safety inspection compared with other food plants because of FDA's staffing and financial constraints [13]. The "Nutrition Facts" label on bottled water usually shows only limited information about the water [1].

Despite the popularity of bottled water in the U.S., there are a number of environmental and social concerns. Plastic bottles are a waste problem adding to landfill overload when not recycled. Water bottling plants have impacts on local groundwater aquifers and streams [5]. Taking too much water can reduce or deplete groundwater reserves and reduce the flow of streams and lakes, causing stress on ecosystems. Although 75% of the world bottled water is produced and distributed on a regional scale, trading and transporting the other 25% bottled water also raises the concern for pollution and carbon dioxide emission [6]. The price of bottled water is on average 500 to 1,000 times higher than that of tap water [6], contributing to concern for affordable access to drinking water. Limited resource populations that use bottled water for drinking are least able to afford the high cost associated with bottled water [1]. Another issue associated with increased consumption of bottled water is that it can erode public tap water revenues and the capacity of governments to provide necessary improvements in basic water infrastructure [7].

Consumers of Bottled Water

Eighty-five million bottles of water are consumed in the United States every day and more than thirty billion bottles a year [14]. The adoption of a health preventive action like drinking bottled water is suggested to be influenced by perception of risk associated with drinking water [15]. The perception of risk is also thought to be closely related to the subjective assessment of drinking water quality [11]. This suggests that perceptions of drinking water safety and beliefs about the ground and surface water quality in a local area might be explanatory factors for a decision to select bottled water over tap water.

Another safety factor influencing consumer decision to select bottled water over tap water is the type of water supply system where the consumer lives. Small water systems (small town, tribal system, rural water district) [16] in the U.S. were found to have problems complying with federal/state quality standards. According to one study, due to inadequate funding and facilities, small water systems reportedly violated federal drinking water regulations more frequently than larger ones [11]. Although the number of public water consumers whose water does not meet current standards has decreased significantly over years, the task of water regulation is still challenging given both the financial limitations and increasing public concern about their drinking water [11].

Socio-economic status is also a factor affecting consumer decisions, particularly given the high cost associated with bottled water. Gender and education differences have been found to affect preference of bottled water over tap water because of their noted differences in perception of environmental risk [11,17].

Risk perception and preventive behaviors are the result of complicated social, cultural, and psychological factors as well as objective information [18]. This suggests that because of the differences in economic, social, and environmental contexts, residents of different regions might have different attitudes towards bottled water. In an earlier study, the findings showed that people in the Pacific region had more per capita consumption of bottled water than in other places of the U.S. [11]. In this article, the regional factor is examined and the popularity of bottled water is mapped across geographic regions.

EXPERIMENTAL SECTION

Hypotheses

Prior studies of bottled water consumption have identified a variety of explanatory factors for consumption behavior. However, these factors have not been considered together in one single model. For example, the regional differences found between the Pacific and the rest parts of the U.S. might be due to confounding factors such as differences in community size, local water quality problems, or water supply systems. Therefore, we propose to test these variables of interest simultaneously using a logistic regression. Hypotheses regarding use of bottled water are as follows:

- H1: Perceptions of poorer groundwater and surface water quality represent higher risk in drinking water and therefore are hypothesized to be associated with higher likelihood of purchasing bottle water as a primary drinking source compared to those reporting perceptions of higher water quality. Related, perceptions that drinking water is not safe are associated with higher likelihood of purchasing bottled water for drinking as a primary water source.
- H2: Based on the observations about small water supply systems, we hypothesize that small water supply (community well and rural district) users are more likely to use bottled water for drinking compared to public municipal water supply users. Community size is used as a control variable.

- H3: Because of the environmental impact associated with bottled water, we test the association between environmental attitudes and bottled water use. The association between the two is hypothesized to be that the more pro-environmental views a person holds, the less likely the person frequently uses bottled water for drinking.
- H4: We hypothesize a regional effect on the use of bottled water, although the specific pattern about such regional differences is not clear at this stage.

Other variables tested in the logistic model include age, education, and gender.

Methodology

Data used for this study were collected from a national stratified random sample mail survey about water issues conducted by Dr. Robert Mahler of University of Idaho. Our analysis used data from twenty-one states, which partially cover five out of the ten U.S. EPA water regions [19]. Data were collected 2004 through 2009 (region 8 and 9, 2004; region 7, 2006; region 6, 2008; and region 4, 2009). Sample sizes for each state were calculated based on the state population and targeted sampling error of four to six percent, with anticipation that the return rate would exceed fifty percent [20]. In each individual state, samples were either randomly selected from phone books or obtained from a professional social sciences survey company (Survey Sampling International, Norwich, Connecticut). The questionnaires were pilot tested, revised, and then mailed to sampled names and addresses. The final sample size was 5,823. Standard mail survey methods [21] were followed in all the regions and institutional review board (IRB) approval was obtained from University of Idaho Office of Research Assurance prior to the survey process. Response rates of each state ranged from 37% to 70%, with median return rates reaching the targeted 50%. The questionnaires, generally about 50 questions, varied in their content and wording due to the regions' differing priorities. However, there were a number of core questions that all states asked. It is these questions in common that make up our data set. These core survey items asked about respondents' perceptions of water quality, use of bottled water, water supply type, general environmental attitudes, and demographic information.

Two sources of drinking water questions were of interest in this study. The first one was "where do you primarily get your drinking water." Possible responses to this question included: private supply (private well, river, pond, lake,*etc.*), public municipal supply, small water supply systems (including rural water district and community well), and purchase bottled water. If respondents chose "purchase bottled water" for this question, they were identified as primary users of bottled water.

The second question asked if the respondent "often use bottled water for drinking purposes." If respondents answered "yes" to this question, they were labeled as regular users of bottled water. The above two questions were not mutually exclusive, which means that a primary bottled water user may be a regular bottled water user.

First, we tested hypotheses one, three and four on the primary bottled water users using a logistic regression model. The independent variables used in this logistic regression were as follows:

Surface and ground water quality perceptions. Respondents were asked to rate the surface and ground water quality in their area. Responses were coded 1 = poor, 2 = fair, 3 = very good/excellent.

Drinking water safety. The original question asked if the respondents felt their home drinking water is safe to drink. Response options were 0 = no, and 1 = yes.

Environmental attitudes. Respondents were asked to indicate where they stand on environmental issues by placing a mark on a line with numbers 1 to 10, where 1 represented preference for total natural resource use and 10 represented preference for total environmental protection.

Community size. Community size was measured by asking respondents to choose from the options which best described their community size, although no strict definition was given to the term "community". Community sizes were measured with five categories. 1 was "less than 3,500 people"; 2 = "3,500 to 7,000"; 3 = "7,000 to 25,000"; 4 = "25,000 to 100,000", and 5 was "more than 100,000."

Age and gender. Age was a continuous variable measuring the ages of respondents, and gender was recorded as 0 = female and 1 = male.

Education. Five categories of formal education levels were provided to choose from, ranging from "less than high school" to "advanced degree."

Residence region. The two bottled water questions of interest were asked in the following regions and states, which include several states of the southeast region (Region 4: Alabama, Florida, Mississippi, Tennessee); the southern region (Region 6: Arkansas, Louisiana, Oklahoma, Texas); the Midwest Heartland region (Region 7: Iowa, Kansas, Missouri, Nebraska); the mountain region (Region 8: Colorado, Montana, North Dakota, South Dakota, Utah, Wyoming); and the southern Pacific region (Region 9: Arizona, California, Nevada [22]). Figure 1 gives a visualization of the above states and regions.

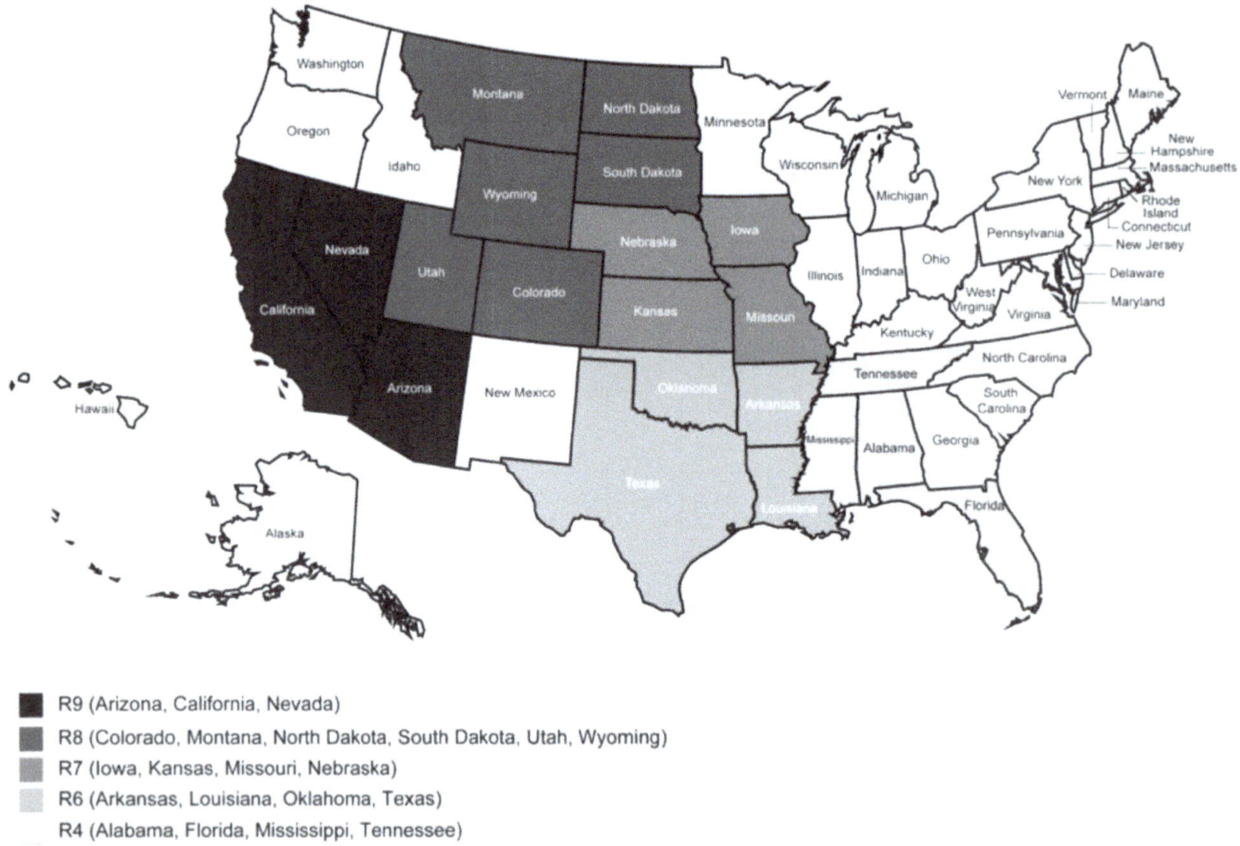

Figure 1: Map of the Sampled Regions and States.

Secondly, we applied a logistic regression on the regular bottled water users. With this part of analysis, we focused on the respondents who used sources other than bottled water for primary drinking purposes but reportedly often used bottled water for drinking. The hypothesis to be tested with this model is the second one, and the independent variable of primary interest is water supply type, which has three categories: 1 = private supply (private well, river, pond, lake, *etc.*), 2 = public municipal supply, and 3 = small water supply systems (including rural water district and community well). All the other independent variables used in the previous model were also included in this logistic regression model.

RESULTS AND DISCUSSION

Descriptive Summary of the Sample

The demographic distribution of survey respondents was similar to that reported for the general adult population based on the 2000 US census data for the demographic factors of community size, age (adult population), and formal education level. The only factor not in line with 2,000 census data was gender. Here, male respondents were much more heavily represented compared to the general population as a whole (about two thirds of the respondents were male, see Table 1). Even though 50% of the mailed surveys were addressed to females, it was apparent that the male adult in the surveyed household was more likely to respond to the survey [20]. The summary of sample statistics is shown in Table 1 below.

Table 1: Summary Statistics

Variable Name	Description	Responses	Mean/%	Std. Dev.
Primary bottled water users	Do you primarily purchase bottled water	0 = no	86.6%	
	for your drinking water? (N = 5,823)	1 = yes	13.4%	
Regular bottled water users	I often use bottled water for drinking	0 = no	54.6%	
	purpose (N = 5,821)	1 = yes	45.4%	

Surface water quality	What is the quality of surface water (rivers, streams, lakes) where you live? (N = 5,142)	1 = poor 2 = fair 3 = excellent	1.99	0.482
Ground water quality	What is the quality of groundwater (sources of well water) in your area? (N = 4,166)	1 = poor 2 = fair 3 = excellent	2.22	0.590
Environmental attitudes	Indicate how you see yourself on environmental issues (N = 5,797)	1 = for total natural resource use 10 = for total environmental protection	5.76	1.580
Age	Age of respondents	Range from 18 to 95	56.39	16.094
Community size	What size do you consider your community to be? (N = 5,749)	1 = less than 3,500 people 2 = 3,500 to 7,000 people 3 = 7,000 to 25,000 people 4 = 25,000 to 100,000 people 5 = more than 100,000 people	3.63	1.386
Education	What level of education you have completed? (N = 5,930)	1 = less than high school or some high school 2 = high school graduate 3 = some college or vocational training 4 = college graduate 5 = advanced college degree	3.40	1.125
Gender	What is your gender? (N = 5,603)	0 = female	33.7%	
		1 = male	66.3%	
Public water supply	Where do you primarily get you drinking water? (N = 5,044)	1 = private water supply (from private well, river, pond, lake)	13.5%	
		2 = public municipal supply	73.1%	
		3 = small water systems (community well or rural water district)	13.4%	
Region	EPA water geographical regions (N = 5,823)	4 (AL, FL, MS, TN)	23.1%	
		6 (AR,LA, OK, TX)	19.3%	
		7 (IA, KS, MO, NE)	16%	
		8 (CO, MT, ND, SD, UT)	17.1%	
		9 (AZ, CA, NV)	24.6%	

Over 13% of all respondents reported that they used bottled water as the primary source for drinking water, while 45.4% of all respondents said they often used bottled water for drinking. The mean for surface water quality perception was 1.99 (fair), and the mean for ground water quality perception was 2.22 (slightly above fair), a little higher than that of surface water. About fifteen percent respondents said they felt their home drinking water was not safe to drink. This percentage corresponded well to the percentage of respondents that used bottled water as their primary drinking source. On a scale of 1 to 10, average environmental attitude score was 5.76, and responses tended to cluster in the middle of the 1 to 10 scale. Thirty-five percent respondents marked their environmental view as 5, midway between totally eco-centric and totally anthropocentric. Other responses with higher percentage are 4 (9%), 6 (15%), and 7 (16%). About 12% respondents responded with higher scores (8–10), and the lower extreme scores (1–3) are only 6% of the total responses. This represents a balanced, somewhat more pro-environmental view towards the relationship between protection of nature and human use of natural resources. Mean age of the survey respondents was 56.8, while average formal educational achievement was between "some college" and "college degree." About two thirds of the respondents were male.

Logistic Regression Model 1: Primary Bottled Water Users

Our first model used a logistic regression model to examine the relationship between primary bottled water users and water quality perceptions (Table 2).

Table 2: Logistic Regression for primary bottled water users (N = 3,232)

Variables	**OR (Odds Ratio)**	**CI 95%**
Surface water quality	0.778 *	0.593–1.021
Ground water quality	0.670 ***	0.529–0.849
Environmental attitudes	0.967	0.897–1.041
Age	0.978 ***	0.970–0.986
Community size	1.116 **	1.014–1.229
Education	0.915	0.817–1.025
Gender (control: male)	1.323 **	1.025–1.707
Drinking water (control: safe)	5.884 ***	4.464–7.757
Region (control: region 9)		
Region 4	0.764	0.550–1.061
Region 6	0.966	0.692–1.348
Region 7	0.191 ***	0.108–0.337
Region 8	0.470 ***	0.317–0.697
Pseudo R^2 (Nagelkerke)	**0.263**	

Note: * P < 0.10; ** P< 0.05; *** < 0.001.

We found that groundwater quality perception was a significant predictor. As the ground water quality perception increased by one ascending-ordered category, the odds of a person using bottled water as primary source of drinking water was reduced by 33%. Compared with a person who feels their home water is safe to drink, a person who does not trust their home drinking water safety was more than 4.8 times more likely to use bottled water as their primary source of drinking water. However, there was no significant difference in bottled water use among respondents with different surface water quality perceptions. Environmental attitudes were not a significant predictor for primary bottled water use.

Age and gender were also found to be significant predictors for bottled water use. When all other conditions were exactly equal, a respondent who was one year older in age was about 2% less likely to use bottled water as the primary source of drinking water. From a gender standpoint, the odds that a female uses bottled water for primary drinking source are 1.32 times as much as the odds for a male, with all other conditions being equal. Education level was not a significant predictor for bottled water use.

Place of residence was found to have important effect on the use of bottled water. For example, community size had a positive relationship with being a primary bottled water user. As the community size increased by one ascending category, the odds of the resident of larger community using bottled water for primary drinking purposes were increased by 0.116 times. The use of bottled water as primary source of drinking water was also closely related to where the respondents lived in the U.S. For example, a respondent in the Midwest (region 7), when compared with a respondent living in the southern Pacific region (region 9), was over 80% less likely to be a primary user of bottled water. Similarly, for a respondent in the mountain region (region 8), the odds of the person using bottled water as primary drinking water source were reduced by 53% compared with a resident in the southern Pacific region (region 9). Similar to the southern Pacific region (region 9), the southern region (region 6) and the southeast region (region 4) also have more residents primarily depending on bottled water for drinking (see appendix for detailed regional bottled water use comparison).

With logistic regression models, there is no equivalent r-squared statistics to show the explained variability in the dependent variable. However, the pseudo R^2 shows that the explanatory variables have moderate strength of associations with consumption of bottled water. The model non-significant chi-square test and likelihood ratio test statistics (1.0), which suggests good model fit [23].

Overall, this model shows that U.S. consumer perceptions about groundwater quality have strong associations on the purchase of bottled water for drinking. This suggests that bottled water use may be considered a substitute for other water sources when groundwater quality is perceived to be poor.

Logistic Regression Model 2: Regular Bottled Water Users

A second logistic regression model was used to predict regular users of bottled water (Table 3).

Table 3: Logistic Regression for regular bottled water users (N = 2,850)

Variables	OR	CI 95%
Surface water quality	1.049	0.874–1.259
Ground water quality	0.661 ***	0.566–0.771
Environmental attitudes	0.973	0.925–1.024
Age	0.984 ***	0.979–0.989
Community size	1.046	0.981–1.116
Education	1.039	0.965–1.120
Gender (control: male)	1.243 **	1.035–1.493
Drinking water (control: safe)	1.742 ***	1.307–2.323
Region (control: region 9)		
Region 4	0.988	0.785–1.243
Region 6	1.144	0.896–1.461
Region 7	0.541 ***	0.413–0.708
Region 8	0.528 ***	0.413–0.676
Small water supply systems		
Private water supply	1.746 **	0.560–0.993
Municipal water supply	0.896	0.709–1.132
Pseudo R^2 (Nagelkerke)	**0.094**	

Note: * P < 0.10; ** P < 0.05; *** P < 0.001.

These results show similar patterns as with primary bottled water users found in Table 2. Groundwater quality perception, safe drinking water perception, age, gender, and region of residence were found to be significant predictors. Community size, however, unlike in the first regression model, was not significant. The likelihood of private water supply users being regular bottled water users was about 25% less than that of small water supply system users. There were no significant differences in bottled water use between municipal water supply users and small water supply system users.

The pseudo r-squared statistics are relatively small compared with our first model, which suggests that the same independent variables do not have particularly strong correlations with or explaining power for regular bottled water usage, although the chi-square test statistic is still non-significant.

DISCUSSION

With findings of both logistic models, we confirmed the hypothesized negative association between perception of ground water quality and bottled water use. Given that an estimate of 49% of the U.S. population depends on groundwater for its drinking water supply from either a public source or private well [24], the groundwater quality perception seems to explain the consumers' behavior regarding bottled water. Perception of drinking water safety is found to be highly associated with bottled water use. The findings about water quality perceptions generally confirmed that when public doubts about the safety of their tap water, they look for alternatives like bottled water [6,14]. No significant relationship, however, was found between surface water quality perception and bottled water use.

Our data do not include actual water quality or safety conditions so it is not known whether consumer's perceptions of the condition of their local drinking water are accurate reflections of the real water quality or not. If perceptions are accurate, then community leadership along with regulatory agencies needs to act to correct the problems for public health to be maintained. However, one might ask why consumers have turned to bottled water purchases rather than voice their concern and pressure public water departments and elected officials for solutions. This is particularly relevant since it is public municipal and rural water system supply users rather than private water supply users that are likely to purchase bottled water. Public water systems are tax supported, regulated and maintained under much more rigorous monitoring and testing conditions than bottled water manufacturers. This suggests that if a large number of consumers purchase drinking water as a substitute for public tap water, they can undermine the water infrastructure investments needed to assure safe public water supplies. This has implications for community capacities to provide low cost, accessible, and safe drinking water for their entire population. Without safe public water supplies, limited income households' health and well-being are at risk.

Our findings show that although municipal water supply users and small water supply users were equally likely to be regular bottled water users when every other condition is held the same, private water supply users (private well or surface water sources) were less likely to use bottled water than small water supply users. Consumers on private wells are often targets of public health campaigns reminding them to have their water tested regularly. To the extent this happens, private water supply users may believe they have more knowledge of and control over the quality of their water supply and thus trust it. Also, media coverage and increased headlines concerning problems with public water systems around the world can lead to high distrust (appropriately) of local water supplies [14]. The poor water conditions also increase the cost of treating water in public systems so that it is safe for consumption. This can lead to changes in water taste despite being safe to drink after treatments. While substituting bottled water for public tap water under these circumstances may be a short term "fix", it does not address long term problems of water quality or the effect it has on escalating the cost of public water as increased treatments become necessary.

Residents of larger communities were found to be more likely to be primary bottled water users, which means that a higher proportion of population in larger communities tend to depend on bottled water rather than their tap water for drinking purpose. Note that this association is established when other conditions are controlled for. That is, for two persons in the same region, with the same perceptions towards their drinking water, surface and ground water quality, and having exactly the same demographic characteristics (age, gender, education), the person from larger community is more likely to depend on bottled water for drinking purpose. As some researchers have suggested, factors like media hype about water supply problems, commercial campaigns on bottled water, or even peer pressure for more fashionable ways of drinking all contribute to bottled water consumption [6,14]. And considering that these factors are usually stronger in larger cities, it is likely that people in larger cities have more negative feelings about their water supply systems and turn to bottled water for solution. However, if respondents were already using some sort of water supply for drinking purpose, then there is no significant association found between their community size and whether or not they regularly consume bottled water. With limited information in our data we were not able to fully explain the associations found between community size and bottled water consumption, and we suggest future research look at community level variables for possible answers.

Our data also show that younger people and females are more likely to purchase bottled water. Young people are generally believed to be more susceptible to marketing and advertising, which are essential keys held by the bottled water companies [6,14]. And the higher likelihood of female drinking bottled water is consistent with previous literature on gender differences in risk, especially health and food related risk perceptions [25,26]. The findings about more consumption in these two groups of people suggests a need to target these audiences with messages about the importance of learning about their local water quality as well as the costs and quality differences between bottled water and public drinking water supplies.

Our hypothesis about environmental attitudes was not supported by the data. The relationship between environmental attitudes and bottled water use was not significant. Consumers with stronger overall concern about the environment do not seem to transfer this concern to pollution and waste problems associated with purchasing bottled drinking water. But again, because of the relatively longer cycle of research using multistate data (data collection in some states were done back in 2004), our data might not be able to reflect the newest trend of national environmental concern on bottled water.

Finally, the hypothesized regional effect regarding bottled water use was confirmed by the data. Residents of the Midwest and west mountain regions were far less likely to use bottled water for either primary drinking purpose or other occasions of regular uses, while residents of the southern pacific, the south, and the southeast were all equally likely to be bottled water users. This suggests that other variables such as culture, actual water quality conditions, media coverage of water issues and other place specific factors may be influencing the decision to use bottled water versus tap water from a private or public system. Water resource quantity and income might also be driving forces for the differences. Further research is needed to better explain regional variations.

CONCLUSIONS

Water is essential to human health and life. Access to safe water supplies and affordability are central concerns of public health and individual consumers. In this study we find that perceptions of ground water quality and local water supply safety are associated with decisions to purchase bottled water versus use public water systems for drinking water. When local water is not considered safe or of high quality U.S. consumers are more likely to use bottled water as a primary water source. Furthermore, negative perceptions of safety increase the likelihood of a consumer frequently purchasing bottled water regardless of whether their primary source of drinking water is a small water system or large municipal water supply system.

Two key implications of our findings are that (1) public health officials and community leaders need to work to assure that public municipal drinking water supplies are safe; in addition, they should find effective ways to communicate to local residents the safety of their water supply; and (2) environmental leaders and activists need to campaign about the long lasting impacts of plastic water bottles. Further the public must be engaged in understanding the relationship of water quality to the capacity of local water systems to maintain safety and good taste standards. Consumer distrust of their groundwater quality should be leveraged to create community action to address legitimate concerns.

ACKNOWLEDGMENTS

This research was partially funded by the National Institute of Food and Agriculture (NIFA), U.S. Department of Agriculture (USDA) under agreement 2008-51130-19526 also known as the Heartland Regional Water Coordination Initiative, the Iowa Agriculture and Home Economics Experiment Station, and USDA agreement 2008-51130-0474, also known as the Pacific Northwest Regional Water Resources Coordination Project.

Appendix 1

A separate analysis, a one-way ANOVA (analysis of variance) was done to compare regional differences in bottled water use for primary drinking purposes. Table 1 shows bottled water use in each region and differences with statistical significance. The variable (primarily purchase bottled water for drinking) is a dichotomous variable with two possible responses 0 (not purchase) and 1 (purchase bottled water for drinking). Therefore the following means reflect proportion of respondents responding with 1 in each region. Post-hoc Bonferroni pair tests were conducted on the means and the last column of the following table shows regions with significant differences (at 0.05 level). For example, the first row shows that region 4 has mean which is significantly different from that of region 6, 7, 8, and 9, respectively.

Table 1: Bottled water use by region

Region	N	Mean	S.D.	Regions with significant differences
R4	1,344	0.136	0.343	R6, R7, R8, R9
R6	1,122	0.185	0.389	R4, R7, R8
R7	932	0.045	0.208	R4, R6, R9
R8	993	0.063	0.244	R4, R6, R9
R9	1,432	0.198	0.398	R4, R7, R8

Region 9 and region 6 have significantly higher percent of primary bottled water users, followed by region 4. Region 7 and region 8 have the least primary bottled water users.

REFERENCES AND NOTES

1. Gleick, PH. The myth and reality of bottled water. In *The World's Water: The Biennial Report on Freshwater Resources 2004–2005*; Island Press: Washington, DC, USA, 2004.
2. *Smaller Categories Still Saw Growth as the U.S. Liquid Refreshment Beverage Market Shrank by 2.0% in 2008*; Beverage Marketing: New York, NY, USA; 31; August; 2009; Available online: http://www.beveragemarketing.com/?section=news&newsID=111 (accessed on 26 September 2010).
3. Lalumandier, JA; Ayers, LW. Fluoride and bacterial content of bottled water *vs*. tap water. *Arch. Fam. Med* 2000,*9*, 246–250.
4. Raj, SD. Bottled water: how safe is it? *Water Environ. Res* 2005, *77*, 3013–3018.
5. Glennon, RJ. *Water Follies: Groundwater Pumping and the Fate of America's Fresh Waters*; Island Press: Washington, DC, USA, 2002; pp. 1–3.
6. Ferrier, C. *Bottled Water: Understanding a Social Phenomenon*; Report Commissioned by World Wild Life: Gland, Switerland; April; 2001; Available online: http://search.panda.org/search?site=panda&client=panda_frontend&proxystylesheet=panda_frontend&output=xml_no_dtd&q=Bottled+water%3A+understanding+a+social+phenomenon (accessed on 26 September 2010).
7. Opel, A. Constructing purity: Bottled water and the commodification of nature. *J. Am. Culture* 1999, *22*, 67–76.
8. Block, B. *Bottled Water Demand May Be Declining*; Worldwatch Institute: Washington, DC, USA; 8; September; 2008; Available online: http://www.worldwatch.org/node/5878 (accessed on 26 September 2010).
9. Laumer, J. *US Bottled Water Sales Are Down Slightly. Why Are So Few Willing to Change?* 8 October 2009. Available online: http://www.treehugger.com/files/2009/10/us_bottled-water-sales-down-slightly-why-few-willing-change.php (accessed on 26 September 2010).
10. Trigaux, R. *Bottled Water Industry Sales Decrease during Recession*, Available online: http://www.tampabay.com/news/business/bottled-water-industry-sales-decrease-during-recession/1028389 (accessed on 12 January 2011).
11. Anadu, EC; Harding, AK. Risk perception and bottled water use. *J. Amer. Water Works Assoc* 2000, *92*, 82–92.
12. Pacific Institute. *Bottled Water Recalls Summary Table*, Available online: http://www.pacinst.org/topics/water_and_sustainability/bottled_water/ (accessed on 14 January 2011).
13. Posnick, LM; Kim, H. *Bottled Water Regulation and the FDA*, Published by Food Safety Magazine, August/September 2002 and reproduced by The Target Group. Available online: http://www.fda.gov/downloads/Food/FoodSafety/Product-SpecificInformation/BottledWaterCarbonatedSoftDrinks/ucm077094.pdf (accessed on 28 September 2010).
14. Gleick, PH. *Bottled and Sold: The Story behind Our Obsession with Bottled Water*; Island Press: Washington, DC, USA, 2010; p. ix.
15. Cleary, PD. Why People Take Health Precautions. In *Taking Care: Understanding and Encouraging Self-Protective Behavior*; Weinstein, ND, Ed.; Cambridge University Press: Cambridge, MA, USA, 1987; pp. 119–149.
16. *Asset Management: A Handbook for Small Water Systems One of the Simple Tools for Effective Performance (STEP) Guide Series*; Office of Water, U.S. EPA: Washington, DC, USA, 2003. Available online: http://www.epa.gov/safewater/smallsystems/pdfs/guide_smallsystems_asset_mgmnt.pdf (accessed on 28 September 2010).
17. Flynn, J; Slovic, P; Mertz, CK. Gender, race, and perception of environmental health risks. *Risk Anal* 1994, *14*, 1101–1108.
18. Glicker, JL. Convincing the public the drinking water is safe. *J. Amer. Water Works Assoc* 1992, *84*, 1–46.
19. The U.S.. EPA groups the fifty states into 10 regions based on geography and regional conditions. Available online: http://www.epa.gov/ow/region.html (accessed on 28 September 2010).
20. Mahler, RL; Gamroth, M; Pearson, P; Sorensen, F; Barber, ME; Simmons, R. Information sources, learning opportunities, and priority water issues in the Pacific Northwest. *J Ext* 2010, *48*, 2. Available online: http://

www.joe.org/joe/2010april/rb2.php (accessed on 28 September 2010).

21. Salant, P; Dillman, DA. *How to Conduct Your Own Survey*; John Wiley & Sons: New York, NY, USA, 1994; pp. 137–147.
22. Hawaii was removed from the study due to its uniqueness in geography and climate, which does not much resemble the rest three states of its region.
23. Hosmer, DW; Lemeshow, S. *Applied Logistic Regression*, 2nd ed; John Wiley & Sons: New York, NY, USA, 2000.
24. Source information from Texas Groundwater Protection Committee. Available online: http://www.tgpc.state.tx.us/DrinkingWater.htm (accessed on 28 September 2010).
25. Gustafson, PE. Gender differences in risk perception: theoretical and methodological perspectives. *Risk Anal*1998, *18*, 805–811.
26. Dosman, DM; Adamowicz, WL; Hrudey, SE. Socioeconomic determinants of health- and food safety-related risk perceptions. *Risk Anal* 2001, *21*, 307–317.

CHAPTER 36

HEALTH BELIEFS ABOUT BOTTLED WATER: A QUALITATIVE STUDY

Lorna A Ward, Owen L Cain, Ryan A Mullally, Kathryn S Holliday, Aaron GH Wernham, Paul D Baillie and Sheila M Greenfield

BACKGROUND

Demand for bottled water has consistently increased during the last decade, making bottled water the fastest growing segment of the non-alcoholic beverage market worldwide [1]. Consumption of bottled water in the UK rose from 1415 to 2275 million litres between 2000 and 2006 [2] and in 2003, UK consumers spent £1 billion on bottled water [3]. Indeed, for some of these consumers, bottled water has become a complete substitute for tap water [4].

However, these increases, in the UK at least, fly in the face of improving tap water quality over the last 10 years [5] and are even more surprising given that bottled water can cost on average 500–1000 times more per litre than tap water [6].

In contrast, recent media reports suggest a new-found scepticism about bottled water. A recent BBC Panorama documentary highlighted the environmental cost of bottled water [7]. There are also reports of the general public's reticence to accept the rising cost of bottled water on the basis of health claims [8, 9].

Citation: Ward LA, Cain OL, Mullally RA, Holliday KS, Wernham AGH, Baillie PD, Greenfield SM (2009) Health beliefs about bottled water: a qualitative study. BMC Public Health 9:196 DOI: 10.1186/1471-2458-9-196

LITERATURE REVIEW

No qualitative studies address public beliefs about bottled water in the English language peer-reviewed literature. A small number of quantitative studies have focused on this general area, although none have been carried out in the UK.

A discussion paper published in 2006 by Doria [1] reviewed the literature and identified dissatisfaction with tap water taste, odour and sight, and health concerns over tap water, as the major motivating factors for choosing bottled water. It also highlighted the lack of peer-reviewed literature on the reasons for choosing bottled water and the need for more research. In another discussion paper published on behalf of the Worldwide Fund for Nature (WWF) in 2001 [6], although there was some reliance on non-peer reviewed literature, similar suggestions were made indicating taste, safety and health issues to be the main motivating factors to buy bottled water.

The most recent original peer-reviewed research was an extensive quantitative study produced by the American Waterworks Association (AWWA) published in 2005 [10], a telephone survey of 2268 American residents. The findings of the study were that bottled water drinkers were satisfied with the quality and safety of tap water and suggested that most bottled water drinkers saw bottled water as a "luxury item", and not something that they purchased due to any perceived problems with their domestic water supply. It also identified taste, safety and healthiness as motivating factors in the choice to purchase bottled water as an alternative to tap water. However, whilst this study highlighted health beliefs as an important factor, it failed to explore what these consumers' health beliefs actually were.

A smaller telephone-interview-based quantitative study of residents of Quebec [11] concluded that taste, rather than safety, was the most important motivating factor for people buying bottled water. This study also noted consumer dissatisfaction with tap water. Research conducted in France in 1989, 1995 and 2000 [12] supports this, showing consistently that taste was a more frequently given reason for drinking bottled water than reasons concerning health and tap water risk.

In contrast, concerns over tap water safety and the consumption of bottled water as a substitute for other beverages were the motivating factors in over 80% of the 1600 Americans interviewed in an older study by the AWWA published in 1993 [13].

A large laboratory-based study by Olson [14] is often referenced in the literature. However, the claim made by the paper that "It is absolutely clear, therefore, that a leading reason for the explosion in bottled water sales is the public perception, fuelled by heavy industry advertising, that bottled water is pure and pristine, and thus a healthier choice than tap water" was not based upon any published research and was merely an expression of the author's opinion.

Existing quantitative literature broadly identifies health, taste, odour and sight as reasons for consumer preferences for bottled water. Health beliefs therefore seem to be an important factor in choosing to drink bottled water. However the existing research is not explicit about what these beliefs are and tends to take a comparative approach in identifying why the consumer might choose to drink bottled water as an alternative to tap water. Hence the health reasons given are generally beliefs concerning unfavorable properties of tap water. Previous research has been based in the USA, Canada and France and although these studies do seem to complement each other in the themes that arise, they concern different populations with different water supplies. It is unclear how generalisable the findings are to the UK.

This study was therefore designed to discover more about individuals' beliefs associated with bottled water and the perceived health effects of bottled water, in a UK setting. We also aimed to determine whether these beliefs were important factors in any motivation to drink bottled water and hence their possible contribution to the growth of the bottled water market.

METHODS

Recruitment of Participants

All participants in this study were users of the Munrow Sports Centre on the University of Birmingham campus. The majority of this sports centre's users are staff and students at the University of Birmingham, with a small number of members from the local community. This cohort was targeted based on the assumption that sports centre users might be more likely to have developed health beliefs than the general public. Recruitment occurred on six separate occasions at varying times of the day on different days of the week in January and February 2008 at the entrance to the sports centre. Every user entering or leaving the building during these times was approached.

Potential participants were briefed about the nature of the study by the researchers and after a provisional agreement to participate they were invited, by a standard e-mail on two occasions, to attend an individual interview at a time and place convenient to them. This ensured a relaxed atmosphere during the interview, with ample time in which to explore the interview themes. Each person who responded to the e-mail invitation was interviewed with no exclusions. Recruitment ended when all of the researchers agreed that no new themes were emerging from the content analysis and that data saturation had been reached.

Data Collection

Semi-structured interviews were deemed the most appropriate method of data collection for this topic for which there is a lack of previous research. Semi-structured interviews use some pre-determined, mainly open ended questions, to help define the area to be explored, but are flexible and allow the interview to diverge from this guide in order to pursue ideas in more detail [15].

A pilot study with four interviews was conducted, which informed the development of a basic interview schedule constructed from previous research findings in this area. The modified interview schedule was subsequently used for all of the interviews in this study. It prompted participants to discuss their habits of bottled water consumption, perceived differences between tap and bottled water, personal beliefs about bottled water and the beliefs that others hold about bottled water. It was emphasised that the term 'bottled water' referred to still bottled water purchased from a supermarket, shop or vending machine, and not sparkling water, or tap water in a bottle.

The interviews were conducted by six of the researchers (LW, OC, RM, AW, KH, PB) who were briefed in the techniques of qualitative interviewing by SG and practiced these techniques informally before data collection began. Two researchers were present at each interview, one conducting the interview and one operating the recording equipment. The interviews were recorded using digital recording equipment and later transcribed verbatim.

In total, 23 interviews were conducted between January and March 2008 in the participants' own homes, places of work, or at the University of Birmingham Medical School. The length of interviews ranged from 5 minutes 52 seconds to 22 minutes 12 seconds, mean length 12 minutes 17 seconds. The variation in length of the interviews was not related to the researcher conducting the interview. Instead the variability is mainly attributable to differences in participant knowledge and convictions about bottled water. The shortest interview resulted from the inclusion of one participant for whom English was not their first-language and hence they found it difficult to give extended answers to the questions.

Data Analysis

A qualitative content analysis using elements of grounded theory was employed to analyse the transcripts. The inductive process of identifying themes from the data collected, rather than applying top down a prioricategories, was appropriate given the paucity of previous research exploring public perceptions of beliefs about bottled water [16].

The data was broken down into meaningful phrases or sentences using open coding. Categories were then identified and used to find the central themes that emerged from the data. In line with the grounded theory approach, these themes were not the same as those in the interview schedule but arose from the coding process. Investigator triangulation ensured trustworthiness of the analysis. The data was divided between the team with two coders for each transcript. Regular meetings of the team gave room for discussion about all levels of coding, particularly about data inconsistent with the evolving categories.

Ethical Considerations

This study did not require ethics committee approval. Participation was voluntary and participants were free to withdraw at any time. Participants were assured that confidentiality would be maintained. Before the interview began, the purpose of the study was explained again and informed verbal consent for participation in this study was obtained.

RESULTS

Socio-demographic information about each of the participants was collected at the beginning of each interview [see Additional file 1]. Of the 23 participants, there were 19 females and 4 males, aged between 18 and 52 years and all were currently living in the West Midlands Region. In terms of ethnicity the majority were White British, although those of White Irish, White Other and Asian British backgrounds were also represented. Eleven of the participants were students at the University of Birmingham, ten were employed by the University of Birmingham and two were employed outside of the University.

Participants were classified into three groups based on their reported average bottled water consumption in litres per week. The majority of participants in this study, 19 participants, were ‹limited consumers› of bottled water and drank 0.5 to 3.5 litres per week. Two participants were ‹consumers›, drinking more than 10 litres each per week. Two participants were ‹non-consumers› and never drank bottled water.

Data analysis revealed two main themes encompassing participants› beliefs about bottled water; health beliefs and environmental concerns. A third theme, factors that motivate participants to purchase bottled water also emerged [see Additional file 2]. Selected quotes are presented to illustrate the different subcategories that arose within these themes.

Health beliefs about Bottled Water

Most participants believed that compared to tap water, bottled water conferred additional health benefits. Bottled water was considered to be a "healthy option" even if participants were unsure as to why:

> *"Um...well I think it's probably better for you than er...it's probably got good minerals and stuff in it... um...probably better for you health-wise."* (P5, Limited consumer)

> *"I mean I know it's good but I'm not sure why it's good."* (P11, Limited consumer)

Not only were the majority unsure as to why bottled water might confer health benefits, the nature of these health benefits was generally not specified either. In cases where participants did indentify a specific reason for the health benefit of bottled water compared with tap water, a belief that the minerals in bottled water conferred a health benefit was most commonly cited:

"It has different minerals added, which I assume to have some benefit." (P19, Limited consumer)

Only three participants gave any other specific health benefits. One participant believed that the bottled water could relieve symptoms of myalgic encephalopathy (M.E.), in the same way that filtering water seemed to provide some benefit to a family member:

> *"Err...it started off with filter water. We used to buy filters and just filter all the tap water and it got too much hassle so we went onto bottled water and we went to filter water because my cousin got M.E. and they found that filtering his water pretty much cured all his symptoms of M.E. So when he...they started doing that we then started doing it as well and then."* (P6, Limited consumer)

A second participant believed that bottled water was «*better for babies*" (P15, Limited consumer), and the third person who ascribed a specific health benefit felt that bottled water could improve bone strength. However, as with many participants, they lacked conviction in their belief and were unable to explain their reasoning:

"Um...er...it makes your bones stronger...I have no idea to be honest." (P5, Limited consumer)

The majority of participants associated bottled water with having fewer impurities than tap water, and were more likely to trust the quality of bottled water than tap water:

> *"They take out all the impurities don't they? And filter it all out."* (P16, Limited consumer)

> *"It's a bit purer than what comes out of the tap."* (P14, Consumer)

Conversely, only two participants believed that tap water in the UK had either higher quality standards than bottled water or fewer impurities:

> *"The quality control on tap water is much higher than on bottled water."* (P7, Limited consumer)

> *"There is all sorts of erm...dubious erm...ingredients in the bottled water."* (P12, Limited consumer)

Despite their beliefs about the purity of bottled water, most participants expressed doubts as to the extent of the health benefits of bottled water compared with tap water:

> *"I really don't think there is much difference between the two to be honest."* (P16, Limited consumer)

> *"If the water quality of the water coming out of the tap is good I'm not sure there are any health claims*

that can be justified with bottled water". (P12, Limited consumer)

These views were not mutually exclusive with beliefs about health benefits, indeed most participants believed that bottled water did have health benefits, but that these benefits were negligible:

"I suppose there must be some benefit I have no idea like you know what percentage of your recommended daily allowance it is...I certainly wouldn't drink it for minerals that otherwise I wouldn't get." (P8, Limited consumer)

A minority of participants consistently denied any health benefits:

"I don't think there is anything inherently better about bottled water than tap water." (P1, Limited consumer)

A minority of participants expressed concerns that drinking bottled water was detrimental to health. One subject believed that the «*actual process of putting it into the plastic bottle*" (P21, Limited consumer) might impair the purity of the water and have a negative impact on consumers' health. Several other participants suggested a link between the plastic used for packaging and cancer. However, these concerns were about the repeated refilling of empty bottles with tap water rather than being about bottled water *per se*:

"I have heard concerns about plastics and the chemicals from plastics getting into water and links to cancer."(P19, Limited consumer)

"What I've heard it did say something about the plastic starting to break down and something about it being able to cause cancer and as soon as anyone says something about cancer you'd think." (P11, Limited consumer)

"Um...it was about like the whole cancer of using the same bottle over and over again, it breaks down or something like that." (P22, Limited consumer)

Environmental Concerns

A number of participants believed that it was «*not eco-friendly*» (P18, Non-consumer) to use bottled water and expressed concerns about bottled water›s «carbon footprint». Some were concerned with the environmental impact of the plastic bottles:

"I prefer to drink tap water because er it's got no packaging." (P15, Limited consumer)

For others, the concern was with the environmental impacts of transporting the water:

"I try to be green (laughs) and buy local mineral water...I try to buy something from the UK so it's not flown over." (P4, Limited consumer)

When asked the reasons that may deter others from drinking bottled water, one person highlighted how topical the environmental issues were, by noting the impact of a recent BBC Panorama documentary:

"Because they'll have watched programmes like the Panorama programme where they see the actual result of all this bottled water coming from all these countries where people can't actually get clean drinking water because we're stealing it all to have in our fancy bottles um and I dunno possibly because like we are here our we have filtered mineral water we don't have bottled water for staff." (P17, Non-consumer)

Motivating Factors

Beliefs about the health benefits of bottled water emerged as a motivating factor influencing participants' decisions to drink bottled water in only a minority of cases, participants 5 and 11, both limited consumers:

"Obviously mineral water has the extra natural goodness in it whereas tap water they probably add things to it to make it cleaned up." (P11, Limited consumer)

Interviewer - *"Ok, and if we gave you a glass of tap water and a glass of mineral water and asked you to drink only one of them, which would you drink and why?"*

Participant- «*Um, the mineral water because it›s gonna have more minerals in, it›s gonna be fresher.*» (P5, Limited consumer)

Others chose to drink bottled water because of concerns over the safety of tap water:

"I'd probably choose the mineral water...because I think I have more confidence that it's been purified than the tap water." (P19, Limited consumer)

However, most people were of the opinion that the tap water was fit for purpose and nobody expressed the view that they had reservations about the safety of tap water that were strong enough to prevent them from using it:

"Tap water in England is fine, perfect" (P9, Limited consumer)

Analysis revealed a number of other motivating factors that were unrelated to health beliefs. Convenience, taste and cost were almost universally important. The most commonly cited reason for purchasing bottled water was convenience:

"When I go away to races I might not have a bottle with me so I'd buy some water..." (P3, Limited consumer)

"The only time I drink bottled water would be if I was out somewhere and wanted some water" (P7, Limited consumer)

Several participants described how they would buy bottled water as they preferred the «taste» of it to tap water:

"Well for me mineral water tastes of nothing...ha... cold nothing, I think tap water has some kind of taste to it..."(P4, Limited consumer)

Participant 16 summarised the general consensus regarding the cost of bottled water:

"If it was a choice to buy [bottled water] and not to buy it and it was drinkable from the tap then [they] would go for the tap water...it is mainly down to money." (P16, Limited consumer)

Other factors included preference over other soft drinks, influence of the media, influence of marketing and advertising, bottled water as a status symbol or as a luxury item and re-use of the bottle as a container for tap water.

DISCUSSION

This study found that most people did hold health beliefs about bottled water, but that in the majority of cases these health beliefs were not strong motivating factors for purchasing bottled water. Other factors such as convenience, cost and taste emerged as far more important reasons for any preference for bottled water. In addition, most participants felt that there was not a significant health benefit in drinking bottled water compared to tap water. From this, it is unlikely that the recent surge in bottled water consumption is due to beliefs about health benefits associated with bottled water.

These results are important because until now, no qualitative studies have been conducted exploring public perceptions about bottled water and the factors that motivate people to buy it. The findings complement previous quantitative studies that have been conducted in this area [10, 11, 12, 13]. The qualitative approach of this study allowed for a deeper exploration of the themes that were used in the quantitative data, and also gave room for new themes, not covered in the top-down approach of the quantitative studies, to emerge.

Convenience was a major motivating factor for buying bottled water, and one that has not been covered in previous quantitative studies. This may be because the ‹top-down› approach of questionnaire design did not include convenience as a category. It seems obvious that people who would normally drink tap water would be motivated to buy bottled water when tap water is unavailable, for example in a shopping centre, or at the cinema. ‹Convenience› is a motivating factor determined by the consumer›s situation, not by the consumer›s beliefs about bottled water.

Participants expressed health beliefs about bottled water that could be categorised as general health benefits or more specific health benefits. Although this is the first study to identify health beliefs about bottled water, the 2006 review by Doria suggests that there is much interest in the subject in both the grey literature and in the peer-reviewed literature, where unsupported claims regarding consumer beliefs are easy to find [1]. For example, Petrie and Wessely claim that bottled water is seen as a «natural antidote» to all the things bad for their health due to modernity [17].

A major emergent health belief was that most people were satisfied with the quality of their tap water supply and that it would not pose an adverse risk to their health. This is consistent with the data from Mackey et al[10], which demonstrated high tap water satisfaction, even in groups who drank bottled water in preference to tap water.

Interestingly, whilst the majority of participants expressed the belief that bottled water has health benefits of some kind, paradoxically these same participants also stated that the health benefits of bottled water are negligible or non-existent. This perhaps reflects confusion in the general public, as suggested by Olson [14], in that they only half-believe the marketing promoting health benefits of bottled water.

Such marketing might also explain why many participants, whilst able to state health beliefs regarding bottled water, were unable to explain or qualify these. The ability of marketing companies to create demand for bottled water "through the skilful use of language and image" has been discussed in a review of American culture [18]. This review suggests that in the public mind, purity, naturalness and healthiness are associated with bottled water through the specific marketing strategies of bottled water companies. The following statements taken from the websites of two leading brands seem to support this suggestion:

"You want the best for your body and we've got it. Taste and feel the volvic difference, pure and natural...courtesy of Mother Earth" [19]

"Replenish your body with the purity of Evian" [20]

There was some discrepancy between the specific health benefits participants believed bottled water to have and reality. Participants often felt that bottled water had an increased mineral content compared to tap water and that this conferred a health benefit. An extensive study conducted in the USA by Azoulay *et al* [21] compared the mineral content of tap water in various areas and a number of commercially available American and European bottled waters. Some brands of mineral water do indeed have a higher mineral content than tap water, which was found to be generally low in minerals, and were recommended as important dietary sources of calcium and magnesium. However, there is a considerable difference between bottled water brands, which no participant in our study seemed to be aware of.

Furthermore, in the USA study, some mineral waters were actually found to have a lower mineral content than the tap water supply, so the belief that all bottled waters are superior to tap water in terms of mineral content is incorrect. Although the study in question was conducted in the USA, the situation is likely to be similar in the UK. Moreover, whilst this study identified that some bottled waters provide a significant amount of the recommended daily intake for magnesium and calcium, none of these brands matched those that our participants drank. Of these, the preferred brands expressed all fell into the low mineral content classification of bottled water except for one, which was classified as moderate mineral content [21, 22].

It is also important to remember that these minerals can be obtained from other sources in the diet, so the health benefits of the minerals contained in bottled water are not exclusive to this source. In addition, research shows that drinking waters with low mineral content does not lead to mineral deficiencies [23].

Where participants were able to give specific health benefits of bottled water we were not able to find supportive evidence in all cases. One participant believed that the symptoms of M.E were ameliorated by bottled water, which is something that existing literature does not appear to support. However, another participant mentioned that they felt that bottled water was especially benficial for babies. Despite finding no recommendations for this practice, we were able to find one study which suggested that choosing a mineral water with low sodium content may be useful in preparing formula milk because a hyperosmolar diet has been linked to hypertension and obesity in later life [23]. Having said this, no evidence could be found to suggest that tap water was unsuitable for this purpose.

Safety has previously been identified as an important motivating factor for buying bottled water [10, 11, 12,13]. Indeed, this was a theme that emerged in this study. It is worth emphasising again that participants did not feel that UK tap water was unsafe. Indeed the quality of this has continued to increase over the last 10 years [5], but participants still felt that bottled water was safer and purer when compared to tap water.

Only one participant correctly stated that tap water was in fact subject to more stringent testing than bottled water in the European Union (EU). The 1980 European Directive on natural mineral water outlines standards for these waters [24]. This became UK law in 1985 [25]. Under these regulations natural mineral water cannot be sterilised or otherwise treated to destroy microorganisms. Bottled water is not free of microorganisms as some might believe and this has been demonstrated by numerous studies [4, 26, 27, 28].

Although European regulations are considered more rigorous than those in the USA [26], natural mineral waters are only tested every two months by independent laboratories, compared to tap water which is tested every two days in urban areas [5]. In addition, quality controls for tap waters are based on 62 parameters, compared to only 26 for mineral waters [5].

Doria notes that whilst there have been outbreaks of disease attributable to tap water, such as in Sydney in 1998, which led to an increase in bottled water sales, bottled water is not without similar events. The well-known brand Perrier was contaminated with benzene in 1990 and in 2004, Coca-Cola withdrew Dasani, its own bottled water, due to concerns about the levels of a potential carcinogen in the water [1, 29].

A number of participants expressed concerns about a link between the plastic container of bottled water and cancer. A carcinogenic substance known as DEHA (di-ethylhexl adipate) is indeed used in the manufacture of PET (polyethylene terephthalate), a plastic used to manufacture most bottled water containers [27]. However, laboratory studies performed by the US Environmental Protection Agency concluded that leaching of DEHA from the bottle is not harmful to human health [30]. Although not true, the concerns held by participants about the plastic bottles are not irrational and replicate concerns that other people seem to have. For example, in 2004 a hoax e-mail circulated in the USA, attributed to Johns Hopkins University, suggesting that the plastic used to manufacture the containers for bottled water contained harmful dioxins, which is untrue [31].

Almost a third of participants expressed concerns over the environmental impact of bottled water. These concerns mirror recent media interest in the subject [7, 8, 9]. These concerns included comments about the ‹carbon footprint› created by the transport of imported bottled water. A 2006 Earth Policy Institute study found that the British bottled water industry generates about 30,000 tonnes of carbon dioxide per year, which was estimated to equal the energy consumption of 6,000 homes per year [32].

The environmental impact of the plastic bottles themselves in their production and disposal was also mentioned by some participants. Packaging is generally made from plastics, either polyvinylchloride (PVC) or PET; the latter is becoming widely used as it is easier to recycle than PVC and does not release chlorine when burnt [5]. In the USA, annual production of PET to meet the needs of the bottled water industry uses around 18 million barrels of oil [32], which is a finite resource. Although the smaller UK market would mean lower oil use, considering that in 2006/2007, only 58.4% of municipal and 28.6% of household waste was recycled in the West Midlands, participants are probably correct in their concerns [33].

LIMITATIONS

This study has several limitations. Selection bias may have occurred in that the participants in the study had both the time available and the inclination to take part. This might mean that those with particularly strong views on the issue were more likely to volunteer, but this does not appear to have been borne out in our results. Availability bias may have occurred in that the issues surrounding bottled water can quite feasibly change over time and certain factors may become transiently important [34]. An example would be negative health beliefs about tap water as a motivating factor to purchase bottled water following media reports of contamination of the supply.

The fact that all of the participants in this study had connections with the Munrow sports centre, and the majority were employed by or studying at the University of Birmingham, has implications for the generalisability of the findings of this study. Hence, the results may not be applicable to people who are unemployed or not in full time education, or people who are not sports centre users. Repetition of the study with a sample more representative of the general population may therefore be of value.

Respondent validation may have proved useful since respondents' reactions to emerging findings can help to refine explanations and can strengthen the rigor of thorough qualitative research [35].

RECOMMENDATIONS

A number of issues arise from this study which may warrant further research. Namely the link between marketing strategies for bottled water and their role in creating health beliefs in the general public. It would also be interesting to see if it is possible to identify people who drink exclusively bottled water and question them about their reasons for this and their health beliefs about bottled water. Such information could then be compared with the results of this study to determine whether people who only drink bottled water are motivated to buy it by the same factors as the participants in this study, and the role of health beliefs within this. Finally, given the lack of knowledge about the purification process and safety of tap water in the UK found by this study, it could be useful to further educate the public with regard to the safety of tap water considering the prevalence of concerns arising that appear to be unfounded.

CONCLUSION

The participants of this study had a range of health beliefs about bottled water which could be classified as general and specific benefits. Participants also held a number of beliefs about the impact the bottled water industry has on the environment.

Although the majority of participants believed that bottled water had some health benefits, these beliefs played a minor role in determining bottled water consumption and are unlikely to be helpful in explaining the recent trends in bottled water consumption. The health beliefs that participants held were supported by scientific evidence to varying extents.

Convenience, cost and taste were more influential factors for participants when deciding whether to buy a bottle of water or not. Most participants did not feel that bottled water conferred significant, if any, health benefits over tap water.

DECLARATIONS

Acknowledgements

The authors would like to thank all of the participants of this research. They would also like to thank the University of Birmingham and the Munrow Sports Centre for allowing the recruitment of participants on the University campus and the University of Birmingham Medical School for allowing the use of their facilities.

REFERENCES

1. Doria MF: Bottled water versus tap water: understanding consumers-preferences. J Water Health. 2006, 4: 271-276.
2. Bottled Water Information: UK Bottled water consumption 2001–2007. [http://www.bottledwaterinformation.co.uk/default.asp?section=2&subsection=24]
3. Petrie KJ, Wessely S: Modern worries, new technology, and medicine. BMJ. 2002, 324: 690-691. 10.1136/bmj.324.7339.690.
4. Armas AB, Sutherland JP: A survey of the microbiological quality of bottled water sold in the UK and changes occurring during storage. Int J Food Microbiol. 1999, 48: 59-65. 10.1016/S0168-1605(99)00027-6.
5. Drinking Water Inspectorate Chief Inspector's Report: Drinking Water for England Annual Report 2002. London. 2002, [http://www.dwi.gov.uk/pubs/annrep02/contents.htm]
6. Ferrier C: Bottled Water: Understanding a social phenomenon. 2001, Report commissioned by the Worldwide Fund for Nature (WWF), [http://assets.panda.org/downloads/bottled_water.pdf]
7. BBC Panorama: Bottled Water – Who needs it?. [http://news.bbc.co.uk/1/hi/programmes/panorama/7247130.stm]
8. The Times online: Why pay the earth for water when it's on tap, ask the French. [http://www.timesonline.co.uk/tol/sport/football/european_football/article1083449.ece]
9. Guardian.co.uk: Is it OK...to drink bottled water?. [http://www.guardian.co.uk/environment/2005/dec/20/water.leohickmanonethicalliving]
10. Mackey E, Davis J, Boulos L, Brown J, Crozes G: Consumer Perceptions of Tap Water, Bottled Water, and Filtration Devices. 2005, London; IWA Publishing
11. Levallois P, Grondin J, Gingras S: Evaluation of consumer attitudes on taste and tap water alternatives in Quebec. J Water Sci Technol. 1999, 40: 135-139. 10.1016/S0273-1223(99)00549-1.
12. Hurd RE: Consumer Attitude Survey on Water Quality Issues. 1993, Denver: AWWA Research Foundation
13. Benevise F: La préoccupation des Français pour la qualité de l'eau. Les données de l'environnement. 2000, 57: 1-4.
14. Olson E: Bottled Water, Pure Drink or Pure Hype?. 1999, New York: Natural Resources Defense Council (NRDC), [http://www.nrdc.org/water/drinking/bw/bwinx.asp]
15. Britten N: Qualitative interviews in medical research. BMJ. 1995, 311: 251-3.
16. Strauss A, Corbin J: Basics of qualitative research. 1990, London: Sage
17. Petrie KJ, Wessely S: Getting Well from Water. BMJ. 2004, 329: 1417-1418. 10.1136/bmj.329.7480.1417.
18. Opel A: Constructing Purity: Bottled Water and the Commodification of Nature. Journal of American Culture. 1999, 22 (4): 67-76. 10.1111/j.1542-734X.1999.2204_67.x.
19. Volvic natural spring water: You and Volvic. [http://www.volvic-na.com/youvolvic/index.html]
20. Evian: Detox with Evian. [http://www.detoxwithevian.co.uk/detox_with_evian-the_benefits_of_detoxing.aspx]
21. Azoulay A, Garzon P, Eisenberg J: Comparison of the mineral content of tap water and bottled waters. J Gen Intern Med. 2001, 16: 168-75. 10.1111/j.1525-1497.2001.04189.x.
22. Green M, Green T: The Good Water Guide. 1994, London: Rosendale Press
23. Petraccia L, Liberati G, Giuseppe Masciullo S, Grassi M, Fraioli A: Water, mineral waters and health. Clinical Nutrition. 2006, 25: 377-385. 10.1016/j.clnu.2005.10.002.
24. European Community Directive: Council Directive No. 80/777/EEC of 15 July 1980 on the approximation of the laws of member states relating to the exploitation and marketing of natural mineral waters. Official Journal of the European Communities. 1980, L229: 1-10.
25. Ministry of Agriculture, Fisheries and Food: The Natural Mineral Waters Regulations 1985: Statutory instruments no. 71. 1985, London: HMSO
26. Rosenberg FA: The Microbiology of Bottled Water. Clinical Microbiology Newsletter. 2003, 25: 41-44. 10.1016/S0196-4399(03)80019-3.
27. Leising L, McCarthy P, Hahn T, Dunford L, McKernon M: Bottled Water Myths: Separating Fact from Fiction. Practical Gastroenterology. 2007, 31: 89-93.
28. Raj SD: Bottled Water: how safe is it?. Water Environ Res. 2005, 77: 3013-3018. 10.2175/106143005X73893.
29. BBC news: "Coke recalls controversial water.". [http://news.bbc.co.uk/1/hi/business/3550063.stm]
30. U.S. Environmental Protection Agency: Di-(2-ethylhexyl) Adipate; Toxic Chemical Release Reporting. Federal Register. 1995, 60: 39132-39134.

31. Johns Hopkins Bloomberg School of Public Health, Public Health News Center: 'Researcher Dispels Myth of Dioxins and Plastic Water Bottles 2006'. [http://www.jhsph.edu/publichealthnews/articles/halden_dioxins2.html]
32. Arnold E, Larsen J: Bottled water: Pouring resources down the drain. Earth Policy Institute. 2006, [http://www.earth-policy.org/Updates/2006/Update51.htm]
33. Department for Environment, Food and Rural Affairs: Key Facts about: Waste and Recycling. Regional household recycling and municipal recovery rates. 2006, [http://www.defra.gov.uk/environment/statistics/waste/kf/wrkf07.htm] /07 Government Office Regions, England
34. Klein JG: Five pitfalls in decisions about diagnosis and prescribing. BMJ. 2005, 330: 781-783. 10.1136/bmj.330.7494.781.
35. Barbour RS: Checklist for improving rigor in qualitative research: a case of tail wagging the dog?. BMJ. 2001, 322: 1115-1117. 10.1136/bmj.322.7294.1115.

INDEX

E

F

G

H

I

J

K

L

M

N

O

P

Q

R

S

T

U

V

W

Y